FORTSCHRITTE DER CHEMIE ORGANISCHER NATURSTOFFE

PROGRESS IN THE CHEMISTRY OF ORGANIC NATURAL PRODUCTS

BEGRÜNDET VON · FOUNDED BY

L. ZECHMEISTER

HERAUSGEGEBEN VON · EDITED BY

W. HERZ   H. GRISEBACH   G. W. KIRBY
TALLAHASSEE, FLA.   FREIBURG i. BR.   GLASGOW

VOL. 31

VERFASSER · AUTHORS

N. H. ANDERSEN · ST. F. BRADY · C. M. HARRIS · TH. M. HARRIS
E. HECKER · K. B. HINDLEY · D. N. McGREGOR · J. A. MARSHALL
J. C. ROBERTS · R. SCHMIDT · G. N. SCHRAUZER · G. A. SWAN
CH. TAMM · H. WAGNER · E. WINTERFELDT

1974

WIEN · SPRINGER-VERLAG · NEW YORK

Mit 60 Abbildungen · With 60 Figures

ISBN-13: 978-3-7091-7096-0   e-ISBN-13: 978-3-7091-7094-6
DOI: 10.1007/ 978-3-7091-7094-6

Inhaltsverzeichnis. Contents

Phorbolesters — the Irritants and Cocarcinogens of Croton Tiglium L. By E. HECKER and
R. SCHMIDT, Biochemisches Institut, Deutsches Krebsforschungszentrum, Heidelberg, BRD 377

Mechanisms of Corrin Dependent Enzymatic Reactions. By G. N. SCHRAUZER, Department of Chemistry, University of California, San Diego, La Jolla, California, USA . 583

Recent Developments in the Chemistry
of Penicillins

By D. N. McGregor, Syracuse, New York, USA

Contents

I. Introduction

The purpose of this review is to summarize the important developments in the chemistry of the penicillin molecule which have been reported in the scientific literature during the approximate period 1964 through 1972. The penicillins were intensively studied from a chemical point of view during the 1940's and this work is discussed in detail in the penicillin monograph (*36*). The isolation of 6-aminopenicillanic acid in 1959 (*16*) led to the preparation of large numbers of penicillin derivatives in which the side chain at the 6-position of the penicillanic acid nucleus was modified. These efforts, which have been successful in introducing a number of important changes in the biological properties of the penicillin molecule, have been reviewed by PRICE (*144*) and others (*55, 175, 72, 145, 2, 91*). This aspect of penicillin chemistry will be dealt with only briefly in this review, and then only with reference to the chemistry involved. Recently, and particularly during the last four years (1969 through 1972), there have been increasing numbers of reports in which the chemistry of the penicillanic acid nucleus itself has been investigated, and it is principally to these studies that this review will address itself (see also *81, 129, 130*, and *66*).

Ever since the structure of the penicillin molecule was elucidated (*36*), there have been continuing efforts directed toward the total synthesis of penicillins and penicillin analogs. These studies are outside the scope of the present review, and the reader is referred to the books by MANHAS and BOSE (*129, 130*) and the chapters by HEUSLER (*89, 86*) for summaries.

II. Nomenclature

The nomenclature which will be utilized throughout this review can be illustrated by referring to the structure of benzylpenicillin (penicillin G) (**1**). This will be designated 6-phenylacetamidopenicillanic acid and

(1)

(2)   (3)

the numbering system shown in (**1**) will be used (note that the term "penicillin" refers to (**2**) except when being used as a general name for the entire class of compounds). Unless otherwise specified either in the name or the structure, the natural penicillin configuration as shown in (**1**) will be assumed (*i.e.*, the 6-substituent β, the 3-substituent α and the 5- and 6-hydrogens *cis*). Other names which, according to other conventions, can be given to (1) are 2,2-dimethyl-6β-phenylacetamidopenam-3α-carboxylic acid, and (2*S*, 5*R*, 6*R*)-3,3-dimethyl-7-oxo-6-(2-phenylacet-amido)-4-thia-1-azabicyclo[3.2.0]heptane-2-carboxylic acid. The penicillanic acid nucleus (3) consists of fused β-lactam and thiazolidine rings—these portions of the nucleus will be referred to by these names.

Several abbreviations will be used throughout the review: Et for ethyl, tBu for *t*-butyl, Ⓞ for phenyl, Ac for acetyl, Ts for *p*-toluenesulfonyl, and Phth for phthaloyl.

III. Reactions at the β-Lactam Ring

Transformations which primarily involve the β-lactam portion of the penicillanic acid nucleus and its substituents will be considered under this heading. Some reactions which involve both the β-lactam and the thiazolidine ring will be dealt with in Section IV C. The chapter by KAISER and KUKOLJA (*106*) includes a discussion of many of these topics.

A. Acylation of the 6-Amino Group

Most of the common methods for forming amide bonds have been applied to the coupling of an organic acid with the amino group of 6-aminopenicillanic acid (4) (henceforth, 6-APA) (*Chart 1*). This aspect

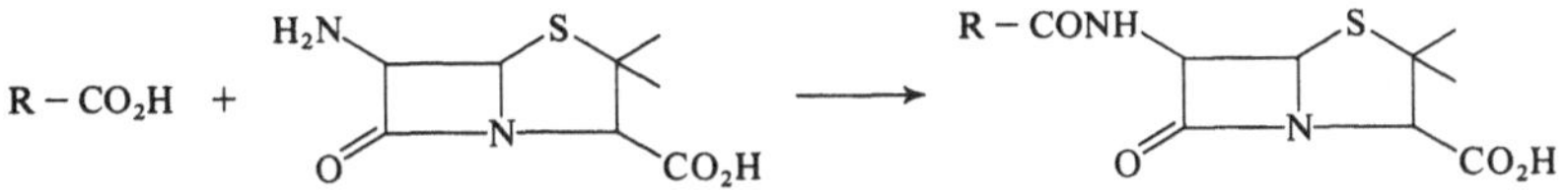

(4)

Chart 1. Acylation of 6-APA

of penicillin chemistry has been reviewed (*135*), and the reader is directed to the references cited in reviews such as the one by PRICE (*144*) for examples of a variety of penicillins and the methods used in their synthesis. Some of the coupling methods which have been described in recent years include the use of coupling reagents such as 1,1′-carbonyldiimidazole (*133*), N,N′-dicyclohexylcarbodiimide (*7*), and N,N-dimethylchloroformiminium chloride (*140*), the use of esters of *p*-nitro- and 2,4-dinitrophenol (*147*), the use of N-carboxy-α-amino acid anhydrides (*76*), and the use of mixed anhydrides with pivalic acid (*53*).

When the acid which is to be coupled to 6-APA also contains an amino group, it is generally necessary to block this amino group to prevent its acylation. A variety of protecting groups which can be removed without destroying the penicillanic acid ring system have been described: the proton (*82*), the carbobenzyloxy group (*126*), the p-nitrocarbobenzyloxy group (*146*), the enamine derived from methyl acetoacetate and other β-dicarbonyl compounds (*159, 53, 126*), the 2-methyl-2-(o-nitrophenoxy)-propionyl group (*103*), the o-nitrophenylsulfenyl group (*59*) and the trityl group (*111*). In addition, an amine precursor such as azide can be substituted for the amine (*58*). Various protecting groups have been employed to block the amine of 6-APA. These have included the trityl group (*20, 21, 138*), Schiff bases such as the p-nitrobenzylidene group (*e.g., 104*), and the 2,2,2-trichloroethoxycarbonyl group (*62*). With the advent of the chemical cleavage of penicillin side chains, however (see *Chart 5*), most secondary amides at the 6-position can serve as blocking groups.

Carboxylic acid blocking groups have frequently been useful in a variety of chemical manipulations involving the penicillanic acid nucleus. These can be applied either to a carboxylic acid in the side chain or to the acid at the 3-position of 6-APA, and have included benzyl esters (*85, 26, 125, 164*), p-nitrobenzyl esters (*1*), methoxymethyl esters (*96*), silyl esters (*187, 68*), stannous esters (*7*), phenacyl esters (*6*), trityl esters (*93*), 3,5-di-*t*-butyl-4-hydroxybenzyl esters (*93*), esters of (*E*)-oximes of benzaldehyde and 2-furaldehyde (*67*), mixed anhydrides with acetic acid (*32*), and amides of N, N'-diisopropylhydrazine (*11*).

In general, it has been necessary to have a carboxamido group at the 6-position of the penicillanic acid nucleus in order to have significant antimicrobial activity. A recent notable exception has been the preparation of several 6-amidino derivatives. One of these, FL-1060 (*Chart 2*), has been reported to have outstanding activity against certain Gram-negative organisms (*128*).

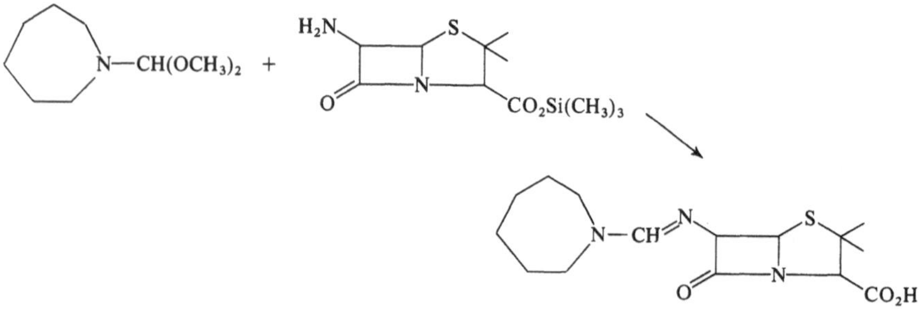

FL-1060

Chart 2. A 6-amidinopenicillin

B. Alkylation of the 6-Amino Group

6-Aminopenicillanic acid has been alkylated either by formation of the Schiff base followed by reduction (122) or by reaction with a diazoalkane (57, 134) (Chart 3).

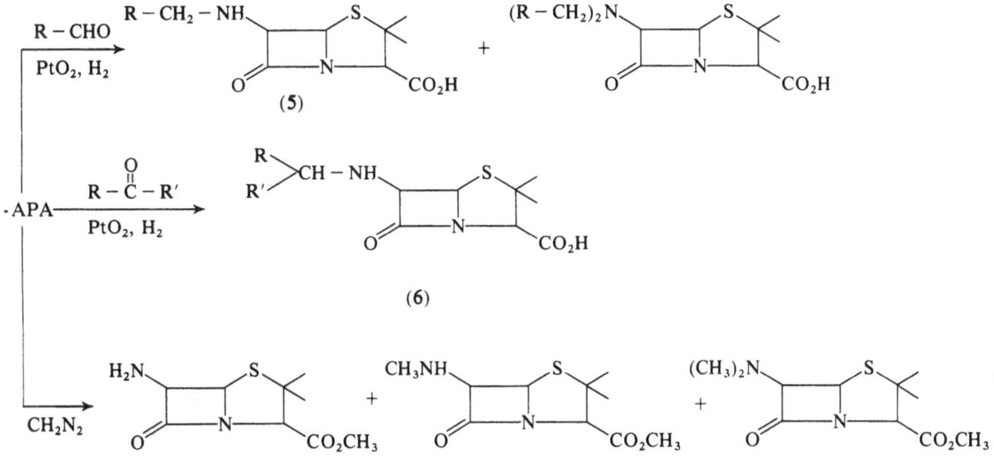

Chart 3. Alkylation of 6-APA

Acylation of compounds of the type (5) did not afford derivatives with interesting biological properties; (6) ($R=R'=CH_3$) could not be acylated.

A special form of N-alkylation is involved in the formation of hetacillin (8) by treatment of ampicillin (7) with acetone under basic conditions (Chart 4) (82). The unique imidazolidinone structure has been confirmed

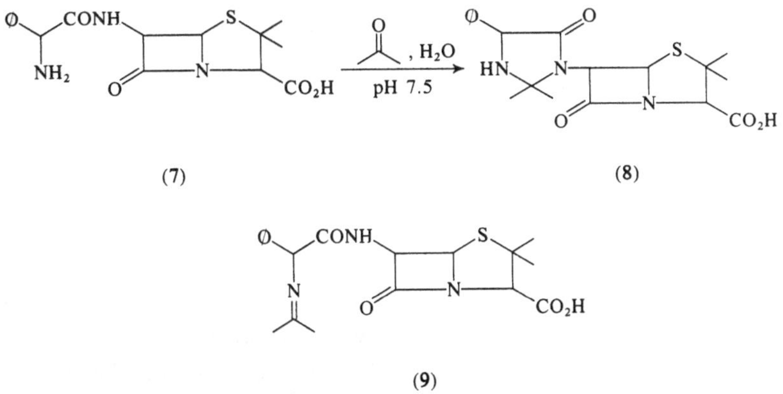

(7)   (8)

(9)

Chart 4. Hetacillin

by X-ray structural analysis. Compared to (7), (8) has a lower isoelectric point, and the β-lactam is less susceptible to opening with acid, 2,4-dinitrophenylhydrazine, and by way of polymerization reactions (*113*). The stability (*162, 113*), microbiological (*179, 171*), and pharmacological (*179*) properties of (8) have been interpreted in terms of an equilibrium between (7) and (8), possibly involving the Schiff base (9) as an intermediate (*56*).

C. Hydrolysis of the 6-Amido Group

Because the cleavage of a penicillin to 6-APA requires the hydrolysis of the more stable 6-carboxamido group in the presence of the labile β-lactam, it would be reasonable to suppose that only the high specificity afforded by an enzymatic reaction would be successful for this conversion (*92*). Weissenberger and van der Hoeven (*187*), however, took advantage of the fact that the 6-carboxamido function is a secondary amide, and were able to convert penicillin G to 6-APA in 91% overall yield by blocking the carboxylic acid as a silyl ester, then converting the 6-carboxamido group to a readily hydrolyzed imino ether by treatment with PCl_5 and l-butanol (*Chart 5*). Similar conversions have been carried out using the

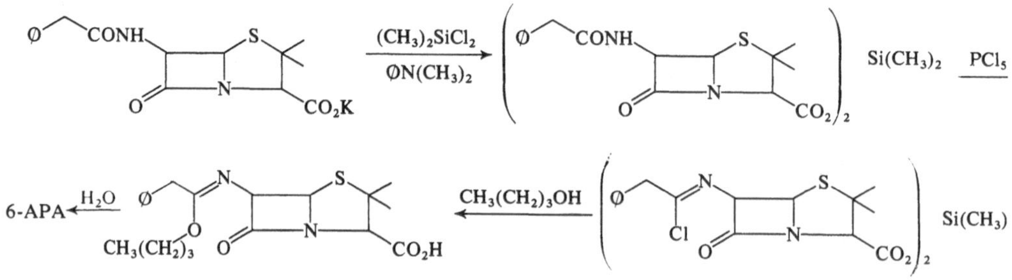

Chart 5. 6-APA from penicillin G

mixed anhydride with acetic acid (*32*) or the ester with the (*E*)-oximes of benzaldehyde or 2-furaldehyde (*67*) to block the carboxylic acid.

A less general method of side chain cleavage was employed to remove the phenylglycyl residue from 6-epihetacillin (*Chart 6*) to afford 6-epiAPA (10) (*100*).

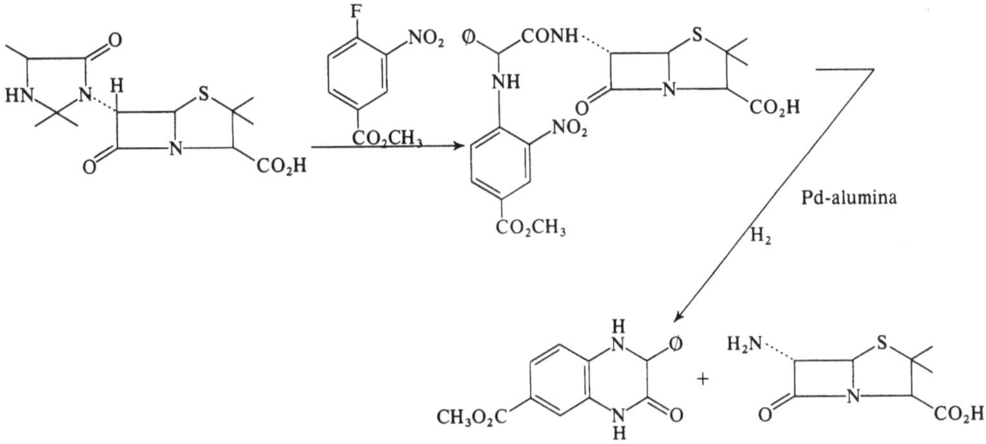

Chart 6. 6-EpiAPA from 6-epihetacillin

Several procedures have been devised which will allow the exchange of one acyl function on the 6-amino group of 6-APA for another, without involving 6-APA as an intermediate. An early example of this type of conversion has been described by SHEEHAN (164) and is shown in Chart 7. A more general exchange reaction is illustrated in Chart 8 (1). Treatment

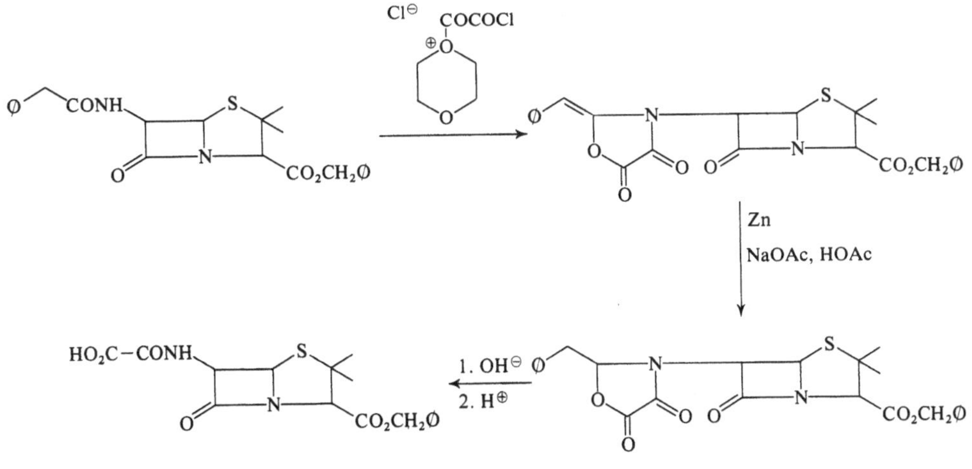

Chart 7. Benzyl 6-oxamidopenicillanate from penicillin G benzyl ester

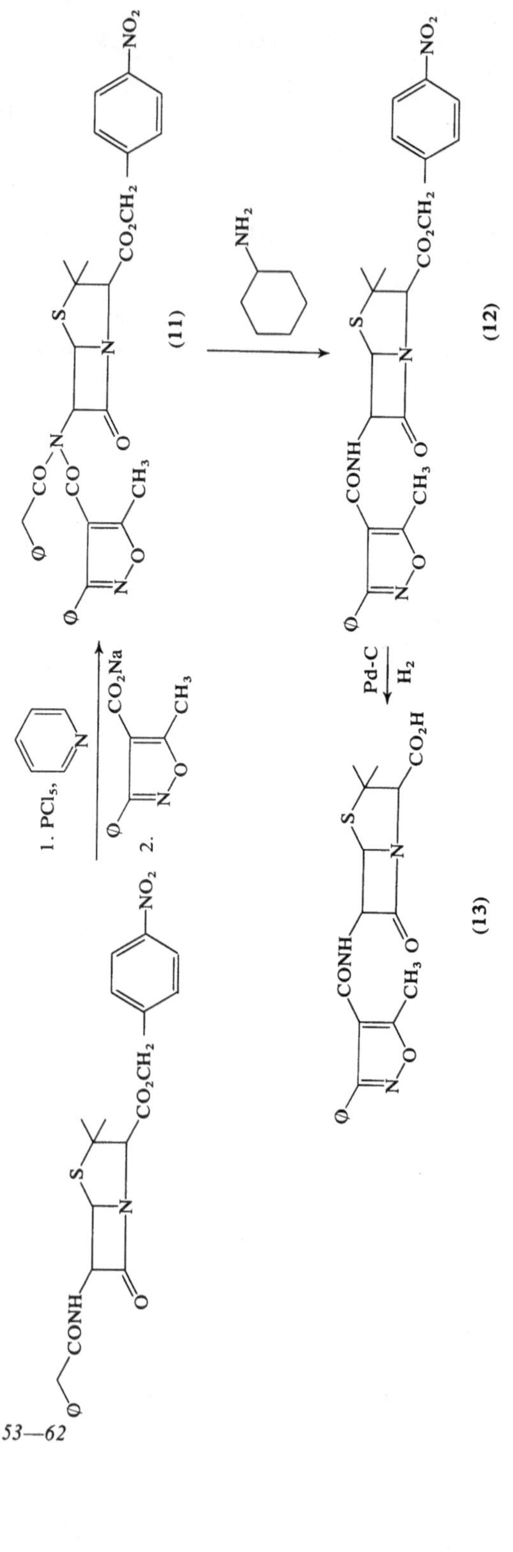

D. N. McGregor :

Chart 8. Oxacillin from penicillin G *p*-nitrobenzyl ester of the penicillin G ester with PCl$_5$ gives the imino chloride which, upon reaction with the carboxylic acid salt, yields an imino ester which re-arranges spontaneously to the diacylamine (11). An acyl acceptor such as cyclohexylamine will react preferentially with the least hindered carbonyl in (11) to afford (12) which can then be deblocked to oxacillin (13). It is interesting to note that this sequence does not require the new acyl group to be converted to an activated acylating species. The success of the exchange, however, depends upon the original acyl group being more readily displaced from the intermediate corresponding to (11) than the new acyl group.

An even more general transacylation reaction is shown in *Chart 9* (*93, 110*). This sequence, which presumably involves an N-acylated imino ether intermediate (14) in the acylation step, is in principle applicable to the exchange of any two acyl functions.

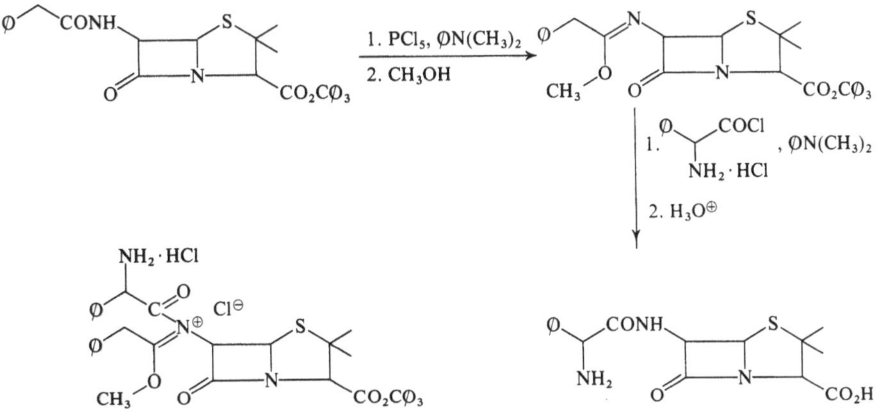

Chart 9. Ampicillin from penicillin G trityl ester

D. Epimerization at the 6-Position

The 6-position of most penicillanic acid derivatives is an optically active center which contains a potentially acidic hydrogen because of its situation α to the β-lactam carbonyl. When the 6-substituent is in the natural (β) configuration, the molecule is in the thermodynamically less stable arrangement. It might be expected, therefore, that epimerization at the 6-position could occur under basic conditions. In fact, simple base treatment of true penicillins (*i.e.*, 6-carboxamidopenicillanic acids, (2)] does not lead to 6-epipenicillins. In recent years, however, a number of

penicillanic acid derivatives have been observed to epimerize at the
6-position. Thus, hetacillin epimerizes either as the salt [(15), R = Na] in
aqueous base or as the methyl ester [(15), R = CH₃] in anhydrous base
(101), and methyl 6-phthalimidopenicillanate (17) was found to epimerize
to (18) by treatment with sodium hydride in tetrahydrofuran, potassium
t-butoxide in t-butyl alcohol, and triethylamine in methylene chloride
(195) (Chart 10). Since these initial observations, epimerizations have

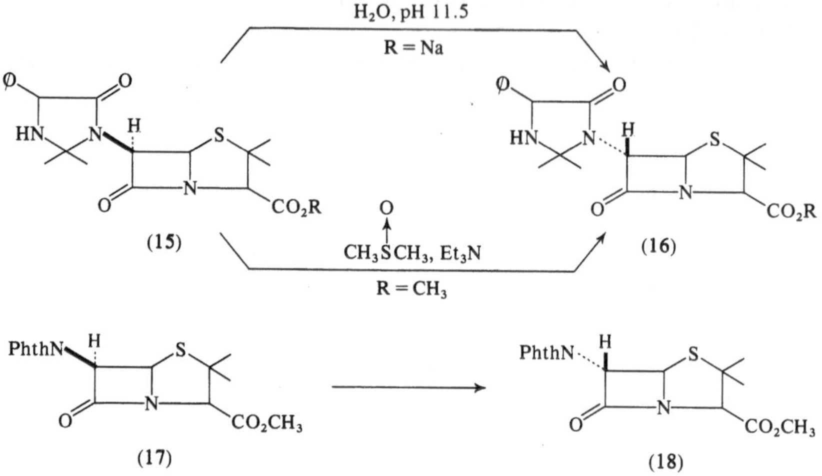

Chart 10. Epimerization of penicillanic acid derivatives

been observed in a variety of other penicillanic acid derivatives: (19) (38),
(20) (150), (21) (198), (22) (97, 95), (23) (95, 96), (24) (95), and (25) (95).

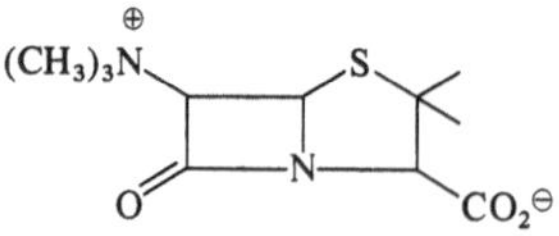

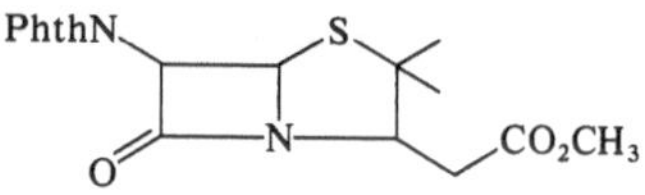

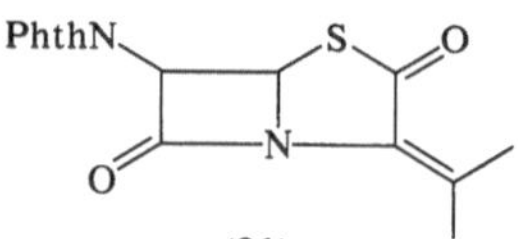

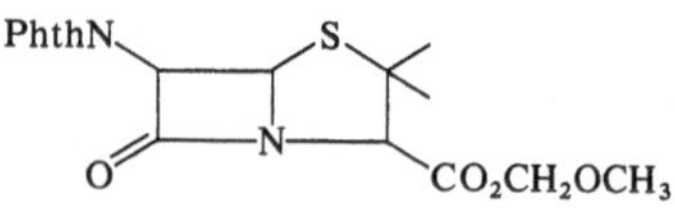

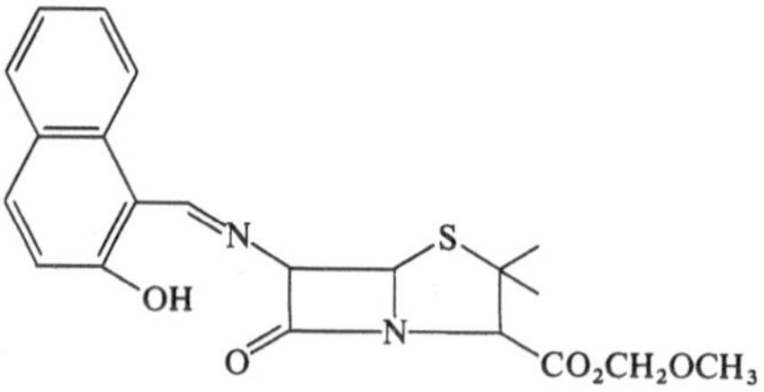

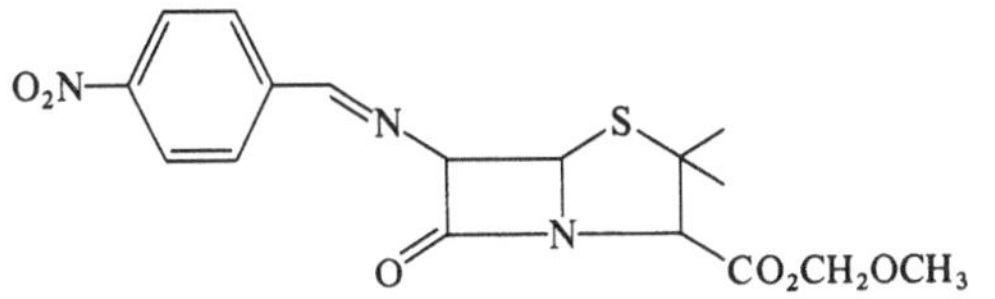

(24) 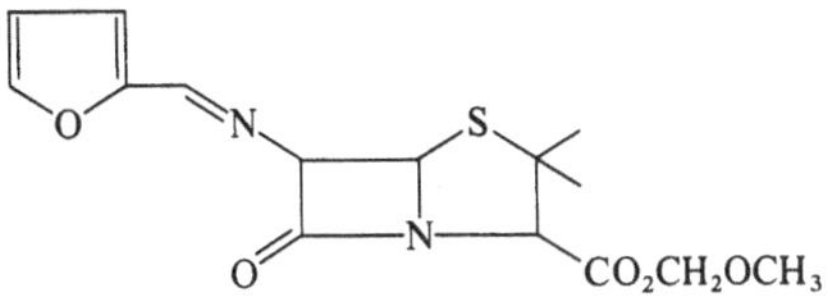   (25)

Just as in the case of the true penicillins mentioned in the previous paragraph, epimerization was not observed with 6-APA itself, 6-dimethylaminopenicillanic acid, nor with 6-tritylaminopenicillanic acid (*38*). If, however, a penicillin such as penicillin V or its ester (**26**) is first reacted with the powerful silylating agent N,O-bis-(trimethylsilyl)-acetamide (BSA) and then treated with a base such as 1,5-diazabicyclo[4.3.0]non-5-ene (DBN) or triethylamine, epimerization to (**27**) is observed (*Chart 11*) (*183*). With the (*S*)-sulfoxide of 2,2,2-trichloroethyl 6-phenoxyacetamidopenicillanate (**28**), BSA alone with no added base is sufficient to establish an equilibrium between the two epimers (*Chart 11*) (*79*).

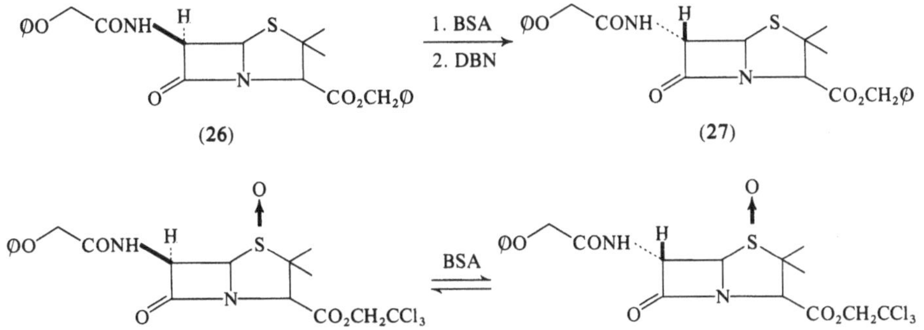

Chart 11. Epimerization of penicillin V derivatives

As might be expected from steric considerations, when the epimerizations described above are allowed to go to completion or until equilibrium is established, the unnatural epimer (with the 6-substituent α) predominates. The effect of solvent and, to a certain extent, the nature of the 6-substituent on the position of this equilibrium has been investigated (*95*).

The incorporation of deuterium during the epimerization reaction has been studied in an effort to gain insight into the mechanism(s) involved. Generally, epimerization in the presence of solvents containing exchangeable deuterium atoms results in the incorporation of deuterium at the 6-position (*101, 195, 38, 150*) (see, however, *198*). When hetacillin [(**15**), R=H] is partially epimerized, the epihetacillin [(**16**),

R = H] produced contains deuterium at the 6-position while the recovered hetacillin does not (101, 38). Deuterium incorporation into epihetacillin [(16), R=H) has been found to be much slower than the rate of incorporation into hetacillin [(15), R=H], indicating that the rate of proton abstraction from the α epimer is slower than from the β epimer (38). Similar rate differences for deuterium incorporation in the two epimers were observed for (20) (150) and (19) (38). The epimerization of methyl 6-phthalimidopenicillanate (17) with potassium t-butoxide in the presence of tBuOD led to deuterium incorporation at C-6 of (18) (195). Deuterium was not incorporated, however, when the weaker base triethylamine was used (198) (see, however, 150).

Under conditions suitable for epimerization, nuclear rearrangement has also been observed (Chart 12). Thus, triethylamine treatment of methyl 6-phthalimidopenicillanate (17) affords not only the 6-epi compound (18) but also the thiazepine (29) (198, 112). A similar re-

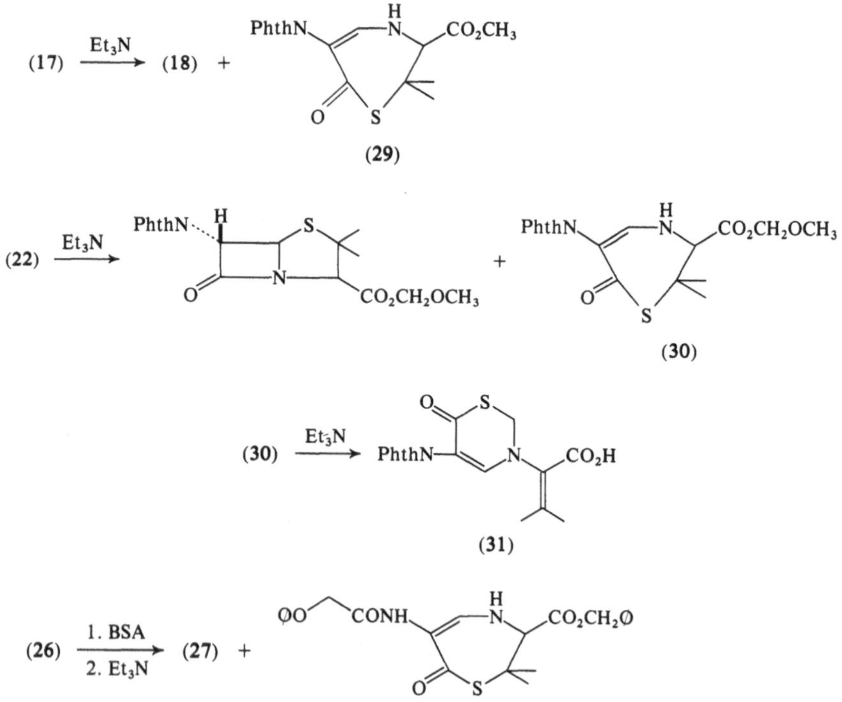

Chart 12. Base-catalyzed rearrangement of penicillanic acid derivatives

arrangement has been observed with (20) (150) and with (26) (183). Those penicillanic acid derivatives in which the carboxyl is blocked as the methoxymethyl ester [e.g., (22) and (24)] afford thiazepines [e.g., (30)] which can undergo further base-catalyzed rearrangement to 1,3-thiazines [*e.g.*, (**31**)] (*94, 97*). The nature of the basic catalyst has been found to be important in determining the ratio of epimer product to rearrangement product. Triethylamine, for example, usually gives more rearrangement product than does DBN (*150, 183*).

The mechanism of the epimerization and the rearrangement has been the subject of some discussion. It has been proposed that both epimerization and rearrangement involve a common intermediate of the type (**32**), but that different mechanisms may be involved depending on the nature of the base and the substrate (*198*). Another proposal suggests that an enolate such as (**33**) is the common intermediate for both epimerization and rearrangement, with (**32**) being a later intermediate for the rearrangement only (*150*).

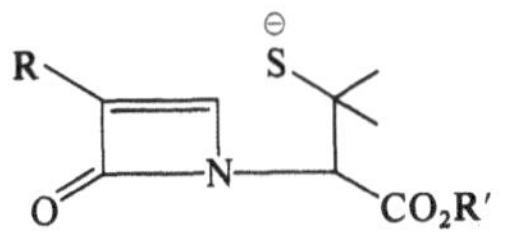

(32) 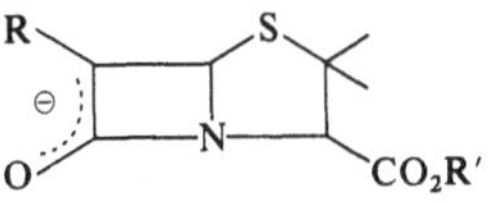   (33)

It has been found that 6-epipenicillins such as 6-epihetacillin [(**16**), R = H] (*101*) and 6-epipenicillin G (prepared from 6-epiAPA) (see *Chart 6*) (*100*) are essentially devoid of antimicrobial activity.

E. Diazotization of the 6-Amino Group

The reaction of 6-APA with nitrous acid in the presence of halide ion has been found to give 6-halopenicillanic acids [(**34**) and (**35**)] (*35*) (*Chart 13*). Nmr studies have indicated that the halogen is in the α con-

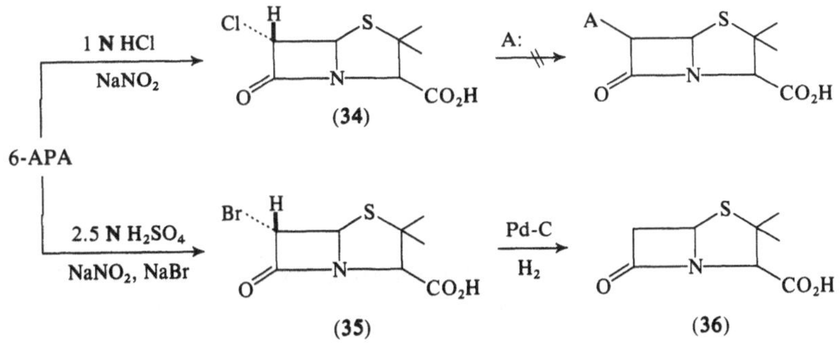

Chart 13. 6-Halopenicillanic acids

figuration (*131*). In spite of the fact that these halogens are adjacent to the β-lactam carbonyl, they have been very resistant to displacement with nucleophiles (A:) (*132*), probably because of steric inhibition of backside attack. It has, however, been possible to hydrogenolyze (**35**) to give penicillanic acid (**36**) (*61*). The intermediate in the diazotization reaction is probably (**37**), since both 6-APA and 6-epiAPA methoxy-methyl esters give (**38**) on treatment with sodium nitrite and hydrochloric acid, and, in a deuterium-containing medium, both incorporate deuterium at the 6-position (*96*). Benzyl 6-diazopenicillanate (**39**) can be isolated from the reaction of 6-APA benzyl ester with nitrous acid in the absence of a reactive nucleophile (*85, 37, 132*).

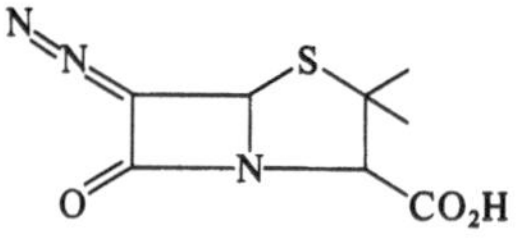

(37) 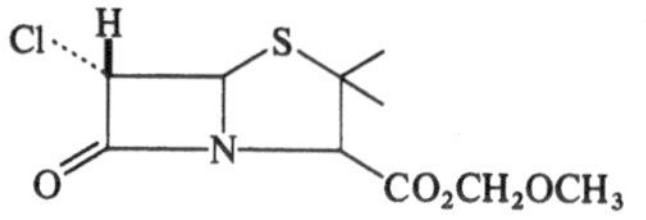　　　　　　　　　　(38) 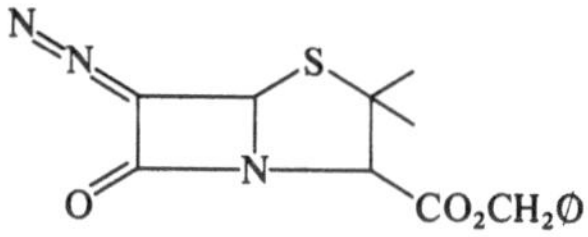　　　　　　　　　　(39)

When the diazotization is carried out in the presence of iodide ion, the major product becomes 6,6-diiodopenicillanic acid (**40**) (*Chart 14*) (*37*).

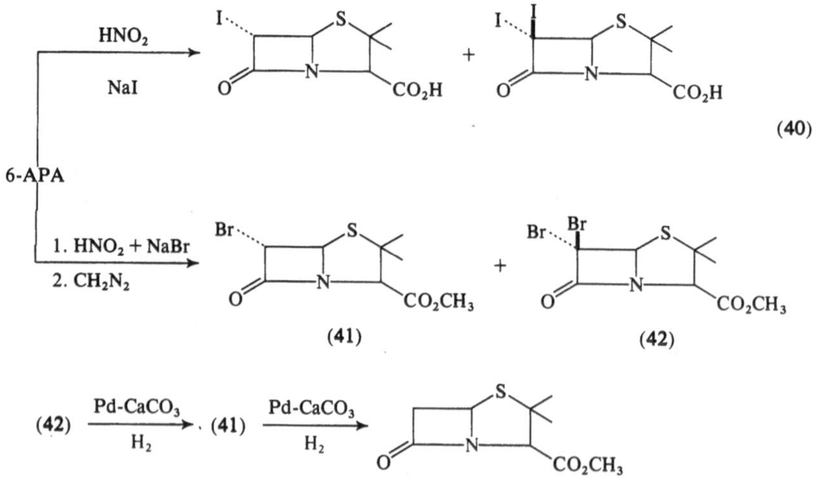

Chart 14. 6,6-Dihalopenicillanic acids

Similarly, the presence of sodium bromide gives rise to the corresponding dibromo compound (**42**), but in smaller proportion. Presumably, these compounds arise from the reaction of the diazo intermediate (**37**) with either iodine or bromine formed by the nitrous acid oxidation of the halide ion. Support for this view was obtained by the formation of (42) by the treatment of 6-APA with nitrous acid and bromine followed by diazomethane. The 6,6-dihalopenicillanic acids were unreactive to nucleophiles either under S_N1 or S_N2 conditions. Methyl 6,6-dibromopenicillanate (42) could, however, be hydrogenolyzed to form initially methyl 6α-bromopenicillanate (41) which could be further hydrogenolyzed to methyl penicillanate.

When 6-APA is treated with sodium nitrite and methanolic hydrochloric acid, a dihydro-1,4-thiazin-3-one (44) is obtained in addition to methyl 6α-chloropenicillanate (43) (Chart 15) (176, 177). An ingenious

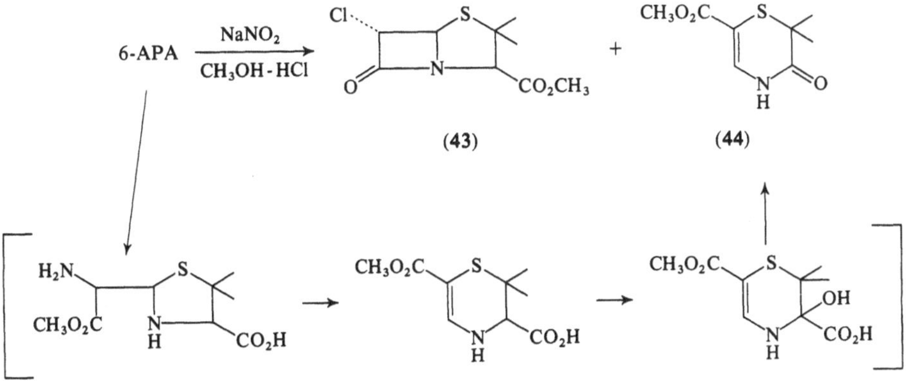

Chart 15. Rearrangement of 6-APA to a dihydro-1,4-thiazin-3-one

series of experiments indicates that the production of (44) from 6-APA probably involves the intermediates shown in Chart 15. Examples of other rearrangements are shown in Chart 16 (131, 132, 40).

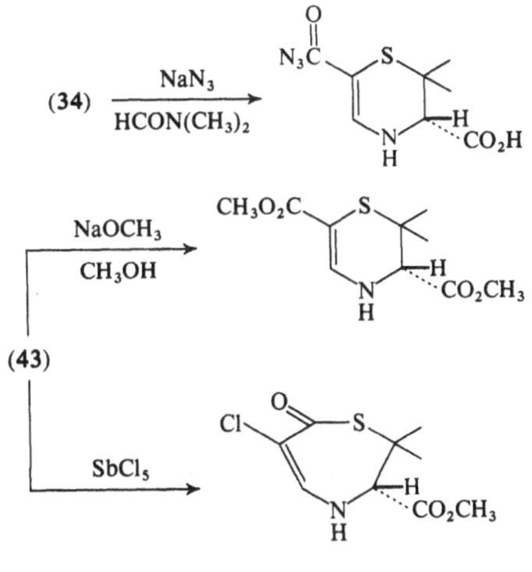

Chart 16. Further rearrangements of 6α-chloropenicillanates

Initial attempts to prepare 6-hydroxypenicillanic acid via the diazotization of 6-APA were unsuccessful (35). Later, however, two methods for the synthesis of 6α-acyloxypenicillanates were devised (85). These are shown in *Chart 17*. The intermediate (45) was also useful for the

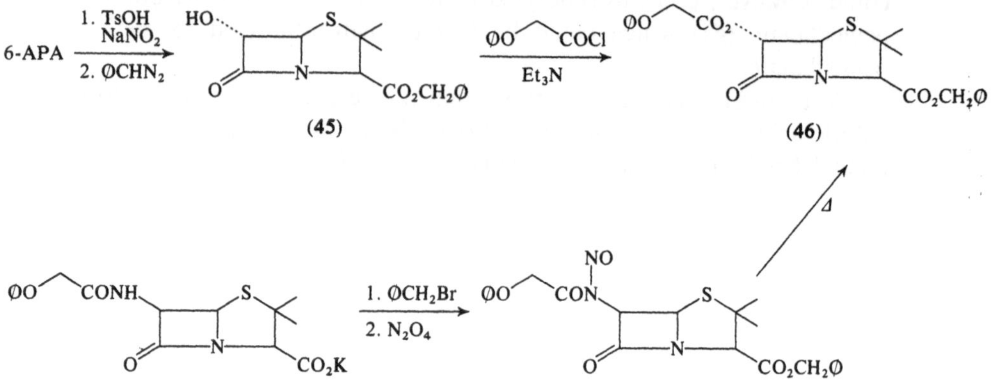

Chart 17. Benzyl 6α-phenoxyacetoxypenicillanate

preparation of 6β-acyloxypenicillanates (125), (*Chart 18*). Interestingly, the acids obtained by the reductive deblocking of (46) and (47) exhibited little or no bioactivity.

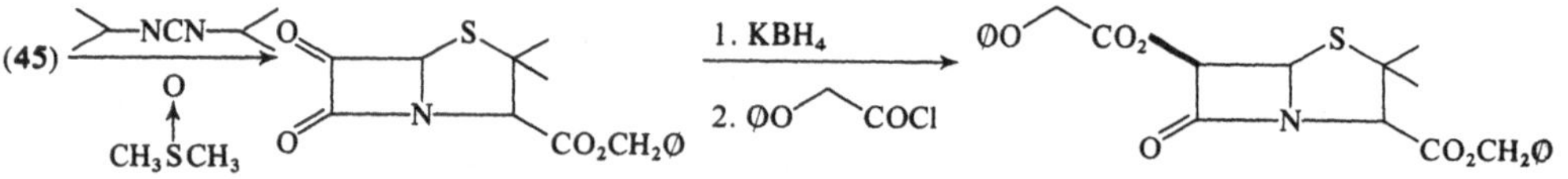

Chart 18. Benzyl 6β-phenoxyacetoxypenicillanate   (47)

Advantage has been taken of the availability of benzyl 6-diazopenicillanate (39) to prepare 6-phenylacetylhydrazono- and 6β-phenylacetylhydrazinopenicillanic acids (*Chart 19*). (26). Both of these derivatives exhibited antimicrobial activity, but at a lower level than that of penicillin G.

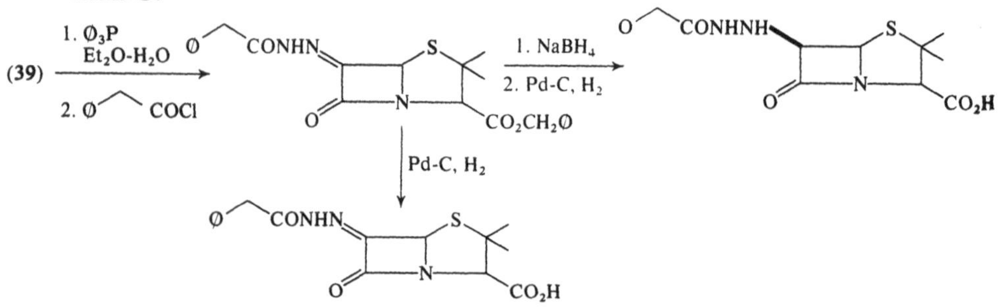

Chart 19. 6-Phenylacetylhydrazono- and 6β-phenyladetylhydrazinopenicillanic acids

References, pp. 53—62

F. Substitution at the 6-Position

On the assumptions that the antimicrobial effect of penicillins is due to inhibition of a transpeptidase-catalyzed step in cell wall peptidoglycan synthesis and that the penicillin mimics the **D**-Ala.**D**-Ala portion of the substrate in its interaction with the enzyme surface, it has been suggested (*178*) that the penicillin molecule might be made to resemble **D**-Ala.**D**-Ala more closely (and hence fit the enzyme surface more closely) if a 6α-methyl substituent could be introduced. One approach to alkylation of the 6-position of the penicillanic acid molecule is shown in *Chart 20* (*105*). This is presumably a [2,3]-sigmatropic rearrangement via the nitrogen ylide.

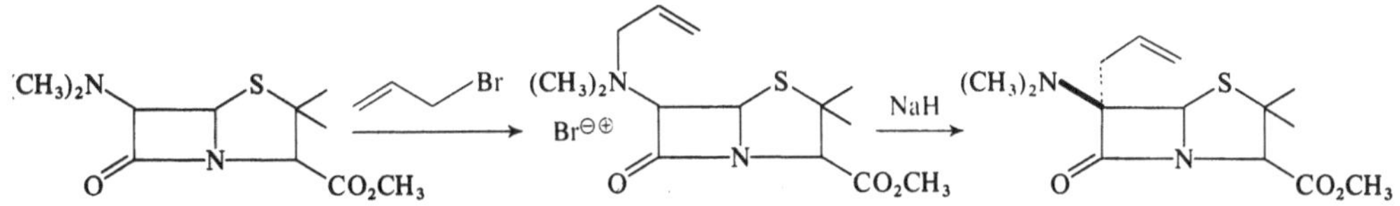

Chart 20. Methyl 6α-allyl-6β-dimethylaminopenicillanate

A more general approach to this alkylation involves preliminary conversion of the amino group of 6-APA to a Schiff base followed by alkylation under basic conditions. Initial efforts in this direction are outlined in *Chart 21* (*154*). Subsequent investigations (*23, 22, 65, 104,*

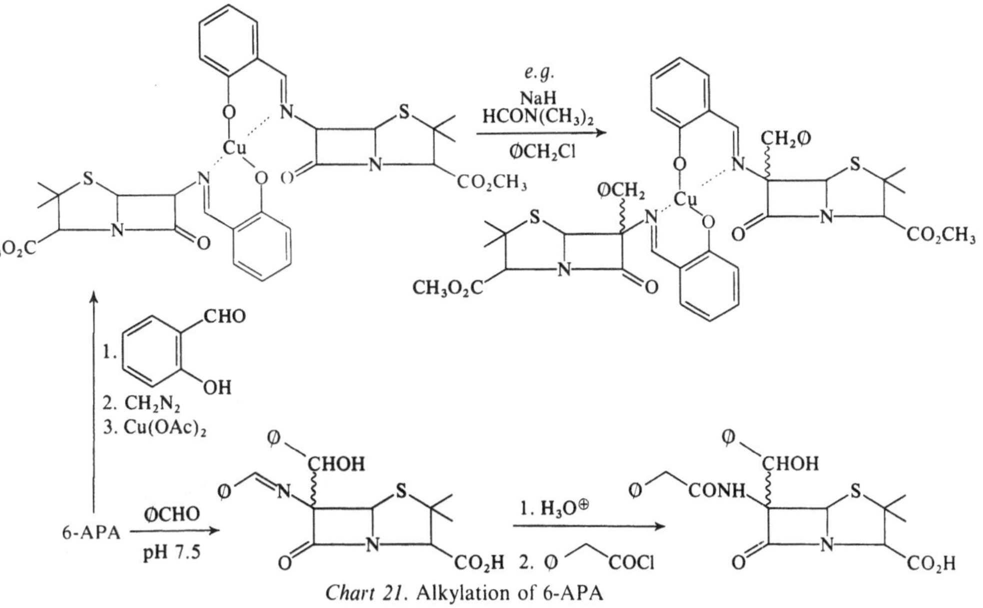

Chart 21. Alkylation of 6-APA

64, 153, 172) have greatly extended the scope of this approach, and some of the transformations achieved are illustrated in *Chart 22*. It is interesting to note that the original compound sought, (49) (R=CH₃), is in fact a poorer inhibitor of the transpeptidase than is the original penicillin (90).

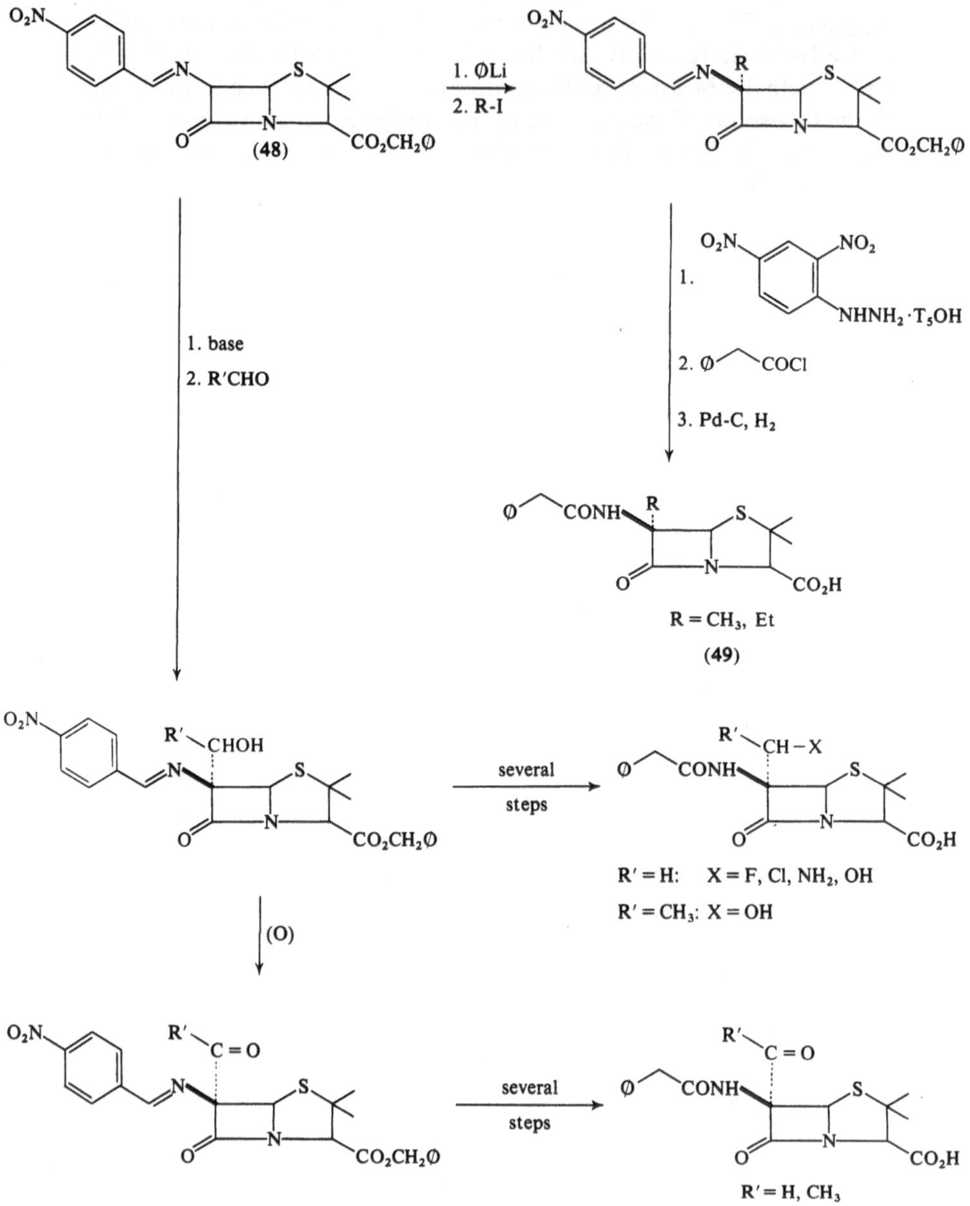

R = CH₃, Et

(49)

R' = H: X = F, Cl, NH₂, OH

R' = CH₃: X = OH

R' = H, CH₃

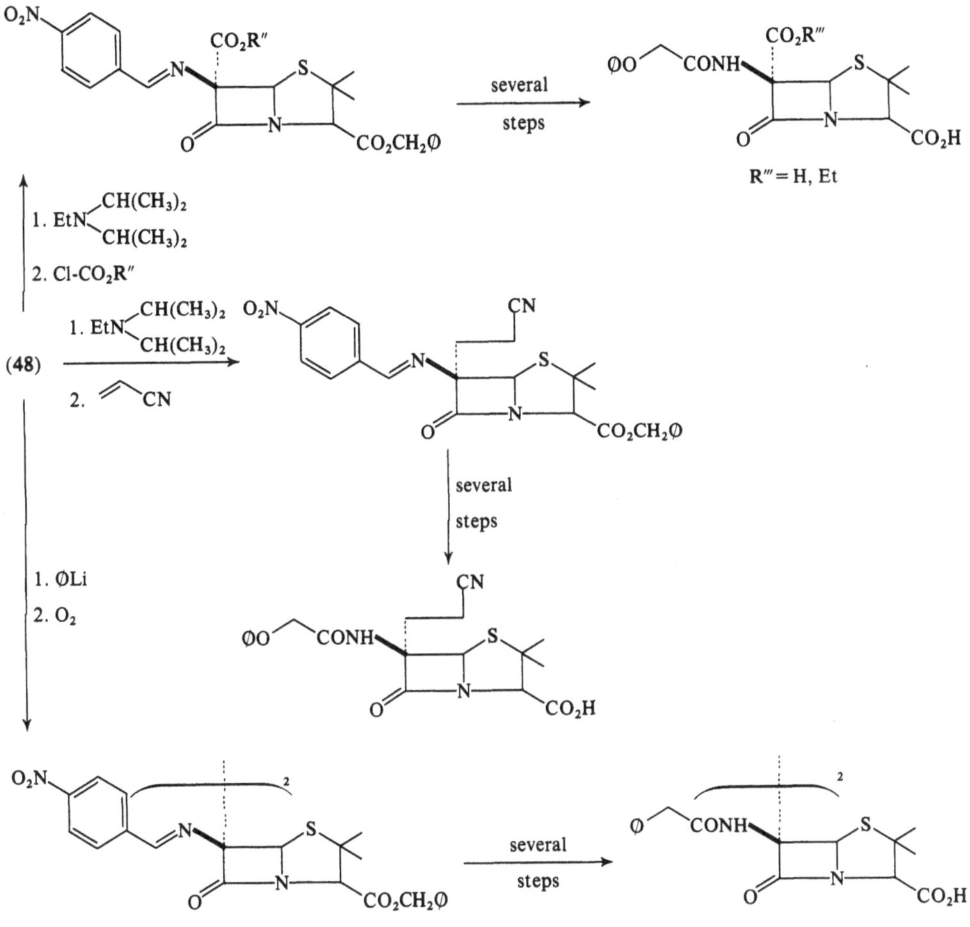

Chart 22. 6-substituted penicillins

A different approach is required for the preparation of 6α-methoxy-penicillin G (50) since this involves a nucleophilic substituent rather than an electrophilic one. *Chart 23* shows the preparation of this compound (*31*). In contrast to the cephalosporin series, (50) shows decreased antimicrobial activity compared to the parent penicillin G.

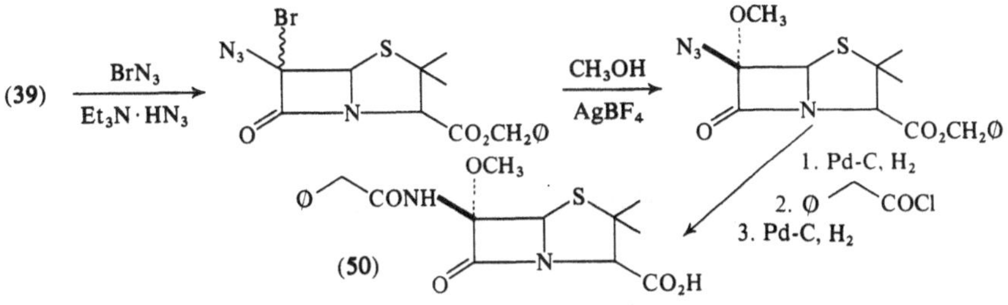

Chart 23. 6α-Methoxypenicillin G

2*

G. Cleavage of the C-7—N-4 Bond

Perhaps the most characteristic reaction of the penicillin molecule is the opening of the β-lactam ring by attack on the carbonyl group by some nucleophile. Penicillins are acylating (penicilloylating) agents with a reactivity roughly comparable to a carboxylic acid anhydride, a fact which accounts for many of the difficulties in their isolation and manipulation, and which also provides the probable mechanism for their ability to inhibit the transpeptidase enzyme. This facet of penicillin chemistry has been extensively studied, both during the early chemical studies (36) and more recently. Of particular interest has been the stability of penicillin antibiotics in solutions for therapeutic use, and the ability of penicillins to form irreversible protein conjugates by penicilloylation of free protein functional groups, thereby providing a mechanism for their allergenicity. This review will not attempt to cover this entire topic, but will make reference to certain selected areas of recent interest.

An example of the products obtained upon degradation of a penicillin under acidic conditions is shown in *Chart 24* (*102*). Thus, methacillin (51), by way of a penicillenic acid intermediate (52), affords a penicilloic acid

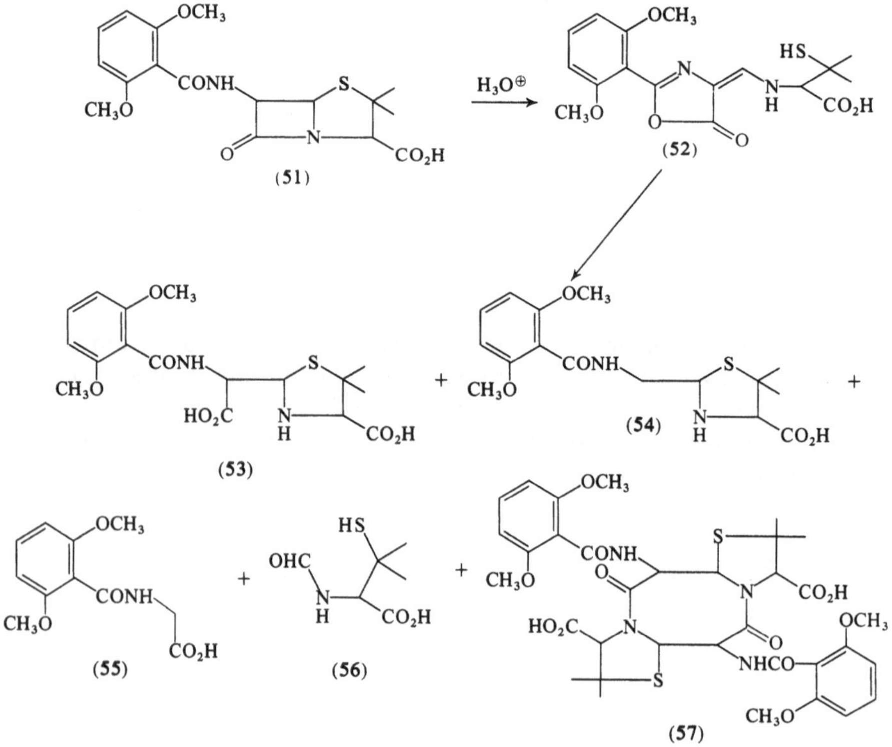

Chart 24. Degradation of methicillin

(53), a penilloic acid (54), a hippuric acid (55), N-formyl-D-penicillamine (56), and a unique dimer (57). The role of penicillenic acids as intermediates in penicillin β-lactam cleavages has been extensively studied (see, for example, *28, 160, 27, 30*). The reactions of penicillenic acids themselves have also been studied. For example, benzylpenicillenic acid (58) reacts with thiols much more rapidly than with amines to give mixtures of benzylpenicilloic acid thioester epimers (*184*). The products of the decomposition of benzylpenicillenic acid (58) as a function of pH have also been studied (*Chart 25*) (*127*). Thus, the penamaldic acid (59) and penillic

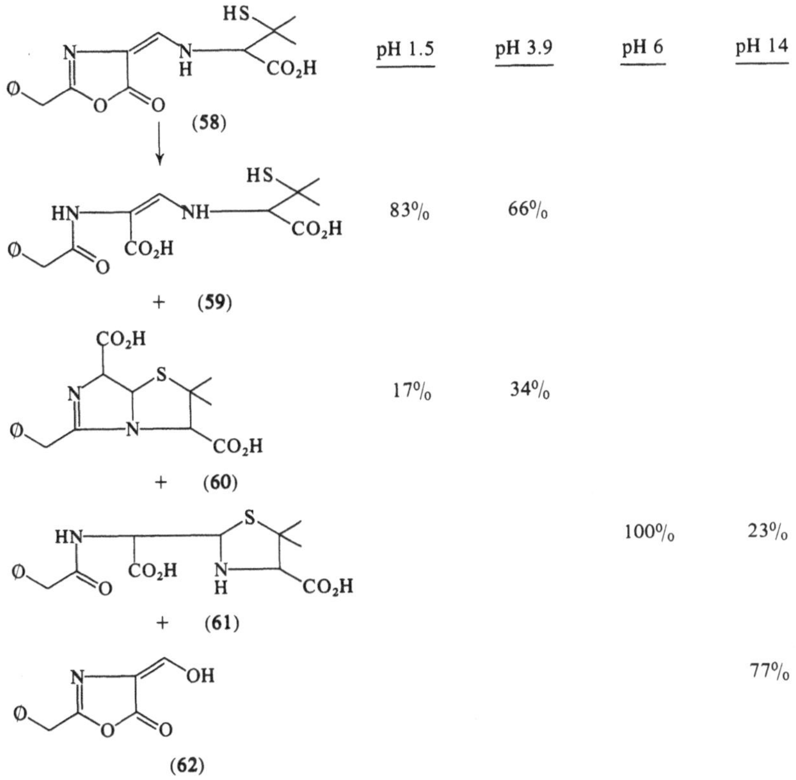

	pH 1.5	pH 3.9	pH 6	pH 14
(59)	83%	66%		
(60)	17%	34%		
(61)			100%	23%
(62)				77%

Chart 25. Decomposition of benzylpenicillenic acid

acid (60) are favored under acidic conditions, while basic conditions favor the penicilloic acid (61) and the oxazolone (62).

Because of the implications with respect to the formation of penicilloyl-specific antigens, the reaction of penicillins with amines, amino acids, and peptides to form N-penicilloyl derivatives has been studied extensively (for examples, see *158, 185, 163, 123*). A special aspect of this is the reaction of a penicillanic acid derivative with itself to form a polymer. Thus, the polymerizations of 6-APA (*77, 54, 169, 170*), ampicillin (**7**) (*29, 169, 170*), and others (*169, 170*) have been studied. The structures suggested (*170*) for several of these polymers are shown in *Chart 26*. The roles of these polymers in penicillin allergies has been discussed (*186*).

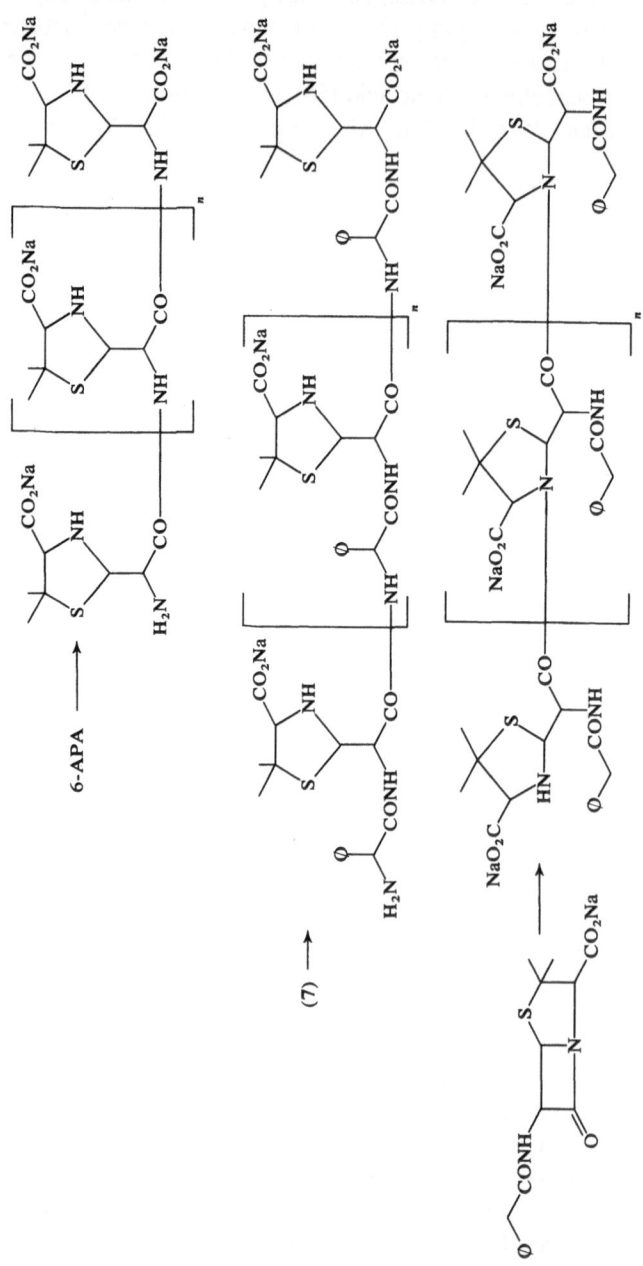

Chart 26. Polymers of penicillanic acid derivatives

The catalysis of β-lactam hydrolysis which leads to loss of antimicrobial activity has also been of interest. Among others, the influence of Cu (II) (*50*), Cu(II)-glycine chelate (*83*), ionic strength (*124*) and buffer salts (*63*) on the rate of hydrolysis has been investigated. A special effect of an inorganic salt, the reaction of 6-APA with sodium bicarbonate and carbon dioxide to form the keto form of 8-hydroxypenillic acid (**63**) (*99, 17*), is shown in *Chart 27*. The most widely studied catalysts of β-lactam hydro-

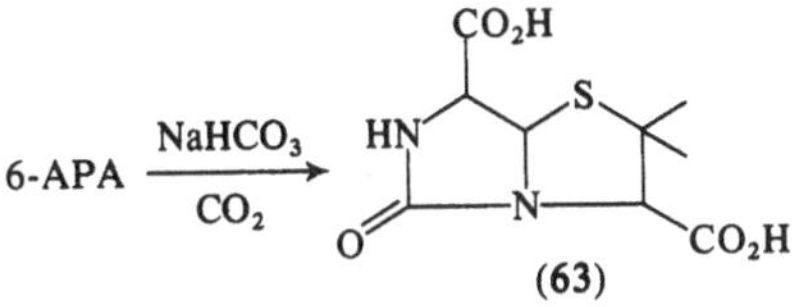

$$6\text{-APA} \xrightarrow[\text{CO}_2]{\text{NaHCO}_3} \text{(63)}$$

Chart 27. Reaction of 6-APA with carbon dioxide

lysis, however, are the β-lactamase enzymes which are produced by various microorganisms. These studies are beyond the scope of this review. Of interest, however, is the discovery (*161, 107, 108*) of several small molecules such as (**64**) which show some of the characteristics of a β-lactamase

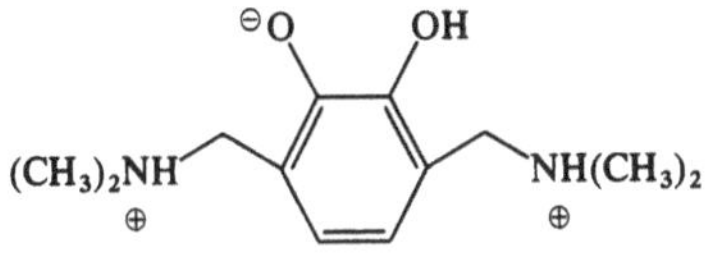

(**64**)

(catalysis of hydrolysis, bell-shaped pH-rate profile). An even closer mimic to a β-lactamase is cycloheptaamylose (BCD, β-cyclodextrin, cyclic 1,4-linked **D**-glucose heptamer), which not only catalyzes β-lactam hydrolysis, but forms a catalyst-substrate complex as a first step, and shows binding specificity based on the penicillin side chain (*181, 182*).

IV. Reactions at the Thiazolidine Ring

For the most part, the reactions to be dealt with in this section will be those involving the thiazolidine ring or a substituent on the thiazolidine ring, while keeping the β-lactam intact. As will be seen, these transformations have provided powerful routes to both known and novel fused ring systems related to penicillanic acid.

A. Rearrangements of Penicillin Sulfoxides

The synthesis, properties, and rearrangement reactions of penicillin sulfoxides have been intensively studied in recent years, especially because of the potentially useful conversion of the penicillanic acid ring system into the cephalosporin ring system (for earlier reviews of this topic, see *14, 46, 42*). Methods for the preparation of penicillin sulfoxides and sulfones have been known for some time (*36, 34, 78, 60*). It has now been shown that, with most oxidizing agents such as peracids, hydrogen peroxide, sodium metaperiodate, N-bromosuccinimide and 1-chlorobenzotriazole, the principal product is the (*S*)-sulfoxide [*e.g.*, (**65**)] (*44, 4, 9*) (*Chart 28*). Nmr and X-ray studies (*44, 5, 4*) have indicated that the (*S*)-

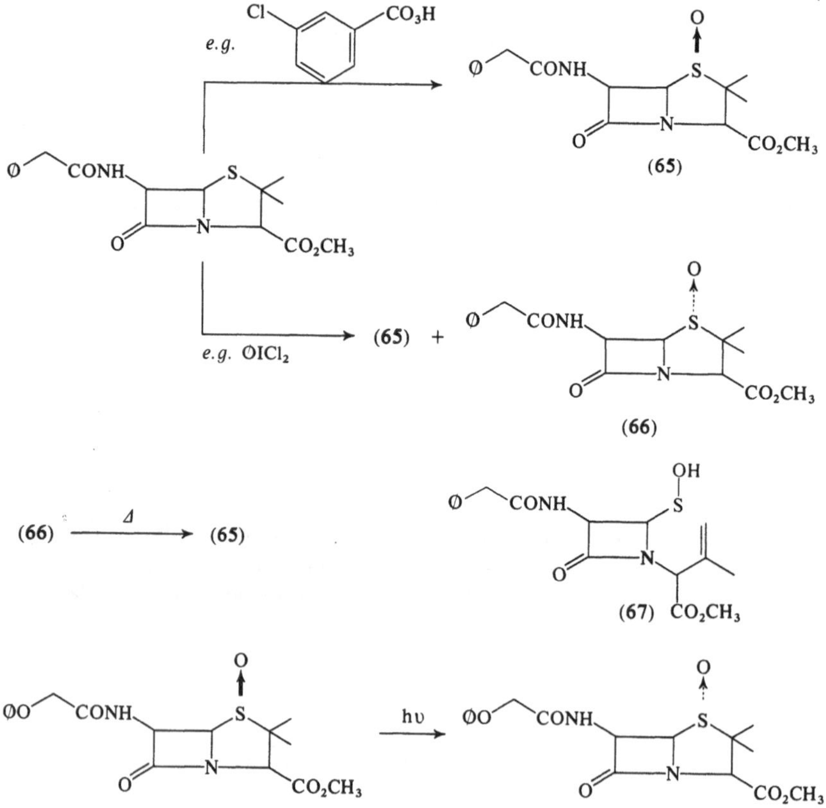

Chart 28. Penicillin sulfoxide preparations and interconversions

sulfoxide adopts a conformation consistent with hydrogen bonding to the 6-NH (a conformation quite different from that of the sulfide). A few oxidants, such as ∅ICl₂ (*9, 10*) and O₃ in H₂O-acetone (*173, 174*), afford a mixture of the (S)- and (R)-sulfoxides [e.g., (65) and (66)], the ratio depending upon the substrate and the reaction conditions. A (R)-sulfoxide such as (66) which has a 6-NH can be completely epimerized to the (S)-sulfoxide with heat, indicating that the latter is thermodynamically more stable. When this epimerization is carried out in the presence of exchangeable deuterium (8, 9), the resultant (S)-sulfoxide contains deuterium in the 2β-methyl group (under these relatively mild conditions, the (S)-sulfoxide does not incorporate deuterium—see 41). To account for these observations, a sulfenic acid intermediate (67) for the epimerization has been proposed. The (S)-sulfoxide can be converted to the (R)-sulfoxide by irradiation with UV light (5). The importance of the 6-NH is illustrated by the oxidation of methyl 6α- and 6β-phthalimidopenicillanate (Chart 29) (45, see also 10). The 6β-epimer (17) affords only the (R)-sulfoxide (68) while the 6α-epimer (18) gives both (R)- and (S)-sulfoxides (70) and (69). The stereoselectivity of these oxidation reactions

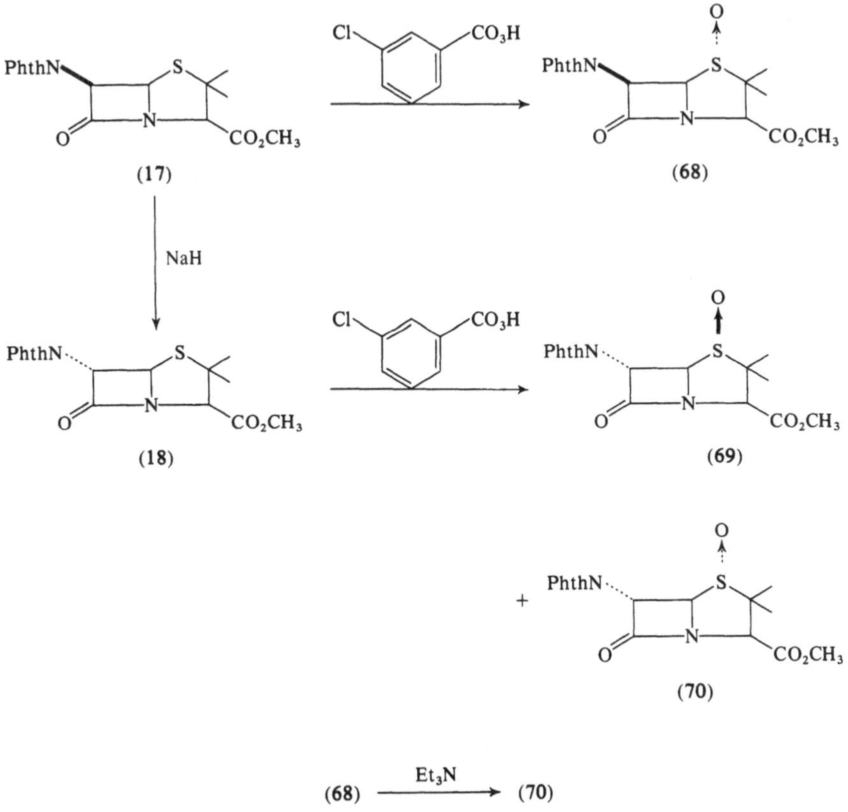

Chart 29. Sulfoxides of methyl 6-phthalimidopenicillanates

has been explained in terms of the 6-NH exerting a strong directing effect on the approaching oxidant and a strong stabilizing effect on the (*S*)-sulfoxide by hydrogen bonding.

MORIN and coworkers (*136, 137*) have examined the rearrangement of penicillin sulfoxides under several sets of conditions. In refluxing acetic

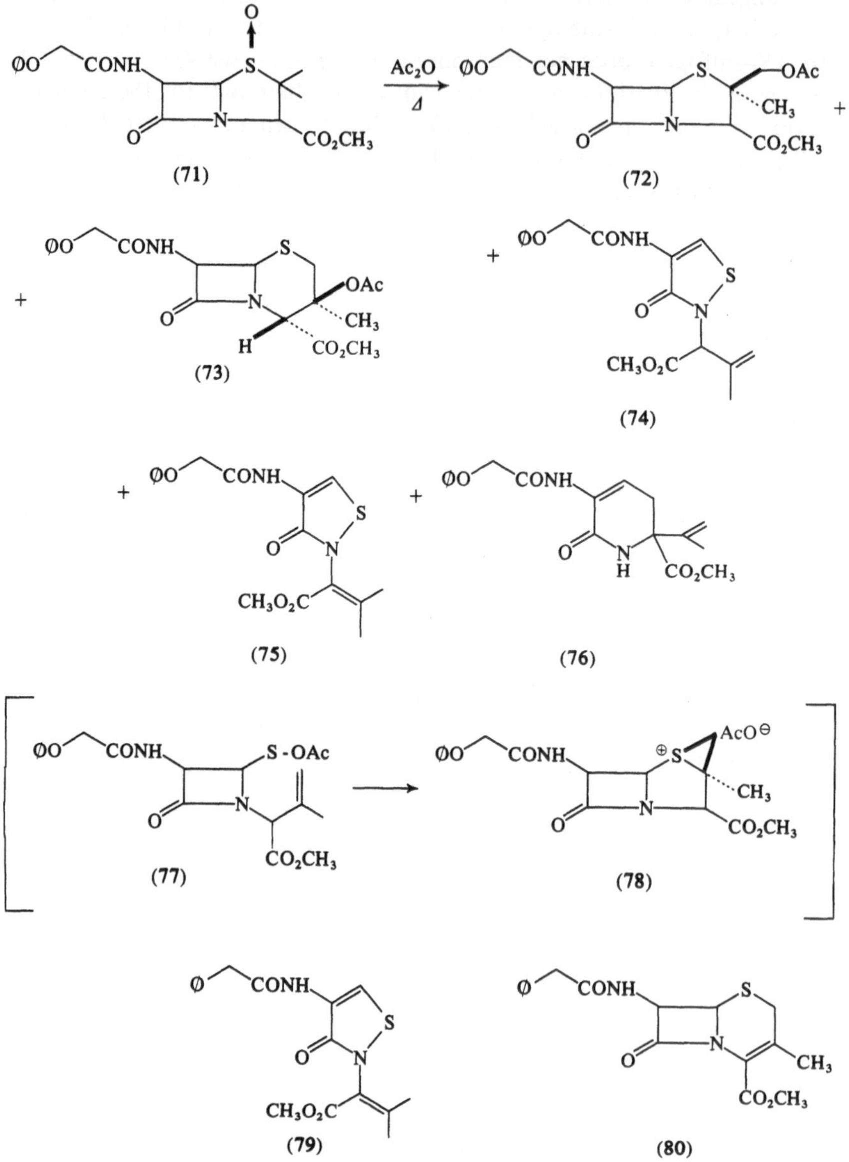

Chart 30. Rearrangement of penicillin ester sulfoxides with acetic anhydride

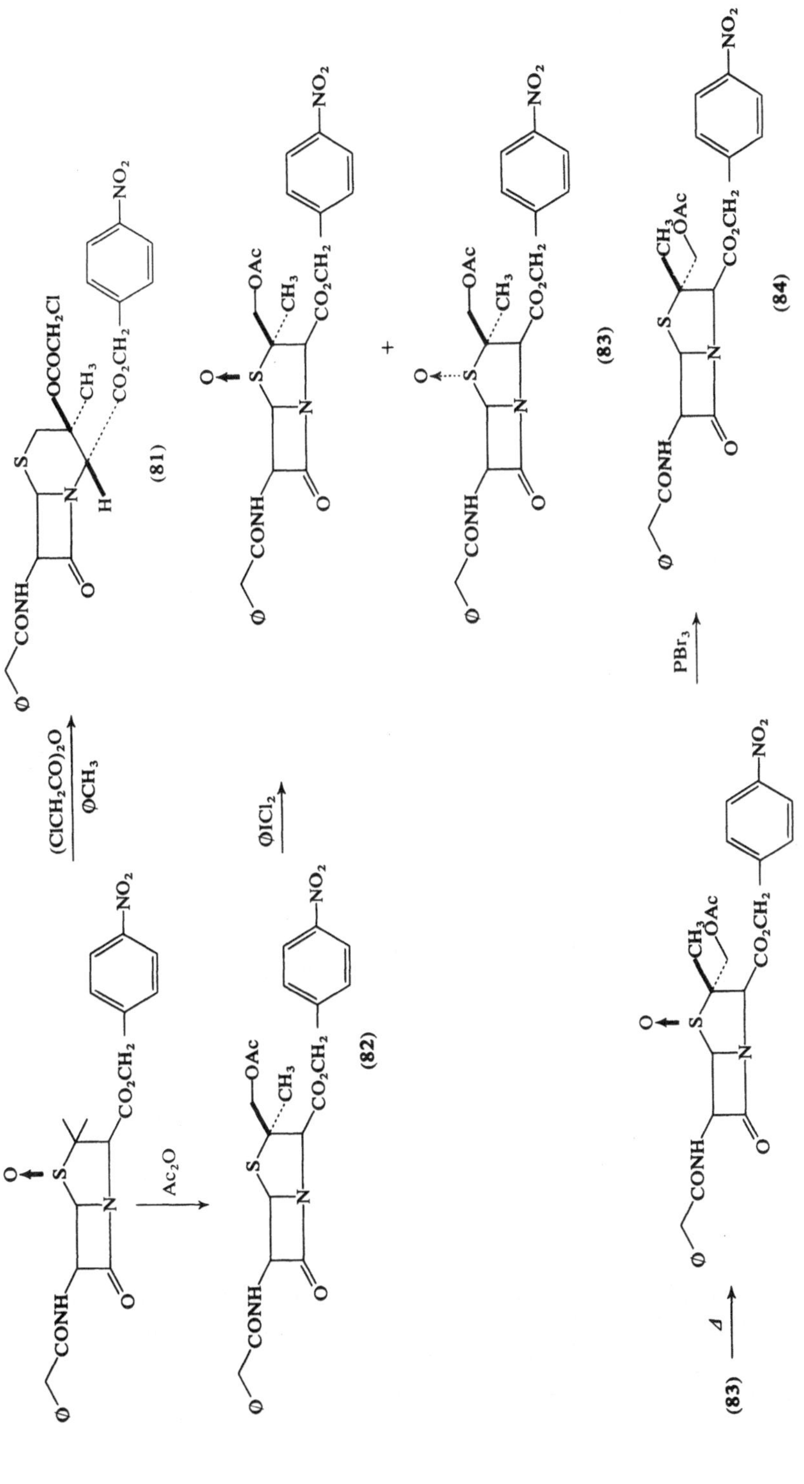

Chart 31. Transformations of penicillin ester sulfoxides

anhydride, (71) rearranged to a mixture of products, the major components of which are the penam (72) followed by the cepham (73) (the term cepham refers to the 8-oxo-5-thia-1-azabicyclo[4.2.0]octane ring system; the term cephem refers to the 8-oxo-5-thia-1-azabicyclo[4.2.0]oct-2-ene ring system) (*Chart 30*). The exact product ratios are a function of the particular reaction conditions. A similar rearrangement was observed with the (*S*)-sulfoxide of 2,2,2-trichloroethyl 6α-phenoxyacetamidopenicillanate (79). For the β-lactam-containing products, it was postulated that the thiazolidine ring opened to form a sulfenic acid, which then reacted with acetic anhydride to form the mixed sulfenic-acetic anhydride (77). Addition of this mixed anhydride to the double bond formed the episulfonium ion (78), from which either (72) or (73) was obtained depending upon the point of acetate attack. The non-β-lactam-containing products were thought to arise from (77), and it is interesting to note that (71), on treatment with refluxing pyridine, gives (75) and (76).

BARTON and coworkers (*9*, see also *168*), working with methyl 6-phenylacetamidopenicillanate-1(*S*)-oxide, obtained essentially the same results, except that no (79) and a small amount of (80) were obtained. In addition, these workers found that the incorporation of NaOAc into the rearrangement medium increased the amounts of (79) and (80) obtained. When chloroacetic anhydride was used instead of acetic anhydride, the only product isolated was the cepham (81) (*Chart 31*) (*cf. Chart 30*). Apparently, therefore, the direction of opening of the episulfonium ion intermediate is dependent upon the nature of the nucleophile. As indicated in *Charts 30* and *31*, the sulfoxide rearrangement yields the 2β-acetoxy-methylpenam [(72) and (82)]. By taking advantage of the facile epimerization of (*R*)-sulfoxides (*Chart 28*), which involves a simultaneous epimerization of the C-2 position, the stereoselective preparation of the 2α-acetoxymethylpenam (84) shown in *Chart 31* becomes possible (see also *12*). The free acids obtained from both (82) and (84) were microbiologically active.

The sulfoxide rearrangement may also be brought about by heating in the presence of an acid catalyst (*136, 137*). Under these conditions, cephems [*e.g.*, (85)] and cephams [*e.g.*, (86)] are the principal products (*Chart 32*). It was found (*80*) that sulfuric acid, sulfate esters, bisulfate salts, and sulfamic acid favored (86) while sulfonic acids favored (85). It appears that (85) and (86) are formed by separate pathways, and that neither one is an intermediate for the other.

Thionyl chloride may also be employed to bring about the sulfoxide rearrangement (*Chart 33*) (*121*). Note that the cepham product (87) can undergo ring contraction to a penam system (88). An episulfonium ion intermediate, analogous to (78), was postulated.

References, pp. 53—62

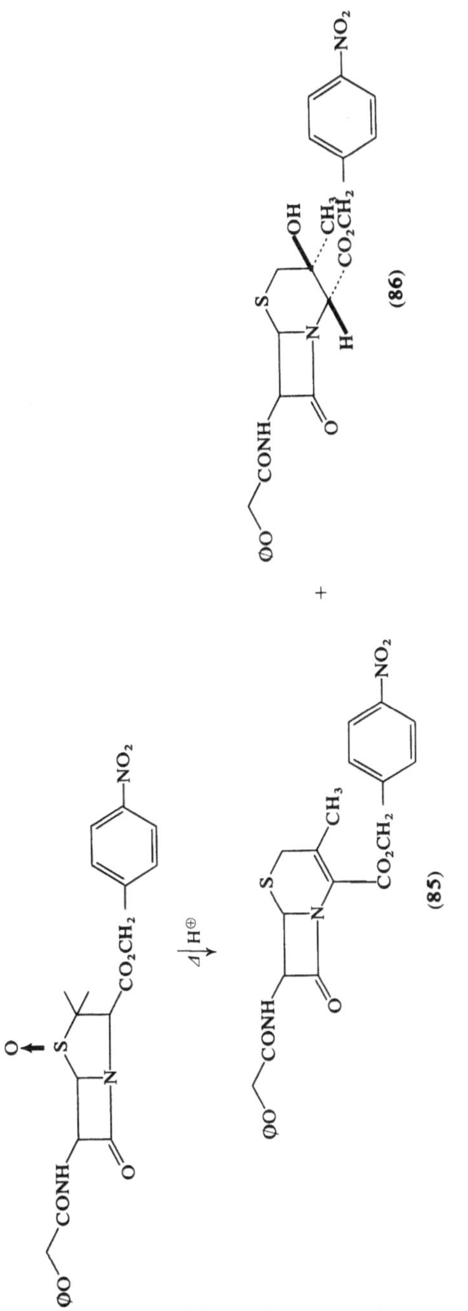

Chart 32. Acid-catalyzed rearrangement of penicillin ester sulfoxides

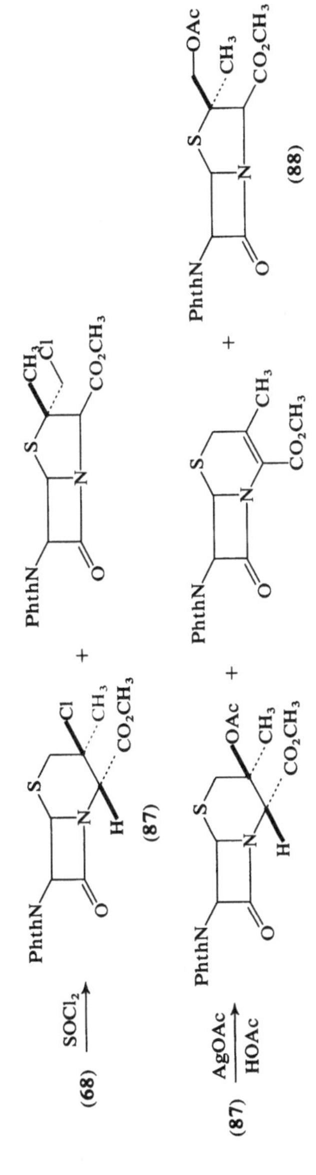

Chart 33. Sulfoxide rearrangement with thionyl chloride

A catalyst of a different sort for effecting the sulfoxide rearrangement is illustrated in *Chart 34 (180)*.

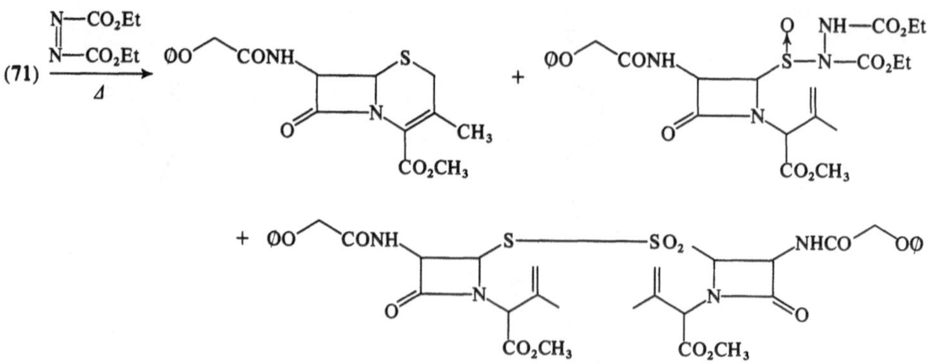

Chart 34. Sulfoxide rearrangement catalyzed by azo-compounds

Two different approaches to the application of the sulfoxide rearrangement for the preparation of the medically important antibiotic cephalexin (89) are shown in Chart 35 (33, 75). It is interesting that the

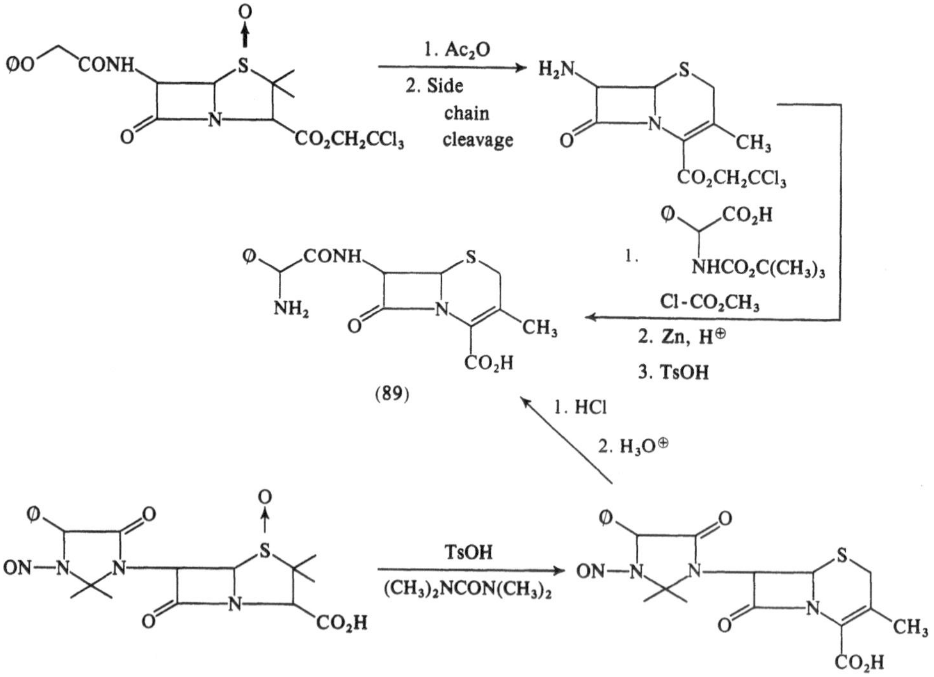

Chart 35. Cephalexin via sulfoxide rearrangement

hetacillin sulfoxide route (75) allows the use of an unblocked carboxyl. Previous attempts to rearrange penicillin sulfoxides as the free acids with acetic anhydride led principally to decarboxylated products (137). Sulfuric acid catalyst, however, does not lead to extensive decarboxylation (46).

Naturally-occurring cephalosporins are substituted at the 3-position of the cephem nucleus with an acetoxymethyl function, not the methyl group found in cephems obtained by penicillin sulfoxide rearrangement. *Chart 36* shows two examples of double sulfoxide rearrangements which

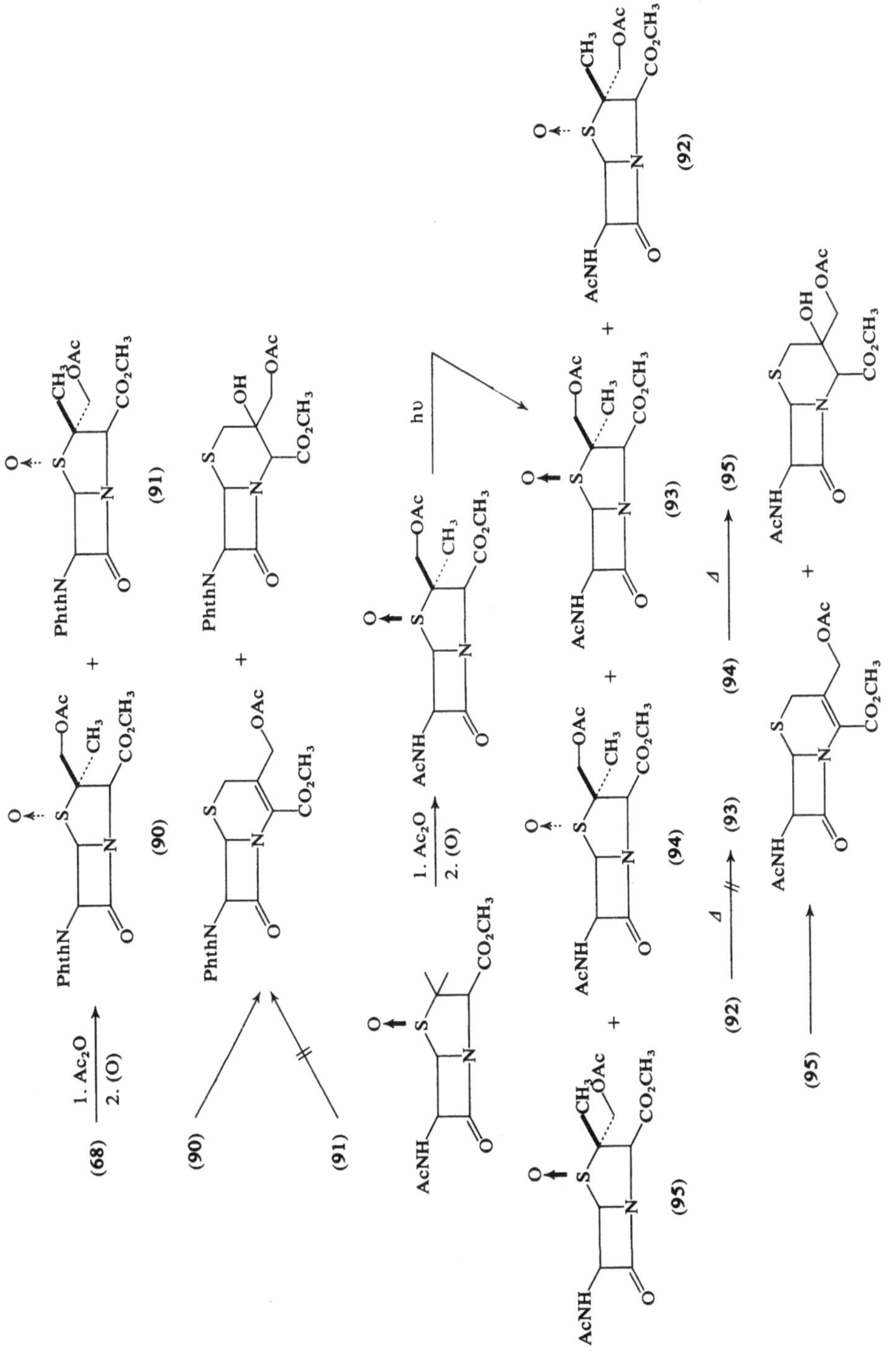

Chart 36. Double sulfoxide rearrangements

afford this type of substitution (*173*). Note that: a) The (*R*)-sulfoxide of methyl 6-phthalimidopenicillanate (**68**), on rearrangement, affords a mixture of 2α- and 2β-acetoxymethylpenams [(**90**) and (**91**)] rather than exclusively 2α (**91**) (see *46* for a rationalization); b) For successful re-arrangement or sulfoxide epimerization, the sulfoxide must be *cis* to an unsubstituted methyl group.

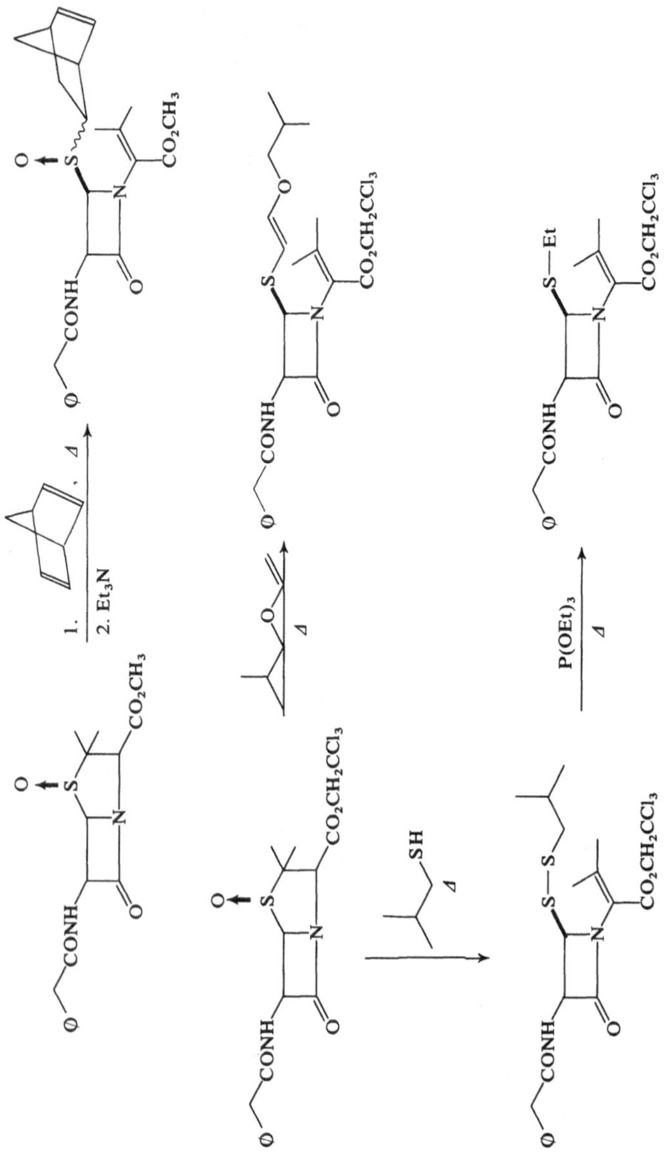

Chart 37. Reactions of sulfenic acid intermediates with olefins and mercaptans

Further evidence for the intermediacy of a sulfenic acid in the sulfoxide rearrangement is to be found in the successful trapping of the sulfenic acid by addition to olefins (*12, 3*), or by reaction with mercaptans (*15*). Some examples are shown in *Chart 37*. Earlier, *Chart 34* provided another example of a trapped sulfenic acid. Still other means of trapping the intermediate sulfenic acid are shown in *Chart 38 (47, 84)*. In these cases,

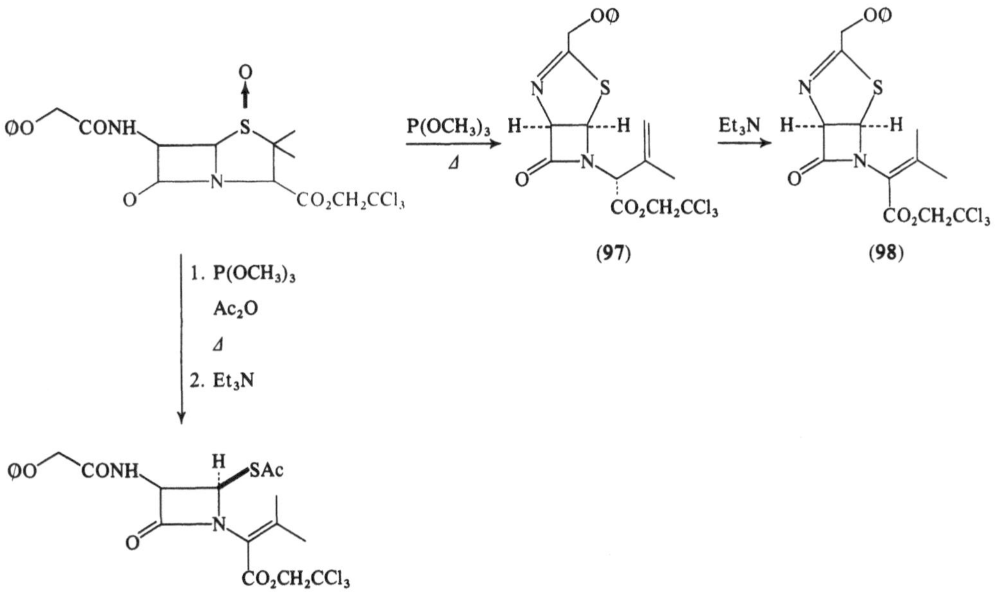

Chart 38. Rearrangement of penicillin sulfoxides with trimethyl phosphite

it has been postulated that the intermediate sulfenic acid is reduced by the phosphite to a mercaptan which can then either react with the side chain amide or be acetylated. It has been further postulated (*43*) that compounds such as (**97**) may be common biosynthetic precursors to both the penicillin and cephalosporin ring system. Support for this possibility was obtained by way of the transformations shown in *Chart 39*. Once again, a sulfenic acid intermediate was postulated.

While they are not strictly sulfoxide rearrangements, the reactions shown in *Chart 40 (141,* see also *200)* are related in that ylide intermediates [(**99**) and (**100**)] somewhat like a sulfoxide are probably involved. In these cases, however, a stable product rather than a labile sulfenic acid results from thiazolidine ring opening.

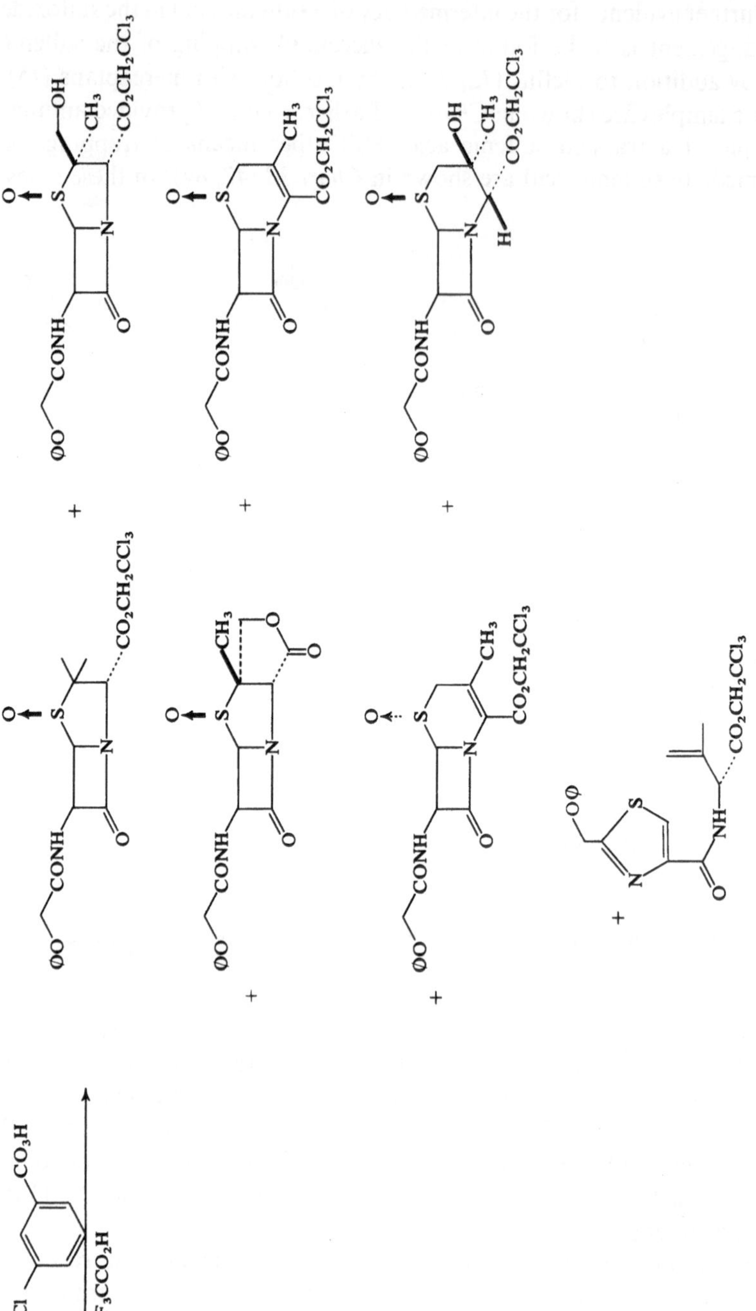

Chart 39. A common precursor for penams and cephams

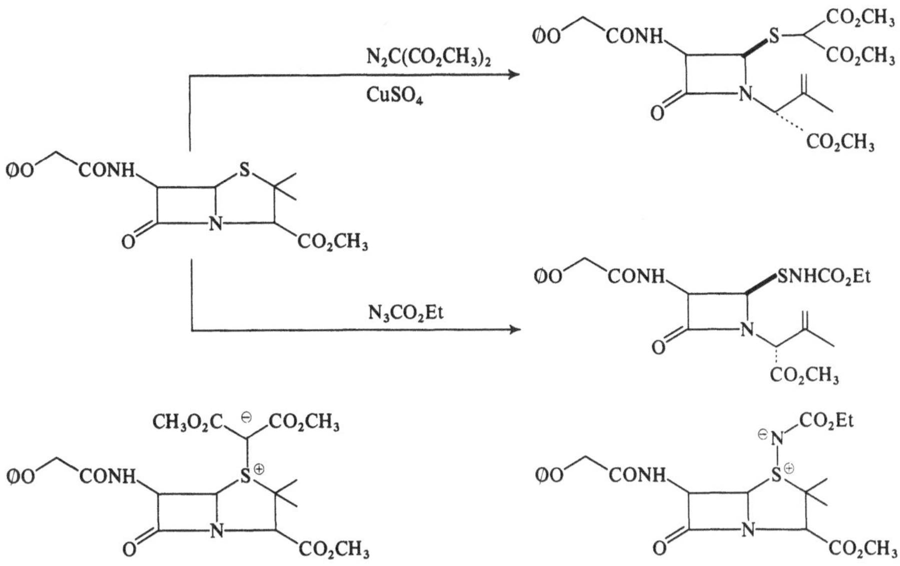

Chart 40. Penicillin rearrangements via diazo and azido compounds

B. Other Thiazolidine Ring Cleavages

The reactions described in the previous section were largely those involving opening of the thiazolidine ring by cleavage of the S-1—C-2 bond. In most cases, the ring-opened intermediates were unstable and reclosed to penam or cepham systems. The last part of the section, however, contained examples of ring-opened products that were stable and did not spontaneously undergo ring reclosure. This type of product is of particular interest because it opens up the possibility of removing the remnants of the original thiazolidine ring, then attaching new residues to the β-lactam which may allow ring closure to systems different from the penam and cepham systems otherwise available. Much of this section will be concerned with transformations of this sort.

One of the first problems encountered when this approach to novel condensed ring systems is applied to S-1—C-2 cleaved thiazolidine intermediates is the removal of the 1-carboxy-2-methyl-1-propenyl ester group which usually remains attached to the β-lactam nitrogen. Several of the methods which have been devised for accomplishing this are shown in Chart 41 (13, 15, 3, 48, 24).

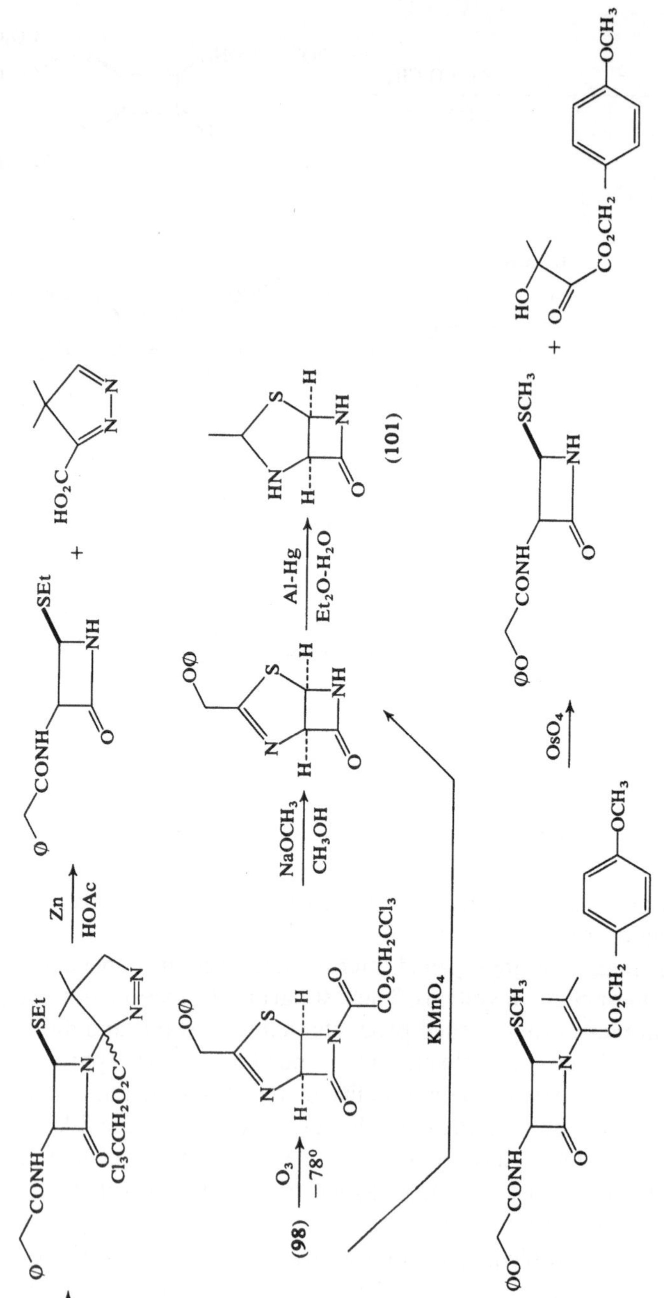

Chart 41. Cleavage of the 1-carboxy-2-methyl-1-propenyl ester residue

The application of this approach to the synthesis of a novel cephem derivative is illustrated in *Chart 42* (*39, 138, 139*). The first step of this

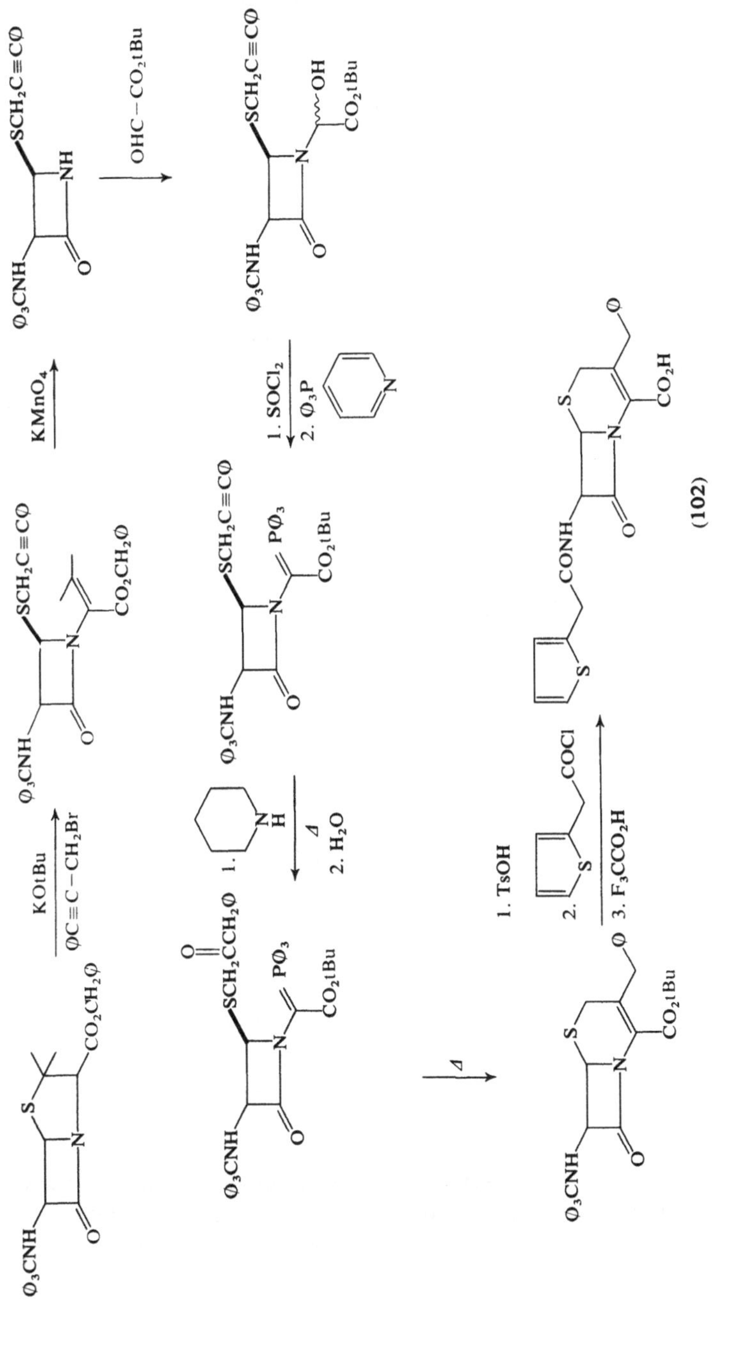

Chart 42. A novel cephem via thiazolidine modification

(102)

sequence, the opening of the thiazolidine ring with a strong anhydrous base and an alkylating agent, represents a novel and potentially very useful thiazolidine cleavage. Once the thiazolidine ring had been opened and the N-substituent removed, application of a synthetic sequence similar to one previously reported (157, 155, 156) permitted the attachment of a new ring component to the β-lactam nitrogen. Hydration of the triple bond followed by ring closure afforded the 3-benzylcephem ring system which, when converted to (102), showed antibacterial activity.

Compound (101), shown in *Chart 41*, also provides potential access to modified fused ring systems (48). Intermediates of this type were central in the total synthesis of cephalosporin C developed by Woodward and coworkers (199), and in the Ciba-Geigy synthesis of cephalosporin analogs to be discussed later.

Application of the modified Curtius reaction to penicillanic acid derivatives has been known for some time (166, 143) (*Chart 43*). It can be

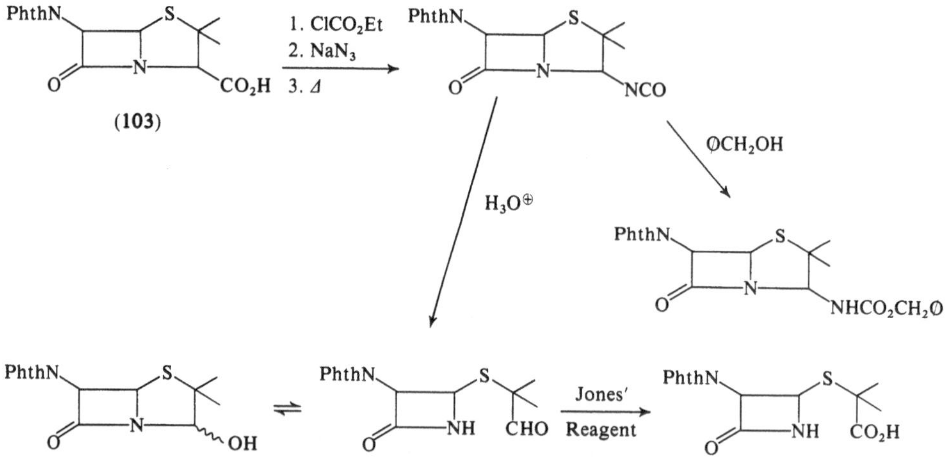

Chart 43. Curtius reaction on 6-phthalimidopenicillanic acid

seen that this sequence represents a method for cleaving the thiazolidine ring at the C-3—N-4 bond. The position of equilibrium between the ring-opened aldehyde and the ring-closed alcohol is a function of the 6-substituent, with bulky groups such as phthalimido favoring the ring-opened form and groups such as phenylacetamido favoring the ring-closed form (88).

The workers at Ciba-Geigy have taken good advantage of the Curtius ring-opening reaction in the preparation of a variety of cephem and related systems (for a summary of this work, see 87). Among many reactions of the 3-hydroxypenam system that were investigated (88). it was found that

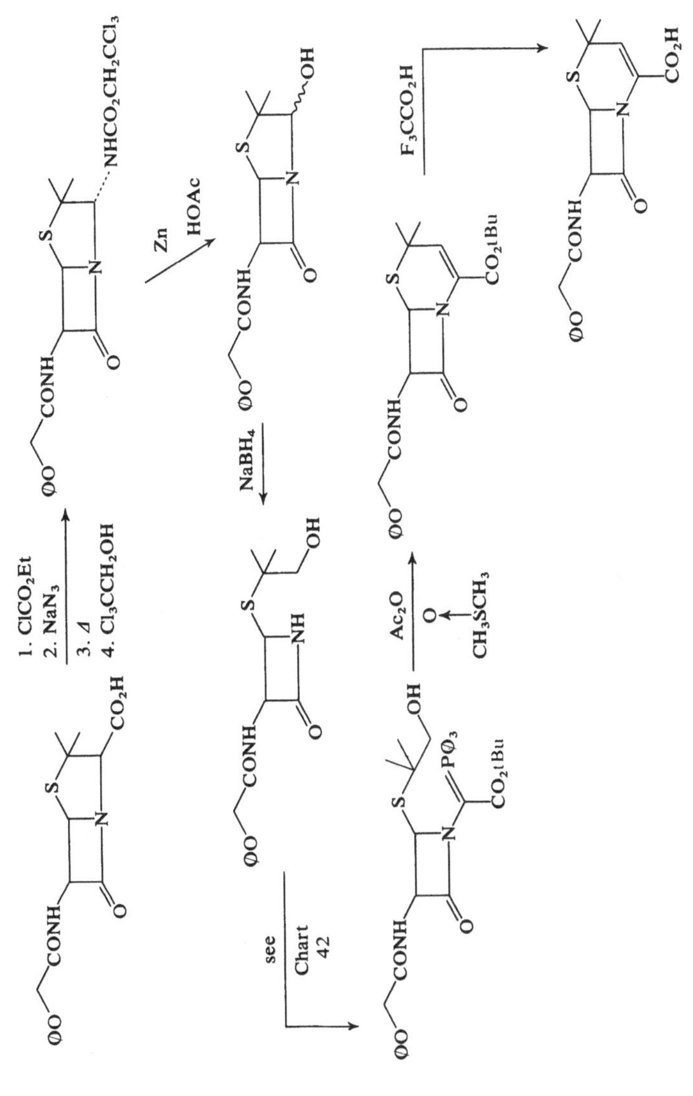

Chart 44. A novel cephem via the Curtius reaction

borohydride reduction afforded the alcohol with no N-substituent (*Chart 44*). The new N-substituent was introduced as shown in *Chart 42*. Oxidation of the alcohol to the aldehyde was followed by spontaneous ring closure, and deblocking of the acid afforded (**104**) (*157*), which had antibacterial activity.

The preparation of an intermediate, (**105**), which is more versatile for the synthesis of novel condensed ring systems is shown in *Chart 45* (*62*).

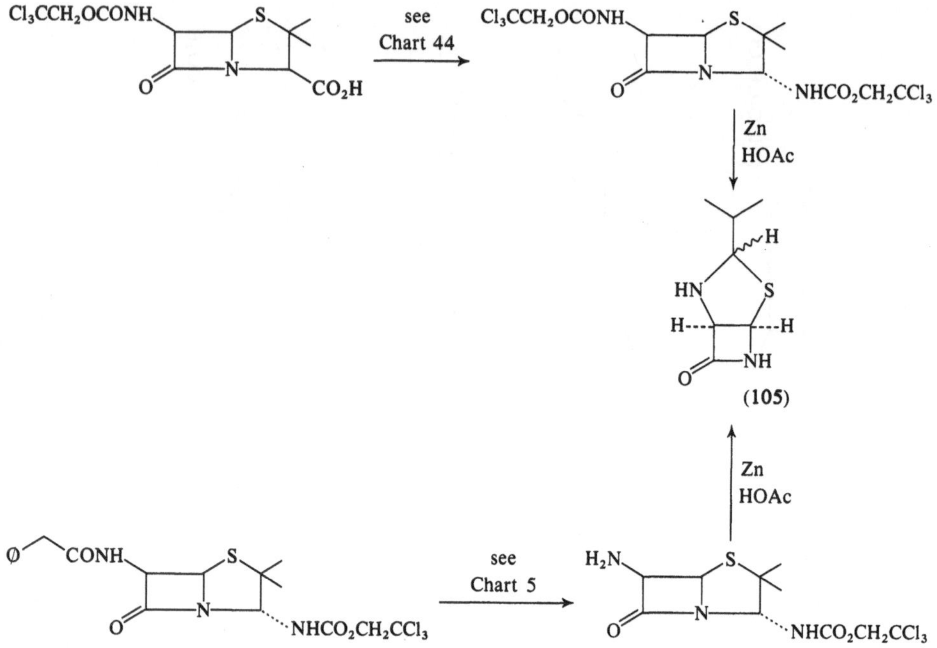

Chart 45. A versatile intermediate for cephems and related compounds

A related compound, **(101)**, has been prepared by an alternate route as described earlier (*Chart 41*). The use of this intermediate for the preparation of a new microbiologically active cephem derivative is shown in *Chart 46* (*155*). Note that, in this sequence, the only vestige of the original thiazolidine ring system retained is the sulfur and those atoms which are also part of the β-lactam, the overall effect being a cleavage at both the S-1—C-2 and C-3—N-4 bonds. Starting with a closely related intermediate, **(106)** (*88*), the synthesis of an antimicrobially inactive homocephem derivative, **(107)**, by a route similar to that shown in *Chart 46* has been described (*156*).

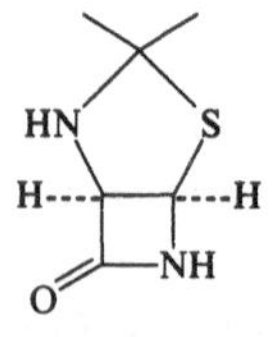

(106) 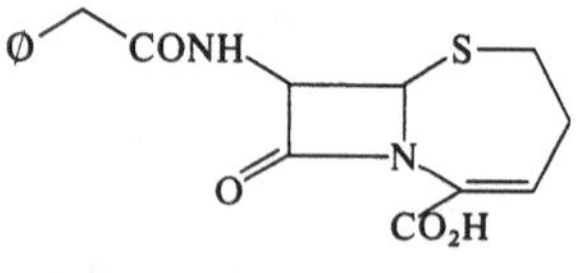   (107)

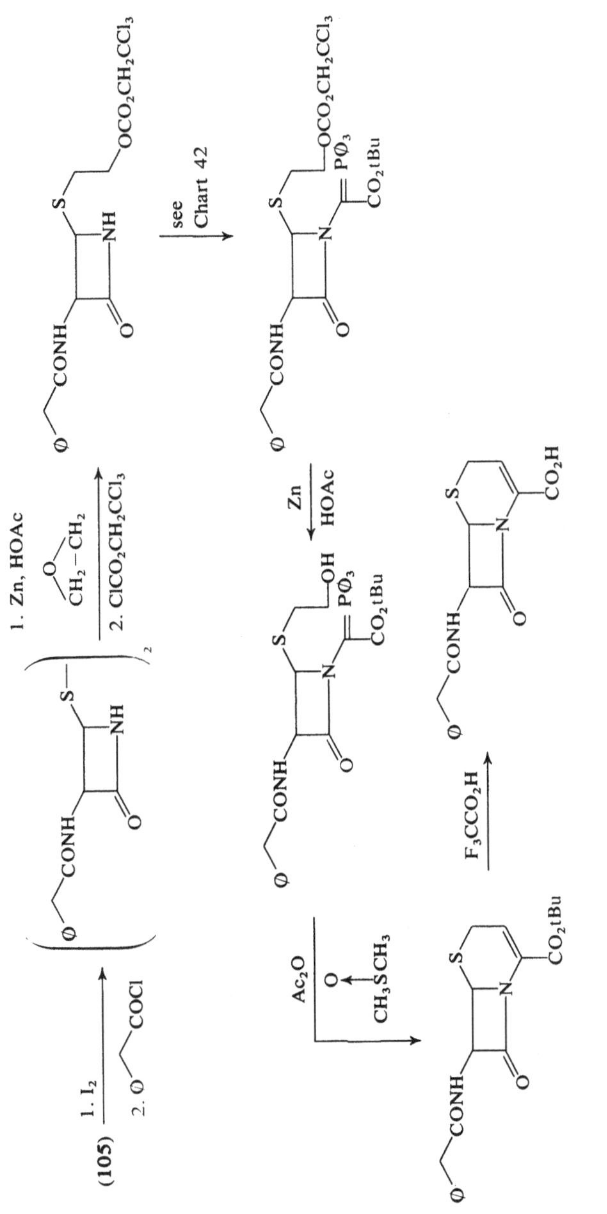

Chart 46. Synthesis of a novel cephem

Some years ago, SHEEHAN (*165*) observed that the treatment of methyl benzylpenicillanate with positive halogen afforded the oxazoline (**108**) (*Chart 47*). Compound (**108**) can also be obtained in small amounts with

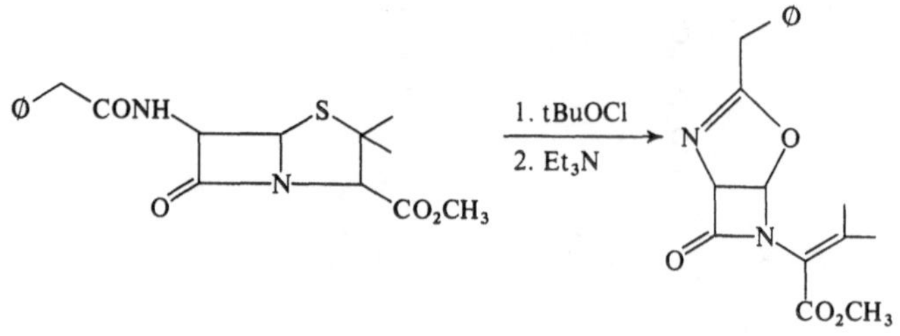

Chart 47. Treatment of methyl benzylpenicillanate with *t*-butyl hypochlorite

iodobenzene dichloride (*10, 9*). This transformation represents still another mode of thiazolidine ring opening: cleavage at both the C-5—S-1 and S-1—C-2 bonds. More recently, KUKOLJA and coworkers have examined the chlorination of penicillanic acid derivatives, and, by control of the

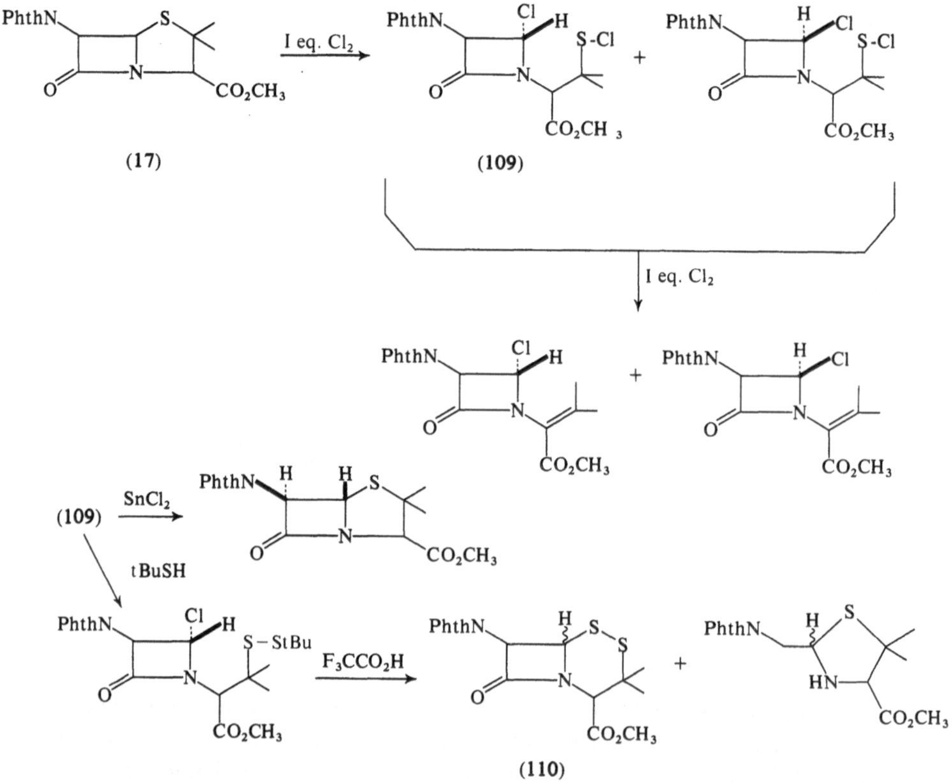

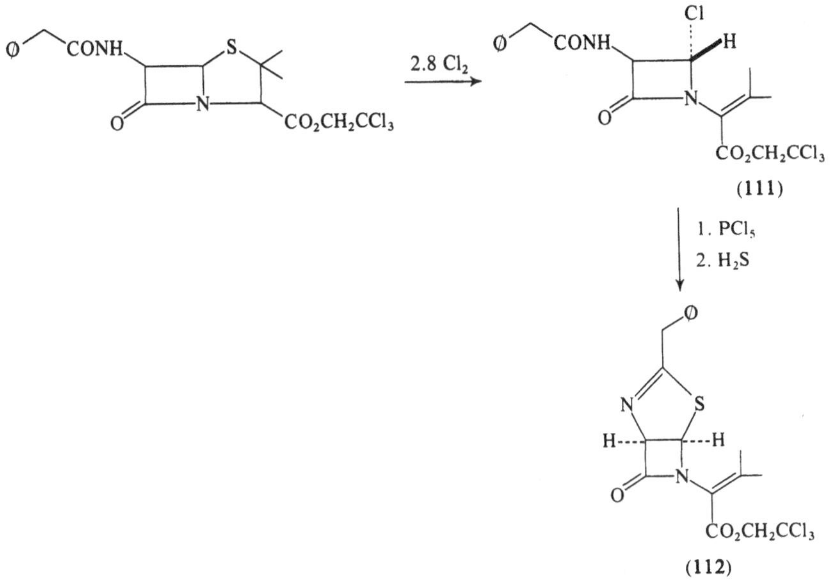

Chart 48. Chlorination of penicillanic acid derivatives and some subsequent reactions

conditions, have been able to separate the C-5—S-1 cleavage from the S-1—C-2 cleavage. Some of the reactions which they have observed are summarized in *Chart 48*. The chlorination (*114*) gives, initially, the mixture of epimeric dichlorides [predominantly (**109**)] (presumably by way of a C-5 carbonium ion) which, on treatment with additional chlorine, form the epimeric olefins. A similar sequence takes place with 1-chlorobenzo-triazole (*120*). Reduction of the sulfenyl halide (**109**) gives the mercaptan which undergoes ring closure with retention (*115*), while reaction with *t*-butyl mercaptan followed by removal of the *t*-butyl group affords the mixture of epimeric disulfides (**110**) (*116, 118*). Conversion of the amide (**111**) to the imino chloride followed by H_2S treatment yielded (**112**) (*119*) (*cf. Chart 38*). The chlorination of anhydropenicillins will be discussed shortly.

The rearrangement of penicillanic acid derivatives to anhydropeni-cillins represents another approach to the S-1—C-2 cleavage of the thiazo-lidine ring (*192, 193, 188*). When the carboxyl group of a penicillin is activated by conversion to an acid chloride or a mixed anhydride, and this derivative is treated with triethylamine, a rearrangement occurs [pre-sumably via the intermediate (**113**)] to an anhydropenicillin (**114**) (*Chart 49*). This rearrangement is successful with a variety of penicillin side chains.

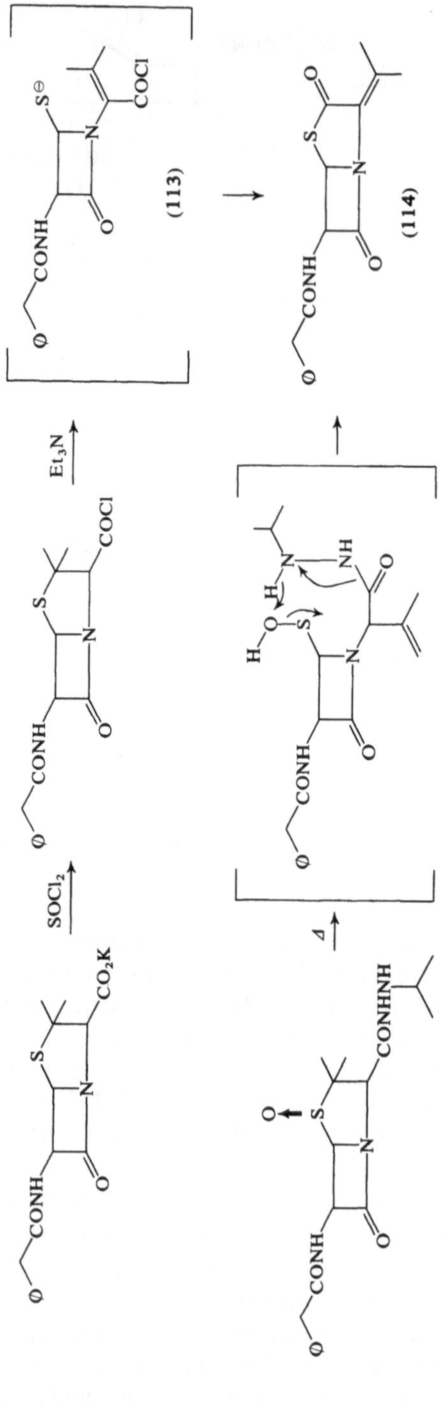

Chart 49. The anhydropenicillin rearrangement

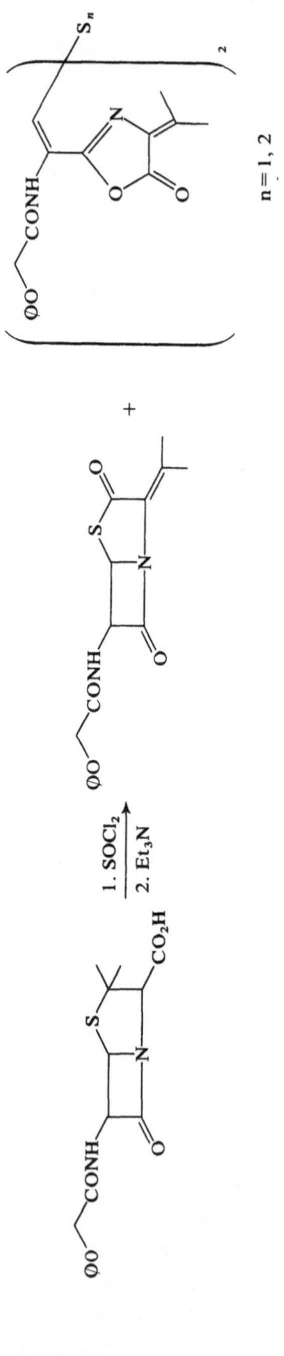

Chart 50. Byproducts of the anhydropenicillin rearrangement

Chart 49 also depicts the novel production of an anhydropenicillin from a penicillin sulfoxide hydrazide (*11*). In addition to an anhydropenicillin, two other products have been identified (*117*) from this rearrangement (*Chart 50*). Perhaps the most interesting property of the anhydropenicillins is their high stability to heat and hydroxylic solvents (in comparison to most penicillanic acid derivatives), despite the fact that the position of the β-lactam carbonyl absorption in the infrared and the lack of planarity of the β-lactam nitrogen in the X-ray crystal structure analysis (*167*) would lead one to expect a highly reactive species.

A variety of reactions of the anhydropenicillins has been investigated. Under carefully controlled conditions, the rearrangement is reversible (*189*) (*Chart 51*). Mercuric acetate oxidation of an anhydropenicillin gave

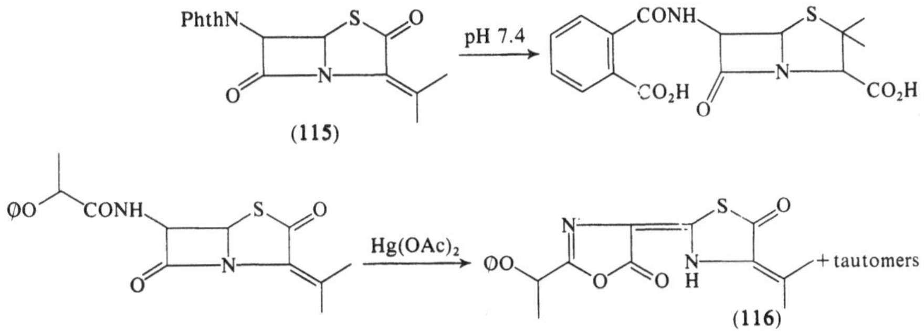

Chart 51. Reactions of anhydropenicillins

a mixture of tautomers of (**116**), termed an anhydropenicillene (*191*) (*Chart 51*). The chlorination of anhydropenicillins is another method for cleaving the thiazolidine ring at the C-5—S-1 and S-1—C-2 positions (*197*) (*Chart 52*) (*cf. Chart 48*). Some of the reactions which these chlorine-containing products undergo (*197, 196, 190*) are also shown in *Chart 52*. Chlorination gives, initially, a mixture of epimeric acid chlorides, which, on treatment with water or an alcohol, gives the acids or esters (*197*). The chlorination is applicable to various 6-substituted anhydropenicillins. Nucleophiles such as azide displace either of the epimeric chlorides with inversion. In contrast to anhydropenicillins themselves, the allylic methyl groups of, for example, (**117**) can be brominated to form the *cis-trans* mixture of monobromides (*196*) (two equivalents of N-bromosuccinimide brominate each methyl group once). The bromides may be displaced with nucleophiles such as azide ion, and the resulting azides can be reduced to the amines. Ring closure of one of the isomers affords (**119**) (*190*). The oxazoline (**108**) (*cf. Chart 47*) can be obtained by heating the

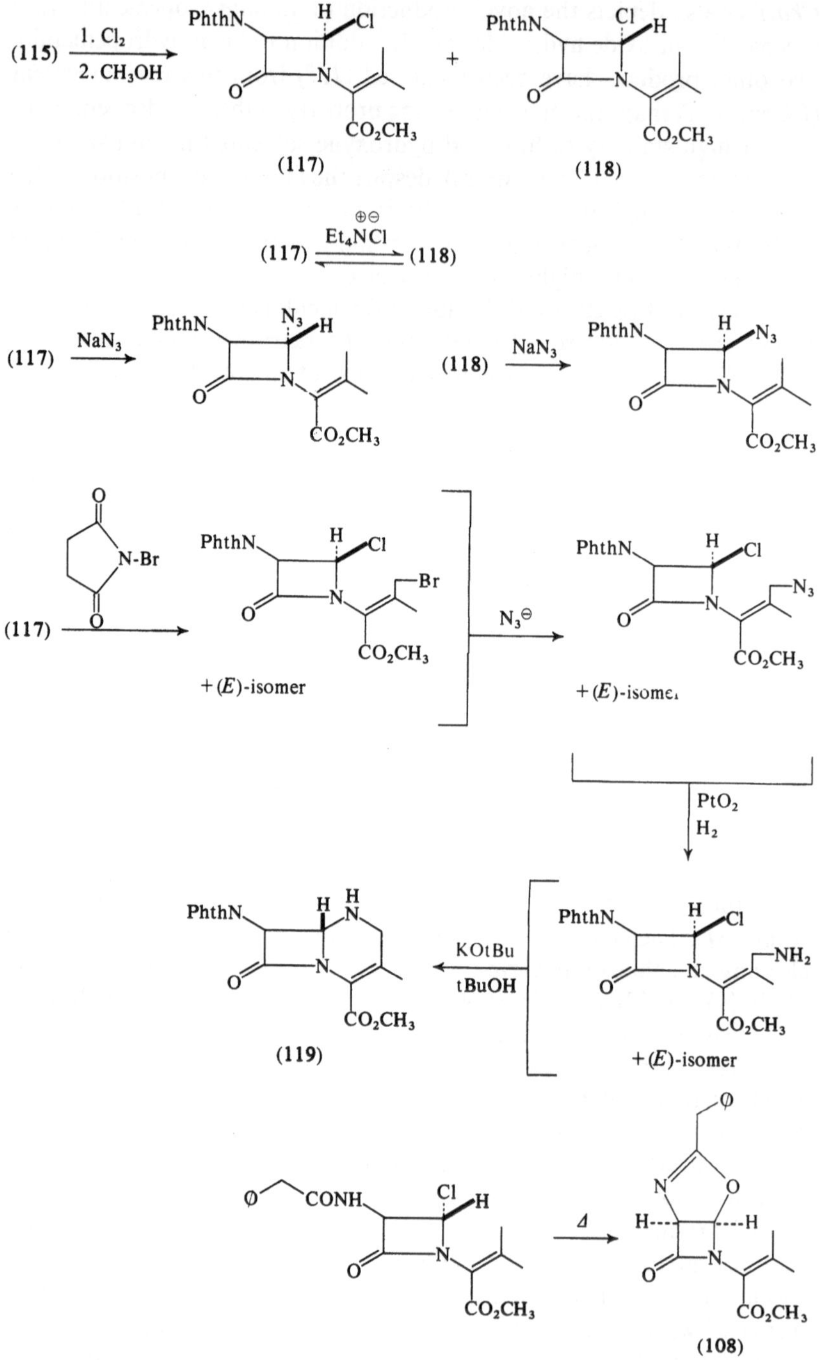

Chart 52. Chlorination and subsequent reactions of anhydropenicillins

appropriate amide-chloride (*190*). The potential which many of these products offer as intermediates for the preparation of novel condensed ring systems should be obvious.

Raney nickel desulfurization of penicillanic acid derivatives is one form of thiazolidine ring cleavage that has been known for some time (*36*). A recent study has shown that, when this reaction is carried out in the presence of D_2O, deuterium is incorporated into one of the *gem*-dimethyl groups (*Chart 53*) (*194*).

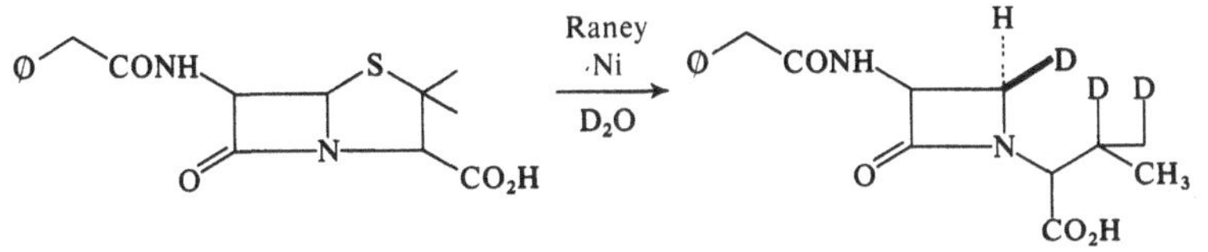

Chart 53. Raney nickel desulfurization of penicillin G

C. Miscellaneous Nuclear Transformations

This section will briefly consider some recently investigated transformations in which both the β-lactam and the thiazolidine ring are cleaved.

The iodometric penicillin determination is based upon the fact that penicilloic acids are susceptible to iodine oxidation while the intact penicillin is not. Recent studies (*69, 142, 70*) have permitted the mechanism shown in *Chart 54* to be postulated for this oxidation.

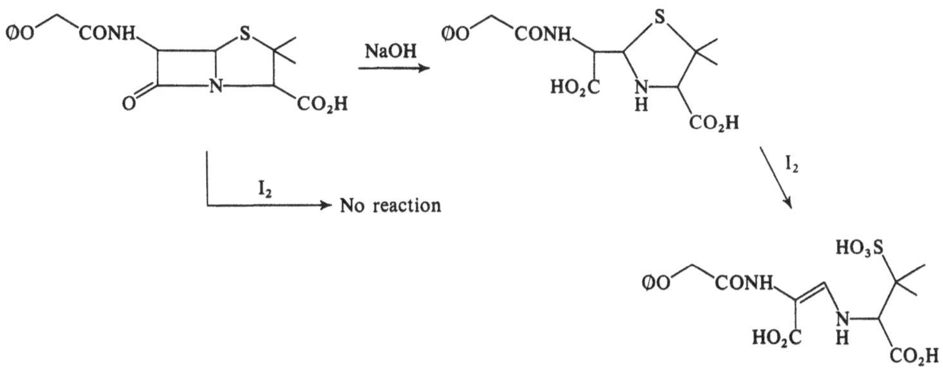

(120)

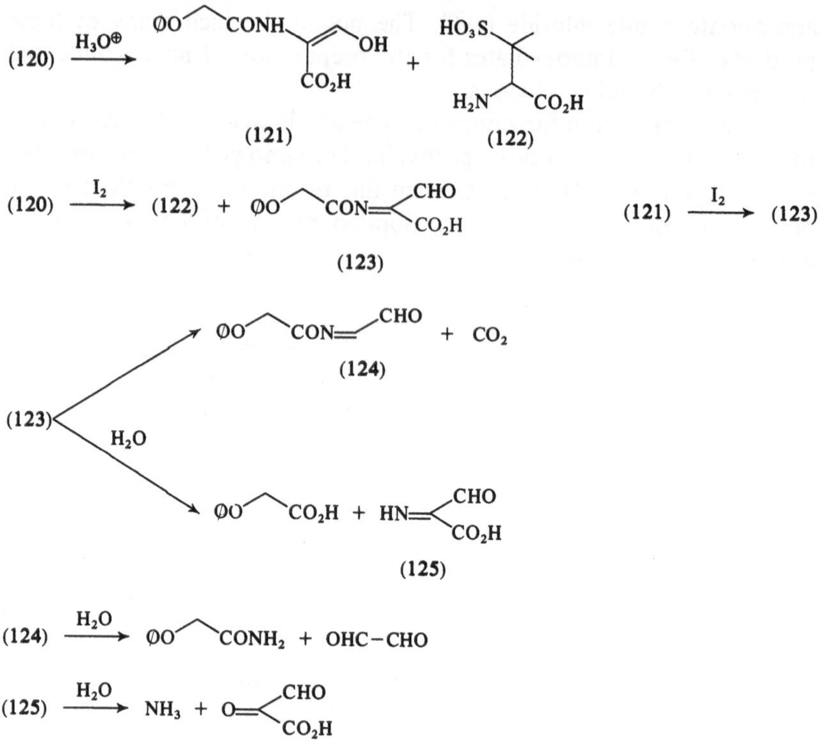

$$(120) \xrightarrow{H_3O^{\oplus}}$$

(121)   (122)

$$(120) \xrightarrow{I_2} (122) + \text{ØO—CON=} \overset{CHO}{\underset{CO_2H}{}}$$

(123)

$$(121) \xrightarrow{I_2} (123)$$

ØO—CON=⟨CHO / CO_2H + CO_2

(124)

(123)

H_2O

ØO—CO_2H + HN=⟨CHO / CO_2H

(125)

$$(124) \xrightarrow{H_2O} \text{ØO—CONH}_2 + OHC-CHO$$

$$(125) \xrightarrow{H_2O} NH_3 + O=\overset{CHO}{\underset{CO_2H}{}}$$

Chart 54. Iodometric penicillin determination

Chart 55 illustrates several diverse transformations which have been observed (*71, 117, 18, 19*).

D. Modification of the 3-Carboxylic Acid

The most thoroughly studied reaction of the 3-carboxylic acid, the Curtius reaction, has already been discussed in Section IV B. Simple esterification of the 3-carboxylic acid has been known for many years (*36*). Recently, however, a particular class of esters, the acyloxymethyl esters [*e.g.*, (**126**)], has aroused special interest because of their enhanced oral absorption over that of the unesterified antibiotic (*98, 51*). A variety of amides (*25, 201*) and hydroxamic acids (*49*) have been prepared. The

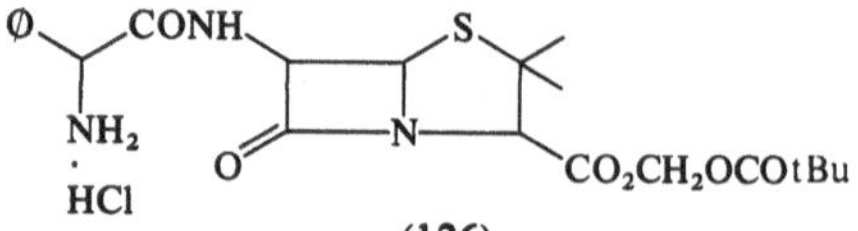

(126)

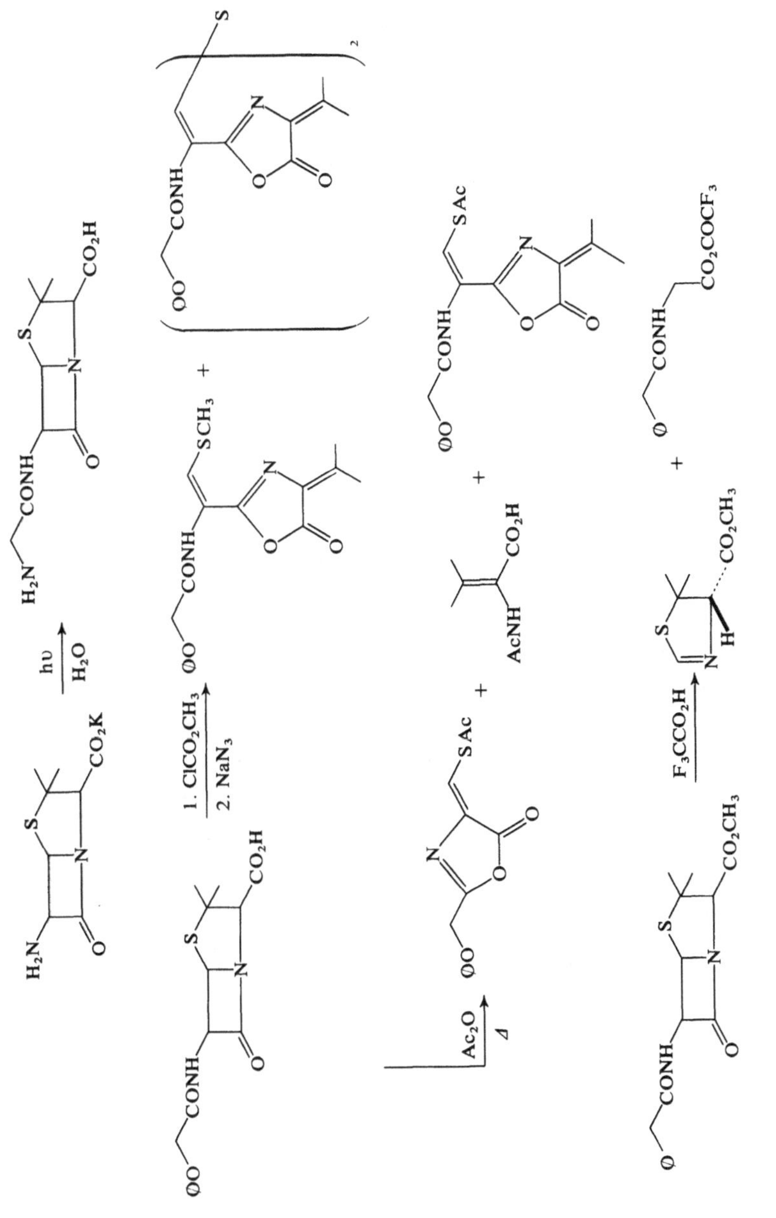

Chart 55. Miscellaneous transformations

conversion of the carboxylic acid to a thiocarboxylic acid (*73*) afforded an intermediate which could be reduced to an aldehyde (*Chart 56*) (*74*).

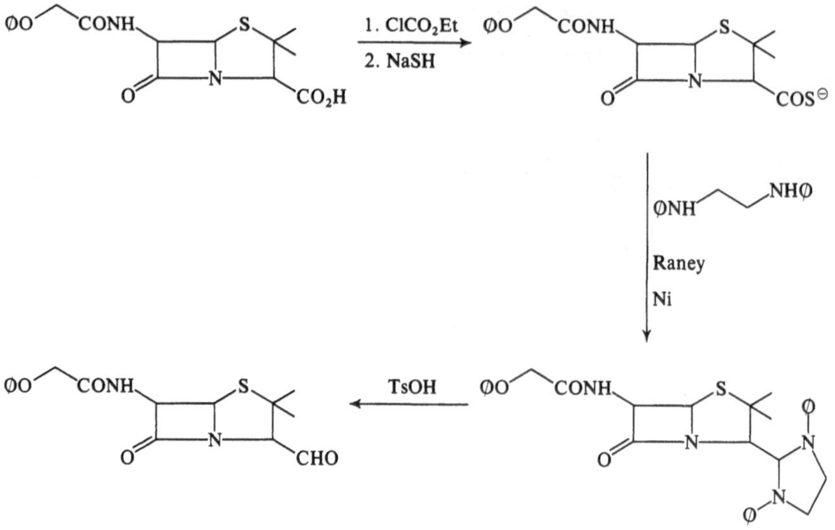

Chart 56. Penicillin thioacids and aldehydes

A further reduction of the 3-carboxylic acid to the penicillanyl alcohol may be achieved by borohydride reduction of the acyl azide (*Chart 57*) (*143*). In addition to being acylated and carbamoylated, these alcohols may be converted to sulfonates which can then undergo intermolecular or intramolecular nucleophilic displacement (*Chart 57*) (*21, 20*).

The conversion of penicillins to the corresponding diazoketones and thence to the 3-acetic acid derivatives has been described (*109, 52*). In addition to homologation, irradiation of the diazoketone (**127**) afforded a novel fused ring system (**128**) (*Chart 58*) (*148, 152*). Treatment of the diazoketone with HCl gave the chloromethyl ketone which, on treatment with base, partially epimerized at C-6 and ring expanded to a cepham derivative (**129**). The epimerization took place on the chloromethyl ketone, not the cepham, and the degree of epimerization was found to be a function of the solvent, the nature of the leaving group (in this example, chlorine) and to a lesser extent, the nature of the base (*149, 151, 152*).

V. Conclusion

Penicillin chemistry, particularly with regard to transformations of the penicillanic acid nucleus, is currently in a very active stage of development. It can be anticipated that the near future will bring a wide and interesting variety of ring modifications that will be of considerable interest from an antimicrobial point of view.

References, pp. 53—62

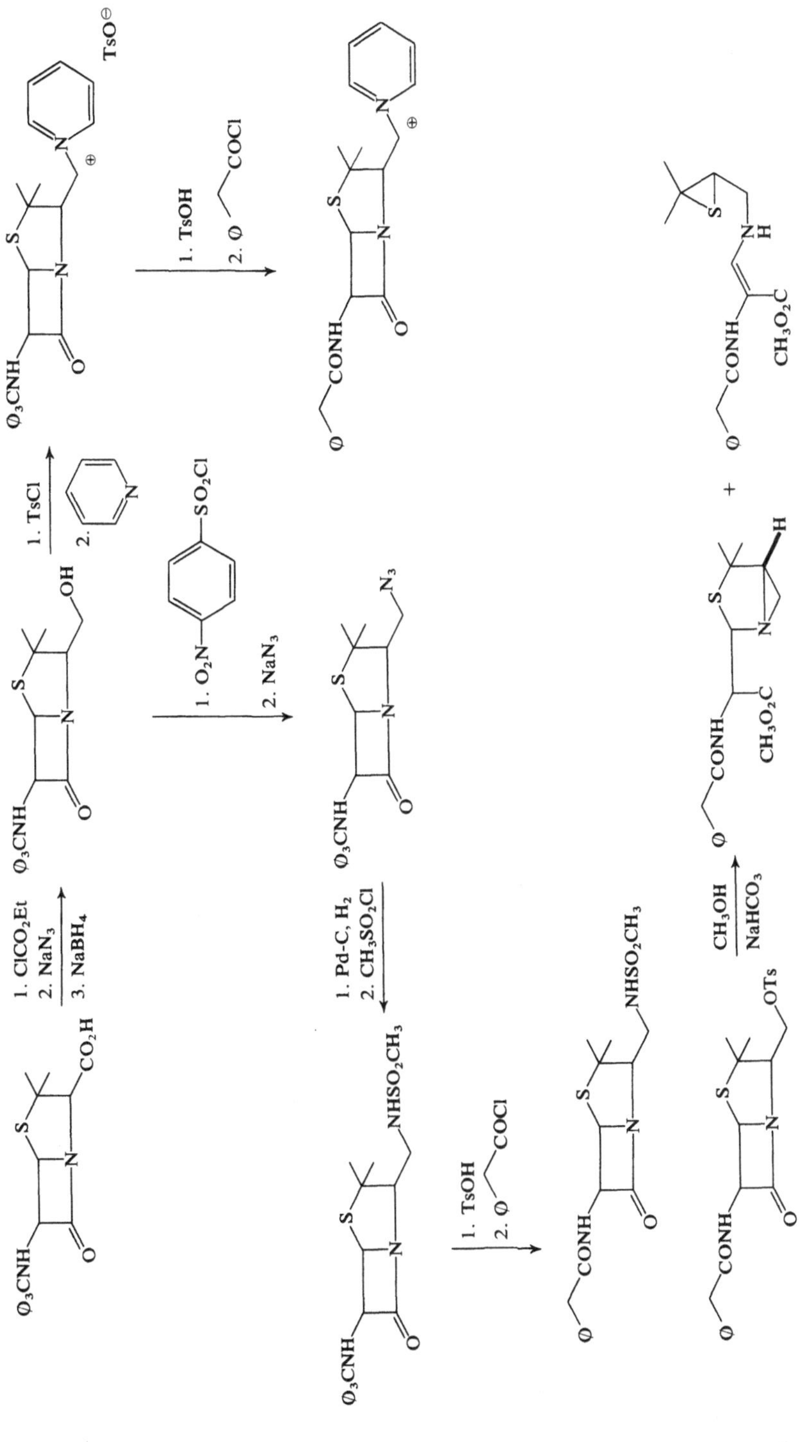

Chart 57. Penicillanyl alcohols and their transformations

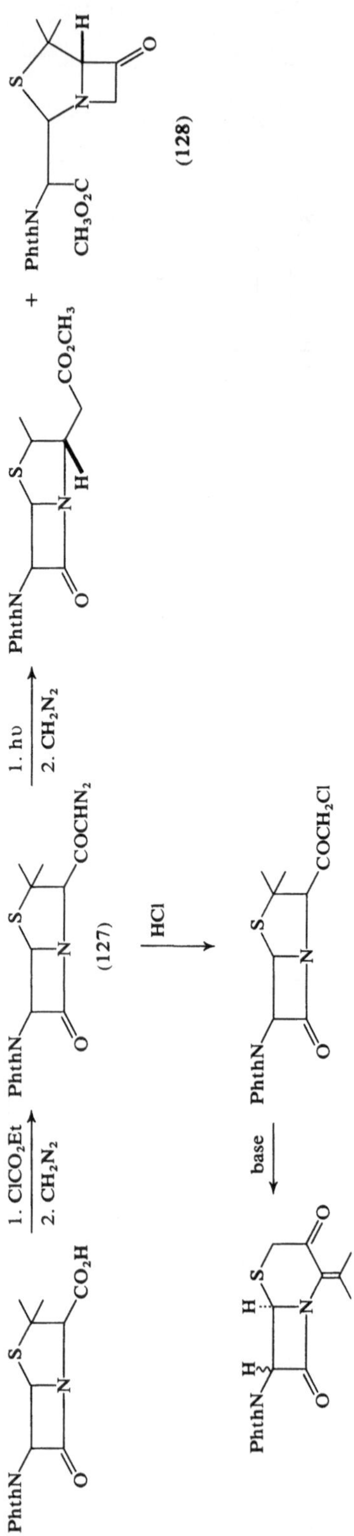

Chart 58. Reactions of 6-phthalimidopenicillanoyldiazomethane

References

1. ABE, J., T. WATANABE, T. TAKE, K. FUJIMOTO, T. FUJII, K. TAKEMURA, K. NISHIIE, S. SATOH, T. KOIDE, and Y. HOTTA: Process for the Production of Isoxazolyl Penicillins. United States Patent 3 668 200 (June 6, 1972).

2. ABRAHAM, E. P.: Penicillins and Cephalosporins – Their Chemistry in Relation to Biological Activity. In Topics in Pharmaceutical Sciences. Vol. 1, D. PERLMAN, Ed., p. 1—31. New York: Interscience Publishers. 1968.

3. AGER, I., D. H. R. BARTON, G. LUCENTE, and P. G. SAMMES: The Reaction of Penicillin Sulphoxides with Vinyl Ethers. Chem. Commun. **1972,** 601.

4. ARCHER, R. A., R. D. G. COOPER, P. V. DeMARCO, and L. F. JOHNSON: Structural Studies on Penicillin Derivatives: ^{13}C Nuclear Magnetic Resonance Studies of Some Penicillins and Related Sulphoxides. Chem. Commun. **1970,** 1291.

5. ARCHER, R. A., and P. V. DeMARCO: Photochemical Preparation and Conformational Analysis by Proton Magnetic Resonance of Penicillin (*R*)-Sulfoxides. J. Amer. Chem. Soc. **91,** 1530 (1969).

6. BAMBERG, P., B. EKSTRÖM, and B. SJÖBERG: Semisynthetic Penicillins. VII. The Use of Phenacyl 6-Aminopenicillanates in Penicillin Synthesis. Acta Chem. Scand. **21,** 2210 (1967).

7. — — — Semisynthetic Penicillins. VIII. The Use of Tributyltin 6-Aminopenicillanate in Penicillin Synthesis. Acta Chem. Scand. **22,** 367 (1968).

8. BARTON, D. H. R., F. COMER, D. G. T. GREIG, G. LUCENTE, P. G. SAMMES, and W. G. E. UNDERWOOD: Rearrangements of Penicillin Sulphoxides. Chem. Commun. **1970,** 1059.

9. BARTON, D. H. R., F. COMER, D. G. T. GREIG, P. G. SAMMES, C. M. COOPER, G. HEWITT, and W. G. E. UNDERWOOD: Transformations of Penicillin. Part I. Preparation and Rearrangements of 6β-Phenylacetamidopenicillanic Sulphoxides. J. Chem. Soc. (London) (C) **1971,** 3540.

10. BARTON, D. H. R., F. COMER, and P. G. SAMMES: Stereoisomerism of Penicillin Sulfoxides. J. Amer. Chem. Soc. **91,** 1529 (1969).

11. BARTON, D. H. R., M. GIRIJAVALLABHAN, and P. G. SAMMES: Transformations of Penicillin. Part II. N,N′-Di-isopropylhydrazine, a New Reagent for Protection of Carboxylic Acids. J. Chem. Soc. (London), Perkin I **1972,** 929.

12. BARTON, D. H. R., D. G. T. GREIG, G. LUCENTE, P. G. SAMMES, M. V. TAYLOR, C. M. COOPER, G. HEWITT, and W. G. E. UNDERWOOD: On the Trapping of Sulphenic Acids from Penicillin Sulphoxides. Chem. Commun. **1970,** 1683.

13. BARTON, D. H. R., D. G. T. GREIG, P. G. SAMMES, and M. V. TAYLOR: Isolation of the β-Lactam Function of Penicillins. Chem. Commun. **1971,** 845

14. BARTON, D. H. R., and P. G. SAMMES: Chemical Relationships between Cephalosporins and Penicillins. Proc. Roy. Soc. (London) **179,** 345 (1971).

15. BARTON, D. H. R., P. G. SAMMES, M. V. TAYLOR, C. M. COOPER, G. HEWITT, B. E. LOOKER, and W. G. E. UNDERWOOD: 4-Alkyl-thio- and -dithio-azetidinones from Penicillins. Chem. Commun. **1971,** 1137.

16. BATCHELOR, F. R., F. P. DOYLE, J. H. C. NAYLER, and G. N. ROLINSON: Synthesis of Penicillin: 6-Aminopenicillanic Acid in Penicillin Fermentations. Nature **183,** 257 (1959).

17. BATCHELOR, F. R., D. GAZZARD, and J. H. C. NAYLER: Action of Carbon Dioxide on 6-Aminopenicillanic Acid. Nature **191,** 910 (1961).

18. BELL, M. R., J. A. CARLSON, and R. OESTERLIN: Degradation of Penicillin G Methyl Ester and Penillonic Acid Methyl Ester to D-5,5-Dimethyl-Δ^2-thiazoline-4-carboxylic Acid Methyl Ester. J. Amer. Chem. Soc. **92,** 2177 (1970).

19. — — — Degradation of Penicillin G Methyl Ester with Trifluoroacetic Acid. J. Organ. Chem. (USA) **37,** 2733 (1972).

20. Bell, M. R., S. D. Clemans, and R. Oesterlin: Transformations of 6-Phenylacetamido- and 6-Tritylaminopenicillanyl p-Toluenesulfonate and p-Nitrobenzenesulfonate. J. Med. Chem. **13,** 389 (1970).

21. Bell, M. R., and R. Oesterlin: Novel Penicillin Transformation Products. Tetrahedron Letters **1968,** 4975.

22. Böhme, E. H. W., H. E. Applegate, J. B. Ewing, P. T. Funke, M. S. Puar, and J. E. Dolfini: 6-Alkyl Penicillins and 7-Alkyl Cephalosporins. J. Organ. Chem. (USA) **38,** 230 (1973).

23. Böhme, E. H. W., H. E. Applegate, B. Toeplitz, J. E. Dolfini, and J. Z. Gougoutas: 6-Methyl Penicillins and 7-Methyl Cephalosporins. J. Amer. Chem. Soc. **93,** 4324 (1971).

24. Brain, E. G., A. J. Eglington, J. H. C. Nayler, M. J. Pearson, and R. Southgate: Oxidation of Some 1,2-Seco-penicillins. Chem. Commun. **1972,** 229.

25. Brown, L. D., W. A. Zygmunt, and H. E. Stavely: Some Active Derivatives of Penicillin. Appl. Microbiol. **17,** 339 (1969).

26. Brunwin, D. M., and G. Lowe: Conversion of Benzyl 6-Diazopenicillanate into 6-Phenylacetylhydrazono- and 6β-Phenylacetylhydrazinopenicillanic Acid. Chem. Commun. **1972,** 192.

27. Bundgaard, H.: Imidazole-catalyzed Isomerization of Penicillins into Penicillenic Acids. Tetrahedron Letters **1971,** 4613.

28. — Kinetic Demonstration of a Metastable Intermediate in Isomerization of Penicillin to Penicillenic Acid in Aqueous Solution. J. Pharm. Sci. **60,** 1273 (1971).

29. Butcher, B. T., M. K. Stanfield, G. T. Stewart, and R. Zemelman: Antibiotic Polymers: α-Aminobenzylpenicillin (Ampicillin). Mol. Cryst. Liq. Cryst. **12,** 321 (1971).

30. Butler, T. C., K. H. Dudley, and D. Johnson: Chemical Studies of Potential Relevance to Penicillin Hypersensitivity: Kinetics of Formation and Disappearance of Benzylpenicillenic Acid and its Derivatives in Solutions of Benzylpenicillin. J. Pharmacol. Exp. Therapeut. **181,** 201 (1972).

31. Cama, L. D., W. J. Leanza, T. R. Beattie, and B. G. Christensen: Substituted Penicillin and Cephalosporin Derivatives. I. Stereospecific Introduction of the C-6(7) Methoxy Group. J. Amer. Chem. Soc. **94,** 1408 (1972).

32. Chauvette, R. R., H. B. Hayes, G. L. Huff, and P. A. Pennington: Preparation of 7-Aminocephalosporanic Acid and 6-Aminopenicillanic Acid. J. Antibiotics **25,** 248 (1972).

33. Chauvette, R. R., P. A. Pennington, C. W. Ryan, R. D. G. Cooper, F. L. José, I. G. Wright, E. M. Van Heyningen, and G. W. Huffman: Chemistry of Cephalosporin Antibiotics. XXI. Conversion of Penicillins to Cephalexin. J. Organ. Chem. (USA) **36,** 1259 (1971).

34. Chow, A. W., N. M. Hall, and J. R. E. Hoover: Penicillin Sulfoxides and Sulfones. J. Organ. Chem. (USA) **27,** 1381 (1962).

35. Cignarella, G., G. Pifferi, and E. Testa: 6-Chloro- and 6-Bromopenicillanic Acids. J. Organ. Chem. (USA) **27,** 2668 (1962).

36. Clarke, H. T., J. R. Johnson, and R. Robinson, Eds.: The Chemistry of Penicillin. Princeton, New Jersey: Princeton University Press. 1949.

37. Clayton, J. P.: The Chemistry of Penicillanic Acids. Part I. 6,6-Dibromo- and 6,6-Diiodo-derivatives. J. Chem. Soc. (London) (C) **1969,** 2123.

38. Clayton, J. P., J. H. C. Nayler, R. Southgate, and E. R. Stove: Penicillanic Acids: Requirements for Epimerization at C-6. Chem. Commun. **1969,** 129.

39. Clayton, J. P., J. H. C. Nayler, R. Southgate, and P. Tolliday: Novel Alkylation of Penicillanates. Chem. Commun. **1971,** 590.

40. Clayton, J. P., R. Southgate, B. G. Ramsay, and R. J. Stoodley: Studies Related to Penicillins. Part IV. The Rearrangement of Penicillanic Acid Derivatives to 1,4-Thiazepines. J. Chem. Soc. (London) (C) **1970,** 2089.

41. COOPER, R. D. G.: Structural Studies on Penicillin Derivatives. V. Penicillin Sulfoxide – Sulfenic Acid Equilibrium. J. Amer. Chem. Soc. **92**, 5010 (1970).
42. — Rearrangements of Cephalosporins and Penicillins. In Cephalosporins and Penicillins, E. H. FLYNN, Ed., p. 183—254. New York: Academic Press. 1972.
43. — Structural Studies on Penicillin Derivatives. VIII. A Possible Model Biosynthetic Route to Penams and Cephems. J. Amer. Chem. Soc. **94**, 1018 (1972).
44. COOPER, R. D. G., P. V. DeMARCO, J. C. CHENG, and N. D. JONES: Structural Studies on Penicillin Derivatives. I. The Configuration of Phenoxymethyl Penicillin Sulfoxide. J. Amer. Chem. Soc. **91**, 1408 (1969).
45. COOPER, R. D. G., P. V. DeMARCO, and D. O. SPRY: Structural Studies on Penicillin Derivatives. II. The Configuration of Phthalimidopenicillin and Epiphthalimidopenicillin Sulfoxides. J. Amer. Chem. Soc. **91**, 1528 (1969).
46. COOPER, R. D. G., L. D. HATFIELD, and D. O. SPRY: Chemical Interconversion of the β-Lactam Antibiotics. Accounts Chem. Res. **6**, 32 (1973).
47. COOPER, R. D. G., and F. L. JOSÉ: Structural Studies on Penicillin Derivatives. IV. A Novel Rearrangement of Penicillin V Sulfoxide. J. Amer. Chem. Soc. **92**, 2575 (1970).
48. — — Structural Studies on Penicillin Derivatives. IX. Synthesis of Thiazolidine-Azetidinones. J. Amer. Chem. Soc. **94**, 1021 (1972).
49. CORONELLI, C., G. C. LANCINI, R. PALLANZA, G. TAMONI, and P. SENSI: Idrossilamidi della Penicillina V. Il Farmaco, Ed. Sci., **21**, 450 (1966).
50. CRESSMAN, W. A., E. T. SUGITA, J. T. DOLUISIO, and P. J. NIEBERGALL: Cupric Ion-Catalyzed Hydrolysis of Penicillins: Mechanism and Site of Complexation. J. Pharm. Sci. **58**, 1471 (1969).
51. DAEHNE, W. V., E. FREDERIKSEN, E. GUNDERSEN, F. LUND, P. MØRCH, H. J. PETERSEN, K. ROHOLT, L. TYBRING, and W. O. GODTFREDSEN: Acyloxymethyl Esters of Ampicillin. J. Med. Chem. **13**, 607 (1970).
52. DAICOVICIU, C., and D. POSTESCU: Penicillin Diazo Ketones. Rev. Chim. (Bucharest) **18**, 179 (1967); Chem. Abstr. **67**, 8538 (1967).
53. DANE, E., and T. DOCKNER: Synthese von 6-[D-α-Amino-phenylacetylamino]-penicillansäure unter Verwendung von β-Dicarbonylverbindungen als Aminoschutzgruppen. Chem. Ber. **98**, 789 (1965).
54. DENNEN, D. W.: Degradation Kinetics of 6-Aminopenicillanic Acid. J. Pharm. Sci. **56**, 1273 (1967).
55. DOYLE, F. P., and J. H. C. NAYLER: Penicillins and Related Structures. Advances in Drug Research **1**, 1 (1964).
56. DURBIN, A. K., and H. N. RYDON: The Equilibrium between the Antibiotics Hetacillin and Ampicillin in Solution. Chem. Commun. **1970**, 1249.
57. DÜRCKHEIMER, W., and M. SCHORR: N-Alkylderivate der 6-Aminopenicillansäure. Liebigs Ann. Chem. **702**, 163 (1967).
58. EKSTRÖM, B., A. GOMÉZ-REVILLA, R. MOLLBERG, H. THELIN, and B. SJÖBERG: Semisynthetic Penicillins. III. Aminopenicillins via Azidopenicillins. Acta Chem. Scand. **19**, 281 (1965).
59. EKSTRÖM, B., and B. SJÖBERG: Semisynthetic Penicillins. VI. The Use of the o-Nitrophenylsulphenyl Protecting Group in the Preparation of Aminopenicillins. Acta Chem. Scand. **19**, 1245 (1965).
60. ESSERY, J. M., K. DADABO, W. J. GOTTSTEIN, A. HALLSTRAND, and L. C. CHENEY: Penicillin Sulfoxides. J. Organ. Chem. (USA) **30**, 4388 (1965).
61. EVRARD, E., M. CLAESEN, and H. VANDERHAEGHE: Gas Chromatography of Penicillin and Penicillanic Acid Esters. Nature **201**, 1124 (1964).
62. FECHTIG, B., H. BICKEL, and K. HEUSLER: Neue β-Lactamsysteme aus Penicillinen. Helv. Chim. Acta **55**, 417 (1972).
63. FINHOLT, P., G. JÜRGENSEN, and H. KRISTIANSEN: Catalytic Effect of Buffers on Degradation of Penicillin G in Aqueous Solution. J. Pharm. Sci. **54**, 387 (1965).

64. Firestone, R. A., N. Schelechow, and B. G. Christensen: Penicillin 6,6'-Dimer. Chem. Commun. **1972**, 1106.

65. Firestone, R. A., N. Schelechow, D. B. R. Johnston, and B. G. Christensen: Substituted Penicillins and Cephalosporins II. C-6(7)-Alkyl Derivatives. Tetrahedron Letters **1972**, 375.

66. Flynn, E. H., Ed.: Cephalosporins and Penicillins. New York: Academic Press. 1972.

67. Fosker, G. R., K. D. Hardy, J. H. C. Nayler, P. Seggery, and E. R. Stove: Derivatives of 6-Aminopenicillanic Acid. Part X. A Non-enzymatic Conversion of Benzylpenicillin into Semi-synthetic Penicillins. J. Chem. Soc. (London) (C) **1971**, 1917.

68. Glombitza, K.-W.: Acylierung der 6-Amino-penicillansäure in wasserfreiem Medium. Liebigs Ann. Chem. **673**, 166 (1964).

69. Glombitza, K.-W., and D. Pallenbach: Die jodometrische Penicillinbestimmung. Arch. Pharmaz. **302**, 695 (1969).

70. — — Die jodometrische Penicillinbestimmung. Arch. Pharmaz. **302**, 985 (1969).

71. Godtfredsen, W. O., W. von Daehne, and S. Vangedal: Photochemical Transformations of 6-Aminopenicillanic Acid and Phenoxymethylpenicillin. Experientia **23**, 280 (1967).

72. Gorman, M., and C. W. Ryan: Structure-Activity Relationships of β-Lactam Antibiotics. In Cephalosporins and Penicillins, E. H. Flynn, Ed., p. 532—582. New York: Academic Press. 1972.

73. Gottstein, W. J., R. B. Babel, L. B. Crast, J. M. Essery, R. R. Fraser, J. C. Godfrey, C. T. Holdrege, W. F. Minor, M. E. Neubert, C. A. Panetta, and L. C. Cheney: Derivatives of 6-Aminopenicillanic Acid. VI. Synthesis of Some Derivatives of 6-Aminothiopenicillanic Acid. J. Med. Chem. **8**, 794 (1965).

74. Gottstein, W. J., G. E. Bocian, L. B. Crast, K. Dadabo, J. M. Essery, J. C. Godfrey, and L. C. Cheney: Derivatives of 6-Aminopenicillanic Acid. VII. Synthesis of Penicillin Aldehydes by a Novel Method. J. Organ. Chem. (USA) **31**, 1922 (1966).

75. Gottstein, W. J., P. F. Misco, and L. C. Cheney: Conversion of Hetacillin into Cephalexin. J. Organ. Chem. (USA) **37**, 2765 (1972).

76. Grant, N. H., and H. E. Alburn: Peptide Synthesis from N-Carboxy-α-amino Acid Anhydrides in Water. Aminoacylpenicillanic Acids. J. Amer. Chem. Soc. **86**, 3870 (1964).

77. Grant, N. H., D. E. Clark, and H. E. Alburn: Poly-6-aminopenicillanic Acid. J. Amer. Chem. Soc. **84**, 876 (1962).

78. Guddal, E., P. Mørch, and L. Tybring: Penicillin Oxides. Tetrahedron Letters **1962**, 381.

79. Gutowski, G. E.: 6-Epi-penicillins and 7-Epi-cephalosporins. Tetrahedron Letters **1970**, 1779.

80. Gutowski, G. E., B. J. Foster, C. J. Daniels, L. D. Hatfield, and J. W. Fisher: Novel 3-Hydroxy, 3-Methyl Cephalosporins from the Rearrangement of Penicillin Sulfoxides. Tetrahedron Letters **1971**, 3433.

81. Hamilton-Miller, J. M. T.: Chemical Manipulations of the Penicillin Nucleus: A Review. Chemotherapia **12**, 73 (1967).

82. Hardcastle, Jr., G. A., D. A. Johnson, C. A. Panetta, A. I. Scott, and S. A. Sutherland: The Preparation and Structure of Hetacillin. J. Organ. Chem. (USA) **31**, 897 (1966).

83. Harwood, A. J., P. J. Niebergall, E. T. Sugita, and R. L. Schnaare: Effect of Copper (II) – Glycine Chelates on Degradation of Penicillin in Mildly Acid Solution. J. Pharm. Sci. **61**, 82 (1972).

84. Hatfield, L. D., J. Fisher, F. L. José, and R. D. G. Cooper: Structural Studies on Penicillin Derivatives: Part VII. Rearrangement of Penicillin Sulfoxides with Trimethylphosphite – Acetic Anhydride. Tetrahedron Letters **1970**, 4897.

85. HAUSER, D., and H. P. SIGG: Desaminierung von 6-Aminopenicillansäure. Helv. Chim. Acta **50**, 1327 (1967).

86. HEUSLER, K.: Advances in the Total Synthesis of β-Lactam Antibiotics. In Topics in Pharmaceutical Sciences, Vol. 1, D. PERLMAN, Ed., p. 33—51. New York: Interscience Publishers. 1968.

87. — Structural Modifications of Penicillins. XXIII rd. International Congress of Pure and Applied Chemistry, Vol. 3, p. 87—109. London: Butterworths. 1971.

88. — Die Umwandlung von Penicillinen in Cephalosporine. Helv. Chim. Acta **55**, 388 (1972).

89. — Total Synthesis of Penicillins and Cephalosporins. In Cephalosporins and Penicillins. E. H. FLYNN, Ed., p. 255—279. New York: Academic Press. 1972.

90. HO, P. P. K., R. D. TOWNER, J. M. INDELICATO, W. A. SPITZER, and G. A. KOPPEL: Biochemical and Microbiological Studies on 6-Substituted Penicillins. J. Antibiotics **25**, 627 (1972).

91. HOOVER, J. R. E., and R. J. STEDMAN: The β-Lactam Antibiotics. In Medicinal Chemistry, Third Edition, A. BURGER, Ed., p. 371—408. New York: Wiley-Interscience. 1970.

92. HUBER, F. M., R. R. CHAUVETTE, and B. G. JACKSON: Preparative Methods for 7-Aminocephalosporanic Acid and 6-Aminopenicillanic Acid. In Cephalosporins and Penicillins. E. H. FLYNN, Ed., p. 27—73. New York: Academic Press. 1972.

93. ISAKA, I., T. KASHIWAGI, K. NAKANO, N. KAWAHARA, A. KODA, Y. NUMASAKI, S. KAWAHARA, and M. MURAKAMI: New Synthetic Procedure of Semisynthetic Penicillin. I. Studies on the Acyl Group Exchange of Natural Penicillin. Yakugaku Zasshi **92**, 454 (1972).

94. JACKSON, J. R., and R. J. STOODLEY: A Novel Rearrangement of a Penicillanic Acid Derivative. Chem. Commun. **1970**, 14.

95. — — Equilibration of Penicillanic Acid Derivatives. Chem. Commun. **1971**, 647.

96. — — Studies Related to Penicillins. Part VII. The Structure of the Epimers Derived from 6β-Substituted Penicillanic Acids. J. Chem. Soc. (London), Perkin I, **1972**, 895.

97. — — Studies Related to Penicillins. Part VIII. The Rearrangement of Penicillanic Acid Derivatives to 1,3-Thiazines. J. Chem. Soc. (London), Perkin I, **1972**, 1063.

98. JANSEN, A. B. A., and T. J. RUSSELL: Some Novel Penicillin Derivatives. J. Chem. Soc. (London) **1965**, 2127.

99. JOHNSON, D. A., and G. A. HARDCASTLE, JR.: Reaction of 6-Aminopenicillanic Acid with Carbon Dioxide. J. Amer. Chem. Soc. **83**, 3534 (1961).

100. JOHNSON, D. A., and D. MANIA: Epi-6-aminopenicillanic Acid and Epi-penicillin G. Tetrahedron Letters **1969**, 267.

101. JOHNSON, D. A., D. MANIA, C. A. PANETTA, and H. H. SILVESTRI: Epi-hetacillin. Tetrahedron Letters **1968**, 1903.

102. JOHNSON, D. A., and C. A. PANETTA: Isomerization and Decomposition Products of Methicillin. J. Organ. Chem. (USA) **29**, 1826 (1964).

103. JOHNSON, D. A., C. A. PANETTA, and R. R. SMITH: Nonenzymatic Conversion of Penicillins to 6-Aminopenicillanic Acid. J. Organ. Chem. (USA) **31**, 2560 (1966).

104. JOHNSTON, D. B. R., S. M. SCHMITT, R. A. FIRESTONE, and B. G. CHRISTENSEN: Substituted Penicillins and Cephalosporins V. 6(7)-Substituted Alkyl Derivatives. Tetrahedron Letters **1972**, 4917.

105. KAISER, G. V., C. W. ASHBROOK, and J. E. BALDWIN: Stereospecific Alkylation of a Penicillin at C-6 Using a Nitrogen Ylide. Methyl-6-α-Allyl-6-β-N,N-dimethylaminopenicillanate. J. Amer. Chem. Soc. **93**, 2342 (1971).

106. KAISER, G. V., and S. KUKOLJA: Modifications of the β-Lactam System. In Cephalosporins and Penicillins, E. H. FLYNN, Ed., p. 74—133. New York: Academic Press. 1972.

107. KINGET, R. D., and M. A. SCHWARTZ: Model Catalysts Which Simulate Penicillinase. III. Structure-Reactivity Relationship in Catalysis of Penicillin Hydrolysis by

Morpholinomethyl Derivatives of Catechol and Pyrogallol. J. Pharm. Sci. **57**, 1916 (1968).

108. KINGET, R. D., and M. A. SCHWARTZ: Model Catalysts Which Simulate Penicillinase. IV. Steric and Electronic Effects in the Catalysis of Hydrolysis of Penicillins and Cephalothin by Aminoalkylcatechols. J. Pharm. Sci. **58**, 1102 (1969).

109. KLEINER, E. M., L. B. SENYAVINA, and A. S. KHOKHLOV: Synthesis and Properties of 6-Benzyl- and 6-Phenoxyacetylaminopenicillan-3-acetic Acids (homopenicillins). Khim. Geterotsikl. Soedin. **1966**, 702; Chem. Abstr. **66**, 75945 (1967).

110. KODA, A., K. TAKANOBU, I. ISAKA, T. KASHIWAGI, K. TAKAHASHI, S. KAWAHARA, and M. MURAKAMI: Studies on the New Synthetic Procedure of Semisynthetic Penicillin. II. The Reaction of Iminoethers and Acid Chlorides. Yakugaku Zasshi **92**, 459 (1972).

111. KOE, B. K.: 6-Aminopenicillanamide and Methyl 6-Aminopenicillanate. Nature **195**, 1200 (1962).

112. KOVACS, O. K. J., B. EKSTRÖM, and B. SJÖBERG: Penicillin Transformations. I. Conversion of a Penicillin into a 7-Oxo-2,3,4,7-tetrahydro-1,4-thiazepine Structure. Tetrahedron Letters **1969**, 1863.

113. KUCHINSKAS, E. J., and G. N. LEVY: Comparative Stabilities of Ampicillin and Hetacillin in Aqueous Solution. J. Pharm. Sci. **61**, 727 (1972).

114. KUKOLJA, S.: Electrophilic Opening of the Thiazolidine Ring in Penicillins. J. Amer. Chem. Soc. **93**, 6267 (1971).

115. — A Stereoselective Synthesis of 6-Phthalimido-5-epipenicillanates. J. Amer. Chem. Soc. **93**, 6269 (1971).

116. — Synthesis of Disulfide Analogs of Penicillins. J. Amer. Chem. Soc. **94**, 7590 (1972).

117. KUKOLJA, S., R. D. G. COOPER, and R. B. MORIN: Structural Studies on Penicillin Derivatives. Part III. Rearrangement and fragmentation of Penicillin V. Tetrahedron Letters **1969**, 3381.

118. KUKOLJA, S., P. V. DeMARCO, N. D. JONES, M. O. CHANEY, and J. W. PASCHAL: Configuration and Conformation of Disulfide Analogs of Penicillins. J. Amer. Chem. Soc. **94**, 7592 (1972).

119. KUKOLJA, S., and S. R. LAMMERT: Thiazabicycloheptenones. Synthesis of Bicyclic Thiazoline Azetidinone Derivatives. Croat. Chem. Acta **44**, 299 (1972).

120. — — Reactions of Penicillin Esters and Related Compounds with 1-Chlorobenzotriazole. Croat. Chem. Acta **44**, 423 (1972).

121. — — Reversible Interconversions of Penam and Cepham Systems via a Common Thiiranium Ion Intermediate. J. Amer. Chem. Soc. **94**, 7169 (1972).

122. LEIGH, T.: N-Alkyl Derivatives of Penicillin V. J. Chem. Soc. (London) **1965**, 3616.

123. LEVINE, B. B.: N(α-D-Penicilloyl) Amines as Univalent Hapten Inhibitors of Antibody-Dependent Allergic Reactions to Penicillin. J. Med. Pharm. Chem. **5**, 1025 (1962).

124. LINDSAY, R. E., and S. L. HEM: Effect of Ionic Strength on Chemical Stability of Potassium Penicillin G. J. Pharm. Sci. **61**, 202 (1972).

125. LO, Y. S., and J. C. SHEEHAN: Synthesis of Benzyl 6-Oxopenicillanate and Derivatives. I. J. Amer. Chem. Soc. **94**, 8253 (1972).

126. LONG, A. A. W., J. H. C. NAYLER, H. SMITH, T. TAYLOR, and N. WARD: Derivatives of 6-Aminopenicillanic Acid. Part XI. α-Amino-p-hydroxybenzylpenicillin. J. Chem. Soc. (London) (C) **1971**, 1920.

127. LONGRIDGE, J. L., and D. TIMMS: Penicillenic Acid, the Mechanism of the Acid and Base Catalysed Hydrolysis Reactions. J. Chem. Soc. (London) (B) **1971**, 852.

128. LUND, F., and L. TYBRING: 6β-Amidinopenicillanic Acids – a New Group of Antibiotics. Nature (New Biology) **236**, 135 (1972).

129. MANHAS, M. S., and A. K. BOSE: Synthesis of Penicillin, Cephalosporin C and Analogs. New York: Dekker. 1969.

130. — — beta-Lactams: Natural and Synthetic. Part 1. New York: Wiley-Interscience. 1971.

131. McMillan, I., and R. J. Stoodley: A Novel Rearrangement of Methyl 6-Chloropenicillanate. Tetrahedron Letters **1966**, 1205.

132. — — Studies Related to Penicillins. Part I. 6-α-Chloropenicillanic Acid and its Reaction with Nucleophiles. J. Chem. Soc. (London) (C) **1968**, 2533.

133. Micetich, R. G., R. Raap, J. Howard, and I. Pushkas: Antibacterial Activity of 6-(5-Membered heteroarylacetamido)penicillanic Acids. J. Med. Chem. **15**, 333 (1972).

134. Moll, F., and M. Hannig: Kondensierte Azetidinone – (2). Arch. Pharmaz. **303**, 321 (1970).

135. Moll, F., and P. Kastenmeier: Die chemische Acylierung der 6-Aminopenicillansäure und 7-Aminocephalosporansäure. Pharmaz. Ztg. **116**, 1345 (1971).

136. Morin, R. B., B. G. Jackson, R. A. Mueller, E. R. Lavagnino, W. B. Scanlon, and S. L. Andrews: Chemistry of Cephalosporin Antibiotics. III. Chemical Correlation of Penicillin and Cephalosporin Antibiotics. J. Amer. Chem. Soc. **85**, 1896 (1963).

137. — — — — — — Chemistry of Cephalosporin Antibiotics. XV. Transformations of Penicillin Sulfoxide. A Synthesis of Cephalosporin Compounds. J. Amer. Chem. Soc. **91**, 1401 (1969).

138. Nayler, J. H. C., M. J. Pearson, and R. Southgate: Hydration of the Triple Bond in Some 4-(Alk-2-ynylthio)azetidin-2-ones. Chem. Commun. **1973**, 57.

139. — — — Novel Conversion of Penicillins into Cephalosporins. Chem. Commun. **1973**, 58.

140. Novak, L., and J. Weichet: Ein neues Verfahren der Partialsynthese der Penicilline. Experientia **21**, 360 (1965).

141. Numata, M., Y. Imashiro, I. Minamida, and M. Yamaoka: Novel Transformations of Penicillins into 2-Azetidinones with Diazo- and Azido-compounds and a Novel Synthesis of Desacetoxycephalosporin. Tetrahedron Letters **1972**, 5097.

142. Pallenbach, D., and K.-W. Glombitza: Die jodometrische Penicillinbestimmung. Arch. Pharmaz. **302**, 863 (1969).

143. Perron, Y. G., L. B. Crast, J. M. Essery, R. R. Fraser, J. C. Godfrey, C. T. Holdrege, W. F. Minor, M. E. Neubert, R. A. Partyka, and L. C. Cheney: Derivatives of 6-Aminopenicillanic Acid V. Synthesis of 6-Aminopenicillanyl Alcohol and Certain Derivatives. J. Med. Chem. **7**, 483 (1964).

144. Price, K. E.: Structure-Activity Relationships of Semisynthetic Penicillins. Adv. Appl. Microbiol. **11**, 17 (1969).

145. Price, K. E., A. Gourevitch, and L. C. Cheney: Biological Properties of Semisynthetic Penicillins: Structure-Activity Relationships. Antimicrobial Agents and Chemotherapy 1966, p. 670—698. G. L. Hobby, Ed. Ann Arbor, Michigan: American Society for Microbiology. 1967.

146. Raap, R.: Synthesis and Antibacterial Activities of Penicillins from (+)- and (−)-α-Amino-4-isothiazolylacetic Acids. J. Antibiotics **24**, 695 (1971).

147. Raap, R., and R. G. Micetich: Penicillins and Cephalosporins from Isothiazolylacetic Acids. J. Med. Chem. **11**, 70 (1968).

148. Ramsay, B. G., and R. J. Stoodley: Studies Related to Penicillins. Part III. 6β-Phthalimidohomopenicillanic Acid. J. Chem. Soc. (London) (C) **1969**, 1319.

149. — — Enlargement of the Thiazolidine Ring of Penicillanic Acid Derivatives. Chem. Commun. **1970**, 1517.

150. — — Epimerization of Penicillanic Acid Derivatives and their Rearrangement to 1,4-Thiazepines: Evidence for an ElcB Mechanism. Chem. Commun. **1971**, 450.

151. — — Studies Related to Penicillins. Part V. The Conversion of 6β-Phthalimidopenicillanic Acid into Cepham Derivatives. J. Chem. Soc. (London) (C) **1971**, 3859.

152. — — Studies Related to Penicillins. Part VI. The Conversion of Penicillin V and 6-Aminopenicillanic Acid into Cepham Derivatives. J. Chem. Soc. (London) (C) **1971**, 3864.

153. Rasmusson, G. H., G. F. Reynolds, and G. E. Arth: 6-Substituted Penicillin Derivatives, VI. Tetrahedron Letters **1973,** 145.

154. Reiner, R., and P. Zeller: Substitution der 6-Aminopenicillansäure am Kohlenstoffatom 6. Helv. Chim. Acta **51,** 1905 (1968).

155. Scartazzini, R., and H. Bickel: Neue β-Lactam-Antibiotika. Über die Darstellung von N-Acylderivaten der 7-Amino-ceph-3-em-4-carbonsäure. Helv. Chim. Acta **55,** 423 (1972).

156. Scartazzini, R., J. Gosteli, H. Bickel, and R. B. Woodward: Neue β-Lactam-Antibiotika. Über die Darstellung der 8-β-Phenylacetamido-homoceph-4-em-5-carbonsäure. Helv. Chim. Acta **55,** 2567 (1972).

157. Scartazzini, R., H. Peter, H. Bickel, K. Heusler, and R. B. Woodward: Neue β-Lactam-Antibiotika. Über die Darstellung der „7-Aminocephalocillansäure". Helv. Chim. Acta **55,** 408 (1972).

158. Schneider, C. H., and A. L. De Weck: Studies of the Direct Neutral Penicilloylation of Functional Groups occuring on Proteins. Biochem. Biophys. Acta **168,** 27 (1968).

159. Schorr, M., and W. Schmitt: Synthese von 6-(Aminomethylphenoxyacetylamino)-penicillansäuren. Arch. Pharmaz. **304,** 325 (1971).

160. Schwartz, M. A.: Mechanism of Degradation of Penicillin G in Acidic Solution. J. Pharm. Sci. **54,** 472 (1965).

161. — Model Catalysts Which Simulate Penicillinase. I. Effect of Ionic Interaction on Catalysis of Penicillin Hydrolysis by Certain Catecholamines. J. Pharm. Sci. **54,** 1308 (1965).

162. Schwartz, M. A., and W. L. Hayton: Relative Stability of Hetacillin and Ampicillin in Solution. J. Pharm. Sci. **61,** 906 (1972).

163. Schwartz, M. A., and G.-M. Wu: Kinetics of Reactions Involved in Penicillin Allergy. I. Mechanism of Reaction of Penicillins and 6-Aminopenicillanic Acid with Glycine in Alkaline Solution. J. Pharm. Sci. **55,** 550 (1966).

164. Sheehan, J. C.: Peptide-type Antibiotics. Pure Appl. Chem. **6,** 297 (1963).

165. — The synthetic Penicillins. In Molecular Modification in Drug Design, Advances in Chemistry Series 45, R. F. Gould, Ed., p. 15—23. Washington, D. C.: American Chemical Society, 1964.

166. Sheehan, J. C., and K. G. Brandt: A Novel Cleavage of the Penicillin Nucleus. J. Amer. Chem. Soc. **87,** 5468 (1965).

167. Simon, G. L., R. B. Morin, and L. F. Dahl: Structural Characterization of an Anhydropenicillin and its Stereochemical Relationship to Penicillins. J. Amer. Chem. Soc. **94,** 8557 (1972).

168. Smart, M. L., and D. Rogers: The Crystal and Molecular Structure of (−)-3β-Acetoxy-4α-t-butylcarbamoyl-3α-methyl-7β-(p-bromophenyl)acetamidocepham 1α-Oxide, $C_{22}H_{28}N_3O_6BrS$. Chem. Commun. **1970,** 1060.

169. Smith, H., J. M. Dewdney, and A. W. Wheeler: A Comparison of the Amounts and the Antigenicity of Polymeric Materials Formed in Aqueous Solution by Some β-Lactam Antibiotics. Immunology **21,** 527 (1971).

170. Smith, H., and A. C. Marshall: Polymers Formed by Some β-Lactam Antibiotics. Nature (New Biology) **232,** 45 (1971).

171. Smith, J. T., and J. M. T. Hamilton-Miller: Hetacillin: A Chemical and Biological Comparison with Ampicillin. Chemotherapy **15,** 366 (1970).

172. Spitzer, W. A., T. Goodson, R. J. Smithey, and I. G. Wright: General Method for the Synthesis of β-Lactam Antibiotics Substituted α to the β-Lactam Carbonyl. Chem. Commun. **1972,** 1138.

173. Spry, D. O.: Conversion of Penicillin to Cephalosporin via a Double Sulfoxide Rearrangement. J. Amer. Chem. Soc. **92,** 5006 (1970).

174. — Oxidation of Penicillin and Dihydrocephalosporin Derivatives with Ozone. J. Organ. Chem. (USA) **37,** 793 (1972).

175. STEWART, G. T.: The Penicillin Group of Drugs. New York: Elsevier Publishing Co. 1965.

176. STOODLEY, R. J.: Deamination of 6-Aminopenicillanic Acid – the Origin of a 2,3-Dihydro-1,4-thiazin-3-one. Tetrahedron Letters **1967**, 941.

177. — Studies Related to Penicillins. Part II. The Rearrangement of 6-β-Amino-penicillanic Acid to 2,3-Dihydro-6-methoxycarbonyl-2,2-dimethyl-1,4-thiazin-3-one. J. Chem. Soc. (London) (C) **1968**, 2891.

178. STROMINGER, J. L., and D. J. TIPPER: Bacterial Cell Wall Synthesis and Structure in Relation to the Mechanism of Action of Penicillins and Other Antibacterial Agents. Amer. J. Med. **39**, 708 (1965).

179. SUTHERLAND, R., and O. P. W. ROBINSON: Laboratory and Pharmacological Studies in Man with Hetacillin and Ampicillin. Brit. Med. J. **2**, 804 (1967).

180. TERAO, S., T. MATSUO, S. TSUSHIMA, N. MATSUMOTO, T. MIYAWAKI, and M. MIYAMOTO: Transformation of Penicillin Sulphoxides into Cephalosporins by Azo-compounds. Chem. Commun. **1972**, 1304.

181. TUTT, D. E., and M. A. SCHWARTZ: Specificity in the Cyclohepta-amylose-catalyzed Hydrolysis of Penicillins. Chem. Commun. **1970**, 113.

182. — — Model Catalysts Which Simulate Penicillinase V. The Cycloheptaamylose – Catalyzed Hydrolysis of Penicillins. J. Amer. Chem. Soc. **93**, 767 (1971).

183. VLIETINCK, A., E. ROETS, P. CLAES, and H. VANDERHAEGHE: A Facile Method for the Preparation of 6-Epi-penicillins. Tetrahedron Letters **1972**, 285.

184. WAGNER, E. S., W. W. DAVIS, and M. GORMAN: The Reaction of Benzylpenicillenic Acid with Thiol-Containing Compounds. The Formation of a Possible Penicillin Antigenic Determinant. J. Med. Chem. **12**, 483 (1969).

185. WAGNER, E. S., and M. GORMAN: The Reaction of Cysteine and Related Compounds with Penicillins and Cephalosporins. J. Antibiotics **24**, 647 (1971).

186. WECK, A. L. DE, C. H. SCHNEIDER, and J. GUTERSOHN: The Role of Penicilloylated Protein Impurities, Penicillin Polymers and Dimers in Penicillin Allergy. Int. Arch. Allergy **33**, 535 (1968).

187. WEISSENBURGER, H. W. O., and M. G. VAN DER HOEVEN: An Efficient Nonenzymatic Conversion of Benzylpenicillin to 6-Aminopenicillanic Acid. Rec. trav. chim. Pays-Bas. **89**, 1081 (1970).

188. WOLFE, S.: Two General Intermediates for the Synthesis of Anhydropenicillins. Canad. J. Chem. **46**, 459 (1968).

189. WOLFE, S., R. N. BASSETT, S. M. CALDWELL, and F. I. WASSON: Reversal of the Anhydropenicillin Rearrangement. J. Amer. Chem. Soc. **91**, 7205 (1969).

190. WOLFE, S., J.-B. DUCEP, G. KANNENGIESSER, and W. S. LEE: Sulfur-free Penicillin Derivatives. III. Regeneration of Fused β-Lactams via Intramolecular Displacement. Canad. J. Chem. **50**, 2902 (1972).

191. WOLFE, S., C. FERRARI, and W. S. LEE: Mercuric Acetate Oxidation of an Anhydro-penicillin. Anhydro-α-phenoxyethylpenicillene, a Novel Antibacterial Agent. Tetrahedron Letters **1969**, 3385.

192. WOLFE, S., J. C. GODFREY, C. T. HOLDREGE, and Y. G. PERRON: Anhydropenicillins: A Novel Rearrangement of the Thiazolidine Ring. J. Amer. Chem. Soc. **85**, 643 (1963).

193. — — — — Rearrangement of Penicillins to Anhydropenicillins. Canad. J. Chem. **46**, 2549 (1968).

194. WOLFE, S., and S. K. HASAN: β-Elimination as a General Process in Penicillin Chemistry. The Stereochemistry and Mechanism of Raney Nickel Desulphurization of Penicillin G and Penicillin V. Chem. Commun. **1970**, 833.

195. WOLFE, S., and W. S. LEE: A Ready C-6 Epimerization of the Penicillin Nucleus. Chem. Commun. **1968**, 242.

196. WOLFE, S., W. S. LEE, J.-B. DUCEP, and G. KANNENGIESSER: Sulfur-free Penicillin Derivatives. II. Functionalization of the Methyl Groups. Canad. J. Chem. **50**, 2898 (1972).

197. Wolfe, S., W. S. Lee, G. Kannengiesser, and J.-B. Ducep: Sulfur-free Penicillin Derivatives. I. Functionalization at C-5. Canad. J. Chem. **50,** 2894 (1972).
198. Wolfe, S., W. S. Lee, and R. Misra: On the Conditions for C-6 Epimerization of the Penicillin Nucleus by a β-Elimination Mechanism. Chem. Commun. **1970,** 1067.
199. Woodward, R. B., K. Heusler, J. Gosteli, P. Naegeli, W. Oppolzer, R. Ramage, S. Ranganathan, and H. Vorbrüggen: The Total Synthesis of Cephalosporin C. J. Amer. Chem. Soc. **88,** 852 (1966).
200. Yoshimoto, M., S. Ishihara, E. Nakayama, E. Shoji, H. Kuwano, and N. Soma: Studies on β-Lactam Antibiotics II. A New Synthesis of 1,2-Secopenicillin and its Conversion to the Cepham Nucleus. Tetrahedron Letters **1972,** 4387.
201. Yurchenko, J. A., M. W. Hopper, T. D. Vince, and G. H. Warren: Substituted Penicillin Amides. Chemotherapy **17,** 405 (1972).

(Received March 27, 1973)

The Antibiotic Complex of the Verrucarins and Roridins

By Ch. Tamm, Basel

With 5 Figures

Contents

I. Introduction

The verrucarins and roridins are secondary metabolites of the soil fungi *Myrothecium verrucaria* (Albertini et Schweinitz) Ditmar ex Fries and *Myrothecium roridum* Tode *ex* Fries. The species *Myrothecium* belongs to the *fungi imperfecti,* order of Moniliales, family Tubercularia-ceae (*30, 17, 32*). The distinction between these and other closely related fungal species is difficult. It has been studied and discussed by various authors (*82, 16, 55, 75, 62, 43*).

Myrothecium species are parasitic on leaves of *Gardenia,* tomatoes, violets, kidney beans, snapdragons and other common plants. They are also found on decaying tissue and in soil. It was known earlier that cultures of the two species mentioned exhibit cellulolytic properties due to the presence of a very active cellulase which was used for the treatment of cellulose during the production of textiles [for leading references see (*42, 66*)].

The first investigation of the secondary metabolites of these micro-organisms was carried out by Brian and McGowan (*16*). They isolated a crystalline compound, designated as glutinosin and assigned the formula $C_{48}H_{60}O_{16}$, from cultures of *Metarrhizium glutinosum* S. Pope, which in fact is a *Myrothecium* species (*82, 55*). Glutinosin exhibited anti-fungal activity. Eight years later Bowden and Schantz (*14*) described the isolation and characterization of a dermatitic or skinirritating crystalline compound melting at 38°, and possessing the formula $C_{15}H_{22}O_4$, from culture filtrates of *Myrothecium verrucaria.* They suspected the presence of additional biologically active substances. In 1961, Nespiak et al. (*55*) reported the antibiotic activity of the mycelium of *Myrothecium roridum* and the isolation of myrothecin to which the formula $C_{18}H_{22}O_6$ was given, from the culture broth. Two additional crystalline compounds, metabolites II and III, were not characterized. Later glutinosin and myrothecin were shown to be a mixture of verrucarins A and B (*43, 36*). Metabolite II proved to be identical with roridin A (*43*). It is probable that compound 379 Y (*49*), isolated from the same organism, represents inhomogeneous material consisting mainly of verrucarin A.

The observation that extracts of *Myrothecium* species exhibit not only antifungal but also a very high cytostatic activity stimulated our research team to undertake a careful chemical examination of the culture filtrate and mycelium of various strains of *Myrothecium verrucaria* and *Myrothecium roridum* about 15 years ago. In 1962 we reported the isolation of ten pure crystalline substances which were designated as ver-rucarins A, B, C, D, E, F, G and roridins A, B and C (*43*). One year later the isolation of muconomycin A from culture filtrates of *Myrothecium verrucaria* was reported (*78*). This is in fact an impure sample of verrucarin A containing the minor metabolites verrucarin B and J (*39*). In 1965 five additional pure metabolites, i.e. verrucarin J and the roridins D, E and H (originally designated as verrucarin H) were isolated from *Myrothecium* species (*11, 83*). A further substance, muconomycin B, (*23*) proved to be identical with verrucarin J. The detection of a very small amount of di-O-acetylverrucarol in the culture of a fluid of a *Myrothecium* sp. isolated from a soil sample originating from Thailand is also to be mentioned (*58*). Very recently we described the isolation of an additional two metabolites myrochromanone and myrochromanol from *Myrothecium roridum* (*79*). Finally the occurence of wortmannin isolated first from *Penicillium* Wortmannii (*15*), in *Myrothecium roridum* has been reported in 1972 (*59*).

II. Production and Isolation

The strains of *Myrothecium* species are grown on culture media which contain the usual nutritional substances. Essentially they are modifications of the CZAPEK-DOX medium, containing yeast extract and peptone with mineral salts and glucose. The cultivation is carried out either by standing surface cultures or by submerged fermentation for 2—4 days at 25°—27° C with aeration and stirring.

The isolation of the metabolites is accomplished either by extraction of the culture filtrate, of the mycelium or of the total culture broth with organic solvents such as ethyl acetate, chloroform, methylene chloride or ethylene chloride. The extracts are washed with dilute hydrochloric acid and sodium hydroxide at 0° and evaporated to dryness. For the treatment of larger batches direct evaporation without washing is more convenient. In this case further purification is achieved by distribution between 80—90 per cent aqueous methanol and petroleum ether in order to remove oily material. The residue obtained after evaporation of the methanolic phase is subjected to chromatography on alumina or silica gel columns or to liquid-liquid partition using 70 to 80 per cent aqueous methanol-carbon tetrachloride (1 : 1). In this manner separation of the antibiotic complex into most components is possible. But often further purification by column chromatography and several recrystallizations are needed.

The total yield of the antibiotic complex and the relative yields of the components vary considerably depending on the strain and culture conditions. For the major metabolites the following maximal yields in mg per liter initial culture broth were reached: verrucarin A: 200 mg; verrucarin B: 52 mg; verrucarin J: 10 mg; 2′-dehydroverrucarin A: 2 mg; roridin A: 300 mg (in one case: 1,3 g); roridin D: 5 mg; roridin E:

Table 1. *Metabolites Isolated from Cultures of Myrothecium verrucaria, Myrothecium roridum and Other Myrothecium Species*

Compound	Molecular Formula	Melting Point (C)	$[\alpha]_D$	Ref.
1. Sterols				
Ergosterol (Roridin B)	$C_{28}H_{44}O$	143—149°[1]	−123° (chloroform)[2]	(43)
			−126° (benzene)	
Metabolite CHT 387	$C_{28}H_{46}O_3$	239—241°	−38° (chloroform-methanol [1:1])	(72)
2. Chromane Derivatives				
Myrochromanone	$C_{13}H_{14}O_2$	46—47°	+53° (chloroform)	(70)
Myrochromanol	$C_{13}H_{16}O_2$	129—131°	+117° (chloroform)	(70)
3. Pyrrole Derivatives				
Verrucarin E	$C_7H_9O_2N$	92—93°	0° (pyridin)	(43)
Verrucarin F	$C_{21}H_{18}O_6N_2(?)$	237—238°	−1° (pyridin)	(43)
Verrucarin G	$C_{15}H_{12}O\ N_2$	131—135°/142—145°	0° (chloroform)	(43)
4. Trichothecane Derivatives				
Verrucarin A	$C_{27}H_{34}O_9$	>330° (dec)	+207° (chloroform)	(43)
			+208° (dioxane)	
2′-Dehydroverrucarin A	$C_{27}H_{32}O_9$	233—240°	+118° (chloroform)	(83)
Verrucarin B	$C_{27}H_{32}O_9$	>330° (dec)	+94° (chloroform)	(43)
			−110° (dioxane)	
Verrucarin J	$C_{27}H_{32}O_8$	>320° (dec)	+147° (benzene)	
			+19° (chloroform)	(11)
			+30° (dioxane)	
			+41° (benzene)	
Roridin A	$C_{29}H_{40}O_9$	198—204°	+130° (chloroform)	(43)
Roridin D	$C_{29}H_{38}O_9$	232—235°	+29° (chloroform)	(11)
Roridin E	$C_{29}H_{38}O_8$	177—178° or 211—212° or 220—221°	−27° (chloroform)	(11)

Roridin H (formerly verrucarin H)	$C_{29}H_{36}O_8$	>320° (dec)	+ 32° (chloroform) + 44° (dioxane) + 60° (benzene)	(11)
Di-O-acetylverrucarol	$C_{19}H_{26}O_6$	148–150°	− 17° (chloroform) − 14° (methanol)	(38, 71) (58)
Roridin C (Trichodermol)	$C_{15}H_{22}O_3$	117–119°	− 33° (chloroform)	(43)
5. Diverse compounds				
Verrucarin C	—	223–224°	—	(43)
Verrucarin D	—	127–128°	—	(43)
Wortmannin	$C_{23}H_{24}O_8$	214°	—	(57)

1 Lit.: 165°
2 Lit.: −135° (chloroform)

5 mg and roridin H: 48 mg. For verrucarin E which is not a trichothecane ester (see below), yields up to 47 mg were observed. The yield of myro-chromanol was 2,1 mg and of myrochromanone 0,15 mg.

III. Structure and Chemical Transformations

The secondary metabolites of *Myrothecium verrucaria* and *Myro-thecium roridum* do not belong to a single chemical class of substances. They are a complex mixture of compounds which can be divided into four structural types: sterols, chromane derivatives, pyrrole derivatives and trichothecane esters. It is not known whether the occurence of the isoprenoid wortmannin is general or incidental. Therefore this metabolite is placed in the group of diverse compounds. The compounds which have been isolated are listed in Table 1. Since the major metabolites are tricho-thecane derivatives, we intend to use in this review the names verrucarins and roridins only for this latter group of substances.

1. Sterols

The minor metabolite originally named roridin B was identified later as ergosterol (1), which is an ubiquitous microbial product.

(1)   (2)

The spectral data of metabolite CHT 387 indicate the presence of three hydroxyl groups and of two isolated olefinic double bonds. CrO_3-oxidation gave an α,β-unsaturated ketone which contained an additional isolated carbonyl group in a cyclohexane ring and a tertiary hydroxyl group. According to the mass spectra the oxygen functions are located in rings A, B and C. The evidence is accommodated by structure (2) for the metabolite whereby the tertiary hydroxyl group is very likely to be attached to C-9.

References, pp. 114—117

2. Chromane Derivatives

Myrochromanol and Myrochromanone

Myrochromanol (**3**) and Myrochromanone (**5**) are crystalline and colourless substances. The molecular formulae, $C_{13}H_{16}O_2$ for (**3**) and $C_{13}H_{14}O_2$ for (**5**), were deduced from the elemental analyses and the high resolution mass spectra. Their structures 2-(1-propenyl)-4-hydroxy-6-methyl-chromane and 2-(1-propenyl)-4-oxo-6-methyl-chromane respectively, were determined using spectroscopic techniques and by chemical degradation (*70*).

The presence of an aromatic chromophoric system was indicated by the UV spectra. The IR spectra revealed the presence of a hydroxyl group in myrochromanol (**3**), of a carbonyl group in myrochromanone (**5**) and of an olefinic double bond with *trans* configuration in both compounds. Myrochromanol (**3**) gave a mono-O-acetyl derivative (**4**) and upon CrO₃-oxidation a monoketone which was identical with the isolated myrochromanone (**5**). The presence of a methyl group attached to the aromatic ring either at C-6 or C-7 was revealed by the NMR-data. Additional structural elements were recognized by double resonance experiments. Oxidative degradation of myrochromanone (**5**) proved the 6-position of the methyl group. Treatment of (**5**) with methanolic NaOH led to the yellow isomer (**6**), which yielded 2-hydroxy-5-methyl benzoic acid (**8**) upon oxidation with H_2O_2-OsO_4 in pyridine.

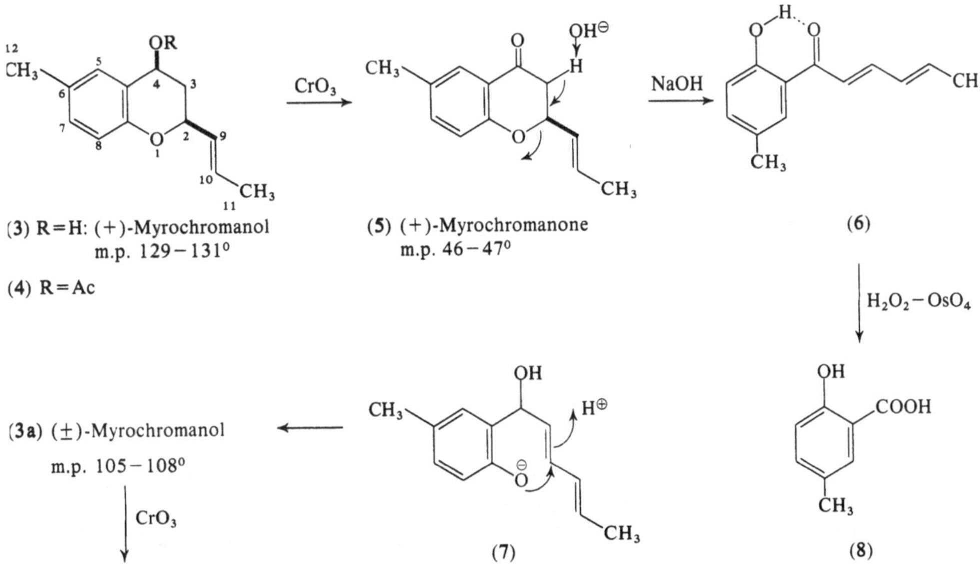

(**3**) R=H: (+)-Myrochromanol
    m.p. 129−131°

(**4**) R=Ac

(**5**) (+)-Myrochromanone
    m.p. 46−47°

(**6**)

(**3a**) (±)-Myrochromanol
    m.p. 105−108°

(**7**)

(**8**)

(**5a**) (±)-Myrochromanone
    m.p. 41−44°

Treatment of myrochromanone (**5**) with NaBH₄ in aqueous dioxane in order to form myrochromanol (**3**) or its epimer, not only resulted in the reduction of the keto group but was accompanied by total racemization. The resulting optically inactive myrochromanol (**3a**) gave upon re-oxidation myrochromanone (**5a**), which was also optically inactive. It was concluded that not only the chiral C-4 atom but also C-2 was racemized. These unexpected results can be explained by assuming a base catalysed ring opening to the dienone (**6**) which is devoid of chirality. The isomer (**6**) is reduced to the alcohol (**7**) which is recyclized to the chromanone system.

The relative configuration of the chiral centres at C-2 and C-4 was established by considering the coupling constants of the protons at C-2, C-3 and C-4 which were determined by double resonance experiments. The conformation of myrochromanol (**3**) shown in Figure 1 is supported by the circular dichroism. The absolute configuration was indicated by the results of partial resolution (dédoublement partiel) using the Horeau method (*44, 45, 46*).

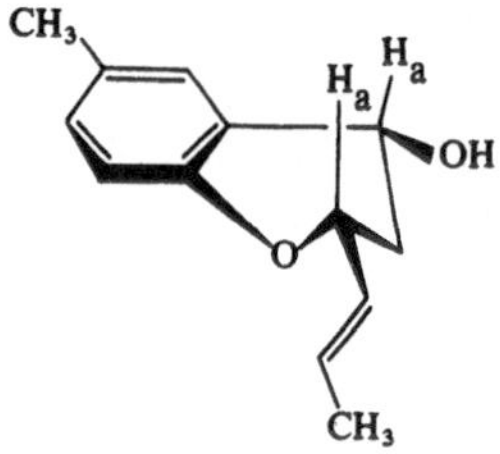

Fig. 1. Conformation of myrochromanol

The chiral C-4 atom probably possesses the (*S*)-configuration.

3. Pyrrole Derivatives

Verrucarin E

So far only the structure of one of the isolated nitrogen containing metabolites, i.e. that of verrucarin E, has been established. However, its structure elucidation was not free of errors. Verrucarin E, a crystalline, colourless substance which decomposes very readily on day light and upon heating, was shown to possess the molecular formula $C_7H_9O_2N$ by elemental analysis and mass spectrometry. It is a neutral compound and very sensitive to acids. The spectral data indicated the presence of a hydroxyl group and a secondary amide, a methyl group attached to an unsaturated carbon atom, and two vinylic protons. Treatment of verrucarin E with acetic anhydride gave a mono-O-acetyl derivative. Catalytic hydrogenation did not lead to defined products. On the other hand, mono-O-acetyl verrucarin E yielded a product which had lost 1 mole of acetic acid. It was concluded that a hydroxymethyl group originally present in verrucarin E had been replaced by a methyl group due to hydrogenolysis. Therefore structures (9) and (10) of a γ-pyridone and of an acetyl pyrrole derivative respectively were considered for verrucarin E (23). There are isoelectronic and vinylogous amides.

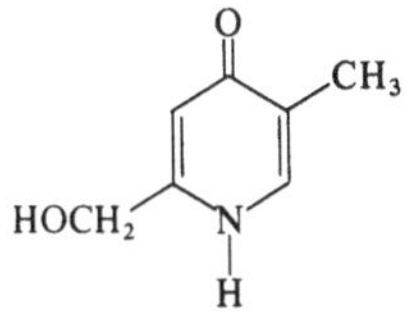

| (9) | (10) Verrucarin E | (11) R=H: Verrucarine E |
| | old structure | revised structure |

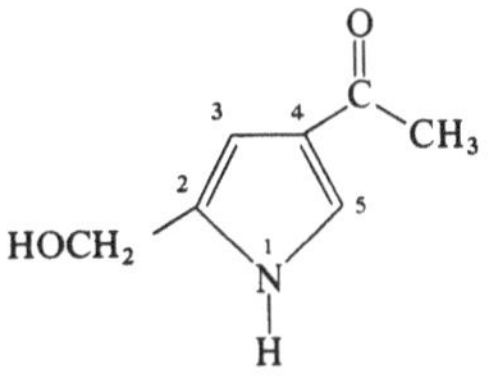

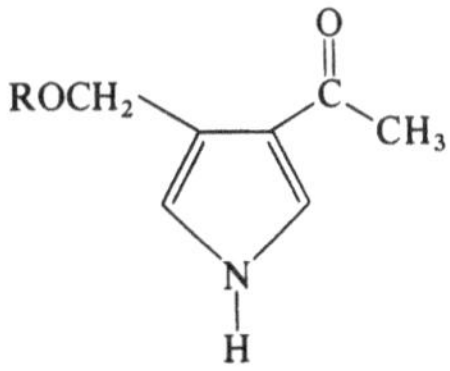

(12) R=Ac

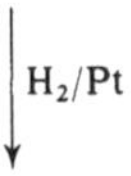

H₂/Pt

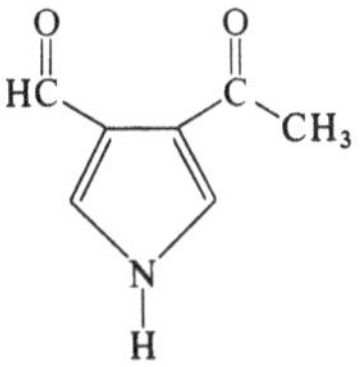

| (13) | (14) | (15) |

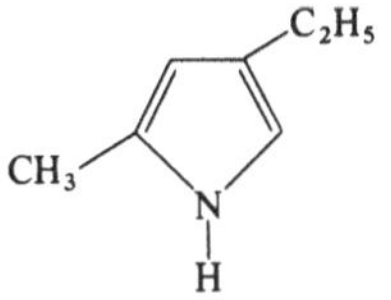

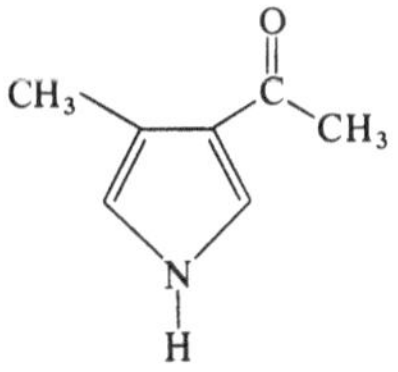

In order to decide between the two possibilities the hydrogenolysis product was treated with LiAlH₄. The resulting product appeared to be identical with the known 2-methyl-4-ethylpyrrole (14) thus excluding the γ-pyridone structure (9). The latter conclusion proved to be correct, but the presumed identity of the LiAlH₄ reduction product with 2-methyl-4-ethylpyrrole (14) could not be confirmed. The circumstance, that such pyrrole derivatives are liquids whose properties are very similar and which are very sensitive to air oxidation, is an understandable source of such an error. The correct structure of 3-acetyl-4-hydroxy-methyl-pyrrole (11) for verrucarin E was finally proved by a careful comparison of the NMR spectra of 3-acetylpyrrole (21) and 3,4-disubstituted pyrroles with those of verrucarin E, and by the non-identity of the hydrogenolysis product with 4-acetyl-2-methyl-pyrrole (22) (6). The NMR studies were extended to 3-acetyl-5-formylpyrrole (13) which was obtained from ver-

rucarin E by oxidation with CrO_3 in pyridine. The two pyrrole derivatives
(21) and (22) were synthetized by condensing amino-acetaldehyde hydrochloride (16) and α-amino-propionaldehyde hydrochloride (17) respectively with ethyl acetylpyruvate (18) to the esters (19) and (20) respectively.
Hydrolysis and decarboxylation gave the desired compounds (21) and
(22) respectively. In addition, a synthetic sample of 3-acetyl-4-methyl-
pyrrole (15) proved to be identical with the hydrogenolysis product of
mono-O-acetyl-verrucarin E (12). The synthesis of verrucarin E (11) was
achieved by reacting 3-acetylpyrrole (21) with formaldehyde in aqueous
Na_2CO_3 solution at 80°. The yield of verrucarin E (11) was low due to
the formation of the four additional products (23) (24) (25) and (26).

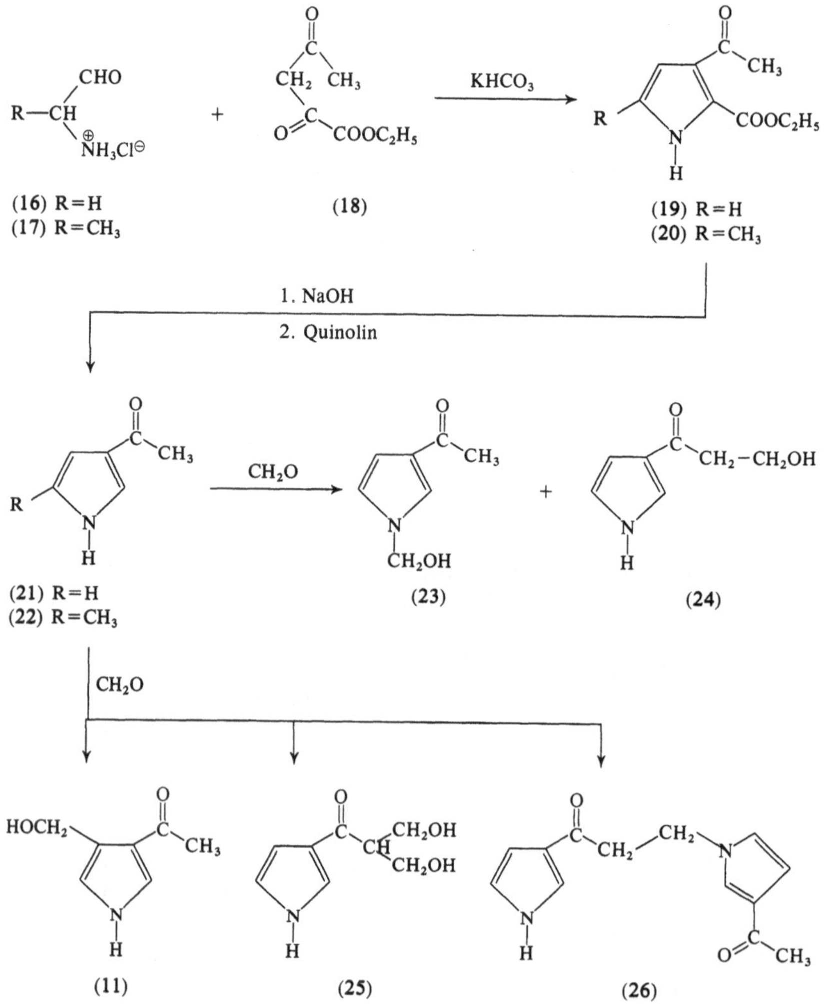

(16) R=H
(17) R=CH₃

(18)

(19) R=H
(20) R=CH₃

1. NaOH
2. Quinolin

(21) R=H
(22) R=CH₃

(23)

(24)

(11)

(25)

(26)

Table 2. *Products from the Base-Catalysed Hydrolysis of the Macrocyclic Verrucarins and Roridins*

Verrucarin A ($C_{27}H_{34}O_9$) $\longrightarrow$	Verrucarol + *cis,trans*-Muconic Acid + Verrucarinolactone
	($C_{15}H_{22}O_4$) ($C_6H_6O_4$) ($C_6H_{10}O_3$)
2'-Dehydroverrucarin A ($C_{27}H_{32}O_9$) $\longrightarrow$	Verrucarol + *cis,trans*-Muconic Acid + 5-Hydroxy-2-oxo-3-methylvalerolactone ($C_6H_8O_3$)
Verrucarin B ($C_{27}H_{32}O_9$) $\longrightarrow$	Verrucarol + *cis,trans*-Muconic Acid + 5-Hydroxy-2,3-epoxy-3-methylvalerolactone ($C_6H_8O_3$)
Verrucarin J ($C_{27}H_{32}O_8$) $\longrightarrow$	Verrucarol + *cis,trans*-Muconic Acid + 5-Hydroxy-3-methyl-2-pentenoic Acid Lactone
	($C_6H_8O_2$)
Roridin A ($C_{29}H_{40}O_9$) $\longrightarrow$	Verrucarol + Roridinic Acid ($C_{14}H_{22}O_7$)
Roridin D ($C_{29}H_{38}O_9$) $\longrightarrow$	Verrucarol + 2-Anhydro-2,3-epoxyroridinic Acid ($C_{14}H_{20}O_7$)
Roridin E ($C_{29}H_{38}O_8$) $\longrightarrow$	Verrucarol + 2-Anhydrororidinic Acid ($C_{14}H_{20}O_6$)
Roridin H ($C_{29}H_{36}O_8$) $\longrightarrow$	Verrucarol + Myrothecinic Acid ($C_{14}H_{18}O_6$)

Verrucarin G

It is uncertain whether verrucarin G also contains a pyrrole ring or whether another ring system is involved.

4. Macrocyclic Trichothecane Esters

This group of metabolites includes the verrucarins A, B, J, 2′-dehydro-verrucarin A, di-O-acetyl-verrucarin A and the roridins A, C, D, E and H. They are the most interesting compounds not only for chemical reasons but also because of their remarkable biological properties. They are colourless, crystalline, optically active solids which are soluble in moderately polar solvents but only very slightly soluble in water. Whereas the verrucarins decompose above 320° without melting, the roridins have definite melting points in the range of 200—235°. They are free of nitrogen, the verrucarin series possessing 27 carbon atoms and 8 to 9 oxygen atoms (di-O-acetyl verrucarol is an exception) and the roridin series 29 carbon atoms and also 8 to 9 oxygen atoms. Roridin C is an exception.

Both the verrucarins and the roridins, with the exception of roridin C, yield the same sesquiterpene alcohol verrucarol (**32**) upon base catalysed hydrolysis. However, they differ from each other by the nature of the acidic hydrolysis products (cf. Table 2). The verrucarins always yield two acidic hydrolysis products whereas the roridins give a single dicarboxylic acid as an acidic hydrolysis product.

Structure elucidation has shown the verrucarins to be macrocyclic triesters and the roridins macrocyclic diesters of verrucarol. Before the actual antibiotics are discussed, however, the structure elucidation and the chemical properties of verrucarol and of the closely related roridin C (trichodermol) will be summarized.

Because in the structural work extensive use was made of ^{1}H-NMR spectroscopy the chemical shifts of the most important protons of the genuine metabolites and their hydrolysis products are compiled in Table 3. We are convinced that the goal of establishing complete structures would never have been reached without the availability of this tool. On the other hand, the usefulness of mass spectrometry was very limited since the interpretation of the fragmentation patterns proved to be extremely difficult. For this reason we never have published our work on the mass spectra.

From *Trichothecium* and *Fusarium* species a series of metabolites has been isolated which are simpler esters (mostly acetyl derivatives) of

other 12,13-epoxy-trichothecenes. The chemistry of these related metabolites has been reviewed recently (*10*) cf. also (*31*).

4.1. Verrucarol and Roridin C (Trichodermol)

Verrucarol is a sesquiterpene alcohol with the molecular formula
$C_{15}H_{22}O_4$. UV, IR and NMR data pointed to the presence of a tri-
substituted olefinic double bond carrying a methyl group, of an additional
tertiary methyl group, of two hydroxyl groups which could be acylated
very readily, of an epoxy group and of a stable cyclic ether bridge (*71*).
The geminal protons of the terminal epoxy group appeared in the NMR
spectrum as double doublet (AB-system) at $\partial = 2.95$—3.00 ppm and
$J = 4$ Hz. One of the hydroxyl groups was shown to be primary, the
other secondary. Catalytic hydrogenation of the double bond produced
a dihydro derivative which, when treated with $LiAlH_4$, was converted
into a triol. The triol contained a new tertiary methyl group and a new
tertiary hydroxyl group.

These results indicated the presence of a terminal epoxy group in
verrucarol (*71*). However, the observation that treatment of verrucarol
with conc. hydrochloric acid gave a chlorohydrin and with dilute sulfuric
acid a product containing four hydroxyl groups which were found to be
stable to sodium periodate and lead tetraacetate led to the conclusion
that a 1,3-diol group originating from an oxetane ring was present instead of a 1,2-diol as would be expected from an oxiran ring. On the
basis of this additional evidence, formula (**27**) was proposed for verrucarol
(*40*) in 1963. This proposal took also in account the similarity to trichothecolone, the sesquiterpene moiety of the mold metabolite trichothecin
(*28, 29*). Later the two compounds were interrelated by a sequence of
chemical reactions (see further below).

(**27**) Verrucarol
old formula

(**28**) Trichothecolone
old formula

(**29**) R=H: Trichodermol
      =Roridin C

(**30**) R=Ac: Trichodermine

In 1964 Danish workers (*1, 33, 34*) isolated a new fungal metabolite,
trichodermin, which was shown to be the acetyl derivative of the sesquiterpene alcohol trichodermol. X-ray analysis of trichodermol p-bromobenzoate established structure (**29**) which is characterized by a terminal

Table 3. *Proton Nuclear Magnetic*

Compound	C-2	C-4	C-10[2]	C-11	C-13[3]	C-14	C-15	C-16[4]	CH₃CO	C-2'
Verrucarol (32)	3.8d (5)	4.7m	5.45d (5)	3.7m	2.95AB (4)	0.92s	3.7AB (12)	1.72s	—	—
Di-O-acetyl-verrucarol (33)	ca. 3.8	5.81dd (3.5; 7.5)	5.50d (5)	ca. 3.8	3.0AB (4)	0.80s	4.15AB (12)	1.72s	2.08	—
Roridin C (Trichodermol) (29)	3.81d (5)	4.3m	5.41d (5)	3.51m	2.95AB (4)	0.80s	0.85s	1.70s	—	—
Verrucarin A (74)	—	5.83dd (5.5; 7.5)	5.46d (5)	—	2.97AB (4)	0.87s	—	1.79s	—	ca. 4.2
2'-Dehydro-verrucarin A (75)	—	5.85dd (5.5; 7.5)	5.50d (5)	—	3.0AB (4)	0.85s	—	1.74s	—	—
Verrucarin B (78)	3.67d (5)	5.90dd	5.47d (5)	3.88m	3.0AB (4)	0.88s	—	1.74s	—	3.41s
Verrucarin J (79)	—	5.8–5.9[5]	5.47d	—	2.98AB (4)	0.83s	4.3	1.72s	—	5.85d (1.5)

Compound	C-2	C-4	C-10	C-11	C-13	C-14	C-15	C-16	–OH	C-2'
Roridin A (84)	—	5.85m	5.44m	—	2.96AB (4)	0.80s	4.44s	1.74s	2.8	4.09d (3)
Dimethyl-roridinate (89)	—	—	—	—	—	—	—	—	ca. 3.1	4.25d (3.5)
Roridin D (105)[7]	—	—	—	—	—	—	—	—	—	3.31s
Dimethyl-2',3'-epoxy-2'-anhydro-roridinate (cf. 108)	—	—	—	—	—	—	—	—	—	3.50s
Roridin E (110)	—	6.22dd (4; 8)	5.5m	—	3.0AB (4)	0.82s	4.15AB (12)	1.75s	—	5.98s[8]
Dimethyl-2'-anhydro-roridinate (113)	—	—	—	—	—	—	—	—	ca. 2.3[5]	5.73d[8] (1.5)[11]
Roridin H (117)	3.8d (5)	ca. 5.9[5]	5.42d (4)	3.64[5]	2.96AB (4)	0.85s	4.15AB (12)	1.69s	—	5.67[11]
Dimethyl-myrothecinate (119)	—	—	—	—	—	—	—	—	—	5.78d[8] (1.5)[11]

[1] Values for chemical shifts are given as ppm (δ) downfield from tetramethyl silane (TMS) used as internal standard. Coupling constants (Hz) are given in parentheses. For unequivocally recognized fine structures the following abbreviations are used: s = singlet; d = doublet; dd = doublet of doublets and m = multiplet. CDCl₃ served as solvent.

[2] The doublet at C-10 often shows a fine structure due to coupling with the 16-methyl group.

[3] For the AB-system of the epoxide the value of the centre is listed. The two doublets are in a distance of ca. 0.15 ppm.

[4] This methyl singlet often appears as relatively broad signal, i.e. as doublet with $J = 1.5$ Hz, due to long range coupling with the proton at C-10[2].

[5] The signal is generally obscured by other signals.

[6] Unequivocal assignment is not possible due to the superposition of signals of protons of the verrucarol moiety. However, the number of protons calculated from the integration curve and after subtraction of the verrucarol protons is in excellent agreement.

Resonance Spectra (NMR spectra)[1]

C-3'	C-4'	C-5'	C-6'	C-2''	C-3''	C-4''	C-5''
—							
—							
—							
—							
—	—	—	0.89d (7)	6.06d (16)	8.08dd (11; 16)	6.70t (11)	6.17d (11)
—	—	—	1.25d (7)	6.04d (16)	7.90dd (11; 16)	6.67t (11)	6.11d (11)
—	—	—	1.56s	6.10d (16)	7.98dd (11; 16)	6.69t (11)	6.19d (11)
—	2.50t (6)	3.82t (6)	2.28d (1.5)	6.05d (16)	8.12dd (11; 16)	6.6t (11)	6.1d (11)

C-3'	C-4'	C-5'	C-12'	C-7'	C-8'	C-9'	C-10'	C-13'	C-14'	C-6'	CH₃O-
[6]	[6]	[6]	1.08d (6.5)	ca. 6.0m (5)	7.68dd (15.5; 11)	6.66t (11)	5.78d	[6]	1.17d (5.5)	[6]	—
ca. 2.1m	1.7dd (6)	3.3–3.7	0.85d (6.5)	5.85dd (15.5; 7.5)	7.58dd (15.5; 11)	6.60t (11)	5.72d	3.3–3.7	1.12d (6)	3.3–3.7	3.75s 3.80s
—	[6]	[6]	1.62s	ca. 6.0m	7.54dd (15.5; 11)	6.61t (11)	5.80d	[6]	1.19d (6)	[6]	—
—	1.7–2.1m [5]	3.6–3.9m	1.38s	ca. 6.0m	7.60dd (15.5; 11)	6.61t (11)	5.74d	3.25m	1.13d (5.5)	3.6–3.9m [5]	3.75s 3.81s
—	2.4–2.7 [5]	3.5–4.0 [5]	2.30d (1.5)[8]	5.7–6.0 [5,9]	7.53dd (11; 15)	6.58t (11)	5.75d	ca. 3.7 [10]	1.22d (6)[10]	3.3—3.5 [5,9]	—
—	2.45t (7)[12]	ca. 3.5m [12]	2.21d (1.5)[8,13]	ca. 5.8dd [9]	7.58dd (11; 15)	6.59t (11)	5.76d	ca. 3.7 [10]	1.13d (6)[10]	3.4–3.7 [9]	3.71s 3.77s
—	2.64d (11)	5.58dd (3.5; 8)	2.27d (1.5)	ca. 5.9m [5]	7.68dd (11; 15.5)	6.55t (11)	5.79d	ca. 3.65m [5,10]	1.32d (6)[10]	4.03	—
—	2.50d (5)	5.2 [11,12]	2.25d[14] (1.5)[8]	5.96dd[9] (7.0; 15.5)	7.60dd (11; 15.5)	6.56t (16)	5.72d (11)[5]	ca. 3.18 [5,10]	1.28d (6)[10]	4.1t (7)[9]	—

[7] Protons of the verrucarol moiety of roridin D (**105**) are not listed.
[8] Spin spin decoupling experiments demonstrated the coupling between the protons at C-2' and C-12'.
[9] Spin spin decoupling experiments demonstrated the coupling between the protons at C-6' and C-7'.
[10] Spin spin decoupling experiments demonstrated the coupling between the protons at C-13' and C-14'.
[11] Unsharp signal.
[12] Spin spin decoupling experiments demonstrated the coupling between the protons at C-4' and C-5'.
[13] During the hydrolysis of roridin H myrothecinic acid isomerized partly at the C-2'—C-3' double bond. Therefore the 12'-methyl group of the *trans*-isomer appears as additional signal at 1.97 ppm. It is a doublet due to long range coupling with C-2'.
[14] For the reasons mentioned [13] the signal of the 12'-methyl group of the *trans*-isomer appears at 1.98 ppm.

epoxy group. Since at the same time trichodermin (**30**) was interrelated with trichothecolone, the structure of the latter compound also had to be revised. Structure (**31**) was assigned to trichothecolone. At the same time it was demonstrated by chemical degradation and interrelation with trichodermin and trichothecolone, respectively, that verrucarol has the same skeleton. These findings established formula (**32**) for verrucarol (*38*). Trichodermol proved to be identical with roridin C (*38*).

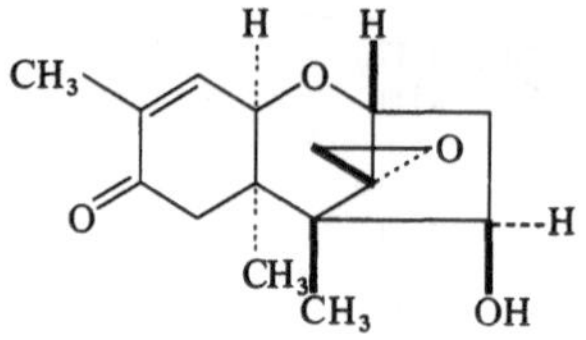

(31) Trichothecolone
revised structure

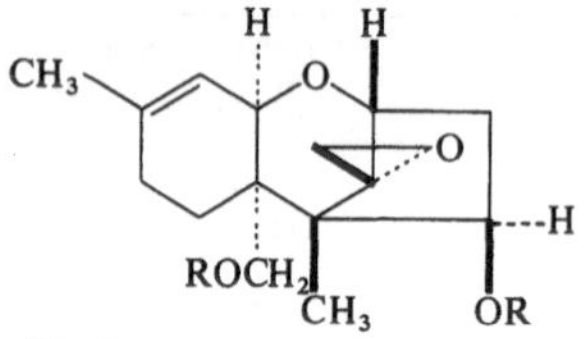

(32) R=H: Verrucarol,
revised structure

(33) R=Ac

In view of the growing number of fungal metabolites (*10*, *31*) which are esters of sesquiterpene alcohols possessing the same tricyclic structure as found in trichodermol (roridin C), trichothecolone and verrucarol, the name trichothecane (**34**) with a particular numbering was introduced for the basic skeleton (*35*). The corresponding rearranged skeleton is called apotrichothecane (**35**). Its atoms retain the numbers they had in (**34**). According to this nomenclature verrucarol is to be named 4β,15-dihydroxy-12,13-epoxy-Δ^9-trichothecene and roridin C (trichodermol) 4β-hydroxy-12,13-epoxy-Δ^9-trichothecene.

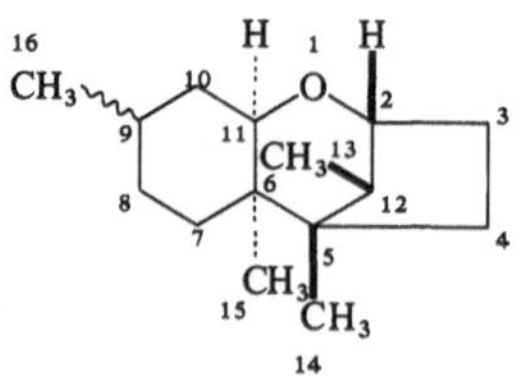

(34) Trichothecane

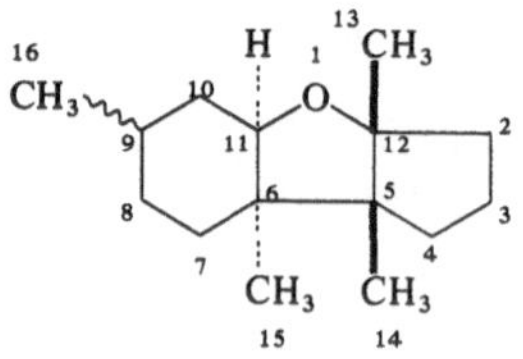

(35) Apotrichothecane

Further details of the relevant reactions used for the elucidation of verrucarol as well as its physical and chemical properties will be discussed in the following paragraphs on the basis of the revised formula (**32**).

Treatment of verrucarol (**32**) with $CrO_3 - H_2SO_4$ in acetone yielded the keto aldehyde (**36**) thus proving the presence of the tertiary hydroxymethyl group and of a secondary hydroxyl group attached to a cyclopentane ring. Catalytic hydrogenation of verrucarol (**32**) gave the dihydro derivative (**39**), whereas treatment with LiAlH₄ led to the isomeric dihydro derivative (**37**) which was devoid of the 12,13-epoxy group. In the NMR spectrum the AB-system of the C(13)-protons had been replaced by a new singlet of a methyl group. Analogous transformation products (**38**) and (**40**) were obtained with dihydroverrucarol (**39**) as starting material. The position of the fourth oxygen atom and the sequence of the remaining carbon and hydrogen atoms of the skeleton were deduced from a detailed analysis of the NMR spectrum of the ketoaldehyde (**36**) using spin spin decoupling experiments. This sequence of atoms was confirmed by the interrelation of verrucarol (**32**) with trichodermin (**30**) and trichothecolone (**31**).

(36)   (32)   (37)

(38)   (39)   (40)

This goal was reached by converting all three compounds into a common product, i.e. 4-O-mesyl trichothecolone (**43**). The 4,15-di-O-mesyl derivative (**41**) of verrucarol (**32**) was treated with NaI in acetone and the resulting 15-monoiodide reduced with zinc to the 4-O-mesyl derivative (**42**). The same compound was obtained by mesylation of trichodermol (roridin C) (**29**). Allylic oxidation of the monomesylate (**42**) with SeO₂ yielded the oxidized mesylate (**43**) which proved to be identical with the mesylation product of trichothecolone (**31**).

After these transformations had been accomplished, the attachment of the primary hydroxyl group of verrucarol (**32**) to C-15 remained to be proved. For this purpose the ketoester (**44**), which was obtained by a rearrangement reaction to be described later, was transformed to the α,β-unsaturated ketone (**45**) by SeO₂-oxidation and the latter hydrogenated to give the tetrahydrodiketo ester (**46**). Retro-Michael cleavage of the diketone (**46**) was achieved by treatment with alkali yielding, *via* the

(32)   (29)   (31)

MsCl/pyridine   MsCl/pyridine   MsCl/pyridine

(41)   (42)   (43)

intermediate (**47**) 1-carboxy-2-methyl-cyclopent-1-ene-3-one (**48**) and, *via* the cyclohexene derivative (**49**), which was treated with diazomethane, methyl-2,5-dihydroxy-4-methylbenzoate (**50**). Hence the hydroxymethyl group of verrucarol must be attached to the cyclohexane ring.

(44)   (45)   (46)

(47)   (48)   (49)   (50)

The stereo formula of verrucarol (**32**) constructed with DREIDING models is reproduced in Fig. 2.

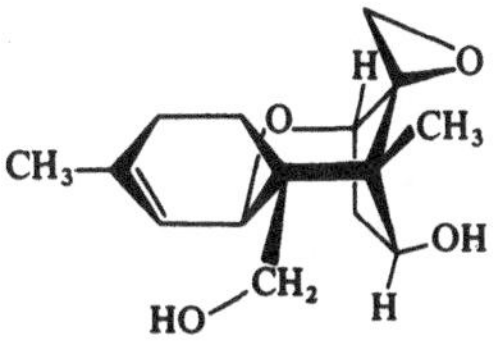

Fig. 2. Conformation of verrucarol

The 12,13-epoxy trichothecenes very readily undergo a rearrangement of the ring system to form the apotrichothecane skeleton (38). This reaction, which is catalysed by mineral acids, is initiated by cleavage of the oxiran ring. Treatment of verrucarol (32) with aqueous sulfuric acid leads to the tetrol (51) and with conc. hydrochloric acid to the chlorohydrin (52) (38). CrO$_3$-oxidation of the chlorohydrin (52) gave a dicarboxylic acid which was converted with diazomethane to the diester (44). The tetrol (51) was stable to NaIO$_4$ and lead tetraacetate. These rearrangements complicated considerably the structural work on the trichothecenes in the earlier stages. As regards the mechanism of these reactions it is assumed that after protonation of the epoxy group the O(1)–C(2) bond is shifted to C-12 with inversion of this centre of chirality and synchronous nucleophilic substitution at C-2. The steric requirements for these transformations are fulfilled since the O(1)–C(2) bond and the C(12)–O (epoxy group) bond are antiparallel and coplanar. The attack of the nucleophile occurs antiparallel to the O(1)–C(2) bond. The postulated reaction mechanism allows the deduction of the stereochemistry of the rearranged products as indicated in the formulae.

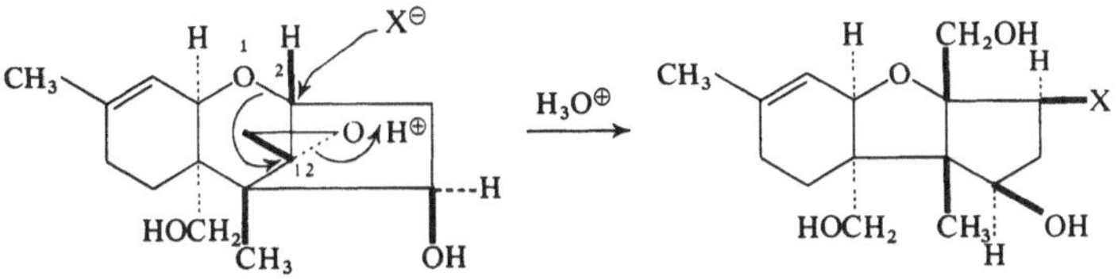

(32)

(51) X=OH
(52) X=Cl

Treatment of the ketoaldehyde (36) with Na$_2$CO$_3$ yielded the hydroxyketoaldehyde (53) (38). It is interesting to note that rearrangement catalysed by acid and by base leads to the same modified skeleton.

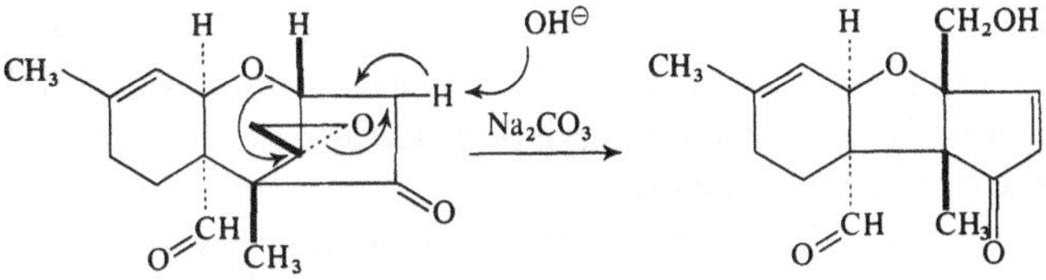

(36)

(53)

When di-O-acetyldihydroverrucarol (**54**) was treated with SOCl$_2$ in pyridine, again substitution with rearrangement to the apotrichothecane derivative (**56**) occurred instead of the expected dehydration (*38*). It is assumed that the reaction proceeds *via* the chlorosulfite (**55**). The reaction of compound (**54**) with trifluoroacetic acid led to the rearrangement products (**59**) and (**60**), probably *via* the carbenium ion (**57**) and the

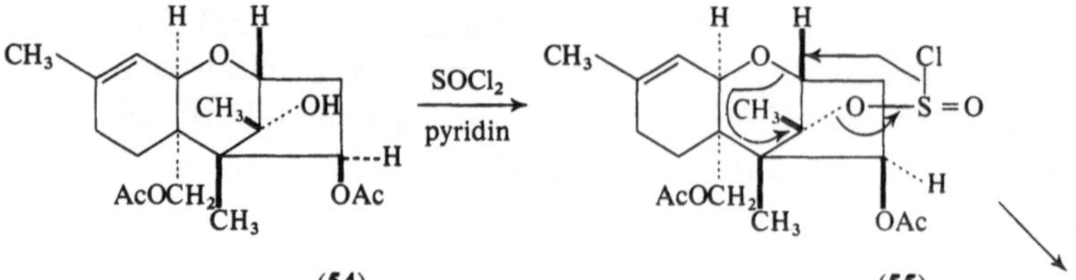

(54)   (55)

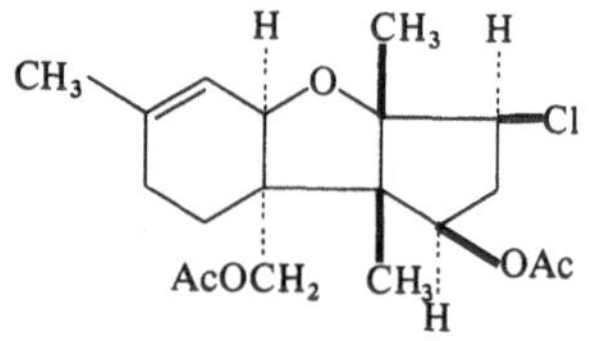

(56)

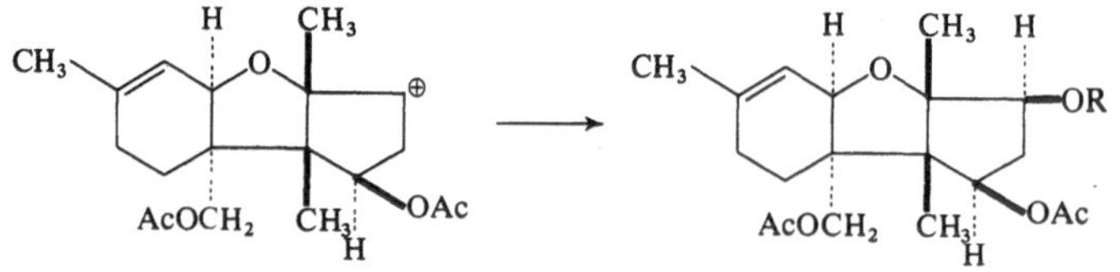

(57)   (58) R = CO · CF$_3$
              (59) R = H

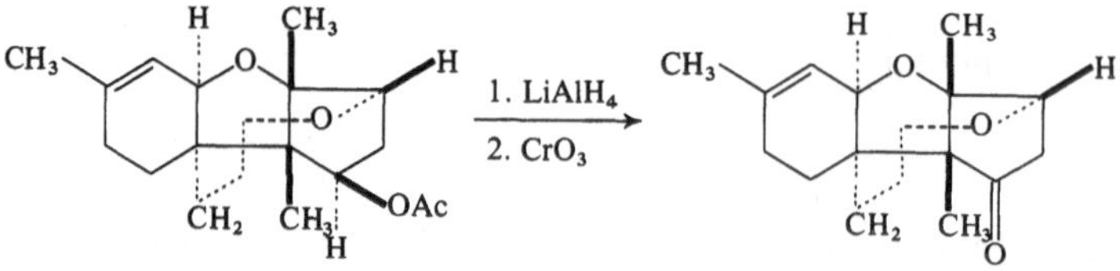

(60)   (61)

trifluoroacetate (**58**) which is hydrolysed during the purification by chromatography on alumina (*38*). The ether bridge was stable to LiAlH$_4$. This reagent effected only deacetylation. CrO$_3$-oxidation of the alcohol yielded the cyclopentanone (**61**).

Finally it should be mentioned that di-O-acetylverrucarol (**33**) was not stable in boiling water. A hydrate of structure (**62**) was formed (*38*).

(33)   (62)

Rearrangements to similar but also to different types of products were observed with the other members of the 12,13-epoxy-trichothecene family. They are reviewed by BAMBURG and STRONG (*10*); cf. also GARDNER et al. (*31*).

The circular dichroism of several trichothecane and apotrichothecane ketones has been studied (*68*). The sign and sometimes even the magnitude of all Cotton effects could be explained by applying the various rules for the cyclohexanone, cyclopentanone and enone chromophores. Thus conformations could also be determined for those compounds which did not yield unambigously to arguments based on conformational analysis.

BAMBURG and STRONG (*10*) have studied the fragmentation patterns of verrucarol and roridin C as well as of other 12,13-epoxy-trichothecenes obtained by mass spectrometry. They also made use of deuterium labelling. However, they concluded that due to the highly variable patterns observed even for very similar compounds in this class, this technique is less useful for analysis of trichothecenes than for many other families of compounds. This conclusion corresponds to our own experience. It is even more true for the macrocyclic verrucarins and roridins.

As already mentioned, in the early phase of the elucidation of the 12,13-epoxy-trichothecanes, formulae containing an oxetane ring instead of the oxiran ring were postulated, because the rearrangement to the apotrichothecane skeleton was not recognized. By ironical and strange coincidence the skeleton postulated in these earlier proposals is in fact identical with that of apotrichothecane. Since at that time very little was known of the chemical and physical properties of oxetane rings fused to a hydrindane system, 2β,13-oxidoapotrichothecanes were synthetized and studied (*69*). 2β,13-dihydroxy-Δ^9-apotrichothecenes which are readily available by the acid catalysed rearrangement of trichothecanes, served as starting materials. In the case of verrucarol, the diacetate (**63**) which was prepared by reacting di-O-acetylverrucarol (**33**) with H_2SO_4 in dioxane-water, was transformed to the monotosylate (**64**).

Ring closure to the oxetane (**65**) was effected quantitatively by treating the tosylate (**64**) with potassium t-butoxide in t-butanol. The 2β,12-oxides starting from roridin C (trichodermol) and trichothecolone were prepared by analogous reactions (*69*). The oxetane ring proved to be

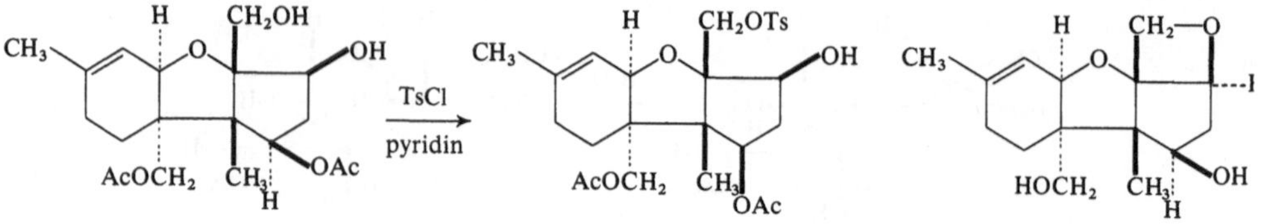

(63)   (64)   (65)

stable to LiAlH₄ and to mineral acids. Cleavage of the ring took place only after the transformation of the secondary hydroxyl group into a keto group. In the NMR spectra of the 2β,12-oxides the AB-system of the C(13)-protons appeared at $\delta = 4.5$ ppm with $J_{AB} = 7$ Hz, whereas in the 12,13-epoxides the AB-system is shifted to $\delta = 3.0$ ppm with $J_{AB} = 4$ Hz. The low field chemical shift in the 2β,12-oxides is unusual.

4.2. Verrucarin A and 2'-Dehydroverrucarin A

Verrucarin A, one of the major metabolites of *Myrothecium verrucaria*, was the first member of the macrocyclic trichothecane ester family whose structure was elucidated (*71, 39*). Formula (*74*) is based on the following evidence. The antibiotic contains one secondary hydroxyl group which gives rise to mono-O-acyl derivatives. The UV spectrum ($\lambda_{max} = 260$ nm; log ε = 4.25) and the IR spectrum indicated the presence of an α,β,γ,δ-unsaturated ester group as chromophoric system. By catalytic hydrogenation verrucarin A was transformed to a hexahydro derivative. Base catalysed hydrolysis of verrucarin A gave, besides verrucarol (*32*), *cis,trans*-muconic acid (*66*) and a neutral substance, C₆H₁₀O₃, which proved to be a lactone of unknown structure. It was named verrucarinolactone and shown to possess structure (*67*). Analogous hydrolysis of hexahydroverrucarin A yielded dihydroverrucarol (*39*), adipic acid and verrucarinolactone (*67*).

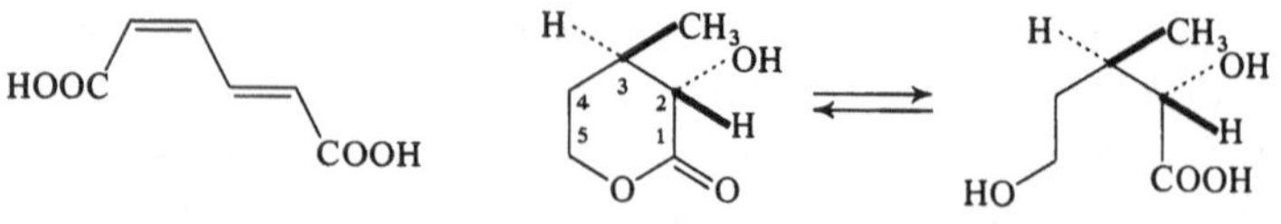

(66)   (67)   (68)

Verrucarinolactone (*67*) showed in the infrared bands indicative of a hydroxyl and a saturated δ-lactone group. It was stable to periodate and was reduced by LiAlH₄ to 3-methyl-pentane-1,2,5-triol. The phenylhydrazide of verrucarinic acid (*68*) was not attacked by periodic acid. It therefore does not contain a 1,2-diol group. These results placed the hydroxyl group of verrucarinolactone at the 2-position. According to the NMR spectrum a secondary methyl group is present which cannot be located at C-4. Consequently it is attached to C-3. The coupling constant of 10 Hz of the vicinal protons at C-2 and C-3 is only compatible with a diaxial *trans*-configuration of the two hydrogen atoms. The lactone therefore exists in the more stable chair form with both the C(2)-hydroxyl and C(3)-methyl in equatorial arrangements.

The absolute configuration of verrucarinolactone (**67**) was established by oxidation with KMnO$_4$ to (*R*)-(+)-methyl succinic acid. Therefore, verrucarinic acid (**68**) has the configuration of (2*S*, 3*R*)-2,5-dihydroxy-3-methyl-valeric acid. Thus, verrucarinolactone (**67**) is a natural isomer of mevalolactone. The biogenetic implications are discussed in Chapter VI. It should be mentioned that verrucarinolactone neither exhibited acetate-replacing activity in cultures of *Lactobacillus acidophilus* nor antagonistic activity to mevalolactone (*39*).

The structure of verrucarinolactone (**67**) was confirmed by synthesis (*4*). β-Methyl-γ-carboxy-γ-butyrolactone (**70**) which is available from β-methyl-glutaric acid (**69**), was reduced by LiAlH$_4$ to the triol (**71**) which is a mixture of diastereoisomers. Separation was achieved by preparing the tris-3,5-dinitrobenzoyl derivatives. One of the two

racemates proved to be identical with the 3,5-dinitrobenzoate of the triol (**71**) derived from natural verrucarinolactone. Treatment of the triol mixture (**71**) with HIO$_4$ gave the hydroxyaldehyde (**72**) which was converted by cyanohydrin synthesis to (±)-verrucarinolactone (**67**) *via* (±)-verrucarinic acid (**73**).

After the structure of the three hydrolysis products of verrucarin A had been established the problem of their interconnection in the actual metabolite remained to be solved. Oxidation of the secondary hydroxyl group of verrucarin A (**74**) gave a monoketone which on hydrolysis yielded free verrucarol (**32**). This result allowed the conclusion that the free hydroxyl group of verrucarin A (**74**) is not located in the verrucarol moiety, but in verrucarinolactone (**67**), which is a secondary product of hydrolysis. It must have existed in verrucarin A in the form of a verrucarinic acid moiety. Therefore verrucarin A is a diester of verrucarol (**32**) with an additional ester bond between a hydroxyl group of verrucarinic acid (**63**) and one of the carboxyl groups of *cis,trans*-muconic acid (**66**).

A distinction between the four possible arrangements of the three building blocks in verrucarin A (*cis,trans*-muconic acid esterified either with the 4-hydroxy or the 15-hydroxy group of verrucarol with the *cis, trans*-sequence being interchangeable) was made possible by NMR studies and by selective oxidative degradation of verrucarin A (**74**) (*84*).

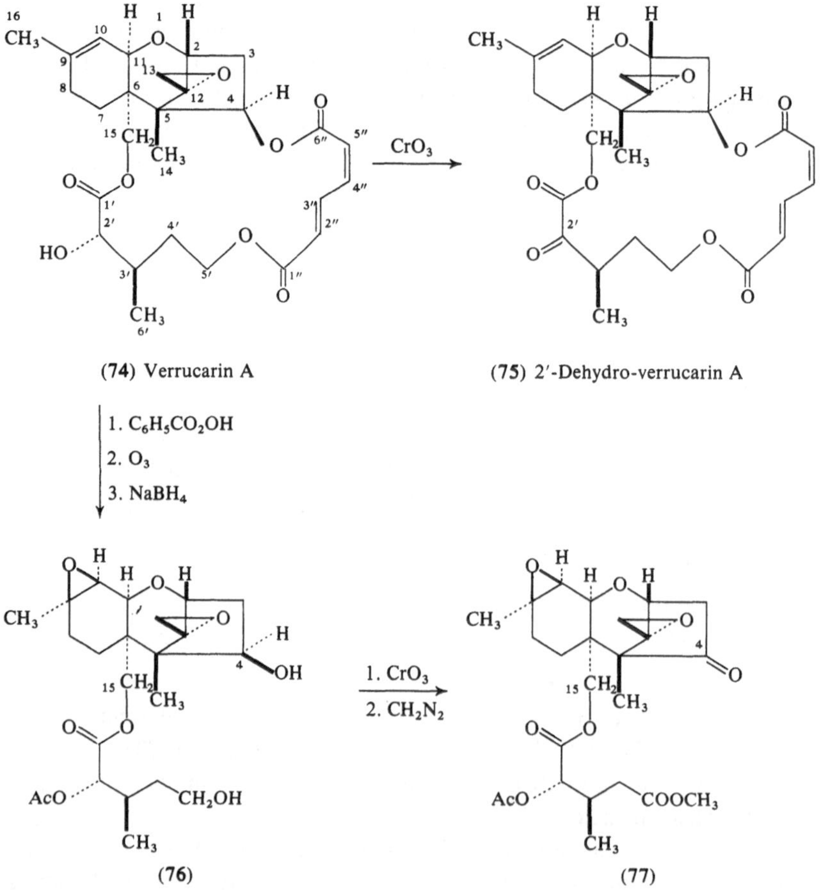

(74) Verrucarin A   (75) 2'-Dehydro-verrucarin A

1. $C_6H_5CO_2OH$
2. O_3
3. $NaBH_4$

(76)   (77)

For this purpose, the 9,10-double bond was protected by epoxidation and the hydroxyl group by acetylation. Ozonolysis of the derivative followed by NaBH₄ reduction and simultaneous partial hydrolysis led to the degradation product (**76**) which was converted to the keto ester (**77**). The NMR spectrum of the diol (**76**) and the formation of a cyclo-pentanone system and not of an aldehyde showed very clearly that it is the primary 15-hydroxyl group of verrucarol, which is esterified with the carboxyl of verrucarinic acid in the original metabolite.

These conclusions as well as the postulated *cis,trans*-sequence of the muconic acid moiety and the absolute configuration of verrucarin A (**74**) were confirmed by an X-ray analysis of the p-iodobenzenesulfonate of the antibiotic (*54*) (cf. Fig. 3).

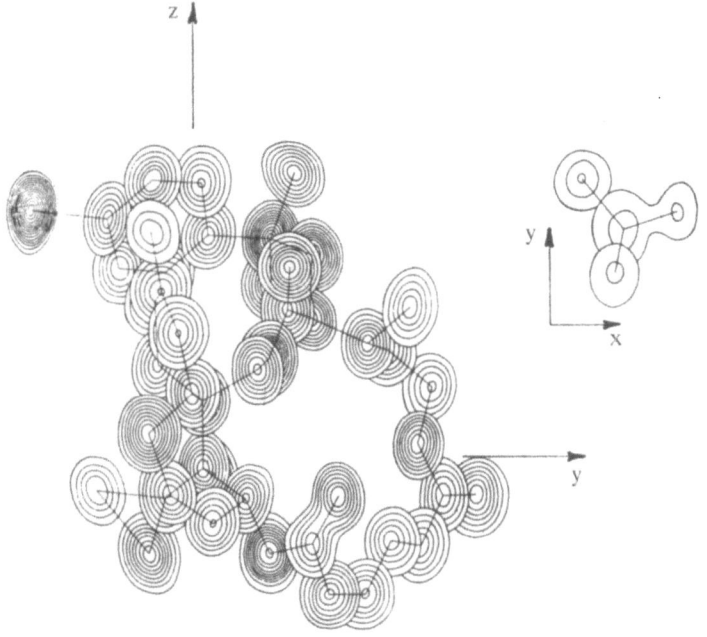

Fig. 3. Three-dimensional electron-density distribution for verrucarin A p-iodobenzene-sulfonate

An investigation of the circular dichroism of the verrucarins and roridins (*68*) revealed that the conformation of the muconic acid moiety of the macrocyclic ring of verrucarin A is the same in dioxane solution as in the solid state. It is to be concluded from this result that the conformation which is most stable in the crystalline state is also the preferred one in solution.

The structure of the minor metabolite *2'-dehydroverrucarin A* (**75**) was established by direct comparison of the natural compound with the oxidation product of verrucarin A (**74**) (*83*).

4.3. Verrucarin B

Verrucarin B is a companion of verrucarin A. The separation and purification is very difficult because of the very similar chemical and physical properties. This slightly less polar metabolite contains two hydrogen atoms less than verrucarin A; it is an isomer of 2'-dehydro-verrucarin A. IR- and NMR-spectra showed that the hydroxyl group of verrucarin A is replaced by an epoxy group at C-2' and C-3' in verrucarin B. Base-catalysed hydrolysis yielded verrucarol (**32**), *cis,trans*-muconic acid (**66**), but not verrucarinolactone (**67**). The corresponding epoxy-lactone could not be isolated. On the basis of this evidence structure (**78**) was assigned to verrucarin B (*41*); however, the configuration of the epoxy group is not yet known.

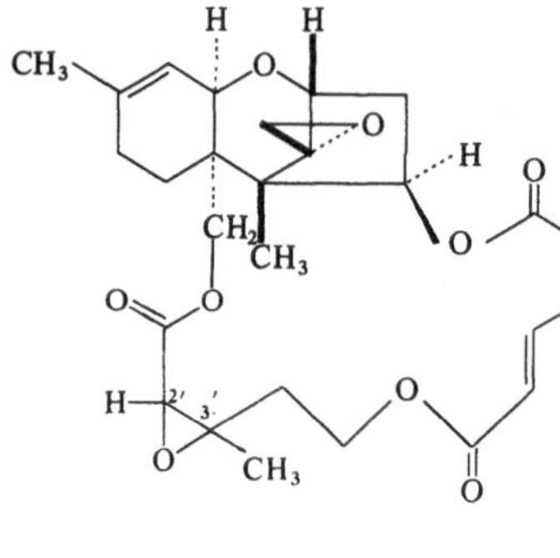

(78) Verrucarin B

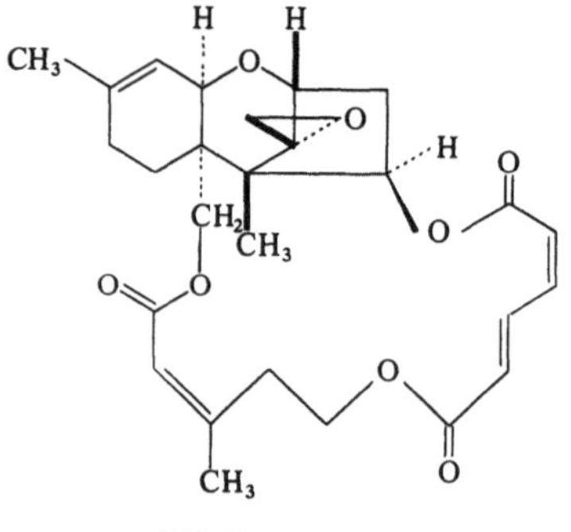

(79) Verrucarin J

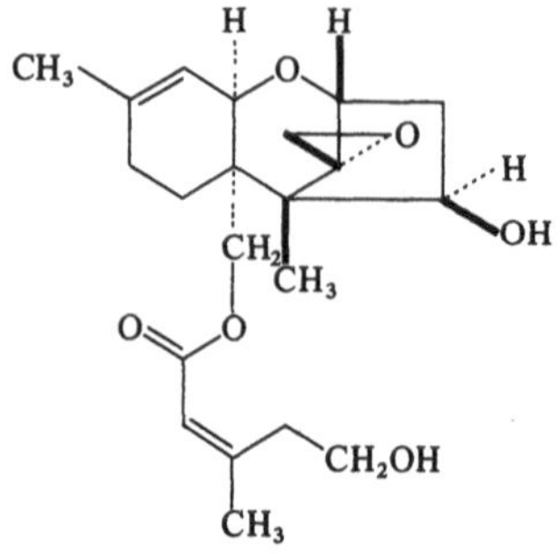

(80)

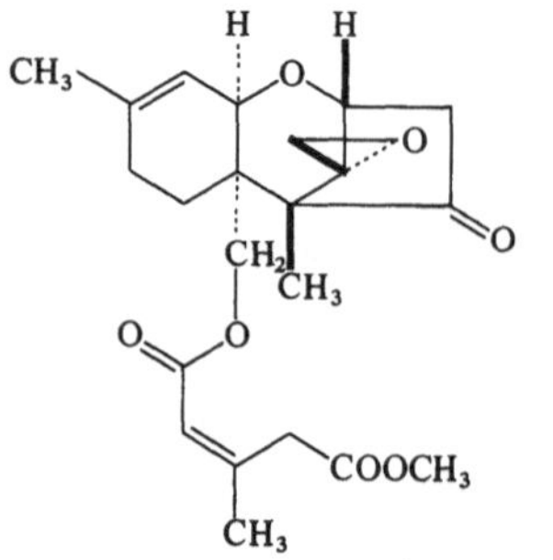

(81)

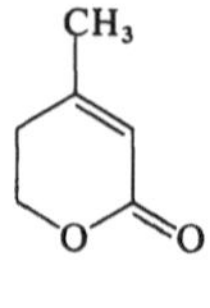

(82)

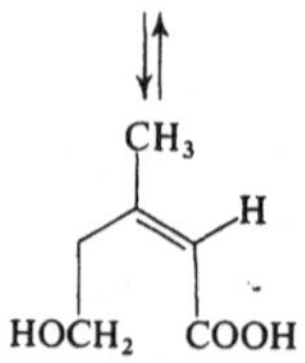

(83)

4.4 Verrucarin J

Like verrucarin B, verrucarin J occurs very often as a companion of verrucarin A. For this reason it is not very easy to separate the three closely related metabolites from each other. Verrucarin J contains one molecule of water less than verrucarin A (**74**). Its structure (**79**) was established by methods similar to those used for the elucidation of verrucarin A and B (*24*). The metabolite is devoid of hydroxyl groups but contains an additional isolated olefinic double bond conjugated to an ester group as demonstrated by the spectral data. Therefore catalytic hydrogenation of verrucarin J yielded an octahydro derivative. Due to the presence of this double bond, total hydrolysis proceeded much slower; it was achieved by treatment with methanolic potassium hydroxide. Verrucarol (**32**), *cis,trans*-muconic acid (**66**) and 5-hydroxy-3-methyl-2-pentenoic acid lactone (**82**) were obtained as hydrolysis products. The isolated oily lactone (**82**) was identical with the product which was obtained by dehydration of mevalolactone with $KHSO_4$ (*19*). It could be transformed to crystalline *cis*-5-hydroxy-3-methyl-*cis*-2-pentenoic acid (**83**). This *cis*-acid has been found also as a building block of the antibiotics ferrirhodin (*48*) and fusigenin (*20, 21*). The isomeric *trans*-acid has also been found in free form in micro-organisms (*22*). The attachment of the carboxyl group of the pentenoic acid (**83**) to the 15-hydroxy group of verrucarol (**32**) in verrucarin J (**79**) could be deduced — in contrast with the other verrucarins — by partial hydrolysis with K_2CO_3 in aqueous methanol. It led to the *cis,trans*-muconic acid (**66**) and to the diol (**80**). The latter was transformed to the keto ester (**81**). The reversed attachment involving the hydroxyl at C-4 of verrucarol would have yielded instead of the keto ester, either a dimethyl ester or a mono methyl ester with an additional aldehyde group. The NMR spectrum of verrucarin J was in agreement with the *cis,trans*-sequence of muconic acid as indicated in formula (**79**).

4.5. Roridin A

The chemical and physical properties of roridin A, the major metabolite of *Myrothecium roridum*, resemble closely those of verrucarin A. The spectral data indicated the presence of the same chromophoric system, i.e. an $\alpha,\beta,\gamma,\delta$-unsaturated ester group with an additional saturated ester group. Roridin A differs, however, in containing two free hydroxyl groups instead of one as in verrucarin A and by the presence of an additional secondary methyl group.

The structural formula (**84**) of roridin A was established on the basis of the following evidence (*12*). Roridin A readily gave di-O-acyl derivatives, e.g. the diacetate (**85**). Treatment with $CrO_3–H_2SO_4$ in acetone led to monodehydrororidin A (**86**) which contained a methyl ketone group as demonstrated by the NMR spectrum and its positive iodoform reaction. Prolonged oxidation gave the diketone (**87**). These results indicated the presence in roridin A of a twocarbon side chain consisting of the CH_3—CH(OH)-group. Catalytic hydrogenation of roridin A led either to tetrahydro- or hexahydrororidin A. Also tetrahydrororidin A gave a monodehydro derivative. Alkaline hydrolysis of the antibiotic gave, besides verrucarol (**32**), only one single additional product, named roridinic acid (**88**), possessing the formula $C_{14}H_{22}O_7$.

Roridinic acid proved to be a very labile substance. It contains two carboxyl and two hydroxyl groups as well as two conjugated olefinic double bonds. Consequently roridinic acid (**88**) was easily transformed to a dimethyl ester (**89**), to a di-O-acetyl dimethyl ester (**90**) and to a tetrahydrodimethyl roridinate (**94**). Selective CrO_3-oxidation of the dimethyl ester produced again a methyl ketone group (**91**). The UV spectra of roridin A, roridinic acid and of verrucarin A were very similar and compatible with the presence of an $\alpha,\beta,\gamma,\delta$-unsaturated carbonyl system. Accordingly in the vinyl proton region of the NMR spectra of the verrucarins, roridin A, as well as the esters of roridinic and *cis,trans*-muconic acid, the chemical shifts of the protons at C-8′, C-9′ and C-10′ were very similar. However, the signal corresponding to the C(7′)-proton of dimethyl roridinate (**89**) differed (quartet instead of a doublet), which means that C-6′ is not a carbonyl group in the roridin series but bears a hydrogen atom. Consequently the following structural elements were recognized in roridinic acid:

$$-\overset{}{\underset{6'}{CH}}-\overset{}{\underset{7'}{CH}}=\overset{}{\underset{8'}{CH}}-\overset{}{\underset{9'}{CH}}=\overset{}{\underset{10'}{CH}}-\overset{}{\underset{11'}{COOH}} \quad \text{and} \quad -\overset{|}{\underset{\underset{6'}{|}}{C}}-\underset{13'}{CH(OH)}-\underset{14'}{CH_3}$$

The NMR-data revealed an additional sequence: HOOC – CH(OH) – – CH(CH$_3$) – CH$_2$ –. The interrelationship of the partial structures together with one methylene group and an inert oxygen atom was established by ozonolysis of di-O-acetyl roridinic acid dimethyl ester (**90**) and subsequent treatment of the ozonide with H_2O_2 – HCl. This resulted in the formation of verrucarinolactone (**67**), propionic acid, α-ketobutyric acid, two moles of acetic acid and two moles of oxalic acid. It is assumed that the formation of these products proceeds *via* the β-hydroxy-

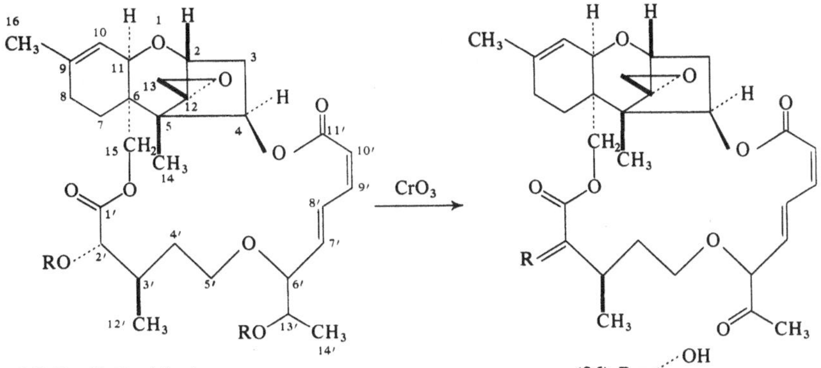

(84) R=H: Roridin A

(85) R=Ac

CrO_3

(86) R= ⟨OH H

(87) R=O

KOH

CrO_3

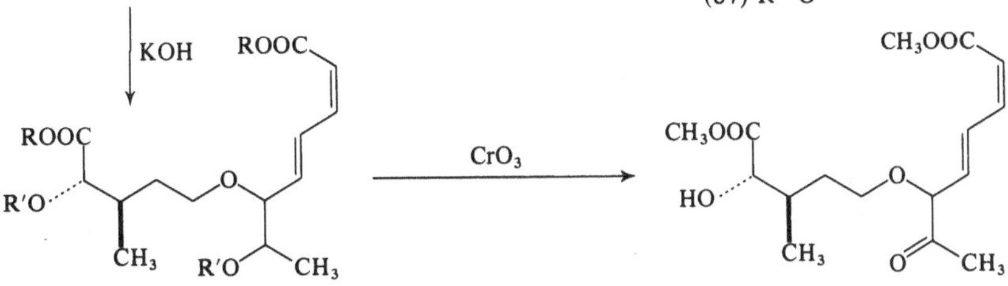

(88) R=R′=H: Roridinic acid

(89) R=CH₃; R=H′

(90) R=CH₃; R′=Ac

(91)

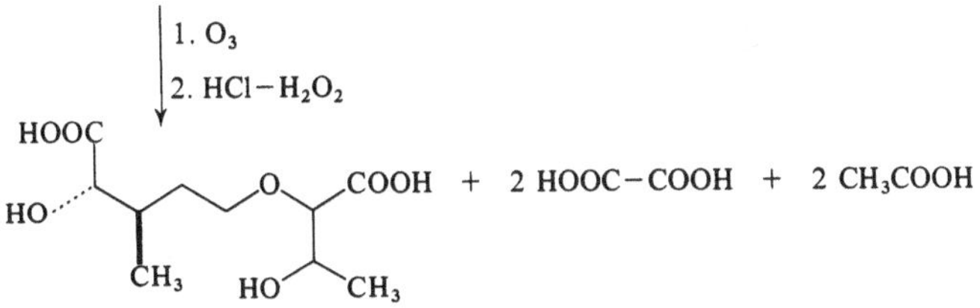

1. O₃

2. HCl−H₂O₂

(92) not isolated   +   2 HOOC−COOH   +   2 CH₃COOH

$H_3O^⊕$

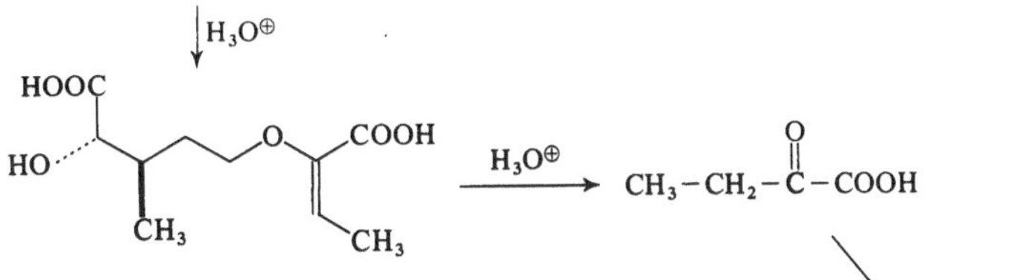

(93) not isolated

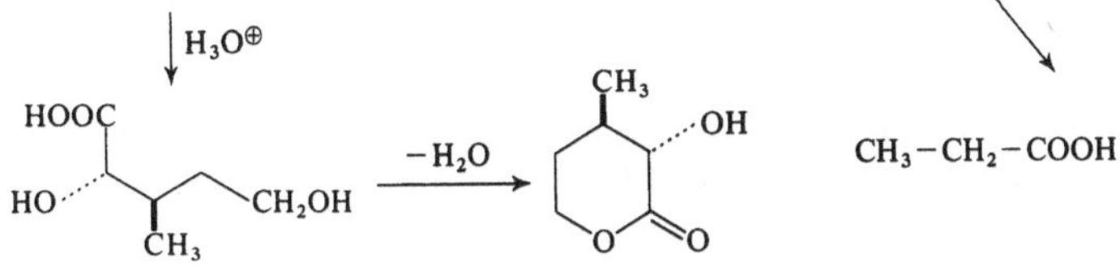

$H_3O^⊕$ → CH₃−CH₂−C−COOH

$H_3O^⊕$

(68) not isolated

−H₂O →

(67)

CH₃−CH₂−COOH

dicarboxylic acid (92) whose β-hydroxyl group is eliminated immediately under the strongly acidic conditions. Product (93) contains a vinyl ether group which is also instable to acid. The cleavage yields verrucarinic acid (68) which is lactonized immediately to stable verrucarinolactone (67).

For the biosynthetic studies to be described in Chapter VI a second degradation scheme was devised for roridinic acid (88) (73).

Selective oxidation of one hydroxyl group of dimethyl tetrahydro-roridinate (94) with $CrO_3 - H_2SO_4$ in acetone and subsequent acetylation led to the ketoester (95) which was transformed to the oxime (96). Treatment of the latter with $SOCl_2$ led to acetonitrile (98) and, after acid hydrolysis of the reaction mixture, to a hydroxyacid (97), which gave after deacetylation verrucarinolactone (67) and monomethyl-adipate semialdehyde (99) as third product. The latter was not isolated but transformed by subsequent reaction with CrO_3 and KOH to adipic acid (100).

alcohol (**97**) and the aldehyde (**99**). In reactions b and c nitrilium ions
are formed which react with water to give amides. Further cleavage of
the amide of reaction c would lead to the same alcohol (**97**) and aldehyde
(**99**) as are obtained by reaction *a* but also requires acetic acid and
ammonia as furthers products. The unequivocal spectroscopic demonstration of acetonitrile in the reaction mixture agrees with the fragmentation and excludes the Beckmann rearrangement path on *b*.

The remaining problem of how verrucarol and roridinic acid are combined in the macrocyclic diester roridin A was solved as follows. The fact that the hydrolysis of the oxidation product (**87**) of roridin A gave verrucarol (**32**) and bis-dehydrororidinic acid demonstrated that both of the free hydroxyl groups of roridin A are located in the roridinic acid moiety. By selective oxidative degradation of the 9,10-epoxide of roridin A with ozone which led to a mixture of the products (**102**) and (**103**) — the tetraacetate (**104**) was obtained in pure form — it was shown that the 1'-carboxy group of roridinic acid (**88**) is esterified with the 15-hydroxyl and the 11'-carboxy group of roridinic acid with the 4-hydroxyl group of verrucarol and not *vice versa* (*5*). This orientation corresponds to that observed in the verrucarin series.

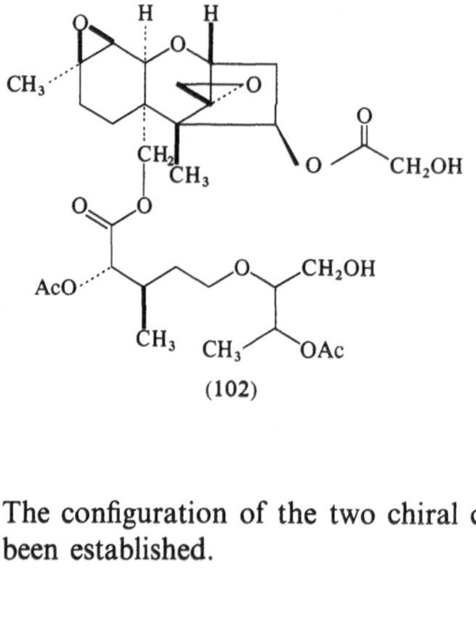

(102)

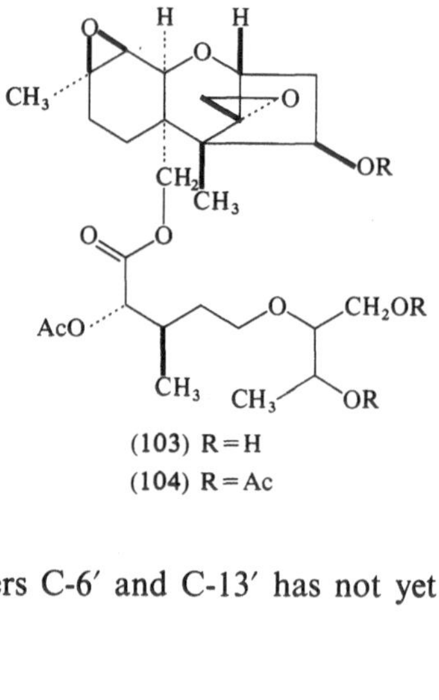

(103) R=H
(104) R=Ac

The configuration of the two chiral centers C-6' and C-13' has not yet been established.

4.6. Roridin D

Roridin D ($C_{29}H_{38}O_9$), one of the minor metabolites of *Myrothecium roridum*, contains 2 hydrogen atoms less than roridin A. According to the spectral data the same chromophoric system as in roridin A is present, but the 2'-hydroxyl group is replaced by a 2',3'-epoxy group. This conclusion, leading to formula (**105**) for roridin D (*13*), was confirmed when it was observed that only a mono-O-acetyl derivative (**106**) was formed and by the results of a base-catalysed hydrolysis which yielded verrucarol (**32**) and 2',3'-epoxy-2'-anhydrororidinic acid (**107**), $C_{14}H_{20}O_7$. Cleavage of the epoxy group by treatment of the mono-O-acetyl dimethyl ester (**108**) with LiAlH$_4$ produced a new

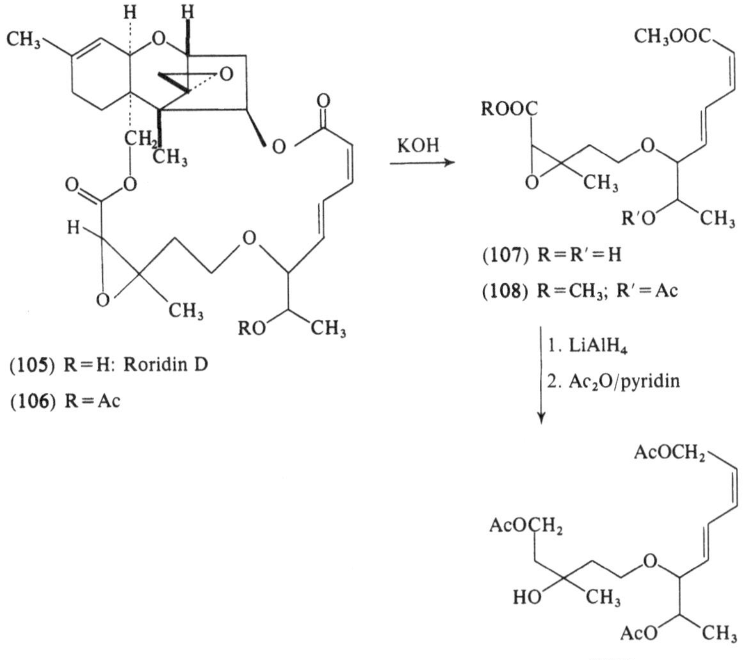

(105) R=H: Roridin D

(106) R=Ac

(107) R=R'=H

(108) R=CH₃; R'=Ac

KOH

1. LiAlH₄
2. Ac₂O/pyridin

(109)

tertiary hydroxyl group as shown by the product (109) which was obtained after subsequent reacetylation.

The relationship between roridin D and roridin A corresponds to that of verrucarin B to verrucarin A. The configurations of the 2′,3′-epoxy group and of the two chiral centres C-6′ and C-13′ are unknown.

4.7. Roridin E

Roridin E, $C_{29}H_{38}O_8$, is an additional companion of both verrucarin A and roridin A isolated from cultures of various strains of *Myrothecium*. Formula (110) was assigned to this metabolite (77). The IR- and NMR-spectra pointed to an $\alpha,\beta,\gamma,\delta$-unsaturated and an additional α,β-unsaturated ester grouping as encountered in verrucarin J (79) with the same *cis,trans*-geometry. The presence of a single secondary hydroxyl group was demonstrated by the formation of a mono-O-acetyl derivative (111). Catalytic hydrogenation of the metabolite yielded octahydrororidin E. Alkaline hydrolysis yielded, besides verrucarol (32), 2′-anhydrororidinic acid (112) which was transformed to the dimethyl ester (113). The product contained a considerable amount of the 2′,3′-*trans*-isomer which is an artefact formed in the

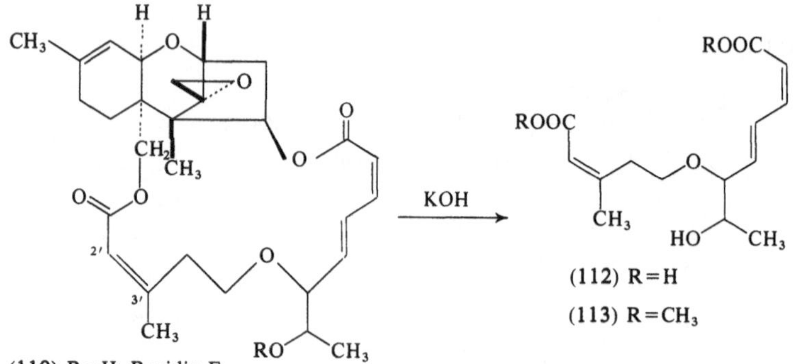

CH₃ ... structures ...

(110) R=H: Roridin E
(111) R=Ac

KOH →

(112) R=H
(113) R=CH₃

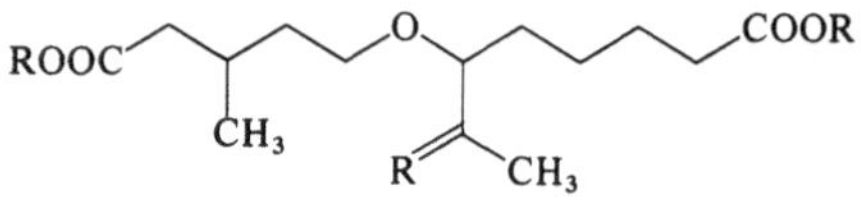

(114) R=H; R'=⟨OH/H
(115) R=CH₃; R'=⟨OH/H
(116) R=CH₃; R'=O

course of the hydrolysis since the 2′,3′-double bond of roridin D (**110**) actually possesses the *cis*-configuration. Hydrolysis of octahydro-roridin D gave dihydroverrucarol (*39*) and hexahydro-2′-anhydro-roridinic acid (**114**). The high resolution mass- and NMR-spectra of the dimethyl ester (**115**) and the oxidation product (**116**) excluded alternative structures. As in roridin A and D the configuration of C-6′ and C-13′ has not yet been established.

4.8. Roridin H

Roridin H is the last member of this family of antibiotics whose structure (**117**) has been elucidated (*76*). Its molecular formula $C_{29}H_{36}O_8$ was established by high resolution mass spectrometry. The metabolite is characterized by the presence of an α,β-unsaturated and an α,β,γ,δ-unsaturated ester group as in the case of verrucarin J and roridin E, but possesses no hydroxyl groups. Detailed analysis of the 220 MHz-NMR-spectrum revealed further structural details. Upon catalytic hydrogenation octahydrororidin H was obtained. Base-catalysed hydrolysis of roridin H (**117**) gave verrucarol (**32**) and a new dicarboxylic acid, $C_{14}H_{18}O_6$, named myrothecinic acid (**118**). Octahydrororidin H yielded dihydroverrucarol (**39**) and hexahydromyrothecinic acid.

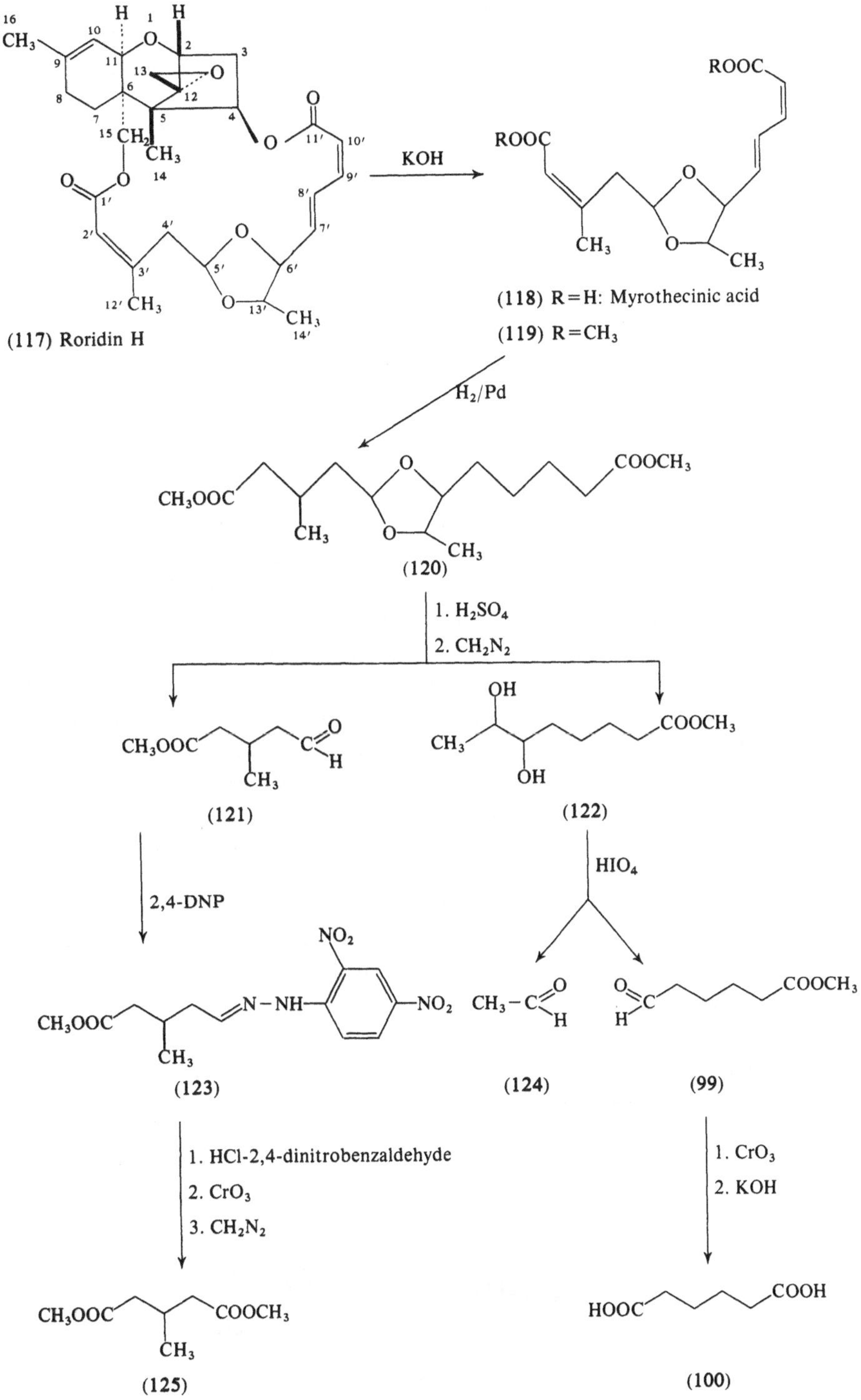
(117) Roridin H
KOH
(118) R=H: Myrothecinic acid
(119) R=CH3
H2/Pd
(120)
1. H2SO4
2. CH2N2
(121)
(122)
2,4-DNP
HIO4
(123)
(124)
(99)
1. HCl-2,4-dinitrobenzaldehyde
2. CrO3
3. CH2N2
1. CrO3
2. KOH
(125)
(100)

Due to the presence of an unusual acetal group the determination of the structure of myrothecinic acid (118) required extended studies. Interpretation of the NMR-spectrum of the dimethyl ester (119) combined with numerous spin-spin decoupling experiments using different solvent systems permitted recognition of the following structural elements:

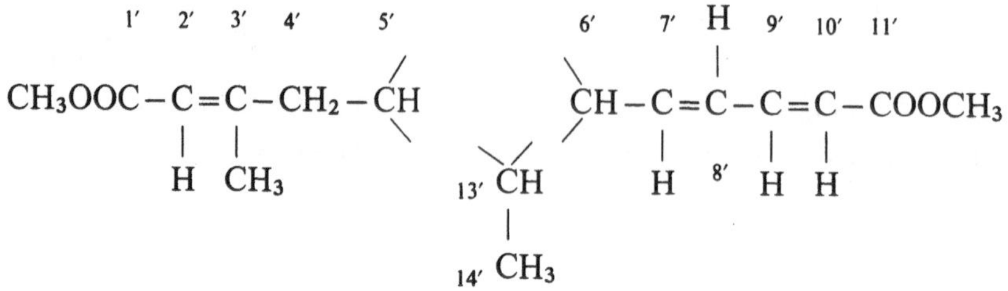

$$
\underset{1'}{} \underset{2'}{} \underset{3'}{} \underset{4'}{} \underset{5'}{} \quad\quad \underset{6'}{} \quad \underset{7'}{} \overset{H}{} \underset{9'}{} \underset{10'}{} \underset{11'}{}
$$

CH$_3$OOC – C = C – CH$_2$ – CH   CH – C = C – C = C – COOCH$_3$
                | | | | | |
                H   CH$_3$ $_{13'}$ CH   H $^{8'}$ H   H
                                                |
                                           $_{14'}$ CH$_3$

On the assumption that the two remaining oxygen atoms form ether linkages, structure (119) was postulated for dimethyl myrothecinate. The fragmentation pattern of the mass spectrum was in agreement with the proposed acetal structure. The latter was confirmed by chemical cleavage of dimethyl hexahydromyrothecinate (120), which was obtained by catalytic hydrogenation of dimethyl myrothecinate (119) or by hydrolysis of octahydrororidin A with conc. H$_2$SO$_4$ in methanol in the presence of 2,4-dinitrophenylhydrazine (DNP). The 2,4-dinitrophenylhydrazone (123) of 3-methylglutaric acid semialdehyde (121) and methyl 6,7-dihydroxy-octanoate (122) were produced. The 2,4-dinitrophenylhydrazone (123) was cleaved with HCl in the presence of 2,4-dinitrobenzaldehyde. The regenerated 3-methylglutaric acid semialdehyde was oxidized with CrO$_3$ and methylated with diazomethane to yield dimethyl 3-methylglutarate (125). Cleavage of methyl 6,7-dihydroxy-octanoate (122) with HIO$_4$ gave acetaldehyde (124) (isolated as acetaldimedone) and methyl adipic acid semialdehyde which was transformed to adipic acid (100).

NMR-data for roridin H indicate a *cis*-arrangement of the H-atoms at C-6′ and C-13′. They do not permit distinction between the two possible diastereoisomers shown in Fig. 4.

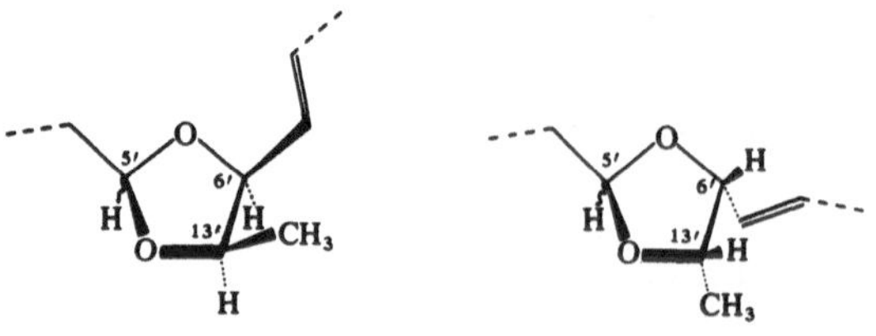

Fig. 4. Diastereoisomers of roridin H

The configuration at C-5′ is unknown. In view of the close relationship with the other roridins it is assumed that in roridin H myrothecinic acid (118) is connected with the verrucarol moiety in an analogous manner as shown in formula (117).

5. Wortmannin

Wortmannin, a neutral solid, was shown to possess structure (126) by chemical and spectroscopic evidence (53). The absolute configuration was established very recently by X-ray analysis (59) and by incorporation experiments with 2-[³H]-lanosterol and 2S-[³H]- and 2R-[³H]-2-[¹⁴C]-mevalonic acid (52).

(126)

IV. Total Synthesis

The first total synthesis of a member of the trichothecane family, (±)-trichodermol (roridin C) (29) and its acetyl derivative trichodermin (30), was reported in 1971 by COLVIN et al. (18). The unsaturated ester (127) served as starting material. It was converted with methylmagnesium chloride to a tertiary hydroxy ester which was hydrolysed to the free acid. The latter underwent an acid-catalysed anionotropic rearrangement to give the cis-fused lactone (128). Monomethylation to the corresponding homolactone and treatment of the latter with the lithium salt of 3,3-diethoxypropyne yielded the hemiacetal (130), which was reduced to the diol (131). The triple bond was reduced selectively to give the corresponding trans-ethylenic acetal which, upon treatment with mild acid, was not only hydrolysed but also cyclized to produce the cis-fused bicyclic hydroxyaldehyde (132). Compound (132) was transformed via the corresponding keto-acid to a mixture of the two possible racemates of the enol-lactone (133). Reduction of this mixture by lithium hydrido tri-t-butoxyaluminate led to the tricyclic ketoalcohol (134). Its acetyl derivative (135) was converted by a WITTIG reaction into the methylene compound (137). Regio- and stereoselective epoxidation of the alcohol (136) with *m*-chloroperbenzoic acid produced exclusively the epoxy-alcohol (±)-trichodermol (29) which was acetylated to give (±)-trichodermin (30). An alternative synthesis of the intermediate (129) has been described (*80, 81*).

For the total synthesis of the pyrrole derivative verrucarin E (11) see Chapter III (p. 70).

(127)   (128) R=H   (130)
         (129) R=CH₃

(131)   (132)   (133)

(134) R=H   (136) R=H   (29) R=H
(135) R=Ac   (137) R=Ac   (30) R=Ac

V. Methods of Assay

No procedure both specific for the metabolites and quantitative has been developed which may be applied without a high degree of purification of the test sample. The characteristic absorption bands in the ultraviolet of the verrucarins and roridins may be used, but they do not allow distinction of the various metabolites. No characteristic colour reactions are known. However, for the identification and control of purity of the metabolites thin-layer chromatography has proved to be most useful. In Table 4 the Rf values obtained in several different solvent systems on both alumina and silica gel G plates are listed. Compounds are arranged in the order of decreasing Rf-values. They are made visible either with iodine vapor or by spraying the plate with conc. sulfuric acid and subsequent heating. To detect the verrucarins and roridins of the trichothecane series ultraviolet light can also be used (43, 83). The pyrrole derivatives, e.g. verrucarin E, are also made visible

Table 4. *Rf-Values on Thin-Layer Chromatograms*

Compound	Rf in solvent system						
	1	2	3	4	5	6	7
Roridin H	—	0.59	—	0.72	0.51	—	—
Verrucarin J	—	0.59	—	0.64	0.42	—	—
Verrucarin B	0.83	0.58	0.69	0.63	0.37	—	—
2′-Dehydroverrucarin A	0.82	0.58	0.68	0.63	—	—	—
Verrucarin A	0.70	0.47	0.59	0.47	0.31	—	—
Roridin E	—	0.40	—	0.35	0.24	—	—
Roridin D	—	0.35	—	0.29	0.18	—	—
Roridin A	0.70	0.21	0.21	0.20	0.14	—	—
Roridin C (Trichodermol)	0.86	0.36	0.41	0.17	—	0.31	0.59
Verrucarol	0.34	0.06	—	0.03	—	0.09	0.51
Ergosterol	0.55	0.26	0.49	—	—	—	—
Verrucarin F	—	0.54	—	—	—	—	—
Verrucarin G	—	0.49	—	—	—	—	—
Verrucarin E[1]	0.0	0.0	0.09	—	—	—	—
Verrucarin C	0.74	0.28	0.52	—	—	—	—
Verrucarin D	0.70	0.28	0.55	—	—	—	—

1: Chloroform-methanol (98:2) on alumina (43)
2: Chloroform-methanol (98:2) on silicagel G (83)
3: Chloroform-methanol (97:3) on silicagel G (83, 43)
4: Benzene-tetrahydrofuran (85:15) on silicagel G (83)
5: Ether, twice (83)
6: Toluene-ethylacetate (1:3) on silicagel G (10)
7: Ethanol-ethylacetate-acetone (1:4:4) (10)

---

[1] Methylene chloride with 3−10% methanol was used as additional solvent system (61).

in the ultraviolet (λ_{max}= 254 nm) in the presence of zinc silicate (61) or with moist HCl vapors.

Gas-liquid chromatography can be used for the detection of verrucarol, roridin C (trichodermol) and the simple esters of the related sesquiterpene alcohols but not for the macrocyclic verrucarins and roridins (10). Trichothecanes containing free hydroxyl groups are converted to their trimethylsilyl ethers.

VI. Biosynthesis

1. Pyrrole Derivatives

Only the biosynthesis of verrucarin E (11) has been investigated since the structures of the other metabolites of this class are not known. The incorporation of nine ^{14}C-labelled, assumed biosynthetic precursors of verrucarin E was measured (61) by feeding them to growing cultures of a strain of *Myrothecium verrucaria* which produces this pyrrole derivative (yield ca. 30 mg/l). A high specific incorporation rate was observed for sodium acetate, whereas sodium mevalonate, L-methionine, glycine, 5-aminolevulinic acid, D,L-proline and D,L-glutamic acid were not incorporated. The significant incorporation of L-serine is due to the fact that this α-amino acid is transformed to acetate. The distribution of radioactivity after incorporation of both [1-^{14}C]- and [2-^{14}C]-acetate was determined by Kuhn-Roth oxidation of verrucarin E (11) and of its transformation product 3-acetyl-4-methylpyrrole (15) (61). The acetic acid obtained was further degraded to methylamine and CO_2. It was found that the skeleton of verrucarin E (11) is built up from four acetate units with loss of one carboxyl C-atom according to the scheme outlined in Fig. 5.

$\longrightarrow$ • = CH$_3$COOH

Fig. 5. Arrangement of the acetate units in verrucarin E

The nature of the N-donor is unknown. The results were rationalized by proposing the following biogenetic pathway:

NH₃ represents the unknown N-donor

2. Trichothecane Esters

The feeding experiments were carried out by adding some likely [14]C- and ^{3}H-labelled precursors to growing cultures of *Myrothecium*. The isolated radioactive metabolites were hydrolysed and subjected to further degradation in order to locate the radioactive atom (73). The incorporation rates did not allow the use of ^{13}C-labelled precursors, ^{13}C being located by NMR-spectroscopy without the necessity of a chemical degradation.

(3 R,S)-[2-^{14}C]-mevalonate as well as the natural (3 R)-enantiomer of mevalonic acid were incorporated into verrucarin A and roridin A. The radioactivities were distributed between verrucarol, verrucarinolactone and roridinic acid respectively; *cis,trans*-muconic acid was radioinactive. It was concluded that 3 molecules of mevalonic acid were incorporated into verrucarol, confirming the sesquiterpene nature of the

trichothecane skeleton, and one molecule of mevalonic acid into verrucarinolactone. Acetate was also incorporated into both antibiotics, the distribution of radioactivity suggesting that of the 12 acetate units of verrucarin A 6 are located in verrucarol and 3 each in verrucarino-lactone and *cis,trans*-muconic acid. Roridinic acid is built up from 6−7 acetate units.

Cis,*trans*-muconic acid and the corresponding structural element of roridinic acid and probably also the C_2-side chain were shown to be formed from acetate units. The alternating labelling of muconic acid and the labelling of C-11′ of roridinic acid after incorporation of [1-^{14}C]-acetate followed from the results of the Schmidt degradation of adipic acid obtained from muconic acid to 1,4-diaminobutane and CO_2 and of 1′-nor-tetrahydro-13′-dehydrororidinic acid (**139**) obtained by $KMnO_4$-oxidation of tetrahydrororidinic acid (**138**). Thus in roridinic acid, C-11′ originates from the carboxyl group of acetic acid. Anal-ogously it was concluded that the carboxyl group of muconic acid (**66**) which is esterified with the 4-hydroxyl group of verrucarol is also derived from C-1 of acetic acid. Catechol may be a biogenetic intermediate of muconic acid.

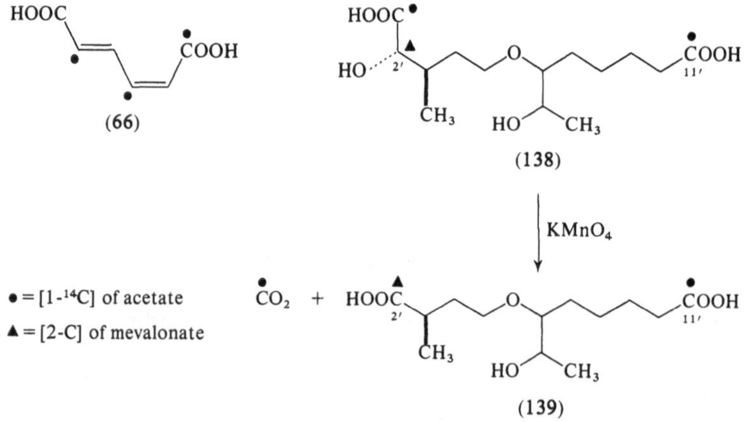

• = [1-^{14}C] of acetate

▲ = [2-C] of mevalonate

Degradation of verrucarinolactone (**67**) derived from [2-^{14}C]- and [5-^{14}C]-mevalonate (**140**) to formaldehyde, 2-methyl-4-hydroxybutanal (**141**) and to methylsuccinic acid (**142**) which gave radioinactive 1,2-diaminopropane (**143**) and radioactive CO_2 demonstrated that the carbon skeleton of mevalonate had not undergone neither a rearrange-ment nor an exchange of the state of oxidation of C-1 and C-5 respectively. Additional proof was adduced by the transformation of roridinic acid (88) to verrucarinolactone (67) on the one hand and to nortetrahydro-13′-dehydrororidinic acid (138) and Schmidt degradation of the latter on the other. The label of C-2 of mevalonate was found in the carboxyl group of the norroridinic acid. Thus it was concluded that C-2 of verrucarinic acid is biogenetically identical with C-2 of the isomeric mevalonic acid.

(140)   (67)   (141)

▲ = [2-^{14}C] of mevalonate
* = [5-^{14}C] of mevalonate

(142)   (143)

To study the mechanism of the later stages of the transformation of mevalonic acid into verrucarinic acid and into verrucarol incorporation of specifically tritiated mevalonates was studied (73, 6). (3R)-[(2S)-2-^{3}H]/(3S)-[(2R)-2-^{3}H]- and (3R)-[(2R)-2-^{3}H]/(3S)-[(2S)-2-^{3}H]-mevalonate in separate experiments gave radioinactive verrucarinolactone from the first experiment and radioactive verrucarinolactone from the second experiment. Since only (3R)-[(2R)-2-^{3}H]-mevalonate was transformed to verrucarinic acid it was concluded that the "pro-2S" hydrogen of mevalonate is lost in the formation of verrucarinolactone. The oxidation of verrucarinate, which had retained the "pro-2R" hydrogen of mevalonate, yielded methyl succinic acid still containing tritium indicating a hydrogen transfer from C-2 of mevalonate to C-3 of verrucarinate. These findings were conclusively proved by the results obtained after the administration of [2-^{3}H, 2-^{14}C]-mevalonate (144) to the mould. The isolated [^{3}H, ^{14}C]-roridin A was hydrolysed to roridinic acid and the latter degraded to verrucarinolactone (145). Subsequent reduction with LiAlH$_4$ and cleavage with HIO$_4$ gave a hydroxyaldehyde which was oxidized with Br$_2$ in water and the 2-methylbutanolide (146) thus formed was treated with phenyl-magnesium iodide. The resulting carbinol (147) gave a 3,5-dinitrobenzoyl derivative (148) with a ^{3}H : ^{14}C ratio which showed that no tritium had been lost during the reaction sequence. Dehydration of the carbinol (148) with SOCl$_2$ yielded the olefin (149) which had retained the ^{14}C-

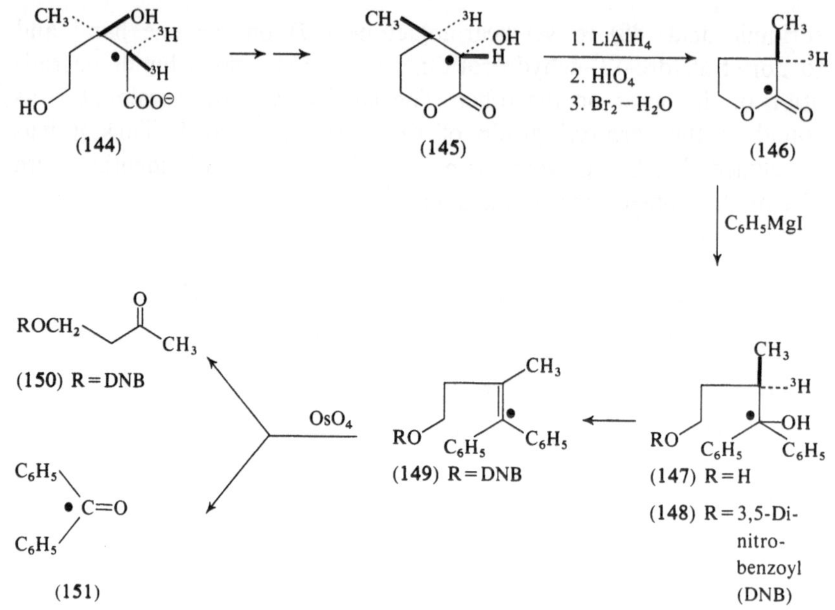

1. LiAlH4
2. HIO4
3. Br2−H2O
(144)
(145)
(146)
C6H5MgI
(150) R=DNB
OsO4
(149) R=DNB
(147) R=H
(148) R=3,5-Dinitrobenzoyl (DNB)
(151)

activity but had completely lost the ^{3}H-activity. The olefin was cleaved with OsO_4-$NaIO_4$ to the butan-1-ol-3-one derivative **(150)** which was radioinactive and to diphenyl ketone **(151)** containing the total ^{14}C-activity.

A possible mechanism for the biogenetic formation of verrucarinate from mevalonate has been proposed and is outlined in the following scheme:

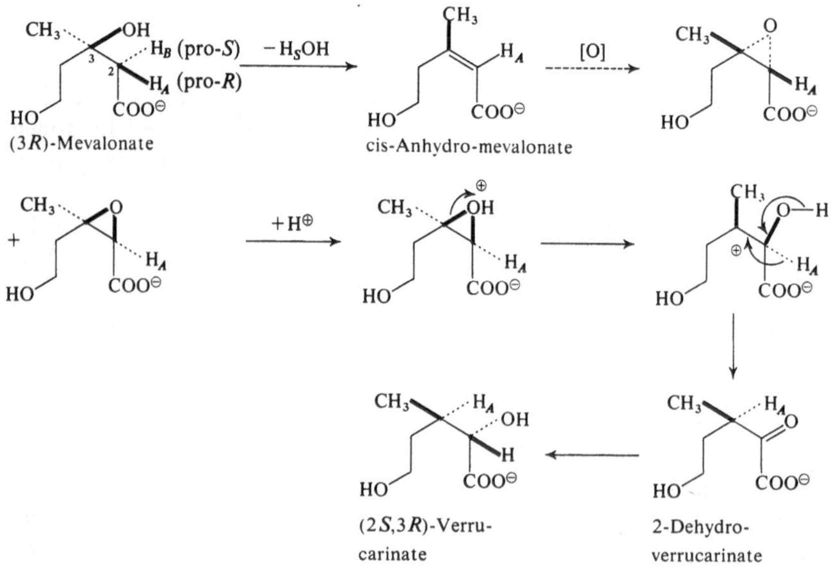

H_B (pro-S)
H_A (pro-R)
−H_SOH
[O]
+H⊕
(3R)-Mevalonate
cis-Anhydro-mevalonate
(2S,3R)-Verrucarinate
2-Dehydro-verrucarinate

The loss of the "pro-2S" hydrogen in mevalonate agrees with the *trans*-elimination of H_2O leading to *cis*-anhydromevalonate. Epoxidation gives rise to two enantiomeric glycidic acids. The "β-epoxide" on the left hand side of the scheme is protonated and cleaved to form a carbenium ion, which is transformed to 2-dehydroverrucarinate by a 1,2-hydride shift involving the "pro-2R" hydrogen (H_A) of mevalonate. In an alternative possibility the hydrogen atom H_A is abstracted from the substrate by an enzyme. The enzyme-bound proton is added to C-3 of the enolic substrate from the same rear side leading again to 2-dehydro-verrucarinate. Another pathway using the "α-epoxide" would require a hydrolytic cleavage of the oxirane ring to a *trans*-diol as intermediate. This possibility could be ruled out since the production of the metabolite in the presence of $H_2^{18}O$ in the medium did not yield an ^{18}O labelled hydroxyl group. — The last step in the formation of (2S,3R)-verru-carinate constitutes the stereospecific reduction of 2'-dehydroverru-carinate. The isolation of metabolites containing *cis*-anhydromevalonic acid, 2,3-epoxy-anhydromevalonic acid and 2'-dehydro-verrucarinic acid provides strong support for the proposed sequence of reactions.

The observation that 3 molecules of mevalonic acid are incorpor-ated into the trichothecane nucleus of verrucarol implies that various rearrangements occur subsequently. To study the distribution of mevalonic acid in the trichothecane skeleton and the mechanism of the rearrangements [^{3}H,^{14}C]-verrucarol (**152**) obtained after the administra-tion of [2-^{3}H,2-^{14}C]-mevalonate was oxidized to the ketoaldehyde (**154**). This reaction resulted in the loss of one fifth of the tritium activ-ity; this is consistent with the attachment of one tritium at C-4. Trans-formation of di-O-acetylverrucarol (**153**) to the 8-oxo product (**155**) was accompanied by the elimination of two tritium atoms from C-8. The second oxidation product, the 16-oxo compound (**157**), and its deacetylation product (**156**) were characterized by a significant increase of the tritium content due to the kinetic isotope effect of tritium. Since C-8 contains tritium this carbon atom is oxidized ca. 20 times more slowly than the competitive carbon atom C-16 which is free of tritium. This observation represents additional evidence for the label-ling of C-8. To confirm that tritium was absent from C-10, verrucarol (**152**) was epoxidized. The 9,10β-epoxide (**158**) could not be isolated due to isomerization to the hydroxy ether (**159**). The latter was converted to the diketo-ether (**160**) with the loss of only the C-4 tritium atom. These results prove that the radioactivity in ring A is located at *C*-8 and not at C-10 as suggested by earlier work for trichothecin (*47*).

The stereochemistry of the hydroxylation step at C-4 of verrucarol was determined by using the tritiated preparations obtained after feeding

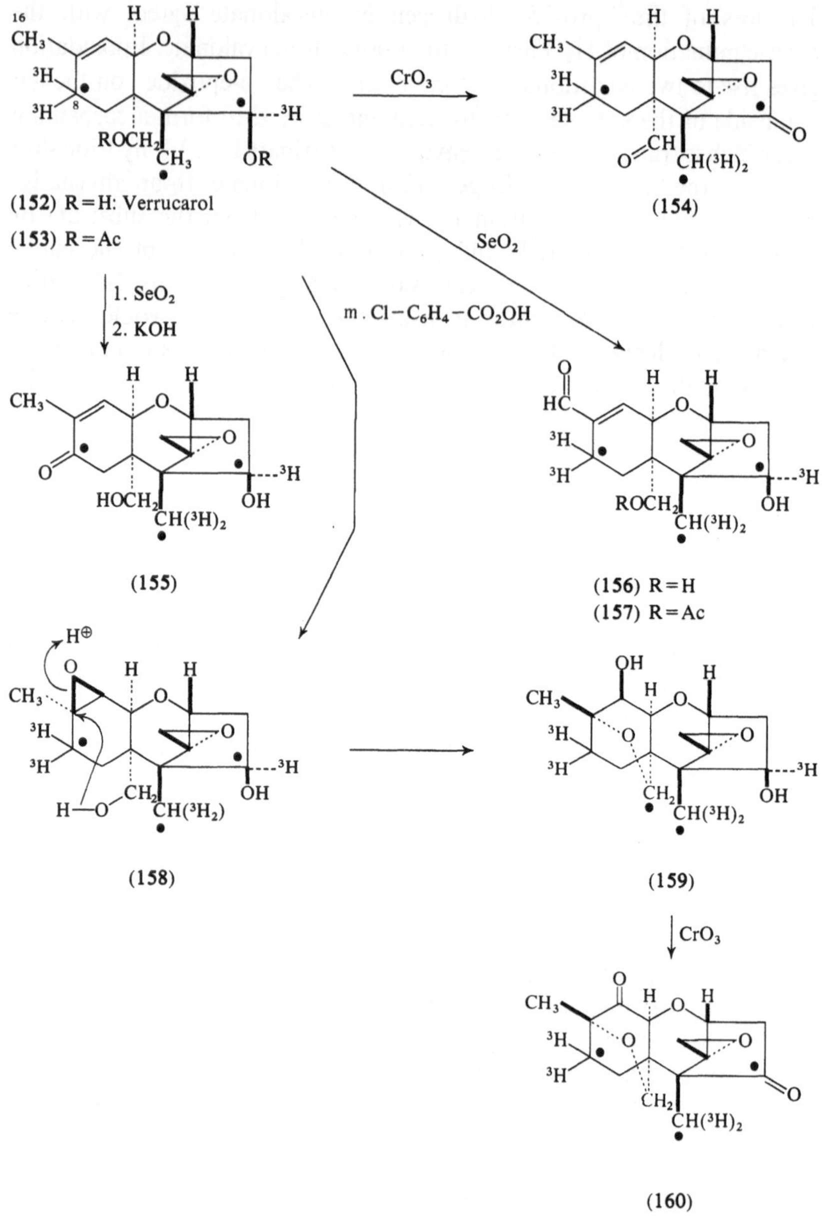
CrO₃
SeO₂
1. SeO₂
2. KOH
m . Cl−C₆H₄−CO₂OH
CrO₃
(152) R=H: Verrucarol
(153) R=Ac
(154)
(155)
(156) R=H
(157) R=Ac
(158)
(159)
(160)

stereospecifically labelled mevalonate preparations. The conversion of verrucarol derived from (3*R*)-[(2*S*)-2-^{3}H]-mevalonate into the keto-aldehyde [structure analogous to formula (**154**)] resulted in ca. 35% loss of tritium activity, consistent with the incorporation of three atoms of tritium into verrucarol, one at C-4. Oxidation of verrucarol derived from $(3R)$-[$(2R)$-2-^{3}H]-mevalonate to the ketoaldehyde produced no change in the specific activity indicating that C-4 carried no tritium. Hence the orientation of the C-4 hydroxy group corresponds to the "pro-$2R$" hydrogen atom of mevalonate, indicating that hydroxylation at C-4 occurs with retention of configuration. These conclusions agreed with those for trichothecin and roridin C (trichodermol) (2, 7).

Mevalonate is transformed to farnesol pyrophosphate (**168**). It has been shown by feeding experiments that this acyclic sesquiterpene acts as precursor of trichothecin (2) and of verrucarol (72). However, the conclusions drawn by ACHILLADELIS et al. (2) do not provide un-equivocal proof for the *trans*-geometry of the 6,7-olefinic linkage of farnesol pyrophosphate, since hydrogen transfers cannot be excluded (73). In fact such a hydrogen migration was demonstrated very recently (8) in case of trichodermol and trichothecin.

In earlier proposals for the biogenetic formation of the tricho-thecane skeleton cyclization of farnesol pyrophosphate to γ-bisabolene (**169**) was postulated (65, 67, 73). Since feeding experiments with cultures of *Myrothecium roridum* and $(+)$-$(1'R)$-$(6R,S)$- and $(-)$-$(1'S)$-$(6R,S)$-α-bisabolol, which was anticipated to undergo elimination in order to form γ-bisabolene, gave negative results, it is also doubtful whether γ-bisabolene is an intermediate in the biosynthesis of the trichothecanes (73). These conclusions agree with those reached recently by FORRESTER and MONEY (26) who demonstrated by administering α-bis-abolol, γ-bisabolene and monocyclofarnesol to cultures of *Trichothecium roseum* that these compounds are not intermediates in the biosynthesis of trichothecin.

The two pairs of diastereoisomers, $(+)$-$(1'R)$-$(6R,S)$- and $(-)$-$(1'S)$-$(6R,S)$-α-bisabolol (**167a**) and (**167b**) respectively were synthetized by reacting [5,6-^{14}C]-4-methyl-pent-3-enemagnesium bromide (**164**) with $(+)$-$(1R)$- and $(-)$-$(1S)$-1-acetyl-4-methyl-cyclohex-3-ene (**166a**) and (**166b**) respectively (72). For the preparation of the Grignard compound (**164**), cyclopropylmethyl ketone (**161**) was treated with [^{14}C]-methyl-magnesium iodide to form the carbinol (**162**). Cleavage of the cyclo-propane ring by HBr yielded radioactive 1-bromo-4-methyl-pent-3-ene (**163**). Treatment of the latter by magnesium gave the first component (**164**). The second components, (**166a**) and (**166b**) respectively, were synthetized starting from $(+)$-(R)- and $(-)$-(S)-limonene (**165a**) and (**165b**) respectively. The endocyclic double bond was protected by selective epoxidation. Subsequent treatment with ozone and reduction with zinc led directly to the desired compounds (**166a**) and (**166b**) respectively.

Feeding [2-^{3}H,2-^{14}C]-geranyl pyrophosphate to *Trichoderma* and *Trichothecium* yielded trichodermol and trichothecin which both had

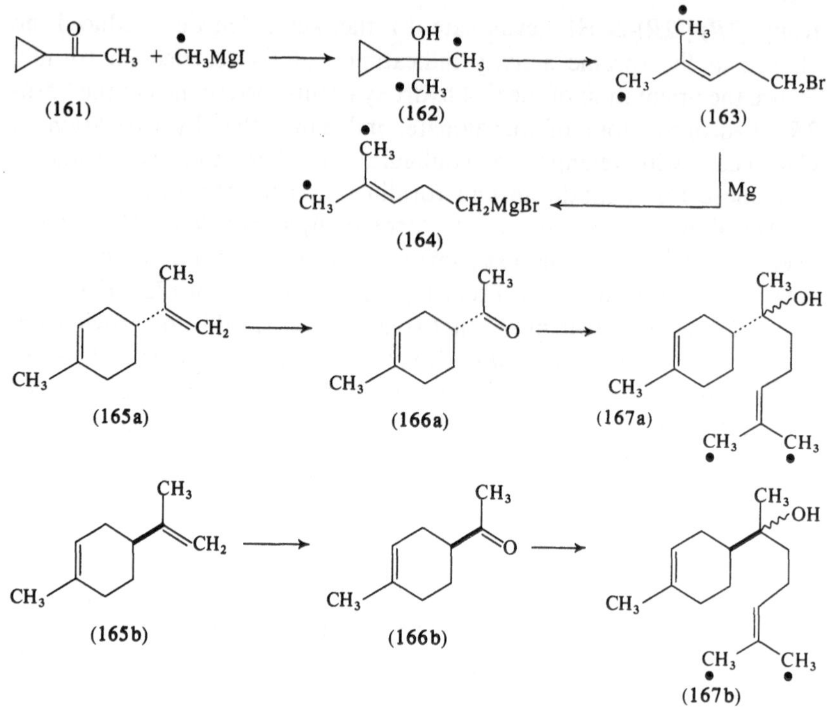

(161)   (162)   (163)

(164)

(165a)   (166a)   (167a)

(165b)   (166b)   (167b)

retained the [2-³H]-geranyl phosphate label and hence the second (4R)-[4-³H]-hydrogen of mevalonate. These results also preclude a bisabolene intermediate in the trichothecane biosynthesis. It is interesting to note that γ-bisabolene (169) had also been excluded as an intermediate in the helicobasidin biosynthesis (9). In order to accomodate these results a concerted sequence in which a hydrogen transfer occurs in the enzyme displacement step had been proposed (3, 8). This transfer could also initiate the two methyl group rearrangements in the carbenium ion (171) which is followed by abstraction of a proton, thus forming the exocyclic methylene group and leading to trichodiene (172) or a related intermediate. This hydrocarbon (172) and the tricyclic hydrocarbon (174) (51) have been isolated from *Trichothecium roseum* and were suggested as precursors of trichothecin (56, 57). Feeding experiments with tritium labelled trichodiene which was incorporated into trichothecolone, 12,13-epoxy-Δ⁹-trichothecene and into trichodiol A have confirmed this suggestion (50).

These and analogous results obtained with verrucarol in *Myrothecium* (72) are summarized in the following scheme which outlines a hypothetical pathway of 12,13-epoxy-Δ⁹-trichothecene biosynthesis. Epoxidation of the 12,13-methylene group and hydroxylation of the tricyclic skeleton lead to trichodermol (roridin C) (29), verrucarol (32) and the other known trichothecane derivatives.

The present knowledge of the biosynthesis of the verrucarins and roridins can be summarized by the statement that they represent a biogenetic combination of an isoprenoid moiety with a polyketide or its biogenetic equivalents. The isoprenoid part containing 20 carbon atoms is built up from a C_{15} unit (trichothecane) and by a C_6-unit (verrucarinate). The details of their formation are known only in part.

The individual macrocyclic metabolites of *Myrothecium* can be arranged in a biogenetic sequence starting with verrucarin J (79) which is followed by verrucarin B (78) and 2'-dehydroverrucarin A (75) and ends with verrucarin A (74). For the roridin series the sequence is roridin E (110) and H (117), roridin D (105) and roridin A (84). However, again a great number of problems remain unresolved, e.g. are the trichothecane moiety and the acids built up separately simultaneously or in subsequent steps before they are joined together? At what stage and by which sequence are the ester linkages formed?

(138)   (168)   (169) γ-Bisabolene

(170)   (171)   (172) Trichodiene

(173)   (174)   (29) R=H: Trichodermol
(32) R=OH: Verrucarol

VII. Biological Activity

The biological activity of the verrucarins, roridins and related trichothecene derivatives has been reviewed recently *in extenso* by BAM-
BURG and STRONG (*10*). We therefore restrict ourselves to summarizing
the significant results of these extended studies. In general the verrucarins and roridins have been recognized as potent cytostatic agents.
In addition they possess some antifungal properties. They are highly
toxic substances.

1. Antibiotic Activity

Only verrucarin A was reported to possess antibacterial activity
(*43*). It showed slight inhibition against gram-negative bacteria (*Escherichia coli, Serratia marcescens, Proteus mirabilis, Pseudomonas aeruginosa,
Salmonella abortus equi*) at 50 μg/ml. Verrucarin A was inactive against
gram-positive bacteria and mycobacteria. Antifungal activity was
exhibited by verrucarin A against *Candida albicans, Saccharomyces cer-
visiae* and *Saccharomyces carlsbergensis* at 0.5 μg/ml. The activity
was much weaker against other fungi *(Aspergillus, Blastomyces)* including human pathogenic fungi *(Trichophyton, Achorion)*. However,
di-O-acetyl-verrucarol showed relatively high activity against *Trichophyton asteroides* (*58*). In addition verrucarin B showed also a fungi-
static effect, but at much higher concentrations while roridin A proved
to be inactive.

Verrucarin A showed antiviral activity. It was effective against *Herpes
simplex* virus strain (HSV) when tested by the agar diffusion plaque
inhibition method (*74*).

2. Cytostatic Activity

The verrucarins A and B, and the roridins A and C (trichodermol)
were found to inhibit the growth of tumor cells of mice (mastocytom
P-815) (*43*). The concentrations causing 50% inhibition (μg/ml) reached
the following values: verrucarin A: 0.0006; verrucarin B: 0.003; roridin
A: 0.001 and roridin C (trichodermol): 0.05. Thus these compounds,
especially verrucarin A, are among the most active cytostatic agents
known. The effect of verrucarin A on chick embryo fibroblasts was
also determined and found to affect dividing cells greatly by 0.01 μg/ml

after about four hours (*64*). Nondividing cells, however, were scarcely affected by 10 µg/ml even after 24 hours of contact. In dividing cells, the most significant effects were on the chromosomes which appeared grossly abnormal and mitosis was completely disrupted. It is possible that verrucarin A interacts with the S-phase, in which the doubling of DNA takes place.

The verrucarins A and B and roridin A also show *in vivo* cytostatic activity against sarcoma 37 and EHRLICH ascites tumors in mice and WALKER carcinoma in rats (*43*). Verrucarin A is also active against YOSHIDA sarcoma (*64*). In higher mammalians such as pigs, dogs and monkeys a pronounced effect on the lymphatic system was observed. Repeated intravenous administration of verrucarin A into these animals resulted in severe leukopenia which was reversible on interruption of treatment (*64*). The animals showed symptoms such as diarrhea, hematuria, vomiting, loss of weight, anorexia, thirst and ataxia. The tumor-specific and embryotoxic effect of verrucarin A on the tumor system of chorio-allantoic membrane were tested (*63*). The mode of action is unknown. It was found that verrucarin A had no significant effect on aerobic or anaerobic glycolysis in cells or tissues suspended in KREBS-RINGER solution (*64*). Rats treated with verrucarin A or J developed creatinurea.

3. Toxicity

The verrucarins and roridins and other trichothecanes cause severe local irritation and inflammation of the skin. The toxicities as expressed in LD_{50} values have been determined for intravenous and intraperitoneal administration. Intravenous application in mice gave the following results: Verrucarin A: 1.5 mg/kg; verrucarin B: 7.0 mg/kg; roridin A: 1.0 mg/kg (*43*). The corresponding values for intraperitoneal applications were found: verrucarin A: 0.5−0.75 mg/kg; verrucarin J: 0.5−0.75 mg/kg. The administration of the verrucarins A or J intraperitoneally in mice did not result in death in less than 18 hours after injection.

These values demonstrate clearly that the verrucarins and roridins are among the most toxic compounds known which do not contain nitrogen.

Verrucarin A and roridin A also proved to have insecticidal activity. They caused the death of Mexican bean beetles (*49*). Symptoms of the toxicity in the insect were ataxia and hyperextention of the legs. The LD_{50} of verrucarin A for the Mexican bean beetle is about 0.5 mg/insect, a value not very different from the mammalian toxicity of verrucarin A.

4. Structure and Biological Activity

So far relatively little work has been carried out to establish the structural requirements for biological activity in the verrucarin and roridin series. From observations on simpler esters of trichothecene alcohols it is apparent that the free sesquiterpene alcohols are not as active as are the esterified compounds. It is quite clear that the high activity of the verrucarins and roridins is due to the presence of the unsaturated macrocyclic ester system. Upon hydrogenation a considerable loss of activity is encountered. Removal of the epoxide ring by reduction with lithium aluminium hydride in a related compound has led to a complete loss of activity (37).

The same was true for products which had undergone rearrangement to the apotrichothecane skeleton. These results indicate that biological activity, i.e. toxicity is associated with the presence of the 12,13-epoxy group with the double bond in the 9-position and with esterification of the hydroxyl groups of the trichothecane nucleus to form the macrocyclic ring system.

Acknowledgement

The author is indebted to the "Swiss National Science Foundation" and to *Sandoz AG.*, Basel, for the generous support of the work reviewed in this article.

References

1. ABRAHAMSSON, S., and B. NILSSON: Direct Determination of the Molecular Structure of Trichodermin. Proc. Chem. Soc. **1964**, 188.

2. ACHILLADELIS, B., P. M. ADAMS, and J. R. HANSON: The Biosynthesis of the Sesquiterpenoid Trichothecane Antibiotics. Chem. Comm. **1970**, 511.

3. — — — Studies in Terpenoid Biosynthesis. Part VIII. The Formation of the Trichothecane Nucleus. J. C. S. Perkin I **1972**, 1425.

4. ACHINI, R., U. MEYER, and CH. TAMM: Synthese des ($\pm$)-Verrucarinsäurelactons. Helv. Chim. Acta **51**, 1702 (1968).

5. ACHINI, R., and CH. TAMM: Der oxydative Abbau von Roridin A, ein weiterer Beweis für die Art der Verknüpfung der Roridinsäure mit Verrucarol. Helv. Chim. Acta **51**, 1712 (1968).

6. ACHINI, R., B. MÜLLER, and CH. TAMM: Biosynthesis of Verrucarol, the Sesquiterpene Moiety of the Verrucarins and Roridins. Chem. Comm. **1971**, 404.

7. ADAMS, P. M., and J. R. HANSON: Biosynthesis of the Sesquiterpenoid Trichothecane Antibiotics. Chem. Comm. **1970**, 1569.

8. — — A Hydrogen Shift in Trichothecane Biosynthesis. Chem. Comm. **1971**, 1414.

9. — — The Biosynthesis of Helicobasidin. J. C. S. Perkin I **1972**, 586.

10. BAMBURG, J. R., and F. M. STRONG: 12,13-Epoxytrichothecenes in "Microbial Toxins", Vol. 7, pg 207. New York: Academic Press, Inc. 1971.

11. BÖHNER, B., E. FETZ, E. HÄRRI, H. P. SIGG, CH. STOLL, and CH. TAMM: Über die Isolierung von Verrucarin H, Verrucarin J, Roridin D und Roridin E aus *Myrothecium*-Arten. Helv. Chim. Acta **48**, 1079 (1965).

12. BÖHNER, B., and CH. TAMM: Die Konstitution von Roridin A. Helv. Chim. Acta **49**, 2527 (1966).

13. — — Die Konstitution von Roridin D. Helv. Chim. Acta **49**, 2547 (1966).

14. BOWDEN, J. P., and E. J. SCHANTZ: The Isolation and Characterization of Dermatitic Compounds Produced by *Myrothecium verrucaria*. J. Biol. Chem. **214**, 365 (1955).

15. BRIAN, P. W., P. J. CURTIS, H. G. HEMMING, and G. L. F. NORRIS: Wortmannin, an Antibiotic Produced by *Penicillium Wortmannii*. Brit. Mycol. Soc. Trans. **40**, 365 (1957).

16. BRIAN, P. W., and J. G. MCGOWAN: Biologically Active Metabolic Products of the Mould *Metarrhizium glutinosum* S. Pope. Nature **157**, 334 (1946).

17. CLEMENTS, F. E.: The Genera of Fungi. New York: Hafner Publishing Co. 1957.

18. COLDIN, E. W., R. A. RAPHAEL, and J. S. ROBERTS: The Total Synthesis of (±)-Trichodermin. Chem. Comm. **1971**, 858.

19. CORNFORTH, J. W., R. H. CORNFORTH, G. POPJAK, and Y. GORE: Studies on the Biosynthesis of Cholesterol. 5. Biosynthesis of Squalene from D,L-3-Hydroxy-3-methyl-[^{14}C]pentano-5-lactone. Biochem. J. **69**, 146 (1958).

20. DIECKMANN, H.: Fusigenin, ein neues Sideramin aus Pilzen. Archiv für Mikrobiologie **58**, 1 (1967).

21. DIECKMANN, H., and H. ZÄHNER: Konstitution von Fusigenin und dessen Abbau zu Δ^2-Anhydromevalonsäurelacton. European J. Biochem. **3**, 213 (1967).

22. DIECKMANN, H.: Die Isolierung und Darstellung von trans-5-Hydroxy-3-methylpenten-(2)-säure. Archiv für Mikrobiologie **62**, 322 (1968).

23. FETZ, E., and CH. TAMM: Die Konstitution von Verrucarin E. Helv. Chim. Acta **49**, 349 (1966).

24. FETZ, E., B. BÖHNER, and CH. TAMM: Die Konstitution von Verrucarin J. Helv. Chim. Acta **48**, 1669 (1965).

25. FISHMAN, J., E. R. H. JONES, G. LOWE, and M. C. WHITING: The Chemistry and Stereochemistry of Trichothecin. J. Chem. Soc. **1960**, 3948.

26. FORRESTER, J. M., and T. MONEY: Sequence Studies in Biosynthesis: Trichothecin. Canad. J. Chem. **50**, 3310 (1972).

27. FREEMAN, G. G., J. E. GILL, and W. S. WARING: The Structure of Trichothecin and its Hydrolysis Products. J. Chem. Soc. **1959**, 1105.

28. FREEMAN, G. G., and R. I. MORRISON: Trichothecin: an Antifungal Metabolic Product of *Trichothecium roseum* Link. Nature **162**, 30 (1948).

29. — — The Isolation and Chemical Properties of Trichothecin, an Antifungal Substance from *Trichothecium roseum* Link. Biochem. J. **44**, 1 (1949).

30. FRIES, E. M.: Systema Mycologicum, Vol. 3¹, 216—218 (1829).

31. GARDNER, D., A. T. GLEN, and W. B. TURNER: Calonectrin and 15-Deacetylcalonectrin, New Trichothecanes from *Calonectria nivalis*. J. C. S. Perkin I **1972**, 2576.

32. GILMAN, J. C.: A Manual of Soil Fungi, 2nd Ed. Ames, Iowa, USA: The Iowa State College Press. 1957.

33. GODTFREDSEN, W. O., and S. VANGEDAL: Trichodermin, a New Antibiotic, Related to Trichothecin. Proc. Chem. Soc. **1964**, 188.

34. — — Trichodermin, a New Sesquiterpene Antibiotic. Acta Chem. Scand. **19**, 1088 (1965).

35. GODTFREDSEN, W. O., J. F. GROVE, and CH. TAMM: Zur Nomenklatur einer neueren Klasse von Sesquiterpenen. Helv. Chim. Acta **50**, 1666 (1967).

36. GROVE, J. F.: The Constituents of Glutinosin. J. Chem. Soc. **1968**, 810.

37. GROVE, J. F., and P. H. MORTIMER: The Cytotoxicity of Some Transformation Products of Diacetoxyscirpenol. Biochem. Pharmacol. **18**, 1473 (1969).

38. GUTZWILLER, J., R. MAULI, H. P. SIGG, and CH. TAMM: Die Konstitution von Verrucarol und Roridin C. Helv. Chim. Acta **47**, 2234 (1964).

39. GUTZWILLER, J., and CH. TAMM: Über die Struktur von Verrucarin A. Helv. Chim. Acta **48,** 157 (1965).

40. — — Über die Verrucarine und Roridine; Struktur von Verrucarol (vorläufige Mitteilung). Helv. Chim. Acta **46,** 1786 (1963).

41. — — Über die Struktur von Verrucarin B. Helv. Chim. Acta **48,** 177 (1965).

42. HALLIWELL, G.: The Action of Cellulolytic Enzymes from *Myrothecium verrucaria.* Biochem. J. **79,** 185 (1961).

43. HÄRRI, E., W. LOEFFLER, H. P. SIGG, H. STÄHELIN, CH. STOLL, CH. TAMM, and D. WIE-SINGER: Über die Verrucarine und Roridine, eine Gruppe von cytostatisch hochwirksamen Antibiotica aus *Myrothecium*-Arten. Helv. Chim. Acta **45,** 839 (1962).

44. HOREAU, A.: Principe et applications d'une nouvelle méthode de détermination des configurations dite "par dédoublement partiel". Tetrahedron Letters **1961,** 506.

45. — Détermination des configurations par "dédoublement partiel" – II. Précisions et compléments. Tetrahedron Letters **1962,** 965.

46. HOREAU, A., and H. B. KAGAN: Détermination des Configurations par "dédoublement partiel" – III. Alcools stéroides. Tetrahedron **20,** 2431 (1964).

47. JONES, E. R. H., and G. LOWE: The Biogenesis of Trichothecin. J. Chem. Soc. **1960,** 3959.

48. KELLER-SCHIERLEIN, W.: Stoffwechselprodukte von Mikroorganismen: Über die Konstitution von Ferrirubin, Ferrirhodin und Ferrichrom A. Helv. Chim. Acta **46,** 1920 (1963).

49. KISHABA, A. N., D. L. SHANKLAND, R. W. CURTIS, and M. C. WILSON: Substances Inhibitory to Insect Feeding with Insecticidal Properties from Fungi. J. Econ. Entomol. **55,** 211 (1962).

50. MACHIDA, Y., and S. NOZOE: Biosynthesis of Trichothecin and Related Compounds. Tetrahedron Letters **1972,** 1969.

51. — — Biosynthesis of Trichothecin and Related Compounds. Tetrahedron **28,** 5113 (1972).

52. MACMILLAN, J., T. J. SIMPSON, and S. K. YEBOAH: Absolute Stereochemistry of the Fungal Product, Wortmannin. J. C. S. Chem. Comm. **1972,** 1063.

53. MACMILLAN, J., A. E. VANSTONE, and S. K. YEBOAH: The Structure of Wortmannin, a Steroidal Fungal Metabolite. Chem. Comm. **1968,** 613.

54. MCPHAIL, A. T., and G. A. SIM: The Structure of Verrucarin A; X-Ray Analysis of Verrucarin A p-Iodobenzenesulphonate. J. Chem. Soc. **1966,** 1394.

55. NESPIAK, A., M. KOCÓR, and A. SIEWINSKI: Antibiotic Properties of Mycelium and Metabolites of *Myrothecium roridum* Tode. Nature **192,** 138 (1961).

56. NOZOE, S., and Y. MACHIDA: Structure of Trichodiene. Tetrahedron Letters **1970,** 2671.

57. — — The Structures of Trichodiol and Trichodiene. Tetrahedron **28,** 5105 (1972).

58. OKUCHI, M., M. ITOH, Y. KANEKO, and S. DOI: A new Antifungal Substance Produced by *Myrothecium.* Agr. Biol. Chem. (Jap.) **32,** 394 (1968).

59. PETCHER, T. J., H.-P. WEBER, and Z. KIS: Crystal Structure and Absolute Configuration of Wortmannin and of Wortmannin p-Bromobenzoate. J. C. S. Chem. Comm. **1972,** 1061.

60. PFÄFFLI, P., and CH. TAMM: Revidierte Struktur von Verrucarin E. Eine Synthese des Antibioticums und verwandter β-Acetyl-Pyrrol-Derivate. Helv. Chim. Acta **52,** 1911 (1969).

61. — — Über die Biosynthese des Antibioticums Verrucarin E. Helv. Chim. Acta **52,** 1921 (1969).

62. PRESTON, N. C.: Observations on the genus *Myrothecium* Tode. I. The three classic species. Brit. Mycol. Soc. Trans. **26,** 158 (1943).

63. RAUEN, H. M., and K. NORPOTH: Tumorspezifische Wirkungen von Substanzen der Podophyllotoxin- und Verrucaria-Reihe am Chorioallantoismembran-Tumor-System. Arzneim.-Forsch. (Drug Res.) **16,** 1594 (1966).

64. RÜSCH, M. E., and H. STÄHELIN: Über einige biologische Wirkungen des Cytostaticums Verrucarin A. Arzneim.-Forsch. (Drug Res.) **15,** 893 (1965).

65. RUZICKA, L.: Perspectives of the Biogenesis and Chemistry of Terpenes. Pure and Appl. Chem. **6**, 493 (1963).

66. SELBY, K.: The Degradation of Cotton Cellulose by the Extracellular Cellulase of *Myrothecium verrucaria*. Biochem. J. **79**, 562 (1961).

67. SIGG, H. P., R. MAULI, E. FLURY, and D. HAUSER: Die Konstitution von Diacetoxyscirpenol. Helv. Chim. Acta **48**, 962 (1965).

68. SNATZKE, G., and CH. TAMM: Konformation und Circulardichroismus der Sesquiterpene vom Trichothecan-Typ und deren makrocyclischen Estern. Helv. Chim. Acta **50**, 1618 (1967).

69. SCHUMACHER, R., J. GUTZWILLER, and CH. TAMM: 2β,13-Oxido-apotrichothecane: Synthese, chemische und physikalische Eigenschaften. Helv. Chim. Acta **54**, 2080 (1971).

70. TAMM, CH., B. BÖHNER, and W. ZÜRCHER: Myrochromanol und Myrochromanon, zwei weitere Metaboliten von *Myrothecium roridum* Tode ex Fr. Helv. Chim. Acta **55**, 510 (1972).

71. TAMM, CH., and J. GUTZWILLER: Über die Verrucarine und Roridine, Partialstruktur von Verrucarin A (Vorläufige Mitteilung) Helv. Chim. Acta **45**, 1726 (1962).

72. TAMM, CH., et al., unpublished results.

73. TAMM, CH.: Biogenetic Pathways of Fungal Metabolites, XXIIIrd Internat. Congress Pure and Appl. Chemistry, Spec. Lectures, Vol. 5, pg 49. London: Butterworths. 1971.

74. TAMURA, G., K. ANDO, S. SUZUKI, A. TAKATSUKI, and K. ARIMA: Antiviral Activity of Brefeldin A and Verrucarin A. J. Antibiotics **21**, 160 (1968).

75. THOMPSON, K. V. A., and ST. C. SIMMENS: Appendages on the Spores of *Myrothecium verrucaria*. Nature **193**, 196 (1962).

76. TRAXLER, P., and CH. TAMM: Die Struktur des Antibioticums Roridin H. Helv. Chim. Acta **53**, 1846 (1970).

77. TRAXLER, P., W. ZÜRCHER, and CH. TAMM: Die Struktur des Antibioticums Roridin E. Helv. Chim. Acta **53**, 2071 (1970).

78. VITTIMBERGA, B. M.: Studies on the Structure of Muconomycin A, a New Biologically Active Compound. J. Org. Chemistry **28**, 1786 (1963).

79. VITTIMBERGA, J. S., and B. M. VITTIMBERGA: Muconomycin B, a New Biologically Active Compound. J. Org. Chemistry **30**, 746 (1965).

80. WELCH, ST. C., and R. Y. WONG: A Synthetic Precursor of (±)-Trichodermin. Tetrahedron Letters **1972**, 1853.

81. — — A Synthetic Intermediate for Trichothecane Phytotoxic Sesquiterpenoids. Synthetic Comm. **2**, 291 (1972).

82. WHITE, W. L., and M. H. DOWNING: The Identity of *"Metarrhizium glutinosum"*. Mycologia **39**, 546 (1947).

83. ZÜRCHER, W., and CH. TAMM: Isolierung von 2′-Dehydroverrucarin A als Metabolit von *Myrothecium roridum* Tode ex Fr. Gattungstyp bei Fries. Helv. Chim. Acta **49**, 2594 (1966).

84. ZÜRCHER, W., J. GUTZWILLER, and CH. TAMM: Der oxydative Abbau von Verrucarin A, ein weiterer Strukturbeweis. Helv. Chim. Acta **48**, 840 (1965).

(Received January 12, 1973)

Aflatoxins and Sterigmatocystins

By John C. Roberts, Nottingham, England

Contents

I. Introduction

In this review the term mycotoxicosis refers to poisoning of man or animals by ingestion of foodstuffs contaminated with certain moulds and/or with their metabolic products (mycotoxins).

From time immemorial, man and his domestic animals have suffered ill-effects from the ingestion of foodstuffs which have been spoiled by moulds or fungi. The earliest recorded reference to man's affliction by moulds is probably that (*1*) to be found in the Book of AMOS (*ca.* 760 B.C.). Ergotism, produced by the consumption of bread made from rye infected with the fungus *Claviceps purpurea,* is a classical instance of mycotoxicosis. The history, incidence and cause of ergotism, which afflicted untold suffering on the peasantry of Europe in medieval times, has been described in a masterly treatise (*5*). A number of other mycotoxicoses (*e.g.,* mouldy rice toxicosis in Japan and alimentary toxic aleukia in Russia) have been recognised and their incidence and cause have been reviewed (*55, 54, 4*).

In the latter part of 1960, some 100,000 young turkeys on farms in England died suddenly from an unidentified disease. A heavy mortality was also discovered among young ducks, partridges, and pheasants. An intensive investigation by veterinary scientists, chemists, and mycologists eventually proved (*73, 75*) that the new disease (which had been called "Turkey X Disease") was a mycotoxicosis caused by the ingestion of ground-nut (pea-nut) meal infected with the mould *Aspergillus flavus* Link *ex* Fries. Further work established (*41*) that this mould, when grown on suitable substrates, produces a number of highly toxic metabolites called, collectively, the aflatoxins (A*spergillus* **flavus toxins**). The individual aflatoxins are colourless, crystalline compounds of rather high melting point. Some eight, different but closely related, aflatoxins are now known (Chart 1), the most notorious of which is aflatoxin-B1 (**1**) (often produced in major proportion). The aflatoxins are highly carcinogenic (*41, 29, 30*), repeated doses of microgram quantities being fatal to small animals such as young poultry, rats, and rainbow trout*. Aflatoxin-B1 is now recognised as the most active hepatocarcinogen yet discovered (*21*). An excellent treatise, covering most aspects of aflatoxicosis and its attendant problems, has been published (*38*).

Sterigmatocystin (**9**) (Chart 2), a yellow, crystalline metabolite of *Aspergillus versicolor* (Vuillemin) Tiraboschi, was isolated and its structure elucidated (1962) (*25, 19*) before its carcinogenic activity had been recognised (*29, 31, 50*). Marked similarity in structure between sterigmatocystin and the aflatoxins became apparent when the structures of the latter were subsequently elucidated (*2, 3*). Sterigmatocystin is, however, considerably less carcinogenic than the aflatoxins. Five other

---

* Disturbing outbreaks of liver cancer in hatchery-raised rainbow trout in various parts of the world had been reported from 1950 onwards. Eventually (1963—65), it became apparent that the problem paralleled that of "Turkey X Disease" and that the aetiological agent was the same (*28, 59*).

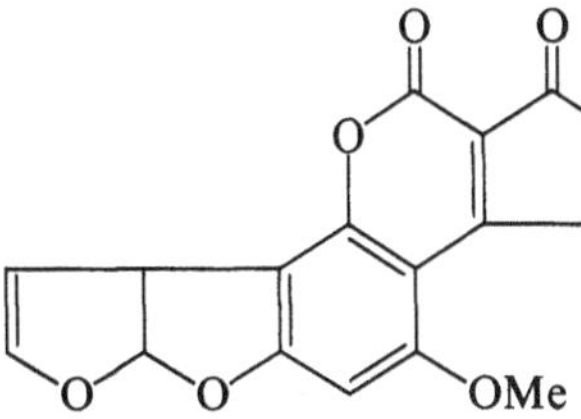

(1) Aflatoxin-B1
m.p. 268−269° (d.), [a]$_D$ −558°

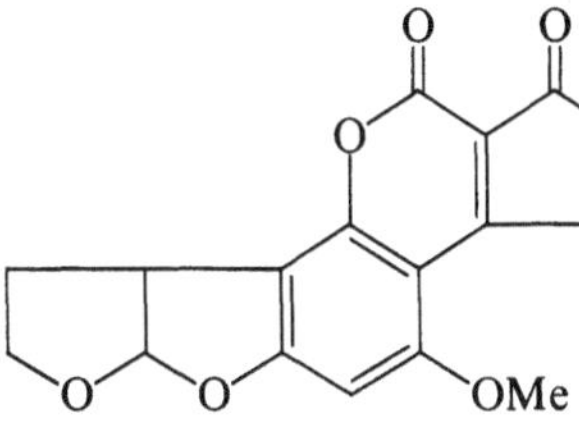

(2) Aflatoxin-B2
m.p. 305−309° (d.), [a]$_D$ −492°

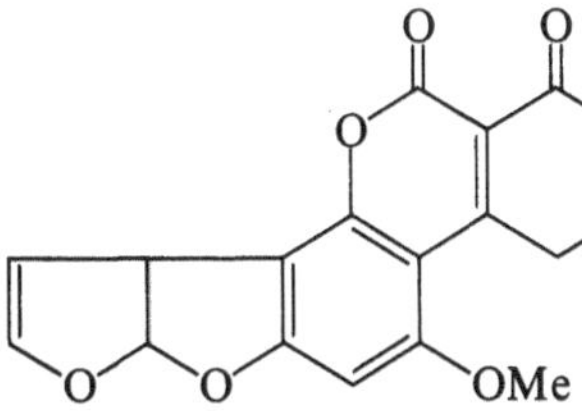

(3) Aflatoxin-G1
m.p. 244−246° (d.), [a]$_D$ −556°

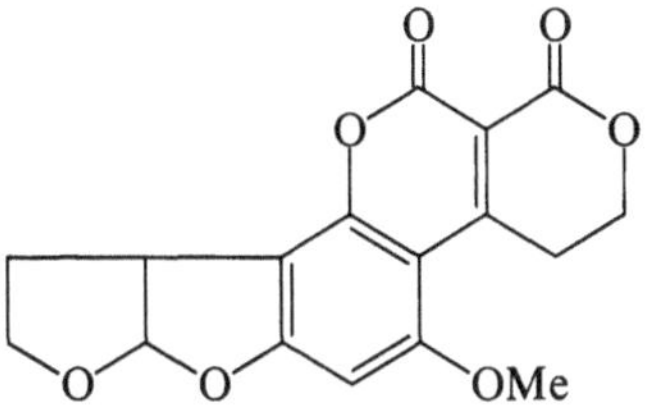

(4) Aflatoxin-G2
m.p. 237−240° (d.), [a]$_D$ −473°

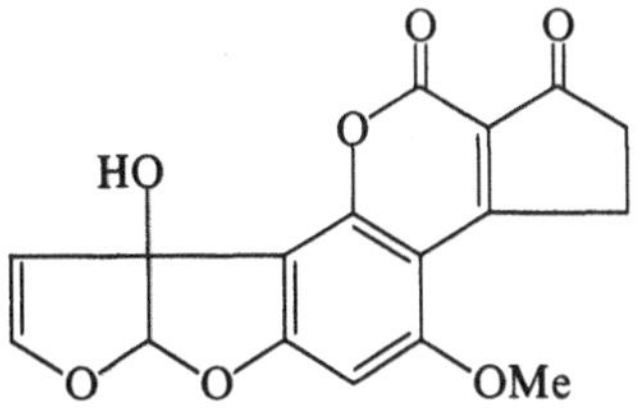

(5) Aflatoxin-M1
m.p. 299° (d.), [a]$_D$ −280°

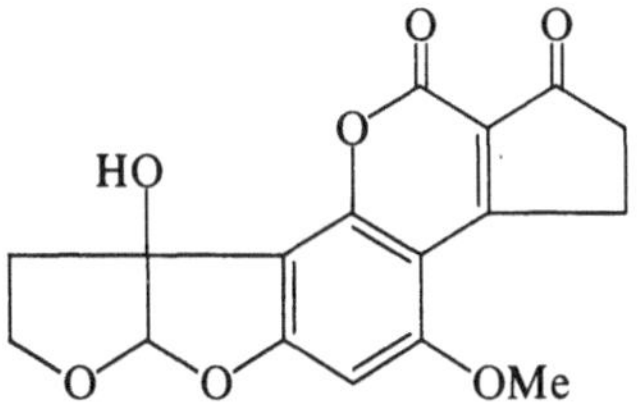

(6) Aflatoxin-M2
m.p. 293° (d.)

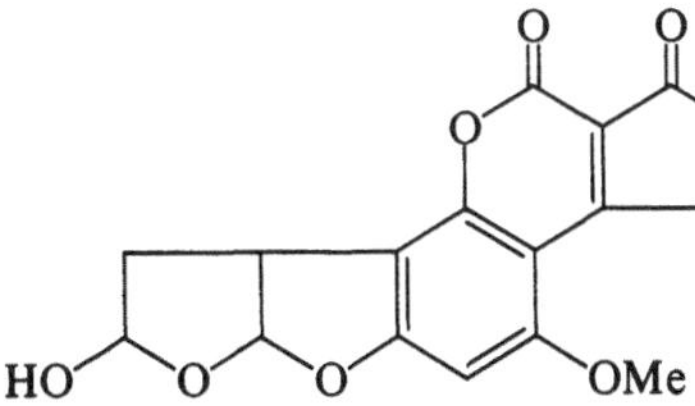

(7) Aflatoxin-B2a
m.p. 240° (d.)

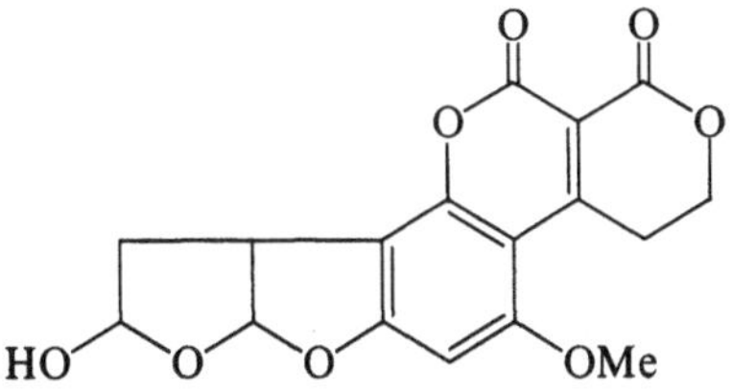

(8) Aflatoxin-G2a
m.p. 190° (d.)

Chart 1. Structures of the aflatoxins. (Physical constants are taken from references given in the text)

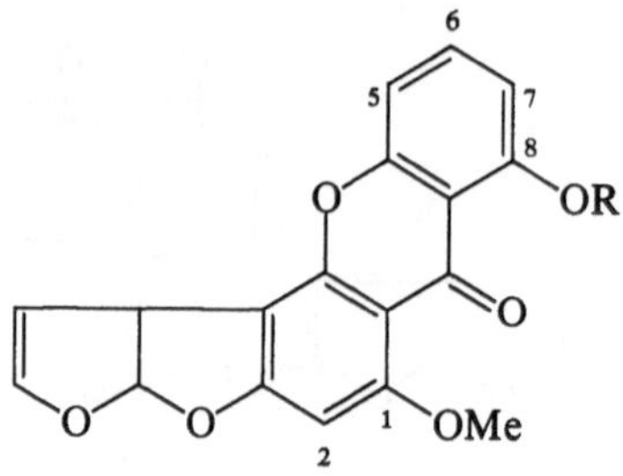

(9) (R=H) Sterigmatocystin,
　　m.p. 246° (d.), $[a]_D -387°$

(9a) (R=Me) O-Methylsterig-
　　matocystin, m.p. 265−267° (d.),
　　$[a]_D - 262°$

(9b) (R=H, OH for OMe at position 1)
　　Demethylsterigmatocystin,
　　m.p. 253−255°, $[a]_D - 483°$

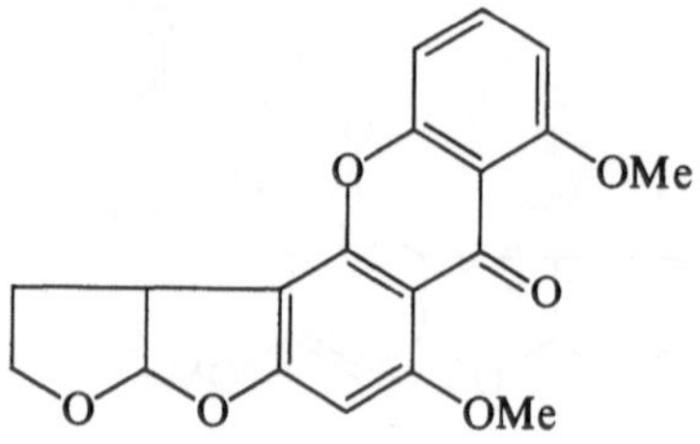

(10) Dihydro-O-methylsterig-
　　matocystin, m.p. 283° (d.),
　　$[a]_D - 255°$

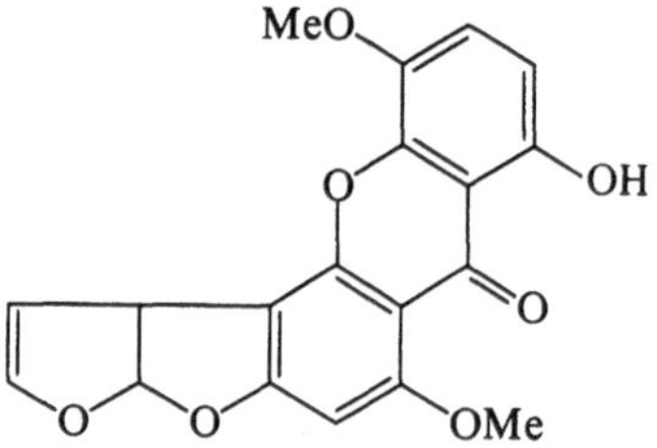

(11) 5-Methoxy-sterigmato-
　　cystin, m.p. 223° (d.),
　　$[a]_D - 360°$

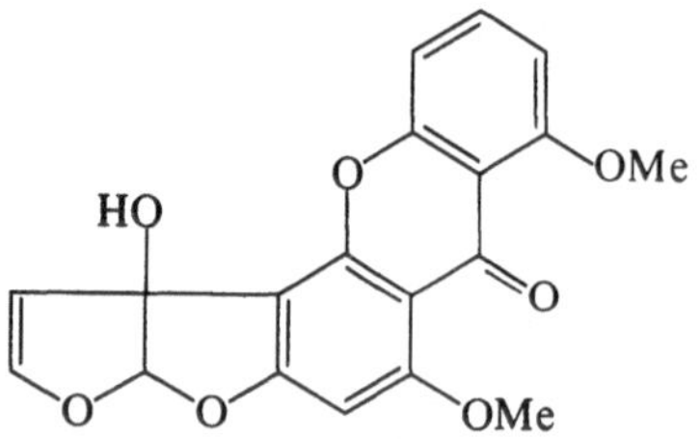

(12) Aspertoxin
　　m.p. 325−327° (d.),
　　$[a]_D - 140°$

Chart 2. Structures of the sterigmatocystins. (Physical constants are taken from references given in the text)

mould metabolites (Chart 2), closely related in structure to sterigmato-cystin, were later discovered.

　　The seriousness of aflatoxicosis in the realms of agriculture and nutrition, coupled with the discovery of new naturally-occurring potent carcinogens, prompted intensive chemical investigations into the nature of the mycotoxins concerned. It is the purpose of this review to consider the chemistry of the aflatoxins together with that of the closely related sterigmatocystins.

II. Isolation and Characterisation of the Mycotoxins

The initial isolates of "aflatoxin" were highly active toxins but, chemically, were crude mixtures (73, 61). A major advance was made in 1963 (41) when workers at the Tropical Products Institute (London) resolved "aflatoxin" into four, pure, crystalline compounds [aflatoxin-B1 (1), -B2 (2), -G1 (3), -G2 (4) in a ratio of *ca.* 40:1:50:1] and determined their molecular formulae and physical, especially spectroscopic, properties. The aflatoxins fluoresce strongly in ultraviolet light (λ *ca.* 365 nm); two of them (-B1 and -B2) produce a blue fluorescence whereas the other two (-G1 and -G2) appear to produce a green fluorescence (41). Four other aflatoxins [-M1 (5), -M2 (6) (51), -B2a (7), and -G2a (8) (35)] which occur in relatively minor proportions, were subsequently isolated from substrates infected with *A. flavus*. A few other metabolites, closely related to those already mentioned, are produced by this same mould (or by its near relative, *A. parasiticus*) in minute amounts (36, 44, 76).

The usual method of isolation of the aflatoxins is to extract the substrate with methanol and to defat the diluted solution with light petroleum; chloroform extraction of the aqueous alcoholic residue then yields a mixture of the mycotoxins which is separated into the various pure compounds by protracted chromatography on alumina or silica followed by preparative thin layer, or paper, chromatography [see *e.g.*, (61, 51)]. The results of a detailed qualitative and quantitative investigation into aflatoxin production, by a number of strains of *A. flavus* on different substrates, have been published from the Northern Regional Research Laboratory, U.S. Department of Agriculture (45). When isolated from mould substrates in small quantities, the aflatoxins may be characterised by their R_F values (t.l.c. on silica gel), fluorescences (in u.v. light), m.p's, and u.v. absorption spectra. Physicochemical methods for the assay of aflatoxins in agricultural commodities have been reviewed (63).

Sterigmatocystin (9) was first obtained from the dried mycelium of *A. versicolor* (42, 25, 19) but was later (50) shown to be produced also by other mould species. An additional five metabolites, closely related to sterigmatocystin, were subsequently described: (I) *O*-methylsterigmatocystin (9a) as a co-metabolite of the aflatoxins in a strain of *A. flavus* (20); (II) demethylsterigmatocystin (9b) from a variant of *A. versicolor* (37); (III) dihydro-*O*-methylsterigmatocystin (10) from a non-aflatoxin producing strain of *A. flavus* (23); (IV) 5-methoxysterigmatocystin (11) from variant strains of *A. versicolor* (18, 46); and (V) aspertoxin (12), a highly toxic metabolite of some strains of *A. flavus* (71, 72, 83). In small quantities, sterigmatocystin may be characterised by methods similar to those used for the aflatoxins. Thin-layer chromatograms, viewed under long-wave u. v. light, reveal the presence of sterigmatocystin as a brick-red fluorescent spot. Detailed analytical methods for the determination of this toxin, in feeding-stuffs spoiled by moulds, have been published (82, 81).

III. Structural Elucidation of the Mycotoxins

Of the mycotoxins considered in this review, sterigmatocystin (9) was the first to be isolated and was also the first to yield to structural elucidation (19).

1. Sterigmatocystin

This compound was originally isolated, in 1954, from the mycelium of A. versicolor (Vuill.) Tiraboschi by Japanese workers (42) who gave it the name by which it is now known. Two independent isolations from the same source were subsequently reported (10, 26, 25). Sterigmatocystin is quite readily available, for the dried mycelial mats of some strains of the mould yield up to 1.3 per cent of the metabolite (25, 19). From the considerable amount of experimental evidence which has been published (42, 43, 10, 26, 25, 19, 70) it is possible to deduce the structure of sterigmatocystin (9) in the following way.

Sterigmatocystin, $C_{18}H_{12}O_6$, forms pale yellow needles, m.p. 246° (d.), and is strongly laevorotatory. Its molecule possesses one methoxyl group (Zeisel), and one free hydroxyl group for it forms a mono-O-methyl ether. Catalytic hydrogenation [to give (10), OH for OMe at position 8] revealed the presence of one easily reducible double-bond, and chemical and spectroscopic (i.r. and u.v.) evidence indicated a xanthone

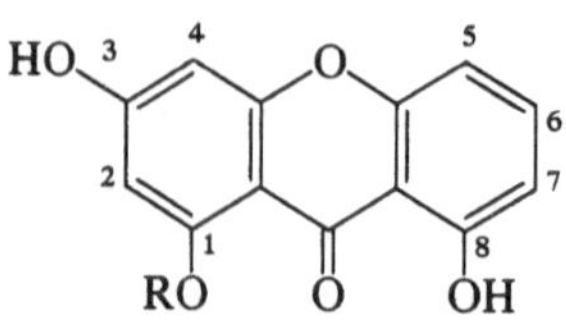

(13) (R=H) 1,3,8-Tri-
hydroxyxanthone

(13a) (R=Me) 3,8-Di-
hydroxy-1-methoxy-
xanthone

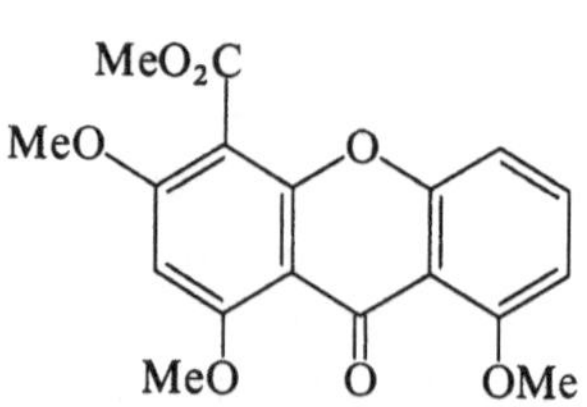

(14) Methyl 1,3,8-tri-
methoxyxanthone-4-
carboxylate

nucleus (67). Degradation of sterigmatocystin with aluminium chloride in chlorobenzene yielded 1,3,8-trihydroxyxanthone (13). When the metabolite was oxidised with a limited quantity of potassium permanganate, there was produced a crystalline acid, $C_{15}H_{10}O_7$, which, on pyrolysis and sublimation, gave 3,8-dihydroxy-1-methoxyxanthone (13a) identical with a specimen synthesised (25) from γ-resorcylic acid and mono-O-methylphloroglucinol. Oxidation of O-methylsterigmatocystin (9a), under similar conditions, gave an acid which, on complete methylation, yielded methyl 1,3,8-trimethoxyxanthone-4-carboxylate (14), identified by comparison with a synthetic sample (70). Chemical and spectroscopic evidence established the presence in the metabolite of a vinyl ether (–CH = CH–O–) group which was not in conjugation with the main chromophore. Taken together, the above lines of evidence indicate the partial structure (15) for sterigmatocystin.

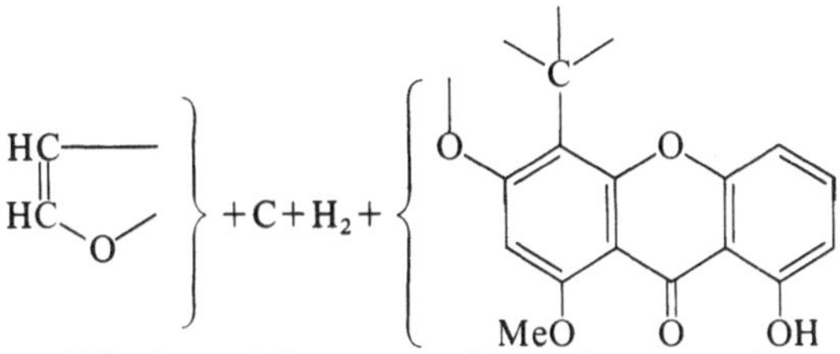

(15) A partial structure for sterigmatocystin

When sterigmatocystin is heated with alcoholic potassium hydroxide, it is converted into an optically inactive isomer (iso-sterigmatocystin) the structure of which (16) may be deduced (19) from the following observations. Iso-sterigmatocystin yielded a di-O-methyl ether which (I) gave a Diels-Alder adduct with maleic anhydride (suggesting the presence of a furan ring), and (II) yielded, on ozonolysis, 1,3,8-trimethoxy-xanthone-4-carboxylic acid [identified as its methyl ester (14)] together with two mol. of formic acid (showing that the furan ring is linked through a β-position to the xanthone nucleus).

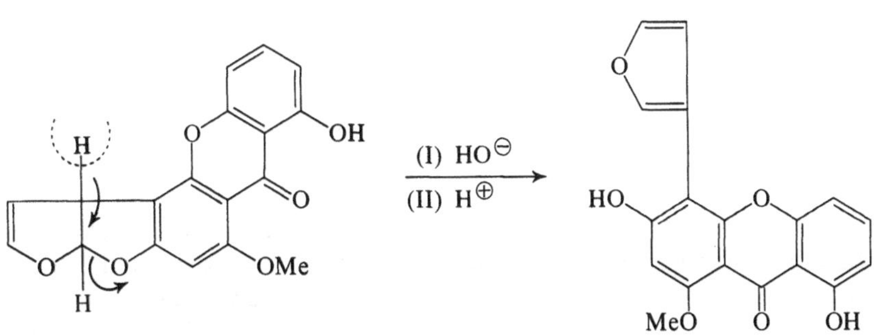

(9) Sterigmatocystin

(16) Iso-sterigmatocystin

Sterigmatocystin may therefore be formulated (*19*) as (**9**) and its isomerisation to iso-sterigmatocystin may be rationalised as shown, (**9**)→(**16**). The correctness of structure (**9**) for sterigmatocystin was confirmed (*19*) by ^{1}H n.m.r. spectroscopy (see Table 1).

Table 1. *Analysis of the ^{1}H n.m.r. Spectrum* of Sterigmatocystin* (**9**)

Proton	OH	H-6	H-5 and H-7	H-2	O.CH$_3$	H$_c$	H$_d$	H$_b$	H$_a$
τ	—3.14	2.55	*ca.* 3.3	3.62	6.06	3.24	5.28	3.52	4.60
Signal**	s	t	m	s	s	d	dt	t	t
Intensity	1	1	2	1	3	1	1	1	1
J (Hz)***		8				7	7 and 2.5	2.5	2.5

* The spectrum was recorded with a Perkin-Elmer R 10, 60 MHz spectrometer using tetramethylsilane as internal reference. The values refer to sterigmatocystin in deuterochloroform solution and are taken from reference (*56*); they differ slightly from those (determined with a different spectrometer on chloroform and methylene dichloride solutions) given in reference (*19*).

** s = singlet; t = triplet; m = multiplet; d = doublet; dt = double triplet.

*** For a discussion on the values of coupling constants in 2,3-dihydrofuran see reference (*53*).

Additional evidence in favour of structure (**9**) for sterigmatocystin was obtained as follows. Submission of dihydrosterigmatocystin [(**10**), OH for OMe at position 8] to a modified Elbs persulphate reaction (*66, 25*) gave dihydro-5-hydroxysterigmatocystin (**17**) which, on oxidation (alkaline hydrogen peroxide) and spontaneous decarboxylation, yielded tetrahydro-4-hydroxy-6-methoxy-furo[2,3-*b*]benzofuran (**18**). The structure of this latter compound was confirmed by comparison with a synthetic (±)-specimen (*57*) (*vide infra*).

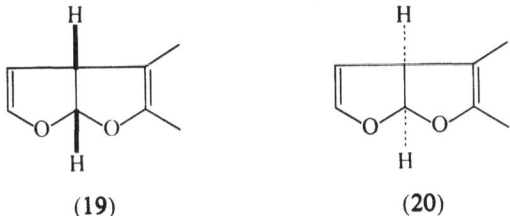

(17) Dihydro-5-hydroxy-
    sterigmatocystin

(18) Tetrahydro-4-hydroxy-
    6-methoxyfuro[2,3-b]-
    benzofuran

Since it is virtually impossible to construct a scale model of sterigmatocystin in which the two dihydrofuran rings are *trans*-fused, it was concluded (19) that the molecule should be represented as (19) or (20). The absolute configuration of sterigmatocystin has been determined (47) as (20) by an adaptation of the method used (12) in the case of the aflatoxins.

(19)   (20)

Stereochemical forms of sterigmatocystin (9)

Finally, the structure of sterigmatocystin (9), including the *cis*-fusion of the two five-membered rings, has been confirmed by X-ray crystallography of the *p*-bromobenzoate derivative (78). The structure (9) has also been proved by a synthesis (65) of the racemic form of the *O*-methyl ether (9a).

2. Other Metabolites Closely Related to Sterigmatocystin

The structures of *O*-methylsterigmatocystin (9a), demethylsterigmatocystin (9b), and dihydro-*O*-methylsterigmatocystin (10) were determined (20, 37, 23) by analysis and spectroscopy, and by the establishment of their identities with authentic sterigmatocystin derivatives.

A methoxy derivative of sterigmatocystin has been isolated from variant strains of *A. versicolor*. It was originally considered (*18*) to be the 6-methoxy derivative but subsequent work (*46*) proved that the metabolite was in fact 5-methoxysterigmatocystin (**11**).

3. Aspertoxin

This compound (**12**) forms colourless needles, m.p. 325—327° (d.) (*83*) [297—304° (d.) (*71, 72*)]. Analysis established the formula, $C_{19}H_{14}O_7$, and u.v. and i.r. spectroscopy indicated a close relationship to *O*-methyl-sterigmatocystin, $C_{19}H_{14}O_6$ (**9a**). Aspertoxin forms a mono-acetate, m.p. 119—121°, and thus contains one hydroxyl group (*72*). The location of this hydroxyl group in aspertoxin (*72, 83*) resulted from an analysis of its ^1H n.m.r. spectrum relative to the spectrum of *O*-methyl-sterigmatocystin. Significant and interpretable differences between the two spectra are produced by replacement of the proton H_d (cf. the formula of sterigmatocystin in Table 1) in structure (**9a**) by the hydroxyl group to give structure (**12**). In the spectrum of aspertoxin, H_c appears as a singlet and both H_a and H_b appear as doublets. These observations leave no doubt as to the position of the hydroxyl group in aspertoxin which therefore has structure (**12**).

4. Aflatoxin-B1, -B2, -G1, and -G2

Initial investigations into the structures of the aflatoxins were hindered by a lack of an adequate amount of material ("aflatoxin") and by the difficulty of separating this crude toxin into the various, pure, individual compounds. After these problems had been overcome (*41, 3*), a number of reports on structural investigations appeared from different laboratories (*33, 41, 60*) but the outstanding contribution (*2, 3*), leading to the complete formulation of the four main constituents, was made by BÜCHI and his collaborators at the Massachusetts Institute of Technology.

The structure of aflatoxin-B1 was deduced (*2, 3*) in the following way. [For clarity in exposition it is desirable to anticipate the deduced structures of aflatoxin-B1 (**1**) and of its reduction product, tetrahydro-deoxo-aflatoxin-B1 (**21**).] The toxin (**1**), $C_{17}H_{12}O_6$, forms colourless needles, m.p. 268—269° (d.), $[\alpha]_D$—558°, λ_{max} 223, 265, and 362 nm (10^{-3} ε 25.6, 13.4, and 21.8), ν_{max} 1760, 1684, 1632, 1598, and 1562 cm^{-1}. Catalytic hydrogenation of the compound over palladised charcoal resulted in the uptake of three mol. of hydrogen and the

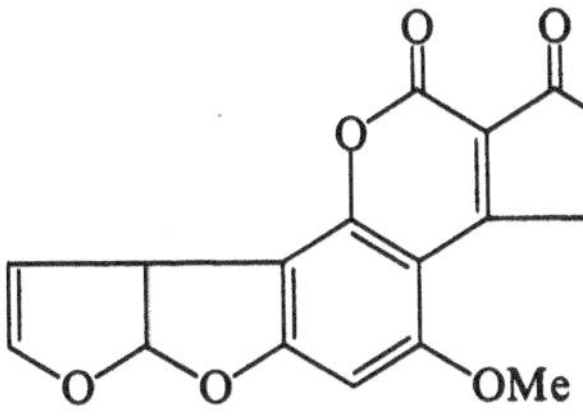

(1) Aflatoxin-B1

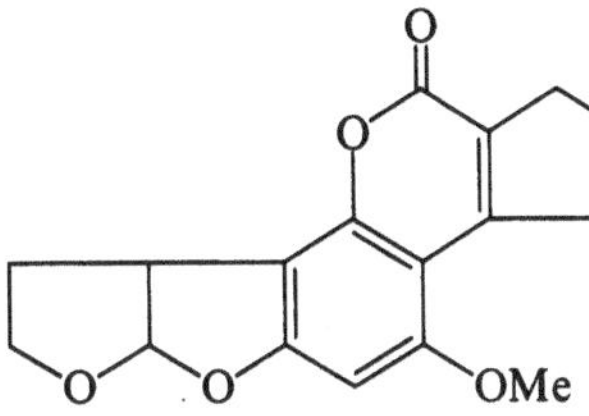

(21) Tetrahydrodeoxo-
aflatoxin-B1

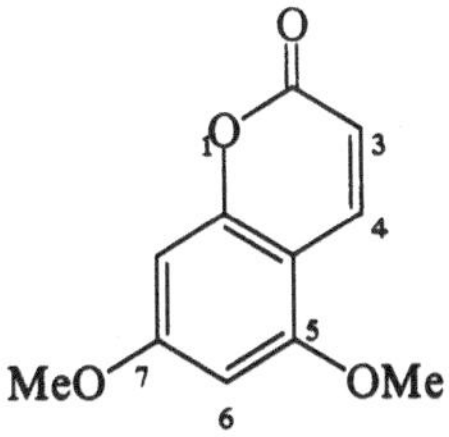

(22) 5,7-Dimethoxy-
coumarin

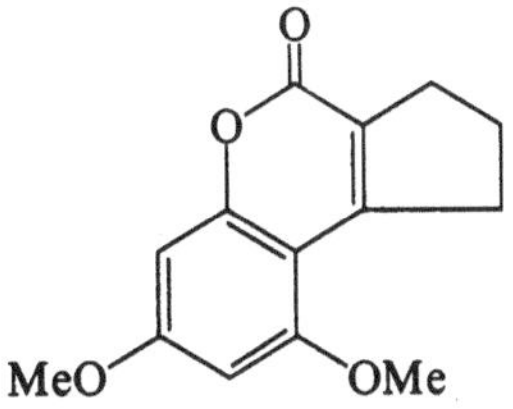

(23) 5,7-Dimethoxy-cyclo-
penteno[c]coumarin

production of tetrahydrodeoxoaflatoxin-B1 **(21)**, $C_{17}H_{16}O_5$, λ_{max} 255, 264, and 332 nm (10^{-3} ε 8.5, 9.2, and 13.9), ν_{max} 1705, 1625, and 1610 cm^{-1}. It was noted that the u.v. spectrum of the reduced compound was similar to that of 5,7-dimethoxycoumarin **(22)** and was even more closely similar to that of a synthetic specimen of 5,7-dimethoxycyclopenteno[c]coumarin **(23)**, λ_{max} 248, 257, and 325 nm (10^{-3} ε 7.7, 7.0, and 16.1). The bathochromic effect (*ca.* 7 nm) in the spectrum of **(21)** relative to that of **(23)** was attributed to an additional alkyl substituent in the former. The change in the molecular composition and in the i.r. absorptions of the compound produced by hydrogenation of aflatoxin-B1 required the presence in the latter of a carbonyl group in conjugation with a double bond, and the two partial structures **(24)** and **(25)** were considered to be possibilities for the toxin.

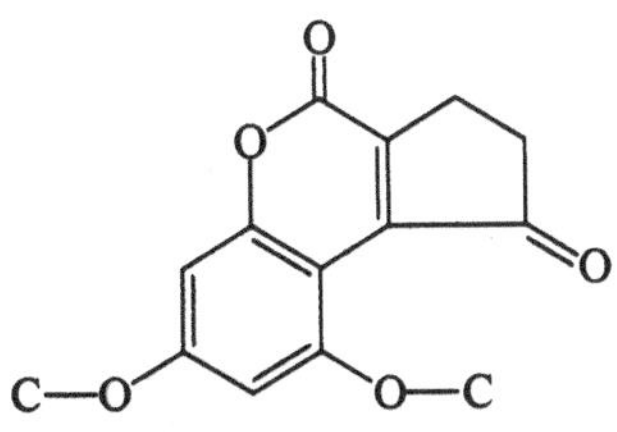

(24)

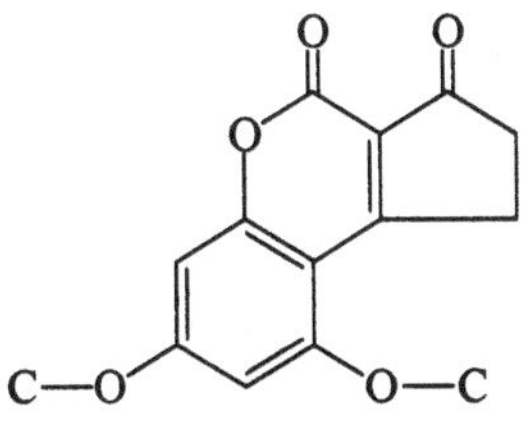

(25)

Partial structures for aflatoxin-B1

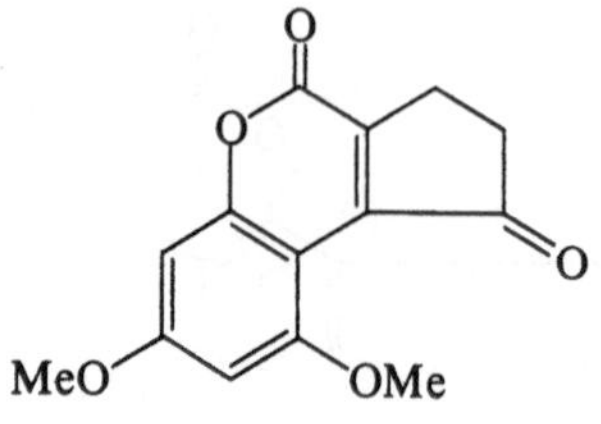

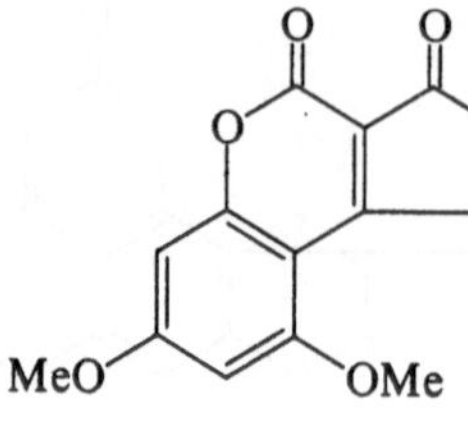

Isomeric 5,7-dimethoxycyclopentenone-coumarins

In pursuance of this idea, the two isomeric coumarins, (26) and (27) were synthesised and it was found that the i.r. spectrum of (27), v_{max} 1759, 1685, 1614, 1594, and 1550 cm^{-1}, was markedly similar to that of aflatoxin-B1. The ^{1}H n.m.r. spectrum of aflatoxin-B1 disclosed the presence of (I) a methoxy group (3H, τ 5.98), (II) four hydrogen atoms in an A_2B_2 system [belonging to the methylene groups in the cyclopentenone ring — see, for example, (27)], and (III) a singlet at τ 3.49 (Ar–H, situated between two alkoxy substituents). The signals for the remaining four protons revealed a pattern (with appropriate τ values and couplings) (2) characteristic of the protons in a fused dihydrofuran ring, and strikingly similar to the pattern which had been observed in the spectrum of sterigmatocystin (19).

The above evidence established with virtual certainty that tetrahydro-deoxoaflatoxin-B1 has structure (21) and that aflatoxin-B1 should be represented as (1). The correctness of both structures was subsequently proved by syntheses (58, 69, 13, 14, 17) of the corresponding ($\pm$)-forms (see later).

The absolute configuration of aflatoxin-B1, on the assumption of a cis-fusion for the two hydrofuran rings, has been determined (12)

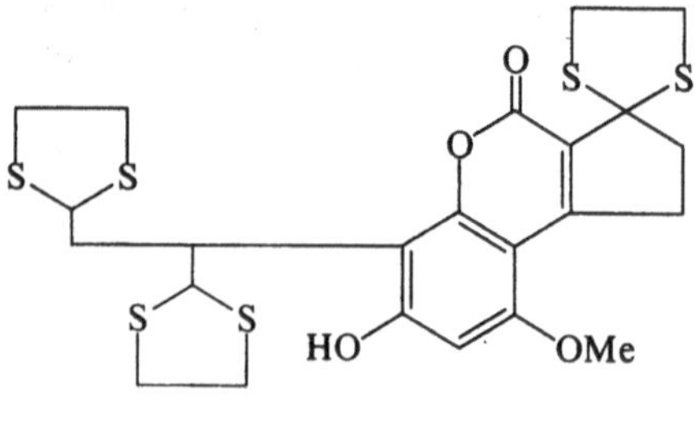

(28)

Trithioketal from
aflatoxin-B1

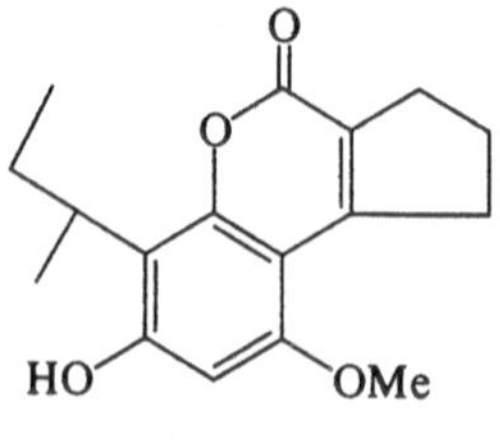

(29)

Desulphurised·product
from (28)

as follows. Aflatoxin-B1 (1), on treatment with ethane dithiol and zinc chloride, gave the trithioketal (28) which was desulphurised by Raney Nickel W2 to give the phenol (29). Chromic acid oxidation of the phenol (29) gave, *inter alia*, (+)-(S)-2-methylbutanoic acid (30) which was converted into the amide (31). The identity of the latter was confirmed by comparison with the amide prepared from authentic (+)-(S)-2-methylbutanoic acid. Aflatoxin-B1 has, therefore, the absolute configuration shown in (1) below.

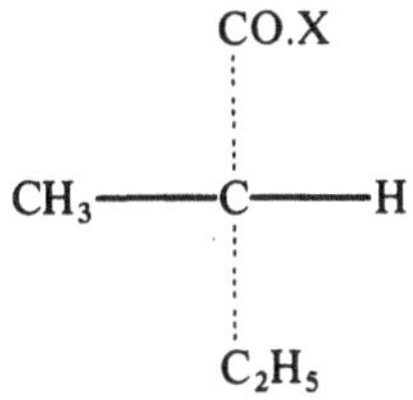

CO.X

CH_3——C——H

C_2H_5

(30) (X=OH) (+)-(S)-2-
Methylbutanoic acid

(31) (X=NH$_2$) (+)-(S)-2-
Methylbutanamide

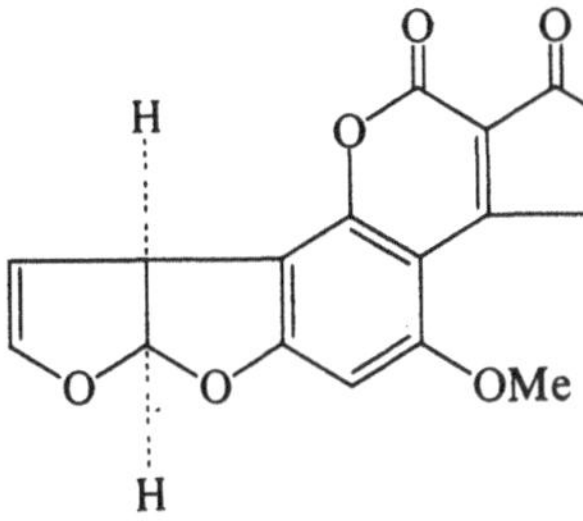

(1)

Absolute configuration of
aflatoxin-B1

The molecule of aflatoxin-G1, $C_{17}H_{12}O_7$, possesses one more oxygen atom than that of -B1 (1) and its u.v., i.r., and ^1H n.m.r spectroscopic properties (3) indicated it to have structure (3). This structure has been confirmed by synthesis (17) and the absolute configuration has been shown (12) to be the same as that of -B1.

Aflatoxin-B2 may be prepared by careful hydrogenation of -B1 (33, 41, 60) and hence possesses structure (2) — a conclusion subsequently verified by an independent synthesis (69). Aflatoxin-G2 may be obtained similarly from -G1 (41, 60) and so possesses structure (4).

It is noteworthy that X-ray crystallography has confirmed the structures of -B2 (74) and of -G1 (22) and has also established the *cis*-fusion of the two hydrofuran rings in both compounds.

5. Aflatoxin-M1 and -M2

Investigations had shown that lactating animals, on ingestion of aflatoxin-B1, secreted a "milk toxin". There was also an indication (52) that this "milk toxin" was present in crude "aflatoxin", and an intensive investigation of the constituents of a large batch of crude "aflatoxin" (51) confirmed this indication. Aflatoxin-M1 was isolated from the latter source as a high-melting, crystalline solid of molecular formula,

9*

$C_{17}H_{12}O_7$, suggesting a near relationship to aflatoxin-B1, $C_{17}H_{12}O_6$. The u.v. and i.r. spectra of the two compounds were very similar but it was noted that the i.r. spectrum of -M1 possessed, exceptionally, a band at 3425 cm^{-1} indicating the presence of a hydroxyl group. This indication was confirmed by the observation that -M1 yielded an acetate derivative. The location of the hydroxyl group was determined by ^{1}H n.m.r. spectroscopy (cf. the solution of a similar problem in the case of aspertoxin) and led to the assignment of structure (5) to aflatoxin-M1. Structure (5) was later confirmed by synthesis (*17*). A second "milk toxin", aflatoxin-M2, was also isolated from the crude "aflatoxin" and was allocated structure (6) on the evidence that it was identical with dihydroaflatoxin-M1 (produced from -M1 by catalytic hydrogenation).

6. Aflatoxin-B2*a* and -G2*a*

These two compounds have been isolated from an extract of an *A. flavus* substrate by a complicated chromatographic procedure (*35*). The structures of the two metabolites were deduced [as (7) and (8) respectively] by analysis and spectroscopy, and by the demonstration (*35*) that each yielded an acetate which was identical with the acetate produced by the catalysed addition of acetic acid across the double bond of the vinyl ether linkage of the appropriate aflatoxin (-B1 or -G1).

IV. Synthesis of the ($\pm$)-Forms of the Mycotoxins and of Related Compounds

All the mycotoxins under review are highly oxygenated, unsaturated compounds and their synthesis has presented a formidable challenge to the organic chemist. One of the difficulties which has always to be surmounted is the construction of a molecule in which the oxygen atoms show an extraordinary variation in functionality. It will be observed, for example, that aflatoxin-B1 (1) or -B2 (2) contains six oxygen atoms every one of which has a different function.

A characteristic of all these toxins is the presence in their molecules of a structural feature which had previously been unknown in natural product chemistry. This feature is the di- (or tetra-)hydrofuro-[2,3-*b*]benzofuran structure (32, 33).

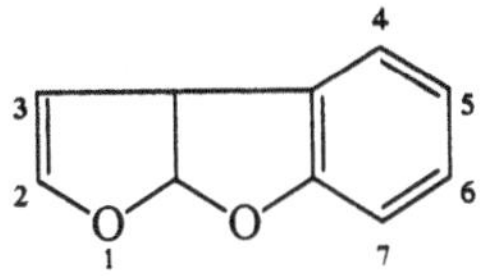

(32) Dihydrofuro[2,3-b]-
benzofuran

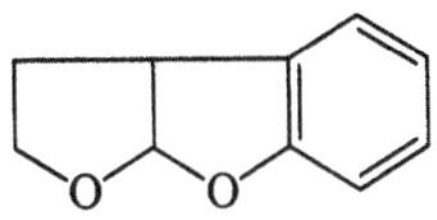

(33) Tetryhydrofuro[2,3-b]-
benzofuran

Although such structures contain two differently constituted asymmetric centres, the stereochemical requirement of *cis*-fusion for the two five-membered rings limits the number of stereoisomers to two*. Chemical syntheses of the mycotoxins therefore lead to single racemates.

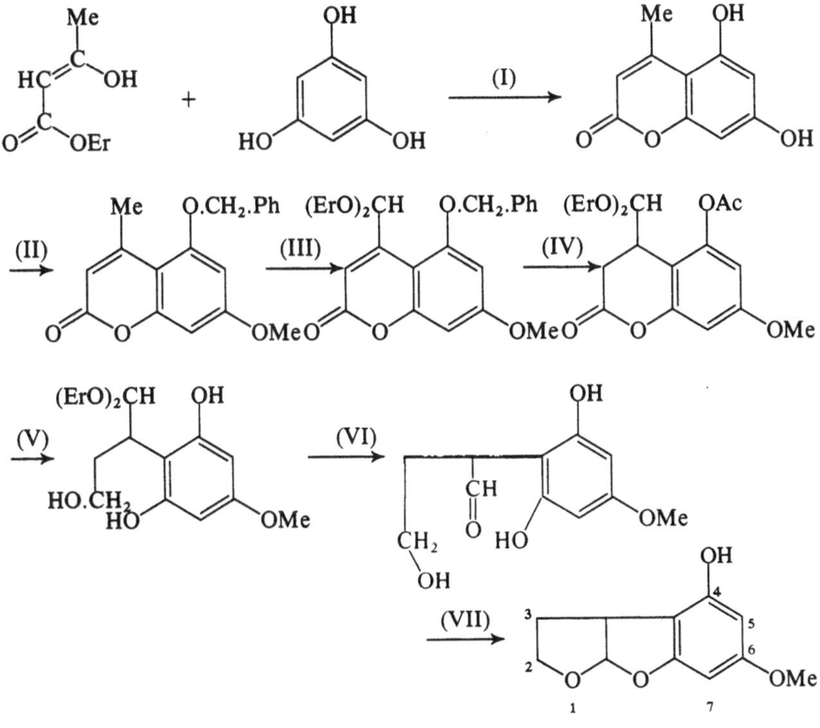

(18)

(I) Pechmann condensation; (II) selective methylation, then benzylation; (III) SeO$_2$ oxidation and acetalisation of the newly-formed aldehyde group; (IV) catalytic hydrogenation to form the dihydrocoumarin (the benzyl group is also removed), then Ac$_2$O; (V) LiAlH$_4$ reduction of the lactone group — the acetate group is also reduced; (VI) aqueous acid to remove the protective acetal group; (VII) spontaneous elimination of the elements of water.

Chart 3. Synthesis of ($\pm$)-tetrahydro-4-hydroxy-6-methoxy-furo[2,3-*b*]-benzofuran (18)

---

* For further consideration of this point, see reference (57). Also, aflatoxins-B2 and -G1 and sterigmatocystin have been shown, by X-ray crystallography (see pp. 127 and 131), to possess the *cis* configuration.

1. Synthesis of (±)-Tetrahydro-4-hydroxy-6-methoxyfuro[2,3-*b*]benzofuran (18)

This compound, in its (–)-form, was originally produced by degradation of dihydro-sterigmatocystin (*25, 57*). Its synthesis, in the (±)-form,
proved of value in two directions. First, it provided confirmation of certain
structural features in sterigmatocystin and, secondly, it proved later to be
a useful intermediate in more sophisticated syntheses — see below.

The compound was synthesised (*57*) by the route shown in Chart 3.

2. Synthesis of (±)-Tetrahydrodeoxoaflatoxin-B1 (21)

The (–)-form of this compound, produced (*3*) by catalytic reduction
of aflatoxin-B1 (**1**), played an important role in the elucidation of the
structure of the aflatoxins. The synthesis of (±)-tetrahydrodeoxoaflatoxin-
B1 (**21**) was the first synthesis of an aflatoxin-type compound — the
reduced toxin retaining intact the carbon skeleton of aflatoxin-B1.

The synthesis was accomplished (*58, 69*) as shown in Chart 4.

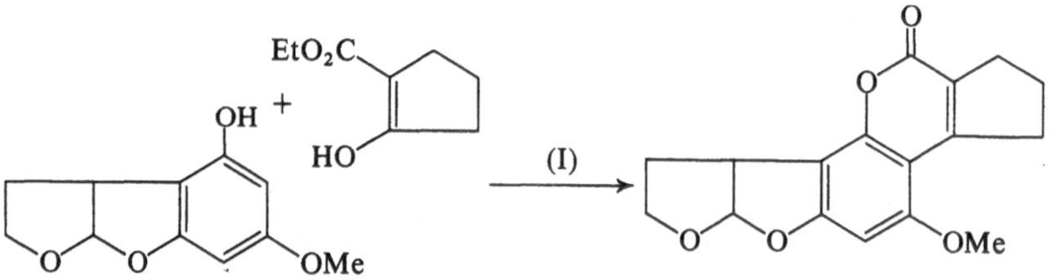

(I) Pechmann condensation

Chart 4. Synthesis of (±)-tetrahydrodeoxoaflatoxin-B1 (**21**)

3. Synthesis of (±)-Aflatoxin-B1 (1)

The presence of the highly reactive vinyl ether group in this molecule (**1**) causes a very real synthetic problem. With this point in mind,
BÜCHI and his collaborators, in their first synthesis, decided to introduce this function as the last step. Other subtleties in this ingenious
synthesis are discussed in the original papers (*13, 14*). The synthesis
is outlined in Chart 5.

References, pp. 148—151

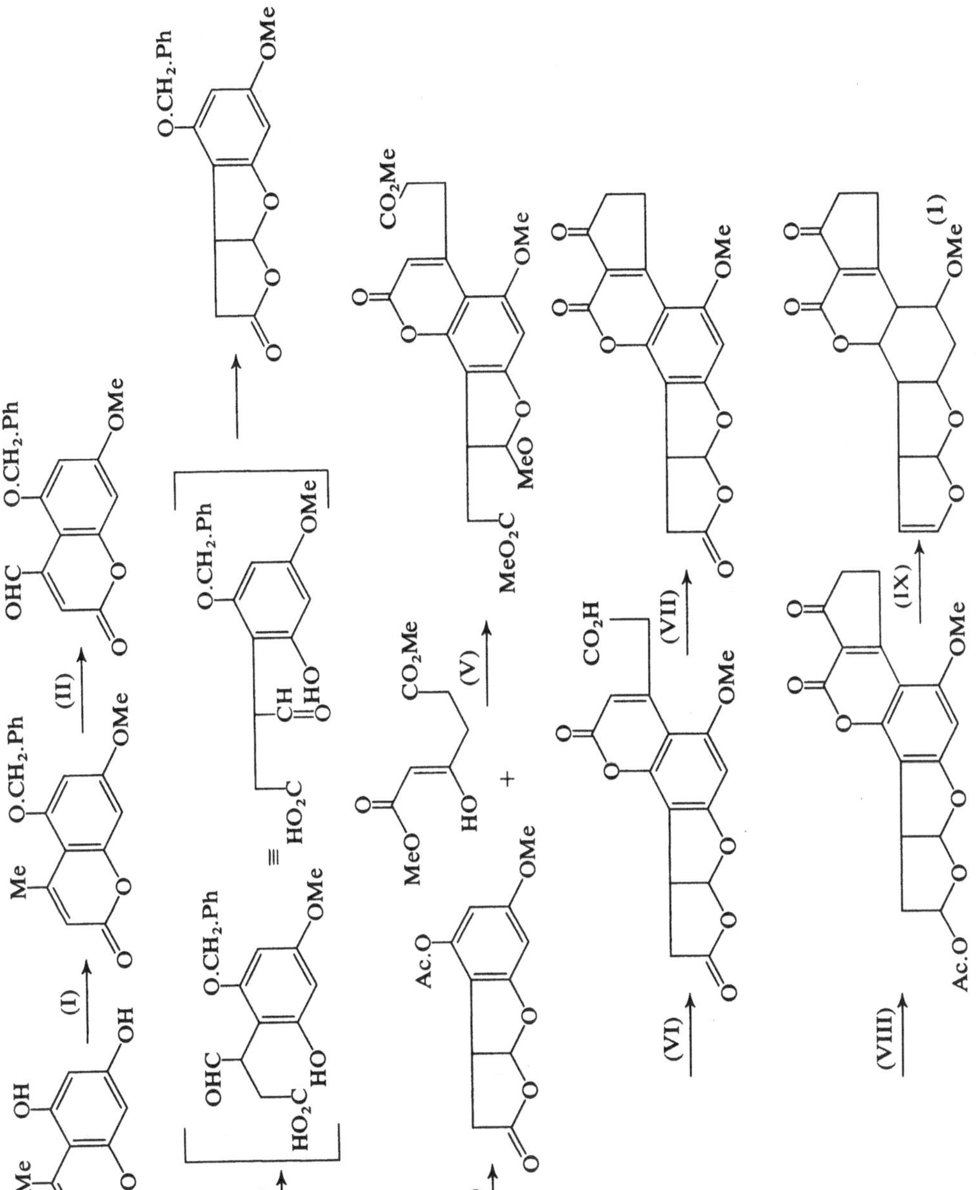

(I) selective methylation, then benzylation; (II) SeO$_2$ oxidation; (III) Zn/HOAc; (IV) removal of benzyl group by hydrogenation, then Ac$_2$O; (V) Pechmann condensation (MeOH/HCl); (VI) H$_2$O/HCl/HOAc; (VII) (COCl)$_2$, then AlCl$_3$/CH$_2$Cl$_2$; (VIII) (iso-C$_5$H$_{11}$)$_2$BH, then Ac$_2$O; (IX) heat-elimination of HOAc.

Chart 5. Büchi's first synthesis of ($\pm$)-aflatoxin-B1 (**1**)

More recently, this same group have developed (*16, 15*) a new type
of coumarin synthesis which has led to a second, and improved synthesis of aflatoxin-B1 (**1**) (*17*) and also (see below) to syntheses of
aflatoxin-G1 and -M1. This new type of coumarin synthesis may prove
of considerable value in natural product chemistry. It accomplishes a
Pechmann-type condensation between a phenol and a vinyl bromide in
presence of zinc carbonate and sodium bicarbonate. The reaction conditions are sufficiently mild for the synthesis to be applied to extremely
sensitive phenols. The second synthesis of aflatoxin-B1 is outlined in
Chart 6.

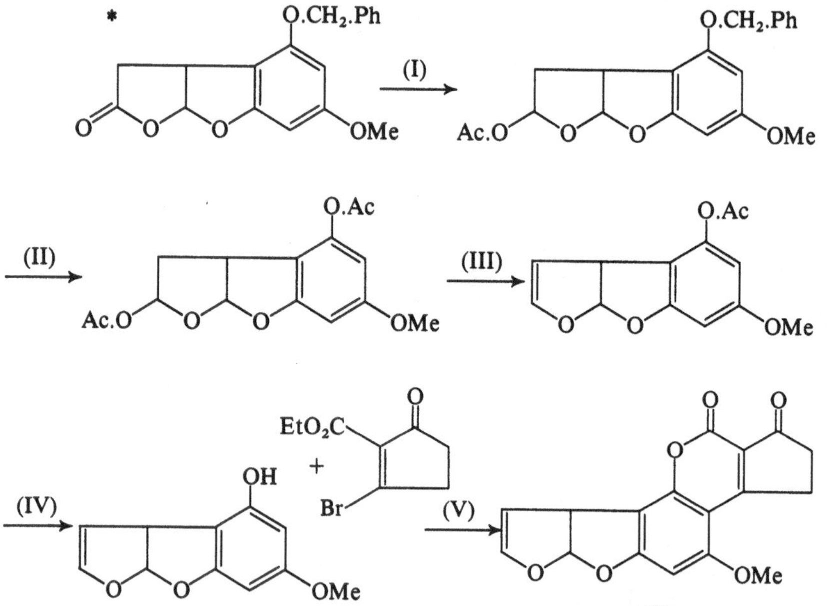

* For preparation of this compound, see Chart 5.

(I) (iso-C₄H₉)₂AlH, then Ac₂O; (II) Pd/H₂, then Ac₂O; (III) pyrolysis (−HOAc);
(IV) basic hydrolysis; (V) coumarin synthesis catalysed by ZnCO₃.

Chart 6. BÜCHI's second synthesis of (±)-Aflatoxin-B1 (**1**)

4. Synthesis of (±)-Aflatoxin-B2 (2)

An independent synthesis of (±)-aflatoxin-B2 was achieved (*69*) by
building the required cyclopentenone-coumarin ring system on to
(±)-tetrahydro-4-hydroxy-6-methoxyfuro[2,3-*b*]benzofuran (**18**). This
synthesis is illustrated in Chart 7.

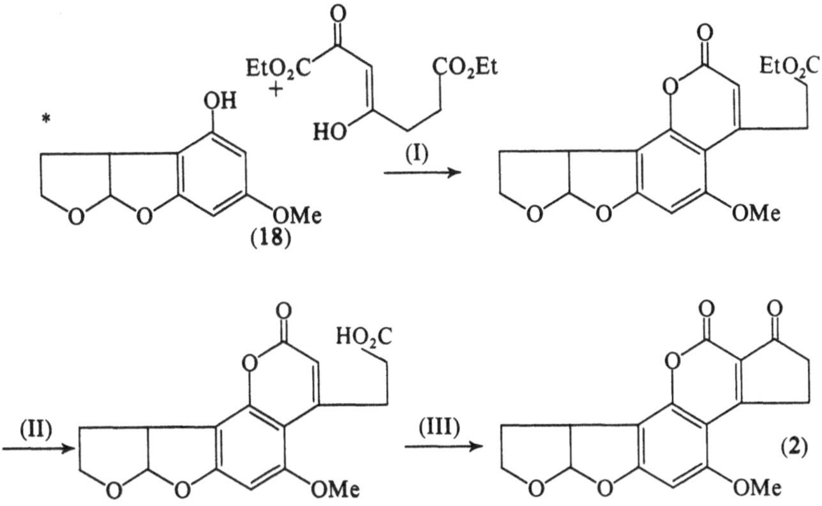

* For preparation of this compound, see Chart 3.

(I) Pechmann condensation; (II) hydrolysis; (III) (COCl)₂ and then AlCl₃.

Chart 7. Synthesis of (±)-aflatoxin-B2 (**2**)

5. Synthesis of (±)-Aflatoxin-G1 (3)

This synthesis was achieved (*17*) by an adaptation of that already described for -B1 (see Chart 6) and is portrayed in Chart 8.

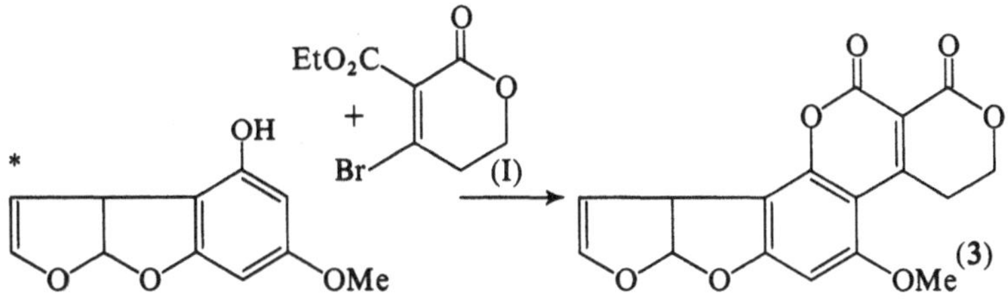

* For preparation of this compound, see Chart 6.

(I) ZnCO₃/LiI catalysed Pechmann condensation.

Chart 8. Synthesis of (±)-aflatoxin-G1 (**3**)

6. Synthesis of (±)-Aflatoxin-M1 (5)

None of the methods yet described is applicable to the synthesis of this "milk toxin" (**5**). Presumably, *in vivo*, this toxin is produced by direct hydroxylation of aflatoxin-B1 (**1**) but no corresponding *in vitro* synthesis has yet been reported.

A practicable synthesis of (±)-aflatoxin-M1 has recently been described (*16, 17*). It involves an adroit, but rather laborious, manipulation of a coumaran-3-one system in order to provide the hydroxylated dihydrofurobenzofuran structure which is found in "milk toxin". This synthesis, summarised in Chart 9, also makes use of the new type of coumarin synthesis on which comment has already been made.

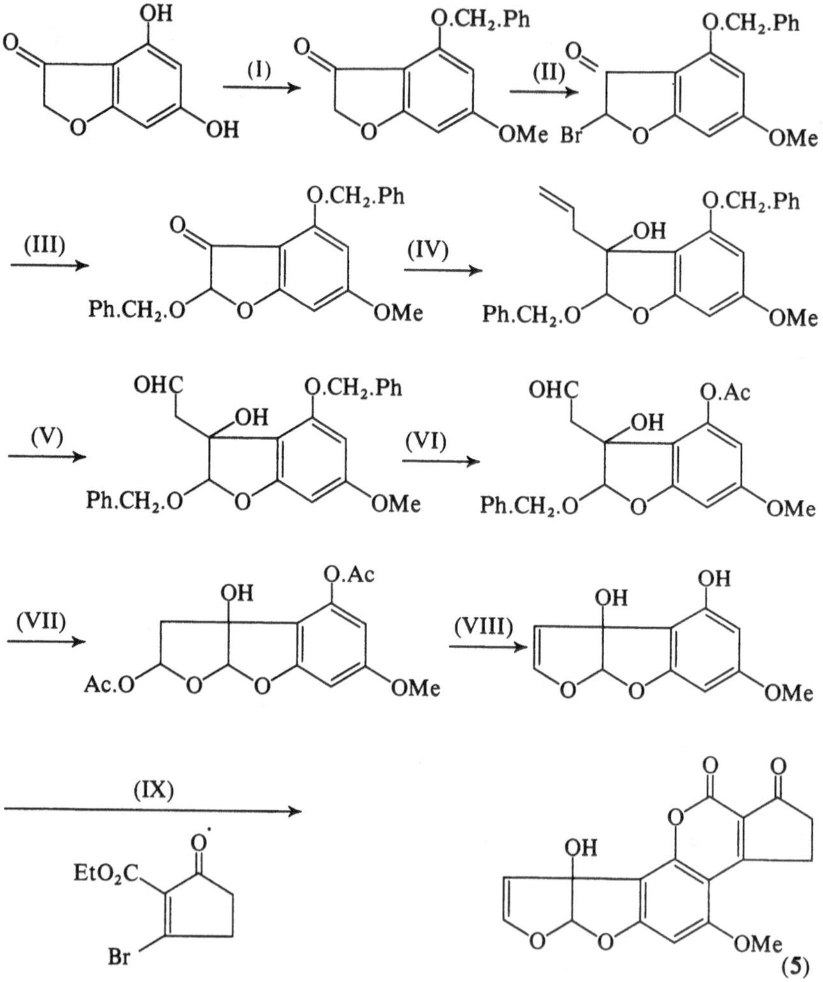

(I) Me_2SO_4 to give the dimethyl ether, then preferential ether cleavage ($AlCl_3$) and, finally, benzylation; (II) $PhN^+Me_3 Br_3^-$; (III) $Ph.CH_2.OH/CaCO_3$; (IV) $CH_2:CH.CH_2$. $Mg.Br$; (V) OsO_4—$NaIO_4$; (VI) $H_2/Pd/C/Ac_2O$ in benzene; (VII) $H_2/Pd/C/EtOAc$, then pyridine/Ac_2O; (VIII) pyrolysis (—HOAc), then hydrolysis; (IX) $ZnCO_3$.

Chart 9. Synthesis of (±)-aflatoxin-M1 (5)

7. Synthesis of (±)-Dihydro-O-methylsterigmatocystin (10)

The synthesis of this metabolite, outlined in Chart 10, was achieved (64, 65) by building the required xanthone system on to the tetrahydrofurobenzofuran (18). A difficulty inherent in this method is the formation of the diphenyl ether linkage (the first stage of an Ullmann xanthone synthesis) under conditions which do not degrade the sensitive phenol (18). This difficulty was eventually overcome by the application of a recently discovered modification (84) of the Ullmann reaction in which the ether linkage is formed under relatively mild conditions.

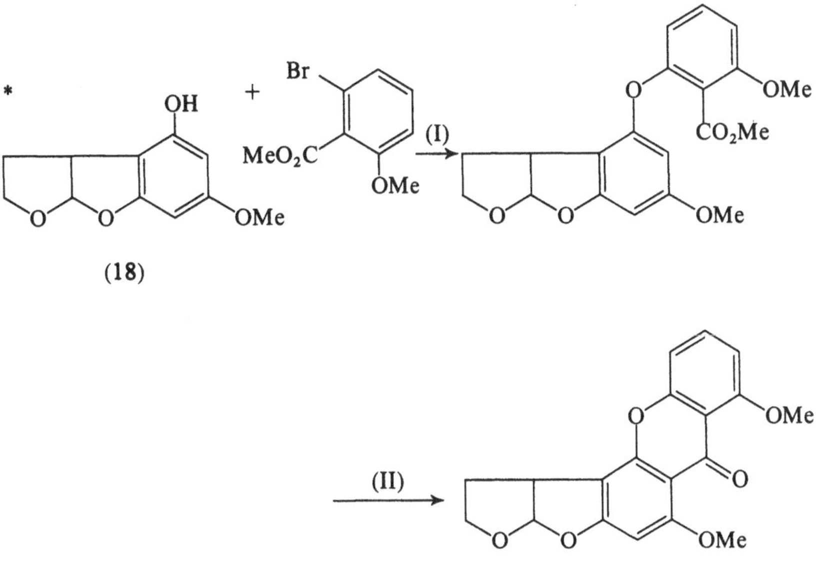

(18)

(10)

* For the synthesis of this compound, see Chart 3.

(I) Cu_2Cl_2/pyridine; (II) hydrolysis, then $(COCl)_2$/CH_2Cl_2.

Chart 10. Synthesis of (±)-dihydro-O-methylsterigmatocystin (10)

8. Synthesis of (±)-O-methylsterigmatocystin (9a)

The procedure is portrayed in Chart 11. The phenolic lactone (34) (14) was converted (65), by several steps including a modified Ullmann reaction (84), into the xanthone-lactone (35). The synthesis was completed as shown in the Chart.

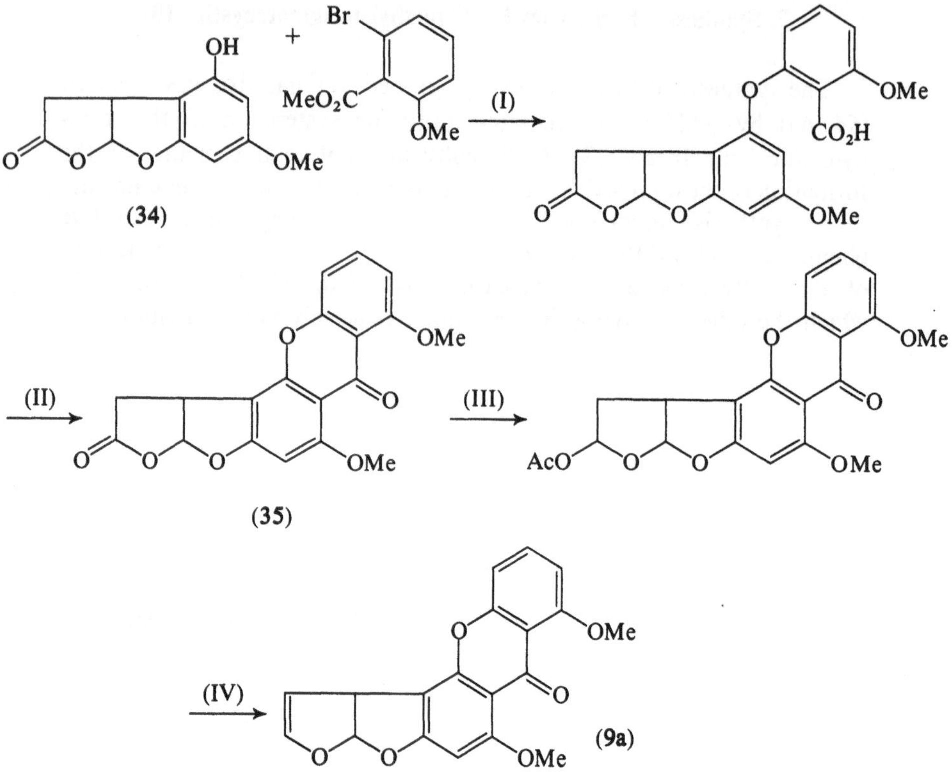

(I) Several steps involving an Ullmann reaction (Cu₂Cl₂/pyridine); (II) (COCl)₂/benzene/
heat; (III) (iso-C₅H₁₁)₂BH, then Ac₂O; (IV) pyrolysis — loss of HOAc.

Chart 11. Synthesis of (±)-O-methylsterigmatocystin (9a)

V. Biogenesis of the Mycotoxins

1. Sterigmatocystin

There now appears to be little doubt that all the seventeen carbon atoms of the sterigmatocystin (9) skeleton (*i.e.*, all carbon atoms in the molecule with the exception of that in the methoxyl group) are of acetate origin. This statement is based on results from (*a*) radio-carbon studies (47) in which the distribution of label was determined by the measurement of the activities of numerous degradation products, and (*b*) ^{13}C n.m.r. spectral analyses (77) of the metabolite which had been grown on a culture medium enriched in *either*

References, pp. 148—151

[1-^{13}C]-acetate *or* [2-^{13}C]-acetate. It also appears that there is an almost constant intensity of labelling. The results, substantiated by both lines of investigation, are expressed diagrammatically in structure (**36**).

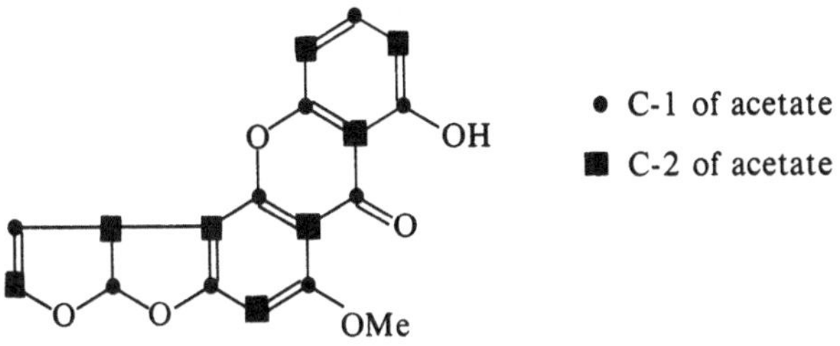

- ● C-1 of acetate
- ■ C-2 of acetate

(**36**) Origin of carbon atoms in sterigmatocystin

One curious feature emerges from an inspection of this diagram; although the usual alternation of C-1 and C-2 acetate carbon atoms (as in polyketides) occurs throughout the xanthone part of the molecule and in the distal dihydrofuran ring, the C–C bond joining the latter function to the xanthone system is formed between two carbon atoms which both originate in the methyl groups (C-2) of acetate. As pointed out in the sequel, two different hypotheses have been advanced which are capable of explaining this unusual phenomenon.

It may be significant that the mould which produces sterigmato-cystin, *A. versicolor,* also produces an array of polyhydroxyanthraquin-ones (Chart 12) some of which carry a six-carbon side chain [*e.g.,* averythrin (**37**) (*68*), averantin (**38**) (*11*)] and some of which possess the hydrofuro[2,3-*b*]benzofuran structure [*e.g.,* aversin (**40**) (*18, 49*), 6-deoxyversicolorin A (**41**) (*37*), and versicolorin C (**42**) (*40*)]. The anthraquinones of type (**37**) and (**38**) undoubtedly appear to be of polyketide origin [see (**39**) (*68*)] and it should be noted that they contain twenty carbon atoms in their molecules. It is possible (*80*) that quinones of the second type [*e.g.,* (**41**)], containing eighteen carbon atoms in their molecules and possessing the hydrofurobenzofuran function, might arise from quinones of the first type by a mechanism illustrated in Chart 13.

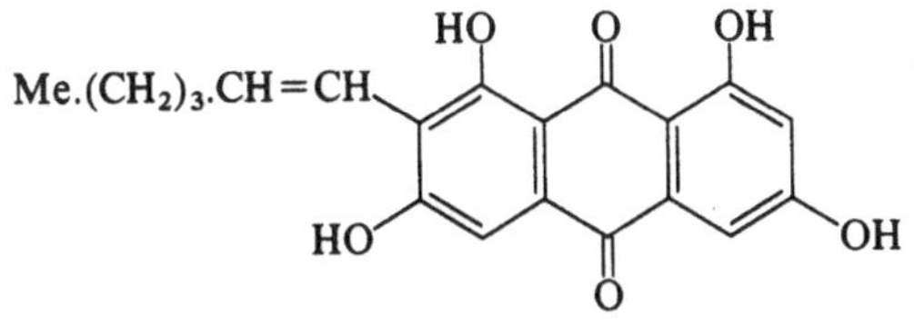

(**37**) Averythrin

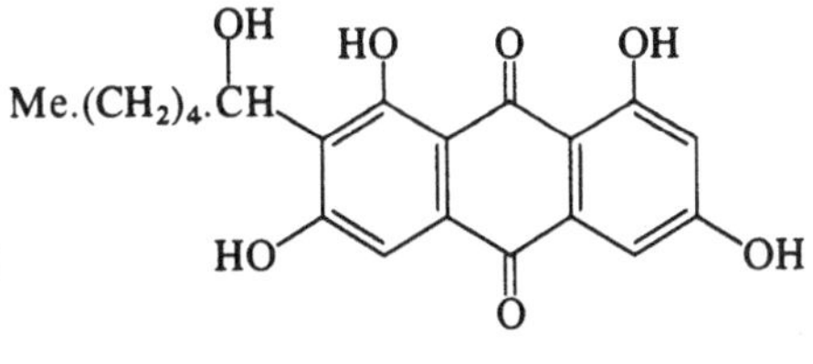

(**38**) Averantin

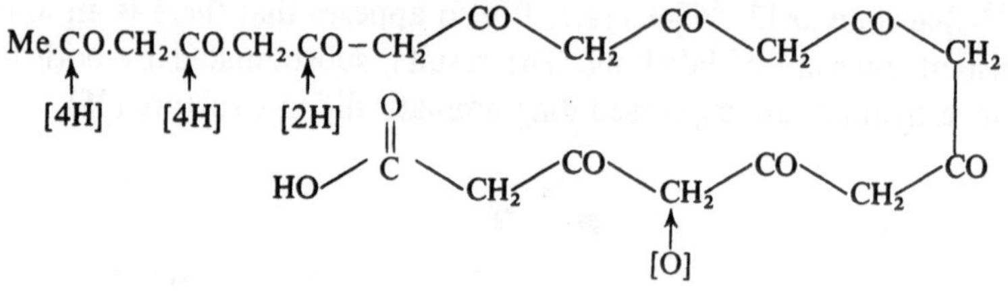
Me.CO.CH2.CO.CH2.CO–CH2
CO
CH2
CO
CH2
CO
CH2
[4H]
[4H]
[2H]
HO
C
O
CH2
CO
CH2
CO
CH2
CO
[O]

(**39**) Suggested polyketide precursor of (**37**) and (**38**)

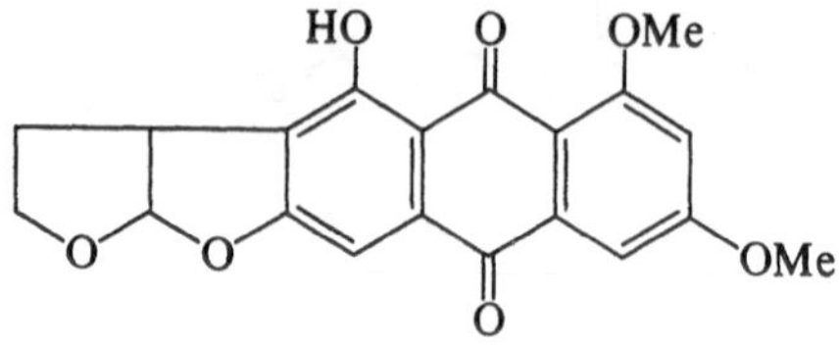
HO
O
OMe
O
O
OMe
O

(**40**) Aversin

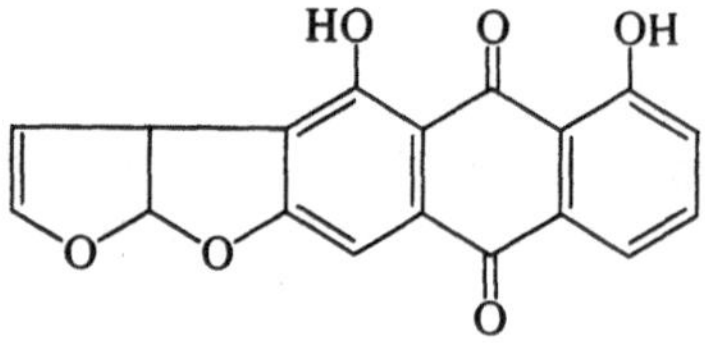
HO
O
OH
O
O
O

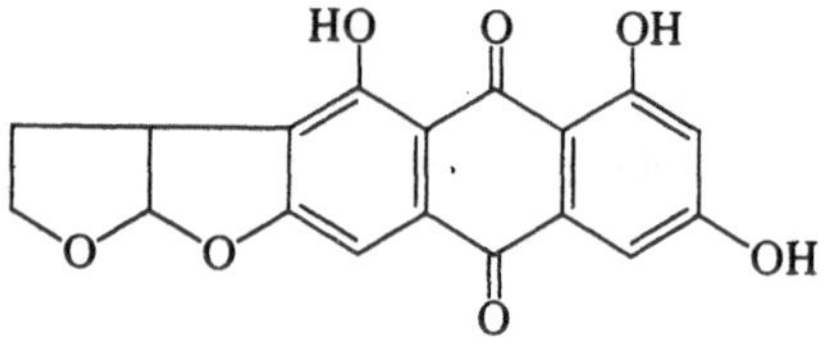
HO
O
OH
O
O
OH
O

(**41**) 6-Deoxyversicolorin A (**42**) Versicolorin C

Chart 12. Some polyhydroxyanthraquinones produced by *A. versicolor* and a suggested polyketide precursor

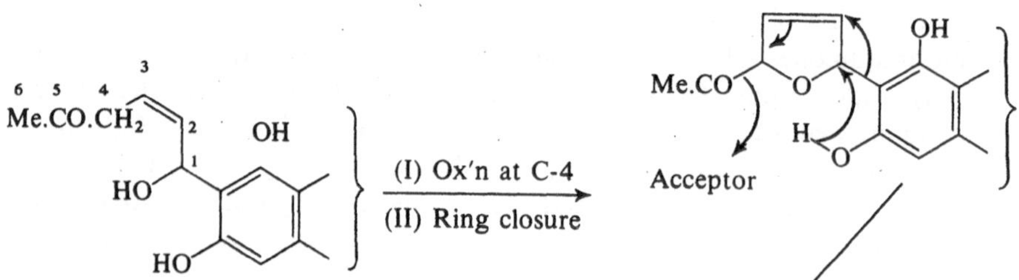
6 5 4
3
Me.CO.CH2
2
1
OH
HO
HO
(I) Ox'n at C-4
(II) Ring closure
Me.CO
O
OH
H
O
Acceptor

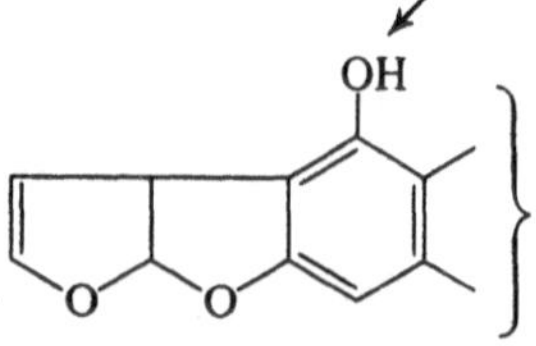
OH
O
O

Chart 13. Suggested biogenesis of the dihydrofuro[2,3-*b*]benzofuran structure

References, pp. 148—151

Now it has been suggested (*79*) that sterigmatocystin, which is known to be acetate derived, might arise *via* an anthraquinone of the second type by an overall process portrayed in Chart 14. An anthraquinone fission of the type shown in Chart 14 is not without precedent in mycological chemistry (*24, 39*).

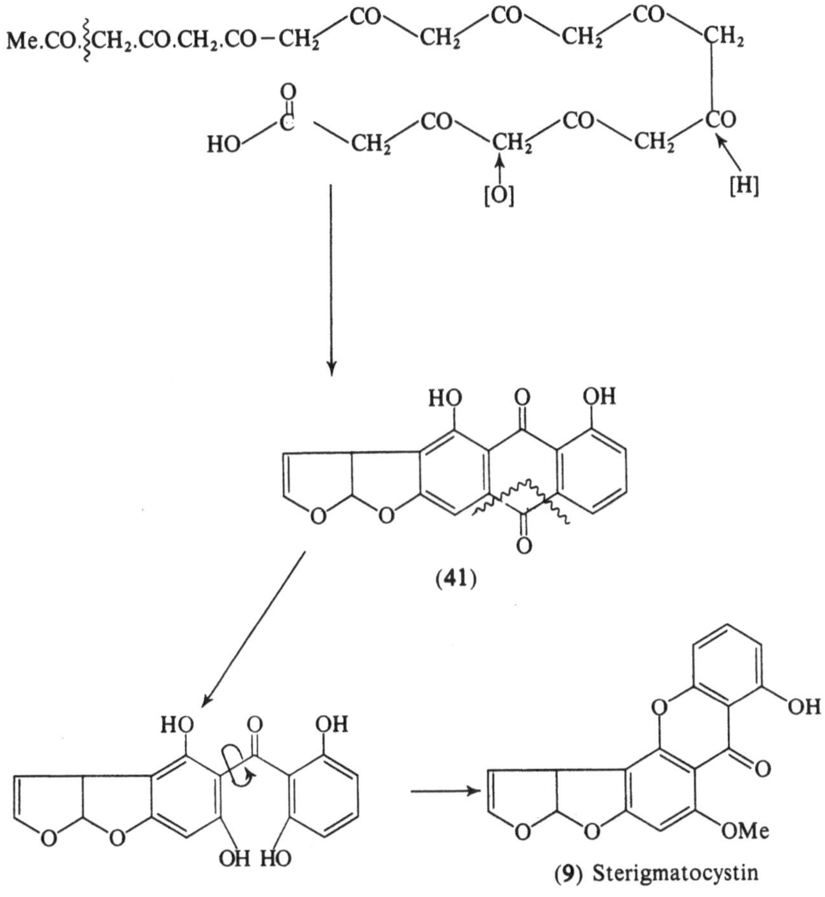

Chart 14. Suggested biogenesis of sterigmatocystin from a polyketide *via* an anthraquinone

The suggested biogenesis of sterigmatocystin outlined in Charts 14 and 13 would account not only for the curious feature in the labelling pattern on which comment has already been made, but also for the *overall* labelling pattern (**36**). An alternative hypothesis for the derivation of sterigmatocystin from a polyketide will be described in connection with the biogenesis of aflatoxin-B1 (see below).

2. Aflatoxin-B1

Fermentation studies (8, 9, 32) have shown that aflatoxin-B1 readily incorporates ^{14}C from methyl-^{14}C-methionine, and from [1-^{14}C]- and [2-^{14}C]-acetate. Both phenylalanine and shikimic acid have been excluded as biogenetic precursors (32). As expected, the radioactivity in aflatoxin-B1, resulting from the administration of methyl-^{14}C-methionine, was found almost exclusively in the methoxyl group. Intensive degradation studies (9) on the radioactive toxin, produced by addition of either [1-^{14}C]-acetate or [2-^{14}C]-acetate to the substrate, established that (I) the carbon skeleton was derived entirely from acetate, (II) the activity of the labelled carbon atoms was virtually equal throughout the molecule, and (III) the origin of the individual carbon atoms was as shown in (43). As in the case of sterigmatocystin, there are two adjacent carbon atoms which both originate from the C-2 of acetate; there are also two adjacent carbon atoms of C-1 acetate origin.

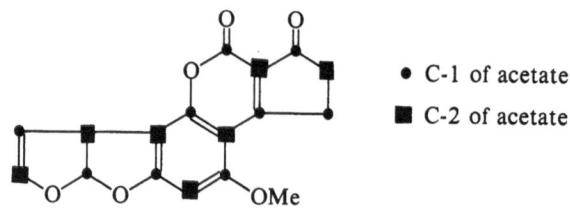

● C-1 of acetate
■ C-2 of acetate

(43) Origin of carbon atoms in aflatoxin-B1

BÜCHI and his collaborators (9) advanced the hypothesis (see Chart 15) that (I) a polyhydroxynaphthacene was first formed from nine acetate residues, (II) oxidation then occurred to yield an anthraquinone with an additional six-membered ring carrying an endoperoxide grouping, and (III) the terminal endoperoxidised ring underwent re-arrangements to give the dihydrofurofuran function. The overall result of the illustrated sequence of operations is the formation of the anthraquinone (41).

The similarity in structure between sterigmatocystin (9) and aflatoxin-B1 (1), and the co-existence of O-methylsterigmatocystin (9a) with aflatoxins in A. flavus (20), lend credibility to the postulate (48, 79) that sterigmatocystin is a precursor of aflatoxin-B1 or that the two metabolites arise from some common intermediate*. Now it has already been pointed

---

* Also, versicolorin C (42), a close relative of a suggested precursor (41) of sterigmatocystin, co-exists with aflatoxins in A. flavus (44).

References, pp. 148—151

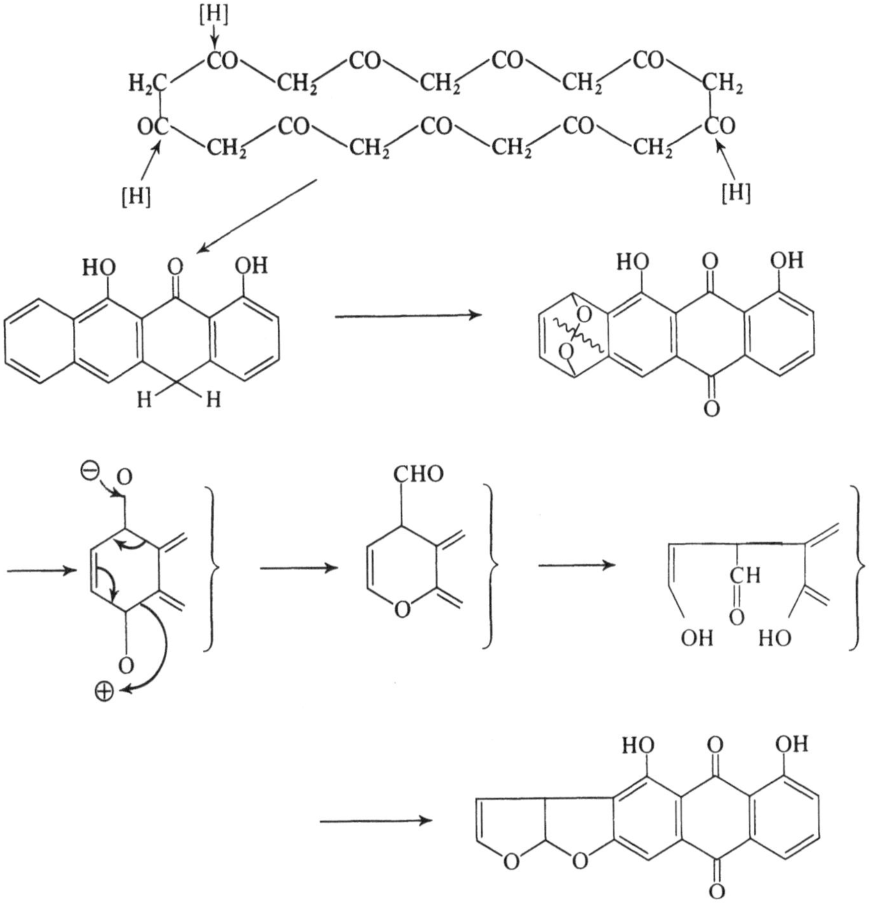

Chart 15. Suggested formation of an anthraquinone carrying a bis-furan side-chain from a polyketide

out that compound **(41)** might be convertible into sterigmatocystin by a route, illustrated in Chart 14, involving loss of one carbon atom of acetate-methyl origin. A scheme for the conversion of sterigmatocystin into aflatoxin-B1 has been advanced by THOMAS (*79*) and is illustrated in Chart 16. This scheme involves oxidative ring cleavage [wavy line in **(9)**], reduction, cyclisation to a cyclopentenone, and the loss (as carbon dioxide) of a second acetate-methyl derived carbon atom.

The combined hypotheses predict (see Charts 15, 14, and 16) that the aflatoxin-B1, produced from a polyketide *via* **(41)** and **(9)**, would

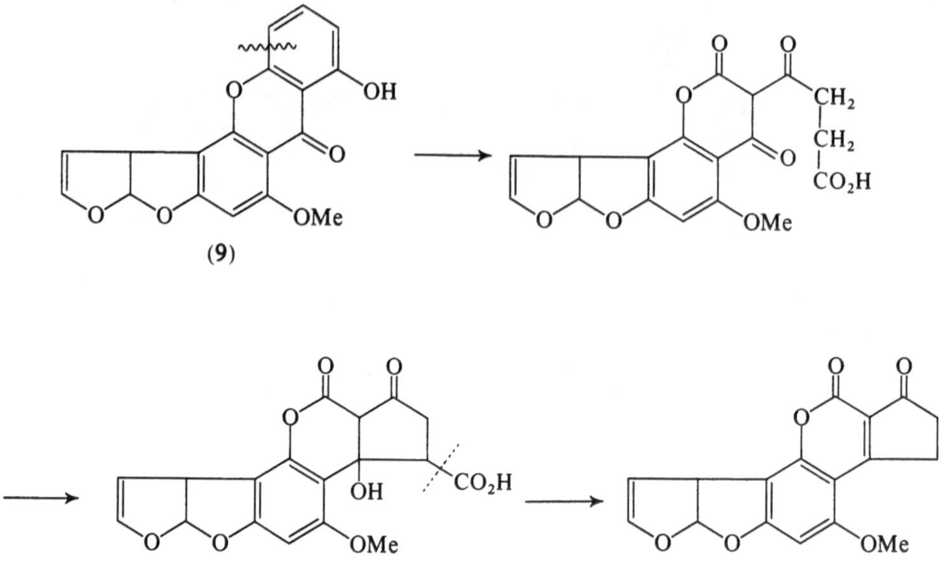

(9) (1)

Chart 16. Suggested biogenesis of aflatoxin-B1 (**1**) from sterigmatocystin (**9**)

have a skeleton containing *nine* atoms of C-1 acetate origin and *seven* of C-2 origin arranged as shown in diagram (**43**). Also a rationalisation is provided for the occurrence in aflatoxin-B1 of two vicinal carbon atoms of acetate-methyl origin and of two other vicinal carbon atoms of acetate C-1 origin.

The overall scheme is attractive and exactly fits the available evidence. Confirmation of such a scheme from feeding experiments with labelled intermediates would, of course, be desirable. A recent preliminary communication (*37*) claims that 5-hydroxy-dihydrosterigmatocystin (**44**), labelled with ^{14}C in the *O*-methyl group, is a biogenetic precursor of aflatoxins-B2 and -G2. Further results will be awaited with interest*.

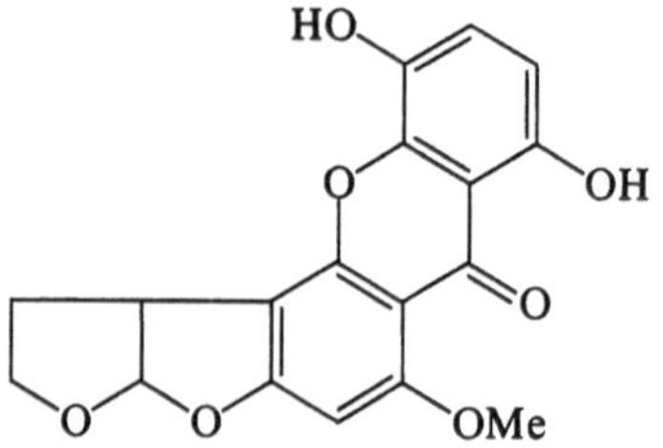

(**44**) 5-Hydroxy-dihydrosterigmatocystin

* For a *Note Added in Proof*, see (*86*).

References, pp. 148—151

VI. Conclusion

The mycotoxins considered in this review have attained notoriety on account of their powerful carcinogenic properties and consequent danger as fortuitous contaminants of feeding-stuffs. Fortunately, analytical methods of control are now well-established and it is to be expected that disastrous outbreaks of aflatoxicosis will not occur in the future. The toxicity of the aflatoxins towards different species of animals varies widely (27). There is little definite knowledge concerning the danger of these mycotoxins to man himself, but it may be significant that there is a high incidence of liver cancer amongst humans (34, 62) in certain tropical or sub-tropical parts of the world where conditions for fungal growth on natural substrates would be very favourable. It has been pointed out (50, 82) that, although sterigmatocystin is considerably less toxic than the aflatoxins, there is a possibility that its importance as a health hazard may still be very significant because of the widespread occurrence of the several moulds responsible for its production, and because of the large quantities in which it is generated.

A detailed review of the pharmacology and toxicology of the aflatoxins is available (27). The relationship between molecular structure and carcinogenicity in compounds of the aflatoxin type has been investigated (85).

For a number of years it was thought that the occurrence, in natural products, of the di- (or tetra-)hydrofurobenzofuran function was peculiar to the aflatoxins, sterigmatocystins, and their polyhydroxyanthraquinone congeners. Recently, however, the fungal toxin implicated in "pine-needle blight", dothistromin (ex Dothistroma pini, a member of the Sphaeropsidales), has been shown (6, 7) to contain this feature and to possess structure (45).

(45) Dothistromin

In conclusion, it may be said that the discovery of aflatoxicosis revealed an additional environmental hazard to world health, and that the ensuing chemical investigations of the aflatoxins and sterigmatocystins have added a new chapter to the organic chemistry of natural products.

References

1. Amos: The Book of Amos. Authorised Version of the Old Testament; Chapter 4; verse 9 (1611).

2. Asao, T., G. Büchi, M. M. Abdel-Kader, S. B. Chang, E. L. Wick, and G. N. Wogan: The Structures of Aflatoxins-B and -G. J. Amer. Chem. Soc. **85,** 1706 (1963).

3. — — — — — — The Structures of Aflatoxins B1 and G1. J. Amer. Chem. Soc. **87,** 882 (1965).

4. Bamburg, J. R., F. M. Strong, and E. B. Smalley: Toxins from Moldy Cereals. J. Agric. Food Chem. **17,** 443 (1969).

5. Barger, G.: Ergot and Ergotism. London: Gurney and Jackson. 1931.

6. Bassett, C., M. Buchanan, R. T. Gallagher, and R. Hodges: A Toxic Difuroanthraquinone from *Dothistroma pini.* Chem. and Ind. **1970,** 1659.

7. Bear, C. A., J. M. Waters, T. N. Waters, R. T. Gallagher, and R. Hodges: X-ray Determination of the Molecular Structure of a Derivative of Dothistromin, a Fungal Toxin Implicated in Pine Needle Blight. Chem. Commun. **1970,** 1705.

8. Biollaz, M., G. Büchi, and G. Milne: The Biogenesis of Bisfuranoids in the Genus *Aspergillus.* J. Amer. Chem. Soc. **90,** 5019 (1968).

9. — — — The Biosynthesis of the Aflatoxins. J. Amer. Chem. Soc. **92,** 1035 (1970).

10. Birkinshaw, J. H., and I. M. M. Hammady: Metabolic Products of *Aspergillus versicolor* (Vuillemin) Tiraboschi: Biochem. J. **65,** 162 (1957).

11. Birkinshaw, J. H., J. C. Roberts, and P. Roffey: "Product B" (Averantin) [(1,3,6,8)-Tetrahydroxy-2-(1-hydroxyhexyl)-anthraquinone], a Pigment from *Aspergillus versicolor* (Vuillemin) Tiraboschi. J. Chem. Soc. (C) (London) **1966,** 855.

12. Brechbühler, S., G. Büchi, and G. Milne: The Absolute Configuration of the Aflatoxins. J. Organ. Chem. (U.S.A.) **32,** 2641 (1967).

13. Büchi, G., D. M. Foulkes, M. Kurono, and G. F. Mitchell: The Total Synthesis of Racemic Aflatoxin-B1. J. Amer. Chem. Soc. **88,** 4534 (1966).

14. Büchi, G., D. M. Foulkes, M. Kurono, G. F. Mitchell, and R. S. Schneider: The Total Synthesis of Racemic Aflatoxin-B1. J. Amer. Chem. Soc. **89,** 6745 (1967).

15. Büchi, G., and E. C. Roberts: Preparation of 2-Carbethoxycyclopentan-1,3-dione. J. Organ. Chem. (U.S.A.) **33,** 460 (1968).

16. Büchi, G., and S. M. Weinreb: The Total Synthesis of Racemic Aflatoxin-M1 (Milk Toxin). J. Amer. Chem. Soc. **91,** 5408 (1969).

17. — — Total Synthesis of Aflatoxin-M1 and -G1, and an Improved Synthesis of Aflatoxin-B1. J. Amer. Chem. Soc. **93,** 746 (1971).

18. Bullock, E., D. Kirkaldy, J. C. Roberts, and J. G. Underwood: Two New Metabolites from a Variant Strain of *Aspergillus versicolor* (Vuillemin) Tiraboschi. J. Chem. Soc. (London) **1963,** 829.

19. Bullock, E., J. C. Roberts, and J. G. Underwood: The Structure of Iso-sterigmatocystin and an Amended Structure for Sterigmatocystin. J. Chem. Soc. (London) **1962,** 4179.

20. Burkhardt, H. J., and J. Forgacs: *O*-Methylsterigmatocystin, a New Metabolite from *Aspergillus flavus* Link *ex* Fries. Tetrahedron **24,** 717 (1968).

21. Butler, W. H.: Acute Toxicity of Aflatoxin-B1 in Rats. Brit. J. Cancer **18,** 756 (1964).

22. Cheung, K. K., and G. A. Sim: Aflatoxin-G1: Direct Determination of the Structure by the method of Isomorphous Replacement. Nature **201,** 1185 (1964).

23. Cole, R. J., J. W. Kirksey, and H. W. Schroeder: Dihydro-*O*-methylsterigmatocystin, a New Metabolite from *Aspergillus flavus.* Tetrahedron Letters **1970,** 3109.

24. Curtis, R. F., C. H. Hassall, and D. R. Parry: The Conversion of the Anthraquinone, Questin, into the Benzophenone, Sulochrin, in Cultures of *Aspergillus terreus*. J. Chem. Soc. (London) Perkin I **1972**, 240.

25. DAVIES, J. E., D. KIRKALDY, and J. C. ROBERTS: Sterigmatocystin, a Metabolite of *Aspergillus versicolor* (Vuillemin) Tiraboschi. J. Chem. Soc. (London) **1960**, 2169.
26. DAVIES, J. E., J. C. ROBERTS, and S. C. WALLWORK: Sterigmatocystin, a Metabolic Product of *Aspergillus versicolor* (Vuillemin) Tiraboschi. Chem. and Ind. **1956**, 178.
27. DETROY, R. W., E. B. LILLEHOJ, and A. CIEGLER: Aflatoxin and Related Compounds. In: Microbial Toxins, Vol. VI (Ed., A. CIEGLER et al.), p. 90 *et seq.* London: Academic Press. 1971.
28. — — — *ibid.*, p. 124.
29. DICKENS, F.: Carcinogenesis: A Broad Critique. Twentieth Annual Symposium on Fundamental Cancer Research, 1966, p. 447. Baltimore: The Williams and Wilkins Co. 1967.
30. DICKENS, F., and H. E. H. JONES: Further Studies on the Carcinogenic Action of Certain Lactones and Related Substances in the Rat and Mouse. Brit. J. Cancer **19**, 392 (1965).
31. DICKENS, F., H. E. H. JONES, and H. B. WAYNFORTH: Oral, Subcutaneous and Intratracheal Administration of Carcinogenic Lactones and Related Substances: The Intratracheal Administration of Cigarette Tar in the Rat. Brit. J. Cancer **20**, 134 (1966).
32. DONKERSLOOT, J. A., D. P. H. HSIEH, and R. I. MATELES: Incorporation of Precursors into Aflatoxin-B1. J. Amer. Chem. Soc. **90**, 5020 (1968).
33. DORP, D. A. VAN, A. S. M. VAN DER ZIJDEN, R. K. BEERTHUIS, S. SPARREBOOM, W. O. ORD, K. DE JØNG, and R. KEUNING: Dihydro-aflatoxin-B1, a Metabolite of *Aspergillus flavus*. Remarks on the Structure of Aflatoxin-B. Rec. Trav. Chim. Pays-Bas **82**, 587 (1963).
34. DUNHAM, L. J., and J. C. BAILAR: World Maps of Cancer Mortality Rates and Frequency Rates. J. Natl. Cancer Inst. **41**, 155 (1968).
35. DUTTON, M. F., and J. G. HEATHCOTE: The Structure, Biochemical Properties and Origin of the Aflatoxins-B2a and -G2a. Chem. and Ind. **1968**, 418.
36. — — O-Alkyl Derivatives of Aflatoxins-B2a and -G2a. Chem. and Ind. **1969**, 983.
37. ELSWORTHY, G. C., J. S. E. HOLKER, J. M. McKEOWN, J. B. ROBINSON, and L. J. MULHEIRN: The Biosynthesis of the Aflatoxins. Chem. Commun. **1970**, 1069.
38. GOLDBLATT, L. A. (Editor): Aflatoxin. New York: Academic Press. 1969.
39. GRÖGER, D., D. ERGE, B. FRANCK, U. OHNSORGE, H. FLASCH, and F. HÜPER: Mutterkorn-Farbstoffe XVIII. Emodin als Biosynthesevorstufe der Ergochrome. Chem. Ber. **101**, 1970 (1968).
40. HAMASAKI, T., Y. HATSUDA, N. TERASHIMA, and M. RENBUTSU: Isolation and Structures of Three Metabolites (from *Aspergillus versicolor*), Versicolorins A, B, and C. Agric. and Biol. Chem. (Japan) **31**, 11 (1967).
41. HARTLEY, R. D., B. F. NESBITT, and J. O'KELLY: Toxic Metabolites of *Aspergillus flavus*. Nature **198**, 1056 (1963).
42. HATSUDA, Y., and S. KUYAMA: Cultivation of *Aspergillus versicolor* and Isolation and Purification of Metabolic Products. J. Agric. Chem. Soc. Japan **28**, 989 (1954).
43. HATSUDA, Y., S. KUYAMA, and N. TERASHIMA: Physical and Chemical Properties and the Chemical Structure of Sterigmatocystin. J. Agric. Chem. Soc. Japan **28**, 992 (1954).
44. HEATHCOTE, J. G., and M. F. DUTTON: New Metabolites of *Aspergillus flavus*. Tetrahedron **25**, 1497 (1969).
45. HESSELTINE, C. W., O. L. SHOTWELL, J. J. ELLIS, and R. D. STUBBLEFIELD: Aflatoxin Formation by *Aspergillus flavus*. Bacteriological Reviews **30**, 795 (1966).
46. HØLKER, J. S. E., and S. A. KAGAL: 5-Methoxysterigmatocystin, a Metabolite from a Mutant Strain of *Aspergillus versicolor*. Chem. Commun. **1968**, 1574.

47. Hølker, J. S. E., and L. J. Mulheirn: The Biosynthesis of Sterigmatocystin. Chem. Commun. **1968**, 1576.

48. Holker, J. S. E., and J. G. Underwood: A Synthesis of a Cyclopentenocoumarin Structurally related to Aflatoxin-B. Chem. and Ind. **1964**, 1865.

49. Holmwood, G. M., and J. C. Roberts: Total Synthesis of (±)-*O*-Methylaversin [(±)-Tri-*O*-methylversicolorin B]: The Structure of Aversin. J. Chem. Soc. (C) (London) **1971**, 3899.

50. Holzapfel, C. W., I. F. H. Purchase, P. S. Steyn, and L. Gouws: The Toxicity and Chemical Assay of Sterigmatocystin, a Carcinogenic Mycotoxin, and its Isolation from Two New Fungal Sources. South African Medical Journal **40**, 1100 (1966).

51. Holzapfel, C. W., P. S. Steyn, and I. F. H. Purchase: Isolation and Structure of Aflatoxins-M1 and -M2. Tetrahedron Letters **1966**, 2799.

52. Iongh, H. de, R. O. Vles, and J. G. van Pelt: Milk of Mammals Fed on Aflatoxin-containing Diet. Nature **202**, 466 (1964).

53. Jackman, L. M.: Applications of Nuclear Magnetic Resonance Spectroscopy in Organic Chemistry, p. 87. London: Pergamon Press. 1959.

54. Joffe, A. Z.: Toxin Production by Cereal Fungi causing Toxic Alimentary Aleukia in Man. In: G. N. Wogan, Mycotoxins in Foodstuffs, p. 77. Cambridge, Mass.: M. I. T. Press. 1965.

55. Kinosita, R., and T. Shikata: On Toxic Mouldy Rice. In: G. N. Wogan, Mycotoxins in Foodstuffs, p. 111. Cambridge, Mass.: M. I. T. Press. 1965.

56. Knight, J. A.: Synthetic Studies in Relation to Mould Metabolites. Ph. D. Thesis, Nottingham (1965).

57. Knight, J. A., J. C. Roberts, and P. Roffey: Synthesis of (±)-Tetrahydro-4-hydroxy-6-methoxyfuro[2,3-*b*]benzofuran, a Racemic Form of a Laevorotatory Degradation Product of Dihydrosterigmatocystin. J. Chem. Soc. (C) (London) **1966**, 1308.

58. Knight, J. A., J. C. Roberts, P. Roffey, and A. H. Sheppard: Synthesis of (±)-Tetrahydrodeoxo-aflatoxin-B1. Chem. Commun. **1966**, 706.

59. Legator, M. S.: Biological Assay for Aflatoxins. In: L. A. Goldblatt, Aflatoxin, p. 107 and p. 125. New York: Academic Press. 1969.

60. Merwe, K. J. van der, L. Fourie, and de B. Scott: On the Structure of the Aflatoxins. Chem. and Ind. **1963**, 1660.

61. Nesbitt, B. F., J. O'Kelly, K. Sargeant, and A. Sheridan: Toxic Metabolites of *Aspergillus flavus*. Nature **195**, 1062 (1962).

62. Oettlé, A. G.: Cancer in Africa, Especially in Regions South of the Sahara. J. Natl. Cancer Inst. **33**, 383 (1964).

63. Pons, W. A., *Jr.*, and L. A. Goldblatt: Physicochemical Assay of Aflatoxins. In: L. A. Goldblatt, Aflatoxin, p. 77. New York: Academic Press. 1969.

64. Rance, M. J., and J. C. Røberts: Total Synthesis of (±)-Dihydro-*O*-methylsterigmatocystin. Tetrahedron Letters **1969**, 277.

65. — — Total Synthesis of (±)-*O*-Methylsterigmatocystin. J. Chem. Soc. (C) (London) **1971**, 1247.

66. Roberts, J. C.: A Novel Method for the Degradation of 1-Hydroxy-xanthones. J. Chem. Soc. (London) **1960**, 785.

67. — Naturally Occurring Xanthones. Chem. Rev. **61**, 591 (1961).

68. Røberts, J. C., and P. Røffey: Averythrin, an Anthraquinonoid Pigment from *Aspergillus versicolor* (Vuillemin) Tiraboschi. J. Chem. Soc. (London) **1965**, 3666.

69. Roberts, J. C., A. H. Sheppard, J. A. Knight, and P. Roffey: Total Synthesis of (±)-Aflatoxin-B2. J. Chem. Soc. (C) (London) **1968**, 22.

70. Roberts, J. C., and J. G. Underwood: Synthesis of 1,3,8-Trimethoxy-xanthone-4-carboxylate, a Degradation Product of Sterigmatocystin. J. Chem. Soc. (London) **1962**, 2060.

71. RODRICKS, J. V., K. R. HENERY-LOGAN, A. D. CAMPBELL, L. STOLOFF, and M. J. VERRETT: Isolation of a New Toxin from Cultures of *Aspergillus flavus*. Nature **217**, 668 (1968).

72. RODRICKS, J. V., E. LUSTIG, A. D. CAMPBELL, and L. STOLOFF: Aspertoxin, a Hydroxy Derivative of O-Methylsterigmatocystin from Aflatoxin-producing Cultures of *Aspergillus flavus*. Tetrahedron Letters **1968**, 2975.

73. SARGEANT, K., A. SHERIDAN, J. O'KELLY, and R. B. A. CARNAGHAN: Toxicity Associated with Certain Samples of Ground-nuts. Nature **192**, 1096 (1961).

74. SØEST, T. C. VAN, and A. F. PEERDEMAN: X-ray Study of Dihydroflatoxin-B1. Kon. Ned. Akad. Wetensch. Proc., Ser. B. **67**, 469 (1964).

75. SPENSLEY, P. C.: Aflatoxin, the Active Principle in Turkey X Disease. Endeavour **22**, 75 (1963).

76. STUBBLEFIELD, R. D., O. L. SHOTWELL, G. M. SHANNON, D. WEISLEDER, and W. K. RØHWEDDER: Parasiticol: A New Metabolite from *Aspergillus parasiticus*. J. Agric. Food Chem. **18**, 391 (1970).

77. TANABE, M., T. HAMASAKI, and H. SETO: Biosynthetic Studies with Carbon-13: ^{13}C Nuclear Magnetic Resonance Spectra of the Metabolite, Sterigmatocystin. Chem. Commun. **1970**, 1539.

78. TANAKA, N., Y. KATSUBE, Y. HATSUDA, T. HAMASAKI, and M. ISHIDA: Structure Analysis of p-Bromobenzoate of Sterigmatocystin by X-Ray Diffraction Method. Bull. Chem. Soc. Japan **11**, 3635 (1970).

79. THOMAS, R.: Biosynthetic Pathways Involving Ring Cleavage. In: Biogenesis of Antibiotic Substances, p. 160 (Z. VANĚK and Z. HOŠŤÁLEK). London: Academic Press. 1965.

80. — Personal communication. See also, MOSS, M. O.: Aflatoxin and Related Mycotoxins. In: Phytochemical Ecology, p. 140 (Ed., J. B. HARBORNE). London: Academic Press. 1972.

81. VORSTER, L. J.: Analysis of Cereals and Ground-nuts for Three Mycotoxins. Analyst **94**, 136 (1969).

82. VORSTER, L. J., and I. F. H. PURCHASE: Determination of Sterigmatocystin in Grain and Oil-seeds. Analyst **93**, 694 (1968).

83. WAISS, A. C. *Jr.*, M. WILEY, D. R. BLACK, and R. E. LUNDIN: 3-Hydroxy-6,7-dimethoxydifuro-xanthone, a New Metabolite from *Aspergillus flavus*. Tetrahedron Letters **1968**, 3207.

84. WILLIAMS, A. L., R. E. KINNEY, and R. F. BRIDGER: Solvent-Assisted Ullmann Ether Synthesis: Reactions of Dihydric Phenols. J. Organ. Chem. (U.S.A.) **32**, 2501 (1967).

85. WOGAN, G. N., G. S. EDWARDS, and P. M. NEWBERNE: Structure-activity Relations in Toxicity and Carcinogenicity of Aflatoxins and Analogs. Cancer Res. **31**, 1936 (1971).

86. *Note Added in Proof.* In a very recent paper [LIN, M. T., and D. P. H. SIEH, J. Amer. Chem. Soc. **95**, 1668 (1973)] it is claimed that the polyhydroxy-anthraquinone, averufin [PUSEY, D. F. G., and J. C. ROBERTS, J. Chem. Soc. (London) **1963**, 3542; ROFFEY, P., and M. V. SARGENT, Chem. Commun. **1966**, 913] is a precursor of aflatoxin-B1 in a strain of *A. parasiticus*.

(Received January 2, 1973)

Flavonoid-Glykoside

Von H. Wagner, München

Inhaltsübersicht

Danksagung: Der Autor dankt Herrn Prof. L. Farkas (Budapest) für Verbesserungs- und Ergänzungsvorschläge.

I. Strukturtypen von Flavonoid-O-Glykosiden und ihre Verbreitung

1. Glykosidierungsmuster (siehe dazu Schema 1)

Von den meisten natürlich vorkommenden Flavonoiden sind O-Glyko-side aufgefunden worden. Zahl und Variationsmöglichkeiten sind bei den Polyhydroxy-Flavonoiden am größten, bei den stark alkylierten Flavonoiden oder solchen mit nicht substituiertem B-Ring am geringsten. Die Häufigkeit, mit der bestimmte Glykosidierungstypen auftreten, korreliert weitgehend mit der von SIMPSON und BETON (*188*) bei der Methylierung von Polyhydroxy-Flavonen und -Flavonolen beobachteten nucleophilen Reaktivität der einzelnen phenolischen OH-Gruppen. Bei *Flavonolen* vom Typ des Quercetins (**1**) mit OH-Substitutionen in C_3-, C_5-, C_7-, $C_{3'}$- und $C_{4'}$-Position stehen z. B. die C_3-O-Glykoside zahlenmäßig an der Spitze. Es folgen mit abnehmender Häufigkeit die C_7-, $C_{4'}$-, $C_{3'}$ und C_5-O-Glykoside. Da sich die C_3- und C_7-OH-Gruppen nur wenig in ihrer Acidität voneinander unterscheiden, findet man auch zahlreiche C_3-, C_7-Di-O-Glykoside, dagegen weniger die Kombination $C_3 C_{4'}$ ($C_{3'}$) oder $C_7 C_{4'}$.

Bei *Flavonen, Isoflavonen* und *Flavanonen* mit fehlender C_3-OH-Gruppe stehen die C_7-O-Glykoside an der Spitze, während die $C_{4'}$-sporadisch, die $C_{3'}$-, $C_{2'}$- und C_5-O-Glykoside sehr selten auftreten. Von den wenigen bisher aufgefundenen Flavanonol-Glykosiden sind die meisten in C_3-, einige auch in $C_{3'}$- und $C_{4'}$-Stellung glykosidiert.

Die als biosynthetische Vorläufer der Flavanone geltenden *Chalkone* und *3,4-Dihydro-chalkone* sind dementsprechend bevorzugt in $C_{4'}$ und C_4- sowie in $C_{2'}$-Stellung glykosidiert. In der Auron (Benzalcumaran-3-on)-Reihe findet man fast nur C_6- und C_4- vereinzelt auch C_6-, $C_{3'}$-O-Glykoside.

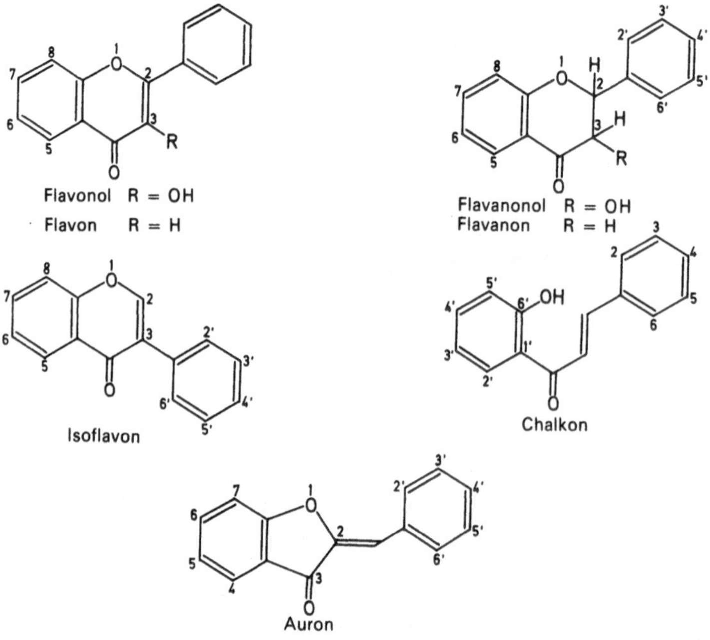

Flavonol R = OH
Flavon R = H

Flavanonol R = OH
Flavanon R = H

Isoflavon

Chalkon

Auron

Schema 1. Flavonoid-Strukturtypen

Während OH- oder Carbonyl-Gruppen in *o*- bzw. *peri*-Stellung zu einer anderen OH-Gruppe im allgemeinen kein Hindernis für eine Glykosidierung darstellen, beobachtet man bei Vorliegen einer vicinalen Trihydroxyl-Substitution (z. B. C_5-, C_6- und C_7-OH im A-Ring) nur Glykosidierungen an der OH-Gruppe mit einer freien Nachbarposition, d. h. nur am C_7-OH. C_6-O-Glykoside sind aus diesem Grunde unbekannt, wenn man von wenig fundierten Angaben absieht. Dagegen existieren C_8-, C_5- und vereinzelt auch $C_{3'}$-O-Glykoside.

Ältere Übersichten über Struktur und Verbreitung von Flavonoid-O-Glykosiden finden sich in den Büchern von GEISSMAN, 1962 (*82*) und KARRER, 1958 (*124*). Die letzten Zusammenstellungen stammen von PRIDHAM, 1965 (Übersicht über Phenol-Kohlenhydrat-Verbindungen) (*171*), HARBORNE, 1967 (alle Flavonoid-Typen) (*74*), FARKAS, 1967 (Aurone) (*51*) und WONG, 1970 (Isoflavone) (*217*).

2. O-Monoside

An der Glykosidbildung von Flavonoiden können folgende Monosaccharide beteiligt sein: D-Glucose, D-Galactose, D-Glucuronsäure, D-Galacturonsäure, D-Xylose, D-Apiose, L-Rhamnose (6-Deoxyhexose) und L-Arabinose. Sie kommen mit Ausnahme der Apiose alle direkt an die Aglykone gebunden als O-Monoside vor. Wenn man von einigen noch unbewiesenen Ausnahmen absieht, sind alle Glykoside mit Zuckern der D-Reihe in Übereinstimmung mit der Klyne-Regel β-konfiguriert, die der L-Serie wie z. B. die Rhamnose und Arabinose α-konfiguriert. Die bisher durchgeführten Synthesen haben außerdem gezeigt, daß sehr wahrscheinlich alle Zucker — mit Ausnahme der Apiose und Arabinose — in der Pyranoseform vorliegen. Während die Apiose nur in der furanoiden Form vorkommt, ist die Arabinose in den Flavonoid-Glykosiden sowohl in der pyranoiden wie in der furanoiden Form aufgefunden worden.

Für das *Guaijaverin* aus *Psidium guaijava* L. (Schmp. = 239° bzw. 256°, $[\alpha]_D^{25} = -97°$ in Äthanol) wurde von EL KHADEM und MOHAMED (*40*) durch Permethylierung des Glykosides, Hydrolyse zu Tetramethylquercetin und 2,3,4-Tri-O-methyl-arabinose und durch Vergleich von natürlichem 5,7,3',4'-Tetramethylguaijaverintriacetat mit synthetischem Material die Struktur eines Quercetin-α-L-arabopyranosids festgelegt. Demgegenüber handelt es sich bei dem erstmals aus *Polygonum aviculare* L. isolierten *Avicularin* (*158*) (Schmp. = 216°—217°, $[\alpha]_D^{25} = -168°$ in Äthanol) um das Quercetin-α-L-arabofuranosid, da bei dessen Permethylierung und Hydrolyse die 2,3,5-Tri-O-methyl-arabinose und durch Weiteroxidation die 2,3,5-Tri-O-methyl-arabonsäure erhalten wurde (*158*). In einem dritten Quercetin-arabinosid, dem *Polystachosid* aus *Polygonum polystachyum* (Schmp. = 246°—247°, $[\alpha]_D = -25,9$ in Methanol), liegt den Enzymhydrolyse-Studien zufolge ein 3-O-β-L-Arabinosid nicht geklärter Struktur vor (*93*). Ein viertes in der Literatur beschriebenes Quercetin-3-arabinosid aus *Foeniculum vulgare* Mill. (*159*), das *Foeniculin* (Schmp. = 256°), ist wiederum verschieden von Avicularin, könnte aber mit Guaijaverin identisch sein.

An zweiter Stelle hinter den O-Monosiden des Flavonols *Quercetin* (**1**) stehen in der Verbreitung die Monoglykoside des *Isoquercetins* (**7**) und *Kämpferols* (**8**). Nur sporadisch kommen die Glykoside des *Myricetins* (**10**), *Quercetagetins* (**11**), *Gossypetins* (**12**) und der zahlreichen Mono-, Di- und Trimethoxy-Flavonole vor. Vom Gossypetin existiert ein 7-O- und ein 8-O-Monoglykosid (Schema 2).

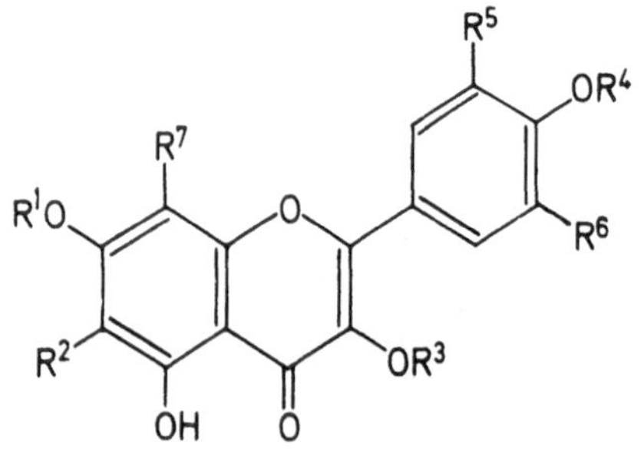

(1)	Quercetin:	R^1, R^2, R^3, R^4, R^6, $R^7 = H$; $R^5 = OH$;
(2)	Isoquercitrin:	R^1, R^2, R^4, R^6, $R^7 = H$; $R^5 = OH$; $R^3 =$ Glucosyl;
(3)	Hyperosid:	R^1, R^2, R^4, R^6, $R^7 = H$; $R^5 = OH$; $R^3 =$ Galactosyl;
(4)	Quercitrin:	R^1, R^2, R^4, R^6, $R^7 = H$; $R^5 = OH$; $R^3 =$ Rhamnosyl;
(5)	Quercimeritrin:	$R^1 =$ Glucosyl; R^2, R^3, R^4, R^6, $R^7 = H$; $R^5 = OH$;
(6)	Spiraeosid:	R^1, R^2, R^3, R^6, $R^7 = H$; $R^5 = OH$; $R^4 =$ Glucosyl;
(7)	Isoquercetin:	R^1, R^2, R^3, R^4, R^6, $R^7 = H$; $R^5 = OCH_3$;
(8)	Kämpferol:	R^1, R^2, R^3, R^4, R^5, R^6, $R^7 = H$;
(9)	Populnin:	$R^1 =$ Glucosyl; R^2, R^3, R^4, R^5, R^6, $R^7 = H$;
(10)	Myricetin:	R^1, R^2, R^3, R^4, $R^7 = H$; R^5, $R^6 = OH$;
(11)	Quercetagetin:	R^1, R^3, R^4, R^6, $R^7 = H$; R^2, $R^5 = OH$;
(12)	Gossypetin:	R^1, R^2, R^3, R^4, $R^6 = H$; R^5, $R^7 = OH$;

Schema 2. Häufig vorkommende Flavonol-Aglykone und ihre Monoglykoside

Die wichtigsten Monoside der Flavon-Reihe leiten sich vom *Apigenin* (**13**) und *Luteolin* (**15**) ab. Relativ selten werden Glykoside des *Acacetins* (**14**), *Chrysoeriols* (**16**), *Diosmetins* (**17**), *Scutellareins* (**18**), *Chrysins* (**19**) oder *Baicaleins* (**20**) angetroffen (Schema 3). Bevorzugte Glykosidierungsstellen sind hier die OH-Gruppen in C_7- und $C_{4'}$-, seltener in C_5-Position. Von den *Isoflavonen* sind nur etwa 30 Glykoside, vorwiegend 7-O-Monoside, davon die meisten in Leguminosen, aufgefunden worden (*217*).

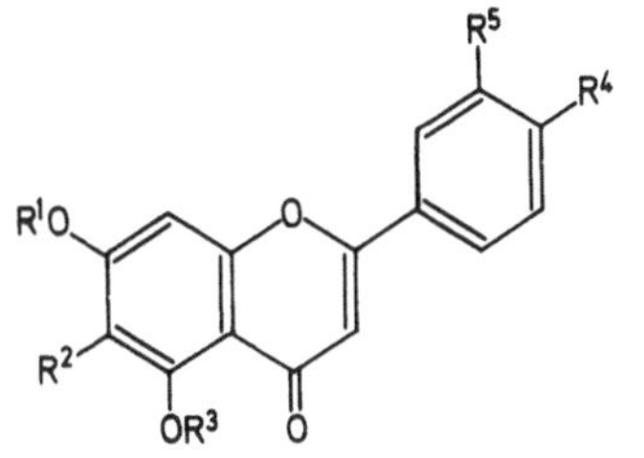

(13) Apigenin: $R^1, R^2, R^3, R^5 = H; R^4 = OH;$
(14) Acacetin: $R^1, R^2, R^3, R^5 = H; R^4 = OCH_3;$
(15) Luteolin: $R^1, R^2, R^3 = H; R^4, R^5 = OH;$
(16) Chrysoeriol: $R^1, R^2, R^3 = H; R^4 = OH; R^5 = OCH_3;$
(17) Diosmetin: $R^1, R^2, R^3 = H; R^4 = OCH_3; R^5 = OH;$
(18) Scutellarein: $R^1, R^3, R^5 = H; R^2, R^4 = OH;$
(19) Chrysin: $R^1, R^2, R^3, R^4, R^5 = H;$
(20) Baicalein: $R^1, R^3, R^4, R^5 = H; R^2 = OH;$

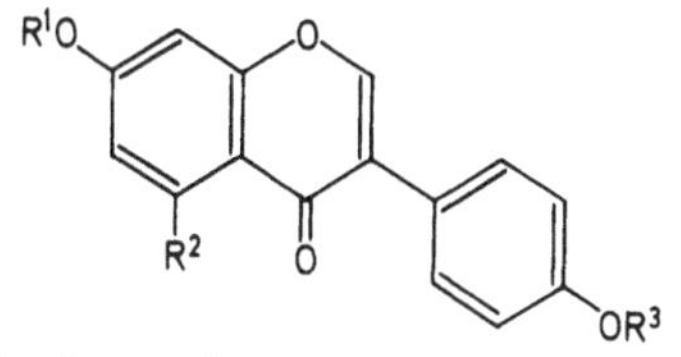

(21) Genistein: $R^1, R^3 = H; R^2 = OH;$
(22) Daidzein: $R^1, R^2, R^3 = H;$

Schema 3. Häufig vorkommende Flavon- und Isoflavon-Aglykone

Die häufigsten Aglukone sind das *Genistein* (21) und *Daidzein* (22) (Schema 3). Wie bei den Isoflavon-Monosiden ist auch bei den Flavanon-Monosiden der Zuckerpartner fast ausschließlich die Glucose. Isoliert wurden zum größten Teil aus *Prunus*-Arten die 7-O-Monoside des *Naringenins* (23), *Eriodictyols* (26), *Hesperetins* (27) und *Isosakuranetins* (28), ein 4′-O-Glucosid des (−) *Liquiritigenins* (30) und je ein 5-O-Glucosid des (+) *Sakuranetins* (29) und (−) bzw (+) Naringenins (23). Mit dem letzten Glykosid verwandt ist das Dihydrochalkon-2′-O-mono-glucosid *Phlorrhizin* (31) aus *Malus*-Arten (Schema 4).

Die ebenfalls vorwiegend als Monoside auftretenden 10 Auron-glykoside sind zu etwa 70% aus Pflanzen der Compositen-Familie (Heliantheae, Coreopsidinae) isoliert worden (51). Häufiger vorkommende Glykoside sind die 4-O- bzw. 6-O-Glucoside des Aureusidins (32), Sulfuretins (33) und Bracteatins (34).

Nahezu alle Monoglykoside der 7 Flavonoidtypen sind heute durch die Synthese in ihrer Struktur bestätigt (siehe Synthese Kapitel). Nur das aus *Polygonum reynoutria* Mak. isolierte und als Quercetin-3-O-mono-xylosid identifizierte Reynoutrin (151) (Schmp. = 203—204°, $[\alpha]_D^{18}$: −175°

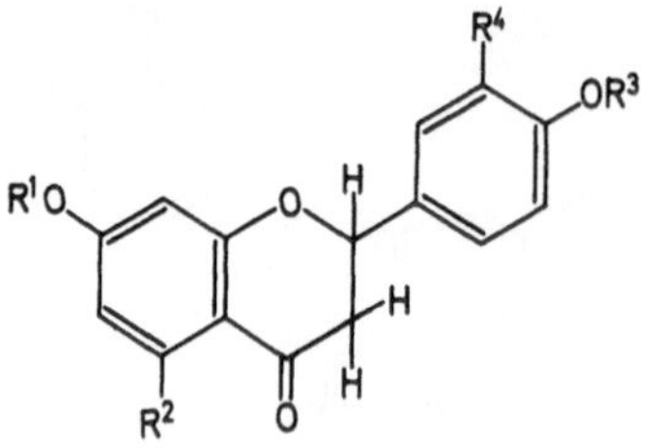

(23)	Naringenin:	$R^1, R^3, R^4 = H; R^2 = OH;$
(24)	Prunin:	$R^1 = Glucosyl; R^2 = OH; R^3, R^4 = H;$
(25)	Salipurposid:	$R^1, R^3, R^4 = H; R^2 = O\text{-}Glucosyl;$
(26)	Eriodictyol:	$R^1, R^3 = H; R^2, R^4 = OH;$
(27)	Hesperetin:	$R^1 = H; R^2, R^4 = OH; R^3 = CH_3;$
(28)	Isosakuranetin:	$R^1, R^4 = H; R^2 = OH; R^3 = CH_3;$
(29)	Sakuranetin:	$R^1 = CH_3; R^2 = OH; R^3, R^4 = H;$
(30)	Liquiritigenin:	$R^1, R^2, R^3, R^4 = H;$

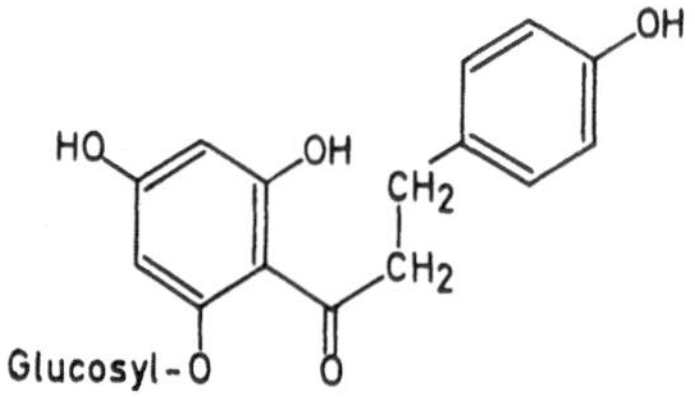

(31)　Phlorrhizin

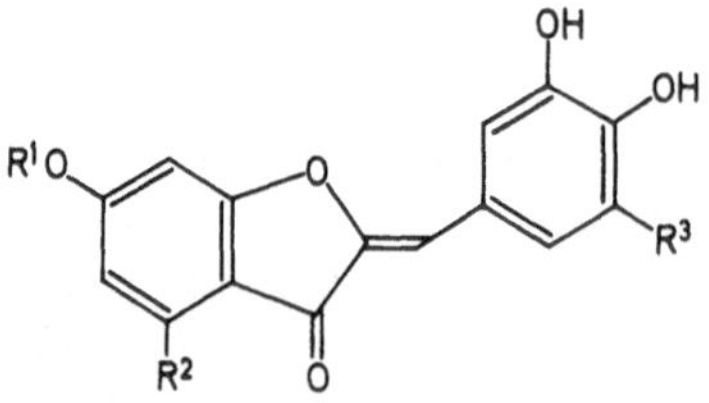

(32)	Aureusidin:	$R^1, R^3 = H; R^2 = OH;$
(33)	Sulfuretin:	$R^1, R^2, R^3 = H;$
(34)	Bracteatin:	$R^1 = H; R^2, R^3 = OH$

Schema 4. Häufig vorkommende Flavanon-, Chalkon- und Auron-Aglykone
und ihre Monoglykoside

in Äthanol) erwies sich nicht mit dem synthetisierten Glykosid identisch
(*98*).

　　Über die wichtigsten aufgefundenen Bis-Glykosid-Kombinationen
informiert die Tabelle 1. Über 90% von ihnen haben die Flavonole
Quercetin, Kämpferol oder Isorhamnetin als Aglykonpartner.

Literaturverzeichnis: SS. 206—216

Tabelle 1. *Bis-Glykosid-Kombinationen in der Monosidreihe*

3,7-Diglucosid	3-Galactosyl-7-glucosid
7,4'-Diglucosid	3-Rhamnosyl-7-glucosid
3,4'-Diglucosid	3-Rhamnosyl-4'-arabinosid
3,3'-Diglucosid	3-Rhamnosyl-3'-glucosid
3,7-Dirhamnosid	7-Rhamnosyl-4'-arabinosid
7,4'-Diglucuronid	7-Rhamnosyl-3'-glucosid
3,7-Diglucuronid	3-Arabinosyl-7-glucosid
3-Glucosyl-7-rhamnosid	3-Arabinosyl-3'-rhamnosid
3-Glucosyl-7-arabinosid	3-Arabinosyl-7-rhamnosid
3-Galactosyl-7-rhamnosid	3-Xylosyl-7-glucosid

3. O-Bioside

Von den in der Tabelle 2 zusammengestellten in Flavonoidglykosiden aufgefundenen 25 Disaccharid-Typen sind 12 strukturell geklärt und davon 9 durch die Synthese bewiesen bzw. mit den Flavonoid-Aglykonen zu den entsprechenden Glykosiden verknüpft worden. Bei allen anderen ist die Verknüpfungsweise ungeklärt oder nicht gesichert.

Tabelle 2. *Disaccharide in Flavonoidglykosiden*

Struktur	Trivialname
6-O-(α-L-rhamnopyranosyl)-β-D-glucopyranose	Rutinose
2-O-(α-L-rhamnopyranosyl)-β-D-glucopyranose	Neohesperidose (Sophorabiose)
3-O-(α-L-rhamnopyranosyl)-β-D-glucopyranose	Rungiose
6-O-(α-L-rhamnopyranosyl)-β-D-galactopyranose	Robinobiose
2-O-(β-D-glucopyranosyl)-β-D-glucopyranose	Sophorose
6-O-(β-D-glucopyranosyl)-β-D-glucopyranose	Gentiobiose
2-O-(α-D-apiofuranosyl)-β-D-glucopyranose	Apiobiose
6-O-(α-L-arabinopyranosyl)-β-D-glucopyranose	Vicianose
2-O-(β-D-xylopyranosyl)-β-D-glucopyranose	Sambubiose
6-O-(β-D-xylopyranosyl)-β-D-glucopyranose	Primverose
2-O-(β-D-xylopyranosyl)-β-D-galactopyranose	Latyrose
2-O-(β-D-glucopyranosyluronsäure)-β-D-glucuronsäure	—

Verknüpfungsweise unbekannt:

Galactosyl-galactose	Rhamnosyl-xylose
Galactosyl-rhamnose	Galactosyl-arabinose
Rhamnosyl-rhamnose	Glucosyl-arabinose
Rhamnosyl-arabinose	Arabinosyl-galactose
Rhamnosyl-galactose	Glucosyl-glucopyranosyluronsäure
Rhamnosyl-glucose	Rhamnosyl-galactopyranosyluronsäure
	Xylosyl-glucopyranosyluronsäure

Tabelle 3. 7-O-Rutinoside und 7-O-Neohesperidoside der Flavanon- und Flavonreihe

Aglykon	R¹ = Rutinosyl	Schmp.	[α]	Vorkommen	Aglykon	R¹ = Rutinosyl	Schmp.	[α]	Vorkommen
Naringenin $R^1R^2=H$, $R^3=OH$ (23)	Narirutin (=Atsinosid) (38)	160—165°	−49,5° (Pyridin)	Citrus sinensis	Apigenin (13)	Isorhoifolin (47)	269—270°	−98,2° (Pyridin) −91,2° (DMF)	Paeonia arab., Dahlia var., Boehmeria niponivea
Isosakuranetin $R^1R^2=H$, $R^3=OCH_3$ (28)	Didymin (39)	209—211°	−72,4° (Pyridin)	Poncirus trifoliata, Monarda didyma, Acinos thymoides	Acacetin (14)	Acaciin (=Linarin) (48)	275—276°	−88,2° (Pyridin)	Robinia pseudacacia, Linaria vulg., Eriodendrum anfractuos., Neviusia alab., Kolkwitzia amnab., Chrysanthemum morifol., Cirsium-, Buddleia-, Tilia-Arten
Eriodictyol $R^1=H$, $R^2R^3=OH$ (26)	Eriocitrin (40)	154—164°	—	Citrus-Arten	Luteolin (15)	Scolymosid (=Lonicerin) (43)	186—189°	−50,9° (Pyridin)	Cynara scolymus, Capsella bursae pastoris, Baptisia leontei, Lonicerajaponica, Citrus-Arten
Hesperetin $R^1=H$, $R^2=OH$ $R^3=OCH_3$ (27)	Hesperidin (41)	257—261°	−80,0° (Methanol)	Citrus-Arten, Verbascum phla., Peucedanum ost.	Diosmetin (17)	Diosmetin-7-O-rutinosid (44)	278—280°	—	Diosma-Arten, Xanthoxylum-Arten, Capsella bursae past., Conium maculat., Hyssopus offic., Menthapulegiumu. crispa, Scrophularia nodos., Dahlia var., Linaria genistifolia, Teucrium mont.

Aglykon	R¹ = Neohesperidosyl	Schmp.	[α]	Vorkommen	Aglykon	R¹ = Neohesperidosyl	Schmp.	[α]	Vorkommen
Pinocembrin R¹R²R³ = H (37)	Pinocembrin-7-O-rutinosid (42)	239—242°	−97,6° (Pyridin)	synth.	Chrysin (19)	Chrysin-7-O-rutinosid (45)	265—266°	−97° (Pyridin)	Dolichandrone falcata
R¹R² = H, R³ = OCH₃ +C₆ − OCH₃ (28)	—	—	—	—	Pectolinarigenin (36)	Pectolinarin (46)	250—257°	−91° bis −93° (Pyridin)	Linaria- und Cirsium-Arten
Naringenin (23)	Naringin (50)	172—174°	−77,1° (Pyridin)	Citrus-Arten	Apigenin (13)	Rhoifolin (55)	202—205°	−118,7° (Methanol)	Paeonia-, Chorisia-, Lupinus-Arten, Rhus succedania, Citrus paradisi
Isosakuranetin (Citrofoliol) (28)	Poncirin (= Neoponcirin) (51)	212—213°	−97,2° (Pyridin)	Citrus paradisi	Acacetin (14)	Fortunellin (56)	219—220°	−102,8° (Pyridin)	Fortunella- u. Mentha-Arten
Eriodictyol (26)	Eriodictyol-7-neohesp. (52)	187—191°	−105° (Methanol) −80° (Methanol)	Fortunella crassifolia	Luteolin (15)	Veronicastrosid (57)	249—251°	−97,55° (Pyridin)	Veronicastrum sibir., Citrus-Arten
Hesperetin (27)	Neohesperidin (53)	239—244°	−95° bis −103° (Pyridin)	Citrus-Arten	Diosmetin (17)	—	—	—	—
Pinocembrin (37)	Sarothanosid (= Isosarothanosid) (54)	277—279°	−104° (Pyridin)	Cytisus comm. Brig. var. merinoii u. Cytisus pyrgas, Sparattosperma vernicosum	Chrysin (19)	Chrysin-7-neohesperidosid (58)	215—218°	−110,3° (Pyridin)	—

(Rutinose (6-O-α-L-Rhamnopyranosyl-β-D-glucopyranose) **(35)**:

Die Rutinose **(35)**, charakteristisches Disaccharid des aus *Ruta gra-veolens* L. erstmals isolierten *Rutins* (Quercetin-3-O-β-rutinosid), wurde von Zemplén und Gerecs *(228)* zunächst für eine 6-O-β-L-Rhamnosyl-Glucose gehalten, später aber von Gorin und Perlin *(69)* durch Perjod-säureoxidation der Rutinose und Isolierung eines C_7-Trialdehydes eindeu-tig als 6-O-α-L-Rhamnosyl-β-D-glucopyranose identifiziert. Da die 1→6 Bindung in der Rutinose relativ stabil ist, kann sie aus den Glykosiden durch milde Essigsäurehydrolyse (10% 6 Stunden) oder Hydrolyse mit Rhamnodiastase intakt erhalten werden. Trotz seines ubiquitären Vor-kommens gibt es nur einige Pflanzen, wie z. B. *Sophora japonica* L. und *Fagopyrum*-Arten in denen die Verbindung in einer auffallend hohen Konzentration (2—5%) gebildet wird.

Häufiger Begleiter des Rutins ist das Kämpferol-3-rutinosid (Nicoti-florin). Das Isorhamnetin-3-rutinosid (Narcissin) kommt nur sporadisch vor und wurde bisher in *Narcissus tazetta* L. var. *chinensis* L., *Lilium auratum* Lindl., *Hippophaë rhamnoides* L., *Bupleurum falcatum* L., *Herniaria glabra* L., *Brassica napus* var. *oleifera* S. *biennis* L. und einigen anderen Pflanzen aufgefunden. Das 7,4'-Dimethyl-quercetin-3-rutinosid (Ombuosid) ist dagegen bisher nur aus *Phytolacca dioica* L., das Patuletin-3-rutinosid nur aus *Plummera ambigens* Blake isoliert worden. In der Flavonreihe sind 7-O-Rutinoside des Apigenins **(13)**, Acacetins **(14)**, Luteolins **(15)**, Diosmetins **(17)**, Chrysins **(19)** und Pectolinarigenins **(36)** bekannt geworden (siehe Tabelle 3).

Ihre homologen 2,3-Dihydroverbindungen, die sich von den in Schema 4 aufgeführten Flavanonen einschließlich Pinocembrin **(37)** (siehe Tabelle 3) ableiten, galten lange Zeit in der Rutaceenfamilie auf die Citrus-arten beschränkt. Mittlerweile sind sie aber auch in Vertretern anderer Pflanzenfamilien gefunden worden. Das *Hesperidin* **(41)**, das 7-O-Rutino-sid des 2S-3',5,7-Trihydroxy-4'-methoxy-flavanons **(27)** ist z. B. aus den Blüten von *Verbascum phloimoides* L. und aus dem Rhizom von *Peuce-danum ostruthium* L. isoliert worden. Die Synthese nahezu aller natürlich vorkommender 7-O-Rutinoside der Flavanon-Reihe und ihrer De-hydrierungsprodukte (siehe Synthesekapitel) führte zur Klärung einer Reihe von strittigen Literaturangaben. Bei dem aus *Poncirus trifoliata* L. isolierten Isosakuranetin-7-O-rhamnoglucosid handelt es sich eindeutig um ein Rutinosid, das mit *Didymin* **(39)** identisch ist. Das Flavonrhamno-glucosid *Acaciin* **(48)** aus *Robinia pseudacacia* L. zeigte mit dem bekannten *Linarin* aus *Linaria vulgaris* Mill. und das *Scolymosid* **(43)** aus *Cynara scolymus* L. mit *Lonicerin* aus *Lonicera japonica* L. Identität.

Die Isoflavonreihe ist mit einigen Rhamnoglucosiden vertreten, die alle in *Baptisia*-Arten gefunden wurden. Durch die Synthese ist nur für ein 7-O-Rhamnoglucosid des Genisteins (21) aus *Baptisia sphaerocarpa* (Sphaerobiosid) (*176*) die Rutinosestruktur bewiesen.

Das einzige bisher beschriebene Auron-rutinosid, das 4,6,3′,4′-Tetrahydroxy-auron-6-rutinosid aus *Citrus limonum* L. (31), erwies sich als ein Kunstprodukt, das bei der Alkoholextraktion aus *Eriocitrin* (**40**) durch Isomerisierung zum entsprechenden Chalkonglykosid und anschließende spontane Oxidation gebildet worden war (*31*).

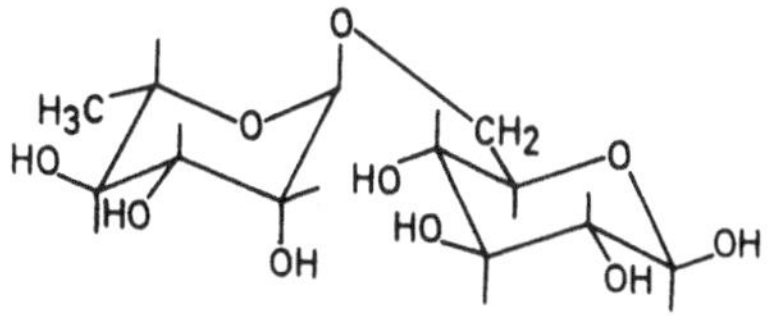

(**35**)   Rutinose

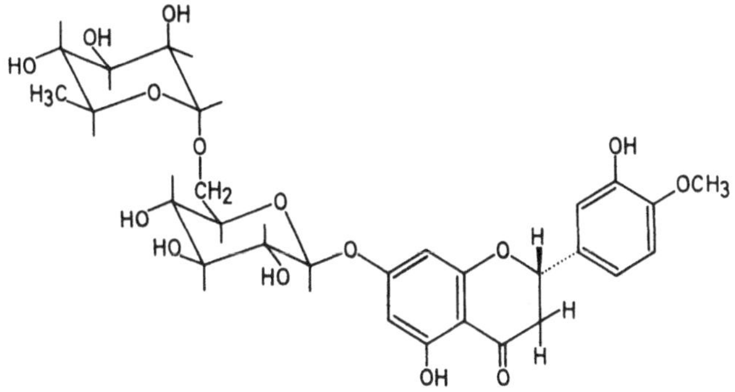

(**41**)   Hesperidin

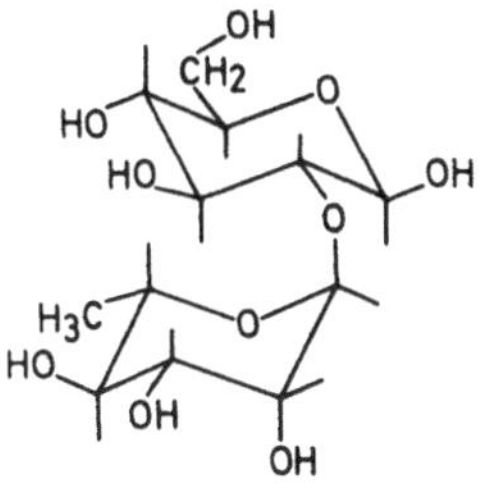

(**49**)   Neohesperidose

Neohesperidose (= Sophorabiose) (2-O-α-L-Rhamno-pyranosyl-β-D-glucopyranose) (**49**):

Glykoside mit Neohesperidose (**49**) als Zuckeranteil sind zunächst ebenfalls nur aus *Citrus* Früchten isoliert worden. Die bekanntesten von ihnen, das *Naringin* (**50**), *Poncirin* (**51**) und *Neohesperidin* (**53**) (siehe Tabelle 3), unterscheiden sich durch ihren stark bitteren Geschmack von den

„geschmacklosen" Rutinosiden. Sie besitzen bei einem Vergleich auf molarer Basis etwa $^1/_5$ der Bitterkeit von Chinin-Hydrochlorid (107). Da die Naringin-Chalkon- und Dihydro-chalkon-neohesperidoside umgekehrt über eine starke Süßkraft verfügen und die Dehydrierung der Flavanon-7-O-neohesperidoside zu einem totalen Verlust der Bitterkeit führt, kann der bittere Geschmack nicht allein an die Neohesperidose-Struktur geknüpft sein (107).

ZEMPLÉN und TETTAMANTI (232) haben für das neue Disaccharid des Neohesperidins zunächst eine 1 → 4 Verknüpfung postuliert. HOROWITZ und GENTILI (109) bewiesen später durch Permethylierung von Naringin (50), Hydrolyse und Identifizierung von 2,3,4-Tri-O-methyl-L-rhamnose und 3,4,6-Tri-O-methyl-β-D-glucose sowie durch Vergleich der optischen Drehwerte von Naringin (50) mit denen von Methyl-α- und β-L-rhamnosid die 2-O-α-L-Verknüpfung. Diese Verknüpfungsweise erklärt, weshalb sich Flavanon-neohesperidoside vom Typ des Naringins (50) im Gegensatz zu den Rutinosiden beim Erhitzen mit 20—25%iger wäßriger Kalilauge ohne Spaltung der Glykosidbindung zu Phloracetophenon-4′-glykosiden (107) abbauen lassen. Bei blockierter C_2,-OH-Gruppe ist eine Ionisierung durch vorübergehende Bildung einer 1,2-Anhydro-Verbindung und Eliminierung des Arylanions wie bei den O-7-Rutinosiden nicht möglich (Schema 5). Aus dem gleichen Grund lassen sich Flavanon-7-neohesperidoside, besonders wenn sie eine freie C_4,-OH-Gruppe besitzen, bereits mit verdünnter Lauge zu den entsprechenden Chalkonglykosiden isomerisieren. Hieraus können durch katalytische Hydrierung mit Palladium-Kohle die heute als Süßersatzmittel in der Diskussion stehenden Dihydrochalkonneohesperidoside hergestellt werden (108). Das Neohesperidindihydrochalkon besitzt verglichen mit Saccharin eine 20fach stärkere Süßwirkung (107).

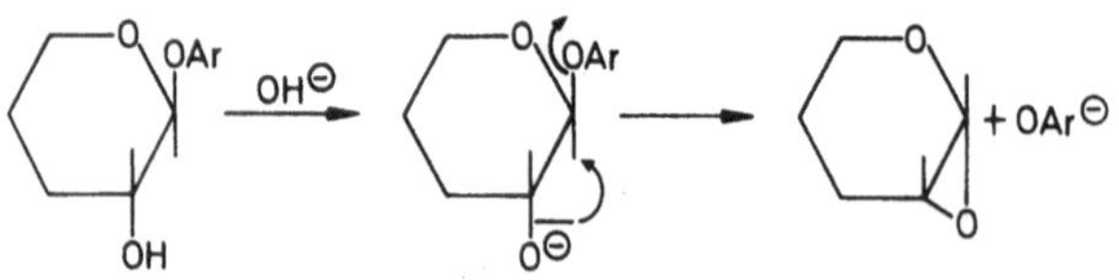

Schema 5. Hydrolytische Spaltung von 7-O-Rutinosiden

Da die 1 → 2 Bindung in der Neohesperidose gegenüber Säuren sehr empfindlich ist, kann die Neohesperidose aus Glykosiden nicht intakt erhalten werden.

Durch die Synthese sind auch hier alle bis heute bekannten Flavanon-7-neohesperidoside einschließlich der Dehydroverbindungen *Rhoifolin* (55), *Fortunellin* (56) und *Veronicastrosid* (57) (siehe Tabelle 3) in ihrer Struktur bestätigt.

Die aus *Cytisus (Sarothamnus) commutatus (175)* und aus *Cytisus purgans (170)* isolierten ebenfalls bitter schmeckenden, optisch aktiven Pinocembrin-7-rhamnoglucoside Sarothanosid bzw. *Isosarothanosid* (**54**) erwiesen sich als miteinander identisch und wurden durch Vergleich mit synthetischen Verbindungen als das 7-O-Neohesperidosid identifiziert (*8*).

Da die Flavanone ein asymmetrisches C_2-Atom besitzen, leiten sich die meisten natürlichen Glykoside von optisch aktiven Aglykonen ab. Für das (−)-Hesperetin (**27**) des Neohesperidins (**53**) ist von HARDEGGER und BRAUNSCHWEIGER (*81*) durch Ozonidabbau des Glykosids und Isolierung des L-Äpfelsäuredimethylesters die 2 S-Konfiguration bestimmt worden. Die gleiche Konfiguration kommt nach ORD-Untersuchungen von GAFFIELD und WAISS (*64*) dem (−)-Naringenin (**23**) von Helichrysin A [(−)-Salipurposid] (**25**) zu, während für ein (+)-Naringenin, das durch Enzymhydrolyse aus einem natürlichen (−)-Naringin (**50**) unbekannter Herkunft erhalten wurde, eine 2R-Konfiguration bestimmt wurde. KAMIYA und Mitarb. (*123*) ermittelten durch ORD-Messung für das Aglykon von natürlichem Naringin die 2 S-Konfiguration. Die gleiche Konfiguration besitzen das (−)-Hesperetin des Hesperidins (**41**) und das (−)-Liquiritigenin (**30**) des 4′-O-Glucosides Liquiritin (*5*). Das (+)-Sakuranetin (**29**) aus Sakuranin, einem 5-O-Glucosid, ließ sich mit Ozon zu D-Äpfelsäure abbauen und hat demnach 2-R-Konfiguration (*5*). Allgemein liefern Glykoside eines links drehenden Flavanons mit 2S-Konfiguration ebenso wie die (−)-Aglykone selbst einen negativen Cotton-Effekt zwischen 280 und 290 nm ($\pi \rightarrow \pi^*$) und einen positiven bei 330 nm ($n \rightarrow \pi^*$) von ähnlicher Intensität. Bei Glykosiden mit einem 2R konfigurierten (+)-Flavanon dagegen erscheint bei 280 bis 290 nm ein positiver Cotton-Effekt. Dieser Regel folgend, wurden für das (−)-Pinocembrin (**37**) des Sarothanosids (**54**) (*8*) und die Aglukone von Poncirin (**51**), Narirutin (**38**) und Didymid (**39**) (*123*) eine 2-S-Konfiguration abgeleitet. Aus natürlichem Eriocitrin (**40**) und Helichrysin B [(+) Salipurposid] (**25**) wurden optisch inaktive Aglukone erhalten.

In der Isoflavonreihe haben FARKAS und Mitarb. (*49*) die „Sophorabiose" des Sophorabiosids aus *Sophora japonica,* als Neohesperidose identifiziert und für das Glykosid die Struktur Genistein-4′-O-neohesperidosid (**59**) durch die Synthese bewiesen. Die Flavonol-Reihe ist mit einem Quercetin-3-O-neohesperidosid aus *Thypa latifolia* L. und *Humulus lupulus* L. vertreten. Von SESHADRI und Mitarb. (*185*) ist in einem als Kämpferol-3-O-β-D-[α-L-rhamnopyranosyl 1→3)]-D-glucopyranosid identifizierten Glykosid aus *Rungia repens* ein von der Rutinose und der Neohesperidose abweichender Disaccharidtyp (Rungiose) aufgefunden worden.

Robinobiose (6-O-α-L-Rhamnopyranosyl-β-D-galactopyranose):
Dieses Disaccharid ist erstmals von ZEMPLÉN und GERECS (*227*) durch Rhamnodiastase-Spaltung von Robinin (**60**), einem Kämpferol-

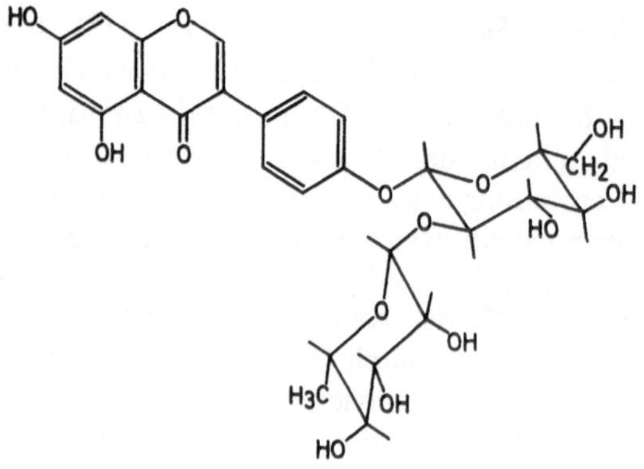

(59) Sophorabiosid

3-O-rhamnosid aus *Robinia pseudacacia* L. erhalten worden. Die von ZEMPLÉN und BOGNÁR (*219*) zunächst angegebene β1 → 6 Verknüpfung ist später von GORIN und PERLIN (*69*) zu α 1→6 korrigiert worden. Robinin selbst existiert in zwei ineinander umwandelbaren Formen, einer α-Form vom Schmp. = 195°—197°, und einer β-Form vom Schmp. = = 249°—250°. Es soll nach MAKSYUTINA und LITVINENKO (*144*) ein Gemisch aus 4 Isomerglykosiden sein, die sich in der Ringgröße der Rhamnose in 7-Position und seiner anomerischen Verknüpfung unterscheiden. Das Robinin ist neben den analogen Quercetin- und Myricetinglykosiden aus *Vinca major* L., *minor* L. var. *alba* und *herbacea*, aus der zu den Vinca-Arten gehörigen *Acocanthera spectabilis* Hook. f., aus *Phaseolus*-Arten, *Pueraria hirsuta* sowie *Azukia angularis* isoliert worden. In einem Kämpferol-3-rhamnogalactosid aus den Blättern von *Atropa belladonna* L. ist die Bindungsweise noch ungeklärt.

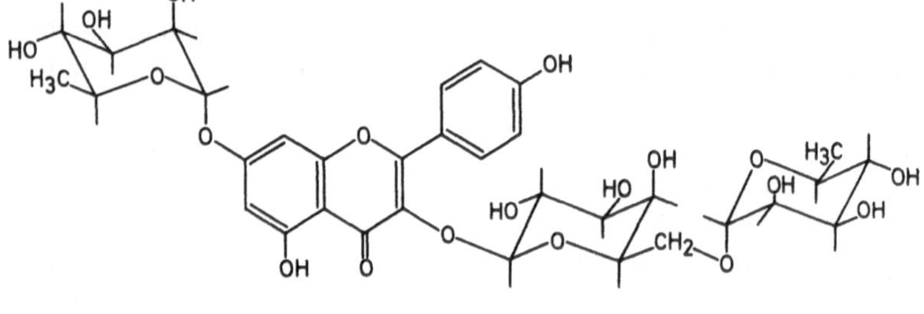

(60) Robinin

Sophorose (2-O-β-D-Glucopyranosyl-β-D-glucopyranose (**61**)
Gentiobiose (6-O-β-D-Glucopyranosyl-β-D-glucopyranose (**62**):
Die Sophorose (**61**) ist erstmals von RABATÉ (*173*) aus einem Kämpferolglykosid von *Sophora japonica* erhalten und mit dem schon

früher von FREUDENBERG (*59*) synthetisierten Disaccharid für identisch befunden worden. Die Sophorose kommt hauptsächlich in Form von Flavonol-3-O-glykosiden vor und ist bisher als Quercetin- bzw. Kämpferol-3-O-sophorosid oder 7-Glucosyl-3-sophorosid z. B. aus *Sorbus aucuparia* L., *Gossypium barb.* L., *Hibiscus mut.* L., *Galanthus niv.* L., *Helleborus niger* L. und verschiedenen *Solanum-*, *Rosa-*, *Pisum-*, *Juglans-*, *Betula-*, *Fagus-*, *Fraxinus-*, *Corylus-* und *Alnus* Arten isoliert worden. Von GIUFFA und Mitarb. (*68*) ist in der Solanacee *Vestia lycioides* Willd. ein Quercetin-3-O-diglucosid mit einem völlig anderen Drehwert ($[\alpha]_D^{20} = -27°$ in Äthanol) entdeckt worden. Nach den durchgeführten Enzymhydrolysen und NMR-Untersuchungen soll es sich um ein Quercetin-3-O-α-D- (2-O-β-D-glucopyranosyl)-glucofuranosid handeln. Über ein Kämpferol-3-α-sophorosyl-7-glucosid aus *Cardamine pratensis* wird von GROUILLER und Mitarb. (*70*) berichtet.

Die mit der Sophorose isomere *Gentiobiose* (**62**) ist bisher als Quercetin- bzw. Kämpferol-3-O-glykosid nur in *Papaver somniferum* L. und einigen *Primula*-Arten nachgewiesen worden.

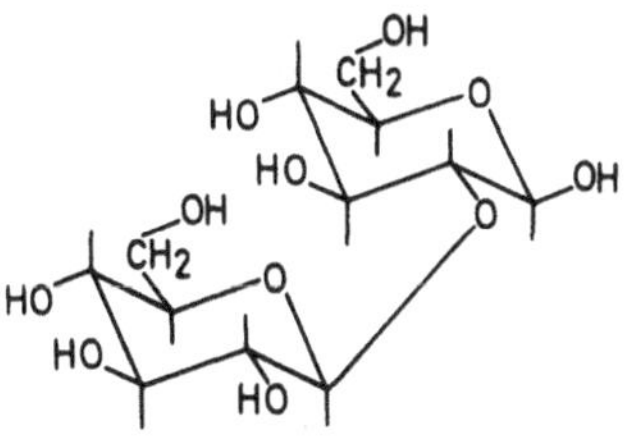

(**61**) Sophorose 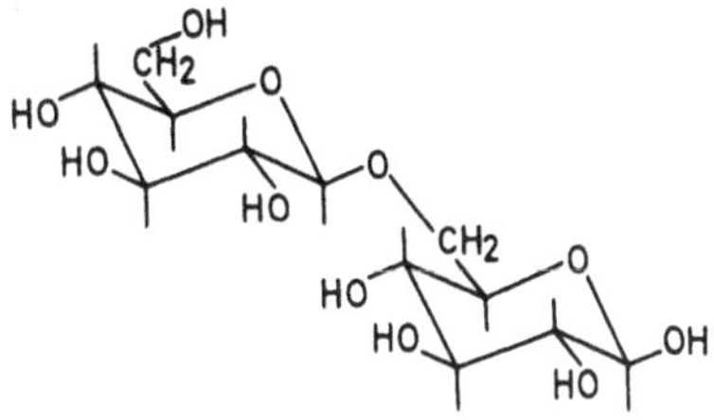   (**62**) Gentiobiose

Sambubiose (2-O-β-D-Xylopyranosyl-β-D-glucopyranose) (**63**)
Latyrose (2-O-β-D-Xylopyranosyl-D-galactopyranose) (**64**)
Primverose (6-O-β-D-Xylopyranosyl-β-D-glucopyranose) (**65**):

Die Sambubiose (**63**) ist erstmals in einem Cyanidin-glykosid aus *Sambucus nigra* aufgefunden worden (*174*). Die Struktur wurde aus den nach Permethylierung und Hydrolyse erhaltenen Methylzuckern und aus der Hydrolysierbarkeit mit Emulsin abgeleitet. Nach dem Ergebnis der partiellen Säurehydrolyse und Enzymspaltung wird für das Xylosylglucose-Disaccharid zweier Flavonol-glykoside aus *Helleborus niger* L. bzw. *Aesculus hippocastanum* (*39, 213*) ebenfalls eine Sambubiose-Struktur vermutet. Bei den zwei Kämpferol-xylosylglucosiden Leucosid und Leucovernid (*99*) aus *Leucojum vernum* L. handelt es sich um Sambubioside, da die für das Leucosid postulierte Kämpferol-3-O-β-D-(2-O-α-D-xylopyranosyl)-glucopyranosid-Struktur durch die Synthese nicht bestätigt werden konnte (*205*).

Die Latyrose (**64**) ist die charakteristische Biose in einem Kämpferol-7-O-rhamnosyl-3-O-xylosylgalactosid aus *Latyrus odoratus* (*75*). Vermutlich das erste Primverosid ist von RABATÉ aus den Blättern von *Salix caesia* Vill. isoliert worden. Bei der Enzymspaltung mit einem Fermentpräparat aus *Gaultheria*-Blättern wurde Luteolin und Primverose (**65**) erhalten. Für ein zweites, sehr wahrscheinlich mit dem ersten identisches Glykosid aus *Salix repens* L., ist die Struktur eines Luteolin-7-O-β-D-(6-O-β-D-Xylopyranosyl)-glucopyranosids ermittelt worden (*195*).

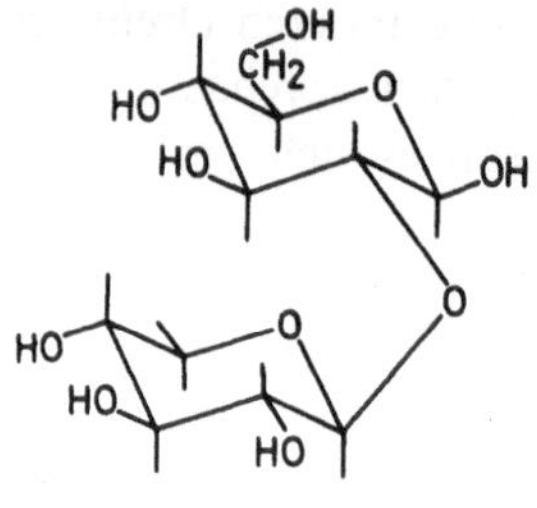

(**63**) Sambubiose 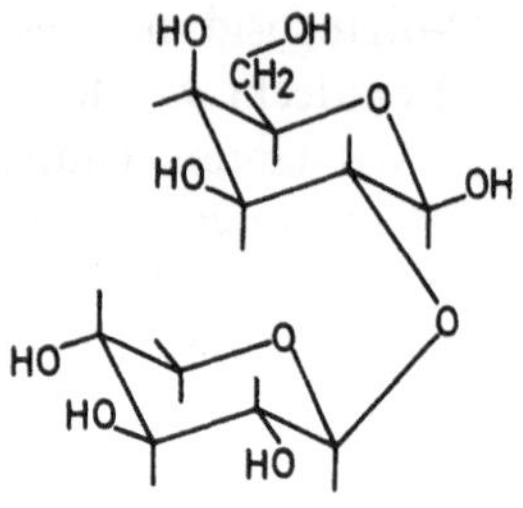   (**64**) Latyrose

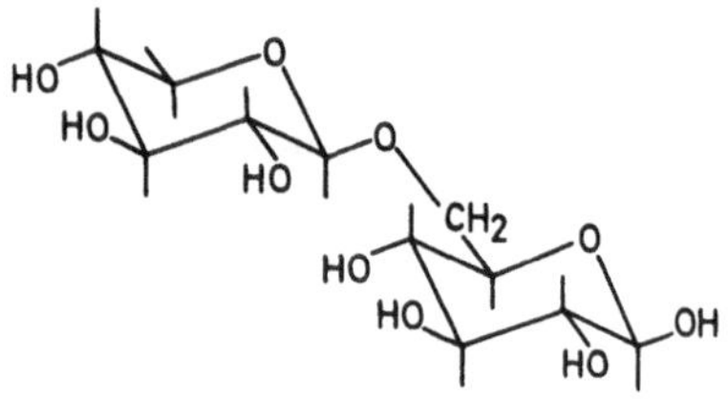

(**65**)   Primverose

Apiobiose (2-O-α-D-Apiofuranosyl-β-D-glucopyranose):

Diese Biose ist der charakteristische Disaccharidanteil von Apiin (**66**), einem Diglykosid aus *Apium petrosilinum* (*112*). Nachdem man schon früher für das Glykosid die Struktur 2-O-(D-Apio-furanosyl)-β-D-glucopyranosyl-7-O-4′,5,7-trihydroxyflavon aufgestellt hatte (*152, 87*), wurde erst später durch Synthese und Periodat-Oxydation (*43*) die D-erythro-Konfiguration der Apiofuranose und durch ORD-Messungen (*113*) die β-Verknüpfung der beiden Zucker ermittelt. Der endgültige Strukturbeweis gelang durch Kupplung von synthetischem 4′-O-Methyl-apigenin-7-O-(3-O-benzyl-4,6-O-benzyliden)-β-D-glucopyranosid mit 2-O-Acetyl-1′, 3-di-O-benzyl-D-apio-D-furanosylbromid, nachfolgende Hydrierung und Deacetylierung zum 4′-O-Methylapiin (*44*).

Das Apiin selbst ist mittlerweile allein oder zusammen mit dem homologen Luteolin-7-apiosylglucosid (Graveobiosid A) noch aus *Cuminum cymium* L., sowie *Vicia hirsuta*, *Petrosilinum-*, *Apium-*, *Chrysanthemum-*, *Matricaria-*, *Bellis-*, *Anthemis-*, *Centaurea-*, und *Capsicum*-Arten isoliert worden. Weitere Apiobiose enthaltende Flavonoidglyko-

(66) Apiin

side sind das Lanceolarin aus *Dalbergia lanceolaria* L. f. (*145*), ein 5,7-
Dihydroxy-4′-methoxy-isoflavon-7-O-apioglucosid, das Chrysoeriol-(3′-
Methylluteolin)-7-O-apioglucosid aus *Luffa echinata* Roxb. (*184*), das
vermutlich mit Graveobiosid B aus den Samen von *Apium graveolens* L.
identisch ist (*55*), und das Homoflavoyadorinin B, ein 7,3′-Di-O-methyl-
luteolin-4′-O-apioglucosid, aus *Viscum album* L. (*160*).

In einem Kämpferol-3-diglykosid aus *Cicer arietinum* L., das bei
totaler Hydrolyse Apiose und Glucose und bei milder Hydrolyse mit
Trifluoressigsäure Kämpferol-3-O-β-D-glucosid (Astragalin) liefert, wird
für die Biose ebenfalls eine Apiobiose-Struktur vermutet (*106*).

Vicianose (6-O-α-L-Arabinopyranosyl-β-D-glucopyranose):
Dieses erstmals aus einem Mandelsäurenitrilglykosid von *Vicia sativa*
L. (Vicianin) und einem Eugenolglykosid von *Geum urbanum* L. (Gein)
erhaltene Disaccharid wird von LEBRETON und BOUCHEZ (*138*) als
Zuckerbestandteil von Peltatosid, einem Quercetin-3-O-arabinoglucosid
aus *Nymphoides peltata* (S. G. Gmel.) Kuntze (Schmp. = 200°—202°,
$[\alpha]_D = -11°$ in Äthanol) angegeben. Der Strukturbeweis basiert auf der
partiellen Enzymhydrolyse zu Isoquercitrin und NMR-spektroskopischen
Untersuchungen. In einem Chrysoeriol-7-O-glucoarabinosid aus *Salix
bakko* Kimura ist dagegen die Verknüpfungsweise der beiden Monosaccharide nicht eindeutig gesichert.

Für alle anderen in der Tabelle 2 aufgeführten Biosetypen, die bisher
in Flavonoidglykosiden aufgefunden wurden, fehlen exakte Untersuchungen über die Verknüpfungsweise.

4. O-Trioside, Tri-, Tetra- und Poly-O-Glykoside

Die wenigen bisher identifizierten Trisaccharide (siehe Tabelle 4)
kann man sich durch Anknüpfung einer Hexose oder Pentose an bereits
bekannte Disaccharide abgeleitet denken. Dies kommt in der Nomenklatur zum Ausdruck. Das Trisaccharid eines Flavonolglykosides aus
Primula sinensis, die *Gentiotriose*, ist aus Gentiobiose und einer $1 \rightarrow 6$

angeknüpften Glucose zusammengesetzt (*79*). Analog bildet die Sophorose mit 1→2 angeknüpfter Glucose die *Sophorotriose*, das Trisaccharid eines Glykosides aus *Pisum sativum* (*76*). Für ein Flavonol-
glucosylrutinosid aus den Blüten von *Solanum tuberosum* wird von HAR-
BORNE (*80*) eine verzweigtkettige Struktur angegeben. Die 1→2 Anknüpfung der Glucose an die Glucose des Rutinoseteils wurde aus dem Hy-
droseverhalten abgeleitet. 3 Flavonoidheteroside, deren Trisaccharide
übereinstimmend aus 1 Mol Galactose und 2 Mol Rhamnose aufgebaut
sind, wurden aus den Früchten von *Rhamnus tinctoria* und *infectorius* L.
(Xanthorhamnin) (*139*), *Rhamnus catharticus* L. (Catharticin) (*167*) und
Rhamnus alaternus L. (Alaternin) (*56*) isoliert.

Das Trisaccharid des Xanthorhamnins erhielt den Namen *Rhamninose*.
SCHMIDT und Mitarb. (*178*) haben durch massenspektroskopische Ana-
lyse der perdeuteromethylierten Glykoside und der hieraus erhaltenen
partiell methylierten Alditolacetate für die beiden Rhamnoseeinheiten
im Alaternin (**67**) eine 1→3 Verknüpfung bewiesen. Beim Xanthorhamnin und Catharticin konnte dagegen, da 1→4 verknüpfte Hexopyrano-
sen und 1→5 verknüpfte Hexofuranosen die gleichen Methylalditol-
acetate liefern, nicht zwischen diesen beiden Alternativen entschieden
werden. In allen drei Glykosiden besitzt die Galactose-Rhamnose-
Einheit Robinobiose-Struktur. Damit ergeben sich für die drei Tri-
glykoside nachstehende Strukturen:

Xanthorhamnin:

7-O-Methylquercetin-3-O-
(Rhamnetin)
$$\begin{cases} \overset{6\leftarrow1}{gal —— rha_p} \overset{4\leftarrow1}{—— rha} \\ \overset{6\leftarrow1}{gal —— rha_f} \overset{5\leftarrow1}{—— rha} \end{cases}$$

Catharticin:

7-O-Methylkämpferol-3-O-
(Rhamnazin)
$$\begin{cases} \overset{6\leftarrow1}{gal —— rha_p} \overset{4\leftarrow1}{—— rha} \\ \overset{6\leftarrow1}{gal —— rha_f} \overset{5\leftarrow1}{—— rha} \end{cases}$$

Alaternin:

7-O-Methylkämpferol-4'-O- $gal \overset{6\leftarrow1}{——} rha \overset{3\leftarrow1}{——} rha$

Ein verzweigtkettiges Trisaccharid enthält das Kämpferol-3-O-rham-
nodiglucosid aus den Pollen von *Populus yunnanensis* Dole (*190*).
Durch Permethylierung des genuinen Glykosides, Permanganatabspal-

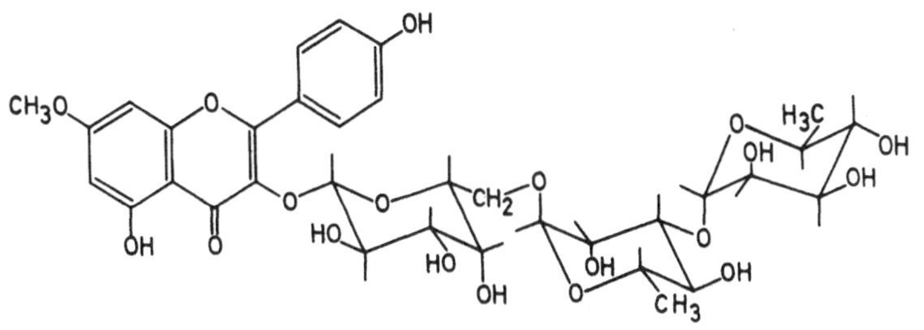

(**67**) Alaternin

tung des intakten Trisaccharids und Darstellung seines Methylproduktes, sowie Identifizierung von Laminariobiose nach β-Glucosidase-Spaltung des Trisaccharides wurde die Struktur als Kämpferol-3-O-β-D-(Gluco-pyranosyl-(1→3)-O-[α-L-rhamnopyranosyl-(1→2)]-D-glucopyranosid (**68**) aufgeklärt. Eine Identität mit einem Kämpferol-3-rhamno-diglucosid aus *Solanum tuberosum* L. und *Camellia sinensis* ist nicht gesichert.

Ein Teil der natürlich vorkommenden Flavonoidtriglykoside liegt in der Bisglykosidform mit einem Disaccharid in der C_3- bzw. C_7- und dem Monosaccharid in der C_7- bzw. $C_{4'}$-Position vor. Aus *Euonymus lanceifolia* wurde z. B. ein Quercetin-3-β-rutinosid-7-β-D-glucosid, aus dem Kraut der *Artischocke* ein Luteolin-7-β-rutinosid-4'-β-D-glucopyranosid (Cynarotriosid) und aus *Citrus paradisi* Macf. die beiden miteinander isomeren (+) Naringenin-7-β-neohesperidosyl- bzw. 7-β-ruti-nosyl-4'-β-D-glucopyranoside (*149*) isoliert. Entsprechende Flavonol-glykoside mit Sophorose, Robinobiose, Sambubiose oder Latyrose in 3- und einem Monosaccharid in 7-Stellung sind in der Papilionaceen-, Primulaceen-, Ranunculaceen- und Solanaceen-Familie mehrfach aufgefunden worden. Über Flavonoltetraglykoside, die sich von C_3-O-Sophorotriosiden und Gentiotriosiden ableiten, ist von HARBORNE (*74*) berichtet worden.

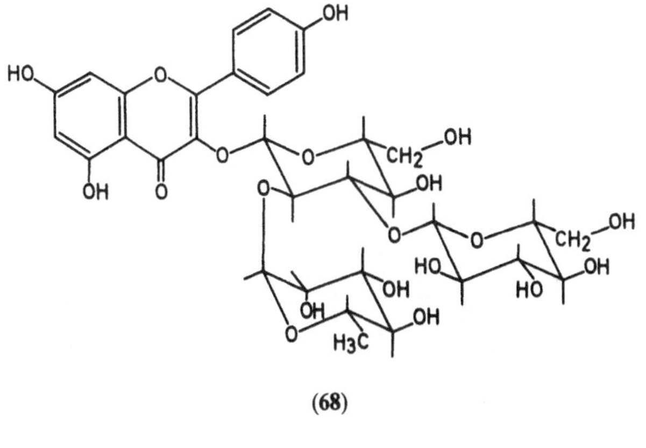

(**68**)

Tabelle 4. *Trisaccharide in Flavonoidglykosiden*

Struktur	Trivialname
O-β-D-glucopyranosyl-(1→6)-O-β-D-glucosyl-(1→6)-β-D-glucopyranose	Gentiotriose
O-β-D-glucopyranosyl-(1→2)-O-β-D-glucosyl-(1→2)-β-D-glucopyranose	Sophorotriose
O-β-D-glucopyranosyl-(1→3)-O-α-L-rhamnopyranosyl-(1→2)-β-D-glucopyranose	2¹-Rhamnosyl-laminaribiose (3ᴳ-Glucosyl-rutinose)
O-β-D-galactopyranosyl-(1→6)-O-α-L-rhamnopyranosyl-(1→4)-α-L-rhamnopyranose	Rhamninose
O-β-D-galactopyranosyl-(1→6)-O-α-L-rhamnopyranosyl-(1→3)-α-L-rhamnopyranose	—
O-β-glucopyranosyl-(1→6)-O-β-D-glucopyranosyl-(1→2)-α-L-rhamnopyranose	2ᴳ-Glucosylrutinose

Verknüpfungsweise unbekannt

Rhamnosyl-glucosyl-glucose

Rhamnosyl-glucosyl-galactose

Rhamnosyl-galactosyl-galactose

Rutinosyl-galactose

Glucosyl-rhamnosyl-xylose

Galactosyl-glucosyl-glucose

Galactosyl-xylosyl-glucose

Glucosyl-glucosyl-glucose

Literaturverzeichnis: SS. 206—216

MARKHAM (*146*) isolierte aus dem Lebermoos *Monoclea fosteri* ein Flavonoidpolysaccharid vom ungefähren MG = 3200. In diesem soll ein 8-Methoxy-5,7,3′,4′-tetrahydroxyflavon an ein aus 18 Zuckereinheiten bestehendes Polysaccharid vom Hemicellulose-Typ gebunden sein. Die Bindung ist glykosidisch und schließt die Galacturonsäure-Reste im Polysaccharid und die 7- und 4′-Hydroxylgruppen im Flavon mit ein. Dieses Glykosid stellt gleichzeitig das erste Phenol-polysaccharid des Pflanzenreichs dar.

5. O-Glykuronide

In der Literatur werden für die glykosidisch gebundenen Glykuron-säuren teils pyranoide, teils furanoide Formen angegeben (*37*). Unter-suchungen an den natürlichen Flavonoidglucuroniden sprechen für das Vorliegen des Pyranotyps (*206*). Stellt man mit Acetanhydrid/Natrium-acetat die Acetate von natürlichem und synthetischem Apigenin-7-[β-D-glucuronid-3,6-lacton] her und vergleicht diese im NMR-Spektrum mit den Modellverbindungen Naphtol-gluco-*furanosiduronsäure*-3,6-lacton bzw. Naphtol-gluco-*pyranosiduronsäure*-3,6-lacton-diacetate, so findet man Übereinstimmung im Bereich der Zuckerprotonen nur mit dem *Glucopyranosiduronsäure*-Glykosid (*206*).

Die leichte Veresterung und Salzbildung der Glykuronide bei der Alkohol-Extraktion bzw. im Isolierungsgang ist die Ursache für die in der Literatur angegebenen unterschied-lichen Schmelzpunkte. Man erhält die reinen kristallinen Glucuronide durch Auflösen des Rohglykosids in 20%iger wäßriger Schwefelsäure, 1 Stunde Stehenlassen bei Raumtempera-tur, tropfenweise Zugabe von 3 n Natronlauge bis pH 3, Ausschütteln mit Äthylacetat und Kristallisation der Glucuronide aus Wasser (*206*).

Die beiden einzigen in Flavonoidglykosiden aufgefundenen Uron-säuren, die Glucuronsäure und die Galacturonsäure, sind entweder direkt oder über eine Hexose oder Pentose in Form eines Di- oder Trisaccharids an das Aglykon gebunden. Die ersten Flavonoid-glucuronide, das Scutel-larein-, Baicalein- und Chrysin-7-O-glucuronid, wurden aus *Scutellaria*-Arten (Lamiaceae) isoliert (*148*). Als weitere Glykosidierungspartner kamen später die Flavone Apigenin (**13**), Acacetin (**14**), Luteolin (**15**), Chrysoeriol (**16**) und Diosmetin (**17**) hinzu (siehe Schema 3).

In der Flavonolreihe sind die Glucuronide nahezu nur auf Quercetin (**1**) und Kämpferol (**8**) beschränkt. Nur einmal wurde ein 3-Glucuronid des 5-Methyl-Quercetins (Azaleatin) und des 6-Methyl-Quercetins (Patu-letin) gefunden. Für ein Acacetin-7-O-diglucuronid aus *Clerodendron trichotomum* Thunb. (Verbenaceae) wurde für die beiden Glucuronsäuren durch Borhydrid-Reduktion des Glykosiddimethylesters, Hakomori-

Permethylierung, nachfolgende Hydrolyse mit 2,5 n Trifluoressigsäure und erneute Borhydrid-Reduktion zu den partiell methylierten Alditolen eine $\beta - 1 \rightarrow 2$ Verknüpfung und damit die Struktur (69) ermittelt (*161*).

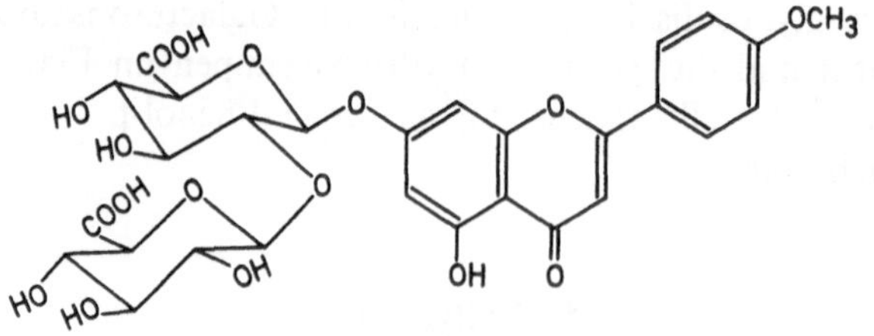

(69)

Für alle anderen Disaccharide mit einem Glucuronsäurepartner ist die Verknüpfungsweise noch ungeklärt. Flavone sind ausschließlich in 7-Stellung, Flavonole in C_3- oder C_3- und C_7-Position mit Glucuronsäure verknüpft.

Eine chemosystematische Bedeutung dürfte dem Vorkommen der Flavonoiduronide nicht zukommen, obgleich eine Anhäufung in den Familien der Asteraceae, Scrophulariaceae, Bignoniaceae, Fabaceae und Lamiaceae, und hier vor allem in *Scutellaria*-Arten (*148*) auffällig ist.

6. Acyl-O-Glykoside

Das erste acylierte Flavonoidglykosid (Tilirosid) wurde von HÖR-HAMMER und WAGNER (*94*) im Jahre 1959 aus den Blättern von *Tilia argentea* Desf. isoliert, nachdem vorher Acylverbindungen nur aus der Anthocyanreihe bekannt waren. Tilirosid wurde zunächst für ein 7-p-Cumaroyl-Kämpferol-3-glucosid gehalten, später aber als ein Kämpferol-3-(p-cumaroylglucosid) (70) identifiziert, da sowohl nach Säurehydrolyse (n-HCl, 20 Min. bei 100°), als auch nach H_2O_2-Oxidation

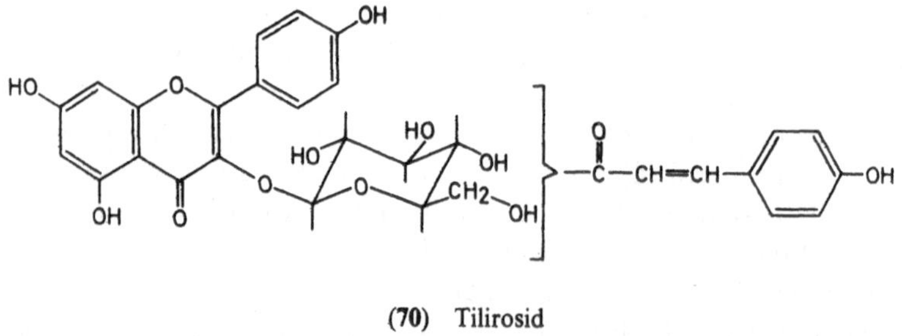

(70) Tilirosid

p-Cumaroylglucose nachgewiesen werden konnte (*77*). Tilirosid ist in der
Zwischenzeit noch in *Planatus acerifolia* (Ait.) Willd., *Fagus silvatica* L.,
Pteridium aquilinum und einer nicht identifizierten *Gnidia*-Art nachge-
wiesen worden.

HARBORNE (*77*) hat in einer systematischen Studie an 15 Anthocyani-
din-3- bzw. 3,5-acylglykosiden gezeigt, daß alle Acylreste an den C_3-
Zucker und nicht an den C_5-Zucker oder an ein phenolisches Hydroxyl
gebunden sind und die intakten Zuckerester durch Säurehydrolyse und
H_2O_2-Oxidation nach der Methode von CHANDLER und HARPER (*25*)
erhalten werden können. Somit dürfte auch dem von KARSTEN (*125*)
aus *Bryophyllum daigremontianum* (R. Hamet et Perr.) Berger isolierten
Kämpferol-cumaroylarabinosid (Bryophyllosid) einem weiteren als
Kämpferol-4'-O-p-cumaroyl-3-diglucosid identifizierten Glykosid aus
Fagus silvatica L. (*34*) sowie einem Kämpferol-3-O-(triglucosidyl)-p-
cumarat aus Erbsenblättern (*63*) eine analoge Struktur zukommen.

Da außerdem keiner der Zuckerester durch β-Glucosidase gespalten
wird (*77*), kann in keinem Fall das reduzierende Ende des Zucker-
moleküls an den Acylsubstituenden gebunden sein, wie dies bei den natürlich vorkommenden Hydroxyzimtsäurezuckerestern der Fall ist. Acylglykoside mit Sophorose als Zuckeranteil liefern daher unter β-Gluco-
sidaseeinwirkung immer den entsprechenden Monoglucoseester. Durch
Vergleich mit verschiedenen synthetischen Feruloyl- bzw. p-Cumaroyl-
glucosen haben BIRKHOFER und Mitarb. bei einem Kämpferol-3-O-
feruloyl-sophorosid (Petunosid) (**71**) aus *Petunia hybrida* Vilm. (*16*) die
2''-O-Verknüpfung und bei dem Bignonosid (**72**), einem Luteolin-7-
(p-Cumaroyl)-O-β-D-glucosid aus *Catalpa bignoioides* Walt. (*17*) die 6''-O-
Verknüpfung bewiesen. Ebenfalls mit der primären alkoholischen Gruppe
verestert ist die p-Cumarsäure in einem Kämpferol-3-O-β-(p-Cumaroyl)-
glucosid (Tribulosid) aus *Tribulus terrestris* L. (*15*).

(**71**)   Petunosid

(72) Bignonosid

Beim Petunosid wurde der intakte Zuckerester durch partielle saure Hydrolyse, beim Bignonosid nur durch β-Glucosidase-Spaltung erhalten. Ungeklärt ist der Ort der Veresterung in einem Quercetin-3-O-(caffeoyl-sophorosid)-7-glucosid aus *Helleborus foetidus* L. *(78)*, einem Apigenin-7-(p-cumaroyl)-β-D-glucosid (Terniflorin) aus *Clematis terniflora* var. *robusta (6)*, das möglicherweise mit einem Acylglykosid aus *Leonurus quinquelobatus* identisch ist und einem Kämpferol-3-O-(p-Oxybenzoyl)-glucosid bzw. Kämpferol-3-O-(benzoyl)-glucosid aus *Narcissus poeticus* L. *(179)*. Beispiele für Acylglykoside mit aliphatischen Säuren sind das Linarin-isovalerianat aus einer *Valeriana wallichii* D. C. Rasse *(196)*, ein Apigenin-7-glucosid-anthemobilinsäureester (Anthemosid) aus *Anthemis nobilis* L. *(88)* und das Biochanin A-7-glucosido-5-maleat *(191)* aus *Trifolium pratense* L. Die Veresterung der Malonsäure am C₅-Hydroxyl wurde aus der leichten Hydrolysierbarkeit zu Biochanin A-7-glucosid in saurem Medium und aus einer IR-Carbonylbande bei 1678 cm^{-1} abgeleitet. Über das Vorkommen von Glykosiden mit Mono- oder Di-acetylzuckern ist mehrfach berichtet worden.

Als innermolekulare Acylglykoside können zwei von Ogiso und Mitarb. *(157)* aus *Leucothoe keiskei* Mig. (Ericaceae) isolierte toxische makrocyclische Flavonoidglykoside vom Biphenyltyp aufgefaßt werden.

Poriolid   **(73)**   R^1=H; R^2=OH;
Isoporiolid **(74)**   R^1=OH; R^2=H;

In den beiden Glykosiden Poriolid (**73**) und Isoporiolid (**74**) ist eine 7-O-gebundene Glucose über das $C_{6'''}$-Hydroxyl als Ester an die Biphenylkomponente eines 3′-substituierten 6-Methyl-5,7,4′-Trihydroxyflavanons gebunden. Die Strukturaufklärung erfolgte vorwiegend durch Abbau zu einer Biphenylcarbonsäure und einem Phenolglykosid, sowie durch Röntgenstrukturanalyse.

II. Flavonoid-C-Glykoside

Die letzten Übersichtsreferate über natürlich vorkommende Flavonoid-C-Glykoside sind von HAYNES (*83*) 1965, CHOPIN (*26*), 1965, WAGNER (*197*) 1966 und ALSTON (*4*) 1967 geschrieben worden. Die bisher aufgefundenen Flavonoid-C-Glykoside lassen sich in 4 Typen unter-· teilen (siehe Schema 6):

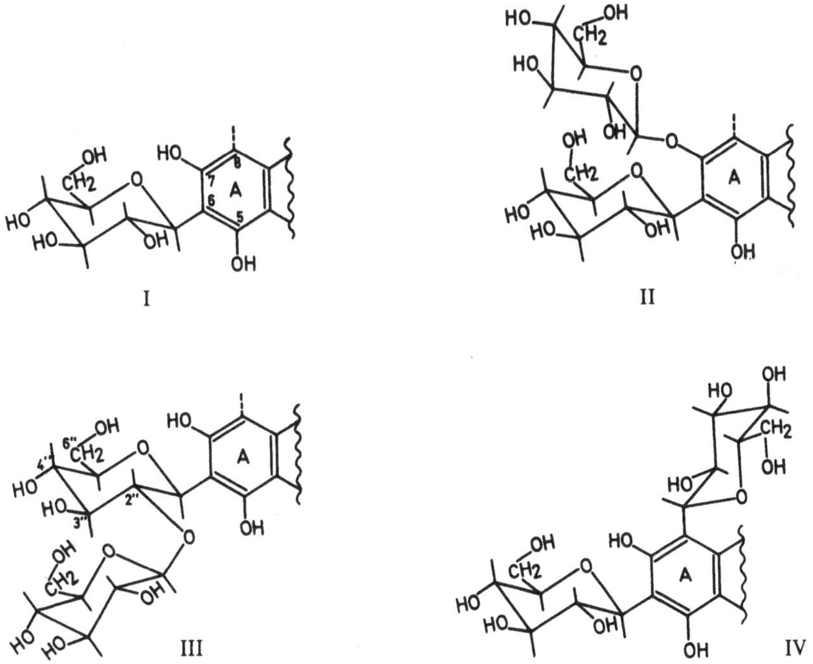

Schema 6. Natürlich vorkommende C-Glykosid-Typen

1. Mono-C-Glykoside, die eine Hexose oder Pentose C-C- an den A-Ring gebunden enthalten (Typ I);
2. Mono-C-Glykoside, die einen zweiten Zucker über O- an das Aglykon gebunden enthalten (Typ II);

Tabelle 5. *Vom Apigenin und Luteolin sich ableitende C-Monolykoside*

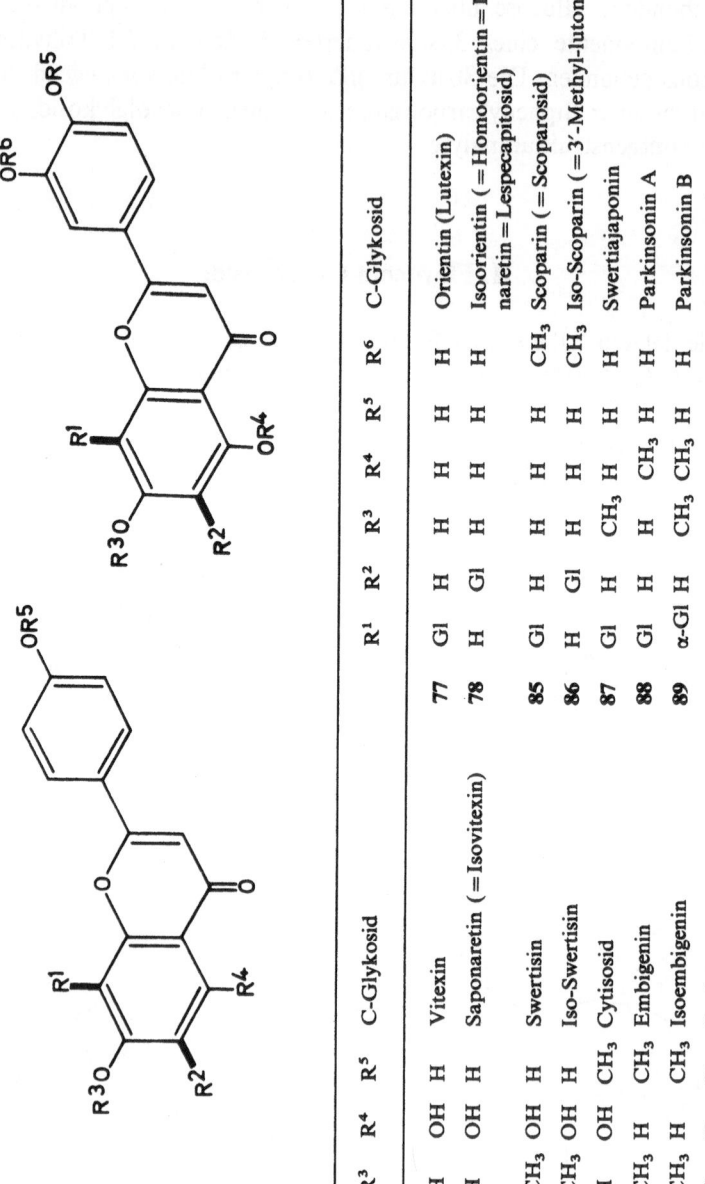

R¹	R²	R³	R⁴	R⁵	C-Glykosid
75	Gl	H	H	OH	H	Vitexin
76	H	Gl	H	OH	H	Saponaretin (=Isovitexin)
79	Gl	H	CH₃	OH	H	Swertisin
80	H	Gl	CH₃	OH	H	Iso-Swertisin
81	Gl	H	H	OH	CH₃	Cytisosid
82	H	Gl	CH₃	H	CH₃	Embigenin
83	Gl	H	CH₃	H	CH₃	Isoembigenin
84	Gl	H	H	H	H	Bayin

	R¹	R²	R³	R⁴	R⁵	R⁶	C-Glykosid
77	Gl	H	H	H	H	H	Orientin (Lutexin)
78	H	Gl	H	H	H	H	Isoorientin (=Homoorientin =Luto-naretin =Lespecapitosid)
85	Gl	H	H	H	H	CH₃	Scoparin (=Scoparosid)
86	H	Gl	H	H	H	CH₃	Iso-Scoparin (=3'-Methyl-lutonarin)
87	Gl	H	CH₃	H	H	H	Swertiajaponin
88	Gl	H	H	CH₃	H	H	Parkinsonin A
89	α-Gl	H	CH₃	CH₃	H	H	Parkinsonin B
90	α-Gl	H	H	H	H	H	Epi-Orientin
91	Gl	H	H	H	CH₃	H	Diosmetin-8-C-glucosid
92	H	Gl	H	H	CH₃	H	Diosmetin-6-C-glucosid

3. Mono-C-Glykoside, die einen zweiten Zucker über O- an den Glykosylrest gebunden enthalten (Typ III);

4. Di-C-Glykoside, die zwei gleich oder verschieden strukturierte Zucker über C an den A-Ring gebunden enthalten (Typ IV).

C-Glykosidierungen wurden bisher nur im A-Ring und zwar an den C-Atomen 6 und/oder 8 bzw. $C_{3'}$ bei Chalkonen beobachtet.

Typ I und II:

Hierzu gehören die weit verbreiteten, sich vom Apigenin (13) bzw. Luteolin (15) ableitenden Isomeren-Paare *Vitexin* (75) / *Saponaretin* (76) und *Orientin* (77)/*Isoorientin* (78) (Tabelle 5).

Vitexin ist bereits im Jahre 1898 von PERKIN aus dem Holz von *Vitex lucens* T. Kirk (littoralis) isoliert worden, aber erst 1957 von EVANS und Mitarb. (*42*) hauptsächlich durch oxidativen Abbau mit Bleitetraacetat und Perjodat als ein in C_8-Stellung mit einem 2,5-Anhydrohexahydroxy-hexylrest substituiertes Apigenin identifiziert worden. HOROWITZ und GENTILI (*110*) bewiesen dann im Jahre 1964 durch Ozonid- und Perjodatabbau sowie NMR-Vergleiche mit verschiedenen Phenylcyclohexylcarbinolen die 8-C-D-Glucopyranosyl-apigenin-Struktur. Das mit Vitexin isomere Saponaretin (76) wurde ebenfalls zum erstenmal aus *Vitex lucens* T. Kirk isoliert. Es wurde später aus dem Saponarin (100) (11) von *Saponaria officinalis* L. durch milde Säurehydrolyse erhalten und ist nunmehr aus dem Vitexin durch Ringisomerisierung nach WESSELY-MOSER (*216*) erhältlich. Durch NMR-Spektroskopie wurde ihm die 6-C-D-Glucopyranosyl-apigenin-Struktur zugeordnet. Die Luteolinhomologen Orientin (77) und Iso-Orientin (78) wurden erstmals im Jahre 1958 von HÖRHAMMER und WAGNER (*101, 105*) in den Blättern von *Polygonum orientale* L. entdeckt. KOEPPEN und Mitarb. (*132*) gelang die Erstisolierung von kristallinem Isoorientin aus *Aspalathus acuminatus* und die Beweisführung, daß Lutexin mit Orientin und Lutonaretin mit Isoorientin identisch sind. Aus den Ergebnissen der vergleichenden NMR-Spektroskopie und Perjodatoxidation leiteten KOEPPEN und ROUX (*133*) für Orientin und Isoorientin die Strukturen 8- bzw. 6-C-β-D-Glucopyranosyl-luteolin ab. HILLIS und HORN (*90*) erweiterten diese NMR-Untersuchungen auf die wichtigsten bis dahin bekannten Flavonoid-C-Glucoside. Aufgrund der bei den NMR-Spektren dieser Glukosidacetate registrierten ax-ax-Kopplungskonstante von 9 Hz für das anomerische Zucker-Proton war für alle C-Glucoside eine β-Verknüpfung gesichert. Die letzte Beweisführung gelang CHOPIN (*27*) durch die Synthese aller natürlich vorkommenden Mono-C-glucoside aus den Aglykonen und α-Acetobromglucose (siehe Kapitel Synthese).

Eine Ausnahme machen zwei C-Glykoside aus *Parkinsonia aculeata* L. (*13*). Die eine Verbindung besitzt gleiches chromatographisches und spektrales Verhalten wie Orientin (Schmp. $= 265° - 267°$, $[\alpha] = + 18°$, in Pyridin), differiert aber von diesem im Schmelzpunkt ($> 300°$) und in der optischen Drehung (um $0°$). Sie wurde Epiorientin (**90**) bezeichnet. Die zweite Verbindung, Parkinsonin B (**89**), ein 8-C-Glucosid des 5,7-Di-O-methyl-luteolins, kann durch Diazomethan-Methylierung in den Epiorientintetramethyläther übergeführt werden. Demnach handelt es sich bei Epiorientin und Parkinsonin B um die entsprechenden 8-C-α-D-Glucopyranoside. Vom Vitexin sich ableitende Mono-Methyläther sind das *Swertisin* (**79**) aus *Swertia japonica* Mak. (*135*) und das *Cytisosid* (**81**) aus *Cytisus laburnum* L. (*165*). Das *Tremasperin* aus *Trema aspera* (*156*), das genuin als Mono-Di-acetat-Gemisch vorliegt, liefert bei milder Säurehydrolyse Cytisosid (**81**) und 1 Mol Glucose, die wie beim Saponarin (**100**) an das C_7-Hydroxyl geknüpft ist. Ein komplex zusammengesetztes C-Glykosid, das sich vom 4′-7, Di-O-methyl-saponaretin *Embigenin* (**82**) ableitet, ist von KAWASE und Mitarb. (*126*) aus den Blüten von *Iris germanica* isoliert worden. Durch Hydrolyse mit 5%iger Schwefelsäure wird das Glykosid Embinin unter Freisetzung von 2 Mol Rhamnose zu dem 8-C-Isomeren *Isoembigenin* (**83**) isomerisiert.

Als ein Orientin-3′-methyläther (8-C-β-D-glucopyranosyl-chrysoeriol) wurde von PARIS und STAMBOULI (*168*) und HÖRHAMMER und WAGNER (*100*) das bereits 1851 isolierte *Scoparin* (**85**) aus *Sarothamnus scoparius* Koch identifiziert. Das entsprechende *Isoscoparin* (**86**) ist zunächst nur als 7-O-Glucosid, später aber auch frei aufgefunden worden (*22*). Mit diesem Scoparin isomer sind die Glykoside Diosmetin-8-β-D- und 6-β-D-Glucopyranosid (**91**), (**92**), die beide von GENTILI und HOROWITZ (*67*) aus Zitronenschalen isoliert wurden. Den 7-O-Methyläther des Isoorientins, das *Swertiajaponin* (**87**), erhielten KOMATSU und TOMINORI (*135*) zusammen mit Swertisin (**79**) aus *Swertia japonica*. Die Nachisolierung und Strukturaufklärung des von PLOUVIER (*169*) aus *Cephalaria leucantha* Schrad, isolierten Leucanthosids ergab Identität mit Swertiajaponin (**87**). Ein 5-O-Methyl-Orientin (Parkinsonin A) (**88**) ist neben dem bereits beschriebenen 5,7-O-Di-methyl-Epiorientin (**89**) in *Parkinsonia aculeata* L. aufgefunden worden (*13*). Ein Flavon-C-Glykosid mit fehlendem C-5-Hydroxyl ist das von EADE und Mitarb. (*38*) aus dem Holz von *Castanospermum australe* A. Cunn et Fraser isolierte *Bayin* (**84**). Seine Struktur wurde durch Vergleich mit einem aus 7,4-Di-O-methyl-tetraacetyl-Vitexin durch Tosylsulfonierung und nachträgliche Hydrogenolyse hergestellten Derivats sowie Abbau zu 2,4-Dihydroxyisophtalsäure und durch NMR-Spektroskopie als 7,4′-Dihydroxy-flavon-8-C-β-D-glucopyranosid bewiesen. Mit Bayin isomer ist Puerarin (**93**), das erste in der Isoflavon-Reihe aufgefundene C-Glykosid. Zusammen mit seinem 7-O-Xylosid wurde Puerarin aus der chinesischen Arzneipflanze *Pueraria Thunbergiana* Benth. isoliert (*150*). Das Glykosid leitet sich vom 7,4′-Dihydroxy-iso-

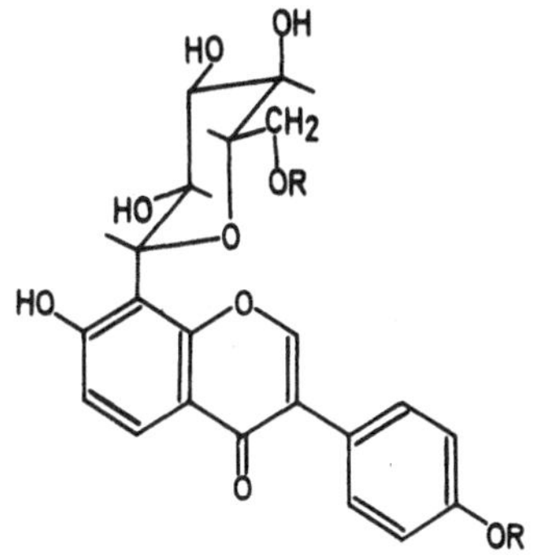

Puerarin (**93**) R=H
Puerarin-4'-6''-diacetat (**94**) R=CH₃CO

flavon Daidzein (**22**) ab und trägt die Glucose in 8-Stellung, wie der Perjodsäureabbau zu 8-Formyldaidzein ergab.

In der Flavonol-Reihe ist bisher nur das C-Glykosid *Keyakinin* bekannt geworden. Keyakinin wurde aus dem Holz von *Zelkowa serrata* (Thunb.) Mak. isoliert (*62*) und von HILLIS und HORN (*91*) als 7-O-Methylkämpferol-6-C-β-D-glucosid identifiziert (**95**). Da man die gleiche Verbindung auch durch Zimtsäure-Natrium katalysierte Luft-Oxidation von Keyakinol (**96**), einem zweiten Flavonoid-C-Glucosid der Pflanze, erhielt, war damit auch dessen Struktur als Dihydro-Keyakinin gesichert. Als Flavanon-C-Glykoside erwiesen sich die beiden von HILLIS und CARLE (*89*) aus dem Harz von *Eucalyptus hemiphloia* F. Muell. entdeckten Glucoside *Hemiphloin* (**97**) und *Isohemiphloin* (**98**). Die Struktur von Isohemiphloin wurde durch NMR-Spektroskopie und Dehydrierung zu Saponaretin bewiesen. Hemiphloin stellt das 8-Isomere dar.

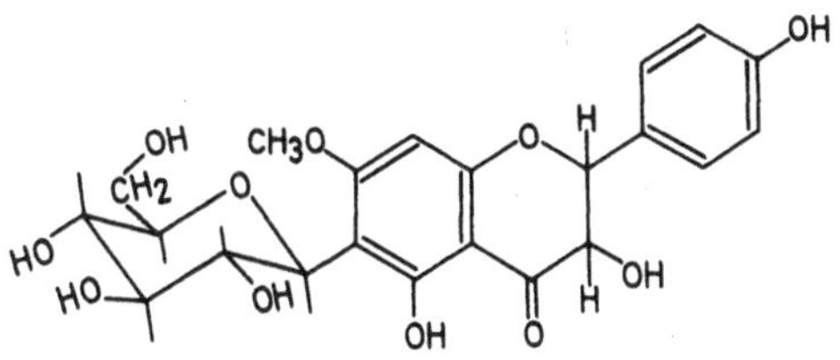

(**95**) Keyakinin

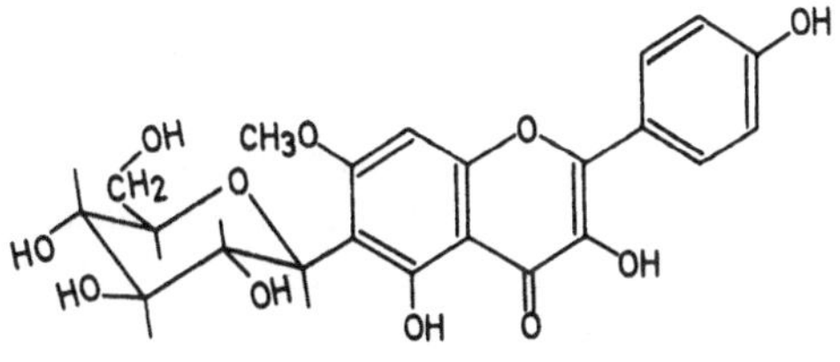

(**96**) Keyakinol

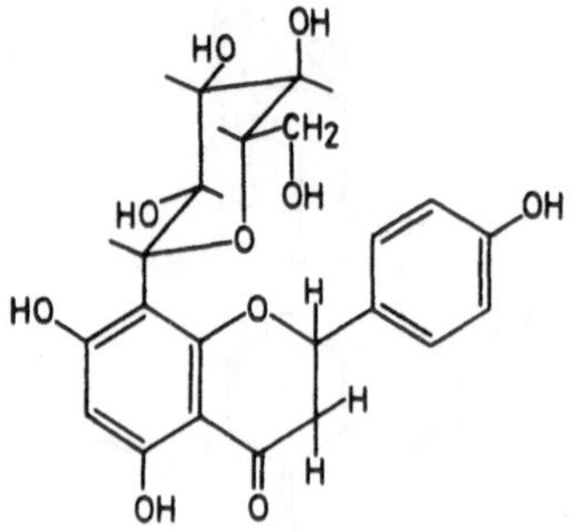

(97) Hemiphloin

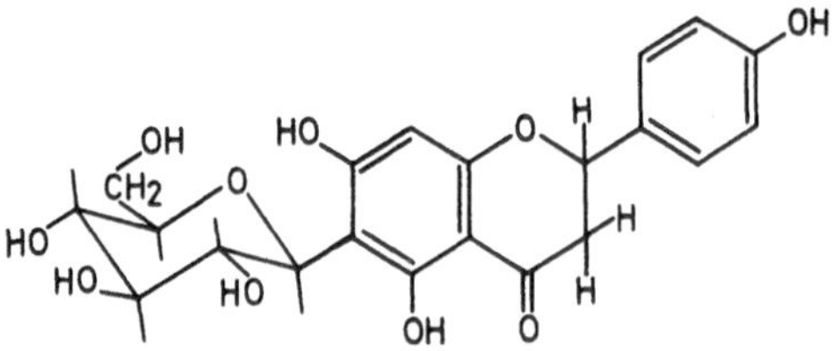

(98) Isohemiphloin

Vom biogenetischen Standpunkt aus ist das Vorkommen eines Dihydrochalkon-C-glucosids interessant. Das Glykosid *Aspalathin* wurde aus den Blättern von *Aspalathus linearis* isoliert und zunächst für ein Flavanon-C-glykosid gehalten. Auf Grund seiner photochemischen Umwandelbarkeit in 2,3-Dihydro-Isoorientin und durch NMR-Untersuchungen wurde später für Aspalathin die Struktur 3′-β-D-Glucopyranosyl-3,4,2′,4′,6′-pentahydroxy-dihydrochalkon ermittelt (**99**) (*134*).

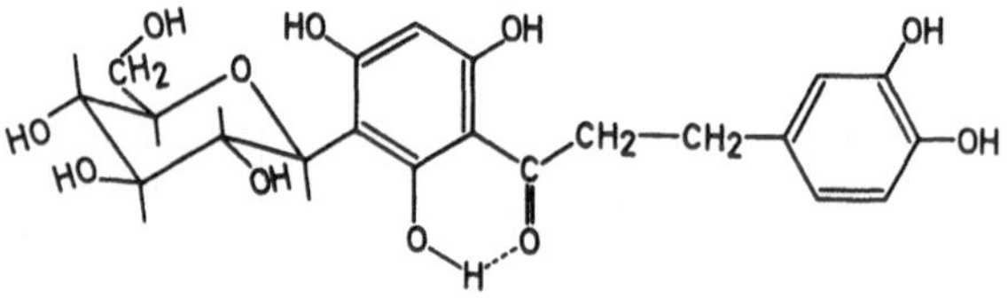

(99) Aspalathin

Vertreter des Glycosidtyps II sind das bereits erwähnte *Saponarin* (**100**) aus *Saponaria officinalis* L., das *Afrosin* (**101**) aus *Achillea fragrantissima* (Forsk.) Sch. Bip. (*186*) und das Lutonarin, ein Isoorientin-7-O-glucosid aus *Hordeum vulgare* L. (*180*). Weitere Beispiele für diesen mittlerweile sehr oft nachgewiesenen Typ sind das Saponaretin-4′-O-glucosid (Isosaponarin) aus *Spirodela oligorrhiza* (*120*), das Vitexin-4′-O-rhamnosid aus *Crataegus oxyacantha* L. (*136*), das Isoorientin-7-O-rutinosid (Wyomin) und Isoswertisin-4′-O-glucosid aus *Triticum aestivum* (*118*).

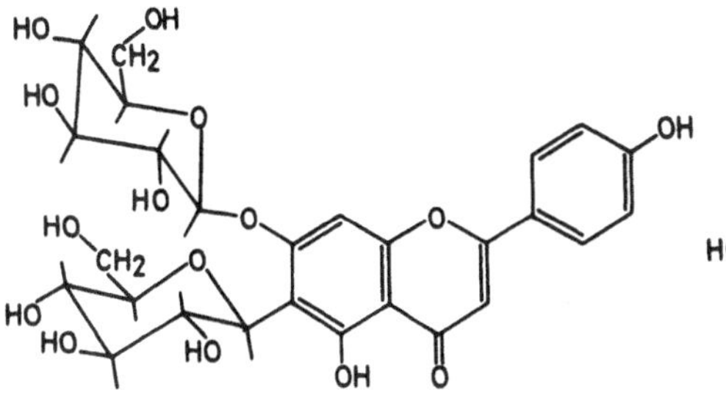

(100) Saponarin 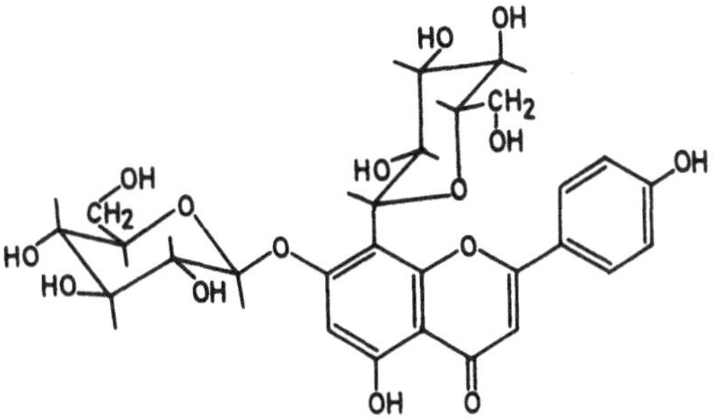   **(101)** Afrosin

Typ III:

Den ersten Hinweis über das Vorkommen eines Flavonoid-Glyko-
sides mit einem C-gebundenen Disaccharid brachten die Untersuchungen
von SEIKEL und Mitarb. (*181*). Zwei der zahlreichen C-Glykoside, die in
Vitex lucens nachgewiesen wurden, wiesen ein unbesetztes C_4- und C_7-
Hydroxyl auf und lieferten bei der sauren Hydrolyse Xylose und Vitexin
bzw. Orientin. Vermutlich das gleiche Vitexin-O-xylosid wurde später
von HOROWITZ und GENTILI (*111*) aus Orangenschalen isoliert. Aus der
beobachteten unterschiedlichen chemischen Verschiebung der Signale für
die 2''- und 6''-O-Acetyl- bzw. Methoxylgruppe in den NMR-Spektren
von verschieden alkylierten und acetylierten Derivaten des Vitexins und
von dem O-Xylosylvitexin-nonamethyläther bzw. seinem Hydrolyse-
produkt leiteten die Autoren für das Glykosid eine 2''-O-β-D-Xylopyra-
nosylvitexin-Struktur (**102**) (*111*) ab. Das Disaccharid hat somit Sambubiose-Struktur. Das gleiche Glykosid ist kürzlich aus *Adonis vernalis* L.
isoliert worden (*65*). Nicht bewiesen ist die Identität mit einem Xylosyl-
Vitexin aus einer *Tragopogon-Art* (*137*).

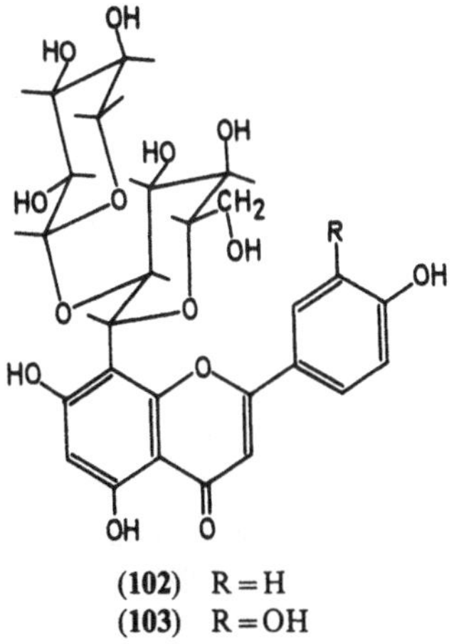

(102) R = H
(103) R = OH

Als Orientinanalogon wurde das schon früher isolierte Adoni-vernith (**103**) (*104*) aus *Adonis vernalis* L. erkannt (*65*). Noch nicht ge-klärt ist die Verknüpfungsstelle der endständigen Rhamnosen im Linosid A und B, einem 4′,7-Di-O-methyl-orientin-O-rhamnosid bzw. (6-O-acetyl)-O-rhamnosid aus *Linum maritimum* L. (*204*), in zwei C-Glyko-siden aus *Phlox drummondii* Hook. (*140*), die bei der Hydrolyse Rhamnose und die bisher noch nicht beschriebenen Apigenin- und Luteolin-6-C-xyloside ergaben und im Embinin (*126*), das als ein 4′,7-Di-O-methyl-apigenin-6-C-(xx′-di-O-L-rhamnosyl-β-D-glucopyranosid identifiziert wurde. Dasselbe gilt für zwei C-Glykoside aus *Trigonella foenum graecum* L., die von ADAMSKA und LUTOMSKI (*1*) als O-Arabinoside des Orientins bzw. Isoorientins beschrieben wurden. Für ein Glykosid aus *Crataegus curvisepala* Lindmann (Cratenacin) ist die Struktur 5,7,4′-Trihydroxy-flavon-8-C-6″-O-acetyl-(4-O-α-L-rhamnopyranosyl)-α-D-glucopyranosid, ermittelt worden (*12*). Außer den Acetyl-Glykosiden Cratenacin (*12*), Linosid A (*204*), Puerarin-4′, 6″-diacetat (**94**) (*14*) und einem Vitexin-4′-O-(O-acetyl) rhamnosid (*58*) ist bisher nur ein Esterglykosid mit einer aromatischen Säure, das p-Hydroxy-benzoyl-vitexin (*111*) bekannt ge-worden. Die NMR-Spektren der Hepta-O-methyl- und Hepta-O-acetyl-Derivate sprechen wie beim Xylosyl-Vitexin für eine Veresterung am 2″-OH des Zuckers (*111*).

Typ IV:

Dem Glykosidierungstyp IV zugehörige Di-C-Glykoside wurden erstmals von SEIKEL (*181*) in dem Holz von *Vitex lucens* T. Kirk entdeckt. Die aufgefundenen 6,8-Di-C-Glykoside des Apigenins wurden als Vice-nine 1, 2 und 3, die Luteolin-di-C-glykoside als Lucenine 1, 2, 3, 4 und 5 bezeichnet (*181*). Nach ihrem Verhalten gegenüber Säuren lassen sie sich in 2 Typen (Typ A und B) unterteilen. Vicenin 2 und Lucenin 2 werden durch Säurebehandlung nicht in Isomerprodukte umgewandelt (Typ B). Sie enthalten daher zwei gleich strukturierte Zucker in C_6- und C_8-Posi-tion. Durch Synthese ausgehend von Vitexin wurde von CHOPIN (*33*) Vicenin 2 als Apigenin-6,8-di-C-glucosid identifiziert. Für das Lucenin 2 ist eine analoge Struktur anzunehmen*. Alle anderen Vicenine und Lu-cenine liefern unter den gleichen Bedingungen Isomerengemische und müssen daher disymmetrisch mit Hexosen bzw. Pentosen glykosidiert sein. CHOPIN fand nach Xylosidierung von Vitexin bzw. einem Orientin-Isoorientin-Gemisch Vicenin 1 mit C_6-Xylosyl-C_8-glucosyl-apigenin und Vicenin 3 mit dem C_6-Glucosyl-C_8-Xylosyl-apigenin identisch (*20*)*.

---

* Da SEIKEL und MABRY (*183*) für das Lucenin-1 aus *Vitex lucens* eine C_6C_8-Di-C-glucosyl-apigenin-Struktur vorgeschlagen haben, dürften die kontroversen Struktur-angaben für Lucenin 1 und 2 auf einer Nr.-Verwechslung beruhen.

Literaturverzeichnis: SS. 206—216

Lucenin 1* erwies sich mit Luteolin-6-xylosyl-8-glucosid, Lucenin 3 mit dem entsprechenden Isomerglykosid identisch (21). Verbindungen vom Vicenin- und Lucenintyp sind mittlerweile in zahlreichen anderen Pflanzen entdeckt worden (siehe Tabelle 6), wobei vor allem ihr Vorkommen in Grünalgen chemotaxonomisches Interesse verdient (147).

Die von KING (127) in Weizenkeimen nachgewiesenen C-Glykoside gehören ebenfalls dem Typ IV zu. Sie wurden adsorptionschromatographisch getrennt und als die Vicenine 1 und 3 und Schaftosid identifiziert. Die gleichen C-Glykoside kommen in der Pflanze außerdem noch in Form der Sinapinsäure-Ester vor (65).

Gleichfalls disymmetrisch glykosidiert ist das Violanthin (104) aus dem Kraut einer Gartenform von *Viola tricolor* L. (212). Violanthin verbraucht gegenüber 10 Mol Perjodat bei Di-C-glucosiden nur 9 Mol Perjodat und liefert beim Viscontini-Abbau neben Glyzerin (von Glucose) das für Rhamnose charakteristische Abbauprodukt 1,2-Propylenglykol. Beweisend für eine C_6-C_8-Rhamnosyl-glucosyl-Apigenin-Struktur waren das NMR-Spektrum (Methyldublett bei $\delta = 1,3$ ppm mit $J = 6$ Hz) und im Massenspektrum die für Di-C-Glykoside typischen 5 bis 6 H_2O-Abspaltungen aus dem Molekül-Ion und das bei ME 283 auftretende Dibenzylkation (212, 172). Die richtige Verteilung der beiden Zucker wurde aus dem chromatographischen Verhalten von Violanthin relativ zu seinem 6-Isomeren Isoviolanthin abgeleitet.

(104) Violanthin

Das Schaftosid aus *Silena Schafta* Gmel. (169) verbraucht ebenfalls 9 Mol Perjodsäure und enthält demnach ebenfalls eine Hexose und Pentose (Arabinose?). Chromatographisch ist es von Violanthin verschieden (65). Mittlerweile ist das Schaftosid zusammen mit einigen Viceninen auch in Weizenkeimen aufgefunden worden (65).

Weitere natürlich vorkommende Di-C-Glykoside leiten sich von den Flavonen Diosmetin, Acacetin und Chrysoeriol ab. Da mittlerweile die Zahl der C-Glykosid-Vorkommen auf etwa 130 gestiegen ist, wird nur eine vereinfachte tabellarische Übersicht ohne Aufzählung der einzelnen Literaturstellen (siehe Tabelle 6) gegeben.

Zum Auffinden von weiteren Flavonoid-C-Glykosiden kann neben der Nichthydrolysierbarkeit mit Säuren die Dufour-Reaktion mit Jodjodkalium (36) herangezogen werden. Diese an sich für lösliche Stärke charakteristische Reaktion gibt auch bei Saponarin (11), Alliarosid (166) und Linosid B bzw. Linosid B (204) einen positiven Ausfall. Saponarin liefert mit Jodwasser oder Jodjodkaliumlösung eine violett gefärbte, kristalline Einschlußverbindung. Nach unseren Erfahrungen ist diese Reaktion ein Hinweis auf das Vorliegen von C-Monoglykosiden, die noch einen 2. Zucker O-glykosidisch an das C_7-Hydroxyl des Aglukons oder an andere OH-Gruppen des C-Zuckers außer $C_{2''}$-OH gebunden enthalten. Von 2''-O-Xylosyl-vitexin oder -Orientin wird die Dufour-Reaktion z. B. nicht gegeben. Die Reaktion kann auch mit Blatt- und Blütenmaterial durchgeführt werden. Nach den systematischen Untersuchungen von SINN (189) ist demnach mit den oben erwähnten Glykosidtypen bei Gramineen, Caryophyllaceen, Linaceen, Liliaceen, Araceen, Lemnaceen, Dipsacaceen, Malvaceen, Cruciferen und Cucurbitaceen zu rechnen (siehe Tabelle 6).

Die weite Verbreitung der Flavonoid-C-glykoside und ihr häufiges gemeinsames Vorkommen mit gleich oder ähnlich strukturierten Flavonoid-O-Glykosiden spricht für enge biosynthetische Beziehungen. Das gehäufte Auftreten von C_6-C_8-Isomeren und das Vorkommen eines Dihydrochalkon-C-glykosides machen eine Glykosidierung auf einer ringoffenen Vorstufe wahrscheinlich. WALLACE, MABRY und ALSTON (215) haben Kulturen der zwei wurzellosen, C-Glykosid-haltigen Wasserpflanzen Lemna und Spirodela mit ^{14}C-Apigenin und ^{14}C-Luteolin inkubiert und neben einer Hydroxylierung und O-Methylierung eine O-Glykosidierung aber keine C-Glykosidierung beobachtet. Analog wiesen nach Inkubation von ^{14}C-Orientin, ^{14}C-Isoorientin und ^{14}C-Isovitexin nur die entsprechenden O-Glykoside, nicht aber die Di-C-Glykoside Vicenin und Lucenin Radioaktivitäten auf. Die Einbaurate von ^{14}C-Phenylalanin-1 in die C-Glykoside war dagegen genau so hoch wie in die O-Glykoside. In Lemna minor scheint darüber hinaus ein Enzym zu existieren, das die Isomerisierung von Vitexin zu Isovitexin katalysiert (214), da nach ^{14}C-Vitexin-Inkubation chromatographisch auch radioaktives Isovitexin nachzuweisen war. Da die umgekehrte Isomerisierung nicht beobachtet wurde, dürfte die Bildung der Isoverbindungen zumindest teilweise über die C_8-C-Glykoside erfolgen.

Nach SEIKELS und GEISSMANS (182) Untersuchungen an Haferkeimlingen wird Saponarin verstärkt im Licht gebildet. Diese erhöhte Saponarinbildung wird nach McCLURE und WILSON (142) durch das Niederenergie-Phytochrom-System kontrolliert, während Pflanzen, die einem hohen

Weißlicht-Niveau oder Blaulicht gefolgt von Rotlicht ausgesetzt werden, nur Spuren eines C-Glykosides, und zwar Lutonarin (78) bilden.
Biosynthetisch verdient das Vorkommen von Flavonoid-C-Glykosiden in den nicht Lignin bildenden Moos- und Farn-Pflanzen Beachtung.

Tabelle 6. *C-Glykosidvorkommen im Pflanzenreich*

Klasse/ Abt. Familie	Art	C-Glykosid Typen*	Autoren/ Veröffentlichungsjahr
ALGAE			
Characeae	*Nitella Hookeri*	IV	MARKHAM (1969)
PTERIDOPHYTA			
Cyatheaceae	*Cyathea fauriei*	Glycflav.	UENO u. a. (1963) SOEDER u. BABB (1972)
Pteridaceae	*Sphenomeris chusana*	Glycflav.	UENO u. a. (1963)
Bryophyta			
Hepaticae	*Hymenophytum flabellatum*	IV	MARKHAM (1969)
	Porella platyphylla (= *Madotheca platyphylla*)	IV	NILSSON (1973)
Musci	*Mnium*-Arten	Glycflav.	MELCHERT u. a. (1965)
	Plagiochila asplenoides	Glycflav.	HARBORNE (1967)
GYMNOSPERMAE			
Pinaceae	*Larix laricina* K. Koch	II	NIEMANN u. a. (1971)
ANGIOSPERMAE/ DICOTYLEDONEAE			
Aceraceae	*Acer palmatum* Thunb.	I	ARITOMI (1963)
Apiaceae	*Anethum graveolens* (Frucht) L. (*Peucedanum graveolens*)	IV	DRANIK (1970)
	Cryptotaenia canadensis DC. Coll.	Glycflav.	CROWDEN u. a. (1969)
	Laretia acaulis Gill. u. Hook	Glycflav.	CROWDEN u. a. (1969)
	Opopanax chironium	Glycflav.	CROWDEN u. a. (1969)
Asteraceae	*Achillea fragrantissima* Sch. Bip.	II	SHALABY (1965)
	Artemisia transsiliensis P. Poljakov	Glycflav.	CHUMBALOV (1970)
	Centaurea cyanus St. Lag.	II	ASEN u. a. (1967)
	Gaillardia pulchella Foug.	I	WAGNER u. a. (1972)
	Helenium-Arten	I	WAGNER u. a. (1972)
Brassicaceae	*Alliaria officinalis* Andrz. (= *Sisymbrium Alliaria*)	II	PARIS (1962), SINN (1967)

---

*Zeichenerklärung: a) I, II, III und IV bezeichnet die vier Flavonoid-C-Glykosid-Typen wie auf Seite 177/179 angegeben; b) Glycflav. = C-Glykoside noch unbekannter Struktur.

Tabelle 6 (Fortsetzung)

Klasse/ Abt. Familie	Art	C-Glykosid Typen	Autoren/ Veröffentlichungsjahr
Caryophyllaceae	*Dianthus*-Arten	II	SINN (1967)
	Elisanthe-Arten	II	SINN (1967)
	Melandrium suaveolens Schischk. (= *Silene suaveolens*)	II	SINN (1967)
	Polycarpon tetraphyllum L.	II	PARIS (1962), SINN (1967)
	Saponaria-Arten	II	PARIS (1962), SINN (1967), BARGER (1906)
	Silene-Arten	II, IV	SINN (1967), PLOUVIER (1967)
Chenopodiaceae	*Beta vulgaris* L.	I	GARDNER (1966)
ANGIOSPERMAE/ DICOTYLEDONEAE			
Cichoriaceae	*Tragopogon*-Arten	I, IV	KROSCHEWSKY (1969)
Combretaceae	*Combretum micranthum* G. Don	I	JENTZSCH u. a. (1962)
Cucurbitaceae	*Bryonia dioica* Jacq.	II	LEIBA (1970), SINN (1967)
Dipsacaceae	*Dipsacus*-Arten	II	PLOUVIER (1966), SINN (1967)
	Succisa pratensis L.	II	PLOUVIER (1966), SINN (1967)
	Cephalaria leucantha Schrad.	I	PLOUVIER (1967)
Ericaceae	*Vaccinium bracteatum* Thunb.	I	MASAITI (1965)
Euphorbiaceae	*Croton zambezicus* Muell.	I, IV	WAGNER u. a. (1970)
	Jatropha gossypifolia L.	I	SUBRAMANIAN (1971)
Fabaceae	*Agryolobium*-Arten	Glycflav.	HARBORNE (1969)
	Aspalathus linearis R. Dahlgr.	I	KOEPPEN (1965)
	Calycotome spinosa L. Link.	Glycflav.	HARBORNE (1969)
	Chamaecytisus-Arten	Glycflav.	HARBORNE (1969)
	Crotalaria-Arten	Glycflav.	HARBORNE (1969)
	Crotalaria-Arten	III	SUBRAMANIAN u. a. (1970)
	Cytisus-Arten	I	HARBORNE (1969)
	Dalbergia paniculata Roxb.	I, IV	ADINARAYANA (19)
	Genista pilosa L.	I	PLOUVIER (1967)
	Lathyrus vernus Bernh.	II	SINN (1967)
	Lespedeza-Arten	I	GLYZIN (1970), PARIS (1962)
	Lupinus-Arten	Glycflav.	HARBORNE (1969)
	Lygos-Arten	Glycflav.	HARBORNE (1969)
	Parkinsonia-Arten	I	BATHIA u. a. (1965)
	Psoralea-Arten	I, II, III, IV	OEKEDON u. a. (1965)
	Pueraria thunbergiana Benth.	I, II	SHIBATA u. a. (1962), BHUTANI u. a. (1969)

Literaturverzeichnis: SS. 206—216

Tabelle 6 (Fortsetzung)

Klasse/ Abt. Familie	Art	C-Glykosid Typen	Autoren/ Veröffentlichungsjahr
	Laburnum anagyroides Medic.	I, II	HARBORNE (1969), HÖRHAMMER u. a. (1962), PARIS (1957), STAMBOULI (1961)
	Sarothamnus scoparius Koch. (= *Cytisus scoparius)*	I	HÖRHAMMER u. a. (1962)
	Spartium junceum L.	I	HÖRHAMMER u. a. (1962)
	Tamarindus indica L.	I	BHATIA u. a. (1967), LEWIS u. a. (1964)
	Teline linifolia Webb u. Berth. (= *Cytisus linifolia)*	Glycflav.	HARBORNE (1969)
	Trigonella foenum-graecum L.	I, II	ADAMSKA (1972)
	Ulex-Arten	Glycflav.	HARBORNE (1969)
Fagaceae	*Castanospermum australe* A. Cunn. u. Fraser.	I	EADE u. a. (1962)
	Nothofagus fusca Oerst.	I	HILLIS (nicht ver.)
GENTIANACEAE	*Swertia japonica* Makino	I	KOMATSU u. a. (1966)
Gesneriaceae	*Cyrtandra pendula* Nadeand.	Glycflav.	HARBORNE (1967)
Linaceae	*Linum*-Arten	II, IV, V	IBRAHIM (1970), SINN (1967)
Malvaceae	*Hibiscus*-Arten	II	NAKAOKI (1944), SINN (1967)
Moraceae	*Humulus japonicus* Sieb. u. Zucc.	I	ARITOMI (1963)
Myrtaceae	*Eucalyptus hemiphloia* F. Muell.	II	HILLIS u. a. (1965), CHOPIN (1966)
Nymphaeaceae	*Nymphaea alba* L.	I	TAKU u. a. (1970)
Oxalidaceae	*Oxalis cernua* Thunb.	I	SHIMOKORYAMA (1962)
Passifloraceae	*Passiflora*-Arten	I, II	GLOTZBACH u. a. (1968)
Polemoniaceae	*Phlox drummondii* Hook.	I, II	CHOPIN (1970), MABRY (1971)
Polygonaceae	*Polygonum orientale* L.	I	HÖRHAMMER u. a. (1962)
Ranunculaceae	*Adonis*-Arten	I, III	HÖRHAMMER u. a. (1960)
	Ranunculus-Arten	I	DROZD (1969)
	Thalictrum-Arten	I	WAGNER u. a. (1971)
	Trollius europaeus L.	I	SACHS (1963)
Rosaceae	*Crataegus*-Arten	II	FIEDLER (1955), FISEL (1966), HRUGASIEWICZ (1964), LEWAK (1966)
Rutaceae	*Citrus*-Arten	I, III, IV	CHOPIN u. a. (1964), HOROWITZ (1966) GENTILI u. a. (1968)

Tabelle 6 (Fortsetzung)

Klasse/ Abt. Familie	Art	C-Glykosid Typen	Autoren/ Veröffentlichungsjahr
Scrophulariaceae	*Gratiola officinalis* L.	I	Litwinenko (1969)
Ulmaceae	*Zelkova serrata* Makino	I	Funaoka (1957)
	Trema aspera Blume	II	Oelrichs (1968)
Verbenaceae	*Vitex*-Arten	I, II, III, IV	Evans u. a. (1957), Hänsel u. a. (1965), Horowitz (1966), Rao u. a. (1956), Seikel u. a. (1966)
Violaceae	*Viola tricolor* L.	IV	Hörhammer u. a. (1965)
Vitaceae	*Vitis cinerea* Noronha	I	Wagner (1967)
ANGIOSPERMAE/ MONOCOTYLEDONEAE			
Araceae	*Arum*-Arten	II	Sinn (1967)
Arecaceae	*Chamaedorea* spec.	Glycflav.	Williams u. a. (1971)
	Chamaerops humilis Mart.	Glycflav.	Williams u. a. (1971)
	Howea forsteriana Becc.	Glycflav.	Williams u. a. (1971)
	Oredoxa regia H. B. u. K.	Glycflav.	Williams u. a. (1971)
	Phoenix dactylifera L.	Glycflav.	Williams u. a. (1971)
ANGIOSPERMAE/ MONOCOTYLEDONEAE			
Commelinaceae	*Commelina communis* L.	I, II	Hayashi u. a. (1958), Takeda u. a. (1966)
Cyperaceae	*Carex*-Arten	Glycflav.	Harborne (1971)
	Cladium mariscus R. Br. Prod.	Glycflav.	Harborne (1971)
	Cyperus longus L.	Glycflav.	Harborne (1971)
	Rhynchospora alba Vahl.	Glycflav.	Harborne (1971)
	Scirpus-Arten	Glycflav.	Harborne (1971)
	Eriphorum-Arten	Glycflav.	Harborne (1971)
Iridaceae	*Iris*-Arten	I, II, IV	Charten (nicht ver.), Kawase (1969), Asen u. a. (1970)
Lemnaceae	*Lemna*-Arten	I, IV	McClure u. a. (1966)
	Spirodela-Arten	I, II, IV	McClure u. a. (1966), Jurd u. a. (1957)
	Wolffia-Arten	I, IV	McClure (1966)
Liliaceae	*Ornithogalum*-Arten	II	Sinn (1967)
Poaceae	*Avena*-Arten	II	Harborne u. a. (1964)
	Agrostis canina L.	I	Harborne u. a. (1964)
	Bromus erectus Huds.	II	Sinn (1967)
	Briza media L.	I	Harborne (1964)

Literaturverzeichnis: SS. 206—216

Tabelle 6 (Fortsetzung)

Klasse/ Abt. Familie	Art	C-Glykosid Typen	Autoren/ Veröffentlichungsjahr
	Hordeum-Arten	II	SEIKEL u. a. (1957) SEIKEL u. a. (1959), SINN (1967)
	Poa compressa Ubw.	I	HARBORNE (1964)
	Stipa-Arten	I, III	HARBORNE (1964), SALEX u. a. (1971)
	Oryza sativa L.	I	HARBORNE (1964)
	Triticum-Arten	I, II, IV	HARBORNE (1964), JULIAN u. a. (1971)
Potamogetonaceae	*Potamogeton natans* L.	I	CHOPIN u. a. (1972)

III. Synthese von Flavonoidglykosiden

1. Darstellung von Acetobromzuckern

Für die Herstellung der zur Glykosidierung notwendigen α-Acetobrom-Derivate von *D-Glucopyranose, D-Galactopyranose* und *D-Xylopyranose* bewährte sich das Verfahren von BARCZAI-MARTOS und KÖRÖSY (*10*). Die Darstellung der *β-L-Acetobromarabopyranose* erfolgt am zweck-mäßigsten nach GEHRKE und AICHNER (*66*), die von α-*Acetobromrhamnose* nach FISCHER und Mitarb. (*57*) und die des α-*Acetobromglucuronsäure-methylesters* nach BOLLENBACK und Mitarb. (*19*).

Die Rutinose kann aus Rutin durch Rhamnodiastase-Spaltung (*228*), durch 6-stündige Säurehydrolyse mit 10%iger Essigsäure (*229*), durch Spaltung mit Dihalomethylmethyläther (*18, 130*) oder synthetisch erhalten werden. Die Synthese erfolgt nach ZEMPLÉN und GERECS (*226*) durch Kupplung von α-Acetobromrhamnose mit α-1-Chlor-2,3,4-tri-acetylglucose in Gegenwart von Quecksilberacetat zu α-Acetochlor-α-L-rhamnosido-6-O-glucose, Austausch von Chlor gegen Acetyl mit Hilfe von Silberacetat in Gegenwart von Essigsäureanhydrid und Natrium-methylat-Verseifung zu Rutinose. Die Darstellung der α-Acetobrom-verbindung erfolgt in üblicher Weise.

KAMIYA und Mitarb. (*121*) synthetisierten Rutinose auf 3 anderen Wegen:

1. durch Kondensation von α-Acetobromrhamnose mit 1,2,3,4-Tetra-O-acetyl-β-D-glucose in Acetonitril mit Quecksilbercyanid und Quecksilberbromid als Katalysatoren (HELFERICH-ZIRNER-Methode),

2. durch Kondensation von 2,3,4-Tri-O-benzoyl-α-L-rhamnosylbromid mit 1,2,3,5-Di-isopropyliden-α-D-glucofuranose (Isodiacetonglucose) unter den gleichen Bedingungen wie bei 1. und

3. durch Kondensation von α-Acetobromrhamnose mit O-Trityl-β-1,2,3,4-tetra-O-acetyl-β-D-glucose in Nitromethan und mit Silberperchlorat als Katalysator (BREDERECK-Verfahren).

Die *Robinobiose* kann nach ZEMPLÉN und Mitarb. (*226*) analog dem Verfahren der Rutinosedarstellung durch Kupplung von 1-Chlor-2,3,4-triacetyl-β-D-galactose mit α-Acetochlorrhamnose in Gegenwart von Quecksilberacetat zu α-1-Chlor-β-6-L-rhamnosido-D-galactose [1,5]-hexaacetat oder nach KAMIYA und Mitarb. (*121*) durch Kondensation von α-Benzobromrhamnose mit Diacetongalactose dargestellt werden.

Neohesperidose wird nach KAMIYA (*121*) durch Kondensation von α-Acetobromrhamnose mit 1,3,4,6-Tetra-O-acetyl-β-D-glucopyranose (*86*) in Acetonitril nach der HELFERICH-Methode erhalten. Einfacher und mit wesentlich größerer Ausbeute (63%) verläuft ein abgeändertes Verfahren von KOEPPEN (*131*), bei dem ebenfalls von der α-anomeren Tetraacetylglucose und α-Acetobromrhamnose ausgegangen und das primär entstehende α-Neohesperidose-hepta-acetat über die Glykosylbromidverbindung mit Quecksilberacetat in das β-Neohesperidoseacetat übergeführt wird. Die Darstellung des α-Acetobromderivates erfolgt in üblicher Weise (Schema 7).

Zur *Sophorose*-Synthese gibt es neben dem älteren Verfahren von FREUDENBERG (*60*) eine Reihe von neueren Methoden mit besseren Aus-

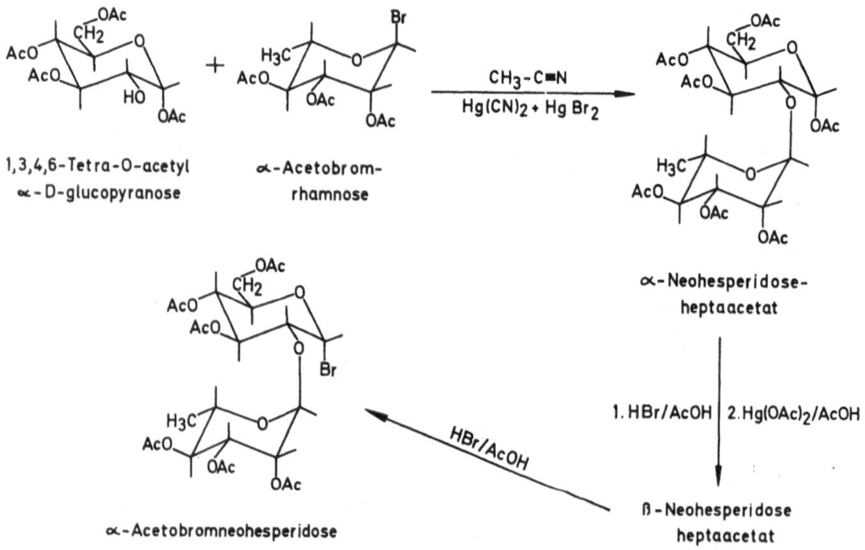

Schema 7. Synthese der α-Acetobrom-Neohesperidose

beuteergebnissen. Die höchsten Ausbeuten (25—30%) liefert ein von KOEPPEN (*129*) ausgearbeitetes kombiniertes Verfahren, bei dem 1,3,4,6-Tetra-O-acetyl-α-D-glucopyranose mit α-Acetobromglucose nach HELFERICH-ZIRNER (*86*) kondensiert wird. Man erhält aus der Reaktionslösung α-Kojibioseoctaacetat und aus der Mutterlauge durch Bromierung direkt die α-Acetobromsophorose in 25%iger Ausbeute.

Die α-*Acetobromgentiobiose* kann aus käuflichem oder nach HELFERICH und KLEIN (*85*) synthetisierbarem β-Gentiobioseacetat, das entsprechende *Vicianosederivat* nach HELFERICH und BREDERECK (*84*) oder McCLOSKEY und COLEMAN (*141*) und die α-Acetobromprimverose in 50%iger Ausbeute nach ZEMPLÉN und BOGNÁR (*218*) erhalten werden. Die Synthese der Sambubiose durch Kondensation von Benzyl-3,5,6-tri-O-benzyl-α-D-glucofuranosid und 2,3,4-Tri-O-acetyl-α-D-xylosylbromid (*41*) ist grundsätzlich gelöst, doch ist die Darstellung des α-Acetobromderivates bisher nur bei der isomeren 2-O-α-D-Xylopyranosyl-D-glucopyranose in guten Ausbeuten gelungen (*205*).

Die *Acetobromapiobiose* selbst ist noch nicht synthetisiert worden. Zur Darstellung des 7-O-Apiobiosids von 4'-Methylapigenin ist man deshalb von partiell kaschiertem 4'-Methylapigenin-7-glucosid ausgegangen und hat dieses mit 2-O-Acetyl-1',3-di-O-benzyl-D-apio-D-furanosylbromid umgesetzt (*44*). Diese Methode stellt eine neue, dem Biosyntheseweg folgende Variante zur Synthese von Flavonbiosiden dar.

Für den Kupplungsverlauf gilt, daß die α-Acetobromzucker der *D*-Reihe unter den Bedingungen der KÖNIGS-KNORR-Synthese (*128*) mit Phenolen unter Walden-Umkehr zu den β-D-Glykosiden umgesetzt werden. Bei Verwendung von Quecksilbersalzen anstelle von Silbersalzen kann die Reaktion auch ohne Walden-Umkehr ablaufen. Dies ist eindeutig der Fall bei der Glykosidierung mit α-Acetobrom-L-rhamnose (*98*). Bei der Kupplung mit β-Acetobrom-L-arabinose tritt dagegen Walden-Umkehr ein und man erhält das α-L-Arabinosid (*40*). NMR-Untersuchungen haben aber gezeigt, daß der Reaktionsverlauf vom Lösungsmittel abhängt (*35*).

In Benzol findet die Kupplung ohne vorherige Anomerisierung unter Konfigurationsumkehr statt. In Pyridin wird jedoch die β-Acetobrom-L-arabinose innerhalb weniger Minuten zur α-Form anomerisiert, so daß die Kupplung selbst ohne Walden-Umkehr verläuft (*35*).

2. Synthese von O-Monosiden und O-Biosiden

Die anzuwendenden Verfahren zur selektiven Glykosidierung einer bestimmten Hydroxylgruppe richten sich nach der Zahl und Stellung bzw. Acidität der anderen in einem Flavonoidtyp noch vorhandenen freien

OH-Gruppen. Liegt nur eine freie, nicht chelierte OH-Gruppe vor, ist die Glykosidierung im Flavonoidmolekül ohne Schwierigkeiten erreichbar, wie FUJISE und Mitarb. (61) sowie JERZMANOWSKA (117) am Beispiel der nicht natürlich vorkommenden 3,5 und 7-Monohydroxy- bzw. 5,7 und 3,5 Dihydroxy-flavone durch Umsetzung nach KÖNIGS-KNORR (128) gezeigt haben.

a) Glykosidierung der C_7-OH-Gruppe

Flavone, Isoflavone, Chalkone oder Flavanone lassen sich entweder nach dem Verfahren von ZEMPLÉN (225) oder nach der KÖNIGS-KNORR-Methode (128) selektiv am C_7-Hydroxyl glykosidieren.

Eine dritte Möglichkeit besteht in der Kondensation eines glykosidierten Acetophenon-Derivates (C_6-C_2) mit einem C_6C_1-Partner zum entsprechenden Chalkonglykosid mit nachfolgender Zyklisierung zum Flavanon-7-O-glykosid.

Nach der ersten Methode erfolgt die Umsetzung des Aglykons mit dem Acetobromzucker in Aceton und in Gegenwart von 1—10%iger wäßriger Kali- oder Natronlauge, wobei das Aglykon des Mono-Kalium-oder Natriumsalzes zur Reaktion kommt: Bei Einsatz von 1 Mol Acetobromzucker pro 1 Mol Aglykon und einer Reaktionsdauer von 14—36 Stunden bei 0° wird nach dem Verseifen des Kupplungsproduktes mit 5—10%iger wäßriger Lauge das Glykosid zumeist direkt ohne weitere Reinigungsverfahren erhalten. Die Ausbeuten liegen zwischen 20% und 40%. Auf diese Weise haben erstmals ZEMPLÉN und FARKAS (225) das 5,7,4'-Trihydroxy-isoflavon-7-O-monoglucosid Genistin hergestellt. In analoger Weise erfolgte z. B. die Darstellung des mit Genistin isomeren Luteolin-7-O-monoglucosids (92) sowie der 7-Glucoside von 5,7,4'-Trihydroxy-3,6,3'-trimethoxyflavon (Jacein) (45) und 5,7,3'-Trihydroxy-3,6,4'-trimethoxyflavon (Centaurein) (46).

Nach der zweiten Methode erfolgt die Umsetzung äquimolarer Mengen von Acetobromzucker und Aglykon in Chinolin und mit Silberoxid oder Silbercarbonat als Katalysator (128). Die Verseifung des Kupplungsproduktes zum freien Glykosid wird in der Regel mit 0.5—2 n Natriummethylatlösung bei 0° durchgeführt. Auf diese Weise haben erstmals ZEMPLÉN und BOGNÁR (221) das in Citrusfrüchten vorkommende Hesperidin (41) synthetisiert. Ebenfalls ohne vorherige Kaschierung der C_4- oder C_3-OH-Gruppe und nach der gleichen Methode wurden die natürlich vorkommenden Flavanon-7-O-rutinoside und 7-O-neohesperidoside Naringin (50), Poncirin (51), Narirutin (38), Didymin (39) und Sarothanosid (54) in Ausbeuten zwischen 40% und 50% als Racemate erhalten (199, 200, 202, 209, 8) (Schema 8). Durch Bromierung und

Dehydrobromierung (*220*) oder Jodierung in Eisessiglösung und in Gegenwart von Kaliumacetat (*143*) lassen sich diese Flavanonglykoside bzw. ihre Vollacetate zu den strukturhomologen Flavon-7-O-glykosiden isomerisieren. Die auf diese Weise erhaltenen natürlich vorkommenden Flavon-O-7-glykoside Fortunellin (**56**), Rhoifolin (**55**), Isorhoifolin (**47**) (*201*), Scolymosid (**43**) (*203*) und Veronicastrosid (**57**) (*115*) wurden in 30—60%iger Ausbeute erhalten. Sie sind in Tabelle 3 aufgeführt. Abweichend von diesen Verfahren synthetisierten TARUSOVA und PREOBRAZHENSKII (*192*) Eriocitrin (**40**) durch Kupplung von α-Acetobromrutinose mit einem partiell in 4'-Stellung benzylierten Eriodictyol (**26**). TEOULE und Mitarb. (*194*) benutzten die in 3'- und 4'-Stellung benzylierten Flavone, um die 7-O-Monoglucoside des Apigenins (**13**), Luteolins (**15**) und Diosmetins (**17**) herzustellen.

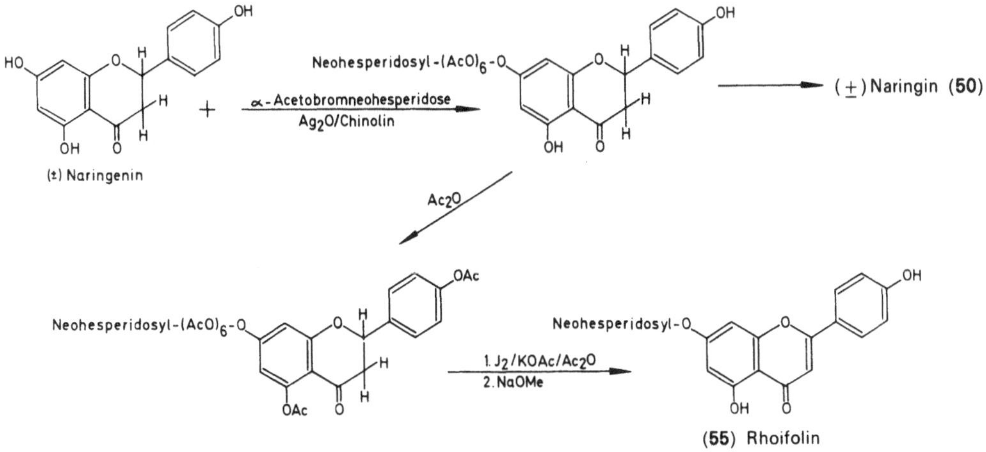

Schema 8. Synthese von Flavanon- und Flavon-7-O-glykoside durch Umsetzung der Aglykone nach KÖNIGS-KNORR (*128*)

Das 3. Verfahren zur Darstellung von 7-O-Glykosiden wurde erstmals von BARGELLINI angewandt. ZEMPLÉN und Mitarb. (*223*) und später TARUSOVA und Mitarb. (*193*) gingen von synthetischen Phloracetophenonglykosiden aus und kondensierten diese in stark alkalischem Medium zu den entsprechenden Chalkonglykosiden. Durch Zyclisierung in schwach saurem Medium wurden hieraus die entsprechenden Flavanonglykoside erhalten. Bei der Umsetzung von freiem Phloracetophenon mit α-Acetobromzuckern nach der Kalilauge oder Chinolin-Silbersalz- Methode entstehen aber keine Monoglykosidacetate, sondern in erster Linie die Phloracetophenon-2,4-diglykosidacetate, die dann bei der Kondensation mit den verschieden substituierten aromatischen Aldehyden die

entsprechenden Chalkon 2',4'-diglykoside liefern (Schema 9) (*164, 103*).
Zyclisierung in Citratpufferlösung oder Natriumacetatlösung führt weiter
zu den nicht natürlich vorkommenden Flavanon-5,7-diglykosiden. Um
nur die 7-O-Glykoside zu erhalten, muß die Zyclisierung der Chalkon-
diglykoside in 0,15%iger Salzsäure durchgeführt werden, oder man spaltet
aus den Flavanon-5,7-diglykosiden nachträglich durch Behandlung mit
0,15%iger Salzsäure selektiv den Zucker in C_5-Stellung (Schema 9) (*103*)
ab. Das Verfahren wurde zur Synthese von Prunin [Naringenin-7-mono-
glucosid](**24**) benützt (*24*). Auf dem gleichen Wege erhielten KAMIYA und
Mitarb. (*122*) erstmals Naringin (**50**) und Neohesperidin (**53**). Die
Autoren führten die Kondensation in 60%iger wäßriger Kalilauge bei
2—3° C bzw. in Natriumäthylatlösung unter Zusatz von p-Toluolsulfo-
säuremethylester bei 80° C durch und zyklisierten die Chalkonglykoside
durch 5-minütiges Erhitzen in einer äthanolischen McIlvain-Pufferlösung
vom pH 7,0.

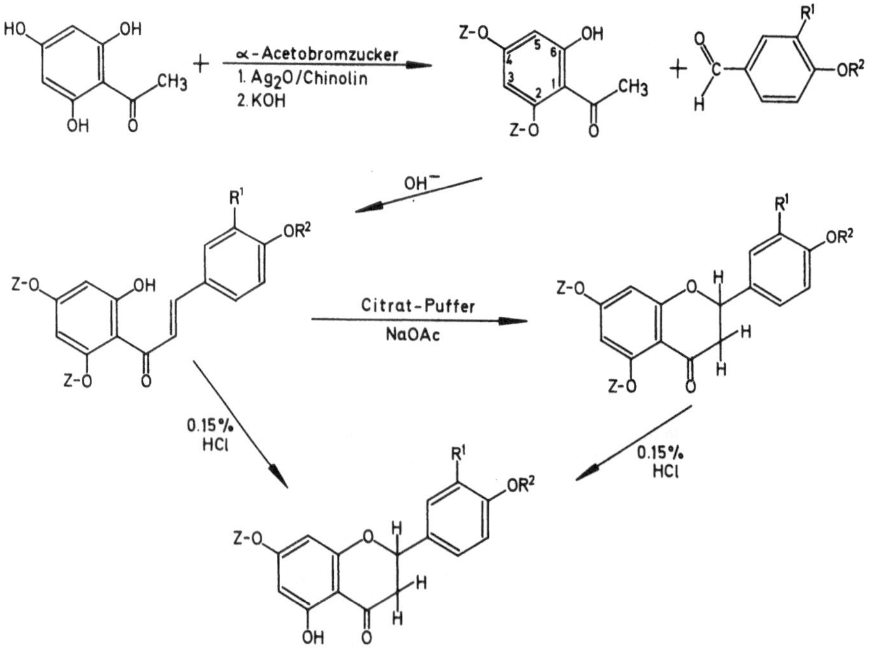

Schema 9. Synthese von Flavanon- und Flavon-7-O-glykosiden nach der
„Kondensations"-Methode I

Höhere Ausbeuten werden erhalten, wenn von den reinen Phloraceto-
phenon-4-O-glykosiden ausgegangen wird. Diese sind zugänglich
durch Umsetzung von 2-Benzoyl-phloracetophenon (*23*) mit Halogen-
zuckern nach der Kalilaugenmethode von ZEMPLÉN (*225*). Phloraceto-

phenon-4-neohesperidosid kann außerdem nach HOROWITZ und GENTILI (*107*) sowie CHOPIN und DELLAMONICA (*30*) durch Alkaliabbau der natürlich vorkommenden Flavanon-7-O-neohesperidoside Naringin oder Neohesperidin erhalten werden.

Nach dieser Methode gelang erstmals die Vollsynthese des Poncirins (**51**) (*202*), (Schema 10). Die Kondensation von synthetischem Phlor-acetophenon-4-neohesperidosid mit Anisaldehyd erfolgte unter Stick-stoff in 25%iger Kalilauge und wasserfreiem Alkohol zum 2′,4′,6′-Tri-hydroxy-4-methoxy-chalkon-4′-neohesperidosid, die nachfolgende Zycli-sierung zum Poncirin in wäßrigem Pyridin. Die nach beiden Verfahren synthetisierten Flavanon-7-glykoside waren miteinander identisch. Die CD-Kurven von synthetischen Flavanonglykosiden besitzen entsprechend dem Razemisierungsgrad geringere oder nur schwache Cotton-Effekte zwischen 250 und 350 nm. Die Tatsache, daß viele synthetische Flavanon-7-O-glykoside ähnliche oder gleiche optische Drehwerte und CD-Kurven wie die natürlichen Glykoside liefern, erklärt sich entweder aus der teilweisen Razemisierung der natürlichen Glykoside bei der Isolierung oder aus der Anreicherung des 2S-Diastomeren Anteils bei der Kristalli-sation der synthetischen Glykoside (**64**) (vgl. hierzu Tabelle 3).

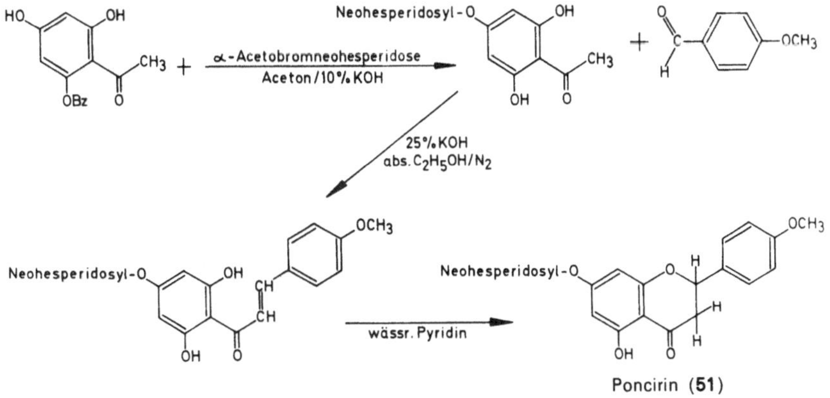

Schema 10. Synthese von Flavanon-7-O-glykosiden nach der Kondensations-Methode II

Da die von Phloracetophenon-4-glykosiden ausgehenden Synthesen über die Chalkonglykoside führen, sind auf diesem Wege gleichzeitig die natürlich vorkommenden Chalkon- bzw. Dihydrochalkon-2′- bzw. 4′-glykoside synthetisierbar (*47*). Durch Selendioxid-Oxidation solcher Chalkon-O-glykoside wurden die 7-O-Glucoside von 7,4′-Dihydroxy-, 7,4′-Dihydroxy-3′-methoxy- und 7,3′,4′-Trihydroxy-Flavon dargestellt (*72*) und damit die Synthesemöglichkeit für Flavon-O-7-glykoside durch eine neue Variante erweitert.

Für die selektive Glykosidierung der C_6-Hydroxyl-Gruppe bei den Auronen, die der C_7-OH-Gruppe bei den Flavonoiden entspricht, arbeiteten FARKAS und Mitarb. (52) ein allgemein anwendbares Verfahren aus, das im Prinzip der „Kondensationsmethode" folgt. Hiernach werden freie oder partiell alkylierte bzw. acylierte Hydroxy-cumaran-3-one mit den Halogenzuckeracetaten nach KÖNIGS-KNORR in Aceton und unter Zugabe von verdünnter Natronlauge umgesetzt. Die Cumaranon-glykoside werden ohne oder nach Entfernung der Schutzgruppen in siedendem Essigsäureanhydrid mit den verschieden substituierten aromatischen Aldehyden zu den Benzalcumaranonglykosidacetaten kondensiert. Verseifung nach ZEMPLÉN liefert die Auronglykoside in guten Ausbeuten (siehe Schema 11). Analog werden Auronglykoside erhalten, die ihre Zucker in 4-Stellung tragen. Zur Synthese des 3′,6-Di-glucosides Palasitrin wurde 6-Hydroxy-cumaranon-6-β-D-glucosidtetraacetat mit Protocatechualdehyd-3-glucosid zu dem entsprechenden Nonaacetat kondensiert (50).

Eine zusammenfassende Darstellung über die bisher synthetisierten Auronglykoside findet sich bei FARKAS und PALLOS (51). Die Möglichkeit, 2′-Hydroxy-chalkonglykoside durch Luftoxydation in Gegenwart von Natriumbicarbonat in Auronglykoside umzuwandeln, nutzten SHIMOKORIYAMA und HATTORI (187) zur Darstellung von Aureusidin-6-rutinosid aus Eriocitrin (40) bzw. seinem Chalkonisomeren.

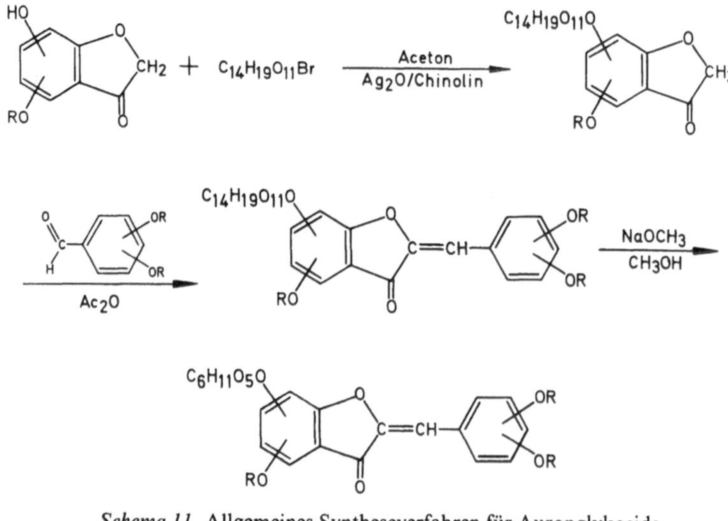

Schema 11. Allgemeines Syntheseverfahren für Auronglykoside

Die Herstellung von Flavonol-7-O-glykosiden in guten Ausbeuten ist nur möglich, wenn die C_3-OH-Gruppe erst nach erfolgter Glykosidierung eingeführt oder wenn die C_3-OH-Gruppe und alle anderen — außer 7-OH — vorher entsprechend kaschiert werden. PACHECO und Mitarb.

(*163*) gingen z. B. von Chalkon-4'- oder Flavanon 7-O-glykosiden aus und oxidierten diese nach ALGAR-FLYNN-OYAMADA (*2, 162*) über die Flavanonolverbindungen oder in einer Stufe direkt zu den Flavonol-7-glykosiden. Auf diese Weise wurde z. B. das 7,5,4'-Trihydroxy-flavanon-7-O-glucosid (Prunin) (**24**) mit H_2O_2 in alkalischer Lösung zu Aromadendrin-7-O-glucosid (Dihydro-kämpferol-7-glucosid) und dieses anschließend mit Natriumbisulfit weiter zum Kämpferol-7-glucosid (Populnin) (**9**) oxidiert. In einem 1-Stufenprozeß gelang die Oxidation von Naringin (**50**) zu Kämpferol-7-O-neohesperidosid und von 2',4'-Dihydroxy-chalkon-4'-glucosid zum 4',7-Dihydroxy-flavon-7-O-glucosid.

Zur Darstellung von Polyhydroxy-flavonolen mit nur einer zur Glykosidierung freien OH-Gruppe eignet sich die von NÓGRÁDI und FARKAS (*155*) entdeckte Transacylierungsmethode. Diese basiert auf der Beobachtung, daß Benzoylgruppen partiell benzoylierter Flavonoide unter den Reaktionsbedingungen einer KÖNIGS-KNORR-Glykosidierung auf phenolische OH-Gruppen niederer Acidität übertragen bzw. umgelagert werden können. Der Austausch dürfte eine basenkatalysierte Reaktion darstellen, wobei das durch Salzbildung mit dem Silbercarbonat entstandene Phenoxid-Anion als Base fungiert.

Da das Silbercarbonat in Pyridin praktisch unlöslich ist, wird das Ionisationsgleichgewicht durch eine Wechselwirkung des Silberkations mit Pyridin auf die rechte Seite verschoben. In Solventien, die mit den Silberkationen in keine Wechselwirkung treten, wie z. B. in Dimethylformamid, wird eine Ionisation der Phenolgruppe nicht beobachtet und

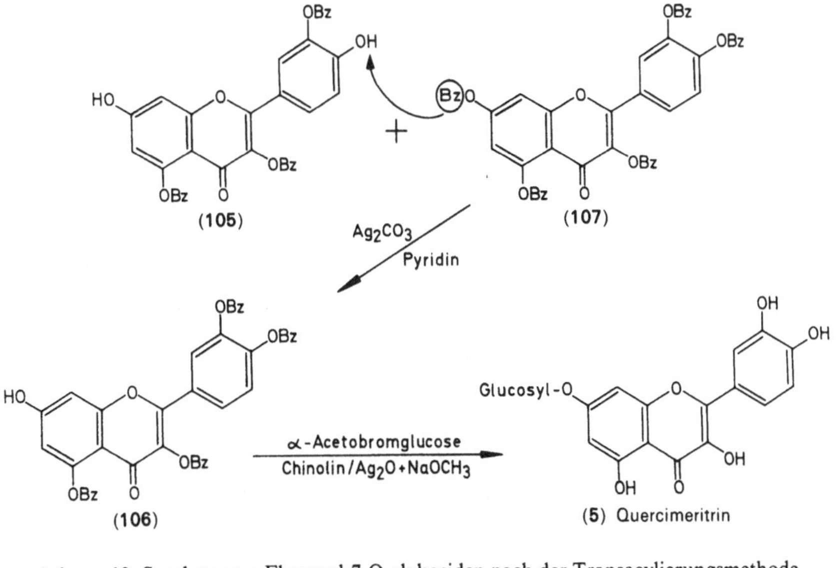

Schema 12. Synthese von Flavonol-7-O-glykosiden nach der Transacylierungsmethode

es findet keine Transacylierungsreaktion statt. Da als Acylfänger auch
reines Phenol geeignet ist, kann die Benzoylwanderung als ein intermole-
kularer Prozeß gedeutet werden. Läßt man z. B. Pentabenzoyl-Quercetin
mit einem Überschuß von Phenol und Silbercarbonat in Pyridin reagieren,
so erhält man 3,5,3′-Tribenzoyl-Quercetin (**105**) oder ein Gemisch aus
dieser Verbindung und 3,5,3′,4′-Tetrabenzoyl-Quercetin (**106**). Dagegen
liefert die von Silbercarbonat katalysierte Konproportionierung von
Pentabenzoyl-Quercetin (**107**) und 3,5,3′-Tribenzoyl-Quercetin (**105**) in
60%iger Ausbeute reines 3,5,3′,4′-Tetrabenzoyl-Quercetin (**106**). Kupplung dieses Derivates mit α-Acetobromglucose unter den üblichen Bedingungen und anschließende Verseifung mit Natriummethylat ergibt
Quercimeritrin (**5**) (*48*) (siehe Schema 12).

    Nach dieser neuen Transacylierungsmethode sind die in der Ta-
belle 7 aufgeführten Flavon-7-O-glykoside dargestellt worden (*203,
155, 48*), obwohl ein Teil von ihnen auch durch Direktkupplung der
freien Aglukone nach ZEMPLÉN (*225*) herstellbar sind.

Tabelle 7. *Synthetisierte natürliche Flavon-7-glykoside*

Cosmosiin	Apigenin-7-glucosid	*Cosmos bipinnatus*
Tilianin	Acacetin-7-glucosid	*Tilia japonica*
Quercimeritrin (**5**)	Quercetin-7-glucosid	*Gossypium herbaceum*
Scutellarin	Scutellarein-7-glucuronid	*Scutellaria altissima*
Plantaginin	Scutellarein-7-glucosid	*Plantago asiatica*
Scolymosid (**43**)	Luteolin-7-rutinosid	*Capsella bursae-pastoris*

b) Glykosidierung der C_3-OH-Gruppe

    Den ersten Versuch einer C_3-OH-Glykosidierung in der Polyhydroxy-
flavonol-Reihe unternahmen ZEMPLÉN und Mitarb. (*224*). Die Autoren
setzten 5,7,3′,4′-Tetraacetyl-Quercetin, das sie durch Hydrolyse eines per-
acetylierten Quercetin-3-rhamnosides erhalten hatten, mit α-Acetobromglucose nach der Chinolin-Silbersalzmethode um. Wegen der dabei ent-
standenen zahlreichen Nebenprodukte konnte das Glucosid nicht erhalten werden. Die Synthese wurde dann im Jahre 1952 von ICE und
WENDER (*114*) durch Kupplung des Quercetin-Kaliumsalzes mit α-
Acetobromglucose in flüssigem Ammoniak realisiert. Analog wurde im
Jahre 1960 von SAMOKHWALOV und Mitarb. (*177*) erstmals Rutin
synthetisiert. Am besten geeignet für die Glykosidierung erweisen sich
Flavonole, die in 7′- und 4′-Stellung benzyliert sind. Nach JURD (*119*)
werden diese durch Benzylierung von Polyhydroxyflavonolperacetaten
und Verseifung der gebildeten 7,4′-Dibenzyl-flavonol-acetate mit Salz-

säure zu den freien 7,4'-Dibenzyläthern erhalten. Durch Umsetzung von 7,4'-Dibenzyl-Quercetin und Isorhamnetin mit den Acetobromzuckern in äquimolarem Verhältnis nach KÖNIGS-KNORR in Pyridin und mit Silberoxid bzw. Silbercarbonat als Katalysator, anschließende Verseifung und hydrogenolytische Entbenzylierung wurden z. B. die 3-O-Glucoside, Galactoside, Rhamnoside, Xyloside, Arabinoside, Rutinoside, Sophoroside und Gentiobioside dargestellt (95, 96, 97, 98, 210, 211). Die Ausbeuten lagen bis auf die 3-O-Rhamnoside zwischen 30% und 70%.

Für die Synthese der Kämpferol-3-O-glykoside (210, 211) mußte das hierfür erforderliche 7,4'-Dibenzyl-Kämpferol vollsynthetisch durch Kondensation von ω-Benzoyloxyphloracetophenon mit p-Benzyloxy-benzoesäureanhydrid nach ALLAN-ROBINSON (3), nachfolgende Acetylierung, partielle Benzylierung auf C_7-OH und Verseifung hergestellt werden. Bei der Synthese des Azaleatin oder 5-Methylquercetin-3-galacto-sides mußte die für die Verbindung charakteristische CH_3-Gruppe erst nachträglich eingeführt werden, da Polyhydroxyflavonole mit einer freien C_5-OH-Gruppe wegen der konkurrierenden Chelatbildung mit der Carbonylgruppe bei der Glykosidierung bessere Ausbeuten ergeben als die total benzylierten Produkte (71). 4',7-Dibenzyl-Quercetin wurde mit α-Acetobromgalactose gekuppelt, das Kupplungsprodukt zunächst partiell auf $C_{3'}$-OH acetyliert und dann partiell auf C_5-OH methyliert. Verseifung und katalytische Entbenzylierung führte zum gewünschten Glykosid (102).

Für die Synthese der Quercetin- und Kämpferol-3-glucuronide (207, 208) wurde von den gleichen 4',7-Dibenzyläthern ausgegangen. Umgesetzt wurde mit dem Acetobromglucuronsäuremethylester nach KÖNIGS-KNORR. Zur Gewinnung der freien Glucuronide wurden die Kupplungs-produkte entbenzyliert, in die Vollacetate übergeführt und dann in der Kälte mit 1 n Natronlauge verseift. Analog erfolgt die Synthese der C_7-O-Glucuronide des Apigenins (206) und Chrysins (198).

c) Glykosidierung der $C_{4'}$-, $C_{2'}$-, $C_{3'}$- und C_5-OH-Gruppen

Vor der Glykosidierung der 4'-OH-Gruppe müssen alle anderen OH-Gruppen des Flavonoids außer der C_5-OH-Gruppe kaschiert sein. Da 7-O-Acyl-Flavonoide unter den Bedingungen der KÖNIGS-KNORR-Glykosidierung zu Transacylierungen Anlaß geben, mußte zur Synthese der 4'-Glykoside des Apigenins bzw. Genisteins von den 7-Benzyl- oder 7-Methyläthern ausgegangen werden. Die Acetobromzucker werden in 2—3 molarem Überschuß zugegeben. Auf diese Weise wurden das Api-genin-4'-glucosid und sein 7-Methyläther (Phegopolin) (54) hergestellt. Für die Synthese des Genistein-4'-neohesperidosids (Sophorabiosid) (59)

(*49*) wurde 7-Benzylgenistein auf zwei Wegen, erstens durch Benzylierung von freiem Genistein mit 1 Mol Benzylchlorid in Dimethylformamid und zweitens durch Benzylierung von natürlichem Genistein-4'-glucosid (Sophoricosid) und nachfolgende Hydrolyse gewonnen. Die Synthese des schon 1946 aus *Spiraea ulmaria* L. isolierten Quercetin-4'-glucosids *Spiraeosid* (**6**), erfolgte unter Ausnutzung der Transacylierungsmethode (*53*). 3,5,5',4'-Tetrabenzoyl-Quercetin (**106**) wurde an seinem freien Hydroxyl benzyliert, anschließend durch Umesterung mit 1 Mol. Phenol die 4'-Benzoylgruppe entfernt und mit diesem Produkt die Glykosidierung durchgeführt. Das einzige bisher bekannt gewordene 2'-O-Glykosid, das Echioidin (5,2'-Dihydroxy-7-methoxy-flavon-2'-O-mono-glucosid) aus *Andrographis echioides* ist durch Umsetzung des Aglykons mit Acetobromglucose nach der Zemplén-Methode synthetisiert worden.

Zur Synthese des aus der Douglastanne, der Roßkastanie und den Baumwollblüten aufgefundenen Quercetin-3'-O-glucosides wurde Quercetin-7,4'-dibenzyläther (*119*) zum 3,7,4'-Tribenzyläther weiter benzyliert und dieser in üblicher Weise umgesetzt und weiter verarbeitet (*53*).

Die Herstellung des isomeren Quercetin-O-5-glucosids ging wiederum vom 3,7,3',4'-Tetrabenzoyl-Quercetin (**106**) aus. Trotz eines Überschusses an Acetobromglucose beträgt die Ausbeute wegen der äußerst schwach aciden chelierten C_5-OH-Gruppe nur etwa 10%.

Nach dem „Kondensationsverfahren" wurde das erste in der Natur aufgefundene Flavonoid-5-O-Glucosid, das Salipurposid (**25**) aus *Salix purpurea* L. synthetisiert (Schema 13). ZEMPLÉN und Mitarb. (*222*) synthe-

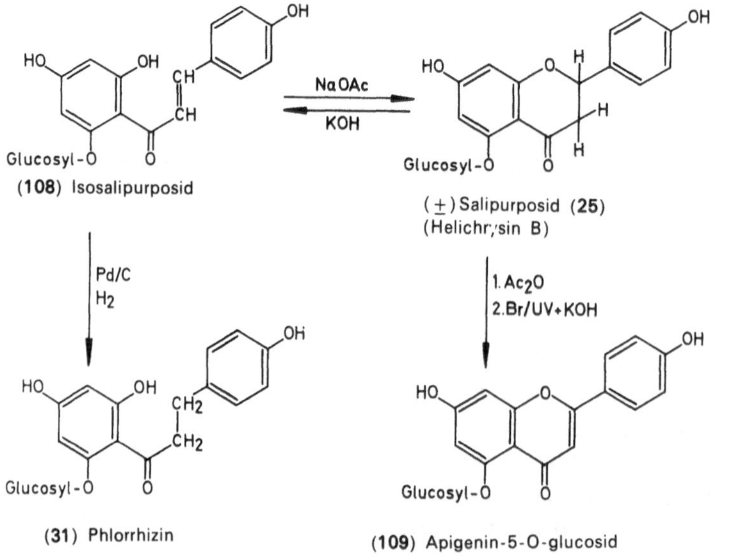

Schema 13. Synthese von Flavanon-5-O-, Flavon-5-O- und Chalkon-2'-O-glykosiden

Literaturverzeichnis: SS. 206—216

tisierten aus 2-(Tetraacetyl-glucosido)-4-benzoyl-phloracetophenon und Oxybenzaldehyd zunächst das 2′,4′,6′,4-Tetrahydroxychalkon-2′-O-glucosid (Isosalipurposid) (108) und isomerisierten dieses durch Natriumacetat zum Naringenin-5-O-glucosid (25). Durch die Isolierung zweier weiterer Naringenin-5-glucoside, Helichrysin A und B, aus den Blüten von *Helichrysum arenarium* (L.) Moench (73), wurde später gezeigt, daß es sich bei dem natürlichen Salipurposid um ein partielles Razemat von (+) und (−) Naringenin-5-β-D-glucosid gehandelt haben muß, das mit Helichrysin B identisch ist (73). Aus Helichrysin *B* wurde unter milden Hydrolysebedingungen optisch inaktives (±) Naringenin erhalten, während Helichrysin A unter den gleichen Bedingungen ein (−) Naringenin ergab (73). Salipurposid läßt sich in Kalilauge zum Isosalipurposid zurückisomerisieren und dann weiter mit Palladium/Kohle zum Phlorrhizin (31) hydrieren (47). Durch Bromierung von Hexaacetyl-Salipurposid unter UV-Bestrahlung und anschließende Dehydrobromierung in alkoholischer Kalilauge wurde das ebenfalls natürlich vorkommende Apigenin-5-glucosid (109) erhalten (231).

3. Synthese von Flavonoid-*O*-Bisglykosiden

Sollen gleich strukturierte Zucker an die C_3- und C_7-OH-Gruppe geknüpft werden, setzt man das freie Aglykon mit 2 Mol Acetobromzucker nach der Kalilaugen-Methode um und reinigt das Kupplungsprodukt von den mitgebildeten 3′- bzw. 7-O-Monoglykosidacetaten durch Polyamidchromatographie. Natriummethylat-Verseifung liefert aus den so gereinigten Kupplungsprodukten Verbindungen, wie z. B. das Quercetin-3,7-diglucosid und das Isorhamnetin-3,7-diglucosid (Brassicosid) (98).

Bisglykoside mit zwei verschiedenen Zuckern in C_3- und C_7-Position sind durch Umsetzung der Flavonol-3-O-glykosidvollacetate mit 2 Mol des zweiten Halogenzuckers nach der Kalilaugenmethode oder durch Kupplung von 7-Hydroxy-flavonol-3-O-mono- oder diglykosidacetat nach KÖNIGS-KNORR herstellbar. Die 7-Monohydroxyverbindung erhält man durch partielle Benzylierung des Glykosidacetates auf C_7 und anschließende katalytische Hydrierung. Die so synthetisierten Glykoside Quercetin-3-α-L-rhamnopyranosyl-7-β-D-glucopyranosid, Quercetin-3-β-D-xylopyranosyl-7-β-D-glucopyranosid (154) und Quercetin-3-β-rutinosyl-7-β-D-glucosid (153) sind natürlich vorkommend in *Euonymus*-Arten aufgefunden worden.

Zur Darstellung des aus *Citrus paradisi* Macf. gewonnenen Naringenin-4′-β-D-glucosyl-7-β-rutinosids (110) wurde Naringenin-7-β-rutinosidhexaacetat (Narirutinhexaacetat) (149) mit einem 2—3 molaren Überschuß an α-Acetobromglucose nach KÖNIGS-KNORR umgesetzt. Das nach dem Verseifen entstandene Gemisch aus gewünschtem Glykosid, isomerem Chalkonglykosid und mitgebildeten Naringenin-5-glykosiden wurde zur Recyclisierung der Chalkonverbindungen mit wäßrigem Pyridin und zur Abtrennung der 5-Glykoside anschließend mit 0,1%iger Salzsäure behandelt. Das Glykosid wurde nach Säulenchromatographie rein erhalten (7). Das hiermit isomere Naringin-4'-glucosid wurde abweichend von (110) durch Kondensation von Phloracetophenon-4-β-neohesperidosid mit dem bereits früher synthetisierten 4-Hydroxy-benzaldehyd-4-β-D-glucopyranosid in 14%iger alkoholischer Kalilauge unter Stickstoff und bei Raumtemperatur, gefolgt von einer Zyclisierung des gebildeten Chalkonglykosides zum Flavanon-tri-glykosid synthetisiert (9).

(110)

4. Synthese von Flavonoid-C-Glykosiden

Methoden zur C-Glykosidierung von Flavonoiden wurden von CHOPIN (28, 27) ausgearbeitet. Sie folgen im Prinzip den Verfahren der C-Methylierung (116). Hiernach werden die durch Synthese oder Isolierung zugänglichen freien oder partiell an den OH-Gruppen kaschierten Aglykone in einem großen Überschuß an Natrium- oder Lithiummethoxid (3—6 Mol) in Methanol oder n-Butanol gelöst und nach Zusatz einer Spur Natriumjodid portionsweise mit dem Halogenzucker bei Raumtemperatur umgesetzt. Die dabei gleichzeitig gebildeten O-Glykoside werden mit 2 n HCl hydrolysiert und die C-Glykoside durch Butanolextraktion und Reinigung über Papier oder Polyamid in Ausbeuten von 0,5% bis 1% erhalten (Schema 14). Bei Einsatz von O-benzylierten Flavonoiden hat der Säurehydrolyse eine katalytische Hydrogenolyse vorauszugehen. Wie bei der C-Methylierung findet die C-Glucosidierung nur am C_6 des A-Ringes statt. Die isomeren 8-C-Glykoside können aus den isolierten oder synthetisierten 6-C-Glykosiden durch Ringisomerisierung nach

Wessely-Moser (216) hergestellt werden (Schema 14). Bei der Glykosidierung mit Acetobromxylose und Acetobromrhamnose erhält man neben den 6-C-Glykosiden auch geringe Mengen an C_6,C_8-Di-C-Glykosiden. Zur Synthese der symmetrischen oder unsymmetrischen C_6,C_8-Di-C-Glykoside werden die natürlich vorkommenden Flavon-8-C-Glykoside weiter mit dem betreffenden Halogenzucker in der beschriebenen Weise umgesetzt.

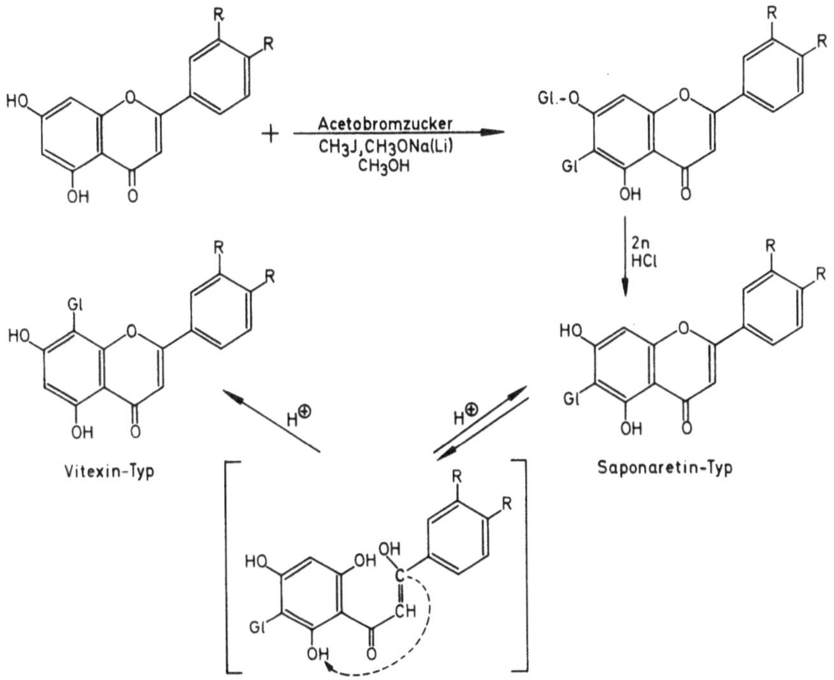

Schema 14. Syntheseverfahren von Flavonoid-C-glykosiden nach Chopin (28)

Von Chopin und Mitarb. wurden die meisten natürlich vorkommenden Mono-C-Glykoside und auch ein Teil der Di-C-Glykoside synthetisiert und damit in ihrer Struktur endgültig bewiesen (27). Die Glykosidierungen wurden durchgeführt mit den Acetobromderivaten von Glucose, Galactose, Xylose, Arabinose und Rhamnose. Die wichtigsten Arbeiten: (28, 32, 33, 20, 21, 29).

Addenda

Die Struktur des Di-C-Glykosides Violanthin (**104**) wurde in der Zwischenzeit durch die Synthese des Isoviolanthins und dessen Isomerisierung zu **104** bestätigt (M. C. BIOL et J. CHOPIN, C. R. Acad. Sci., Ser. C, **275**, 1523, 1972).

Eine neue Synthesemethode für Flavonoid-C-Glykoside, die Grignardierung von substituiertem Benzol, Darstellung des Phenyl-C-glykosides, Kondensation des entsprechenden 2-Hydroxy-Acetophenon-C-glykosides zum Chalkon-'C-glykosid und SeO_2-Oxidation zum Flavon-C-8-glykosid, wurde von EADE und MCDONALD zur Synthese des 7,4'-Di-O-Methylbayins ausgearbeitet (Private Mitteilung 1973).

Literaturverzeichnis

1. ADAMSKA, M., und J. LUTOMSKI: C-Flavonoidglykoside in den Samen von *Trigonella Foenum-graecum*. Planta med. **20**, 224 (1972).
2. ALGAR, J., und J. P. FLYNN: A new Synthesis of Flavonols. Proc. Roy. Irish. Acad. **42 B**, 1 (1934).
3. ALLAN, J., and R. ROBINSON: CCXC. An Accessible Derivative of Chromonol. J. chem. Soc. [London] **125**, 2192 (1924).
4. ALSTON, R. E.: C-Glycosyl Flavonoids. In: T. J. MABRY, Recent Advances in Phytochemistry Vol. 1, S. 305. New York: Appleton-Century-Crofts. 1968.
5. ARAKAWA, H., und M. NAKAZAKI: Die absolute Konfiguration der optisch aktiven Flavanone. Ann. **636**, 111 (1960).
6. ARITOMI, M.: Terniflorin, a New Flavonoid Compound in Flowers of *Clematis terniflora* var. *robusta*. Chem. Pharm. Bull. (Tokyo) **11**, 1225 (1963).
7. AURNHAMMER, G., H. WAGNER, L. HÖRHAMMER und L. FARKAS: Erste Synthese eines Flavanontriglykosides, des 4'-β-D-glucosyl-β-rutinosyl-7-naringenins. Chem. Ber. **103**, 1578 (1970).
8. — — — — Über die Identität der Flavanonrhamnoglucoside Sarotanosid und Isosarotanosid, Synthese von ($\pm$) Pinocembrin-7-β-neohesperidosid. Chem. Ber. **103**, 3667 (1970).
9. — — — — Synthese des 7-β-Neohesperidosyl-4'-β-D-glucopyranosyl-naringenins, eines Flavanontriglykosids aus Citrusfrüchten. Chem. Ber. **104**, 473 (1971).
10. BARCZAI-MARTOS, M., and F. KÖRÖSY: Preparation of Acetobromsugars. Nature (London) **165**, 369 (1950).
11. BARGER, B.: Saponarin, a New Glucoside, Coloured Blue with Iodine. J. Chem. Soc. (London) **89**, 1210 (1906).
12. BATYUK, V. S., N. V. CHERNOBROVAYA und D. G. KOESNIKOV: Flavonoids of the Hawthorn Structure of Cratenacin. Khim. Priv. Soedin. **5**, 234 (1969).
13. BHATIA, V. K., S. R. GUPTA, and T. R. SESHADRI: C- Glycosides of the Leaves of *Parkinsonia aculeata*. Tetrahedron (London) **22**, 1147 (1966).
14. BHUTANI, S. P., S. S. CHIBBER, and T. R. SESHADRI: Components of the Roots of *Pueraria tuberosa*, Isolation of a New Isoflavone-C-glycoside (Di-O-acetyl-puerarin). Indian J. Chem. **7**, 210 (1969).
15. — — — Flavonoids of the Fruits and Leaves of *Tribulus terrestris*: Constitution of Tribuloside. Phytochemistry **8**, 299 (1969).
16. BIRKHOFER, L., C. KAISER und H. KOSMOL: Konstitution des Petunosids. Z. Naturforsch. **20 b**, 605 (1965).
17. BIRKHOFER, L., C. KAISER und F. BECKER: Bignonosid, ein acyliertes Flavon aus *Catalpa bignonioides*. Z. Naturforsch. **20 b**, 923 (1965).

18. BOGNÁR, R., I. FARKAS SZABO, I. FARKAS and H. GROSS: Cleavage of Oligosaccharide Glykosides by Means of Dihalomethyl-Methylethers: a Novel Preparation of Rutinose. Carbohyd. Res. **5**, 241 (1967).

19. BOLLENBACK, G. N., J. W. LONG, D. G. BENJAMIN, and J. A. LINDQUIST: The Synthesis of Aryl-D-glucopyranosiduronic Acids. J. Am. Chem. Soc. **77**, 3310 (1955).

20. BOUILLANT, M. L., et J. CHOPIN: C-Xylosylation de la Vitexin, Synthèse de la C-Xylosyl-6-C-glucosyl-8-apigenine et comparaison avec les vicènines de *Vitex lucens*. C. R. Acad Sci. Ser. C **273**, 1759 (1971).

21. — — C-Xylosylation. D'un melange orientine u. homoorientine: Comparaison des produits obtenus avec les lucénines de *Vitex lucens*. C. R. Acad. Sci. Ser. C **274**, 193 (1972).

22. BOUTARD, B., M. L. BOUILLANT, J. CHOPIN et PH. LEBRETON: Isolement de l'isoscoparine (C-glucosyl-6-chrysoériol) de *Potamogeton natans* L. C. R. Acad. Sci. Ser. D **274**, 1099 (1972).

23. CANTER, F. W., F. H. CURD, and A. ROBERTSON: CLXVI. Hydroxycarbonyl Compounds. III. The Preparation of Coumarins and 1:4-Benzo-pyrones from Phloroglucinol and Resorcinol. J. Chem. Soc. (London) 1245 (1931).

24. CHABANNES, B., A. VILLE, A. GROUILLER et H. PACHECO: Chimie des composés flavoniques. V. Synthèse de la Prunine ^{14}C-2. Bull. soc. chim. France **1967**, 223.

25. CHANDLER, B. V., and K. A. HARPER: Identification of Saccharides in Anthocyanins and other Flavonoids. Austral. J. Chem. **14**, 586 (1961).

26. CHOPIN, J.: Les C-Glycoflavonoides. In: Actualités de Phytochemie, S. 44, Paris: Masson. 1966.

27. — Synthesis of C-Glycoflavonoids. In: Pharmacognosy and Phytochemistry ed. H. WAGNER and L. HÖRHAMMER, S. 110. Berlin-Heidelberg-NewYork: Springer. 1970.

28. CHOPIN, J., M. L. BOUILLANT et A. DURIX: Structure et synthèse du cytisoside. C. R. Acad. Sci. **260**, 4850 (1965).

29. CHOPIN, J., M. CHADENSON et M. HAUTEVILLE: C-glucosylation des dihydroxy-5,7-flavonols. C. R. Acad Sci. Ser. C **270**, 733 (1970).

30. CHOPIN, J., et G. DELLAMONICA: Synthese de quelques neohesperidosides-7 de flavanones. C. R. Acad. Sci. Ser. C **262**, 1712 (1966).

31. CHOPIN, J., G. DELLAMONICA et P. LEBRETON: Isolement d'un glucoside-6 de l'aureusidine à partir des extraits d'ecorce de citron. C. R. Acad. Sci. **257**, 534 (1963).

32. CHOPIN, J., et A. DURIX: C-glucosylation de la naringénine Synthèse de l'hemiphloine. C. R. Acad. Sci. Ser. C **263**, 951 (1966).

33. CHOPIN, J., B. ROUX, M. L. BOUILLANT, A. DURIX, A. D. D'ARCY, T. MABRY, et H. YOSHIOKA: Structure et synthèse de la di-C-glucosyl-6,8-apigénine du Citron. C. R. Acad. Sci. Ser. C 980 (1969).

34. DIEDRICHS, H. H., und E. SCHAICH: Flavonolglucoside in den Knospen und Blättern der Rotbuche *Fagus sylvatica* Linn. Naturwiss. **50**, 478 (1963).

35. DIRSCHERL, R.: Thesis, München (1968).

36. DUFOUR, J: Recherches sur l'amidon soluble et son role physiologique chez les vegetaux: Extrait du Bulletin de la Soc. vand. des sciences natur. Vol. XXI, Nr. 93.

37. DUTTON, G. J.: Glucuronic Acid, S. 4. New York and London: Academic Press. 1966.

38. EADE, R., A. I. SALASOO, and J. J. H. SIMES: Bayin (5-Deoxyvitexin): A New Glycoflavone. Chem. and Ind. 1720 (1962).

39. EGGER, K., und M. KEIL: Flavonolglykoside in Ranunculaceen. Ber. dtsch. bot. Ges. **78**, 153 (1965).

40. EL KHADEM, H., and Y. S. MOHAMED: Constituents of the Leaves of Psidium guaijava L. II. Quercetin, Avicularin and Guaijaverin. J. Chem. Soc. (London) **1958**, 3320.

41. ERBING, B., and B. LINDBERG: Synthesis of Sambubiose. Acta Chem. Scand. **23**, 2213 (1969).

42. Evans, W. H., A. McGookin, L. Jurd, A. Robertson, and W. R. N. Williamson: Vitexin I. J. Chem. Soc (London) 1957, 3510.

43. Ezekiel, A. D., W. G. Overend, and N. R. Williams: Synthesis of 1,2-O-Isopropylidene-α-D-apio-D-furanose and D-Apiose. Tetrahedron Letters 1969, 1635.

44. — — — Branched chain Sugars, XIII. Synthesis of 4'-O-Methylapiin. J. Chem. Soc. (London) 1971, 2907.

45. Farkas, L., L. Hörhammer, H. Wagner, H. Rösler und R. Gurniak: Die Struktur des Jaceins und dessen Synthese aus dem Aglukon und Acetobromglucose. Chem. Ber. 97, 610 (1964).

46. — — — — — Die Struktur des Centaureins und dessen Synthese aus dem Aglukon und Acetobromglucose. Chem. Ber. 97, 1666 (1964).

47. Farkas, L., M. Nogradi und A. Major: Synthese zweier Dihydrochalkon-glucoside aus *Malus trilobata* L. und *Malus sieboldii arborescens*. Chem. Ber. 98, 2926 (1965).

48. Farkas, L., M. Nógrádi, B. Vermes, A. Wolfner, H. Wagner, L. Hörhammer und H. Krämer: Transacylierungsreaktionen in der Flavonoid-Reihe. V: Synthese Quercimeritrins und Quercetin-3,7-diglucosids. Chem. Ber. 102, 2583 (1969).

49. Farkas, L., M. Nógrádi, H. Wagner und L. Hörhammer: Endgültige Strukturaufklärung und vollständige Synthese des Sophorabiosids, eines Glykosids aus *Sophora japonica* L. Chem. Ber. 101, 2758 (1968).

50. Farkas, L., und L. Pallos: Synthese des Palasitrins, eines Glucosids von *Butea frondosa*. Chem. Ber. 93, 1272 (1960).

51. — — Natürlich vorkommende Auronglykoside. In: Fortschritte der Chemie organischer Naturstoffe, Bd. XXV, S. 150. Wien-New York: Springer. 1967.

52. Farkas, L., L. Pallos und Z. Paal: Synthese und endgültige Strukturaufklärung des Sulfureins. Chem. Ber. 92, 2847 (1959).

53. Farkas, L., B. Vermes und M. Nógrádi: Die erste Synthese von drei natürlichen Quercetin-glucosiden Spiraeosid, Quercetin-5- und 3'-O-β-D-glucopyranosid. Chem. Ber. 105, 3505 (1972).

54. Farkas, L., A. Wolfner, M. Nógrádi, H. Wagner und L. Hörhammer: Transacylierungsreaktionen in der Flavonoid-Reihe. III: Synthese des Apigenin-4'-β-D-glucosids und des Phegopolins. Chem. Ber. 101, 1630 (1968).

55. Faroog, M. O., S. R. Gupia, M. Kiarmuddin, W. Rahman, and T. R. Seshadri: Chemical Examination of Celery Seeds. J. Sci. Ind. Res. (India) 12 B, 400 (1953).

56. Faugeras, G., et R. Paris: Heterosides flavoniques du *Rhamnus alaternus* L. Ann. pharm. franc. 20, 217 (1962).

57. Fischer, E., M. Bergmann und A. Rabe: Über Acetobromrhamnose und ihre Verwendung zur Synthese von Rhamnosiden. Chem. Ber. 53, 2362 (1920).

58. Fisel, J.: Neue Flavonoide aus Crataegus. 1. Die Isolierung eines acetylierten Vitexin-4'-rhamnosides aus *Crataegus*. Arzneimittelforsch. 15, 1417.

59. Freudenberg, K., und H. Knauber: Die Identität der Sophorose mit synthetischer 2-(β-glucosido)-Glucose. Naturwiss. 34, 344 (1947).

60. Freudenberg, K., und K. Soff: Biose aus Methylbiosid. Synthese der 2-(β-Glucosido)-Glucose. Ber. dtsch. chem. Ges. 69, 1245 (1936).

61. Fujise, S., und S. Mitui: Versuche zur Synthese von Oxyflavanon-glucosiden. Ber. dtsch. chem. Ges. 71, 912 (1938).

62. Funaoka, K.: Flavonoids of *Zelkova serrata* Wood. Mokuzai Gakkaishi 3, 218 (1957). [Chem. Abstr. 52, 12395 (1958)].

63. Furuya, M., A. Galston, and B. B. Stowe: Isolation from Peas of Cofactors and Inhibitors of Indolyl-3-acetic acid Oxidase. Nature (London) 193, 456 (1962).

64. Gaffield, W., and A. C. Waiss jr.: Optical Rotatory Dispersion and Absolute Configuration of Flavanones, 3-Hydroxyflavanones, and Their Glycosides. Chem. Commun. 1968, 29 und W. Gaffield, Circular Dichroism, Optical Rotatory Dispersion and

Absolute Configuration of Flavanones, 3-Hydroxyflavanones and Their Glycosides, Tetrahedron **26**, 4093 (1970).

65. GALLE, K.: Thesis (München) in Vorbereitung.

66. GEHRKE, M., und F. X. AICHNER: Über das Arabinal. Ber. dtsch. chem. Ges. **60**, 918 (1927).

67. GENTILI, B., und R. M. HOROWITZ: Flavonoids of Citrus IX. Some New C-Glycosyl-flavones and a NMR-Method for Differentiating 6- and 8-C-Glycosyl-Isomers. J. Org. Chem. **33**, 1571 (1968).

68. GIUFFA, E., C. GALEFFI, M. C. RENDINA e E. M. DELLE MONACHE: 3-α-(2-O-β-D-glucopyranosyl)-D-glucofuranoside della quercetina: nuovo glucoside flavonico della *Vestia lycioides*. Ann. Ist. Super Sanitá **7**, 23 (1971).

69. GORIN, P. A. J., and A. S. PERLIN: Configuration of Glycosidic Linkages in Oligosaccharides, VIII. Synthesis of α-D-Mannopyranosyl- and α-L-Rhamno-pyranosyl-Disaccharides by the Königs-Knorr Reaction. Canad. J. Chem. **37**, 1930 (1959)*8*

70. GROUILLER, A., K. EGGER et H. PACHECO: Sur un nouveau glycoside du Kaempferol. extract de *Cardamine pratensis*. C. R. Acad. Sci., Ser. C, **271**, 769 (1970).

71. GROUILLER, A., et H. PACHECO: Chimie des composés flavoniques. VII. Problèmes posés par la glucosylation en 3 des flavanols. Bull. Soc. chim. (France) **1968**, 4981.

72. GUPTA, S. R., B. RAVINDRANATH, and T. R. SESHADRI: Synthesis of some Flavonoid Glucosides of *Trifolium subt*. Phytochem. **10**, 877 (1971).

73. HÄNSEL, R., und D. HEISE: Zwei diastereomere Naringenin-β-D-glucoside aus *Flores Stoechados*. Arch. Pharmaz. **292**, 398 (1959).

74. HARBORNE, J. B.: Comparative Biochemistry of the Flavonoids. London: Academic Press. 1967.

75. — Flavonoid Pigments of *Lathyrus odoratus*. Nature (London) **187**, 240 (1960).

76. — Flavonoid Sophorosides. Experientia **19**, 7 (1963).

77. — Plant Polyphenols XI. The Structure of Acylated Anthocyanins. Phytochemistry **3**, 151 (1964).

78. — Plant Polyphenols XIV. Characterization of Flavonoidglycosides by Acidic and Enzymic Hydrolyses. Phytochemistry **4**, 107 (1965).

79. — Plant Polyphenols XV. Flavonols as Yellow Flower Pigments. Phytochemistry **4**, 647 (1965).

80. — Plant Polyphenols XVI. The Flavonol Glycosides of Wild and Cultivated Potatoes. Biochem. J. **84**, 100 (1962).

81. HARDEGGER, E., und H. BRAUNSCHWEIGER: Die absolute Konfiguration des Aglukons im Neohesperidin. Helv. Chim. Acta **34**, 1413 (1961).

82. HATTORI, S.: Glycosides of Flavones and Flavonols. In: T. A. GEISSMAN, The Chemistry of Flavonoid Compounds, S. 317. Oxford: Pergamon Press. 1962.

83. HAYNES, L. J.: Naturally Occurring C-Glycosyl Compounds. In: Advances in Carbohydrate Chemistry Vol. 20, S. 357. New York: Academic Press. 1965.

84. HELFERICH, B., und H. BREDERECK: Zuckersynthese VIII. Ann. **465**, 166 (1928).

85. HELFERICH, B., und W. KLEIN: Zur Synthese von Disacchariden IV. Zwei Tetra-acetyl-β-d-glucosen. Ann. **450**, 219 (1926).

86. HELFERICH, B., und J. ZIRNER: Synthese einiger Disaccharide. Chem. Ber. **95**, 2604 (1962).

87. HEMMING, R., and W. D. OLLIS: Flavone Glykosides of Parsley. II. The Structure of Apiin. Chem. and Ind. **1953**, 85.

88. HERRISSET, A., R. PARIS et J. P. CHAUMONT: Flavonoides de Romain camomile (*Anthemis nobilis*), Plant Med. Phytother. **5**, 234 (1971).

89. HILLIS, W. E., and A. CARLE: The Chemistry of Eucalypt Kinos, IV. *Eucalyptus hemiphloia* Kino. Aust. J. Chem. **16**, 147 (1963).

90. Hillis, W. E., and D. H. S. Horn: Nuclear Magnetic Resonance Spectra and Structures of Some C-Glycosylflavonoids. Aust. J. Chem. **18**, 531 (1965).

91. — — The Structure of Keyakinin. Aust. J. Chem. **19**, 705 (1966).

92. Hörhammer, L., L. Farkas, H. Wagner und J. Ostermayer: Isolierung des Luteolin-7-glucosids aus *Achillea millefolium* L. und eine neue Synthese des Glucosids. Acta chim. **40**, 463 (1964).

93. Hörhammer, L., R. Hänsel, G. Kriesmair und W. Endres: Zur Kenntnis der Polygonaceenflavone, I. Über ein mittels α-Glykosidasen spaltbares Quercetinarabinosid aus *Polygonum polystachyum*. Arch. Pharmaz. **268**, 419 (1955).

94. Hörhammer, L., L. Stich und H. Wagner: Über die Flavonglykoside der Lindenblüten, I. Arch. Pharmaz. **294**, 687 (1961).

95. Hörhammer, L., H. Wagner, H. G. Arndt und L. Farkas: Über die Synthese von Isorhamnetin-glykosiden I., Isolierung und Synthese zweier Flavonolglykoside von *Cereus grandiflorus*. Chem. Ber. **99**, 1384 (1966).

96. Hörhammer, L., H. Wagner, H. G. Arndt, R. Dirscherl und L. Farkas: Über die Synthese von Quercetin-3-glykosiden, I. Synthese und Strukturbeweis von Isoquercitrin, Hyperosid und Quercitrin. Chem. Ber. **101**, 450 (1968).

97. Hörhammer, L., H. Wagner, H. G. Arndt, G. Hitzler und L. Farkas: Über die Synthese von Quercetin-3-glykosiden. II. Synthese des Ombuosids und eine rationelle Synthese des Rutins. Chem. Ber. **101**, 1183 (1968).

98. Hörhammer, L., H. Wagner, H. G. Arndt und H. Kraemer: Synthese natürlich vorkommender Polyhydroxy-Flavonol-Glykoside. Tetrahedron Letters **1966**, 567.

99. Hörhammer, L., H. Wagner und K. Beck: Isolierung neuer Flavonololigosaccharide aus den Blüten von *Leucojum vernum* L. und *Galanthus nivalis* L. Z. Naturforsch. **22 b**, 896 (1967).

100. Hörhammer, L., H. Wagner und P. Beyersdorff: Über die Struktur des Scoparins und weitere Vorkommen von C-Glykosiden der Flavon-Reihe. Naturwiss. **48**, 392 (1962).

101. Hörhammer, L., H. Wagner und F. Gloggengiesser: Über einen neuen Glykosidtyp der Flavonreihe. I. Isolierung eines Luteolin- und Apigeninglykosides aus *Polygonum orientale* L. Arch. Pharmaz. **291**, 126 (1958).

102. Hörhammer, L., H. Wagner, G. Hitzler, L. Farkas, A. Wolfner und M. Nógrádi: Synthese von Thalictiin und Azaleatin-3-mono-β-D-galactosid, zwei natürlichen Flavongalactosiden. Chem. Ber. **102**, 792 (1969).

103. Hörhammer, L., H. Wagner, H. Krämer und L. Farkas: Über die Struktur synthetischer und natürlicher Eriodictyol-Glykoside. Tetrahedron Letters **1966**, 5133.

104. Hörhammer, L., H. Wagner und W. Leeb: Über einen neuen Glykosidtyp der Flavonreihe, IV. Adonivernith, ein Luteolin-8-hexosyl-monoxylosid aus *Adonis vernalis* L. Arch. Pharmaz. **293**, 264 (1960).

105. Hörhammer, L., H. Wagner, H. Nieschlag und G. Wildi: Über einen neuen Glykosidtyp der Flavonreihe, III. Zur Konstitution von Orientosid und Orientin. Arch. Pharmaz. **292**, 380 (1959).

106. Hösel, W., und W. Barz: Flavonoide aus *Cicer arietinum* L. Phytochem. **9**, 2053 (1970).

107. Horowitz, R. M.: Relations between the Taste and Structure of some Phenolic Glycosides. In: J. B. Harborne, Biochemistry of Phenolic Compounds. S. 545. London-New York: Academic Press. 1964.

108. Horowitz, R. M., and Gentili, B.: Dihydrochalcone Derivatives and their Use as Sweeting Agents. U. S. Patent. 3,087,82: (1963).

109. — — Flavonoids of Citrus, VI. The Structure of Neohesperidose. Tetrahedron **19**, 773 (1963).

110. — — Structure of Vitexin and Isovitexin. Chem. and Ind. (London) **1964**, 498.

111. — — Long Range Shielding in C-Glycosyl Compounds: Structure of some New C-Glycosylflavones. Chem. and Ind. **1966**, 625.

112. HUDSON, C. S.: Apiose and the Glycosides of the Parsley Plant. Adv. Carbohydrate Chem. **4**, 57 (1949).

113. HULYALKAR, R. K., J. K. N. JONES, and M. B. PERRY: Chemistry of D-Apiose, II. The Configuration of D-Apiose in Apiin. Canad. J. Chem. **73**, 2085 (1965).

114. ICE, C. H., and S. H. WENDER: The Synthesis of Isoquercitrin. J. Amer. chem. Soc. **74**, 4606 (1952).

115. INOUYE, H., Y. AOKI, H. WAGNER, L. HÖRHAMMER, G. AURNHAMMER und W. BUDWEG: Die Flavonglykoside von *Veronicastrum sibiricum* Pennell var. *japonicum* Hara und Synthese des Luteolin-7-neohesperidosids (Veronicastrosid). Chem. Ber. **102**, 3009 (1969).

116. JAIN, A. C., and T. R. SESHADRI: Nuclear Methylation of Flavonoids. J. Sci. Ind. Res. India **14 A**, 227 (1955).

117. JERZMANOWSKA, Z. I., and M. J. MICHALSKA: Synthese von Polyhydroxyflavone Glucosides. Chem. and Ind. (London) **1952**, 1318.

118. JULIAN, E. A., G. JOHNSON, D. K. JOHNSON, and BR. J. DONNELLY: The Glycoflavonoid Pigments of Wheat, *Triticum aestivum* leaves. Phytochemistry **10**, 3185 (1971).

119. JURD, L.: The Selective Alkylation of Polyphenols. II. Methylation of 7-4'- and 3'-Hydroxyl Groups in Flavonols. J. Org. Chem. **27**, 1294 (1962).

120. JURD, L., T. A. GEISSMAN, and M. K. SEIKEL: The Flavonoid Constituents of *Spirodela oligorrhiza*, II. The Flavone Constituents. Arch. Biochem. Biophys. **67**, 284 (1957).

121. KAMIYA, S., S. ESAKI, and M. HAMA: Glycosides and Oligosaccharides in the L-Rhamnose Series, II. Syntheses of Certain α-L-Rhamnosyl-Disaccharides. Agric. Biol. Chem. **31**, 261 (1967).

122. — — — Glycosides and Oligosaccharides in the α-L-Rhamnose Series, IV: Synthesis of Naringin and Neohesperidin. Agric. Biol. Chem. **31**, 402 (1967).

123. KAMIYA, S., E. SACHIKO, and K. FUKUKO: Flavonoids in Citrus and related Genera. IV. Narirutin, Neoeriocitrin and Veronicastroside in Citrus. Agric. Biol. Chem. **36**, 1461 (1972).

124. KARRER, W.: Konstitution und Vorkommen der organischen Pflanzenstoffe. Basel: Birkhäuser. 1958.

125. KARSTEN, U.: Bryophyllosid, ein neues Kämpferolglykosid aus *Bryophyllum daigremontianum* (R. HAMET et PERR.) Berg. Naturwissenschaften **52**, 84 (1965).

126. KAWASE, A., and K. YAGISHITA: Flavonoid of Iridaceae, I. The Structure of a New C-Glycosyl Flavone Embinin, isolated from the petals of *Iris germanica*. Agric. Biol. Chem. (Tokyo) **32**, 537 (1968). Chem. Abstr. **69**, 44173 (1968).

127. KING, H. G. C.: Phenolic Compounds of Commercial Wheat Germ. J. Food Science **27**, 446 (1962).

128. KÖNIGS, W., und E. KNORR: Über einige Derivate des Traubenzuckers und der Galactose. Ber. dtsch. chem. Ges. **34**, 957 (1901).

129. KOEPPEN, B. H.: Some Observations on the Synthesis of Sophorose. Carbohyd. Res. **7**, 410 (1968).

130. — Reaction of Acetates of Neohesperidose, Rutinose and some Flavonoid-L-Rhamnosyl-D-Glucosides with Di-Halomethyl-Methylethers. Carbohyd. Res. **10**, 105 (1969).

131. — Synthesis of Neohesperidose. Tetrahedron **24**, 4963 (1968).

132. KOEPPEN, B. H., C. J. B. SMIT, and D. G. ROUX: The Flavone-C-Glycosides and Flavonol-O-Glycosides of *Aspalathus acuminatus* (Rooibos Tea). Biochem. J. **83**, 507 (1962).

133. KOEPPEN, B. H., and D. G. ROUX: C-Glycosylflavonoids. The Chemistry of Orientin and Iso-Orientin. Biochem. J. **97**, 444 (1965).

134. — — C-Glycosylflavonoids, the Chemistry of Aspalathin. Biochem. J. **99**, 604 (1966).

135. KOMATSU, M., and T. TOMINORI: Studies on the Constituents of *Swertia japonica*:

Isolation and Structure of New Flavonoids, Swertisin and Swertiajaponin. Tetrahedron Letters **1966**, 1611.

136. KRANEN-FIEDLER, U.: Neue Inhaltsstoffe aus Crataegus oxyacantha. Die Flavonderivate. Arzneimittel-Forsch. **5**, 609 (1955).
137. KROSCHEWSKY, J. R., T. J. MABRY, K. R. MARKHAM, and R. E. ALSTON: Flavonoids from the Genus *Tragopogon*. Phytochemistry **8**, 1495 (1969).
138. LEBRETON, P., et M. P. BOUCHEZ: Sur un nouvel heteroside naturel de la quercétine, extrait de *Nymphoides peltata* (Gmel.) Kuntze. C. R. Acad. Sci. Ser. D **268**, 1661 (1969).
139. LIEBERMANN, C., und O. HÖRMANN: Über das Glykosid der Gelbbeeren und den Rhamnodulcit. Ber. dtsch. chem. Ges. **11**, 952 (1878).
140. MABRY, T. J., H. YOSHIOKA, S. SUTHERLAND, S. WOODLAND, W. RAHMAN, M. ILYAS, I. N. USMANI, N. HAMEED, J. CHOPIN, and M. L. BOUILLANT: New C-Glycosylflavones from *Phlox Drummondii*. Phytochemistry **10**, 677 (1971).
141. McCLOSKEY, CH. M., and G. H. COLEMAN: The Preparation of β-Primverose, Heptaacetate and β-Vicianose Heptaacetate. J. Amer. Chem. Soc. **65**, 1778 (1943).
142. McCLURE, J. W., and K. G. WILSON: Photocontrol of C-Glycosylflavones in Barley seedlings. Phytochemistry **9**, 763 (1970).
143. MAHESH, V. B., and T. S. SESHADRI: Iodine Oxidation (dehydrogenation) of Hydroxyflavanones to Hydroxyflavones. J. Sci. Ind. Res. **14 B**, 608 (1955).
144. MAKSYUTINA, N. P., und V. I. LITVINENKO: Robinin. Dopov. Acad. Nauk. UKr. RSR ser. B 29 (5) 443 (1967). [Chem. Abstr. **68**, 49983 (1968)].
145. MALHOTRA, A., V. V. S. MURTI, and T. R. SESHADRI: Lanceolarin, a New Isoflavone Glycoside of *Dalbergia Lanceolaria*. Tetrahedron **1967**, 405.
146. MARKHAM, K. R.: A Novel Flavone-Polysaccharide Compound From *Monoclea Forsteri*. Phytochemistry, **11**, 2047 (1972).
147. MARKHAM, K. R., and L. J. PORTER: Flavonoids in the Green Algae (Chlorophyta). Phytochemistry **8**, 1777 (1969).
148. MARSH, C. A.: Glucuronide Metabolism in Plants. 2. The Isolation of Flavoneglucosiduronic acids from Plants. Biochem. J. **59**, 58 (1955).
149. MIZELLE, J., W. DUNLAP, and S. WENDER: Isolation and Identification of two Isomeric Naringenin Rhamnodiglucosides from Grapefruit. Phytochemistry **6**, 1305 (1967).
150. MURAKAMI, T., Y. NISHIKAWA, and T. ANDO: Constituents of Japanese and Chinese Crude Drugs, IV. Constituents of *Pueraria* root. Chem. Pharm. Bull. (Tokyo) **8**, 688 (1960).
151. NAKAOKI, T., and N. MORITA: Components of the Leaves of *Polygonum reynoutria* and *Polygonum sachalinense*. J. Pharm. Soc. Japan **76**, 323 (1956).
152. NARASIMHACHARI, N., and T. R. SESHADRI: A New Synthesis of Flavones. Proc. Indian Acad. Sci. **30 A**, 151 (1949).
153. NÓGRÁDI, M., L. FARKAS und V. OLECHNOWICZ-STEPIEN: Die Synthese des Quercetin-3-β-rutinosid-7-β-D-glucosids. Chem. Ber. **103**, 3414 (1970).
154. — — — Synthese von Flavonoid-bis-glykosiden. IV: Die Synthese von Quercetin-3-α-L-rhamnopyranosid-7-β-D-glucopyranosid und Quercetin-3-β-D-xylopyranosid-7-β-D-glucopyranosid, zwei Flavon-bis-glycosiden aus *Evonymus*-Arten. Chem. Ber. **104**, 3618 (1971).
155. NÓGRÁDI, M., L. FARKAS, H. WAGNER und L. HÖRHAMMER: Transacylierungsreaktion in der Flavonoid-Reihe, II: Eine neue Synthese des Cosmosiins und des Tilianins. Chem. Ber. **100**, 2783 (1967).
156. OELRICHS, P. J., T. B. MARHALL, and D. H. WILLIAMS: 7-O-β-D-Glucosyl-8-C-D-glucosyl-4'-O-methylapigenin, A New Flavone from Trema aspera. J. Chem. Soc. (London) C. **1968**, 941.

157. OGISO, A., A. SATO, S. SATO, and CH. TARNURA: Novel Macrocyclic Flavonoid Glycosides, Toxic Components from *Leucothoe keiskei*. Tetrahedron Letters **1972**, 3071.

158. OHTA, T.: Über ein Glykosid der *Polygonum aviculare* L. var. *buxifolium* Ledeb. Hoppe Seyler's Z. physiol. Chem. **263**, 221 (1940).

159. OHTA, T., and T. MIYAZAKI: Fenicularin, a Quercetin-3-arabinoside from the Leaves of *Foeniculum vulgare*. J. Pharm. Soc. Japan, **79**, 986 (1959).

160. OHTA, N., and K. TAGISHITA: Isolation and Structure of New Flavonoids, Flavoyadorinin B and Homoflavoyadorinin B in the Leaves of *Viscum album varcoloratum* epiphyting to *Pyrus communis* (pear). Agr. Biol. Chem. (Japan) **34**, 900 (1970). [Chem. Abstr. **74**, 983 a (1971)].

161. OKIGAWA, M., H. HATANAKE, and N. KAWANO: A New Glycoside, Acacetin-7-glucuro-1→2-glucuronide from the Leaves of *Chlerodendron trichotomum*. Tetrahedron **33**, 2935 (1970).

162. OYAMADA, T.: Eine neue allgemeine Methode zur Synthese von Derivaten des Flavonols. Bull. Chem. Soc. Japan **10**, 182 (1935).

163. PACHECO, H., et A. GROUILLER: Conversion de β-D-glucosyloxy-4'-hydroxy-2'-chalcones en β-D-glucosyloxy-flavonols. Bull. soc. chim. France **1966**, 3212.

164. PACHECO, A., A. GROUILLER et A. HOURFAR: Chimie des composés flavoniques, II: Synthese de di-(β-D-glucopyranosyloxy)-5,7-flavanones et de β-D-glucopyranosyloxy-7, hydroxy-5, flavanones (1,2). Bull. soc. chim. France **1965**, 2937.

165. PARIS, R. R.: Sur la cytoside flavonoide des fleurs et de feuilles de *Cytisus laburnum* L. C. R. Acad. Sci. **245**, 443 (1957).

166. PARIS, R. R., et P. DELAVEAU: Isolement d'un nouvel heteroside flavonique. L' „alliaroside" de feuilles de *l'Alliaria officinalis* Andrz. (Cruciferes). C. R. Acad. Sci. **254**, 928 (1962).

167. PARIS, R. R., et M. QUIRIN: Sur le catharticoside, heteroside flavonique des fruits du Nerprun (*Rhamnus cathartica* L.) C. R. Acad. Sci. **256**, 2448 (1960).

168. PARIS, R. R., et A. STAMBOULI: Sur quelques flavonosides difficilement hydrolysai .es (scoparoside, cytisioside et aphloioside). C. R. Acad. Sci. **252**, 1659 (1961).

169. PLOUVIER, V.: Sur trois C-glycosylflavonoides nouveaux. C. R. Acad. Sci. Ser. D **265**, 516 (1967).

170. — Structure de trois flavonoids, isosarotanoside de Cytisus purgans, catharticoside de *Rhamnus utilus* et rhodotyposide de *Rhodotypos kerrioides*. C. R. Acad. Sci. Ser. D **265**, 2120 (1967).

171. PRIDHAM, J. B.: Phenol-Carbohydrate Derivatives in Higher Plants. In: M. WOLFROM and R. STUART TIPSON, Advances in Carbohydrate Chemistry Vol. 20, S. 371. New York-London: Academic Press. 1965.

172. PROX, A.: Massenspektrometrische Untersuchung einiger natürlicher C-Glucosyl-Verbindungen. Tetrahedron **24**, 3697 (1968).

173. RABATÉ, J.: Etude du sophorose, Bull. soc. chim. France **1940**, 565.

174. REICHEL, L., und W. REICHWALD: Über die Farbstoffe der schwarzen Hollunderbeere. Naturwiss. **47**, 40 (1960).

175. RIBAS, J., D. LOZANO, F. REFOJO, and S. PEREIRA: An Investigation of the Glucoside isolated from the Leaves of *Sarothamnus commutatus* from Galicia. An. Real. Soc. espan. Fisica. Quim. Sev. B **52**, 271 (1956) Chem. Abstr. **51**, 386 c (1957).

176. ROESLER, H., T. J. MABRY, and J. KAGAN: Sphaerobiosid, ein Isoflavonglykosid aus *Baptisia sphaerocarpa*. Chem. Ber. **98**, 2193 (1965).

177. SAMOKHVALOV, G. T., M. K. SHAKHOVA und N. A. PREOBRAZHENSKII: Synthese von Rutin. Pokl. Acad. Nauk. S.S.S.R. **123**, 305 (1958).

178. SCHMIDT, R. D., P. VARENNE, and R. PARIS: Mass Spectrometry of Flavonoid Trisaccharides. The Structure of Xanthorhamnin, Alaternin and Catharticin. Tetrahedron **28**, 5037 (1972).

214 H. WAGNER:

179. SCHÖNSIEGEL, I., K. EGGER und M. KEIL: Acylierte Kämpferolglykoside in *Narcissus poeticus* L. Z. Naturforsch. **24 b**, 1213 (1969).

180. SEIKEL, M. K., and A. J. BUSHNELL: The Flavonoid Constituents of Barley *(Hordeum vulgare)* II. Lutonarin. J. Organ. Chem. **24**, 1995 (1959).

181. SEIKEL, M. K., J. H. S. CHOW, and L. FELDMAN: The Glycoflavonoid Pigments of *Vitex lucens* Wood. Phytochemistry **5**, 439 (1966).

182. SEIKEL, M. K., and T. A. GEISSMAN: The Flavonoid Constituents of Barley *(Hordeum vulgare)* I. Saponarin. Arch. Biochem. Biophys. **71**, 17 (1957).

183. SEIKEL, M. K., and T. J. MABRY: A New Type of Glycoflavonoid from *Vitex lucens*. Tetrahedron Letters **1965**, 1105.

184. SESHADRI, T. R., and S. VYDEESWARAN: Chemical Examination of *Luffa Echinata*. Phytochem. **10**, 667 (1971).

185. — — Chrysoeriol-Glycosides and other Flavonoids of *Rungia repens* Flowers. Phytochem. **11**, 803 (1972).

186. SHALABY, A. F., K. TSINGARIDAS und E. STEINEGGER: Zur Kenntnis der Flavonoide von *Achillea fragrantissima* (Forsk) Sch. Bip. Pharm. Act. Helv. **40**, 19 (1965).

187. SHIMOKORIYAMA, M., and S. HATTORI: Anthochlor Pigments of *Cosmos sulfureus*, *Coreopsis lanceolata* and *C. saxicola*. J. Amer. Chem. Soc. **75**, 1900 (1953).

188. SIMPSON, T. H., and J. L. BETON: Anthoxanthins, I. Selective Methylation and De-methylation, J. Chem. Soc. (London) **1954**, 4065.

189. SINN, M.: Thesis. Würzburg 1968.

190. SOSA, F., et F. PERCHERON: Le Kempferol-3-O-rhamnodiglucoside, Nouvel Heteroside du pollen de *Populus yunnanensis*. Phytochem. **9**, 441 (1970).

191. TAMURA, S., C. F. CHANG, A. SUZUKI, and S. KUMAI: Chemical Studies on "Clover Sickness", I. Isolation and Structural Elucidation of Two New Isoflavonoids in Red Clover. Agric. Biol. Chem. **33**, 391 (1969).

192. TARUSOVA, N., and N. A. PREOBRAZHENSKII: Studies in the Field of the Group P Vitamins. VI: Synthesis of 4'-Benzyloxy-3',5,7-trihydroxy-flavanone. Biol. Activn. Soedin. Akad. Nauk. S. S. S. R., 210 (1965).

193. TARUSOVA, N. T., A. N. VETVOV, and N. A. PREOBRAZHENSKII: Synthetic Studies of Flavonoids, V. Synthesis of Eryodictiol-7-β-D-glucoside and its Methylether. Zh. Obsheh. Khim. **34**, 3300 (1964).

194. TEOULE, R., J. CHOPIN et C. MENTZER: Nouvelle methode de synthèse des glucosides flavoniques. C. R. Acad. Sci. **250**, 3669 (1960).

195. THIEME, H.: Über die Flavonglykoside der Blätter von *Salix repens* L., Isolierung und Konstitutionsaufklärung des Caesiosids. Tetrahedron Letters **1968**, 2781.

196. THIES, P. W.: Linearin-isovalerianat, ein bisher unbekanntes Flavonoid aus *Valeriana wallichii* D. C. Planta medica **16**, 361 (1968).

197. WAGNER, H.: Flavonoid-C-Glycosides in Comparative Phytochemistry, S. 309 I. New York: Academic Press. 1966.

198. WAGNER, H., G. AURNHAMMER, H. DANNINGER, O. SELIGMANN, L. PALLOS und L. FARKAS: Synthese von Glucuroniden der Flavonoid-Reihe. IV: Synthese von Chrysin-7-β-D-glucopyranuronid, -7-β-D-neohesperidosid und 7-β-D-rutinosid. Chem. Ber. **105**, 257 (1972).

199. WAGNER, H., G. AURNHAMMER und L. HÖRHAMMER: Endgültige Konstitutions-aufklärung und Synthese von Narirutin, Didymin, Rhoifolin, Poncirin und Fortunellin. Tetrahedron Letters **1968**, 1635.

200. WAGNER, H., G. AURNHAMMER, L. HÖRHAMMER und L. FARKAS: Synthese und Strukturbeweis von Narirutin, einem 5,7,4'-Trihydroxy-flavanon-7-rutinosid aus *Citrus sinensis* (L.) Osb. Chem. Ber. **102**, 2089 (1969).

201. — — — — Endgültige Konstitutionsaufklärung und Synthese der Flavonglykoside Fortunellin, Rhoifolin und Isorhoifolin. Chem. Ber. **102**, 2083 (1969).

202. WAGNER, H., G. AURNHAMMER, L. HÖRHAMMER, L. FARKAS und M. NÓGRÁDI:

Synthese von Poncirin, einem Isosakuranetin-7-β-neohesperidosid aus *Poncirus trifoliata* Raf. Chem. Ber. **102**, 785 (1969).

203. WAGNER, H., W. BUDWEG, L. HÖRHAMMER, B. VERMES und L. FARKAS: Transacylierungsreaktionen in der Flavonoid-Reihe, VI. Synthese von Luteolin-7-β-rutinosid (Scolymosid) sowie 3',4'-Di-O-methyl-luteolin-7-β-neohesperidosid und 7-β-rutinosid. Chem. Ber. **104**, 2118 (1971).

204. WAGNER, H., W. BUDWEG, M. A. IYENGAR, O. VOLK und M. SINN: Linosid A und B, zwei neue Flavon-C-glycoside aus Linum maritimum L. Z. Naturforsch. **27 b**, 809 (1972).

205. WAGNER, H., und V. M. CHARI: Synthese von Flavonol-3-O-xylosyl-glucosiden. I. Struktur von Leucosid aus *Leucojum vernum*. Chem. Ber. (im Druck).

206. WAGNER, H., H. DANNINGER, M. A. IYENGAR, O. SELIGMANN, L. Farkas, S. S. SUBRAMANIAN und A. G. R. NAIR: Synthese von Glucuroniden der Flavonoid-Reihe. III. Isolierung von Apigenin-7-β-D-glucuronid aus *Ruellia tuberosa* und seine Synthese. Chem. Ber. **104**, 2681 (1971).

207. WAGNER, H., H. DANNINGER, O. SELIGMANN und L. FARKAS: Synthese von Glucuroniden der Flavonoid-Reihe. I. Erste Synthese eines natürlich vorkommenden Flavonoidglucuronids (Quercetin-3-β-D-glucuronid). Chem. Ber. **103**, 3674 (1970).

208. WAGNER, H., H. DANNINGER, O. SELIGMANN, M. NOGRADI, L. FARKAS und N. FARNSWORTH: Synthese von Glucuroniden der Flavonoid-Reihe. II. Isolierung von Kämpferol-3-β-D-glucuronid aus *Euphorbia esula* L. und seine Synthese. Chem. Ber. **103**, 3678 (1970).

209. WAGNER, H., L. HÖRHAMMER, G. AURNHAMMER und L. FARKAS: Strukturaufklärung und Synthese des Didymins, eines Isosakuranetin-7-β-rutinosids aus *Monarda didyma* L. Chem. Ber. **101**, 445 (1968).

210. WAGNER, H., L. HÖRHAMMER, R. DIRSCHERL, G. HITZLER, L. FARKAS und M. NÓGRÁDI: Über die Synthese von Quercetin-3-glykosiden, IV. Synthese des Quercetin- und Kämpferol-3-β-gentiobiosids. Chem. Ber. **101**, 3419 (1968).

211. WAGNER, H., L. HÖRHAMMER, R. DIRSCHERL, L. FARKAS und M. NÓGRÁDI: Über die Synthese von Quercetin-3-glykosiden. III. Synthese des Quercetin-3- und Kämpferol-3-β-sophorosids. Chem. Ber. **101**, 1186 (1968).

212. WAGNER, H., L. ROSPRIM und P. DÜLL: Die Flavon-C-glykoside von *Viola tricolor* L. Z. Naturforsch. **27 b**, 954 (1972).

213. WAGNER. J.: Über die Inhaltsstoffe des Roßkastaniensamens. III. Untersuchungen an Flavonolen. Hoppe Seyler's Z. physiol. Chem. **335**, 232 (1964).

214. WALLACE, J. W., and T. J. MABRY: The Conversion of the 8-C-Glycosylflavone Vitexin to the 6-Isomer, Isovitexin in *Lemna minor*. Phytochem. **9**, 2133 (1970).

215. WALLACE, J. W., T. J. MABRY, and R. E. ALSTON: On the Biogenesis of Flavone-O-glycosides and C-glycosides in the Lemnaceae. Phytochem. **8**, 93 (1969).

216. WESSELY, F., und G. H. MOSER: Synthese und Konstitution des Scutellareins. Monatshefte Chem. **56**, 97 (1930).

217. WONG, E.: Structural and Biogenetic Relationships of Isoflavonoids. In: Fortschritte der Chemie organischer Naturstoffe, Bd. 28, S. 1. Wien-New York: Springer-Verlag. 1970.

218. ZEMPLÉN, G., und R. BOGNÁR: Einwirkung von Quecksilbersalzen auf Acetohalogenzucker. XII. Eine neue, ausgiebige Synthese der Primverose-Derivate und der Primverose. Ber. dtsch. chem. Ges. **72**, 47 (1939).

219. — — Endgültige Konstitutionsaufklärung des Robinins. Ber. dtsch. chem. Ges. **74**, 1783 (1941).

220. — — Umwandlung des Hesperetins in Diosmetin, des Hesperidins in Diosmin und des Isosakuranetins in Acacetin. Ber. dtsch. chem. Ges. **76**, 452 (1943).

221. — — Synthese des Hesperidins. Ber. dtsch. chem. Ges. **76**, 773 (1943).

222. ZEMPLÉN, G., R. BOGNÁR und I. SZÉKELY: Synthese des Salipurposids und des Isosalipurposids. Ber. dtsch. chem. Ges. **76**, 386 (1943).
223. ZEMPLÉN, G., R. BOGNÁR und L. SZEZÖ: Synthese des Eriodictyols und des 3-Oxy-p-phlorrhizins. Ber. dtsch. chem. Ges. **76**, 1112 (1943).
224. ZEMPLÉN, G., Z. CRÜRÖS, A. GEVERZ und ST. ACZEL: Beiträge zur Kenntnis des Phlorrhizins und des Quercitrins. Ber. dtsch. chem. Ges. **61**, 2486 (1928).
225. ZEMPLÉN, G., und L. FARKAS: Synthese des Genistins. Ber. dtsch. chem. Ges. **76**, 1110 (1943).
226. ZEMPLÉN, G., und A. GERECS: Einwirkung von Quecksilbersalzen auf Aceto-halogen-zucker. IX. Synthese von Derivaten der β-1-L-Rhamnosido-6-D-glykose. Ber. dtsch. chem. Ges. **67**, 2049 (1934).
227. — — Über Robinobiose und Kämpferol-rhamnosid. Ber. dtsch. chem. Ges. **68**, 2054 (1935).
228. — — Konstitution und Synthese der Rutinose, der Biose des Rutins. Ber. dtsch. chem. Ges. **68**, 1318 (1935).
229. — — Synthese des Lusitanicosids (Chavicol-β-rutinosids) des Glykosids aus *Cerasus lusitanica* Lois. Ber. dtsch. chem. Ges. **70**, 1099 (1937).
230. ZEMPLÉN, G., A. GERECS und H. FLESCH: Einwirkung von Quecksilbersalzen auf Aceto-halogen-zucker. XI. Synthese einiger Derivate der β-1-L-Rhamnosido-6-D-galactose. Ber. dtsch. chem. Ges. **71**, 774 (1938).
231. ZEMPLÉN, G., und L. MESTER: Synthese des Apigeninglykosids-(5) der *Amorpha fruticosa* L. Chem. Ber. **76**, 776 (1943).
232. ZEMPLÉN, G., und A. K. TETTAMANTI: Über die Biose des Hesperidins und des Neohesperidins. Ber. dtsch. chem. Ges. **71 B**, 2511 (1938).

(Eingelaufen am 18. März 1973)

Biogenetic-Type Syntheses of Polyketide Metabolites

By Th. M. Harris, C. M. Harris, and K. B. Hindley,
Nashville, Tennessee, USA

Contents

Acknowledgement. The authors acknowledge the generous support by the United States Public Health Service (Grant No. GM-12848).

I. Introduction

The term "polyketide" was coined in 1907 by Collie (*37*) and later defined more explicitly by Birch (*13*) to describe the aromatic natural products which are formed in nature from acetic acid *via* β-polycarbonyl intermediates. Biosynthetically, the polyketides arise by the acyl poly-malonate route, which also produces the fatty acids. The latter are formed from linear arrays of acetate units but not *via* polycarbonyl intermediates.

Biogenetic-type syntheses are preparative procedures which are modelled upon proven or speculative biosynthetic pathways. In practice, the starting materials employed in these reactions are not always identical with the natural ones and non-physiological conditions are often required. Moreover, the products are not always identical with the natural products, but may merely contain some of their important structural features.

If a valid model is chosen, the biogenetic-type synthesis can provide information about the biosynthetic processes that would be difficult to obtain by direct study of the latter. However, one must be careful not to overvalue this information because the models can never be completely accurate representations of the enzymic processes. Thus, these models can never *prove* anything, but they will often provide insight and sometimes suggest decisive biological experiments. Many biogenetic-type syntheses are undertaken for their preparative value alone, perhaps on the assumption that in natural processes enzymes facilitate reactions which are already favored chemically. In some cases this approach has been very successful.

Biogenetic-type syntheses of natural products were reviewed by van Tamelen in an earlier volume in this series (*162*). That review antedates most of the studies in the polyketide series. However, the reader's attention is directed to its introduction which contains an excellent essay on the requirements, uses, and limitations of biogenetic-type syntheses.

Studies of the biogenetic-type synthesis of polyketide metabolites as well as of their biosynthesis, have been hampered by the inaccessibility and instability of the appropriate β-polycarbonyl compounds. Recent work has provided access to polycarbonyl compounds having part or all of the carbonyl groups protected with masking groups. More significant, however, have been the syntheses of unprotected β-poly-ketones and β-polyketo acids and the finding that the free polycarbonyl compounds are sufficiently stable in many cases to permit isolation and characterization. Cyclizations of both the protected and the free poly-carbonyl compounds have given phenols identical to or resembling natural products. In some cases these syntheses have been accomplished with a high degree of specificity. Hopefully, the progress that has been made with biogenetic-type syntheses with these compounds can eventually be matched in the biosynthetic studies.

The synthetic approaches to both protected and unprotected poly-carbonyl compounds are surveyed in this review, as well as the use of these compounds in biogenetic-type syntheses of phenols. It should be noted that few, if any, of the syntheses of polycarbonyl compounds can be construed to be biogenetically modelled. The synthesis of these precursors has, on the whole, been more difficult than their trans-formation to aromatic products and consequently has dominated the efforts of most of the workers in this field.

This review will be limited to consideration of the formation of phenolic compounds (and to a lesser extent pyrone derivatives), specifically excluding models of subsequent transformations of the aromatic metabolites, unless these reactions have led to additional aromatic rings. Oxidations, alkylations and other electrophilic reactions of the aromatic rings will not be reviewed, although a number of inter-esting models of these biosynthetic processes have been investigated. The literature published through late 1972 has been surveyed.

II. Early Studies

In 1866 GEUTHER observed the formation of dehydroacetic acid by self-condensation of ethyl acetoacetate (65). Controversy arose con-cerning its structure with FEIST (57) advocating pyrone (1) and COLLIE (35) pyrone (2). The disagreement was ultimately resolved in favor of (1) in 1924 by RASSWEILER and ADAMS (138), but prior to that date COLLIE carried out extensive studies of the chemistry of dehydroacetic acid in an attempt to establish his structure. From this unsuccessful venture, Collie obtained information about the reactivity of polycarbonyl compounds, providing the basis for the polyketide hypothesis of biogenesis, which he announced in 1907 (*37, 38*).

Collie noted that treatment of dehydroacetic acid with hot, concentrated sodium hydroxide gave orcinol (**3**) but that treatment with

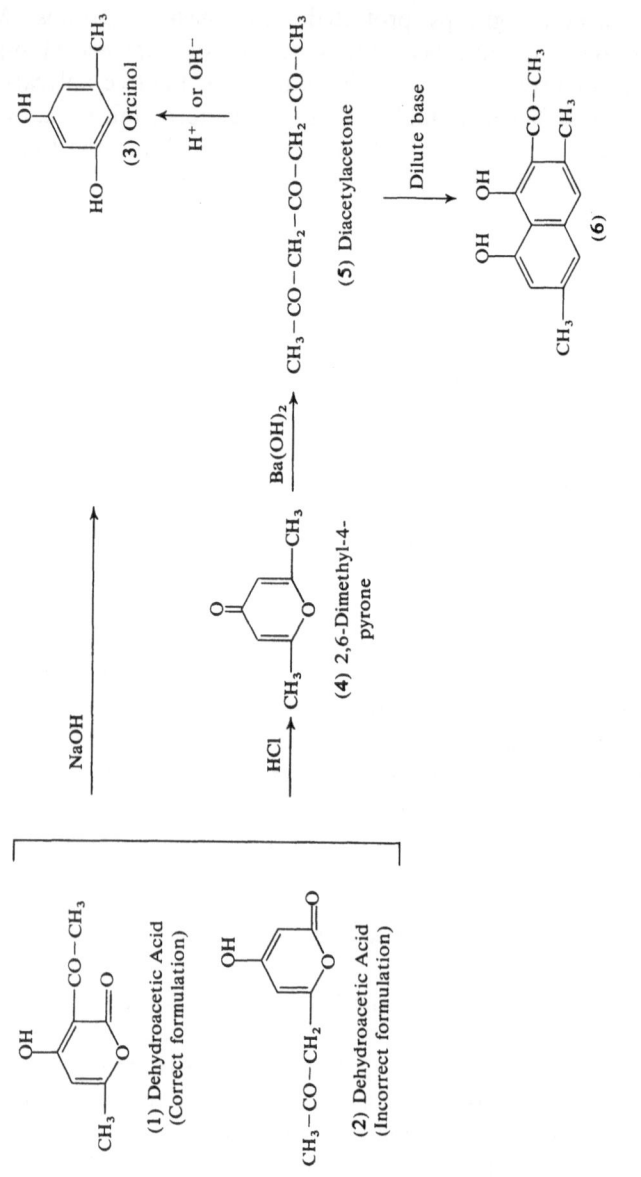

Chart 1. Transformations of dehydroacetic acid

hot, concentrated hydrochloric acid gave 2,6-dimethyl-4-pyrone (**4**) (*39, 57*). Treatment with barium hydroxide transformed the latter into diacetylacetone (**5**), which also gave orcinol in acid or strong base. In dilute base diacetylacetone underwent a complex dimeric condensation to give naphthalenediol (**6**) (Chart 1).

COLLIE was aware that orcinol (**3**) is a natural product and that other transformation products of dehydroacetic acid, including naphthalenediol (**6**), resemble natural products. He visualized the natural compounds as polymers of ketene, hence "polyketides", and proposed that they were derived from multiple condensations of acetic acid or a similar C_2 moiety (*38*). Self-condensation of the basic building block would give β-polyketone intermediates, from which aromatic metabolites would arise by cyclizations (Chart 2).

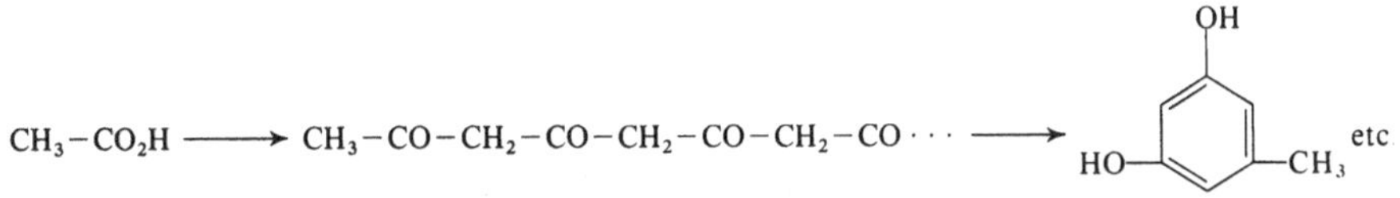

$$CH_3-CO_2H \longrightarrow CH_3-CO-CH_2-CO-CH_2-CO-CH_2-CO\cdots \longrightarrow$$

(3) Orcinol

Chart 2. COLLIE's polyketide hypothesis

The limitation of COLLIE's hypothesis is that it was based solely on structural arguments, metabolic information being unavailable. A. W. STEWART, a student of COLLIE, provided a detailed description of COLLIE's conceptualization of biogenesis in the textbook, "Recent Advances in Organic Chemistry" (*157*). Unfortunately, the book emphasized the erroneous proposals that carbohydrates and polycarbonyl compounds might be directly interconvertible and that isoquinoline alkaloids could be of polyketide origin. Despite the promising start, interest in polyketide biogenesis waned for lack of experimental methods to test these theories.

In 1948 ROBINSON re-opened the question, proposing a biosynthetic scheme for formation of orsellinic acid (**7**) from two molecules of acetoacetic acid and suggesting that, in lichens, the latter is probably derived from erythrose (*141*). Following the same line of reasoning, he suggested that dehydration of a hexose could give triacetic acid and further dehydration could give phloroglucinol (**8**) (Chart 3). Although these comments re-awakened interest in the polyketide hypothesis, they were, as before, based on structural considerations rather than metabolic information.

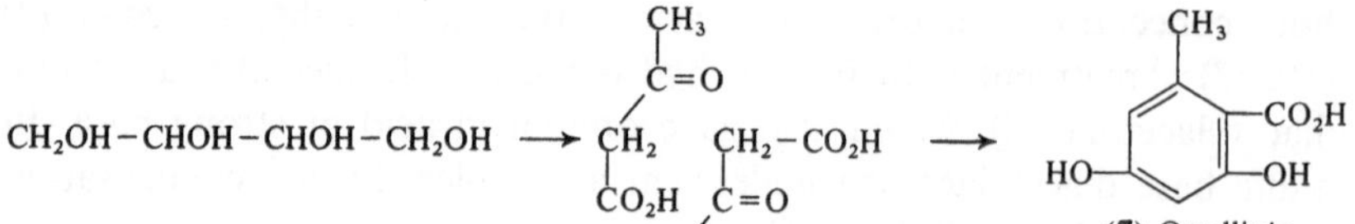

$$CH_2OH-CHOH-CHOH-CH_2OH \longrightarrow$$

$$\begin{matrix} & CH_3 \\ & | \\ & C=O \\ & | \\ CH_2 & CH_2-CO_2H \\ | & | \\ CO_2H & C=O \\ & | \\ & CH_3 \end{matrix} \longrightarrow$$

(7) Orsellinic Acid

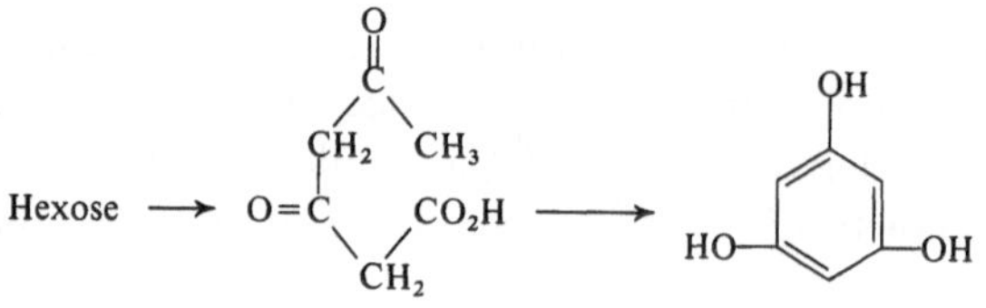

$$Hexose \longrightarrow$$

$$\begin{matrix} & O \\ & || \\ & C \\ & / \ \\ CH_2 & CH_3 \\ | & \\ O=C & CO_2H \\ \ & / \\ & CH_2 \end{matrix} \longrightarrow$$

(8) Phloroglucinol

Chart 3. ROBINSON's polyketide hypothesis

III. Modern Concepts of Polyketide Biosynthesis

The modern era can be said to have begun in 1953 when BIRCH and DONOVAN restated COLLIE's early ideas about the biosynthesis of phenolic compounds, placing the polyketide pathways in accord with new biochemical data (15). Investigations by RITTENBERG and BLOCH using isotopic tracers had shown that fatty acids are derived from multiple molecules of acetic acid coupled together in a head-to-tail fashion (140). BIRCH and DONOVAN suggested a possible relationship between the biosynthesis of the fatty acids and the aromatic metabolites. The key difference in the two pathways would be that the keto groups along the chain are reduced in the formation of fatty acids but are retained and undergo intramolecular condensations in the formation of aromatic metabolites. Continuing the analogy with fatty acid biosynthesis, they proposed that acetyl coenzyme A, rather than free acetic acid, is the C_2 moiety undergoing condensations (114. 153).

A review by BIRCH, appearing in this series in 1957, expounded this theory more fully and provided numerous examples of the manner in which it could be used to explain the formation of aromatic natural products (12). An initial Claisen-type condensation between acetate units (i.e., acetyl coenzyme A) would give acetoacetate, further condensations with acetate giving in sequence linear triacetic acid, tetra-acetic acid, penta-acetic acid, etc. At the triketo acid stage four cyclizations could occur, two giving phenols and the others pyrones (Chart 4). Aldol closure between positions 2 and 7 followed by dehydration would give orsellinic acid (**7**), while Claisen-type closure between positions 6 and 1 would give acetylphloroglucinol (**9**), methyl ethers of which are known metabolites (*148, 150*). Heterocyclic closures could give either tetra-acetic lactone (**10**), which is a 2-pyrone, or 4-pyrone (**11**). Neither of the pyrones was known to BIRCH, but (**10**) and decarboxylated (**11**) have since been obtained from fungal sources (*10, 61*).

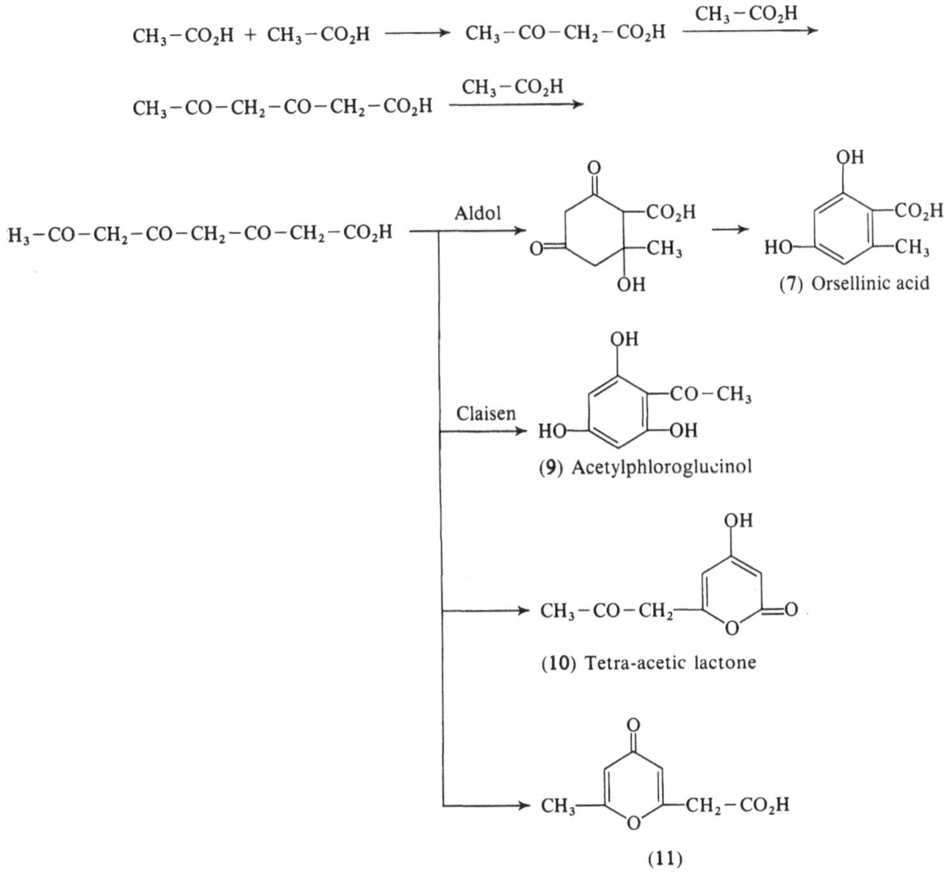

$$CH_3-CO_2H + CH_3-CO_2H \longrightarrow CH_3-CO-CH_2-CO_2H \xrightarrow{CH_3-CO_2H}$$

$$CH_3-CO-CH_2-CO-CH_2-CO_2H \xrightarrow{CH_3-CO_2H}$$

Chart 4. BIRCH's polyketide hypothesis

An increasing number and variety of cyclization products can arise from the higher polyketo acids. For example, a tetraketo acid can, at least in theory, undergo seven primary cyclizations, a pentaketo acid ten, a hexaketo acid thirteen, etc. With compounds containing five or more carbonyl groups, some of the initial cyclization products can undergo secondary ring closures to give polycyclic compounds. Among the many

cyclization products of hepta-acetic acid are the well-known metabolites, alternariol (12), lichexanthone (13), griseofulvin (14), rubrofusarin (15) (158) and flavasperone (16) (27), as well as the more
obscure ones, 6-hydroxymusizin (17) (24) and barakol (18) (28)
(Chart 5)*.

$$CH_3-CO-CH_2-CO-CH_2-CO-CH_2-CO-CH_2-CO-CH_2-CO-CH_2-CO_2H \longrightarrow$$

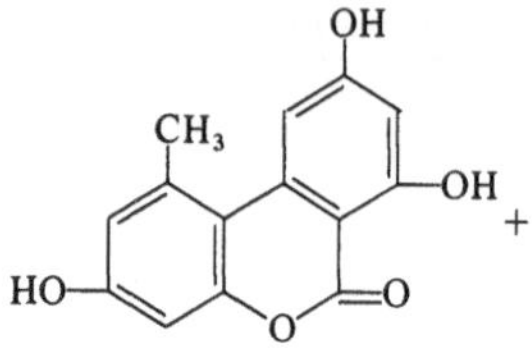

(12) Alternariol 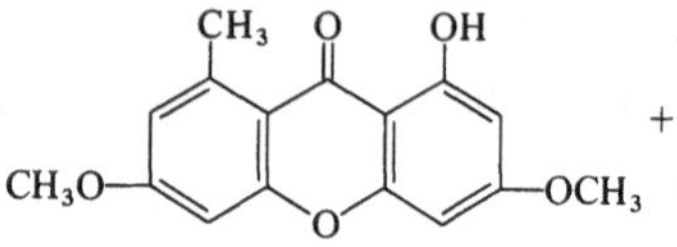   (13) Lichexanthone 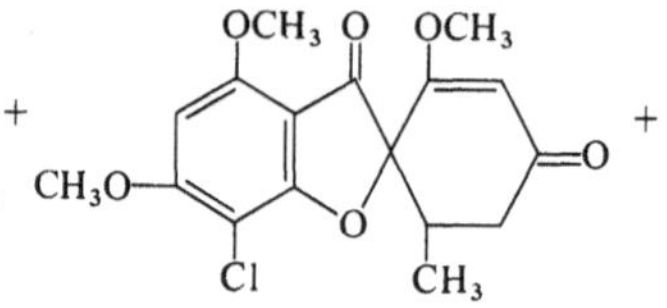   (14) Griseofulvin

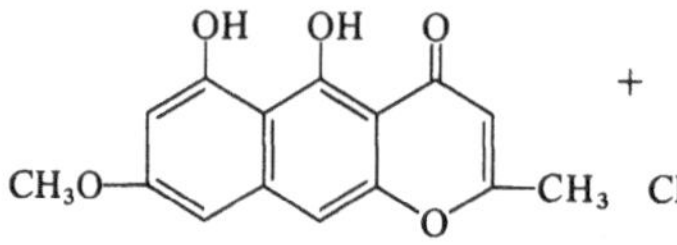

(15) Rubrofusarin 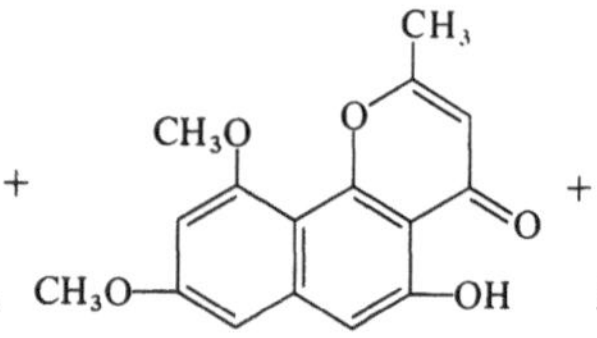   (16) Flavasperone 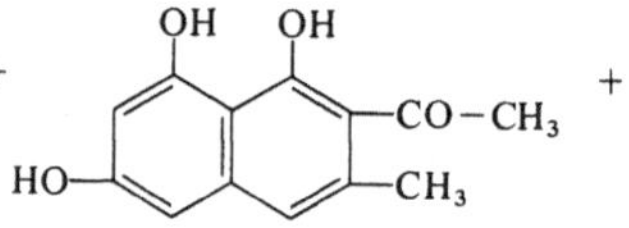   (17) 6-Hydroxymusizin

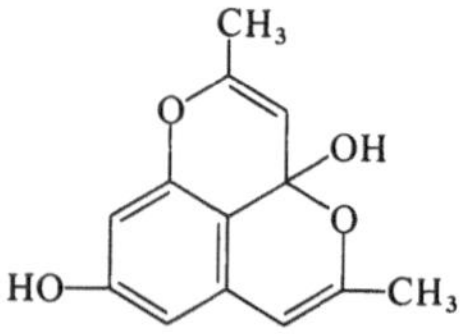

(18) Barakol

Chart 5. Natural products derived from hepta-acetic acid

In order to account for many of the natural products of probable
polyketide origin, BIRCH found it necessary to make several additional
hypotheses. Other carboxylic acids must sometimes serve as initiators
of polyketo acid chains; both aliphatic and aromatic acids could serve
in this capacity. Chain extension by acetate (more accurately, malonate)
is the rule with aromatic metabolites. This is not always true with

---

* The formation of some of these metabolites requires additional modifications of
the aromatic nuclei.

fatty acids and macrolides, but polyketone intermediates may not be involved in these cases.

Deletion of one or more of the aromatic hydroxyl groups is not uncommon; examples range from 6-methylsalicylic acid (19) to pretetramid (20). BIRCH proposed that deoxygenation occurred by reduction prior to aromatization. The alternate proposition appears to be less likely, at least with monocyclic compounds, in view of the difficulty found in hydrogenolysis of phenolic hydroxyl groups in the laboratory.

Extra oxygens are often found in metabolites, anthraquinones like endocrocin (21) representing important examples of this. Oxidations are viewed as normally occurring after aromatization. In some cases, as with patulin (22), the oxidative processes cause ring cleavage of the initially formed phenols, obscuring the polyketide origin of the final products. In other cases, oxidative coupling occurs; griseofulvin (14) is an intramolecular example of this phenomenon.

Other irregularities in the structures of aromatic metabolites include the presence of alkyl substituents. These include both C- and O-derivatives and involve methyl groups, alkyl groups of mevalonate origin and carbohydrates. Introduction of the alkyl groups could occur either at the polyketide stage or after aromatization. Metabolites (13), (14) and (16) contain extra methyl groups on oxygen. Chlorination, as in (14), is also observed in a few instances. Electrophilic mechanisms for introduction of halogen are readily accomodated.

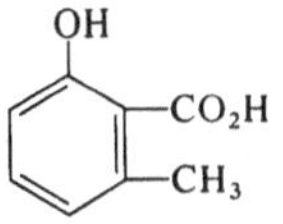

(19) 6-Methylsalicylic acid 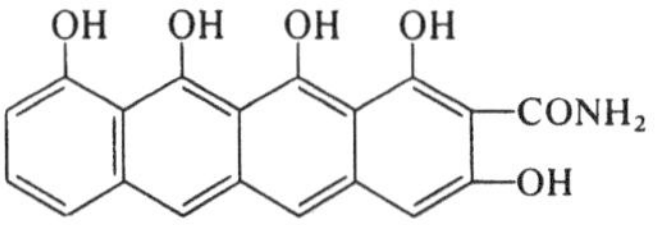   (20) Pretetramid

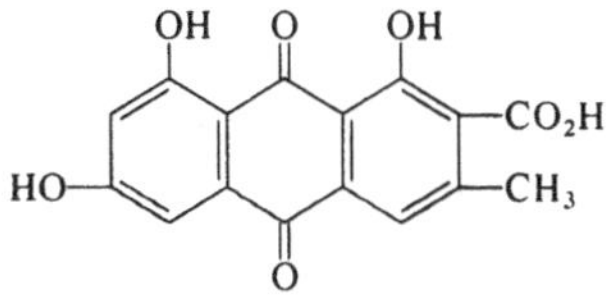

(21) Endocrocin 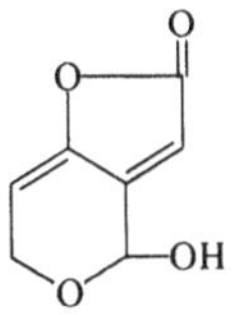   (22) Patulin

Chart 6. Polyketide metabolites having modified oxidation states

IV. Experimental Support for the Polyketide Theory

Experimental studies have provided substantial support for Birch's hypotheses, the first evidence being obtained by Birch's own group (17). 1-^{14}C-labelled acetic acid fed to the fungus, *Penicillium griseofulvum* (Dierckx), was converted to labelled 6-methylsalicylic acid (19), and a systematic degradation showed that labelling had occurred preferentially at the sites predicted by the polyketide hypothesis. Similar investigations of a number of other aromatic metabolites, suspected of being of polyketide origin, have given the same result (2, 18, 19, 20, 59, 63, 133, 159). In a few cases, the enzymes responsible for syntheses of the aromatic metabolites have been isolated. These include the synthetase systems for 6-methylsalicylic acid (19) (111, 115), alternariol (12) (62, 149), orsellinic acid (7) (64) and 5-methylorsellinic acid (61). Acetyl coenzyme A has been shown to be the chain initiator with malonyl coenzyme A as the acetate moiety involved in chain extension. This is a direct analogy with the biosynthesis of fatty acids (165).

Although many details of the polyketide pathways are now known, the growing polyketo acid chains have never been isolated or detected, nor has the biological incorporation of polycarbonyl compounds into aromatic metabolites ever been demonstrated*. The interception of intermediates has met with some success. Triacetic lactone (23) and its 3-methyl derivative (24), as well as tetra-acetic lactone (10), have been identified in fungal cultures (1, 10, 82, 112). Gatenbeck and co-workers have described an interesting experiment involving denaturation of 5-methylorsellinate synthetase from *Aspergillus flaviceps* after incubation with 1-^{14}C-acetic acid (61). Small quantities of abnormal products were isolated, including orcinol (3), 2,6-dimethyl-4-pyrone (4) and tetra-acetic lactone (10). These products are believed to arise by cyclization (and decarboxylation) of tetra-acetic acid either before or after release from the binding site of the enzyme.

(23) R=H, Triacetic lactone
(24) R=CH$_3$

---

* The report by Hillis of biosynthesis of stilbenes in eucalyptus from 9-phenyl-3,5,7-trioxo-8-nonenoic acid was subsequently retracted (92, 93).

References, pp. 274—282

The paucity of detailed biosynthetic information that has been obtained thus far reflects in part difficulties associated with isolation of enzymes from organisms in which they are not involved in a major pathway. Moreover, the enzymes are often unstable, frustrating attempts to purify them and to study them systematically. Except in the case of colored metabolites, only limited use has been made of mutant organisms, since useful mutants are not readily recognized (60, 111, 118). Generally, polyketide metabolites are non-essential to the organisms and deprivation in mutants does not affect their well-being.

Undoubtedly, a further problem associated with biosynthetic studies of polyketides has been the inaccessability of β-polycarbonyl compounds. Attempts to isolate polycarbonyl compounds from biological systems have been thwarted by lack of information on the chemical and physical properties of the intermediates and particularly by lack of comparison samples. Moreover, polycarbonyl compounds suitable for enzymic incorporation have not been available. The combined difficulties in the biosynthetic studies have provided an impetus for chemists to prepare polycarbonyl compounds and to attempt biogenetically modelled syntheses of the aromatic metabolites.

V. Syntheses of β-Polycarbonyl Compounds

The important preparative routes for linear β-polyketones and β-polyketo acids are described in this section. Some syntheses are included of derivatives of polycarbonyl compounds in which one or more of the carbonyl groups have been protected. These derivatives have generally been prepared indirectly, rather than from the polycarbonyl compounds themselves. In some cases, the protective groups have been or conceivably could be removed to furnish syntheses of the free polycarbonyl compounds. In others, the conditions required for unmasking would also cause cyclizations and other reactions to occur. The original interest in protective groups stemmed from concern that free β-polycarbonyl compounds would be too unstable to permit isolation and manipulation. Future interest will center upon their use in facilitating syntheses and directing cyclizations of polycarbonyl compounds.

1. 3,5-Diketo Acids and Esters

The most direct approach to 3,5-diketo acids is from β-diketones. Diketones, when treated with two equivalents of alkali amides, undergo two-fold ionization giving dianions. The external anion of these intermediates is much more reactive than the internal one and treatment of the dianions with carbon dioxide has given exclusively the 3,5-diketo acids (*77, 78, 88, 134*). Triacetic acid (**25**) has been prepared from acetylacetone in 44% yield using sodium amide as the base (*78*) (Chart 7). A complication of this procedure has been that the dianions are prepared in liquid ammonia, which must be replaced with an unreactive solvent prior to addition of carbon dioxide. Although no examples have been reported, the use of lithium diisopropylamide in tetrahydrofuran would be more convenient and would probably give superior results. This base-solvent combination has been employed in carboxylations of higher polyketones; removal of diisopropylamine before the addition of carbon dioxide has not been required (*85, 96, 130*).

3,5-Diketo esters can be prepared from the acids by treatment with diazomethane (*22*); they can also be prepared directly from the diketones. Excellent yields of the diketo esters have been obtained when dilithio-acetylacetone and benzoylacetone were acylated with dimethyl carbonate (*168*). Diketo esters can also be prepared by acylation of the dianion of acetoacetic ester. The initial procedure (*171*), employing sodium hydride as the condensing agent, has been superseded by one involving *n*-butyl-lithium in conjunction with sodium hydride (*97*). Methyl acetoacetate was converted to the monoanion with sodium hydride and then to the dianion with *n*-butyllithium. Acylation with methyl acetate gave 71% of the methyl ester (**26**) of triacetic acid (Chart 7). Comparable results were obtained in acylations with other esters.

The oldest method for synthesis of diketo acids and esters is by opening the lactone ring of 4-methoxy- and 4-hydroxy-2-pyrones. Ring cleavage by hydroxide ion has given the diketo acids, when care was taken to avoid complete degradation (*7, 22, 56, 67, 127, 139, 167*). The ethyl ester (**27**) of triacetic acid has been prepared by thermal ethanolysis of triacetic lactone (**23**) (*152, 170*) (Chart 7). A photochemical methanolysis of tri-acetic lactone has also been reported (*9*).

In view of these transformations, 4-hydroxy-2-pyrones can be considered to be masked diketo acids. They can be prepared by cyclization of diketo acids with acidic catalysts or acetic anhydride (*7, 77, 78, 88, 139*), and also by spontaneous cyclization of diketo thiolacids (*81, 83*). Independent methods are more important for the synthesis of 4-hydroxy-2-pyrones, if the latter are then to be used for preparation of diketo acids. Chief among these is the deacylation of 3-acyl-4-hydroxy-2-pyrones obtained from the self-condensation of β-keto esters (*3, 36, 107, 108, 170*). This method is exemplified by the conversion of dehydroacetic acid (**1**) into triacetic lactone (**23**) in 90% sulfuric acid (Chart 7). Another important method is the condensation of ketones with malonates, the best results being obtained with α-substituted malonates (*1, 21, 110, 177, 178*). The 4-

hydroxy-2-pyrone ring has been constructed by several other condensation reactions (*102, 103, 104, 116, 173*), but the methods have not been
explored so extensively. A final approach to 4-hydroxy-2-pyrones is
by condensations of anions of triacetic lactone and its methyl ether with
electrophilic reagents; the method is useful for placing complex substituents at the 6 position (*26, 55, 56, 164*).

BRAM has reported a novel synthesis of a protected diketo acid (*23*).
Condensation between ketal-protected acetoacetylimidazole (**28**) and the
magnesium salt of the dianion of monoethyl malonate gave protected
triacetic ester (**29**) (Chart 7).

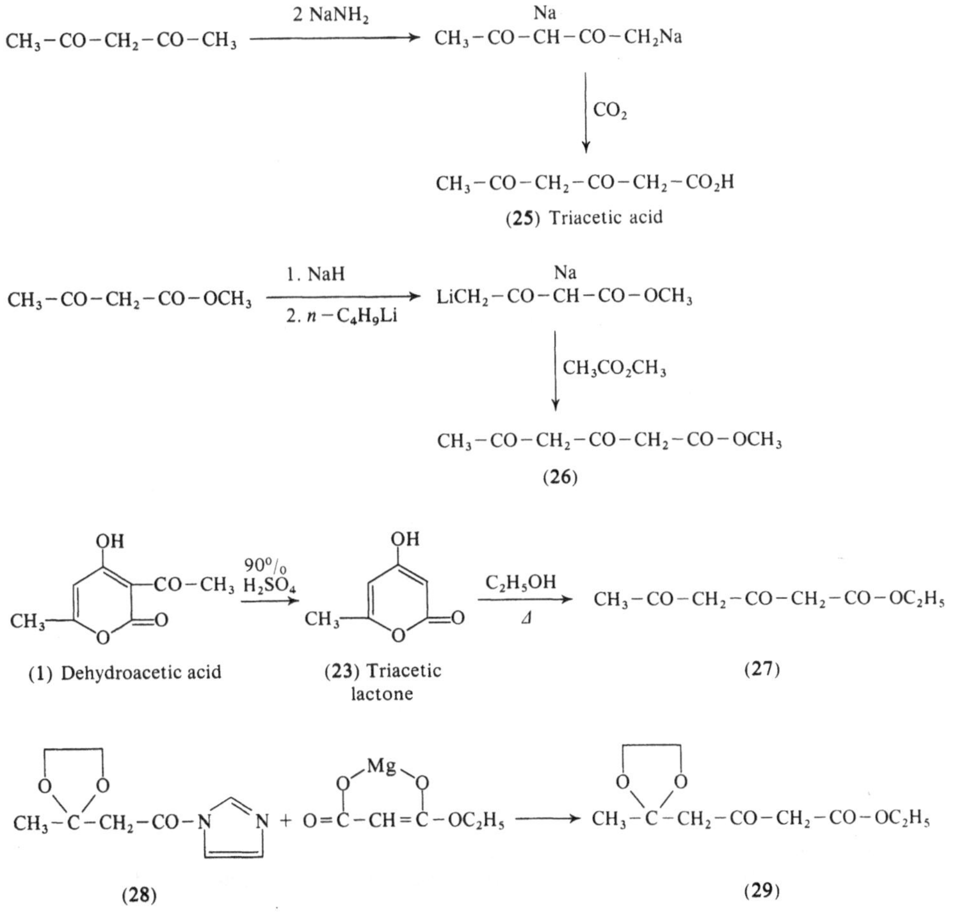

$$CH_3-CO-CH_2-CO-CH_3 \xrightarrow{\text{2 NaNH}_2} CH_3-CO-\overset{\text{Na}}{\underset{}{CH}}-CO-CH_2Na$$

$$\downarrow CO_2$$

$$CH_3-CO-CH_2-CO-CH_2-CO_2H$$

(**25**) Triacetic acid

$$CH_3-CO-CH_2-CO-OCH_3 \xrightarrow[\text{2. } n-C_4H_9Li]{\text{1. NaH}} LiCH_2-CO-\overset{\text{Na}}{\underset{}{CH}}-CO-OCH_3$$

$$\downarrow CH_3CO_2CH_3$$

$$CH_3-CO-CH_2-CO-CH_2-CO-OCH_3$$

(**26**)

(**1**) Dehydroacetic acid   (**23**) Triacetic lactone   (**27**)

(**28**)   (**29**)

Chart 7. Syntheses of diketo acids and esters

2. 1,3,5-Triketones

Acylation of the dianions of β-diketones is the most generally useful method for preparation of 1,3,5-triketones (Chart 8, Reaction 1).

$$R-CO-CH_2-CO-CH_3 \xrightarrow{\text{Base}} R-CO-\bar{C}H-CO-\bar{C}H_2 \xrightarrow{R'CO_2CH_3} \quad (1)$$

$$R-CO-CH_2-CO-CH_2-CO-R'$$

$$C_6H_5-CO-CH_2-CO-CH_2-CO-CH_3 \quad (2)$$

$$\downarrow \text{Base}$$

$$C_6H_5-CO-\bar{C}H-CO-\bar{C}H-CO-CH_3 \xrightarrow[\text{CH}_3\text{I}]{\text{Excess}} C_6H_5-CO-\underset{\underset{CH_3}{|}}{C}H-CO-\underset{\underset{CH_3}{|}}{C}H-CO-CH_3$$

$$\downarrow \text{Base}$$

$$C_6H_5-CO-\bar{C}H-CO-\bar{C}H-CO-\bar{C}H_2 \xrightarrow{C_6H_5CH_2Cl}$$

$$C_6H_5-CO-CH_2-CO-CH_2-CO-CH_2-CH_2-C_6H_5$$

Chart 8. Syntheses of triketones

The dianions, formed by treatment of the diketones with two equivalents of strong bases, undergo acylation exclusively at the more nucleophilic external anion. Various bases have been employed in this procedure; potassium amide was the original one but was found to be limited to acylations with aromatic esters (*88, 106, 113, 134*). Sodium hydride, a more easily manipulated reagent than amide bases, has been used to effect aroylations (*70, 122, 169*). Some unanswered questions remain concerning the mechanism of action of sodium hydride; dianions may not be involved in acylation reactions. Aliphatic esters initially gave disappointing results because ionization competed with acylation. The problem is minimized with lithium amide, the nucleophilicity of dilithio diketones being enhanced relative to their basicity (*76, 169, 174*). Recent results with lithium diisopropylamide suggest that it may be the reagent of choice for most acylations (*130, 168*).

A few examples have been reported of the preparation of triketones by base-catalyzed alkylation of simpler homologs. The site of alkylation is dependent upon the state of ionization, the mono- and dianions alkylating at the internal methylene positions (*29, 40, 169*) and the trianion alkylating at the terminal methyl position (*70*) (Chart 8, Reaction 2).

FEIST and COLLIE explored the opening of 4-pyrones by barium hydroxide as a route to 1,3,5-triketones (*39, 57*) (Chart 1). The method has since been applied in a number of other cases (*6, 14, 52, 143, 151*) and is limited primarily by the availability of 4-pyrones. These have usually been prepared by rearrangement-decarboxylation of 3-acyl-4-hydroxy-2-pyrones (*14, 57, 166*), for which a variety of syntheses are known (*6, 65, 69, 83, 99, 107, 108, 166, 170, 172*). Several other methods for synthesis of 4-pyrones have been reported, but their generality has not been established (*33, 98, 142, 163*).

3. 3,5,7-Triketo Acids and Esters

3,5,7-Triketo acids have, without exception, been synthesized by carboxylation of the trianions of triketones (Chart 9, Reaction 1). At least eight triketo acids have been prepared in sufficiently pure state to obtain satisfactory elemental analyses; others have been obtained in crude form for use as intermediates. The crystalline members of this class are relatively stable; however, they undergo cyclization reactions in polar media and decarboxylate at their melting points.

$$R-CO-CH_2-CO-CH_2-CO-CH_3 \xrightarrow{\text{Base}} R-CO-\bar{C}H-CO-\bar{C}H-CO-\bar{C}H_2 \quad (1)$$

$$\downarrow CO_2$$

$$R-CO-CH_2-CO-CH_2-CO-CH_2-CO_2H$$

(30) $R = C_6H_5$

(31) $R = n-C_7H_{15}$

(32) $R = CH_3$, Tetra-acetic acid

$$C_6H_5-CO-CH_2-CO-CH_2-CO-CH_3 \xrightarrow[\text{2. } CH_3OCO_2CH_3]{\text{1. } 3\,LiN(iso-C_3H_7)_2} \quad (2)$$

$$C_6H_5-CO-CH_2-CO-CH_2-CO-CH_2-CO-OCH_3$$

(33)

$$2\,CH_3-CO-CH_2-CO-OCH_3 \xrightarrow{\text{1. } 9\,LiN(iso-C_3H_7)_2} \quad (3)$$

$$CH_3-CO-CH_2-CO-CH_2-CO-CH_2-CO-OCH_3$$

(34)

Chart 9. Syntheses of unprotected triketo acids and esters

A major consideration in the synthesis of triketo acids has been the choice of ionizing base. Sodium amide in liquid ammonia was the first investigated but has given variable results (*70, 74, 76, 81*). Satisfactory yields of some triketo acids have been obtained with it, for example, 46% of (**30**), but poor yields have been obtained of others, particularly aliphatic acids. Thus, sodium amide gave < 2% of triketo acids (**31**) and (**32**). Potassium amide gave marginally better results in the carboxylation

of aliphatic triketones, the yield of (31) being raised to 30% but the yield
of (32) remaining very low (76). Lithium amide was completely ineffective
in carboxylation reactions. The alkali metal amides appear not to be
strong enough bases to effect complete three-fold ionization of triketones;
this problem is compounded by low solubilities of sodium and lithium
amides and of the salts of some triketones. Uniformly better yields have
been obtained using lithium diisopropylamide in tetrahydrofuran (89,
90, 96, 130). This is undoubtedly a much stronger base and the lithium
salts of the triketones are relatively soluble in tetrahydrofuran. A further
advantage of this reagent is that carbon dioxide can be introduced directly
into the reaction mixture; whereas with sodium and potassium amides
the solvent used for the ionization process, liquid ammonia, must be
replaced with an ethereal solvent prior to treatment with carbon dioxide.

Triketo esters have been prepared from the acids by treatment with
diazomethane (75, 76, 89, 90, 96, 130). When the acids are liquids or
otherwise difficult to purify, esterification of the crude acids can be
advantageous because the esters can usually be purified by chromatography on silica gel. During treatment with diazomethane, care must be
exercised to avoid alkylation of the enolic hydroxyl groups (84). Under
the conditions of acid catalysis required for esterification with alcohols,
the triketo acids cyclize.

In one case a triketo ester has been prepared directly from a triketone.
Condensation of the trilithium salt of 1-phenyl-1,3,5-hexanetrione with
dimethyl carbonate has given 30% of triketo ester (33) (168) (Chart 9,
Reaction 2). This yield, although not as high as the combined yield (68%)
for carboxylating the triketone and esterifying the resulting acid (30)
with diazomethane, is sufficiently good to make the reaction an attractive
alternative to the two-step process.

The methyl ester (34) of tetra-acetic acid has been prepared in 59%
yield by condensation between the mono- and dianions of methyl acetoacetate (132) (Chart 9, Reaction 3). This is a highly attractive procedure
and similar self-condensations of other β-keto esters should be possible.

Most of the examples of protected triketo acids are pyrones; only
two cases having been reported in which the carbonyl groups were protected by other means. BRAM has prepared ester (35) of tetra-acetic acid,
which has the 5 carbonyl group protected as an enol ether and the 7 carbonyl group protected as a ketal (23) (Chart 10). The synthesis started
with ketal-protected triacetic ester (29), the preparation of which is
outlined in Chart 7. The remaining keto group was converted into an
enol ether and the carboxyl group activated as the imidazolide prior to
condensation with the magnesium salt of the dianion of monoethyl
malonate. SCHMIDT and SCHWOCHAU have prepared ester (36) of tetra-
acetic acid, which has the 3 keto group protected as a hemithioketal (144).

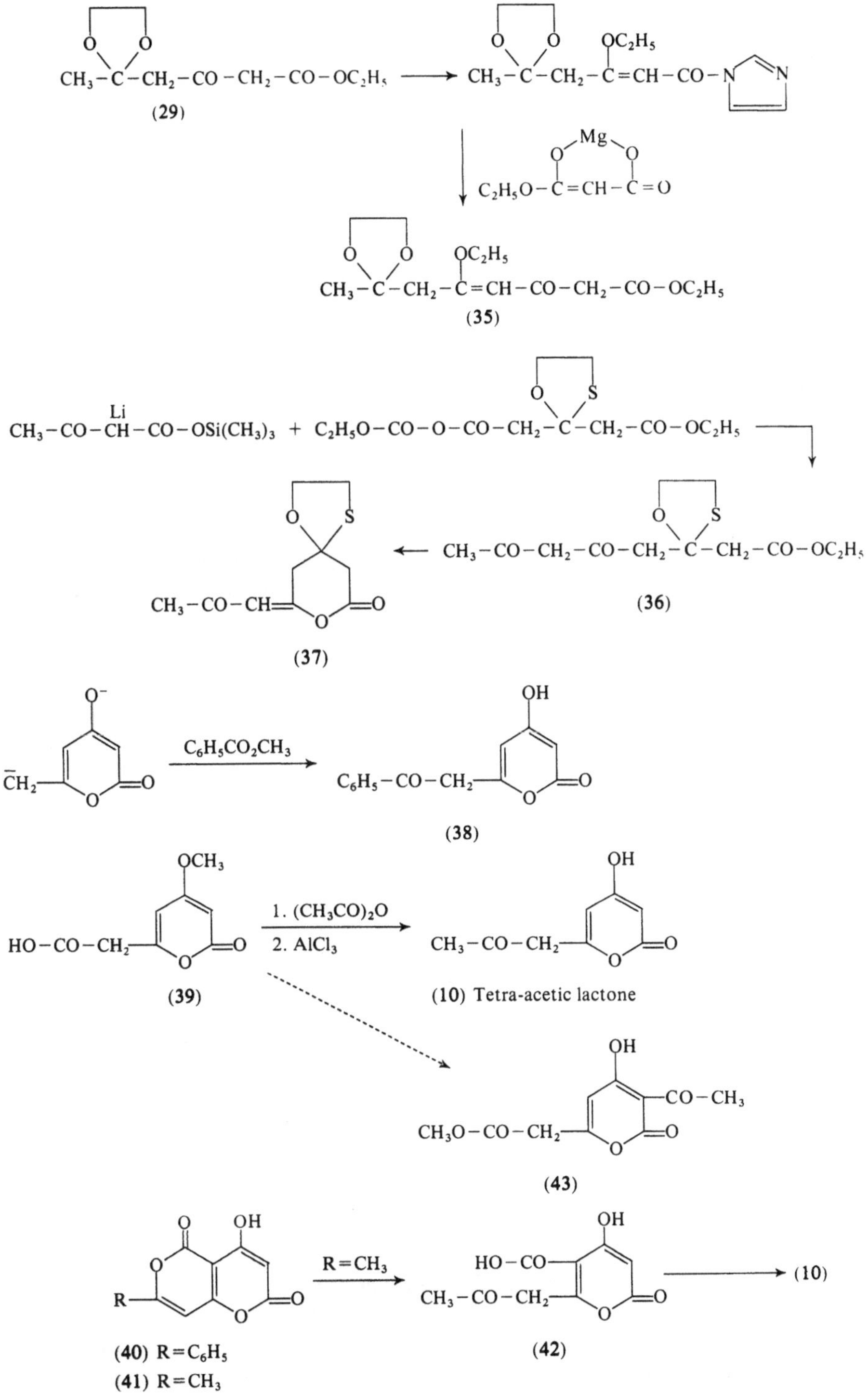

$$CH_3-\overset{O\diagdown\diagup O}{\underset{|}{C}}-CH_2-CO-CH_2-CO-OC_2H_5 \longrightarrow CH_3-\overset{O\diagdown\diagup O}{\underset{|}{C}}-CH_2-\overset{OC_2H_5}{\underset{|}{C}}=CH-CO-N\diagup\diagup N$$

(29)

$$C_2H_5O-\overset{O-Mg-O}{\underset{\diagdown\diagup}{C}}=CH-\overset{}{C}=O$$

$$CH_3-\overset{O\diagdown\diagup O}{\underset{|}{C}}-CH_2-\overset{OC_2H_5}{\underset{|}{C}}=CH-CO-CH_2-CO-OC_2H_5$$

(35)

$$\overset{Li}{CH_3-CO-CH-CO-OSi(CH_3)_3} \;+\; C_2H_5O-CO-O-CO-CH_2-\overset{O\diagdown\diagup S}{\underset{|}{C}}-CH_2-CO-OC_2H_5 \longrightarrow$$

$$\overset{O\diagdown\diagup S}{\underset{CH_3-CO-CH=\!\!\!\diagdown\!\!\!O\!\!\!\diagup\!=O}{\bigcirc}} \longleftarrow CH_3-CO-CH_2-CO-CH_2-\overset{O\diagdown\diagup S}{\underset{|}{C}}-CH_2-CO-OC_2H_5$$

(37)   (36)

$$\overset{O^-}{\overset{|}{CH_2}}\!\!\diagup\!\!\bigcirc\!\!\diagup\!\!=O \quad \overset{C_6H_5CO_2CH_3}{\longrightarrow} \quad C_6H_5-CO-CH_2\!\!\diagup\!\!\overset{OH}{\bigcirc}\!\!\diagup\!\!=O$$

(38)

$$HO-CO-CH_2\!\!\diagup\!\!\overset{OCH_3}{\bigcirc}\!\!\diagup\!\!=O \quad \overset{1.\;(CH_3CO)_2O}{\underset{2.\;AlCl_3}{\longrightarrow}} \quad CH_3-CO-CH_2\!\!\diagup\!\!\overset{OH}{\bigcirc}\!\!\diagup\!\!=O$$

(39)   (10) Tetra-acetic lactone

$$CH_3O-CO-CH_2\!\!\diagup\!\!\overset{OH}{\bigcirc}\!\!\overset{CO-CH_3}{\underset{O}{\diagdown}}=O$$

(43)

$$\overset{O}{\underset{R}{\diagup\bigcirc\diagdown}}\overset{OH}{\underset{O}{\diagup\bigcirc\diagdown}}=O \quad \overset{R=CH_3}{\longrightarrow} \quad \overset{HO-CO}{\underset{CH_3-CO-CH_2}{}}\!\!\diagup\!\!\overset{OH}{\bigcirc}\!\!\diagup\!\!=O \quad \longrightarrow \quad (10)$$

(40) R=C$_6$H$_5$ (42)
(41) R=CH$_3$

Chart 10. Syntheses of protected triketo acids and esters

The synthesis involved acylation of the lithium salt of trimethylsilyl acetoacetate by a mixed carbonic anhydride of protected acetonedicarboxylic
acid. Lactone (37) has also been prepared.

Pyrones, such as (10) and (38), can be considered to be protected
forms of triketo acids, since they are enol lactones. Members of this
class have been prepared in a variety of ways, the most direct being by
cyclization of the triketo acids (81, 89, 130). The acids lactonize in
acetic anhydride by a pathway apparently involving mixed anhydrides.
Pyrones (10), (38) and others have been prepared in this way (Chart 10).
Pyrone (38) has also been obtained from an attempted synthesis of a
thiol analog of triketo acid (30); the thiolacid was unstable, giving the
pyrone by spontaneous cyclization (81).

Another approach has been acylation of triacetic lactone (23) via
its dianion (68, 87, 145, 164). Acylation with methyl benzoate gave a good
yield of (38) but acetylating agents failed to give tetra-acetic lactone (10).
Yamamura and coworkers have prepared tetra-acetic lactone (10) by
acylation of pyrone-acid (39) with acetic anhydride followed by demethylation with aluminum chloride (175, 176). Pyrone-acid (39) is a useful
intermediate for synthesis of polyketide compounds; its preparation, as
developed by Douglas and Money, involves a condensation between the
ether of triacetic lactone (23) and ethyl oxalate (55).

7-Substituted 4-hydroxypyrano[4,3-b]pyran-2,5-diones, including (40)
and (41), have been exploited extensively as masked triketo acids. They
have been synthesized by condensations between ketones and two equivalents of malonate derivatives but better results are obtained by condensations of 4-hydroxy-2-pyrones with stoichiometric amounts of the malonate (66, 177). For the synthesis of (40), the 2,4-dichlorophenyl ester of
malonic acid has been employed as the acylating agent; whereas malonyl
chloride in trifluoroacetic acid has been used in the preparation of (41)
(53, 54, 124). Pyranopyrandione (41) was used by Scott and coworkers
in their synthesis of tetra-acetic lactone (10) (68, 145). The procedure
involved selective hydrolysis to pyrone-acid (42) followed by decarboxylation. The decarboxylation step is a troublesome one and these workers
have since reported a modified procedure involving the methyl ether of
(42) (146).

Douglas and Money have synthesized still another pyrone-type,
protected triketo acid; a transformation of pyrone-acid (39) gave ester
(43) (55) (Chart 10).

4. 1,3,5,7-Tetraketones

The first synthesis of an acyclic β-tetracarbonyl compound occurred
in 1963 when Hauser's group synthesized tetraketone (44) in 52% yield by

acylation of 1-phenyl-1,3,5-hexanetrione with methyl benzoate in the presence of sodium hydride (*121*) (Chart 11, Reaction 1). Tetraketone (**44**) has also been prepared by a comparable condensation using sodium amide as the base but the yield was much lower (*70*). Lithium diisopropylamide has been used to effect the condensation of diacetylacetone (**5**) with methyl benzoate, which gave 19% of tetraketone (**45**) (*85*). No other examples of acylations of triketones have been reported; lithium diisopropylamide would be more likely than the other two bases to give satisfactory acylations with aliphatic esters.

A better method for synthesis of tetraketones is the acylation of diketones with β-keto esters (Chart 11, Reaction 2). MURRAY and HARRIS

$$R-CO-CH_2-CO-CH_2-CO-CH_3 \xrightarrow[\text{Base}]{R'CO_2CH_3} R-CO-CH_2-CO-CH_2-CO-CH_2-CO-R' \quad (1)$$

$$R-CO-CH_2-CO-CH_3 \xrightarrow[\text{Base}]{R'COCH_2CO_2R''} R-CO-CH_2-CO-CH_2-CO-CH_2-CO-R' \quad (2)$$

R−CO−CH₂−CO−CH₂−CO−CH₂−CO−R′

(**44**) R=R′=C₆H₅

(**45**) R=CH₃; R′=C₆H₅

(**46**) R=R′=CH₃

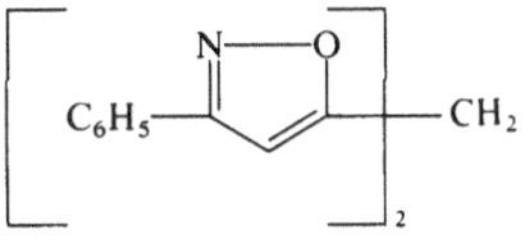

(**47**)

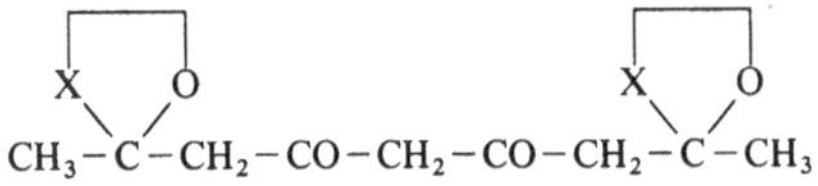

(**48**) X=O
(**49**) X=S

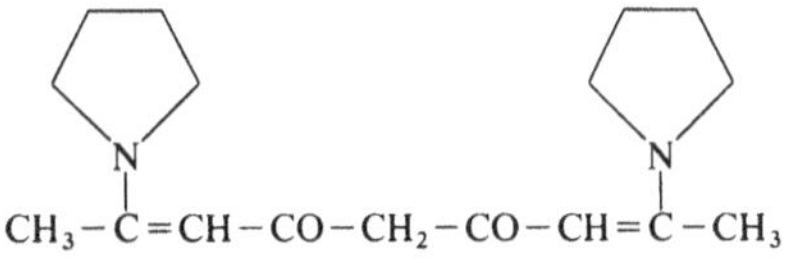

(**50**)

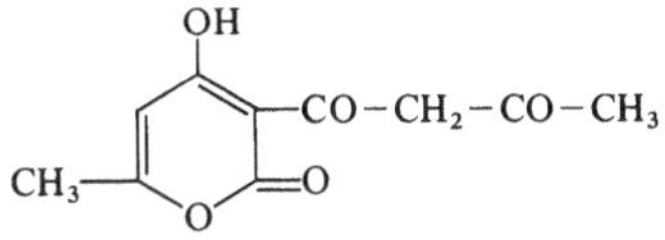

(**51**)

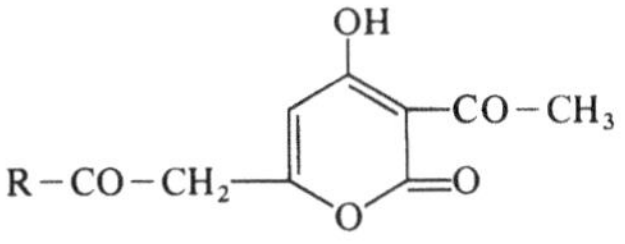

(**52**) R=C₆H₅
(**53**) R=CH₃

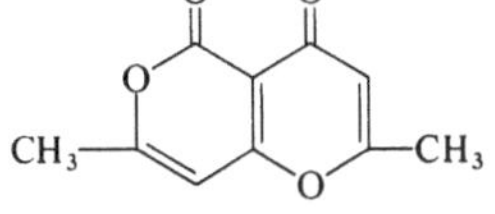

(**54**)

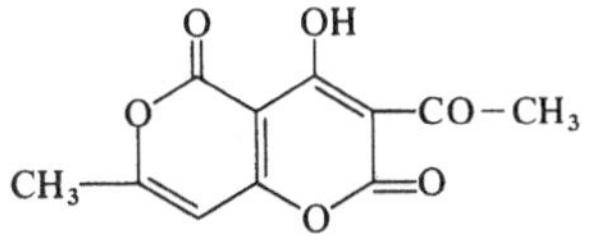

(**55**)

Chart 11. Syntheses of tetraketones

have reported the acylation of the dilithium salt of benzoylacetone by ethyl sodiobenzoylacetate, which gave tetraketone (44) in 51% yield (131). The procedure is applicable with methyl acetoacetate as well; condensations with benzoylacetone and acetylacetone gave tetraketones (45) and (46) in yields of 30 and 31% respectively. The availability of starting materials and the apparent general applicability make this method an attractive one.

Casnati and coworkers have described a synthesis of tetraketone (44) from bis-isoxazole (47), which was prepared by double 1,3-dipolar additions of benzonitrile oxide to 1,4-pentadiene (30, 31). The applicability of this method to the synthesis of other tetraketones has not been demonstrated, although other bis-isoxazoles have been synthesized (32).

Protected tetraketones have received attention from a number of groups, resulting, in part, from the apparent lack of generality of the syntheses of free tetraketones based on triketones and bis-isoxazoles. The studies of protected compounds have concentrated on derivatives of tetraketone (46), which is the least stable of the three known tetraketones.

Stetter and Vestner have synthesized bis-ketal (48) by a Claisen condensation between a ketal of methyl acetoacetate and a monoketal of acetylacetone (156). Schmidt and Schwochau have prepared bis-hemithioketal (49) by two-fold acylation of a malonate ester with a hemi-thioketal-protected acetoacetate (144). Stephen and Marcus have prepared bis-enamine (50) from pyrone (51) (155).

Both 3β and 6α acylation products of dehydroacetic acid (1) are known. Pyrone (51) has been prepared by acetylation of dehydroacetic acid (1) and by acetoacetylation of triacetic lactone (23) (117, 137). Pyrone (52), but not (53), has been obtained by 6α acylation of dehydroacetic acid (79, 83). However, pyrone (53) has been synthesized by acylation of tetra-acetic lactone (10) (145). Pyranopyrandiones equivalent to tetra-ketone (46) include (54) and (55), which have been prepared from pyrone (51) and pyranopyrandione (41), respectively (124, 137).

5. 3,5,7,9-Tetraketo Acids and Esters

Only one unprotected tetraketo acid has been synthesized; carboxylation of the tetra-anion of tetraketone (45) gave 56% of tetraketo acid (56) (85) (Chart 12). The corresponding ester has been prepared by treatment of acid (56) with diazomethane. Presumably, the same sequence could be used with tetraketone (46) for synthesis of penta-acetic acid (57) and its ester.

Three protected tetraketo acids have been prepared. Enol lactone (58) has been formed by treatment of acid (56) with acetic anhydride (85).

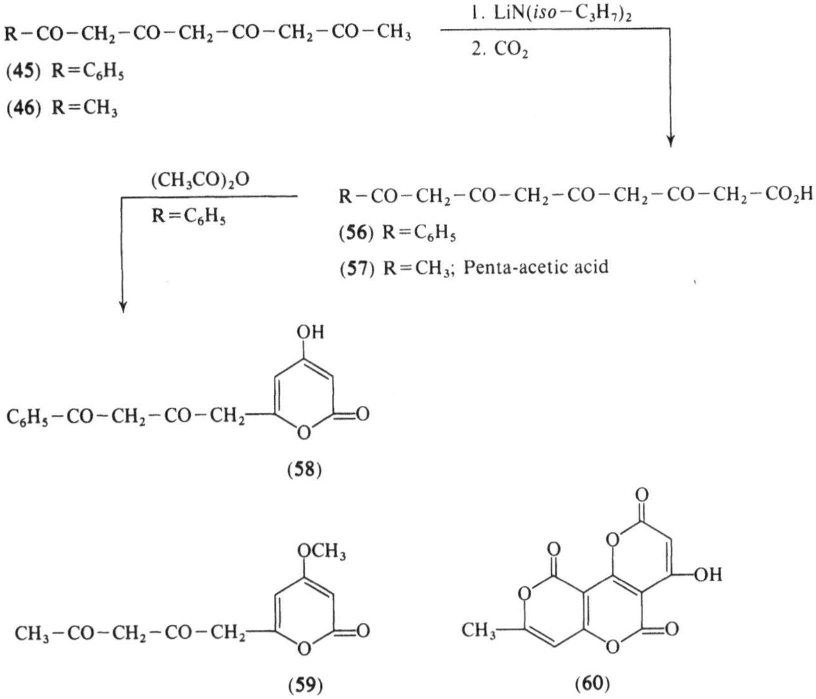

$R-CO-CH_2-CO-CH_2-CO-CH_2-CO-CH_3$ $\dfrac{1.\ LiN(iso-C_3H_7)_2}{2.\ CO_2}$

(45) R=C₆H₅

(46) R=CH₃

$\dfrac{(CH_3CO)_2O}{R=C_6H_5}$ $R-CO-CH_2-CO-CH_2-CO-CH_2-CO-CH_2-CO_2H$

(56) R=C₆H₅

(57) R=CH₃; Penta-acetic acid

$C_6H_5-CO-CH_2-CO-CH_2-$

(58)

$CH_3-CO-CH_2-CO-CH_2-$

(59)

(60)

Chart 12. Syntheses of tetraketo acids and esters

Pyrone (59) and dipyranopyrantrione (60), which are masked penta-acetic acids, have been synthesized by acylation of pyrone-acid (39) and pyrano-pyrandione (41) with isopropenyl acetate and malonyl chloride, respectively (*124, 147*).

6. 1,3,5,7,9-Pentaketones

Two methods for the synthesis of pentaketones have been developed by HARRIS and coworkers. The first is acylation of a tetraketone (Chart 13, Reaction 1). Pentaketone (61) was prepared by benzoylation of tetraketone (45) via its tetra-anion. A slightly more convenient procedure to effect this reaction is a two-fold benzoylation of diacetyl-acetone (5). Both methods gave approximately 20% of the pentaketone (*85*). The use of aliphatic esters to synthesize pentaketones, such as (62) and (63), has not been demonstrated and may be difficult to effect. The second approach to pentaketones is the acylation of triketones with β-keto esters (Chart 13, Reaction 2). The trianion of the triketone and the monoanion of the keto ester are employed (*131, 132*). The condensation of ethyl benzoylacetate with 1-phenyl-1,3,5-hexanetrione and with diacetylacetone (5) gave pentaketones (61) and (62) in yields of 41 and 26%, respectively. Acylation of diacetylacetone with methyl acetoacetate gave pentaketone (63); a poor yield in this case resulted from instability of the pentaketone.

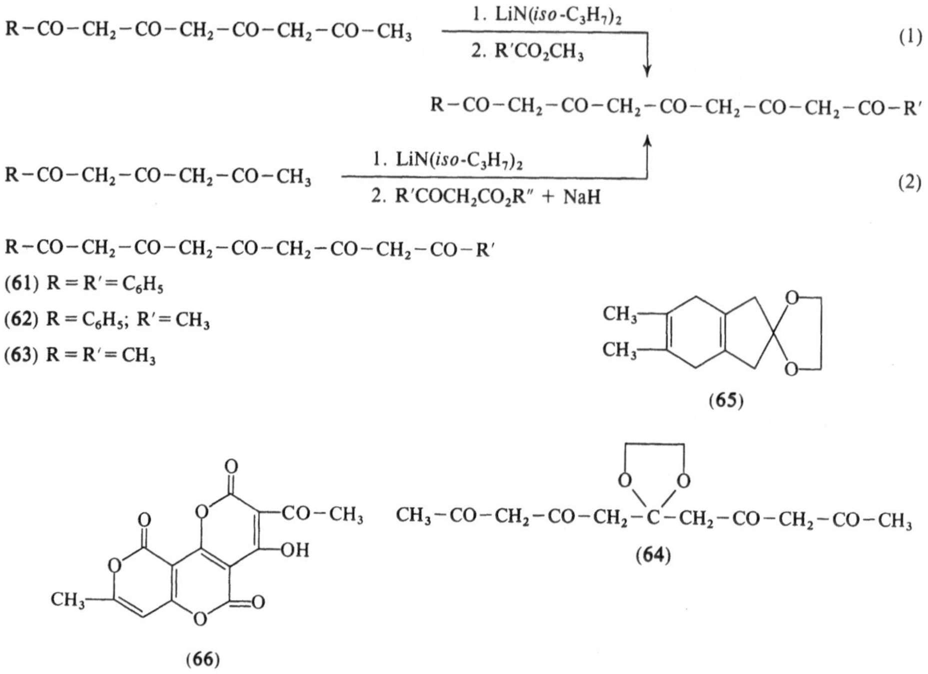

$$R-CO-CH_2-CO-CH_2-CO-CH_2-CO-CH_3 \quad \xrightarrow[\text{2. } R'CO_2CH_3]{\text{1. } LiN(iso\text{-}C_3H_7)_2}$$

(1)

$$R-CO-CH_2-CO-CH_2-CO-CH_2-CO-CH_2-CO-R'$$

$$R-CO-CH_2-CO-CH_2-CO-CH_3 \quad \xrightarrow[\text{2. } R'COCH_2CO_2R'' + NaH]{\text{1. } LiN(iso\text{-}C_3H_7)_2}$$

(2)

$$R-CO-CH_2-CO-CH_2-CO-CH_2-CO-R'$$

(61) R = R' = C₆H₅
(62) R = C₆H₅; R' = CH₃
(63) R = R' = CH₃

(65)

(66)

$$CH_3-CO-CH_2-CO-CH_2-\overset{\displaystyle O-\hspace{-0.3em}O}{\underset{|}{C}}-CH_2-CO-CH_2-CO-CH_3$$

(64)

Chart 13. Syntheses of pentaketones

Two protected pentaketones are known. BIRCH and coworkers have synthesized monoketal (64) of pentaketone (63) by ozonolysis of dihydro-indane (65) (16). The latter had been prepared by a Birch reduction of the corresponding indane. No further examples of this approach have appeared but the method deserves additional attention in view of its potential applicability to higher polyketides. Dipyranopyrantrione (66), which is formally a protected pentaketone, has been synthesized by MONEY and coworkers by acetylation of dipyranopyrantrione (60) (124).

7. 3,5,7,9,11-Pentaketo Acids and Esters

Free pentaketo acids are unknown but two derivatives of hexa-acetic acid have been synthesized (Chart 14). Dipyrone (67) has been obtained in 10% yield by acylation of triacetic lactone (23) with pyrone-acid (39) in the presence of aluminum trichloride (55). Tripyranopyrantetraone (68) has been prepared in 31% yield by treatment of dipyranopyrantrione (60) with 2,4-dichlorophenyl malonate at high temperature (124).

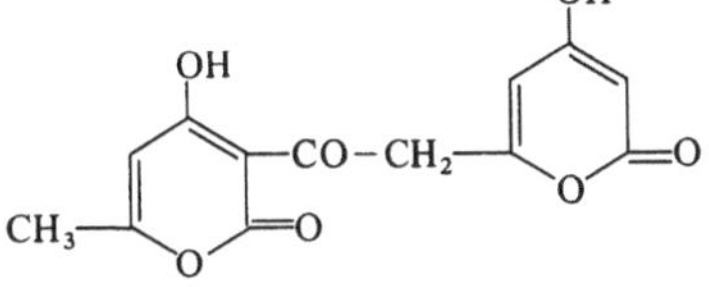

(67) 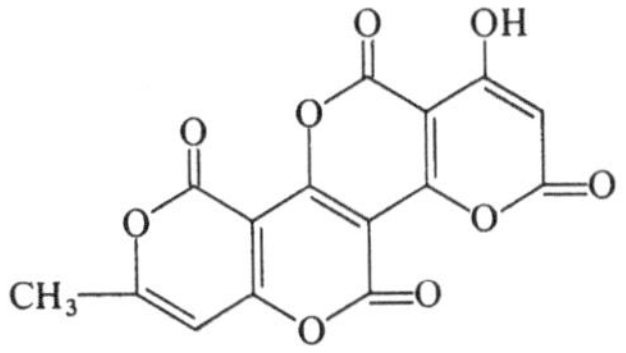   (68)

Chart 14. Syntheses of protected pentaketo acids

8. 1,3,5,7,9,11-Hexaketones

Only one example is known, hexaketone (69) (168). The compound was prepared by WITTEK in 19% yield by a condensation between tetraketone (45) and ethyl benzoylacetate (Chart 15, Reaction 1). The reaction involves the tetra-anion of the tetraketone and the monoanion of the keto ester. Hexaketone (69) was also synthesized by a two-fold condensation of ethyl benzoylacetate and acetylacetone (Chart 15, Reaction 2). The reaction undoubtedly involves anions of tetraketone (45) as intermediates; it is interesting that a better yield (40%) was obtained by the second procedure than by the first.

$$C_6H_5-CO-CH_2-CO-CH_2-CO-CH_2-CO-CH_3 \quad \xrightarrow[\substack{2.\ C_6H_5COCH_2CO_2C_2H_5 \\ +\ NaH}]{1.\ LiN(iso\text{-}C_3H_7)_2} \quad (1)$$

(45)

$$C_6H_5-CO-CH_2-CO-CH_2-CO-CH_2-CO-CH_2-CO-CH_2-CO-C_6H_5$$

(69)

$$CH_3-CO-CH_2-CO-CH_3 \quad \xrightarrow[\substack{2.\ 2\,C_6H_5COCH_2CO_2C_2H_5 +\ NaH \\ 3.\ LiN(iso\text{-}C_3H_7)_2}]{1.\ LiN(iso\text{-}C_3H_7)_2} \quad (2)$$

Chart 15. Syntheses of hexaketone (69)

9. β-Heptacarbonyl Compounds

Only one unprotected example is known, heptaketone (70); WITTEK's preparation was modelled after the synthesis of hexaketone (69) (Chart 16).

A two-step condensation of diacetylacetone (5) with ethyl benzoylacetate gave 15% of the heptaketone (*168*). Pentaketone (62), although undoubtedly an intermediate, was not detected in the product mixture.

$$CH_3-CO-CH_2-CO-CH_2-CO-CH_3 \xrightarrow{\begin{array}{l}1.\ LiN(iso\text{-}C_3H_7)_2\\2.\ 2\,C_6H_5COCH_2CO_2C_2H_5 + NaH\\3.\ LiN(iso\text{-}C_3H_7)_2\end{array}}$$

(5) Diacetylacetone

$$C_6H_5-CO-CH_2-CO-CH_2-CO-CH_2-CO-CH_2-CO-CH_2-CO-CH_2-CO-C_6H_5$$

(70)

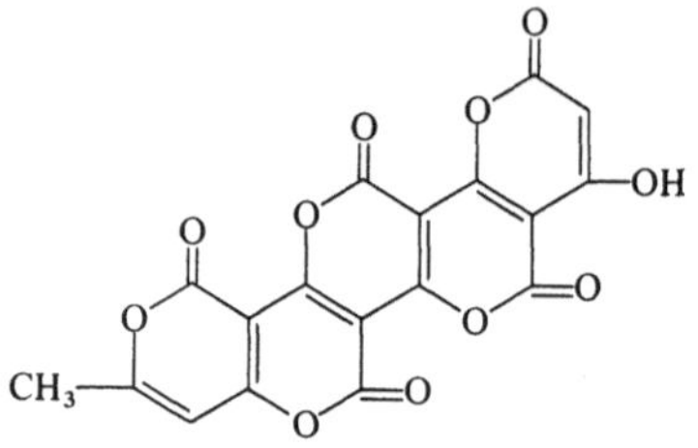

(72)

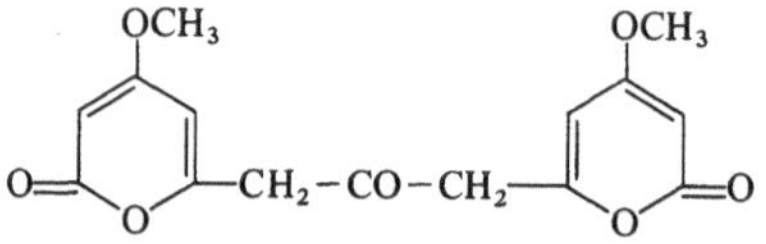

(71)

Chart 16. Syntheses of heptacarbonyl compounds

Dipyrone (71), which is a protected pentaketo diacid, has been synthesized by Scott and coworkers by decarboxylative dimerization of pyrone-acid (39) in the presence of acetic anhydride (*135, 147*). A 48% yield was obtained. Yamamura and coworkers have also obtained (71) as a byproduct in their synthesis of tetra-acetic lactone (10) from (39) (*175, 176*). The choice of whether pyrone-acid (39) is converted into tetra-acetic lactone (10) or dipyrone (71) depends upon the amount of acetic anhydride employed.

Tetrapyranopyranpentaone (72) has been obtained by Scott's group along with other polypyrones from the condensation of 2,4-dichlorophenyl malonate with pyranopyrandione (41) (*145*). The compound contains the backbone of hepta-acetic acid, but the yield of the reaction was too low (0.04%) to be of practical value.

10. β-Octacarbonyl Compounds

Wittek has prepared octaketone (73) by two-fold β-ketoacylation of tetraketone (46) with ethyl benzoylacetate by a polyanion procedure (*168*) (Chart 17). The octaketone, although reasonably stable, was obtained in only a 3% yield. No other free or protected octacarbonyl compounds have been prepared.

$$CH_3-CO-CH_2-CO-CH_2-CO-CH_2-CO-CH_3$$

(46)

1. LiN(iso-C$_3$H$_7$)$_2$
⟶
2. 2 C$_6$H$_5$COCH$_2$CO$_2$C$_2$H$_5$ + NaH
3. LiN(iso-C$_3$H$_7$)$_2$

$$C_6H_5-CO-CH_2-CO-CH_2-CO-CH_2-CO-CH_2-CO-CH_2-CO-CH_2-CO-CH_2-CO-C_6H_5$$

(73)

Chart 17. Synthesis of octaketone (73)

11. β-Nonacarbonyl Compounds

Free nonacarbonyl compounds have not been synthesized. However, Scott's group have prepared bis-pyranopyrandione (74), which is a protected heptaketo diacid (147) (Chart 18). Pyranopyrandione (75), prepared from pyrone-acid (39) by an annelation with 2,4-dichlorophenyl malonate, was self-condensed in the presence of acetic anhydride and sodium acetate to give bis-pyranopyrandione (74). This is the largest of the ketide systems that have been synthesized.

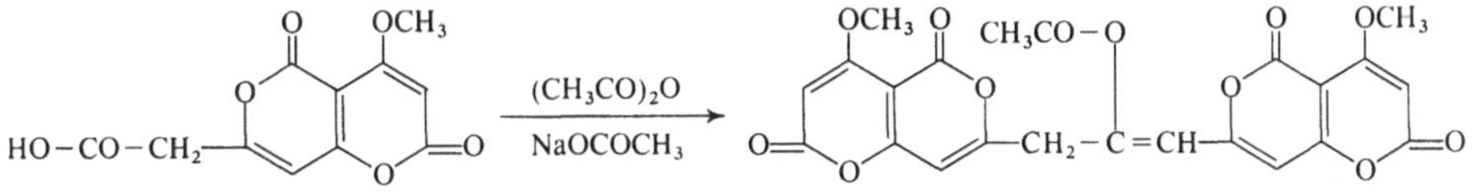

Chart 18. Synthesis of protected heptaketo acid (74)

12. Reduced Polycarbonyl Compounds

Many polyketide metabolites, such as 6-methylsalicylic acid (19) and pretetramid (20), lack hydroxyl groups at positions predicted by the acetate hypothesis. The inherent difficulty of reducing the phenolic hydroxyl function non-enzymically suggests that in nature reduction, at least in monocyclic cases, probably occurs prior to aromatization (13, 115). β-Polycarbonyl compounds in which one or more keto groups have been reduced to carbinols, as in (76), would be suitable model compounds for study of this group of metabolites. Equally useful would be the dehydrated compounds containing the 2-pentene-1,5-dione moiety (77). Unsaturation introduces an additional consideration into cyclization reactions since most ring closures require the cis configuration. An examination of the literature reveals that neither class has been studied extensively; there are a few examples of unsaturated compounds, but most of them are encumbered with extraneous substituents on the double bond [for example, see (80)]. At the simplest level, 5-oxohexenoic acid (78) has been synthesized (71), but heptene-2,6-dione (79) is unknown (Chart 19).

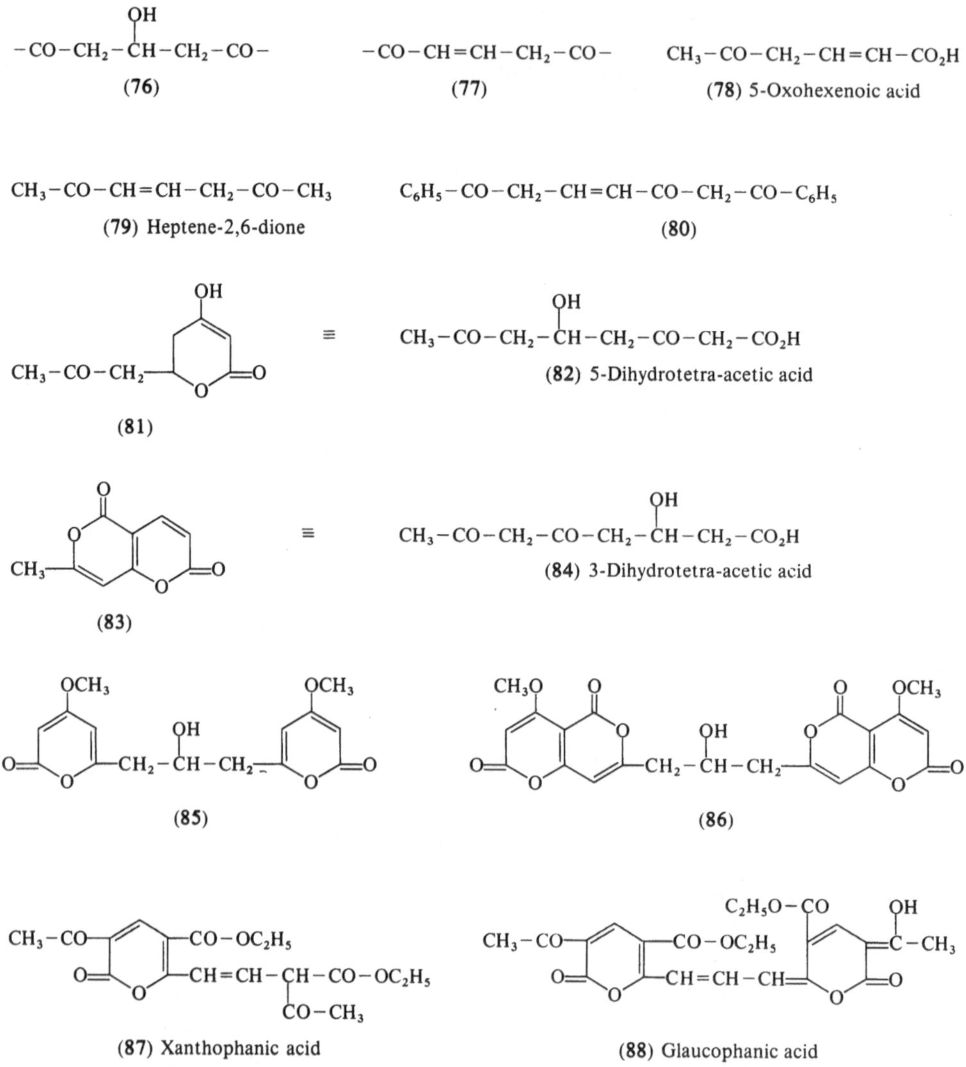

$$-CO-CH_2-\overset{\overset{\textstyle OH}{|}}{CH}-CH_2-CO-$$

(76)

$$-CO-CH=CH-CH_2-CO-$$

(77)

$$CH_3-CO-CH_2-CH=CH-CO_2H$$

(78) 5-Oxohexenoic acid

$$CH_3-CO-CH=CH-CH_2-CO-CH_3$$

(79) Heptene-2,6-dione

$$C_6H_5-CO-CH_2-CH=CH-CO-CH_2-CO-C_6H_5$$

(80)

(81) $\equiv$ $$CH_3-CO-CH_2-\overset{\overset{\textstyle OH}{|}}{CH}-CH_2-CO-CH_2-CO_2H$$

(82) 5-Dihydrotetra-acetic acid

(83) $\equiv$ $$CH_3-CO-CH_2-CO-CH_2-\overset{\overset{\textstyle OH}{|}}{CH}-CH_2-CO_2H$$

(84) 3-Dihydrotetra-acetic acid

(85)

(86)

(87) Xanthophanic acid

(88) Glaucophanic acid

Chart 19. Syntheses of reduced polycarbonyl compounds

One very promising approach to reduced polycarbonyl compounds is the aldol condensation of β-ketoaldehydes with polycarbonyl compounds. HARRIS has condensed the anion of benzoylacetaldehyde with the dianion of benzoylacetone to obtain unsaturated triketone (**80**) and tautomers thereof (*72*). For this reaction to be of significant value in the preparation of reduced polyketides, the practicability must be demonstrated with aliphatic β-ketoaldehydes and additional polyketone nucleophiles.

SCOTT and coworkers have synthesized dihydropyrone (**81**) by catalytic reduction of tetra-acetic lactone (**10**) (*68, 146*). The compound is equivalent to 5-dihydrotetra-acetic acid (**82**). An attempt to synthesize the phenyl analog from pyrone (**38**) was unsuccessful, reduction of the phenacyl carbonyl group occurring preferentially.

CHENG and TAN have prepared pyranopyrandione (**83**) by the acid-catalyzed condensation of triacetic lactone (**23**) with formylacetic acid (*34*). The compound can be regarded as a protected form of 3-dihydrotetra-acetic acid (**84**).

Carbinol (**85**) has been prepared by SCOTT's group by reduction of keto-dipyrone (**71**) with sodium borohydride (*147*). The same procedure should be applicable with bis-(pyranopyrandione) (**74**) to form carbinol (**86**), which is only one acetate short of the precursor required for synthesis of pretetramid (**20**).

Finally, xanthophanic acid (**87**) and glaucophanic acid (**88**) are β-polycarbonyl compounds having several carbonyl omissions. They also have considerable branching, making them of limited value for model studies. CROMBIE and coworkers recently established the structures of these condensation products of diethyl ethoxymethylenemalonate and have put the compounds to good use in studies of the effects of metal chelation on aromatization reactions (*43, 44, 45, 46, 47, 48, 49*).

VI. Cyclizations of β-Polycarbonyl Compounds

This section contains examples of polyketide annelation reactions resembling biosynthetic processes by which the same or related structures arise in nature. The emphasis is on carbocyclic, aromatic structures but appropriate heterocyclic examples are included. The section is divided into three parts. The first contains examples in which multiple ketide fragments condense to give aromatic products. In the second are examples of compounds arising from cyclizations of single polycarbonyl chains. The final one reviews cases in which polycyclic structures have been synthesized from polyketide precursors having one or more pre-formed aromatic rings.

1. Using Several Ketide Fragments

The biosynthesis of phenolic compounds by condensations between two polycarbonyl compounds appears to be relatively rare; this minimizes the significance of biogenetically-modelled syntheses involving such processes. Several examples having historical significance in connection with Collie's theories were described in Section II. These include the self-condensation of ethyl acetoacetate to give dehydroacetic acid (1), which Collie erroneously concluded had the structure of tetra-acetic lactone (2), and the self-condensation of diacetylacetone (5) to give naphthalene-diol (6) (Chart 20). The latter reaction has been re-investigated by Birch and coworkers and by Bethell and Maitland (11, 14). By careful control of reaction conditions, two intermediates, phenol (89) and its aldol cyclization product (90), have been isolated. Birch has remarked on the similarities between condensation products (90) and (6) and the *Campnosperma brevipetiolata* metabolites (91) and (92), (91) being the apparent precursor of (92) (13).

Recently, Kato and Hozumi reported the formation of methyl orselli-nate (93) by acylation of methyl acetoacetate with ketene dimer in the presence of sodium hydride (105). The reaction probably involves the branched triketo ester (94) rather than the linear one.

Mühlemann has synthesized emodin (97) from diketone (95) by a condensation with dimethyl acetonedicarboxylate to give diarylmethane (96) followed by closure of the central ring and removal of the carbo-methoxyl group (129). Endocrocin has been prepared by a similar procedure (154).

Polycyclic, phenolic natural products containing 4-pyrone rings have been synthesized by annelation of phenols with β-keto esters, usually without isolation of intermediates. Numerous examples of the use of this reaction to synthesize chromones, flavones and xanthones can be found in Dean's monograph "Naturally Occurring Oxygen Ring Compounds" (51). A typical example is the thermal condensation of α-naphthol (98) with ethyl acetoacetate to give dimethyl ether (99) of eleutherinol (58). The relationship of these condensations to physiological processes is slight, although it is true that the heterocyclic ring of many of these metabolites is probably the last one formed.

Phloroglucinol derivatives have been synthesized by trimerization of esters. The diethyl esters of malonic and acetonedicarboxylic acids have both been self-condensed in this manner (4, 42, 100).

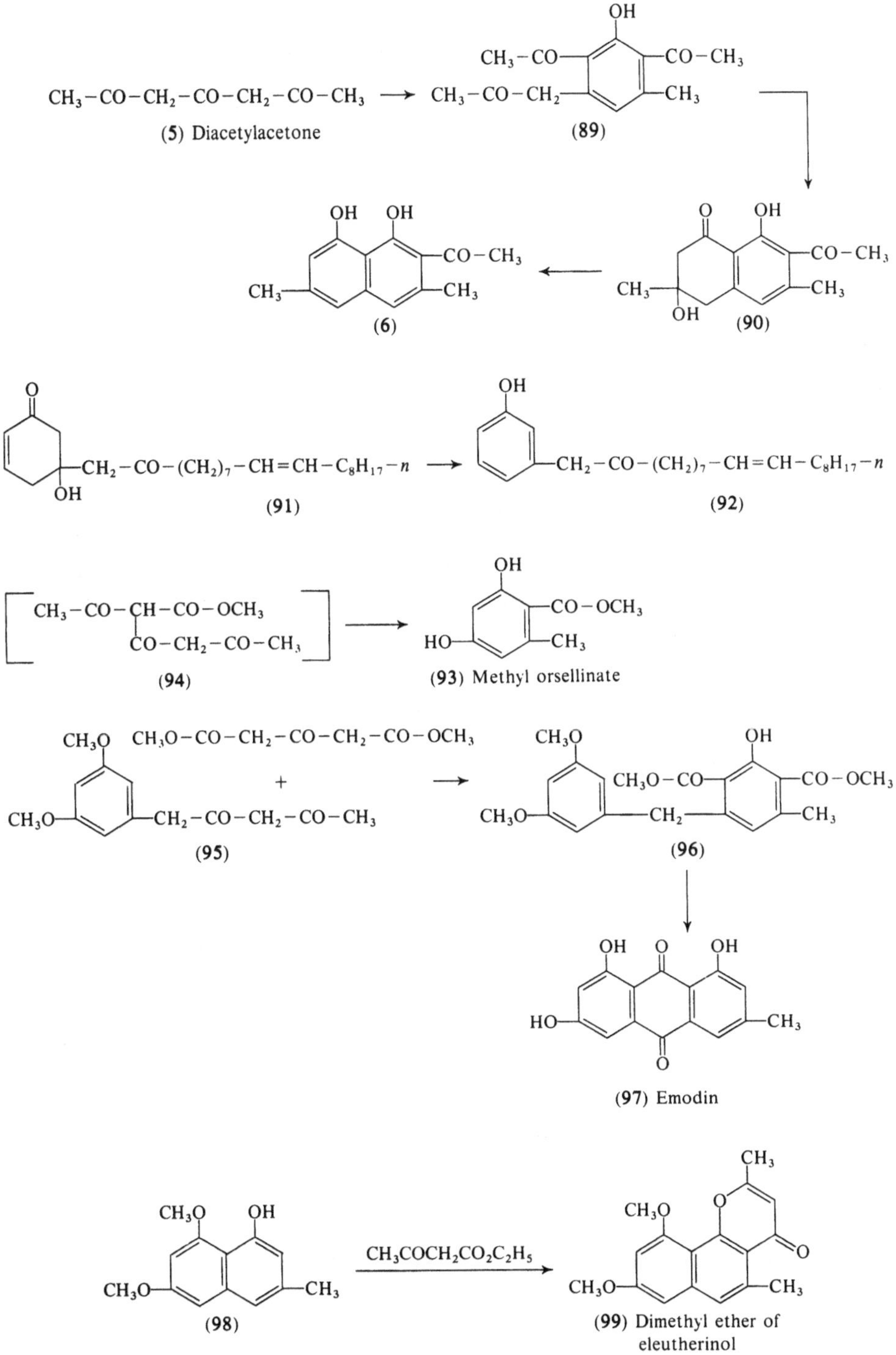

$$CH_3-CO-CH_2-CO-CH_2-CO-CH_3 \longrightarrow$$

(5) Diacetylacetone

(89)

(6) ← **(90)**

(91) → **(92)**

(94) → **(93)** Methyl orsellinate

(95) → **(96)**

(97) Emodin

(98) $\xrightarrow{CH_3COCH_2CO_2C_2H_5}$ **(99)** Dimethyl ether of eleutherinol

Chart 20. Syntheses of phenols from two ketide fragments

2. Using a Single Polycarbonyl Compound

a) With 3,5-Diketo Acids

3,5-Diketo acids are capable of two dehydrative cyclizations; these are closure to phloroglucinols and to δ-enol lactones, i.e., 4-hydroxy-2-pyrones. Enzymic formation of phloroglucinol from triacetic acid appears unlikely because it would require acylation of the relatively unreactive terminal methyl group. Phloroglucinol occurs in nature; but it probably arises by degradation of flavanoid metabolites, rather than as a direct product of the cyclization of triacetic acid (*161*). 4-Hydroxy-2-pyrones are well known metabolites; their chemistry has been reviewed in this series (*128*).

A non-enzymic phloroglucinol cyclization has been claimed by KOMNINOS (*109*). The process involved the acid chloride of triacetic acid, which was prepared by a condensation of acetone and malonyl chloride. The yield of phloroglucinol was very low; the cyclizations to form 4-hydroxy-2-pyrones, which were described in Section V.1, are a much more general process. Similar processes are probably involved in the biosynthesis of 4-hydroxy-2-pyrones.

b) With 1,3,5-Triketones

Triketones also have limited possibilities for cyclization. Their closure to 4-pyrones is well known (*70, 88, 174*), but the 4-pyrones represent a rare class of polyketide metabolite (*61*). Triketones can also cyclize to give resorcinols. COLLIE and MYERS observed the first example of this reaction; formation of orcinol (**3**) from diacetylacetone (**5**) in acid or strong base (*39*). BIRCH and coworkers attempted, without success, to cyclize triketone (**100**) to pinosylvin (**102**); however, they were able to obtain dihydropinosylvin (**103**) from triketone (**101**) (*14*) (Chart 21). In general, the cyclizations of triketones to resorcinols do not go well because they require the unreactive terminal methyl group to enter into aldol condensations. BIRCH concluded that the cyclization of triketones was a poor model of the biosynthesis of resorcinols and correctly predicted that triketo acids would in the laboratory, as in the cell, undergo more facile cyclizations.

$R-CO-CH_2-CO-CH_2-CO-CH_3 \longrightarrow$

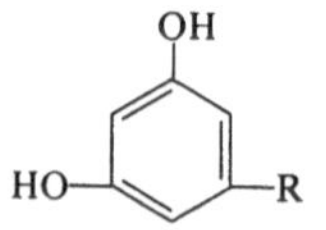

(**100**) $R=C_6H_5-CH=CH$

(**101**) $R=C_6H_5-CH_2-CH_2$

(**102**) $R=C_6H_5-CH=CH$, Pinosylvin

(**103**) $R=C_6H_5-CH_2-CH_2$, Dihydropinosylvin

Chart 21. Cyclizations of triketones to resorcinols

c) With 3,5,7-Triketo Acids and Esters

Triketo acids are capable of four cyclizations and examples are available for at least three, and possibly all, occurring in nature (see Section III and Chart 4)*. The reactions of one member of this class, 7-phenyl-3,5,7-trioxoheptanoic acid (30), have been studied extensively by HARRIS and coworkers. The ease of synthesis and physical properties of this acid make it more convenient to use than many of its homologs. Some more limited investigations have been made with other triketo acids. No significant differences in the chemistry of the simple triketo acids have been found other than that acids with small substituents at the 7 position, in particular, tetra-acetic acid (32), are much more reactive.

Acid (30), although relatively stable in most organic solvents, was found to be highly reactive in aqueous solutions. Aldol cyclization was observed over a range of acidity from pH 4 to strong base (74, 76, 95). In alkaline solution, dianion (104) was obtained quantitatively. The dianion has been characterized by its n.m.r. and u.v. spectra and apparently exists as a single stereoisomeric form. Free acid (106) has eluded isolation; acidification of solutions of dianion (104) has given resorcylic acid (107). Even at pH 5, dehydration occurs, suggesting that acid (106) is able to catalyze its own dehydration. With acidic media (pH 4—5), resorcylic acid (107) is the only product that has been obtained from (30). The intermediate acid (106) has not been detected; presumably, aromatization occurs faster than cyclization under these conditions. The specificity of aldol cyclization is very high; resorcylic acid (107) was isolated in 86% yield from the cyclization of (30) at pH 5 (Chart 22).

Cyclization to 4-pyrone (108) occurred quantitatively when triketo acid (30) was treated with anhydrous, liquid hydrogen fluoride (70). Pyrone (108) decarboxylated readily to give 2-methyl-6-phenyl-4-pyrone. The lactonization of triketo acid (30) in acetic anhydride to give pyrone (38) was described in Section V.3. No evidence of Claisen cyclization of (30) to give benzoylphloroglucinol (109) has been found. This reflects the fact that in base anionic attack on the carboxylate anion is unlikely. Internal O-acylation to give pyrone (38) is the preferred reaction pathway for the mixed acetic anhydride.

Interesting results have been obtained with the methyl ester (33) of triketo acid (30) (Chart 23). In methanolic sodium acetate and in aqueous sodium bicarbonate solutions, cyclization of ester (33) occurred rapidly

---

* This excludes the possibility that the relatively unreactive 8 position of an aliphatic 3,5,7-triketo acid will enter into an aldol condensation. In subsequent sections of this chapter, a similar assumption is made with higher homologs.

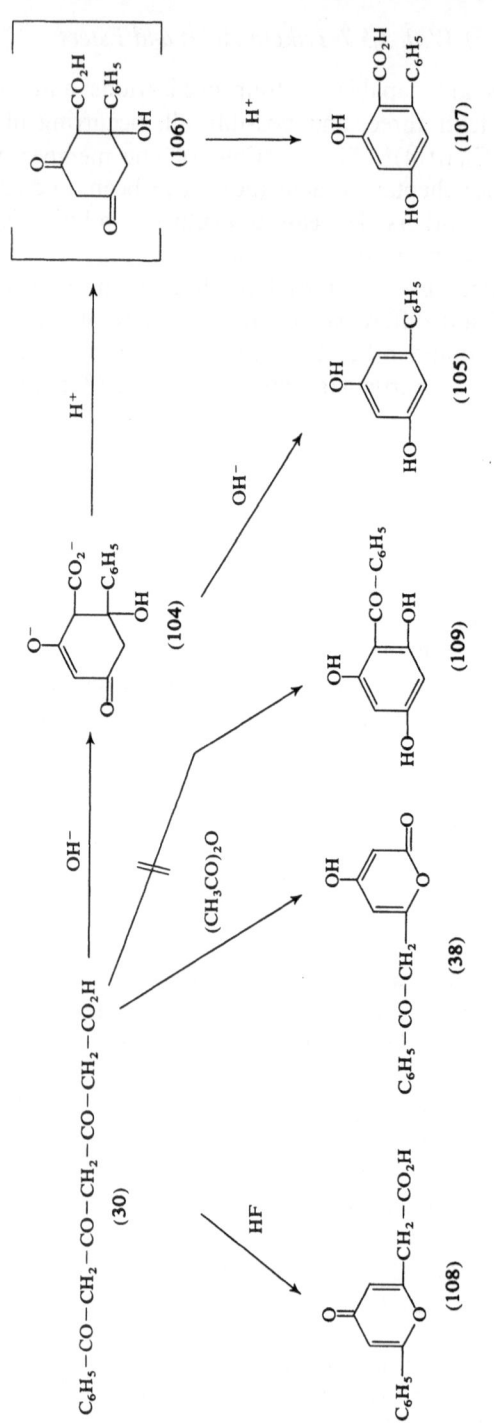

Chart 22. Cyclizations of triketo acid (**30**)

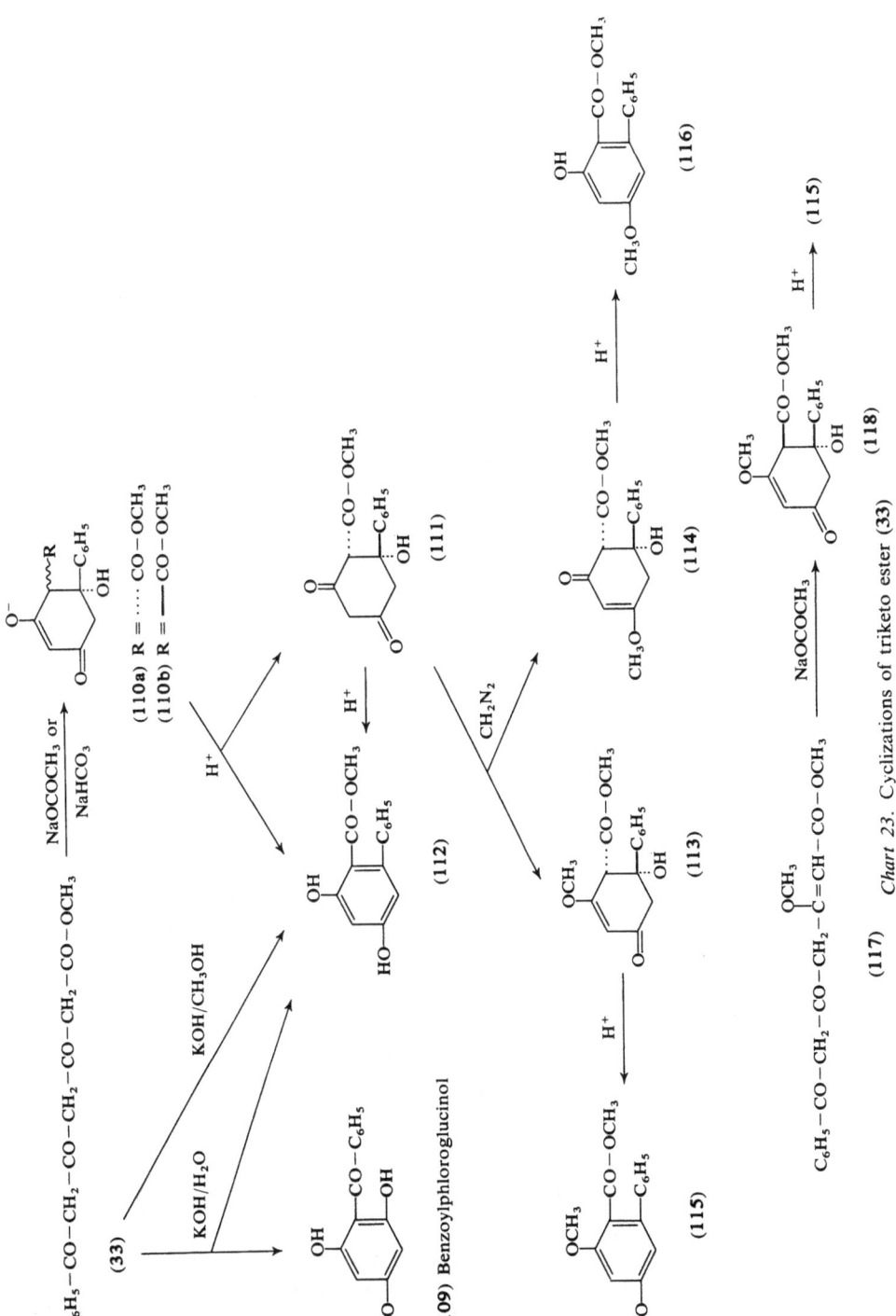

Chart 23. Cyclizations of triketo ester (33)

giving anion (110) (95). The anion was characterized by ultraviolet and n.m.r. spectra; the latter indicated that both epimers (110a) and (110b) were present. The initial ratio of the two was 2.3:1, but this dropped to 1.3:1 after equilibration. Careful neutralization of the anions gave 53% of aldol product (111), the remainder going to resorcylic ester (112). The epimer of (111) was not detected. In acidic solution, quantitative conversion of (111) to (112) was observed. Dissolution of (111) in aqueous sodium bicarbonate gave back anion (110a), which gradually epimerized to an equilibrium mixture with (110b). Deuterium exchange studies have shown that the mechanism of equilibration involves ionization of the 1-methinyl proton rather than reversal of the cyclization process.

The relative configuration of aldol product (111) was established by correlation with enol ether (113) for which stereochemical assignments could be made by spectroscopic comparisons with its epimer (118). Methylation of (111) with diazomethane gave a mixture of enol ethers (113) and (114), the structures of which were assigned by acid-catalyzed aromatization to O-methylated resorcylic esters (115) and (116), respectively. Facile aromatization of enol ether (113) thwarted attempts to epimerize it but epimer (118) was ultimately synthesized by an independent method involving cyclization of ester (117), formed by treatment of triketo acid (30) with diazomethane (84). The key differences observed in the spectra of the epimeric enol ethers are that the infrared spectrum of (113) showed intramolecular hydrogen bonding between the hydroxyl and carbomethoxyl groups and the n.m.r. spectrum showed allylic coupling (J = 2 Hz) between the 1-methinyl and 3-vinyl protons. These data establish the carbomethoxyl group as equatorial in (113) and axial in (118) with the phenyl group equatorial in both. Therefore, the structure of the aldol product is (111), because epimerization is unlikely to have occurred during treatment with diazomethane. These results lend credence to BIRCH's prediction that aldol metabolites, like (111), might be relatively stable, requiring enzymes to catalyze dehydration (13).

Treatment of triketo ester (33) with methanolic potassium hydroxide gave resorcylic ester (112) directly in 92% yield, the aldol product (111) being unstable in strong base. Remarkably, the last remaining cyclization product was formed when ester (33) was treated with aqueous potassium hydroxide. The Claisen product, benzoylphloroglucinol (109), was obtained along with a lesser amount of resorcylic ester (112). The ratio of the two was approximately 2:1; the isolated yield of benzoylphloroglucinol was 47% (75, 76).

The mechanistic basis for the change of course of the cyclization of triketo ester (33) between alcoholic and aqueous base is not known. Attempts to accomplish Claisen cyclization included the use of magnesium hydroxide, magnesium methoxide and a pH 9.5 buffer containing magnesium ion. In all these cases aldol closure was the only cyclization observed; moreover, the rate of cyclization was reduced by precipitation of magnesium chelates (76, 86).

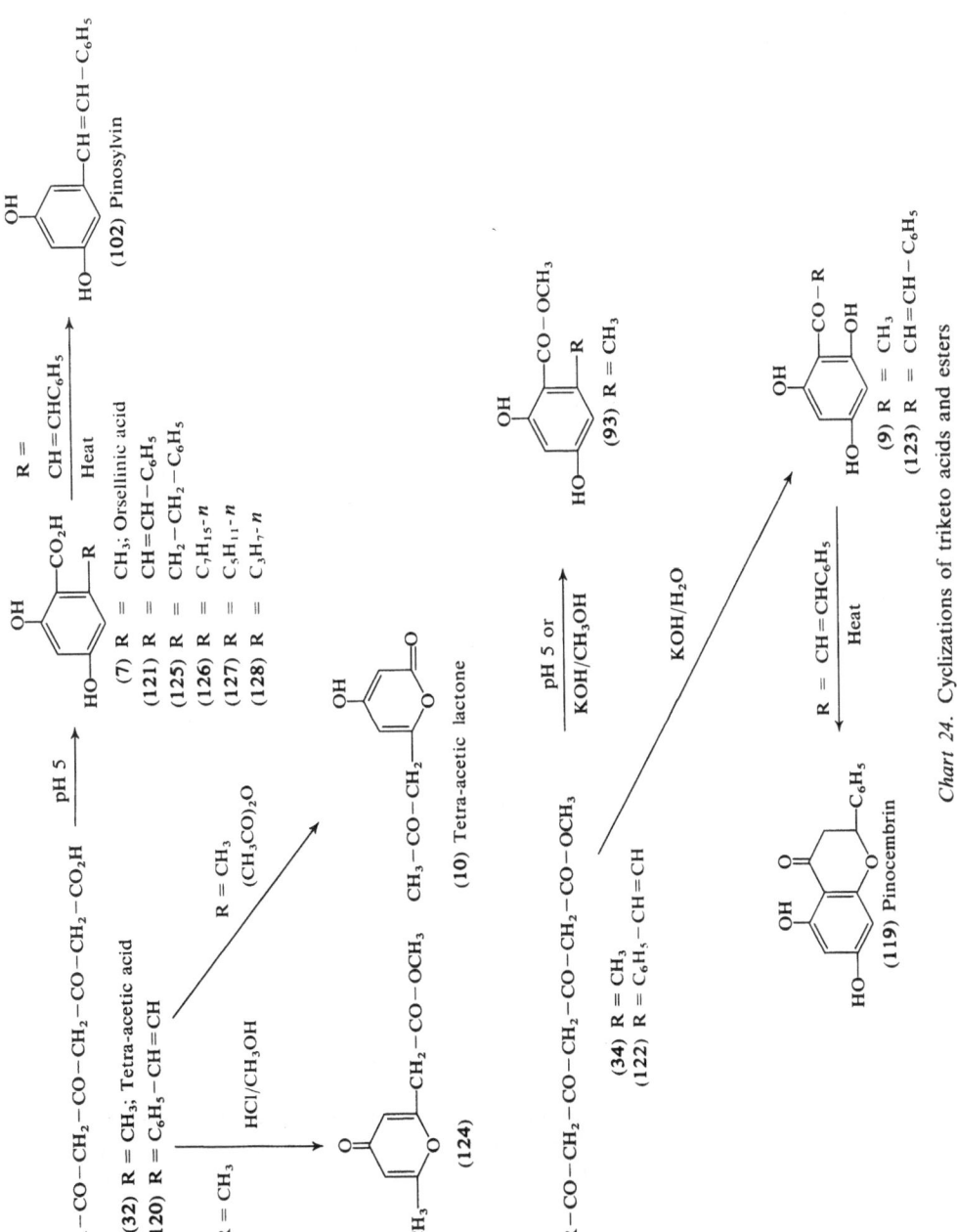

Chart 24. Cyclizations of triketo acids and esters

Benzoyl initiation of polycarbonyl chains is rare in nature; cotoin, which is a methyl ether of phloroglucinol (109), is the principal example of this class of metabolites (*101, 136*). By starting with other triketo acids a number of tetraketide natural products have been synthesized. One of the most interesting cases is the synthesis of pinosylvin (102) and pinocembrin (119) (*74, 75, 76*). Both compounds are found in the heart-wood of pine trees and Birch has suggested that they arise from a single precursor, the coenzyme A ester of triketo acid (120) (*13*). Aldol cyclization of triketo acid (120) occurred readily (88%) in pH 5 buffer; the resulting resorcylic acid (121) decarboxylated on heating to give pinosylvin (102). Triketo ester (122) underwent aldol cyclization in weak base (69%) but gave predominantly (82%) the Claisen product, acyl-phloroglucinol (123), in aqueous potassium hydroxide. The acylphloro-glucinol cyclized, when heated, to give pinocembrin (119) (Chart 24).

Tetra-acetic acid (32) is also of interest because its cyclization products are well known. The acid undergoes extremely facile aldol cyclization to orsellinic acid (7) (53%) in pH 5 buffer and lactonization to tetra-acetic lactone (10) (66%) in acetic anhydride (*96, 130*). Treatment with methanolic hydrogen chloride gave 4-pyrone-ester (124). Methyl ester (34) of tetra-acetic acid cyclized under most conditions to methyl orsellinate (93) but gave (39%) of the Claisen product, acetylphloro-glucinol (9), on treatment with aqueous potassium hydroxide (Chart 24).

Aldol cyclizations of several other triketo acids have been reported (*74, 76*). Resorcylic acids (125)—(128) have been synthesized in this manner; they or the related resorcinols are known metabolites (Chart 24).

The simplest of the protected triketo esters is enol ether (117), the aldol cyclization of which is described above and depicted in Chart 23. The initial aldol adduct (118) readily aromatized under acidic and more strongly basic conditions, giving methylated resorcylic ester (115) (*84*). Treatment of enol ether (117) with pH 9.5 buffer gave a mixture of ester (115) and the Claisen product (129), identical with natural cotoin (Chart 25). These reactions of enol ether (117) have been proposed as models of the hypothetical biosynthetic sequence in which O-methyl groups are introduced into phenolic natural products prior to formation of the aromatic rings (*84*).

Bram has obtained aldol cyclization concurrent with removal of the two protective groups from ester (35). Ethyl orsellinate was not obtained; hydrolysis of the ester and decarboxylation occurred as well and orcinol (3) was the only product identified (*23*) (Chart 25).

4-Hydroxypyrano[4,3-b]pyran-2,5-diones can be viewed as stabilized 3,5,7-triketo acids. Although pyranopyrandiones are stable under most conditions, hydrolysis of the lactone rings occurs in base and the resulting triketo diacids (or diesters) can undergo the same cyclization

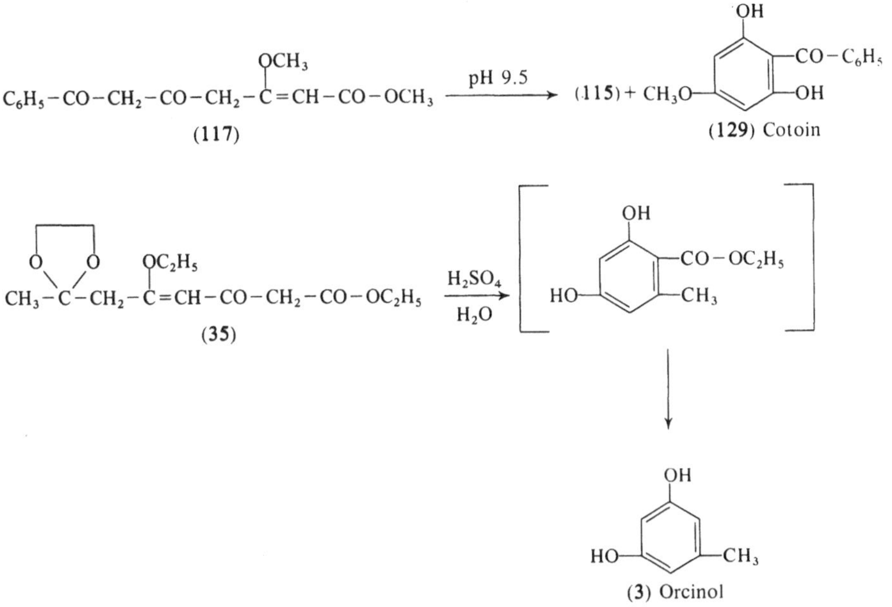

$$C_6H_5-CO-CH_2-CO-CH_2-\overset{\overset{\displaystyle OCH_3}{|}}{C}=CH-CO-OCH_3 \xrightarrow{\text{pH 9.5}} (115)+CH_3O-$$

(117)

(129) Cotoin

$$CH_3-\overset{\overset{\displaystyle O\diagdown\diagup O}{|}}{C}-CH_2-\overset{\overset{\displaystyle OC_2H_5}{|}}{C}=CH-CO-CH_2-CO-OC_2H_5 \xrightarrow[H_2O]{H_2SO_4}$$

(35)

(3) Orcinol

Chart 25. Cyclizations of acyclic protected triketo esters (**35**) and (**117**)

reactions as their monocarboxylic counterparts. As models, pyrano-pyrandiones suffer from the potential shortcoming that the "extra" carboxyl group can alter the reactivity of the remainder of the molecule. MONEY, SCOTT and their coworkers have carried out detailed investigations of four pyranopyrandiones; these studies were undertaken prior to publication of a synthesis of unprotected 3,5,7-triketo acids in 1965 (*53, 54, 124, 125, 126*).

With 7-methyl pyranopyrandione (**41**), cleavage and aldol-type recyclizations were observed with potassium hydroxide (Chart 26). With aqueous potassium hydroxide, the sole isolable (6%) product was orsellinic acid (**7**). Better recoveries were obtained with methanolic potassium hydroxide; the reaction gave 18% of methyl orsellinate (**93**), 5.4% of diester (**130**) and 10% of an ester identified as (**131**). With methanolic magnesium methoxide, recyclization followed a Claisen pathway and phloroglucinol (**132**) was obtained in 8.5% yield. The low recovery of products in all three cases reflects a fundamental problem; namely, the acyclic intermediates are substantially less stable in base than their precursors. These acyclic intermediates have not been detected, nor is it likely that they could be, except possibly in the magnesium methoxide reaction where magnesium chelates might precipitate.

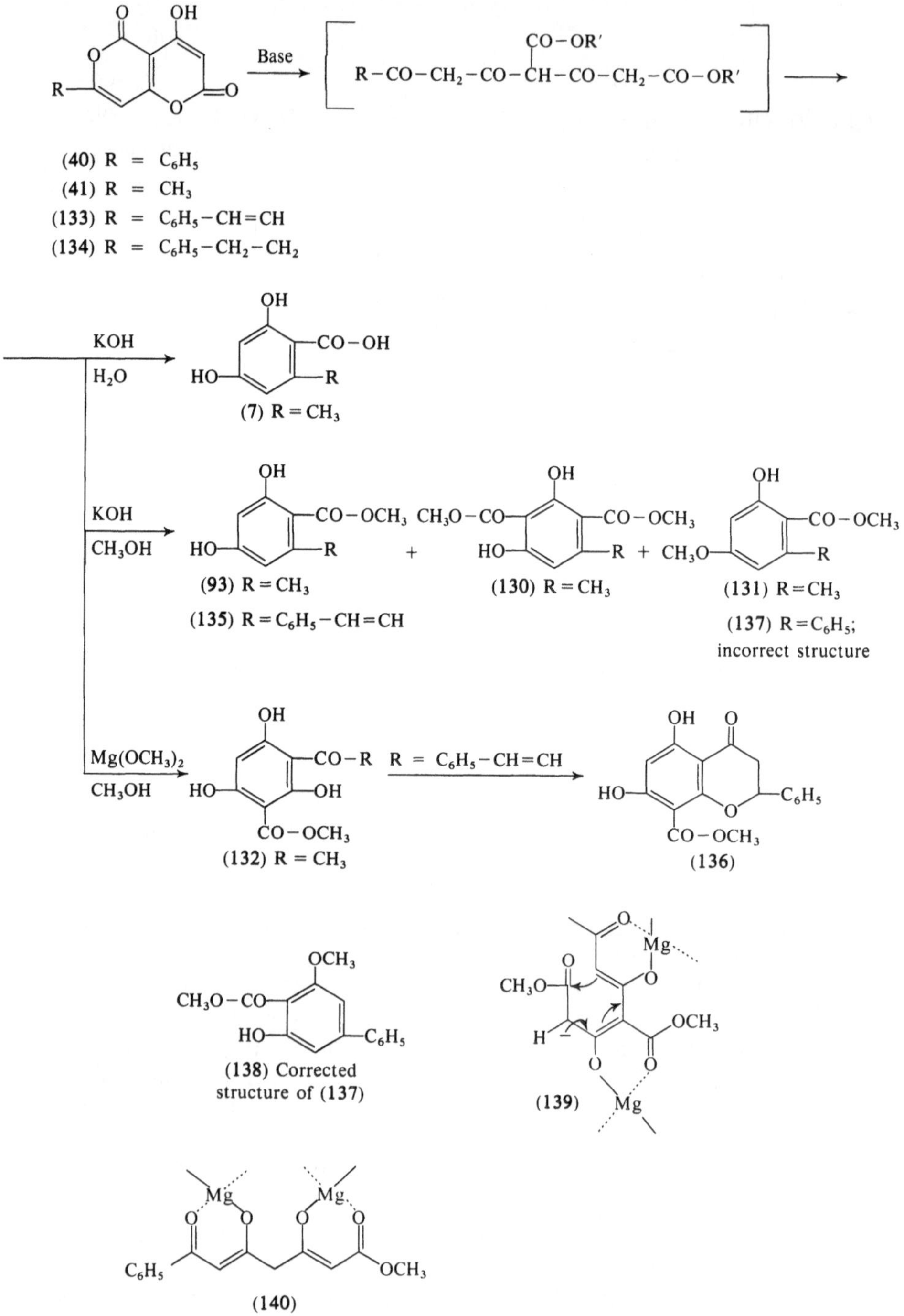
Base
R–CO–CH2–CO–CH(CO–OR′)–CO–CH2–CO–OR′
(40) R = C6H5
(41) R = CH3
(133) R = C6H5–CH=CH
(134) R = C6H5–CH2–CH2
KOH
H2O
(7) R = CH3
KOH
CH3OH
(93) R = CH3
(135) R = C6H5–CH=CH
(130) R = CH3
(131) R = CH3
(137) R = C6H5; incorrect structure
Mg(OCH3)2
CH3OH
(132) R = CH3
R = C6H5–CH=CH
(136)
(138) Corrected structure of (137)
(139)
(140)

Chart 26. Cleavage-recyclization of pyranopyrandiones

7-Phenyl (**40**), 7-styryl (**133**) and 7-phenethyl (**134**) pyranopyrandiones have given similar results. With the styryl compound (**133**), methanolic potassium hydroxide gave resorcylic ester (**135**) and magnesium methoxide gave flavanone (**136**). The latter presumably arose via the corresponding cinnamoylphloroglucinol. Resorcylic ester (**135**) and flavanone (**136**) were subsequently transformed into pinosylvin (**118**) and pinocembrin (**119**), respectively. With phenyl pyranopyrandione (**40**), the structure (**137**) of the methyl ether obtained from the reaction with methanolic potassium hydroxide has been challenged by HOWARTH and HARRIS on the basis of an independent synthesis and has been reassigned as isomeric ether (**138**) (*94*). The latter is more reasonable on mechanistic grounds, and it seems probable that similar reassignments should be made for the corresponding methyl ethers obtained from the other pyranopyrandiones.

The effect of magnesium ion is interesting; CROMBIE and JAMES have put forward the suggestion that two magnesium ions simultaneously chelate with the triketo diester intermediate, holding it in a conformation (**139**) favoring Claisen cyclization (*50*). This proposal is compatible with the observation that magnesium ion does not direct Claisen cyclization of triketo monoester (**33**), where, with one of the chelation sites of (**139**) absent, chelation with two magnesium ions would hold the carbonyl chain of (**33**) in conformation (**140**) from which cyclization could not occur (*76*). As a consequence, although chelation undoubtedly plays a major role in some biological cyclization reactions of polycarbonyl compounds, the effect seen here is unlikely to have a direct biological counterpart.

Monocyclic pyrones containing the 3,5,7-triketo acid skeleton have also been investigated. The simplest members of this group are the enol lactones of the triketo acids. Tetra-acetic lactone has been investigated by BENTLEY and ZWITKOWITS, who found that the lactone rearranged to orsellinic acid (**7**) and orcinol (**3**) on treatment with potassium hydroxide and with sulfuric acid and that uncatalyzed ethanolysis of the lactone (110° for 48 hours) gave ethyl orsellinate (*10*). The Claisen rearrangement product, acetylphloroglucinol (**9**), was not detected.

Phenacyl pyrone (**38**) has been studied in both the SCOTT and the HARRIS laboratories (*68, 86, 87, 145*). The compound is somewhat more stable than tetra-acetic lactone (**10**). Treatment of (**38**) with aqueous potassium hydroxide gave the aldol products, resorcinol (**105**) and resorcylic acid (**107**), in low yields. Better results were obtained with methanolic potassium hydroxide; 55—67% of resorcylic ester (**112**) was isolated. Another product, present in trace quantities, was detected and tentatively identified as the Claisen product, phloroglucinol (**109**) (Chart 27).

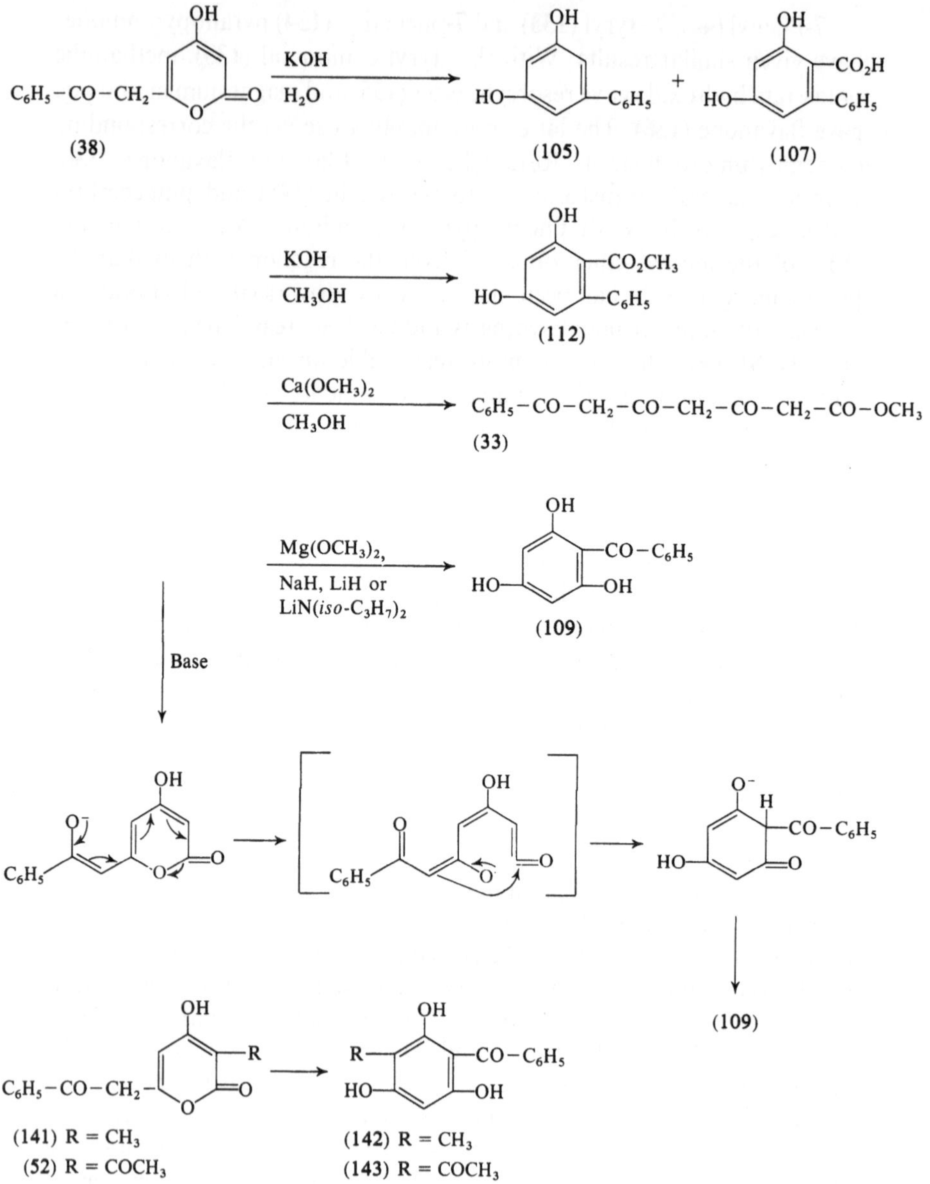

$$C_6H_5-CO-CH_2- \quad (38) \quad \xrightarrow[\text{H}_2\text{O}]{\text{KOH}} \quad (105) \quad + \quad (107)$$

$$\xrightarrow[\text{CH}_3\text{OH}]{\text{KOH}} \quad (112)$$

$$\xrightarrow[\text{CH}_3\text{OH}]{\text{Ca(OCH}_3)_2} \quad C_6H_5-CO-CH_2-CO-CH_2-CO-CH_2-CO-OCH_3 \quad (33)$$

$$\xrightarrow[\substack{\text{NaH, LiH or} \\ \text{LiN(}iso\text{-C}_3\text{H}_7)_2}]{\text{Mg(OCH}_3)_2,} \quad (109)$$

Base

(109)

(141) R = CH$_3$
(52) R = COCH$_3$

$$C_6H_5-CO-CH_2- \quad \longrightarrow \quad (142) \quad R = CH_3 \quad (143) \quad R = COCH_3$$

(142) R = CH$_3$
(143) R = COCH$_3$

Chart 27. Cleavage and recyclization of phenacyl pyrones (**38**), (**52**) and (**141**)

Harris and coworkers investigated the use of other bases in a search for conditions under which phloroglucinol (**109**) would become the major product (*86, 87*) Magnesium methoxide and calcium methoxide were found to be highly effective for this purpose. With one equivalent of magnesium methoxide in methanol, 44% of phloroglucinol (109) and 15% of resorcylic ester (112) were obtained. A trace of triketo ester (33) was also detected in the product mixture. The yield of triketo ester (33) was optimized (58%) by the use of a large excess of calcium methoxide in methanol. Apparently, the calcium salt of (33) is only sparingly soluble in methanol and consequently survives the reaction conditions. This represents the only example of successful unmasking of a protected triketo acid. An excellent yield (87%) of phloroglucinol (109) was obtained using 0.27 equivalents of magnesium methoxide in hot dimethylformamide.

The formation of phloroglucinol (109) under these conditions was surprising because triketo ester (33) was thought to be an intermediate in the formation of (109) and magnesium methoxide was known to catalyze only aldol cyclization of the triketo ester. HARRIS and coworkers initially proposed that the chelate obtained when magnesium methoxide and triketo ester are mixed differs from the chelate derived from cleavage of pyrone (38) (87). The structural differences direct aldol cyclization in the first case and Claisen cyclization in the second. This proposal became unattractive when non-nucleophilic bases were also found to convert pyrone (38) into the Claisen product (109) (86). Equimolar lithium hydride in tetrahydrofuran gave 86% of phloroglucinol (109). A similar result was obtained with lithium diisopropylamide in tetrahydrofuran. It is interesting that these reactions required no more than one equivalent of base and were actually slowed by excess base. In the light of these results, HARRIS and WACHTER proposed (86) that Claisen product (109) arises via a ketene, this intermediate being formed by a spontaneous cleavage of the monoanion of the pyrone. Recyclization of this ketene gives an anion of phloroglucinol (109). Triketo ester (33) and the aldol products (105), (107) and (112) may arise by attack of nucleophilic bases and solvents on the ketene intermediate or directly on pyrone (38). The Claisen rearrangement reaction has been employed successfully with pyrones (141) and (52), giving phloroglucinols (142) and (143), respectively.

The ketene-mediated acylphloroglucinol synthesis is of value in its own right but its relevance to biosynthetic processes is unknown. Ketene intermediates have been implicated in related reactions of β-keto esters (25, 160) and it is possible that they play an, as yet, unrecognized part in biological reaction of the carboxyl group of β-polyketo esters.

Cleavage-recyclization reactions of the methyl ethers of pyrones (10) and (38) have been investigated (Chart 28). SCOTT and coworkers have transformed methyl ether (144) into methylated resorcylic ester (146) using an unspecified base but failed to obtain the comparable conversion of (145) into (115) with potassium hydroxide or sodium methoxide (146).

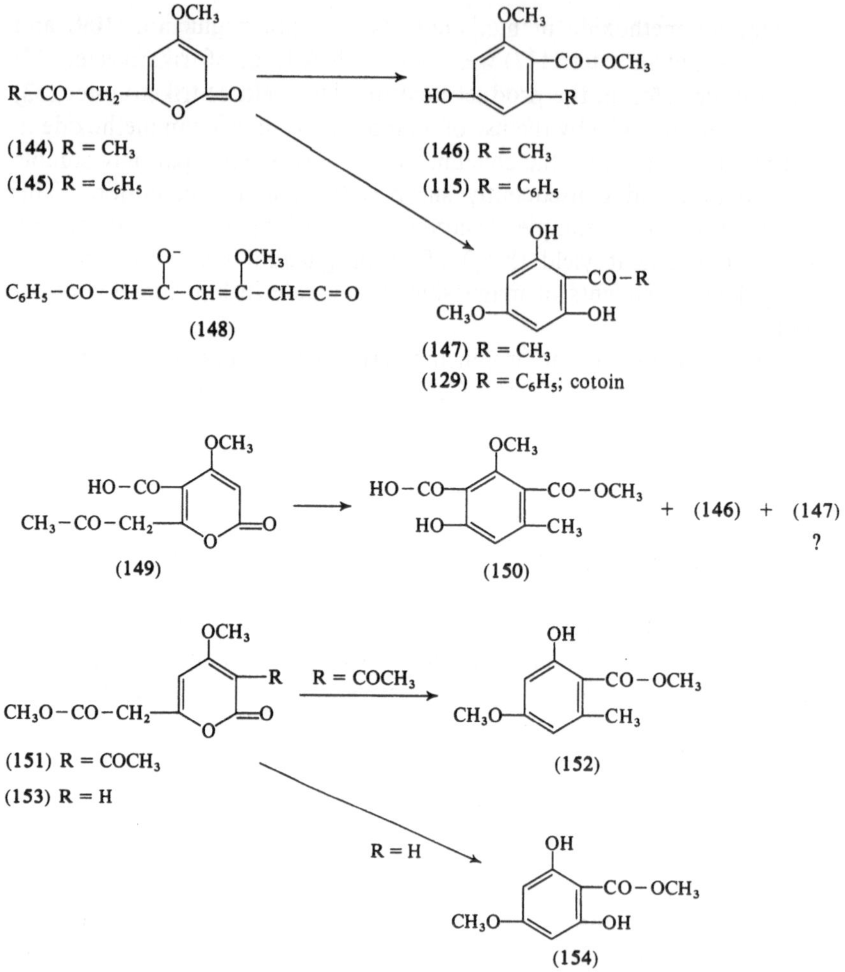

Chart 28. Cleavage and recyclization of additional pyrones

Claisen-type rearrangement of methyl ether (144) into methylated phloro-glucinol (147) was achieved by YAMAMURA and coworkers, who also did not specify their conditions (175); HARRIS and WACHTER have reported the Claisen-type rearrangement of (145) into cotoin (129) via ketene (148) in 89% yield using lithium hydride (86).

Pyrone-acid (42), which is an intermediate in SCOTT's first synthesis of tetra-acetic lactone (10) from pyranopyrandione (41) (68), has been transformed into aldol products. SCOTT and coworkers obtained orcinol (3) and orsellinic acid (7) from treatment of pyrone-acid (42) with potassium hydroxide but failed to obtain reaction with magnesium methoxide, apparently because of precipitation of the magnesium salt of (42) (*145*). The methyl ether (149) of pyrone-acid (42) did react with magnesium methoxide and gave methylated resorcylic ester (146) (28%) along with a lesser amount of ester-acid (150) and a trace of a third component, provisionally identified as the Claisen product, phloroglucinol (147) (*146*). Inexplicably, CARNDUFF *et al.*, during an attempt to repeat SCOTT's synthesis of tetra-acetic lactone (10), obtained the Claisen product, acetylphloroglucinol (9), from pyrolysis of (42) in the presence of a copper-bronze catalyst (*29*).

Pyrone-ester (151) contains the tetra-acetic acid skeleton in a different configuration from that found in pyranopyrandione (41) and in pyrone (10). MONEY and his group have investigated the cleavage-recyclization of (151) with the hope that it might preferentially rearrange in base to Claisen product(s). The results were disappointing; only the aldol product, ester (152), was obtained (*8*). However, pyrone-ester (153), rearranged in good (~ 50%) yield to Claisen product (154) on treatment with base. Pyrone-ester (153) might alternatively have given an isomeric Claisen product but it could not give a resorcylate derivative.

d) With 1,3,5,7-Tetraketones

β-Tetraketones can cyclize to give 4-acylresorcinols. Diphenyl tetraketone (44) has been synthesized in several different ways but it has never been cyclized. The compound is relatively stable but would probably give resorcinol (155) in base. Phenyl methyl tetraketone (45) has been cyclized (*130*). The reaction occurred readily in base, giving predominantly (62%) resorcinol (156), although a minor amount (7%) of isomer (157) was also isolated. Dimethyl tetraketone (46) is the most reactive of the three, cyclizing on silica gel to give resorcinol (158) (*131*). These results point to a strong steric preference for nucleophilic attack at acetyl rather than benzoyl groups (Chart 29).

A variety of protective groups for tetraketones have been investigated. Bis-ketal (48) and bis-enamine (50) of tetraketone (46) both gave resorcinol (158) under conditions of acid hydrolysis (*155, 156*). Pyranopyrandione (54) has been transformed into the same product by treatment with acid (*91*). Base treatment of (54) gave products in which one of the pyrone rings had not been opened. Pyranopyrandione (55) gave a low yield of resorcinol (158) when treated with methanolic potassium hydroxide; acidic conditions have not been investigated (*124, 126*). Pyrone (53) was found to be stable to methanolic sodium hydroxide at room temperature but degraded within two hours in refluxing base (*145*).

Pyrone (52), which is a protected form of tetraketone (45), failed to rearrange to any phenolic products on treatment with methanolic

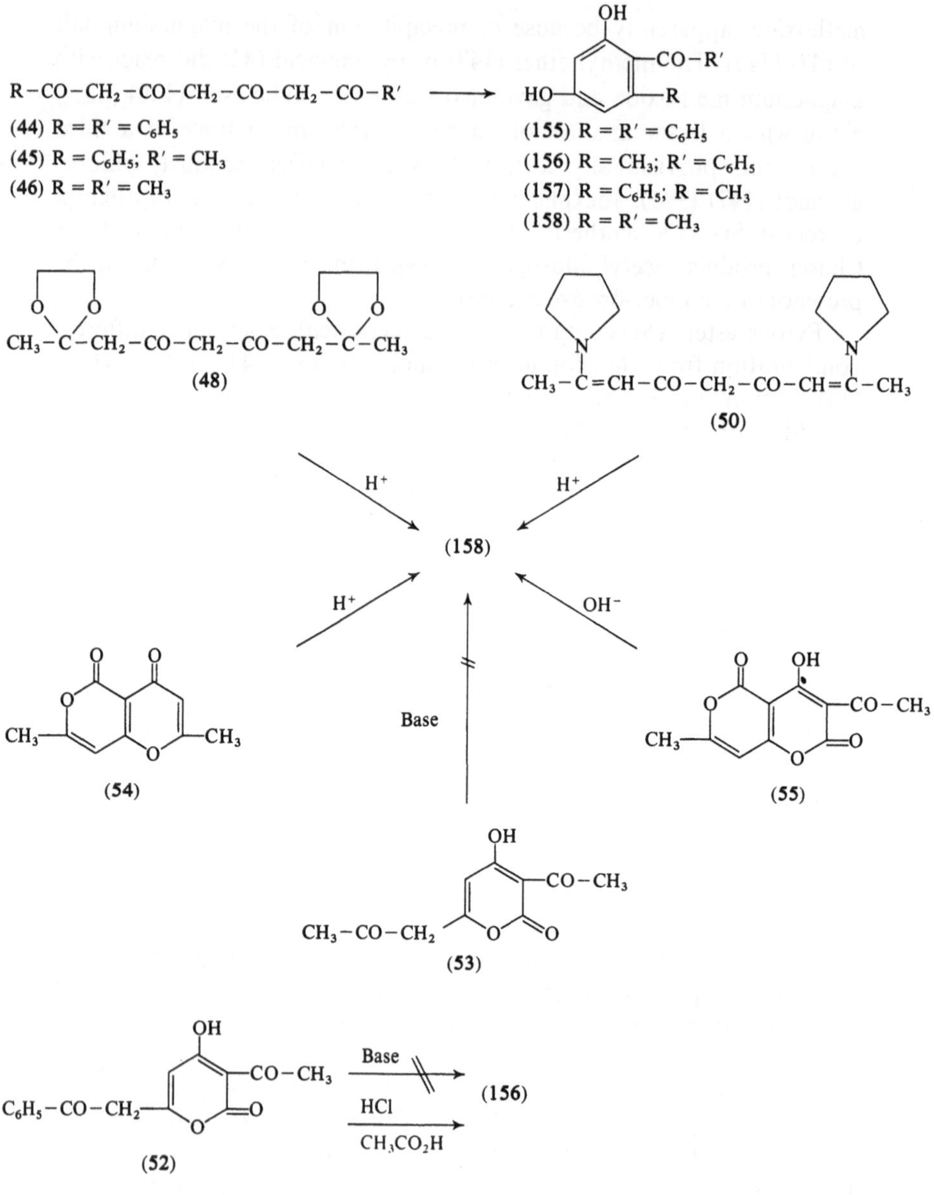

$R-CO-CH_2-CO-CH_2-CO-CH_2-CO-R' \longrightarrow$

(44) $R = R' = C_6H_5$

(45) $R = C_6H_5$; $R' = CH_3$

(46) $R = R' = CH_3$

(155) $R = R' = C_6H_5$

(156) $R = CH_3$; $R' = C_6H_5$

(157) $R = C_6H_5$; $R = CH_3$

(158) $R = R' = CH_3$

$CH_3-C-CH_2-CO-CH_2-CO-CH_2-C-CH_3$

(48)

$CH_3-C=CH-CO-CH_2-CO-CH=C-CH_3$

(50)

(158)

(54)

(55)

Base

(53)

(52) $\xrightarrow[\text{CH}_3\text{CO}_2\text{H}]{\text{HCl}}$ (156)

Chart 29. Cyclizations of free and protected tetraketones

potassium hydroxide (*145*), but gave 24% of resorcinol (**156**) in acid (*83*). The isomeric resorcinol (**157**) was not detected in the latter reaction, indicating that the acid-catalyzed cleavage-recyclization reaction had the same steric preference as the basic cyclization of tetraketone (**45**).

e) With 3,5,7,9-Tetraketo Acids and Esters

β-Tetraketo acids are capable of undergoing four separate carbocyclic ring closures; one is a Claisen-type reaction yielding a phloroglucinol and the other three are aldol types giving resorcinols. Heterocyclic ring closures are also possible; moreover, the carbocyclic products can undergo secondary cyclizations to give benzopyran derivatives. Pentaketide natural products are known exemplifying all of the homocyclic ring closures.

HARRIS and MURPHY have studied the cyclizations of tetraketo acid (56), which is the only known unprotected tetraketo acid (85, 130) (Chart 30). Treatment of acid (56) with aqueous sodium bicarbonate gave 84% of resorcinol (159), an unstable compound which readily cyclized to coumarin (160). Aqueous potassium hydroxide gave 28% of resorcinol (159) but the major product (67%) was resorcinol (161). Weakly acidic conditions, pH 5, were investigated briefly and appeared to give results similar to aqueous bicarbonate. The cyclizations of the methyl ester of tetraketo acid (56) in base followed the same pattern as the free acid. Neither with the ester nor with the free acid was the third aldol product, resorcinol (162), or the Claisen product, phloroglucinol (163), or derivatives thereof observed. The results with the tetraketo ester must be contrasted with the cyclizations of triketo esters in aqueous potassium hydroxide which give predominantly the Claisen products.

MONEY and his coworkers have investigated the reactions of dipyrano-pyrantrione (60), which is a masked penta-acetic acid (124, 126) (Chart 30). Treatment of (60) with aqueous potassium hydroxide gave a low yield of resorcinol (158) as the only product. Methanolic potassium hydroxide gave more interesting results. With it eight products, totalling a 12—15% yield, were identified. Resorcinols (130) and (158) resulted from cleavage of the penta-acetate chain and are aldol products. The other six, (164)—(169), are aldol-type products arising from ring closure between the 2 methylene and 7 carbonyl groups of the intermediate 3,5,7,9-tetraketo acid. This result provides a provocative contrast with the reactions of tetraketo acid (56) and its ester in which the corresponding cyclization product (162) was not observed. Treatment of dipyrano-pyrantrione (60) with magnesium methoxide gave a different group of products, the total yield being 10—15% (41). Four of these, (93), (130), (170) and (171), are degraded species. The remaining two are derived from penta-acetic acid, resorcinol (172) resulting from a 3:8 aldol cyclization of the penta-acetate chain and chromone (173) arising from Claisen-type cyclization followed by ring closure. Apparently, regio-specificity of magnesium chelation is incomplete in the cleavage of this dipyranopyrantrione.

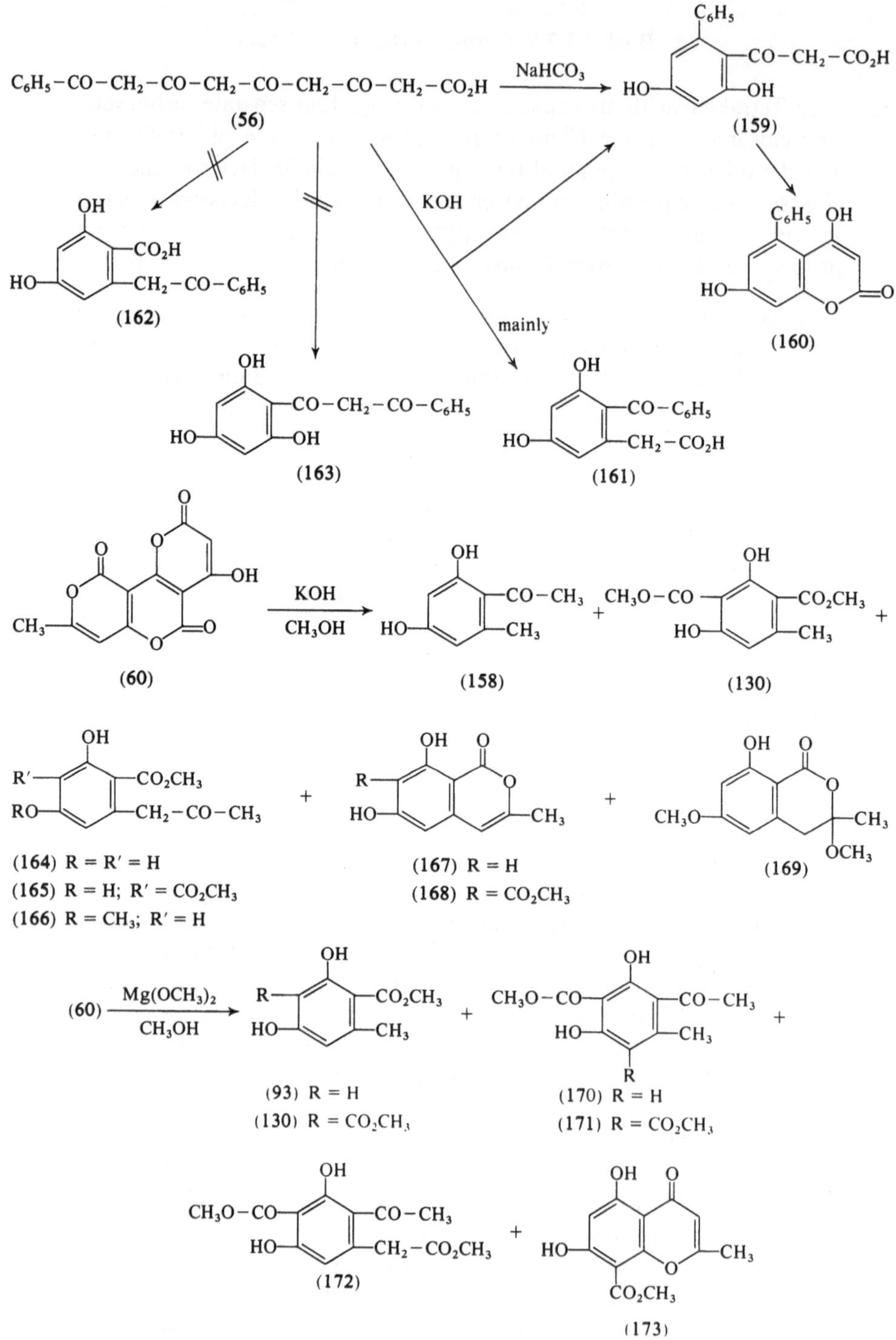

$$C_6H_5-CO-CH_2-CO-CH_2-CO-CH_2-CO-CH_2-CO_2H \xrightarrow{\text{NaHCO}_3}$$

(56)

(159)

(162)

KOH

(160)

mainly

(163)

(161)

(60) $\xrightarrow[\text{CH}_3\text{OH}]{\text{KOH}}$ (158) + (130) +

(164) R = R' = H
(165) R = H; R' = CO$_2$CH$_3$
(166) R = CH$_3$; R' = H

(167) R = H
(168) R = CO$_2$CH$_3$

(169)

(60) $\xrightarrow[\text{CH}_3\text{OH}]{\text{Mg(OCH}_3)_2}$

(93) R = H
(130) R = CO$_2$CH$_3$

(170) R = H
(171) R = CO$_2$CH$_3$

(172)

(173)

Chart 30. Cyclizations of free and protected tetraketo acids

References, pp. 274—282

In summary, all four of the carbocyclic ring closures have been observed with the two compounds, (**56**) and (**60**), but the results are incomplete in both cases. It must be concluded that the procedures for control of the course of aromatizations, which had been developed with triketo acids and pyranopyrandiones, work only poorly with tetraketo acids and dipyranopyrantriones. Further study, particularly with tetraketo acids and esters, should be fruitful.

f) With 1,3,5,7,9-Pentaketones

Pentaketones can give two types of aldol cyclization products and both have been obtained selectively from pentaketone (**61**) (*132, 168*) (Chart 31). With aqueous potassium hydroxide, resorcinol (**174**) was obtained in 53% yield. The other cyclization occurred on activated silica gel; the reaction gave 73% of resorcinol (**175**) plus a trace of resorcinol (**174**). The cyclization reactions of acetyl-terminated pentaketones, such as compounds (**62**) and (**63**), can potentially give naphthalene derivatives but have not been studied yet.

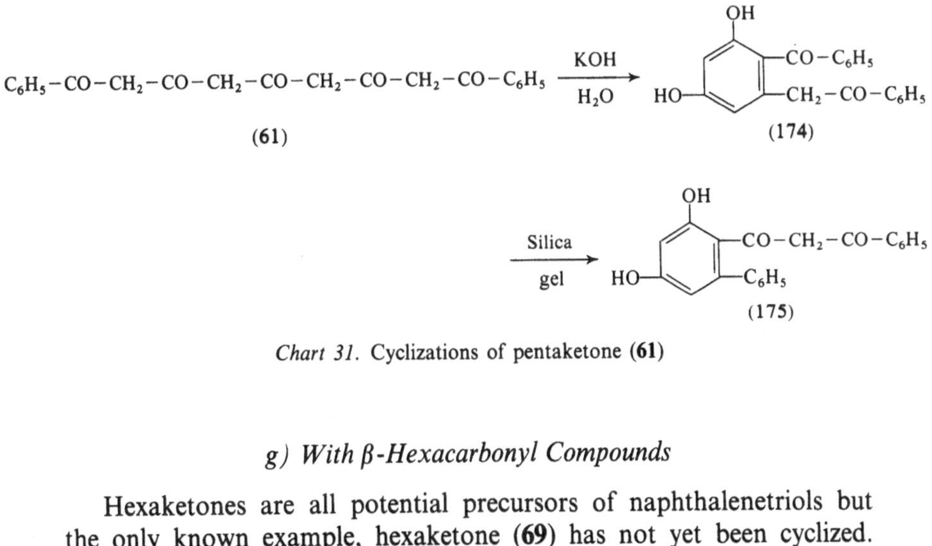
$C_6H_5-CO-CH_2-CO-CH_2-CO-CH_2-CO-CH_2-CO-C_6H_5$
KOH
H_2O
OH
$CO-C_6H_5$
HO
$CH_2-CO-C_6H_5$
(61)
(174)
Silica
gel
OH
$CO-CH_2-CO-C_6H_5$
HO
C_6H_5
(175)

Chart 31. Cyclizations of pentaketone (**61**)

g) With β-Hexacarbonyl Compounds

Hexaketones are all potential precursors of naphthalenetriols but the only known example, hexaketone (**69**) has not yet been cyclized. Tripyranopyrantetraone (**68**), which is a protected hexa-acetic acid, has been studied by Scott and coworkers (*145*). Partial cleavage occurred in aqueous potassium hydroxide giving a dipyrone, identified as (**176**) (Chart 32). In spite of numerous attempts, no conditions were found under which (**68**) or (**176**) could be transformed into carbocyclic products. The fundamental problem here would appear to be that cleavage of the polyketide chain occurs more readily than opening of the pyrone rings. Dipyrone (**67**) is currently under investigation (*123*). Preliminary results

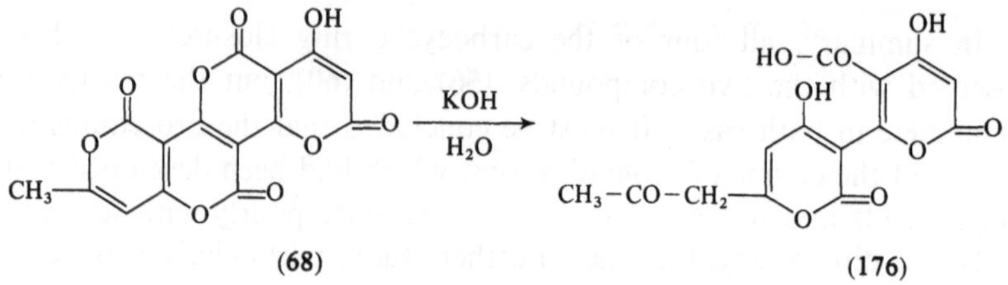

Chart 32. Attempted aromatization of tripyranopyrantetraone (68)

suggest that carbocyclic products containing the undegraded hexaketide chain can be obtained from the dimethyl ether of (67) in basic media.

h) With β-Heptacarbonyl Compounds

The synthesis of phenolic compounds from heptaketone (70) or from tetrapyranopyranpentaone (72) has not been studied. The former should produce a number of interesting products but use of the latter does not appear promising. Scott and coworkers have studied the reactions of dipyrone (71), which is a latent, linear pentaketo diacid (135, 147) (Chart 33). Treatment of this compound with methanolic potassium hydroxide gave xanthone (178) in 15% yield. This transformation requires cleavage of both pyrone rings, Claisen-type recyclization to give benzophenone (177) and finally closure of the heterocyclic ring. An acyclic heptacarbonyl compound could be an intermediate in this process but sequential

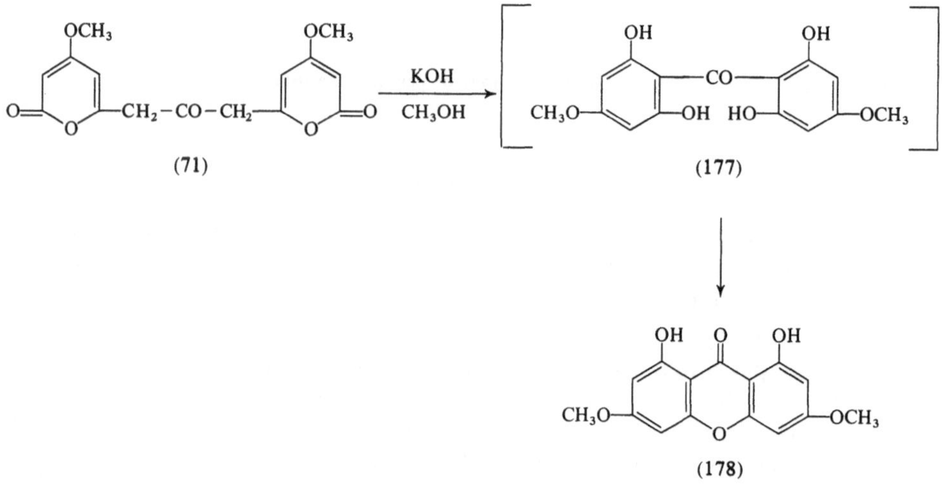

Chart 33. Aromatization of dipyrone (71)

cleavage-recyclization of the pyrone rings seems more probable since an acyclic intermediate would be expected to undergo competing aldol cyclizations. The pattern of functionalization of xanthone (**178**) does not conform precisely to that of naturally occurring polyketide xanthones in which one ring normally arises by aldol cyclization and the other by Claisen cyclization of a hexaketo acid; however, the results are of sufficient interest to encourage further exploration along these lines.

i) With β-Octacarbonyl Compounds

The cyclization reactions of octaketone (**73**) have not been studied; but the reactions should give anthracene derivatives.

j) With β-Nonacarbonyl Compounds

Bis(pyranopyrandione) (**74**), a protected heptaketo diacid, is the largest polyketide system to have been studied (*147*). The compound has promise as a precursor of anthracene derivatives and falls short by only one carbonyl group of being large enough to give tetracyclic products. The acetyl group is readily removed by base, giving a deep blue anion believed to be the dianion of bis (pyranopyrandione) (**179**). SCOTT chose this compound for study because earlier studies had shown that the A rings of similar pyranopyrandiones cleave in base more readily than the B rings (*68, 146*). Thus, base should convert (**179**) into dipyrone (**180**) (Chart 34). Only one aldol cyclization would be possible with (**180**), that giving resorcinol (**181**). Slow cleavage of the pyrone rings followed by rapid aldol condensations might give anthracene dicarboxylic acid (**182**). This sequence, when attempted, was only partially successful. A product, identified as resorcinol (**181**) or a lactone thereof, was isolated but it resisted all attempts to transform it into a naphthalene or anthracene derivative.

k) With Reduced β-Polycarbonyl Compounds

Polyketide metabolites lacking one or more of the expected phenolic hydroxyl groups are common in nature but only a limited number of biogenetic-type syntheses of this class of phenolic compound have been reported. Polycarbonyl compounds with one or more keto groups reduced to carbinols are appropriate starting materials for these syntheses as are the olefinic derivatives resulting from dehydration of the carbinols. The paucity of examples stems mainly from difficulties associated with the synthesis of both the polyketo carbinols and the polyketo olefins.

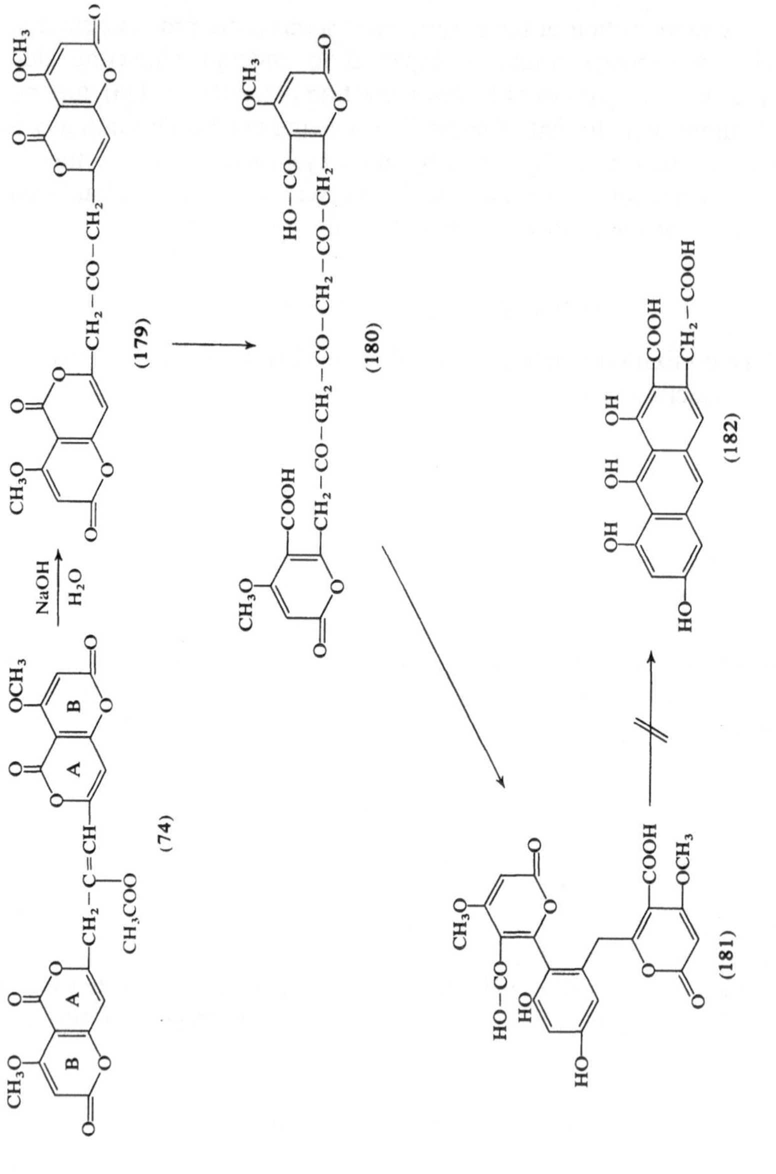

Chart 34. Cleavage-recyclization of bis-(pyranopyrandione) (74)

The most impressive biogenetic-type synthesis to appear thus far is the preparation of 6-methylsalicylic acid (19) from dihydropyrone (81) by SCOTT's group (68, 146). Aqueous potassium hydroxide cleaved dihydro-pyrone (81) to give an orange anion believed to be trianion (183). Un-saturated diketo acid (184) could be obtained by protonation but the acid was too unstable to permit purification and complete character-ization. Cyclization of the acid occurred readily giving 6-methylsalicylic acid (19) (Chart 35). A 28% yield of (19) was obtained by more extended treatment of dihydropyrone (81) with base. Methanolic potassium hydroxide gave comparable results showing that ring opening in (81) occurred by β-elimination rather than by nucleophilic attack on the lactone carbonyl group.

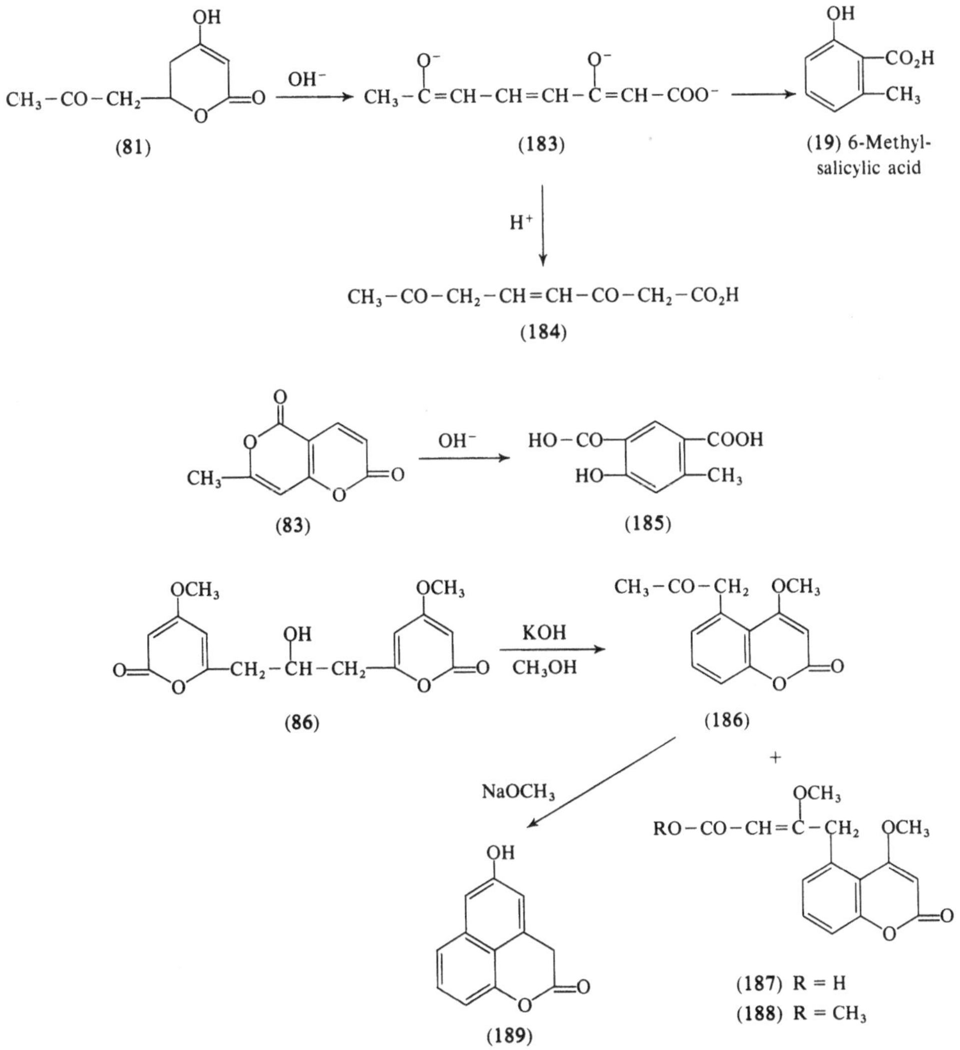

Chart 35. Syntheses with dihydropolyketides

Pyranopyrandione (**83**) behaves similarly. With it, ring cleavage is slower than recyclization and intermediates have not been observed. CHENG and TAN obtained acid (**185**) from base treatment of (**83**) (*34*). This represents an aldol process; it is noteworthy that the "extra" carboxyl group has been retained.

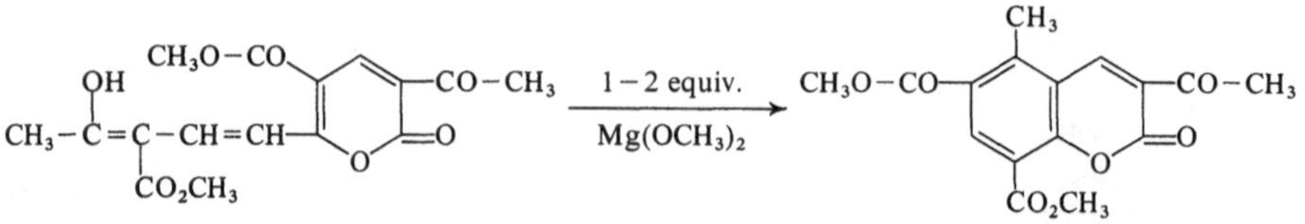

(**87**) Xanthophanic enol

(**190**)

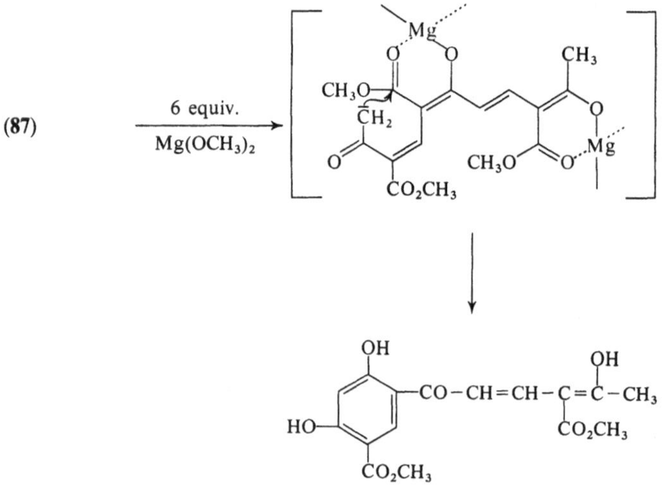

(**191**)

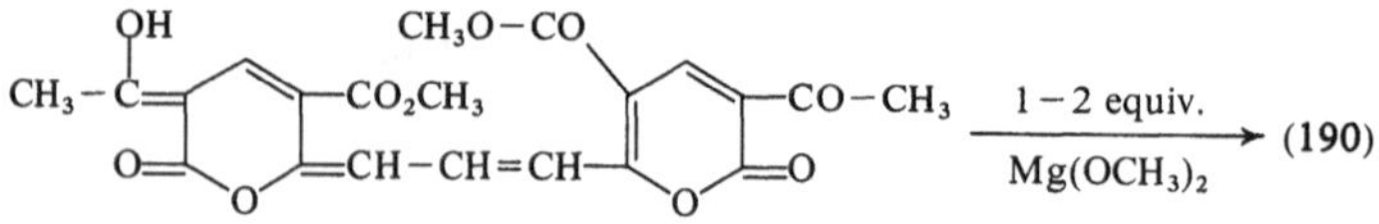

(**88**) Glaucophanic enol

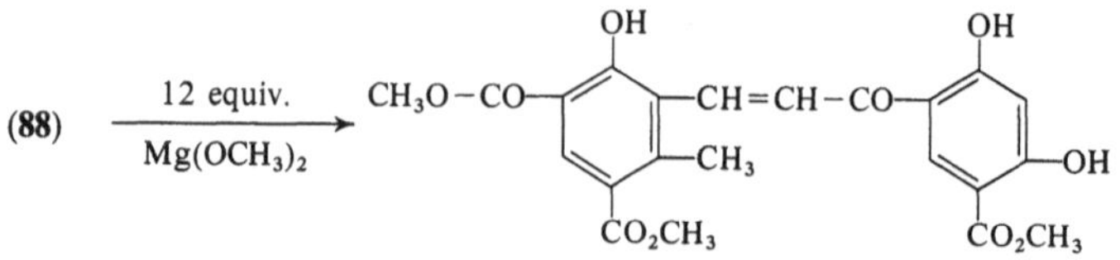

(**192**)

Chart 36. Cleavage-recyclization reactions of xanthophanic enol (**87**) and glaucophanic enol (**88**)

Carbinol-dipyrone (86) is a masked dihydropentaketo diacid and a potential precursor of naphthalenediols. Treatment of (86) with methanolic potassium hydroxide by SCOTT and coworkers gave a complex mixture from which three coumarins (186)—(188) were isolated (147). These coumarins arise by cleavage of the pyrone rings followed by aldol cyclization and lactonization. The coumarin ring structure is very stable but treatment of coumarin (186) with sodium methoxide in refluxing methanol gave a small amount of naphthol (189). The cleavage-recyclization reaction of carbinol-dipyrone (86) provides an interesting contrast with the cleavage-recyclization of the corresponding keto-dipyrone. The latter compound undergoes Claisen recyclizations to give a xanthone. The pathway is repressed in (86) by the reduced acidity of the methylene groups adjacent to the pyrone rings.

Xanthophanic enol (87) and glaucophanic enol (88) are masked poly-carbonyl compounds which have several keto groups deleted. They differ from natural polycarbonyl compounds by being branched but they react similarly and undergo cleavages and recyclizations in base to give phenolic compounds. In a detailed study of the chemistry of (87), (88) and related compounds, CROMBIE and coworkers discovered that the course of the recyclization reactions could be controlled by chelation with metallic cations (43, 44, 45, 46, 47, 48, 49). Xanthophanic enol (87), when treated with sodium methoxide or a limited amount (1—2 equivalents) of magnesium methoxide underwent cleavage and aldol-type recyclization to give coumarin (190) and related species; whereas a large excess (6 equivalents) of magnesium methoxide gave resorcinol (191), which is a Claisen product. Aldol cyclization appears to be the normal course of aromatization of the acyclic intermediate from (87); but, in the presence of high concentrations of magnesium ion, simultaneous chelation of magnesium ions at two sites holds the molecule in configurations directing Claisen cyclization but barring the aldol closure (Chart 36). A similar result has been obtained with glaucophanic enol (88). Excess magnesium methoxide (12 equivalents) gave chalcone (192) via two Claisen cyclizations. With limited amounts of magnesium methoxide, aldol cyclization and partial degradation gave coumarin (190).

3. Using Partially Cyclized Polycarbonyl Compounds

The biosynthesis of polycyclic aromatic metabolites requires multiple intramolecular condensations of polycarbonyl compounds. The natural processes must involve specific sequences for these condensations, the order of which can sometimes be discerned from consideration of cometabolites or on structural grounds alone. Many advances have been made in the synthesis and manipulation of free and protected polycarbonyl compounds, but, as seen in the previous section, adequate control of cyclization does not yet extend beyond the synthesis of monocyclic systems and of a few bicyclic compounds in which one of the rings is heterocyclic.

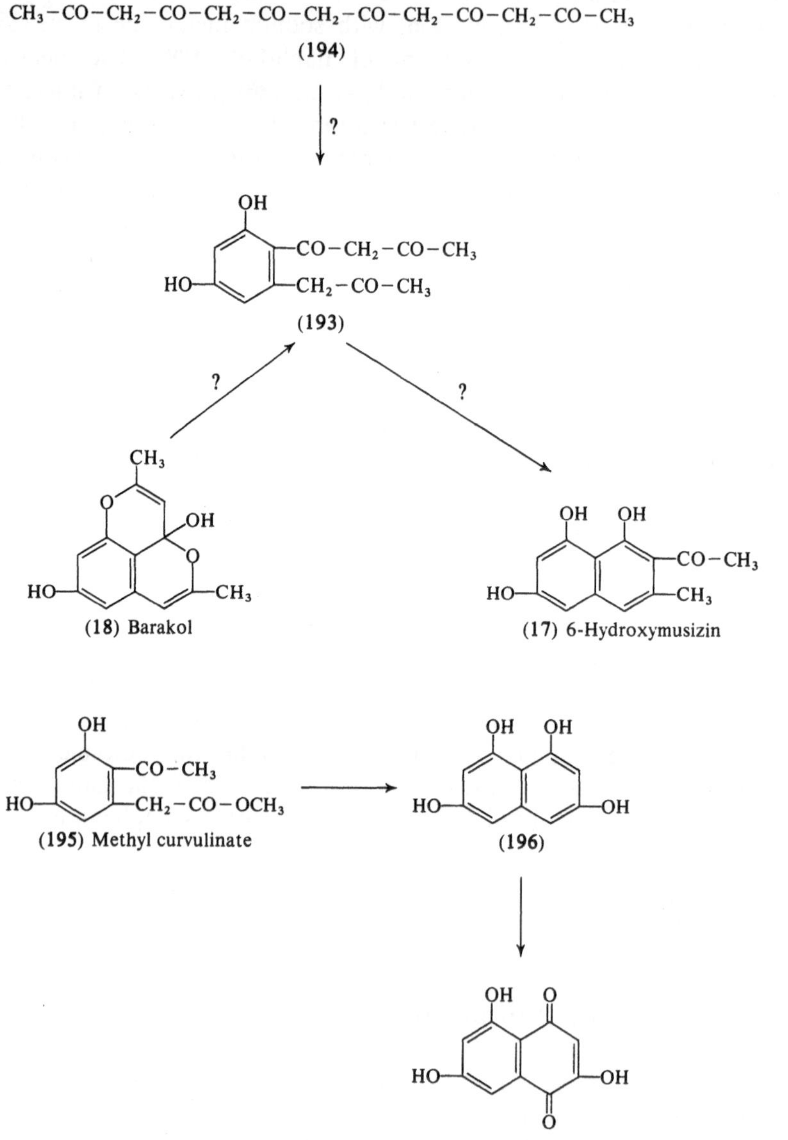

$$CH_3-CO-CH_2-CO-CH_2-CO-CH_2-CO-CH_2-CO-CH_2-CO-CH_3$$

(194)

(193)

(18) Barakol

(17) 6-Hydroxymusizin

(195) Methyl curvulinate

(196)

(197) Flaviolin

Chart 37. Biogenetic-type syntheses of 6-hydroxymusizin (17) and flaviolin (197)

References, pp. 274—282

Another approach to the synthesis of polycyclic compounds along biogenetic lines is their preparation from the species that would result from closure of the first ring. For example, the naphthalenoid metabolite, 6-hydroxymusizin (17), might be synthesized from resorcinol (193) instead of from hexaketone (194) or the corresponding hexaketo acid. A study of this reaction of (193) could provide information that would ultimately be required for a *de novo* synthesis of 6-hydroxymusizin from hexaketone (194). Resorcinol (193) has not, as yet, been synthesized, but barakol (18) is a potential source of it (*28*) (Chart 37).

Flaviolin has been synthesized in this manner by BAKER and BYCROFT (*5*). The methyl ester (195) of curvulinic acid cyclized on treatment with sodium methoxide to give the unstable tetrahydroxynaphthalene (196). Flaviolin was obtained by air oxidation of (196) (Chart 37). Naphthalenoid metabolites having other patterns of oxygenation have been obtained by these workers using related procedures.

Turning to tricyclic examples, the dimethyl ether (201) of eleutherinol has been synthesized by HARRIS (*73*). The condensation of isocoumarin (198) with the dianion of acetylacetone gave adduct (199), which was not isolable, cyclizing to naphthalene (200). The final cyclization was accomplished with trifluoroacetic acid to give naphthopyrone (201) (Chart 38).

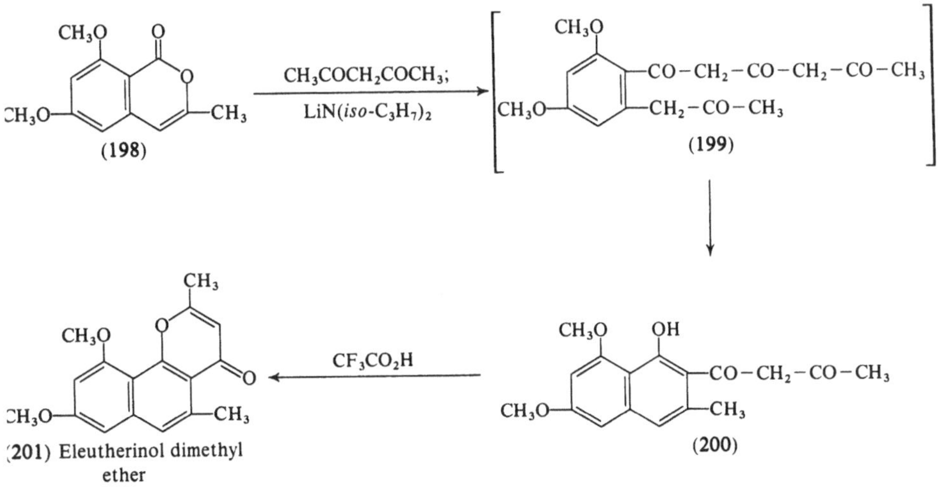

Chart 38. Biogenetic-type synthesis of eleutherinol dimethyl ether (201)

The preparation of alternariol (12) and lichexanthone (13) from orcinyl triketo esters by HAY and HARRIS is an important example of the use of partially cyclized polycarbonyl compounds in biogenetic-type syntheses of polycyclic metabolites (*89, 90*). Triketo ester (202), which has the phenolic hydroxyl groups protected as methyl ethers, was cyclized in aqueous potassium hydroxide to give 25% of benzophenone (203) and 5% of xanthone (204). Methanolic potassium hydroxide converted both the triketo ester and the benzophenone into xanthone (204), methylation of which gave lichexanthone (13). Benzophenone (203) is a Claisen product; aldol products were not detected (Chart 39).

Unprotected orcinyl triketo ester (205), which was prepared by hydrogenolysis of the dibenzyl analog of triketo ester (202) and existed in equilibrium mainly as cyclic hemiketal (206), underwent aldol but not Claisen cyclization. Treatment of ester (205) with equimolar sodium acetate and acetic acid gave 52% of alternariol (12) and 46% of coumarin (207); the latter is a cleavage product and may be directly derived from

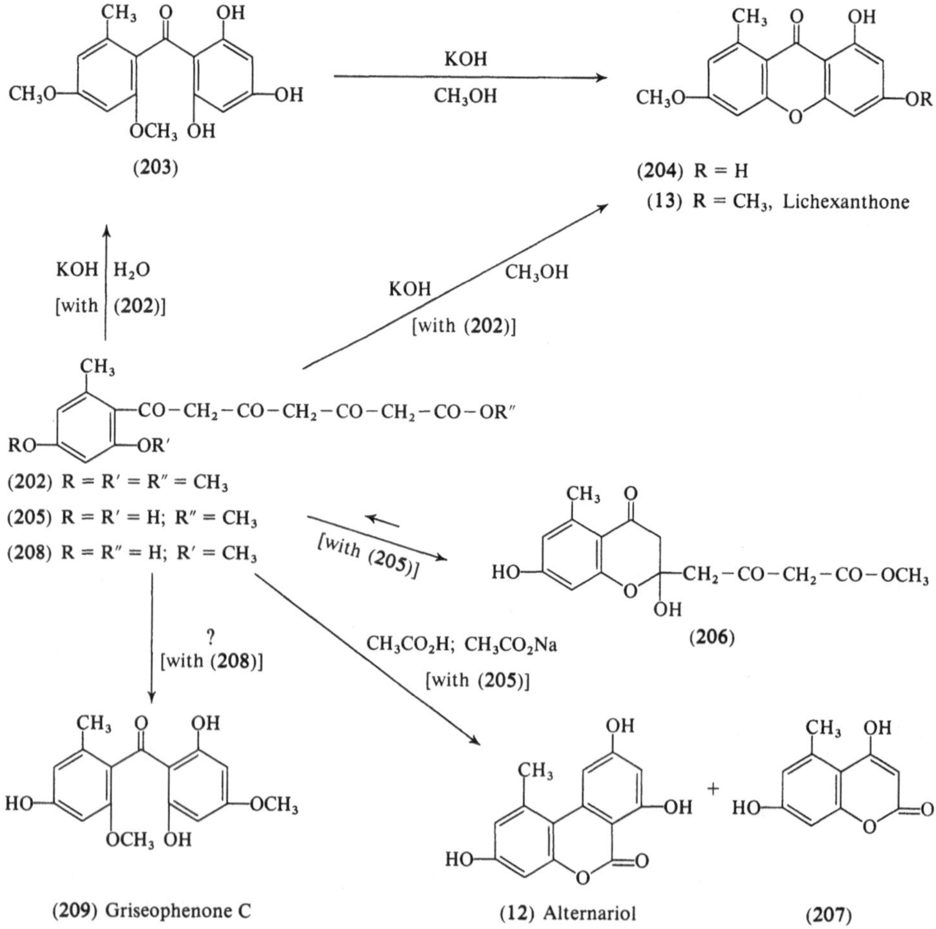

(203)

(204) R = H
(13) R = CH$_3$, Lichexanthone

KOH | H$_2$O
[with (202)]

KOH   CH$_3$OH
[with (202)]

$-CO-CH_2-CO-CH_2-CO-CH_2-CO-OR''$

(202) R = R' = R'' = CH$_3$
(205) R = R' = H; R'' = CH$_3$
(208) R = R'' = H; R' = CH$_3$

[with (205)]

(206)

?
[with (208)]

CH$_3$CO$_2$H; CH$_3$CO$_2$Na
[with (205)]

(209) Griseophenone C

(12) Alternariol

(207)

Chart 39. Biogenetic-type syntheses of alternariol (12) and lichexanthone (13)

References, pp. 274—282

hemiketal (**206**). Under more basic conditions, the yield of coumarin (**207**) increased at the expense of alternariol (**12**). No intermediates in the formation of (**12**) nor Claisen products were detected.

The difference in cyclization preferences of the two triketo esters has been accounted for in terms of participation of the *ortho*-hydroxyl and methoxyl substituents. With triketo ester (**202**), the methoxyl group provides steric hindrance to nucleophilic attack on the 7-carbonyl group and Claisen cyclization becomes the most facile alternative to aldol cyclization. With triketo ester (**205**), the *ortho*-hydroxyl group hydrogen bonds with the 7-carbonyl group, holding it coplanar with the aromatic ring. As a consequence, nucleophilic attack, i.e., aldol cyclization, is both sterically and electronically facilitated.

In view of these results, it would be interesting to know the stage at which each of the O-methyl groups is introduced during the biosynthesis of griseophenone C (**209**); in particular, whether they are introduced after closure of both aromatic rings. This information would provide insight into the more difficult question of the sequence of formation of the rings. Should the orcinyl ring be formed first, the enzymic closure of

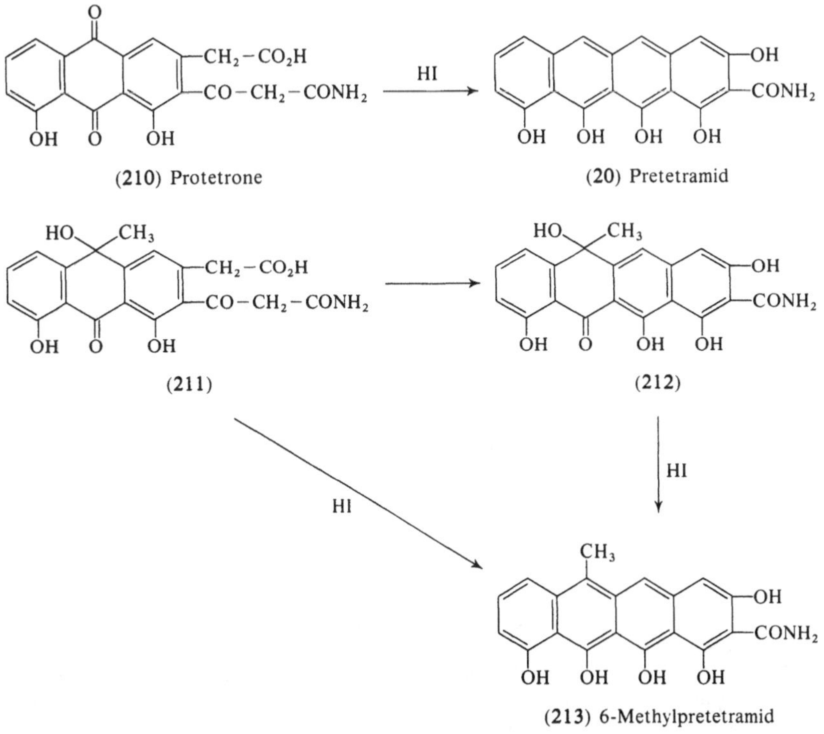

Chart 40. Biogenetic-type syntheses of pretetramid (**20**) and 6-methylpretetramid (**213**)

the phloroglucinol ring may, as in its non-enzymic counterpart, require methylation of the *ortho*-hydroxyl group [giving (**208**)] in order to block aldol cyclization.

The final examples are drawn from the tetracycline studies of McCORMICK and collaborators at Lederle Laboratories (*119, 120*). Protetrone (**210**) is a shunt metabolite elaborated by a mutant of *Streptomyces aureofaciens*. In all probability, the compound is similar to one of the intermediates in the tetracycline biosynthetic pathway. Treatment of protetrone (**210**) with hydrogen iodide caused reduction of the quinone and closure of the fourth ring to give pretetramid (**20**), identical with natural material (Chart 40). From another mutant, these workers have isolated methylanthrone (**211**), which again is a shunt metabolite. The compound was found to be unstable, undergoing spontaneous cyclization to (**212**). Both methylanthrone (**211**) and cyclization product (**212**) gave 6-methylpretetramid (**213**) on treatment with hydrogen iodide. These cyclizations show the facility with which final cyclizations can occur once the proper folding pattern has been established by the initial cyclizations of the polyketide precursor.

VII. Conclusions

During the past decade, expanding activity has been seen in biogenetic-type syntheses of polyketide metabolites. Monocyclic compounds have been studied most systematically and with the best success, but a few good results and a number of promising leads have been obtained with more complex, polycyclic systems. The recent availability of larger polycarbonyl compounds, i.e., those having six or more carbonyl groups, both in masked and free states suggests that the future will see increased success with syntheses of the polycyclic metabolites. The large polycarbonyl compounds can undergo many different cyclization reactions and further efforts will be required to define conditions under which these reactions can be achieved selectively. Solutions to these problems will place the direct syntheses of compounds as complex as pretetramid (**20**) within the reach of workers in this field.

References

1. ACKER, T. E., P. E. BRENNEISEN, and S. W. TANENBAUM: Isolation, Structure, and Radiochemical Synthesis of 3,6-Dimethyl-4-hydroxy-2-pyrone. J. Amer. Chem. Soc. **88**, 834 (1966).

2. ALLPORT, D. C., and J. D. BU'LOCK: Biosynthetic Pathways in *Daldinia concentrica*. J. Chem. Soc. (London) **1960**, 654.

3. ARNDT, F., B. EISTERT, H. SCHOLZ und E. ARON: Zur Synthese der Dehydracetsäure aus Acetessigester. Ber. Dtsch. chem. Ges. **69**, 2373 (1936).

4. BAEYER, A.: Über die Synthese des Acetessigäthers und des Phloroglucins. Ber. Dtsch. chem. Ges. **18**, 3454 (1885).

5. BAKER, P. M., and B. W. BYCROFT: A Convenient Route to Some Naturally Occurring Hydroxynaphthaquinones. Chem. Commun. **1968**, 71.

6. BALENOVIĆ, K., and R. MUNK: Polyoxo Compounds. III. sym-Dibenzoylacetone (1,3,5-Trioxo-1,5-diphenylpentane). Arhiv Kem. **18**, 41 (1946); [Chem. Abstr. **42**, 2926 (1948)].

7. BALENOVIĆ, K., und D. SUNKO: Über γ-Benzoyl-acetessigsäure. Monatsh. Chem. **79**, 1 (1948).

8. BEDFORD, C. T., J. L. DOUGLAS, B. E. MCCARRY, and T. MONEY: Pyrone Studies: Conversion of 2-Pyrones into Aromatic Compounds. Chem. Commun. **1968**, 1091.

9. BEDFORD, C. T., and T. MONEY: Photochemistry of 4-Hydroxy-6-methyl-(2H)-pyran-2-one (Triacetic Acid Lactone). Chem. Commun. **1969**, 685.

10. BENTLEY, R., and P. M. ZWITKOWITS: Biosynthesis of Tropolones in *Penicillium stipitatum*. VII. The Formation of Polyketide Lactones and Other Nontropolone Compounds as a Result of Ethionine Inhibition. J. Amer. Chem. Soc. **89**, 676 (1967).

11. BETHELL, J. R., and P. MAITLAND: Organic Reactions in Aqueous Solution at Room Temperature. Part III. The Influence of pH on the Self-condensation of Diacetyl-acetone: Constitution of Collie's Naphthalene Derivative. J. Chem. Soc. (London) **1962**, 3751.

12. BIRCH, A. J.: Biosynthetic Relations of Some Natural Phenolic and Enolic Compounds. Fortschr. Chem. organ. Naturstoffe **14**, 186 (1957).

13. — Some Pathways in Biosynthesis. Proc. Chem. Soc. (London) **1962**, 3.

14. BIRCH, A. J., D. W. CAMERON, and R. W. RICKARDS: Studies in Relation to Bio-synthesis. Part XXIII. The Formation of Aromatic Compounds from β-Polyketones. J. Chem. Soc. (London) **1960**, 4395.

15. BIRCH, A. J., and F. W. DONOVAN: Studies in Relation to Biosynthesis. I. Some Possible Routes to Derivatives of Orcinol and Phloroglucinol. Austral. J. Chem. **6**, 360 (1953).

16. BIRCH, A. J., P. FITTON, D. C. C. SMITH, D. E. STEERE, and A. R. STELFOX: Studies in Relation to Biosynthesis. Part XXXII. Preparation, Spectra, and Hydrolysis of Poly-β-carbonyl Compounds. J. Chem. Soc. (London) **1963**, 2209.

17. BIRCH, A. J., R. A. MASSY-WESTROPP, and C. J. MOYE: Studies in Relation to Bio-synthesis. VII. 2-Hydroxy-6-methylbenzoic Acid in *Penicillium griseofulvum* Dierckx. Austral. J. Chem. **8**, 539 (1955).

18. BIRCH, A. J., R. A. MASSY-WESTROPP, R. W. RICKARDS, and H. SMITH: Studies in Relation to Biosynthesis. Part XIII. Griseofulvin. J. Chem. Soc. (London) **1958**, 360.

19. BIRCH, A. J., O. C. MUSGRAVE, R. W. RICKARDS, and H. SMITH: Studies in Relation to Biosynthesis. Part XX. The Structure and Biosynthesis of Curvularin. J. Chem. Soc. (London) **1959**, 3146.

20. BIRCH, A. J., J. F. SNELL, and P. J. THOMSON: Studies in Relation to Biosynthesis. Part XXVIII. Oxytetracycline (Terramycin). J. Chem. Soc. (London) **1962**, 425.

21. BOLTZE, K.-H., und K. HEIDENBLUTH: Zur Synthese 3-substituierter 4-Hydroxy-pyrone-(2), I. Ringschlüsse mit Malonsäure-dichloriden. Chem. Ber. **91**, 2849 (1958).

22. BORSCHE, W., und C. K. BODENSTEIN: Untersuchungen über die Bestandteile der Kawawurzel, IX.: Die Synthese des Yangonins. Ber. Dtsch. chem. Ges. **62**, 2515 (1929).

23. BRAM, G.: Synthese d'un derive du trioxo-3,5,7 octanoate d'ethyle modele chimique de biosynthese de l'orcinol. Tetrahedron Letters **1967**, 4069.

24. BROWN, K. S., D. W. CAMERON, and U. WEISS: Chemical Constituents of the Bright Orange Aphid, *Aphis Nerii* Fonscolombe. I. Neriaphin and 6-Hydroxymusizin 8-O-β-*D*-Glucoside. Tetrahedron Letters **1969**, 471.

25. BRUICE, T. C., and B. HOLMQUIST: The Establishment of a Carbanion Mechanism for Ester Hydrolysis and the Unimportance of Electrostatic Effects of α Substituents on the Rates of Hydroxide Ion Attack at the Ester Carbonyl Group. J. Amer. Chem. Soc. **90**, 7136 (1968).

26. BU'LOCK, J. D., and H. G. SMITH: Pyrones. Part I. Methyl Ethers of Tautomeric Hydroxypyrones and the Structure of Yangonin. J. Chem. Soc. (London) **1960**, 502.

27. BYCROFT, B. W., T. A. DOBSON, and J. C. ROBERTS: Studies in Mycological Chemistry. Part VIII. The Structure of Flavasperone ("Asperxanthone"), a Metabolite of *Aspergillus niger*. J. Chem. Soc. (London) **1962**, 40.

28. BYCROFT, B. W., A. HASSANIALI-WALJI, A. W. JOHNSON, and T. J. KING: The Structure and Synthesis of Barakol; a Novel Dioxaphenalene Derivative from *Cassia siamea*. J. Chem. Soc. C (London) **1970**, 1686.

29. CARNDUFF, J., J. A. MILLER, B. R. STOCKDALE, J. LARKIN, D. C. NONHEBEL, and H. C. S. WOOD: Synthesis and Reactions of 3,3-Dimethylallyl Derivatives of Acetylacetone and Other Poly-β-carbonyl Compounds. J. Chem. Soc. (London) Perkin I, **1972**, 692.

30. CASNATI, G., A. QUILICO, A. RICCA, and P. VITA-FINZI: New Synthesis of polyketones: 1,5-dibenzoylacetylacetone. Chim. et ind. **47**, 993 (1965); [Chem. Abstr. **64**, 8068 (1966)].

31. — — — — Some Synthetic Applications of the Reaction of Reductive Opening of the Isoxazole Ring. Tetrahedron Letters **1966**, 233.

32. — — — — Synthesis of 5,3'- and 5,5'-Methylenediisoxazoles, Intermediates in the Preparation of β-Tetraketones. Gazz. chim. ital. **96**, 1064 (1966); [Chem. Abstr. **66**, 37811 (1967)].

33. CHAUVELIER, J.: Sur une nouvelle synthèse des pyrones. C. R. hebd. séances Acad. Sci. **226**, 927 (1948).

34. CHENG, F. C., and S. F. TAN: Synthesis and Reactions of 7-Methylpyrano[4,3-b]-pyran-2,5-dione. J. Chem. Soc. (London) C **1968**, 543.

35. COLLIE, J. N.: On the Constitution of Dehydracetic Acid. J. Chem. Soc. (London) **59**, 179 (1891).

36. — The Lactone of Triacetic Acid. J. Chem. Soc. (London) **59**, 607 (1891).

37. — Derivatives of the Multiple Ketene Group. Proc. Chem. Soc. (London) **23**, 230 (1907).

38. — Derivatives of the Multiple Keten Group. J. Chem. Soc. (London) **91**, 1806 (1907).

39. COLLIE, J. N., and W. S. MYERS: The Formation of Orcinol and other Condensation Products from Dehydracetic acid. J. Chem. Soc. (London) **63**, 122 (1893).

40. COLLIE, J. N., and B. D. STEELE: Dimethyldiacetylacetone, Tetramethylpyrone, and Orcinol Derivatives from Diacetylacetone. J. Chem. Soc. (London) **77**, 961 (1900).

41. COMER, F. W., T. MONEY, and A. I. SCOTT: Conversion of a Polypyrone into Phenolic Compounds. Chem. Commun. **1967**, 231.

42. CORNELIUS, H., und H. VON PECHMANN: Über die Synthese des Orcins aus Acetondicarbonsäureäther. Ber. Dtsch. chem. Ges. **19**, 1446 (1886).

43. CROMBIE, L., M. ESKINS, and D. E. GAMES: Ratio-dependent Products from Xanthophanic Enol and Magnesium Methoxide; Reaction Control by Substrate-chelation. Chem. Commun. **1968**, 1015.

44. CROMBIE, L., D. E. GAMES, and M. H. KNIGHT: Structures of Xanthophanic and Glaucophanic Acid. Tetrahedron Letters **1964**, 2313.

45. — — — Base-catalysed Cyclisation of Highly Enolisable Systems: Diversion of Pathway by Magnesium Chelation. Chem. Commun. **1966**, 355.

46. — — — Polyketo-enols and Chelates. Part I. The Formation and Constitution of Xanthophanic Enol and the Xanthyrones. J. Chem. Soc. (London) C **1967**, 757.

47. CROMBIE, L., D. E. GAMES, and M. H. KNIGHT: Polyketo-enols and Chelates. Part II. The Chemistry of the Xanthophanic Enols. J. Chem. Soc. (London) C 1967, 763.

48. — — — Polyketo-enols and Chelates. Part III. The Constitution and Chemistry of the Glaucophanic Enols. J. Chem. Soc. (London) C 1967, 773.

49. — — — Polyketo-enols and Chelates. Part IV. The By-Product in the Xantho-phanic-Glaucophanic Enol Reaction: 2-Acetyl-4,7-alkoxycarbonyl-1,6-dimethyl-naphthalene. J. Chem. Soc. (London) C 1967, 777.

50. CROMBIE, L., and A. W. G. JAMES: The Control of Pyrone and Aromatic Cyclisation in Polyketonic-Polyenolic Systems by Magnesium Alkoxide Concentration. Chem. Commun. 1966, 357.

51. DEAN, F. M.: Naturally Occurring Oxygen Ring Compounds. London: Butterworth & Co. 1963.

52. DESHAPANDE, S. S., Y. V. DINGANKAR, and D. N. KOPIL: Synthesis and Structure of Dipropionylacetone and of Di-n-butyrylacetone. J. Indian Chem. Soc. 11, 595 (1934).

53. DOUGLAS, J. L., and T. MONEY: Biogenetic-type Synthesis of Benzophenones and Biphenyls. Canad. J. Chem. 45, 1990 (1967).

54. — — Pyrone Studies. II. Biogenetic-type Synthesis of Phenolic Compounds. Tetrahedron 23, 3545 (1967).

55. — — Pyrone Studies. Linear α-Pyrone Route to Protected β-Polyketones. Canad. J. Chem. 46, 695 (1968).

56. EDWARDS, R. L., and D. V. WILSON: Constituents of the Higher Fungi. Part II. The Synthesis of Hispidin. J. Chem. Soc. (London) 1961, 5003.

57. FEIST, F.: Über Dehydracetsäure. Liebigs Ann. Chem. 257, 253 (1890).

58. FREI, H., und H. SCHMID: Inhaltstoffe aus Eleutherine bulbosa (Mill.) Urb. VIII. Synthese des Eleutherinols. Liebigs Ann. Chem. 603, 169 (1957).

59. GATENBECK, S.: On the Biosynthesis of the Pigments of Penicillium islandicum. II. Acta Chem. Scand. 14, 296 (1960).

60. GATENBECK, S., and P. BARBESGÅRD: On the Biosynthesis of the Pigments of Penicillium islandicum. III. Acta Chem. Scand. 14, 230 (1960).

61. GATENBECK, S., P. O. ERIKSSON, and Y. HANSSON: Cell-free C-Methylation in Relation to Aromatic Biosynthesis. Acta Chem. Scand. 23, 699 (1969).

62. GATENBECK, S., and S. HERMODSSON: Enzymic Synthesis of the Aromatic Product Alternariol. Acta Chem. Scand. 19, 65 (1965).

63. GATENBECK, S., and K. MOSBACH: Acetate Carboxyl Oxygen (^{18}O) as Donor for Phenolic Hydroxy Groups of Orsellinic Acid Produced by Fungi. Acta Chem. Scand. 13, 1561 (1959).

64. GAUCHER, G. M., and M. G. SHEPHERD: Isolation of Orsellinic Acid Synthase. Biochem. Biophys. Res. Comm. 32, 664 (1968).

65. GEUTHER, A.: Untersuchungen über einbasische Kohlenstoffsäuren: 1. Über die Essigsäure. Z. Chem. (Jena) 2, 8 (1866); [Chem. Zbl. 37, 801 (1866)].

66. GOETSCHEL, C., et C. MENTZER: Synthèse de nouveaux dérivés α-pyroniques par condensation thermique de quelques esters maloniques avec des cétones. Bull. soc. chim. France 1962, 365.

67. GOTTLIEB, O. R., and W. B. MORS: The Chemistry of Rosewood. III. Isolation of 5,6-Dehydrokavain and 4-Methoxyparacotoin from Aniba firmula Mez. J. Organ. Chem. (USA) 24, 17 (1959).

68. GUILFORD, H., A. I. SCOTT, D. SKINGLE, and M. YALPANI: The Synthesis of Tetraacetic Acid Lactone and a Model for the Biosynthesis of 6-Methylsalicylic Acid. Chem. Commun. 1968, 1127.

69. HALE, W. J.: The Constitution of Dehydroacetic Acid. J. Amer. Chem. Soc. 33, 1119 (1911).

70. HAMPTON, K. G., T. M. HARRIS, C. M. HARRIS, and C. R. HAUSER: Condensations

at the Terminal Methyl Group of 1,3,5-Triketones by Means of Sodamide in Liquid Ammonia. J. Organ. Chem. (USA) **30**, 4263 (1965).

71. HARRIS, C. M., and T. M. HARRIS: Synthesis of 5-Oxohexenoic Acid. J. Organ. Chem. (USA) **36**, 2181 (1971).

72. — — Unpublished results.

73. HARRIS, T. M.: Unpublished results.

74. HARRIS, T. M., and R. L. CARNEY: Biogenetically Modeled Synthesis of β-Resorcylic Acids. J. Amer. Chem. Soc. **88**, 2053 (1966).

75. — — Biogenetic-Type Synthesis of an Acylphloroglucinol. J. Amer. Chem. Soc. **88**, 5686 (1966).

76. — — Synthesis of 3,5,7-Triketo Acids and Esters and Their Cyclizations to Resorcinol and Phloroglucinol Derivatives. Models of Biosynthesis of Phenolic Compounds. J. Amer. Chem. Soc. **89**, 6734 (1967).

77. HARRIS, T. M., and C. S. COMBS, JR.: Synthesis of Certain Naturally Occurring 2-Pyrones via 3,5-Diketo Acids. J. Organ. Chem. (USA) **33**, 2399 (1968).

78. HARRIS, T. M., and C. M. HARRIS: Carboxylation of β-Dicarbonyl Compounds through Dicarbanions. Cyclizations to 4-Hydroxy-2-pyrones. J. Organ. Chem. (USA) **31**, 1032 (1966).

79. — — Condensations at the 6-Methyl Position of Dehydracetic Acid. A Novel Site of Reactivity. Chem. Commun. **1966**, 699.

80. — — Condensations of 2,6-Diphenyl-4-pyrone with Carbanions. J. Organ. Chem. (USA) **32**, 970 (1967).

81. — — Lactonization of 3,5,7-Trioxo-7-phenylheptanoic Acid and its 2-Methyl Homolog. Tetrahedron **25**, 2687 (1969).

82. HARRIS, T. M., C. M. HARRIS, and R. J. LIGHT: The Isolation of Triacetic Acid Lactone from Cultures of *Penicillium patulum*. Biochim. Biophys. Acta **121**, 420 (1966).

83. HARRIS, T. M., C. M. HARRIS, and M. P. WACHTER: Condensations of Dehydroacetic Acid at the 6-Methyl Position. Tetrahedron **24**, 6897 (1968).

84. HARRIS, T. M., T. T. HOWARTH, and R. L. CARNEY: Models of the Biogenesis of Polyketide-Type Phenolic Ethers. J. Amer. Chem. Soc. **93**, 2511 (1971).

85. HARRIS, T. M., and G. P. MURPHY: Synthesis of 1,3,5,7,9-Pentacarbonyl Compounds. J. Amer. Chem. Soc. **93**, 6708 (1971).

86. HARRIS, T. M., and M. P. WACHTER: Aromatization Reactions of 4-Hydroxy-6-phenacyl-2-pyrone and Related Compounds. Tetrahedron **26**, 5255 (1970).

87. HARRIS, T. M., M. P. WACHTER, and G. A. WISEMAN: The Cleavage and Aromatic Recyclization of 4-Hydroxy-6-phenacyl-2-pyrone. A Novel Metallic Cation Effect. Chem. Commun. **1969**, 177.

88. HAUSER, C. R., and T. M. HARRIS: Condensations at the Methyl Group Rather than the Methylene Group of Benzoyl- and Acetylacetone Through Intermediate Dipotassio Salts. J. Amer. Chem. Soc. **80**, 6360 (1958).

89. HAY, J. V.: Biogenetic-type Synthesis of Heptaketide Phenolic Natural Products: Alternariol and Lichexanthone. Dissert., Vanderbilt Univ. 1972.

90. HAY, J. V., and T. M. HARRIS: Biogenetic-type Syntheses of Heptaketide Natural Products: Alternariol and Lichexanthone. Chem. Commun. **1972**, 953.

91. HEDGECOCK, P. F., P. F. G. PRAILL, and A. L. WHITEAR: Acid-catalysed Conversion of Pyronopyrones into Phenolic Ketones. Chem. and Ind. **1966**, 1268.

92. HILLIS, W. E., and N. ISHIKURA: An Enzyme from *Eucalyptus* which Converts Cinnamoyl Triacetic Acid into Pinosylvin. Phytochem. **8**, 1079 (1969).

93. HILLIS, W. E., and Y. YAZAKI: The Biosynthesis of Stilbenes in Eucalypt Leaves. Phytochem. **10**, 1051 (1971).

94. HOWARTH, T. T., and T. M. HARRIS: Methyl 2-Hydroxy-4-methoxy-6-phenylbenzoate and Methyl 2-Hydroxy-6-methoxy-4-phenylbenzoate; Reassignment of a Structure. Canad. J. Chem. **46**, 3739 (1968).

95. HOWARTH, T. T., and T. M. HARRIS: Biogenetic-type Synthesis of β-Resorcylic Acids. Isolation and Characterization of the Aldol Intermediate. J. Amer. Chem. Soc. 93, 2506 (1971).

96. HOWARTH, T. T., G. P. MURPHY, and T. M. HARRIS: Preparation and Biogenetic-Type Aromatizations of Tetraacetic Acid (3,5,7-Trioxooctanoic Acid). J. Amer. Chem. Soc. 91, 517 (1969).

97. HUCKIN, S. N., and L. WEILER: Claisen Condensation of the Dianion of β-Keto Esters. Tetrahedron Letters 1972, 2405.

98. HÜNIG, S., E. BENZING und K. HÜBNER: Synthesen mit Enaminen, VI. Reaktionen mit Diketen zu γ-Pyronen. Chem. Ber. 94, 486 (1961).

99. IGUCHI, S., and K. HISATSUNE: Pyrone Derivatives. I. Syntheses and Antibacterial Properties of Ethyl 4-Hydroxy-1-oxo-6-methyl-2-pyrone-3-decanoate and Related Compounds. J. Pharmac. Soc. Japan 77, 94 (1957); [Chem. Abstr. 51, 8733 (1957)].

100. JERDAN, D. S.: The Condensation of Ethylic Acetonedicarboxylate and Constitution of Triethylic Orcinoltricarboxylate. J. Chem. Soc. (London) 75, 808 (1899).

101. JOBST, J., und O. HESSE: Über die Cotorinden und ihre charakteristischen Bestand-teile. Liebigs Ann. Chem. 199, 17 (1879).

102. JULIA, M., et J. BULLOT: Synthèses de la paracotoine et de quelques α-pyrones apparentées. Bull. soc. chim. France 1959, 1689.

103. — — Sur quelques alcools dichlorovinyliques et leur transformation en acide éthyléniques. Bull. soc. chim. France 1959, 1828.

104. JULIA, M., et C. B. DU JASSONNEIX: Synthèse de la méthoxy-4 paracotoine et de quelques composés apparentés. C. R. hebd. séances Acad. Sci. 253, 872 (1961).

105. KATO, T., and T. HOZUMI: Studies on Ketene and Its Derivatives. XLIX. Reaction of Diketene with β-Ketoesters to Give Ethyl Orsellinate, Divarate, Olivetol Carboxylate, and Sphaeropherol Carboxylate. Chem. Pharm. Bull. (Japan) 20, 1574 (1972).

106. KIRBY, F. B., T. M. HARRIS, and C. R. HAUSER: Acylation vs. Conjugate Addition of Dipotassio β-Diketones with Cinnamic Esters. Synthesis of Unsaturated 1,3,5-Triketones and t-Butyl 5,7-Dioxoalkanoates. J. Organ. Chem. (USA) 28, 2266 (1963).

107. KÖGL, F., und C. A. SALEMINK: Untersuchungen über Derivate von 2,3-Dihydro-pyran-2,4-dion. (1. Mitteilung.) Rec. trav. chim. Pays-Bas 71, 779 (1952).

108. — — Untersuchungen über Derivate von 2,3-Dihydro-pyran-2,4-dion. (2. Mit-teilung.) Rec. trav. chim. Pays-Bas 74, 221 (1955).

109. KOMNINOS, T.: Nouveau mode de passage de la série grasse a la série aromatique. Bull. soc. chim. France [iv] 23, 449 (1918).

110. LEFEUVRE, A., et C. MENTZER: Applications du phényl-thiomalonate d'éthyle en synthèse hétérocyclique. Bull. soc. chim. France 1964, 623.

111. LIGHT, R. J.: The Biosynthesis of 6-Methylsalicylic Acid. J. Biol. Chem. 242, 1880 (1967).

112. LIGHT, R. J., T. M. HARRIS, and C. M. HARRIS: Metabolism of Triacetic Acid and Triacetic Acid Lactone. Biochemistry 5, 4037 (1966).

113. LIGHT, R. J., and C. R. HAUSER: Aroylations of β-Diketones at the Terminal Methyl Group to Form 1,3,5-Triketones. Cyclizations to 4-Pyrones and 4-Pyridones. J. Organ. Chem. (USA) 25, 538 (1960).

114. LYNEN, F., E. REICHERT und L. RUEFF: Zum biologischen Abbau der Essigsäure. VI. („Aktivierte Essigsäure", ihre Isolierung aus Hefe und ihre chemische Natur.) Liebigs Ann. Chem. 574, 1 (1951).

115. LYNEN, F., und M. TADA: Die biochemischen Grundlagen der „Polyacetat-Regel". Angew. Chem. 73, 513 (1961).

116. MACIEREWICZ, Z.: Synthesis of the Lactone of the Mother Substance of Yangonin. Roczniki Chem. 24, 144 (1950); [Chem. Abstr. 48, 10013 (1954)].

117. MARCUS, E., J. F. STEPHEN, and J. K. CHAN: A Study of the Acylation of 4-Hydroxy-6-methyl-2-pyrone and 4-Hydroxy-6-phenyl-2-pyrone. J. Heterocycl. Chem. 6, 13 (1969).

118. McCORMICK, J. R. D.: Biosynthesis of the Tetracyclines. In: Z. Vaněk and Z. Hošťálek, Biogenesis of Antibiotic Substances, p. 73. Prague: Publishing House of the Czechoslovak Academy of Sciences 1965.

119. McCORMICK, J. R. D., and E. R. JENSEN: Biosynthesis of Tetracyclines. X. Protetrone. J. Amer. Chem. Soc. **90,** 7126 (1968).

120. McCORMICK, J. R. D., E. R. JENSEN, N. H. ARNOLD, H. S. COREY, U. H. JOACHIM, S. JOHNSON, P. A. MILLER, and N. O. SJOLANDER: Biosynthesis of Tetracyclines. XI. The Methylanthrone Analog of Protetrone. J. Amer. Chem. Soc. **90,** 7127 (1968).

121. MILES, M. L., T. M. HARRIS, and C. R. HAUSER: Aroylation at the Terminal Methyl Group of a 1,3,5-Triketone to Form a 1,3,5,7-Tetraketone. J. Amer. Chem. Soc. **85,** 3884 (1963).

122. — — — Aroylations at the Methyl Group of Benzoylacetone and Related β-Diketones with Esters to Form 1,3,5-Triketones by Sodium Hydride. Other Terminal Condensations. J. Organ. Chem. (USA) **30,** 1007 (1965).

123. MONEY, T.: Biogenetic-type Synthesis of Phenolic Compounds. Chem. Rev. **70,** 553 (1970).

124. MONEY, T., F. W. COMER, G. R. B. WEBSTER, I. G. WRIGHT, and A. I. SCOTT: Pyrone Studies. I. Biogenetic-type Synthesis of Phenolic Compounds. Tetrahedron **23,** 3435 (1967).

125. MONEY, T., J. L. DOUGLAS, and A. I. SCOTT: Biogenetic-type Synthesis of Phenolic Compounds. J. Amer. Chem. Soc. **88,** 624 (1966).

126. MONEY, T., I. H. QURESHI, G. B. WEBSTER, and A. I. SCOTT: Chemistry of Polypyrones. A Model for Acetogenin Biosynthesis. J. Amer. Chem. Soc. **87,** 3004 (1965).

127. MORS, W. B., O. R. GOTTLIEB, and C. DJERASSI: The Chemistry of Rosewood. Isolation and Structure of Anibine and 4-Methoxyparacotoin. J. Amer. Chem. Soc. **79,** 4507 (1957).

128. MORS, W. B., M. T. MAGALHÃES, and O. R. GOTTLIEB: Naturally Occurring Aromatic Derivatives of Monocyclic α-Pyrones. Fortschr. Chem. organ. Naturstoffe **20,** 131 (1962).

129. MÜHLEMANN, H.: Anthraquinones and Anthraquinone Glycosides. XII. Nuclear Synthesis of the Emodins of *Frangula* and of a Chrysophanic Acid Isomer (1,6-Dihydroxy-3-methylanthraquinone). Pharm. Acta Helv. **26,** 195 (1951). [Chem. Abstr. **46,** 8078 (1952)].

130. MURPHY, G. P.: Synthesis of Tetra- and Pentacarbonyl Compounds Using Lithium Diisopropylamide. Dissert., Vanderbilt Univ. 1971.

131. MURRAY, T. P., and T. M. HARRIS: Negatively Charged Electrophiles. Acylation of Strong Nucleophiles by Enolate Salts of β-Keto Esters. J. Amer. Chem. Soc. **94,** 8253 (1972).

132. — — Unpublished results.

133. OLLIS, W. D., I. O. SUTHERLAND, R. C. CODNER, J. J. GORDON, and G. A. MILLER: The Incorporation of Propionate in the Biosynthesis of ε-Pyrromycinone (Rutilantinone). Proc. Chem. Soc. (London) **1960,** 347.

134. O'SULLIVAN, W. I., and C. R. HAUSER: Certain Condensations at the Terminal Methyl Group of 3-Phenylpentane-2,4-dione through Its Dipotassio Derivative. Cyclizations. J. Organ. Chem. (USA) **25,** 1110 (1960).

135. PIKE, D. G., J. J. RYAN, and A. I. SCOTT: Synthesis and Aromatisation of the Linear Hepta-β-carbonyl System. Chem. Commun. **1968,** 629.

136. POLLAK, J.: Notiz über das Cotoin. Monatsh. Chem. **22,** 996 (1901).

137. PRAILL, P. F. G., and A. L. WHITEAR: The Acid-catalysed Self-condensation of Acetic Anhydride and Analogous Compounds. Proc. Chem. Soc. (London) **1961,** 112.

138. RASSWEILER, C. F., and R. ADAMS: The Structure of Dehydro-acetic Acid. J. Amer. Chem. Soc. **46,** 2758 (1924).

139. RESPLANDY, A.: Synthèse et propriétés de deux α-pyrones permettant d'accéder à la méthoxy-4, paracotoine. Étude de quelques substances apparentées. Bull. soc. chim. France **1962**, 1332.

140. RITTENBERG, D., and K. BLOCH: The Utilization of Acetic Acid for Fatty Acid Synthesis. J. Biol. Chem. **154**, 311 (1944).

141. ROBINSON, SIR ROBERT: The Structural Relations of Some Plant Products. J. Roy. Soc. Arts **96**, 795 (1948).

142. RUHEMANN, S.: The Formation of 4-Pyrone Compounds from Acetylenic Acids. Part I. J. Chem. Soc. (London) **93**, 431 (1908).

143. — The Formation of 4-Pyrone Compounds from Acetylenic Acids. Part II. J. Chem. Soc. (London) **93**, 1281 (1908).

144. SCHMIDT, U., und M. SCHWOCHAU: β-Polycarbonylverbindungen, 3. Mitt.: Über Synthesen mit den Trimethylsilylestern der Acetessigsäure und Malonsäure. Ein neuer Weg zu Diacyl-methanen und Diacyl-essigsäureestern. Monatsh. Chem. **98**, 1492 (1967).

145. SCOTT, A. I., H. GUILFORD, J. J. RYAN, and D. SKINGLE: Biogenetic-type Synthesis of Polyketides. Part VIII. Experiments with the Tetra- and Hexa-acetate Systems. Tetrahedron **27**, 3025 (1971).

146. SCOTT, A. I., H. GUILFORD, and D. SKINGLE: Biogenetic-type Synthesis of Polyketides. Part IX. A Model for the Biosynthesis of 6-Methyl Salicylic Acid. Tetrahedron **27**, 3039 (1971).

147. SCOTT, A. I., D. G. PIKE, J. J. RYAN, and H. GUILFORD: Biogenetic-type Synthesis of Polyketides. Part X. Synthesis and Reactions of Hepta- and Nona-β-Carbonyl Chains as Substrate Models. Tetrahedron **27**, 3051 (1971).

148. SEMMLER, F. W., und E. SCHOSSBERGER: Zur Kenntnis der Bestandteile ätherischer Öle. (Zusammensetzung des ätherischen Öles von *Xanthoxylum aubertia* Cordemoy [*Evodia aubertia* Cordemoy] und *Xanthoxylum alatum* Roxb.) Ber. Dtsch. chem. Ges. **44**, 2885 (1912).

149. SJÖLAND, S., and S. GATENBECK: Studies on the Enzyme Synthesizing the Aromatic Product Alternariol. Acta Chem. Scand. **20**, 1053 (1966).

150. SMITH LABORATORIES, T. and H.: Further Note on Brevifolin. Pharm. J. **123**, 604 and 611 (1929); [Chem. Abstr. **24**, 2547 (1930)].

151. SOLIMAN, G., and I. E. EL-KHOLY: The Pyrone Series. Part I. 2:6-Diaryl-4-pyrones. J. Chem. Soc. (London) **1954**, 1755.

152. SPROXTON, F.: The Esters of Triacetic Lactone and Triacetic Acid. J. Chem. Soc. (London) **89**, 1186 (1906).

153. STADTMAN, E. R., M. DOUDOROFF, and F. LIPMANN: The Mechanism of Acetoacetate Synthesis. J. Biol. Chem. **191**, 377 (1951).

154. STEGLICH, W., and W. REININGER: A Synthesis of Endocrocin, Endocrocin-9-anthrone and Related Compounds. Chem. Commun. **1970**, 178.

155. STEPHEN, J. F., and E. MARCUS: A Facile Route to a Novel Derivative of 2,4,6,8-Nonanetetraone. J. Organ. Chem. (USA) **35**, 258 (1970).

156. STETTER, H., und S. VESTNER: Synthese des 2,4,6-Trioxa-adamantan-Ringsystems. Chem. Ber. **97**, 169 (1964).

157. STEWART, A. W.: Recent Advances in Organic Chemistry, vol. II, 5th edition, chapt. IX. London: Longmans, Green & Co. 1927.

158. STOUT, G. H., D. L. DREYER, and L. H. JENSEN: Structure of Rubrofusarin. Chem. and Ind. **1961**, 289.

159. THOMAS, R.: The Biosynthesis of Alternariol. Proc. Chem. Soc. (London) **1959**, 88.

160. TOBIAS, P. S., and F. J. KÉZDY: The Alkaline Hydrolysis of 5-Nitrocoumaranone. A Method for Determining the Intermediacy of Carbanions in the Hydrolysis of Esters with Labile α Protons. J. Amer. Chem. Soc. **91**, 5171 (1969).

161. TOWERS, G. H. N.: Metabolism of Phenolics in Higher Plants and Micro-organisms.

In: J. B. HARBORNE, Biochemistry of Phenolic Compounds, p. 249. New York: Academic Press, Inc. 1964.

162. VAN TAMELEN, E. E.: Biogenetic-type Syntheses of Natural Products. Fortschr. Chem. Organ. Naturstoffe 19, 242 (1961).

163. VORLÄNDER, D., and G. A. MEYER: Überführung des Dibenzal-acetons in α,α-Diphenyl-pyron. Ber.Dtsch. chem. Ges. 45, 3355 (1912).

164. WACHTER, M. P., and T. M. HARRIS: Condensations at the 6α-Position of Triacetic Lactone Via the Dianion. Tetrahedron 26, 1685 (1970).

165. WAKIL, S. J., and J. GANGULY: On the Mechanism of Fatty Acid Synthesis. J. Amer. Chem. Soc. 81, 2597 (1959).

166. WILEY, R. H., C. H. JARBOE, and H. G. ELLERT: 2-Pyrones. XV. Substituted 3-Cinnamoyl-4-hydroxy-6-methyl-2-pyrones from Dehydroacetic Acid. J. Amer. Chem. Soc. 77, 5102 (1955).

167. WINZHEIMER, E.: Investigation of Kava Root. Arch. Pharmaz. 246, 338 (1908); [Chem. Abstr. 3, 429 (1909)].

168. WITTEK, P. J., and T. M. HARRIS: Unpublished results.

169. WITTEK, P. J., K. B. HINDLEY, and T. M. HARRIS: Synthesis of C-Methyl Derivatives of 1-Phenyl-1,3,5-hexanetrione. J. Organ. Chem. (USA) 38, 896 (1973).

170. WITTER, R. F., and E. STOTZ: Synthesis and Properties of Triacetic Acid. J. Biol. Chem. 176, 485 (1948).

171. WOLFE, J. F., T. M. HARRIS, and C. R. HAUSER: Condensations at the Methyl Group of Ethyl Acetoacetate by Means of Potassium Amide or Sodium Hydride. J. Organ. Chem. (USA) 29, 3249 (1964).

172. WOODS, L. L., and P. A. DIX: Acylation, Bromination and Oxidation of 4-Pyrones and Pyronones. J. Organ. Chem. (USA) 26, 2588 (1961).

173. WOODS, L. L., D. JOHNSON, and F. THOMAS: 6-Aryl-5-carboxy-4-hydroxy-2-pyrones. Texas J. Science 19, 227 (1967); [Chem. Abstr. 68, 21777 (1968)].

174. WORK, S. D., and C. R. HAUSER: Acylations of Dilithio β-Diketones with Aliphatic Esters to Form 1,3,5-Triketones. Cyclizations to 4-Pyrones and 4-Pyridones. J. Organ. Chem. (USA) 28, 725 (1963).

175. YAMAMURA, S., K. KATO, and Y. HIRATA: The Reaction of 6-Carboxymethyl-4-methoxy-2-pyrone with Acetic Anhydride. Chem. Commun. 1968, 1580.

176. — — — Reactions of 4-Methoxy-2-oxopyran-6-ylacetic Acid With Acid Anhydrides. J. Chem. Soc. (London) C 1969, 2461.

177. ZIEGLER, E., und H. JUNEK: Synthesen von Heterocyclen. XI. Mitteilung: 4-Hydroxy-2-pyrone. Monatsh. Chem. 89, 323 (1958).

178. ZIEGLER, E., und E. NÖLKEN: Synthesen von Heterocyclen. XII. Mitteilung: Über das Anibin. Monatsh. Chem. 89, 391 (1958).

(Received March 30, 1973)

The Chemistry of Spiro[4.5]Decane Sesquiterpenes

By J. A. MARSHALL, Evanston, Illinois, USA, ST. F. BRADY, Rahway, New Jersey, USA, and N. H. ANDERSEN, Seattle, Washington, USA

Contents

A. Introduction

The sesquiterpenes provide a rich storehouse of diverse carbocyclic structural types whose perverse nature has taxed the talent and imagination of organic chemists since the beginning of modern organic chemistry. Studies on this widely varied group of natural products have contributed significantly to the advancement of spectroscopic techniques, mechanistic insight, and synthesis methodology, and the trend along these lines shows no sign of letup, as new and increasingly complex structures join the list of known members. Indeed, almost every conceivable carbon skeleton derivable via rational chemical transformations of farnesol (1), the biological precursor of the sesquiterpenes (80), seems to have found its way into one or another species of plant life.

The possibility of farnesol-derived spiro[4.5]decanes was suggested in 1953 by Ruzicka who postulated the cation (2) as an intermediate in the biosynthesis of cedrene (3) (87). However, the actual occurrence

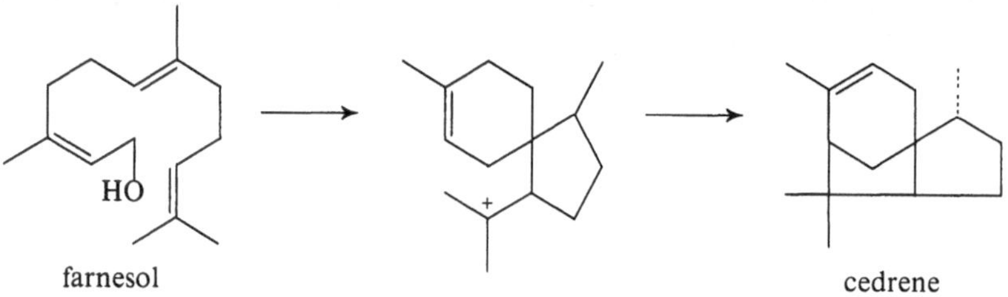

farnesol   cedrene

(1)   (2)   (3)

Chart 1. Proposed biogenesis of cedrene from farnesol

of sesquiterpenes possessing a spiro[4.5]decane carbon framework was not actually established until 1956 when Šorm and his co-workers (97) published their interpretation of degradative experiments on acorone (4), a constituent of sweet flag oil (91). Further studies led to the characterization of a number of related spiro[4.5]decane derivatives from the same oil (106, 107). Recent work has uncovered a series of skeletally related enantiomeric compounds, the alaskenes (11, 13, 17).

A second type of spiro[4.5]decane-derived sesquiterpene was formulated by Bhattacharyya and co-workers in the course of their studies on constituents of the essential oil derived from fungus-infected agarwood (104). This new sesquiterpene, agarospirol (5), differs from the acorane type in the placement of methyl and isopropyl groupings. Recently the number of known natural sesquiterpenes possessing this particular skeletal arrangement was increased considerably by the discovery that β-vetivone (6) and related vetivane sesquiterpenes, previously thought to be hydroazulenes (84), possess the same carbon skeleton as that proposed for agarospirol (64).

A third type of spiro[4.5]decane sesquiterpene has been isolated from marine sources. The lone member of this class known to date, spiroaurenone (7), is one of the few known natural products containing bromine (96).

References, pp. 371—376

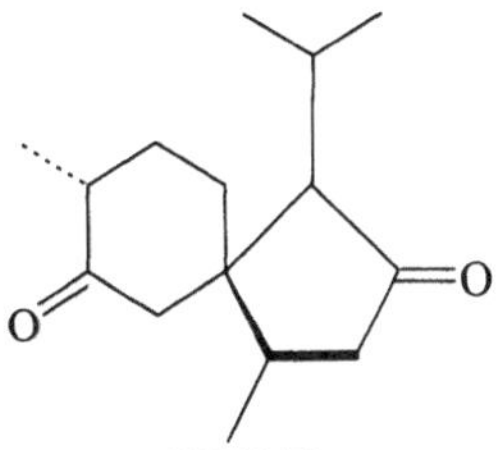

acorone
(4)

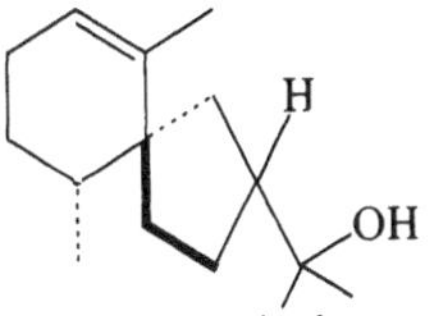

agarospirol
(5)

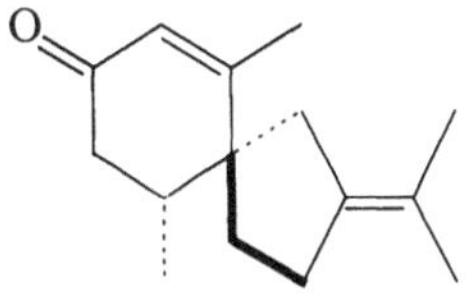

β-vetivone
(6)

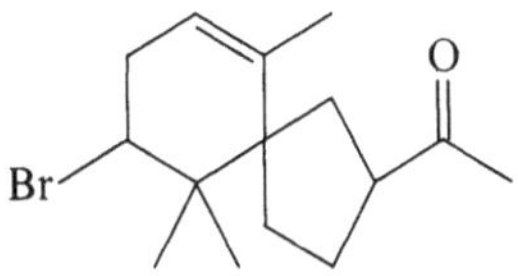

spiroaurenone
(7)

Chart 2. Representative spiro[4.5]decane sesquiterpenes

This chapter will review the important aspects of the structure elucidation, synthesis and biogenesis of the naturally occurring spiro[4.5]-decanes.

B. The Acoranes and Alaskanes

1. Structure Elucidation of the *Acorus* Spiranes

ŠORM and HEROUT in connection with their studies on the constitution of sweet flag *(Acorus calamus* L.) isolated two ketones, acorone (m.p. 100−101°) and isoacorone (m.p. 96−97°) whose interconversion in basic solution suggested an epimeric relationship *(91, 92)*. The infrared spectrum of acorone showed two carbonyl bands which, together with chemical evidence, indicated the presence of cyclohexanone and cyclopentanone moieties *(97)*. Condensation of acorone with excess benzaldehyde afforded only the monobenzylidene derivative (**10**) *(Chart 3)*. Thus the cyclopentanone ring must possess a free α-methylene grouping. Actually, the cyclohexanone ring also contains this structural feature, but steric factors apparently retard benzylidene formation at that center.

The cyclohexanone carbonyl grouping could be selectively reduced to give a hydroxy ketone (acorolone) which yielded the enone (**8**) as a mixture of isomers upon dehydration. The infrared spectrum of this mixture confirmed the presence of the cyclopentanone ring. Dehydrogenation of this

material afforded a product (11) whose spectral properties indicated
the presence of a benzene ring and a cyclohexanone grouping. Thus,

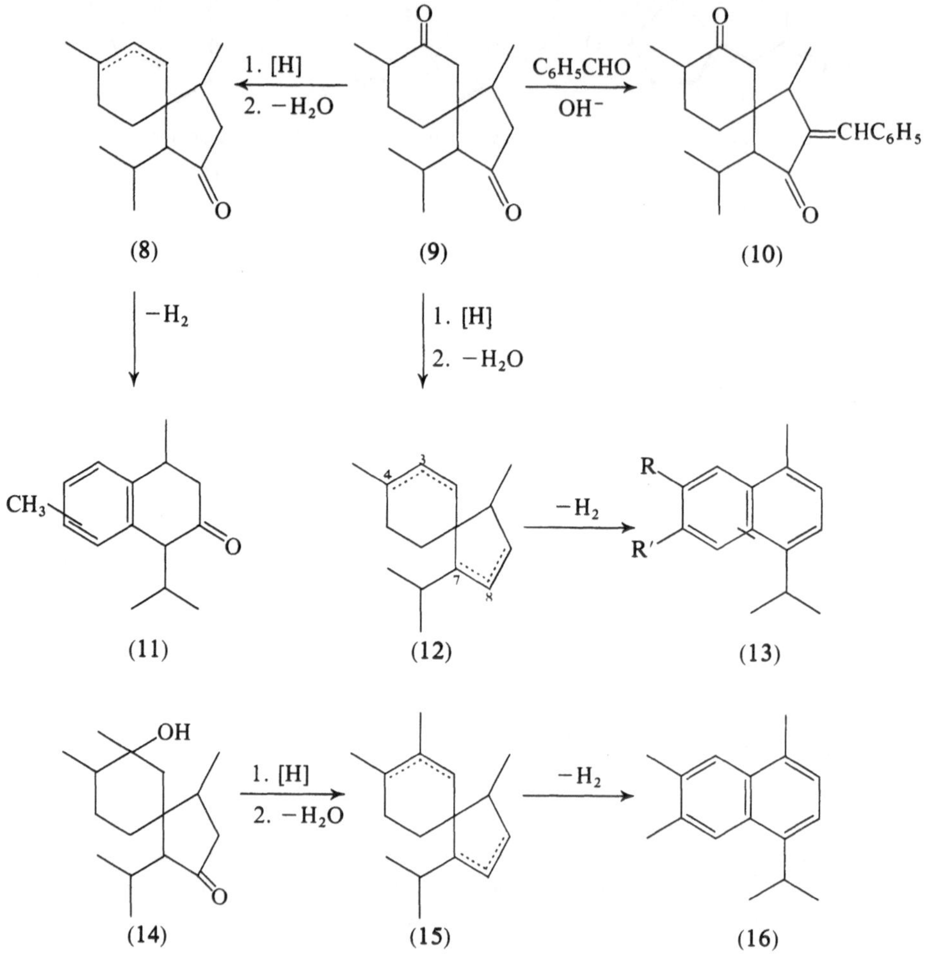

(8)   (9)   (10)

(11)   (12)   (13)

(14)   (15)   (16)

Chart 3. Degradation of the acorones

expansion of the cyclopentane ring must have taken place during the
dehydrogenation reaction. This finding indicated that a carbon atom
common to both rings must be quaternary, and suggested a spiro[4.5]decane
skeleton for acorone. An analogous isomerization took place upon
dehydrogenation of acoradiene (12), secured via reduction and dehydration of acorone. In this case, the well-known hydrocarbon cadalene
[(13): R=H, R′=CH₃] was obtained. Similarly, the isomeric diene
mixture, isoacoradiene (obtained from isoacorone in a like manner)
afforded the isomeric hydrocarbon 1,7-dimethyl-4-isopropylnaphthalene

[(13): R = CH$_3$, R' = H] upon dehydrogenation. A small amount of azulenic material was also detected in these experiments, thus indicating that expansion of the six membered ring can also occur during the aromatization reactions. The apparent divergence in the pathways of dehydrogenation of the two stereoisomers represented by (12) to give isomeric naphthalenes is noteworthy. A recent reinvestigation by ANDERSEN has indicated that the hydrocarbon mixture secured via dehydration of the acoradiols derived from acorone and isoacorone is principally the 3,7-diene [(12), double bonds at 3,4 and 7,8]. (1).

Acorone afforded only the keto alcohol (14), the product of addition to the cyclohexanone carbonyl, upon treatment with excess methylmagnesium iodide (92). The crowded steric environment of the cyclopentanone carbonyl grouping coupled with the intrinsically greater reactivity of cyclohexanone vs. cyclopentanone carbonyls presumably renders conversion to the ditertiary diol difficult in this case. However, reduction of keto alcohol (14) to the corresponding secondary tertiary diol could be effected easily with lithium aluminium hydride. Subsequent dehydration then afforded the diene mixture (15). Dehydrogenation of this mixture yielded 1,6,7-trimethyl-4-isopropylnaphthalene (16). A comparison of the naphthalenes (13) (R = H, R' = CH$_3$) and (16) provided the basis for assigning the location of the cyclohexanone carbonyl grouping.

The relationship of the cyclohexanone ring substituents was confirmed via dehydrogenation of the diacid (19) (Chart 4). This acid was secured through ozonolysis of the hydroxymethylene derivative of acoranone (18) prepared by desulfurization of the mono ethylenethioketal derivative (17) of acorone. The formation of p-isobutyltoluene and p-ethyltoluene

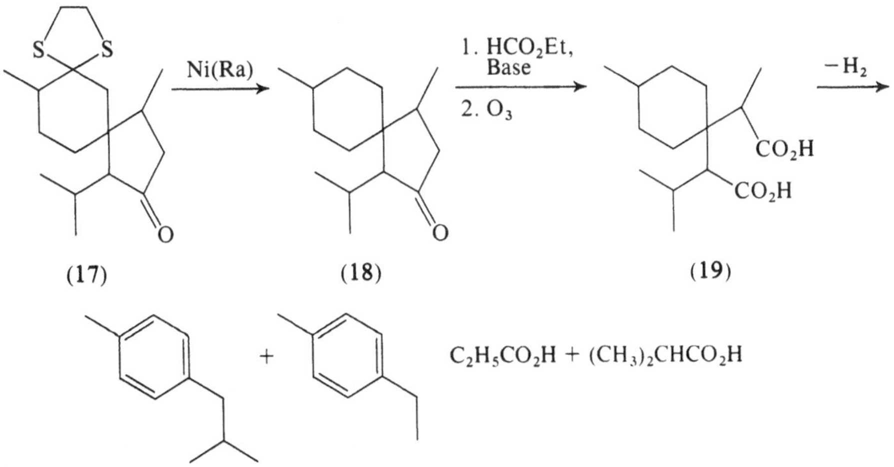

Chart 4. Location of acorone ring substituents

along with propionic and isobutyric acids was entirely consistent with the earlier assigned structure.

The location of the cyclopentanone carbonyl grouping was unequivocally established through the degradative sequence outlined in *Chart 5.* (*109*). Thus acoronol (**20**) obtained by selective reduction of acorone with lithium tri-*t*-butoxyaluminohydride, yielded the lactone (**21**) upon Baeyer-Villiger oxidation. Further reduction with lithium aluminum hydride afforded the triol (**22**). This substance gave rise to a diketo acid (**23**) which lacked an infrared band in the region of 7.35 μ characteristic of methyl ketones. Such a product (**24**) would have

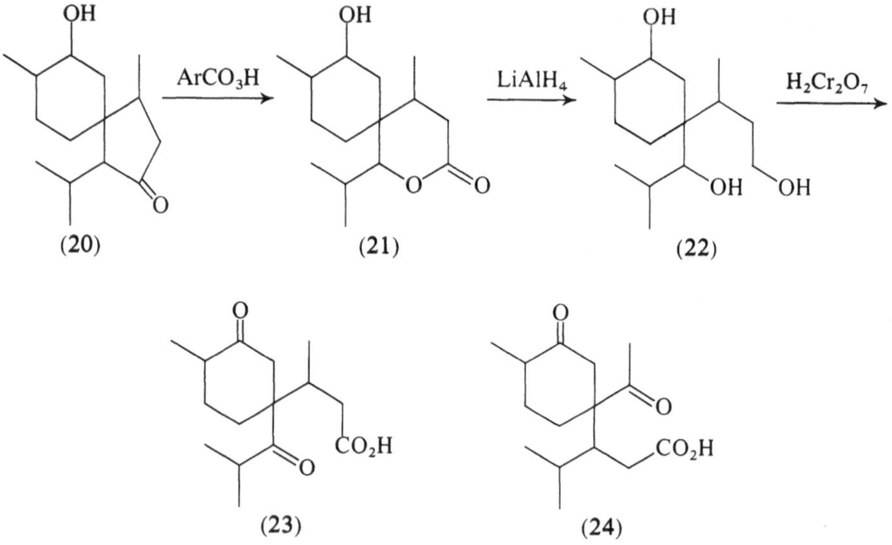

Chart 5. Location of acorone cyclopentanone carbonyl group

resulted had the cyclopentanone carbonyl grouping of acorone been adjacent to the methyl rather than the isopropyl grouping. Additional support for structure (**23**) (*vs.* **24**) came from pyrolysis of the barium salt which afforded butyric acid. The isomeric diketo acid (**24**) would have given rise to γ-methylvaleric acid.

In their work on the stereochemistry of acorone, ŠORM and his group made perceptive use of a variety of physical measurements (*109*). The absolute stereochemistry of the cyclohexane methyl group was deduced by application of the Hudson-Klyne rule to the keto lactone (**25**). Since the cyclohexanone carbonyl grouping of acorone undergoes addition reactions much more readily than the cyclopentanone carbonyl, the lactone (**25**) could be prepared directly by treatment of acorone with peroxyphthalic acid. In a like manner, isoacorone and cryptoacorone

(*108*), two naturally occurring stereoisomers of acorone, were found to possess a cyclohexyl methyl grouping of opposite configuration to that of acorone.

The Hudson-Klyne rule also proved instrumental in elucidating the orientation of the isopropyl grouping of acorone and isoacorone. In these cases the lactones (**26**), obtained via Baeyer-Villiger oxidation of the keto alcohols (**20**), were shown to have identically oriented iso-propyl groupings. This approach could not be employed for the third isomer, cryptoacorone, since the requisite acid catalysis of the oxi-dation step effected its isomerization to isoacorone. Even so, this obser-vation is still significant in that it reveals the epimeric relationship of cryptoacorone and isoacorone.

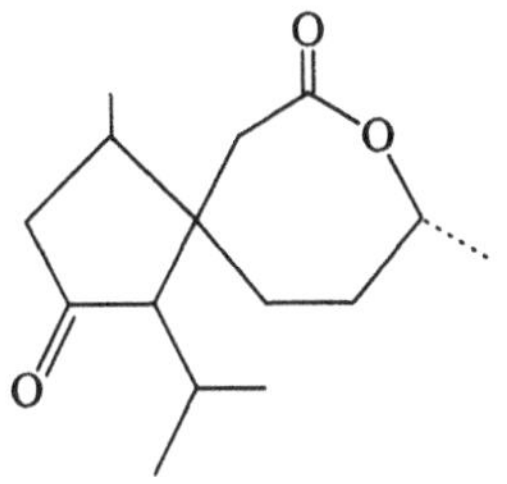

(25) 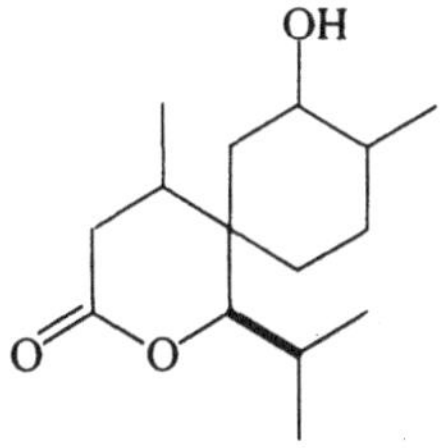   (26)

On the basis of the optical rotatory dispersion curves, Šorm and co-workers concluded that acorone and isoacorone adopt differing chair conformations in solution, as depicted below by (**27**) and (**28**), respect-ively (*99*). The data for cryptoacorone, on the other hand, seemed best accommodated by the twist boat conformer (**29**). The final remaining stereochemical ambiguity of the acorones, the relative configuration of the spiro carbon atom, was resolved on the basis of dipole moment studies. An exact correspondence between calculated and observed dipole moments for the three isomers could not be made owing to uncertainties in the conformation of the cyclopentane ring. Nonetheless, fairly close agreement was reached through careful analysis, and this conclusion, together with the aforementioned optical rotational properties of the

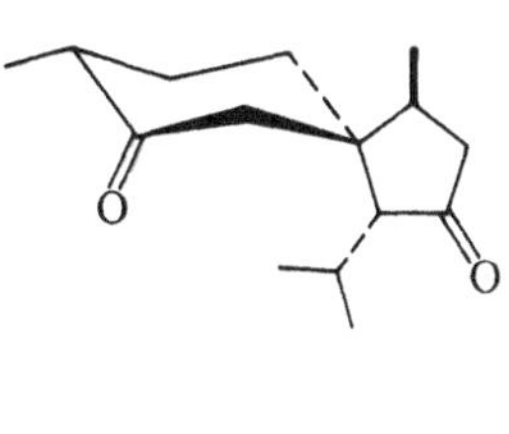

acorone   isoacorone   cryptoacorone
(27)   (28)   (29)

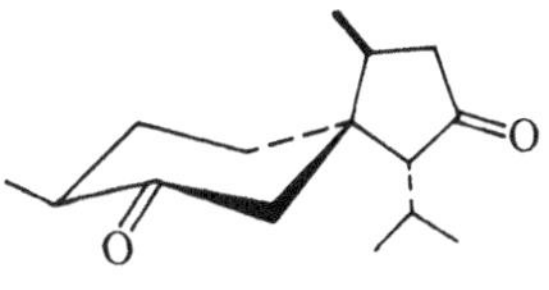

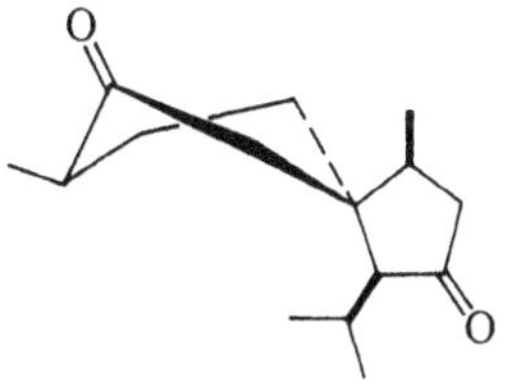

ketones and lactones, and the relative stability of the various epimers, served as the basis for the indicated structure assignments. However, a certain ambiguity still remained as these assignments presupposed that *trans*-1,3-disubstituted cyclopentanones are more stable than their *cis*-counterparts. If this assumption proved invalid, then the assignment of relative configuration to the cyclopentane methyl group would need to be reversed. This possibility was suggested by the work of JACQUES who found that *cis*-1,3-dialkylcyclopentanones are indeed more stable than the corresponding *trans*-isomers (*43*).

The problem was resolved by single crystal X-ray structure analysis of the *p*-bromophenylhydrazone derivatives (**30**) of acorone (*55*). This analysis confirmed the stereochemistry depicted in formula (**27**) initially proposed by ŠORM and co-workers. Interestingly, the derivative employed in the X-ray work adopts the alternative chair conformation in which the cyclohexane methyl grouping is axially oriented and the iso-propyl-bearing cyclopentane carbon is equatorial as shown in (**30b**). Conceivably, this situation is brought about by the unfavorable $A^{(1,3)}$ interaction between the phenylhydrazone NH grouping and the equatorial methyl as indicated below in **30a** (*47*). Of course, as noted by McEACHAN et al., crystal packing forces can favor conformations in the solid phase that may be present to only a small extent in solution (*56*).

(**30a**)   (**30b**)

The absolute stereochemistry of the acorones was recently confirmed through a correlation with cedrene (*14*). Cyclization of the diol (**31**) derived from isoacorone yielded (+)-α-cedrene (**32**), the enantiomer of natural α-cedrene, along with other products.

(**31**)   (**32**)

Vrkoč, Herout and Šorm were able to isolate a fourth acorane sesquiterpene, acorenone (**33**), from the oil of sweet flag (*107*). The infrared spectra of acorenone and its hydrogenation product (**34**) showed them to be substituted cyclohexanones. Further reduction of (**34**) via desulfurization of the thioketal derivative (**35**) afforded acorane (**38**) (*Chart 6*). The choice in favor of structure (**33**) over (**37**) for acorenone could be made on the basis of the ozonolysis product, a C_{13} dicarboxylic acid (**36**). The enone (**37**) would have afforded a C_{14} keto acid upon ozonolysis.

Zalkow and co-workers subsequently characterized a related acorane sesquiterpene, acorenone-B (**39**) (*51*). Hydrogenation of this new acorane afforded two dihydro derivatives, neither of which corresponded to the dihydroacorenone (**34**) of Vrkoč et al. (*107*). However, upon ozonolysis acorenone-B gave the diacid (**36**) previously obtained from acorenone. Accordingly, the two acorenones must differ only in the relative configuration of their spiro carbon atoms. The complete structure of acorenone-B (**39**) was unambiguously established

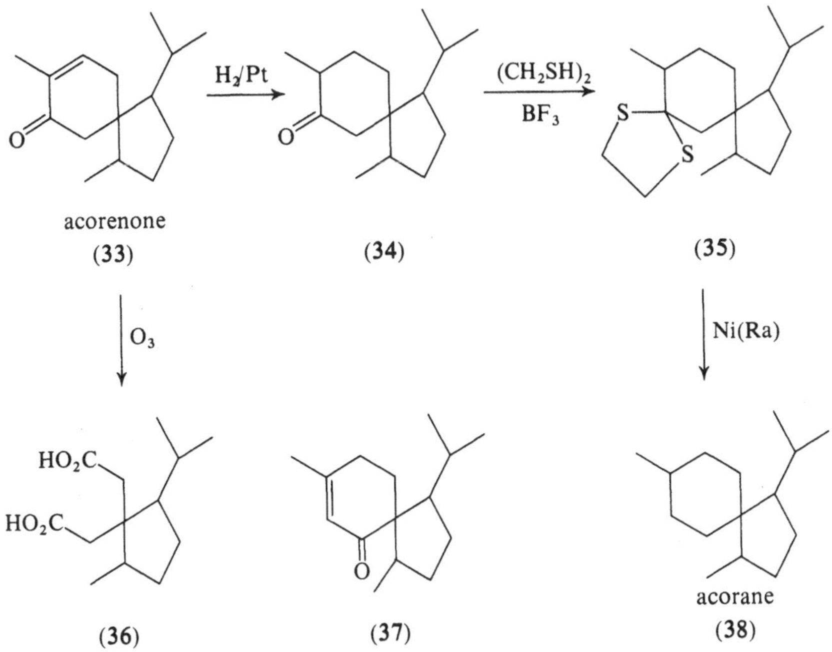

acorenone (33) (34) (35)

(36) (37) acorane (38)

Chart 6. Characterization of acorenone

by X-ray crystallographic analysis of the *p*-iodo-*o*-nitrophenylhydrazone derivative. Acorenone, therefore, can now be formulated as the spiro epimer (**33**) (*51*).

acorenone-B	acorenone
(39)	**(33)**

MINATO and co-workers have recently described a new acorane sesquiterpene, "acoronene" (**40**) which differs from ŠORM's "acorenone" by possessing an additional carbonyl grouping in the cyclopentane ring (*68*). The proof of structure entailed hydrogenation to a mixture of isoacorone and acorone. Spectral data indicated the presence of a conjugated cyclohexenone moiety.

acoronene	acorone + isoacorone
(40)	

2. The Alaskanes, Precursors of Cedrene

Although the acoranes have been postulated as the biogenetic precursors of cedrene, the relative and absolute stereochemistry of the two classes of sesquiterpenes do not correlate (*80*). This postulate must therefore be regarded as unlikely. However, spiranes with the cedrene absolute stereochemistry have recently been discovered in conifer essential oils by two groups. TOMITA et al. reported the isolation of α-acoradiene (**41**) and α-acorenol (**42**) in *Juniperus rigida,* a cedrene producing species

acorane	α-acoradiene	α-acorenol
(38)	**(41)**	**(42)**

(*101*). The structure proof involved hydrogenation to a saturated hydrocarbon identified as acorane by its infrared spectrum. The relationship between the two substances follows from the basic alumina dehydration of the alcohol (**42**) to give the diene (**41**). The stereochemical assignments were based upon the facile conversion of both substances to natural (−)-α-cedrene *(Chart 7)*.

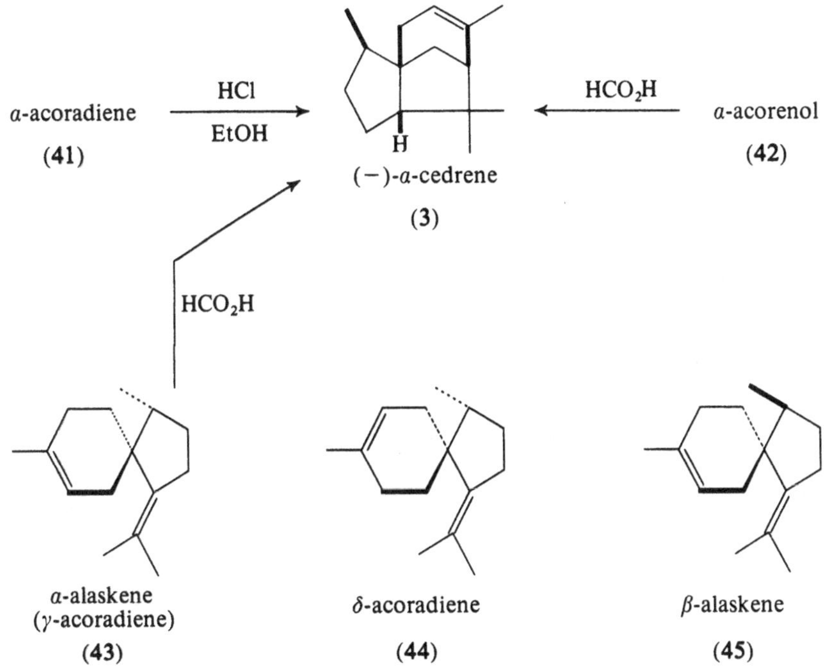

Chart 7. Correlation of acoradiene with cedrene

Andersen and Syrdal obtained two acorane-type sesquiterpenes from *Chamaecyparis nootkatensis* (*12, 13*), a conifer that is viewed as a close relative of genus *Juniperus* (*40*). These two substances, α-alaskene (**43**) and its epimer β-alaskene (now known to be **45**), were assigned spirane structures based on an infrared correlation with acorane and

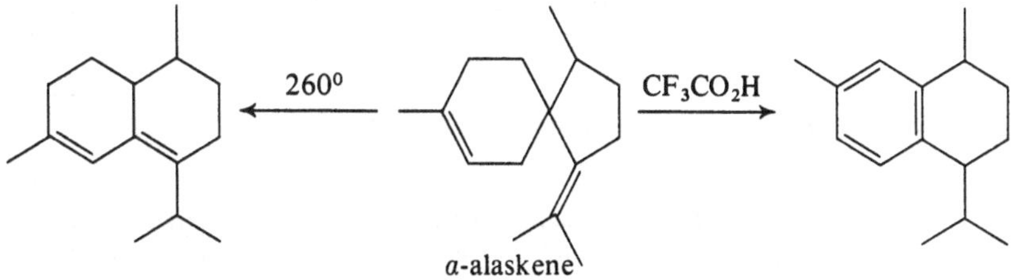

α-alaskene
β-alaskene

rearrangement to both of the two possible ortho-fused hydronaphthalene skeletons *(13)*. On milder acid treatment (2 : 1 $HCO_2H - THF$) α-alaskene afforded cedrene in greater than 90% yield supporting the relative configuration shown *(Chart 7)*. β-Alaskene proved less reactive and gave only uncharacterized mixtures.

Further study of *Juniperus rigida* uncovered dienes (**43**) (designated γ-acoradiene) and (**44**) (designated δ-acoradiene) as constituents of this

| β-acorenol | "β-acoradiene" | δ-acoradiene |
| (46) | (47) | (44) |

oil *(103)*. In addition another tertiary alcohol and its dehydration product were found. These were designated β-acoradiene (**47**) and β-acorenol (**46**) and unlike the α-isomers, they did not cyclize readily. In fact β-acoradiene afforded δ-acoradiene rather than tricyclic sesquiterpenes on acid treatment.

The remaining stereochemical questions were resolved through chemical correlations with isoacorone *(Chart 8)* *(11)*. Reduction of isoacorone (**28**) followed by dehydration afforded a mixture of olefins containing dienes (**48**) and (**49**) with minor amounts of epimeric cedrenes (**50**), whose CD spectra establish the absolute configuration at the 10-position of the acorones.

The absolute stereochemistry of (−)-α-alaskene was clear from its high yield conversion to natural (−)-α-cedrene. The first evidence that β-alaskene was in fact an acorane* came from an ORD comparison of the products of CF_3CO_2H treatment of β-alaskene and acoradiene (**48**) *(18)*. Both spiranes afforded a 2:3 mixture of *cis-* and *trans-*isocalamenenes (**51**) with identical ORD spectra *(Chart 8)*. α-Alaskene in contrast, produced primarily the *cis-*isomer (**51a**) *(Chart 9)*.

---

* It has been suggested that the name "acorane" be used for the spiranes having the C-10 absolute configuration of acorones. The enantiomeric series, which yields natural cedrene, is designated "alaskane".

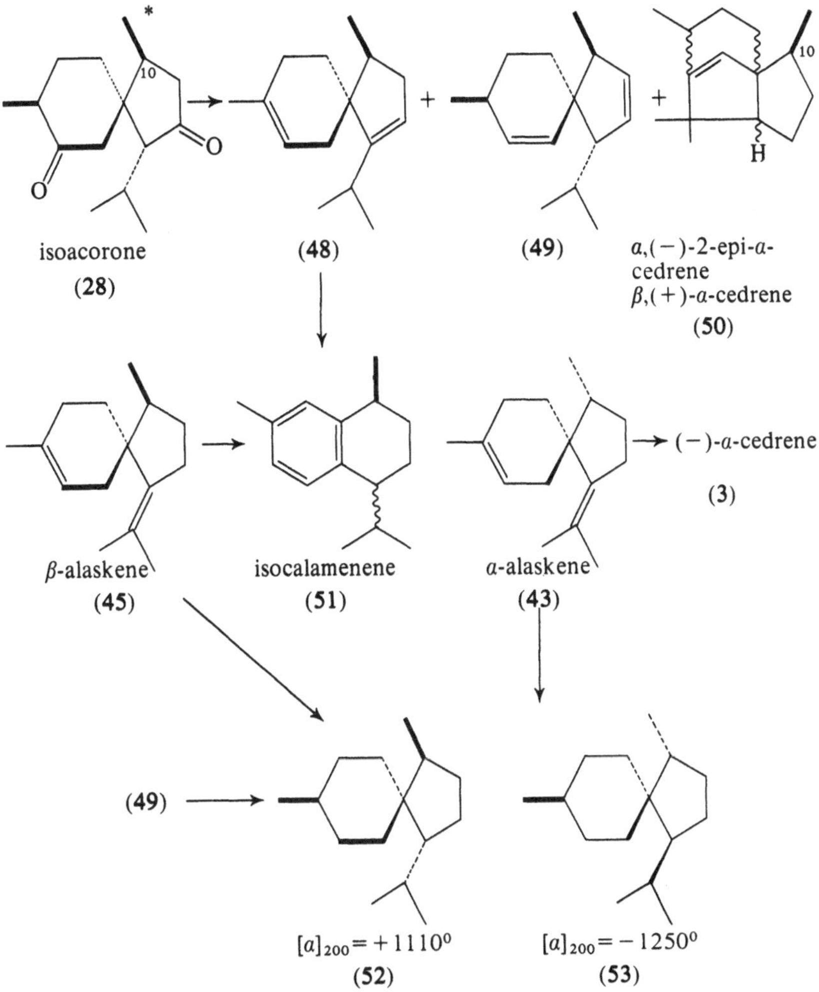

isoacorone
(28)

(48)

(49)

a,(−)-2-epi-a-
cedrene
β,(+)-a-cedrene
(50)

β-alaskene
(45)

isocalamenene
(51)

a-alaskene
(43)

(−)-a-cedrene
(3)

(49) →

$[a]_{200} = +1110°$
(52)

$[a]_{200} = -1250°$
(53)

Chart 8. Correlation of isoacorone with alaskenes

The assignment of enantiomeric skeletons to the alaskenes was confirmed
by the ORD spectra of one of the diastereomeric hydrogenation products
(52) and (53) *(Chart 8)* (*11*). The relative stereochemistry was based on
the fact that diene (49) afforded only (52) on hydrogenation using a homogeneous rhodium catalyst.

A recent reinvestigation of vetiver oil has established that the
acoradiene (48) is a natural product (*48*). In addition the acoradiene (54)*

* The structure shown is enantiomeric to that which appears in (*48*). Recent unpublished work indicates that this revision should be made (*70*).

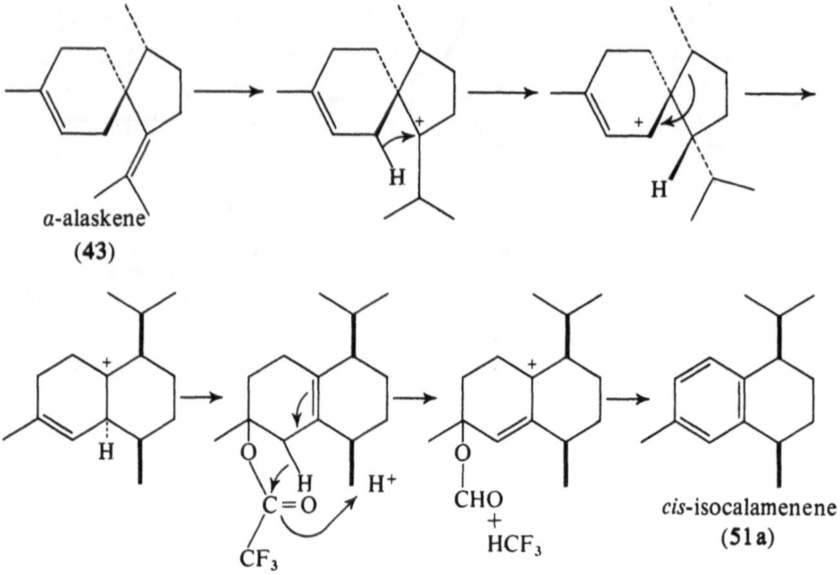

Chart 9. Conversion of α-alaskene to *cis*-isocalamenene

was found together with a series of enantiomeric cedrenes (**55**) of unusual functionalization. The biogenetic significance of these findings will be considered in a later section.

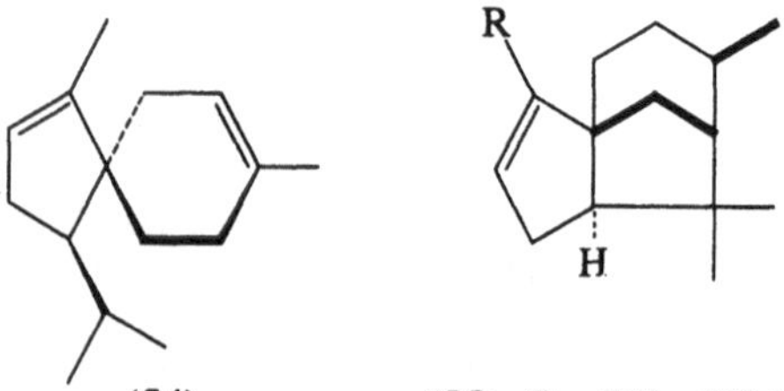

(**54**)　　　　　(**55**); R=CH₃, CH₂OH,
　　　　　　　　　　　　　　CHO, CO₂H

3. Mass Spectral Considerations

The mass spectra of various acorane and related spiro[4.5]decane derivatives have proved useful in structure elucidation. The spectra typically reveal an otherwise uncommon tendency to produce prominent stable radical cations (even m/e) rather than cations (odd m/e). This was first noted for the chamigrenes which are spiro[5.5]undecanes (79). Other examples that have appeared are also given.

Fragmentation pathways for Spiro Compounds

Reference

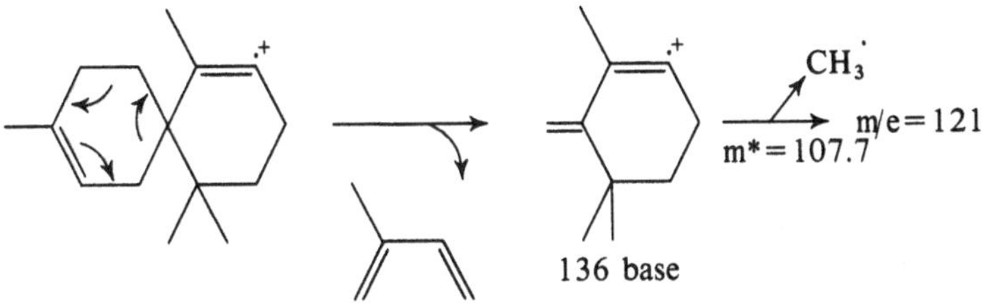

$$m^* = 107.7 \xrightarrow{\quad CH_3^. \quad} m/e = 121 \qquad (79)$$

136 base

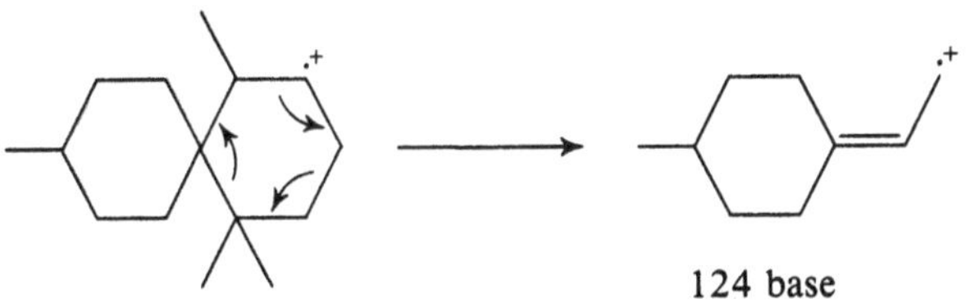

124 base   (79)

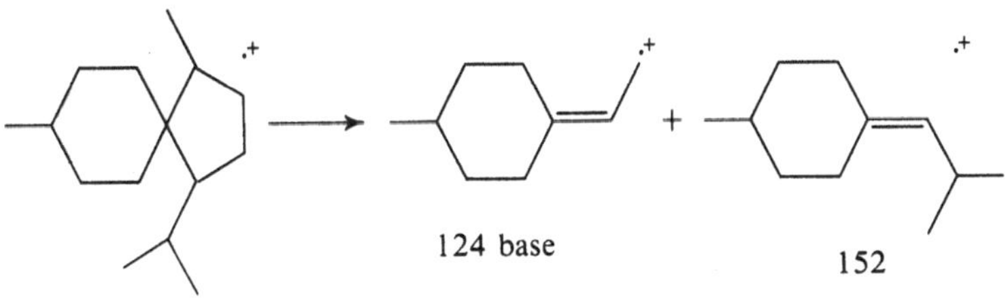

124 base   152   (13)

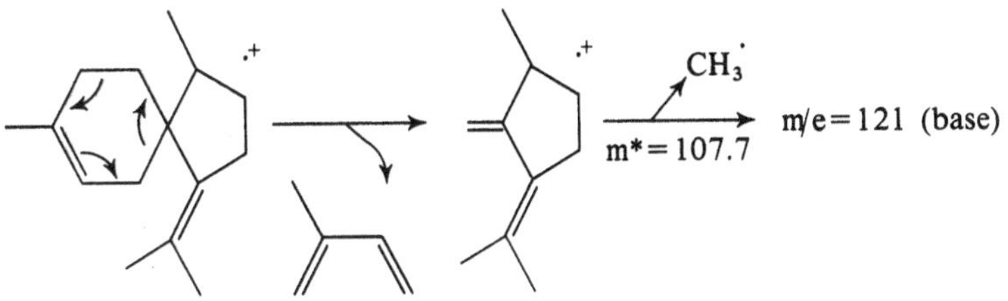

$$m^* = 107.7 \xrightarrow{\quad CH_3^. \quad} m/e = 121 \ (base) \qquad (13)$$

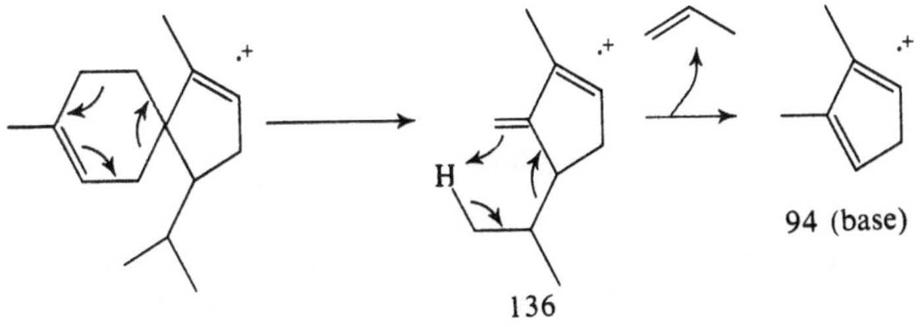

136   94 (base)   (35, 48)

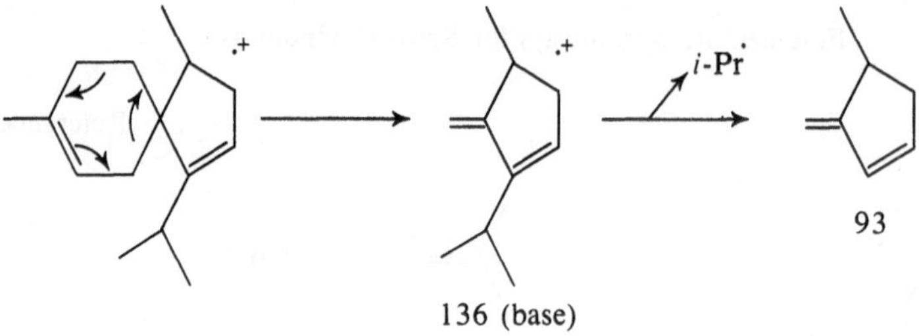
.+
.+
i-Pr
136 (base)
93

(*48, 100*)

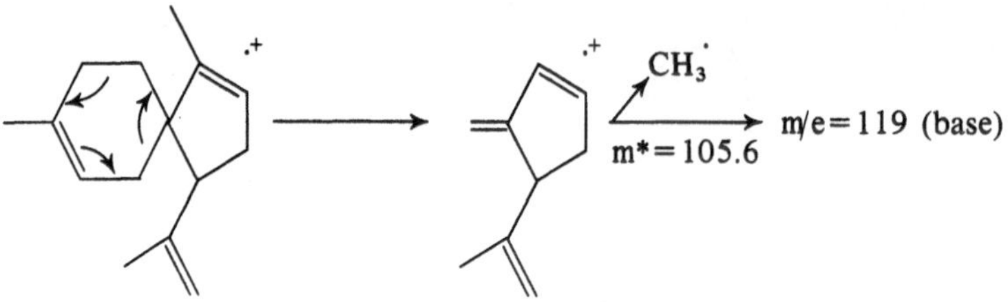
.+
.+
CH3
m*=105.6
m/e=119 (base)

(*35*)

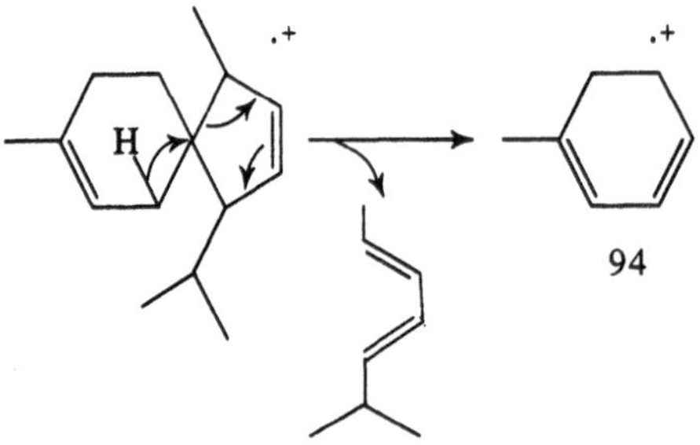
.+
H
.+
94

(*15*)

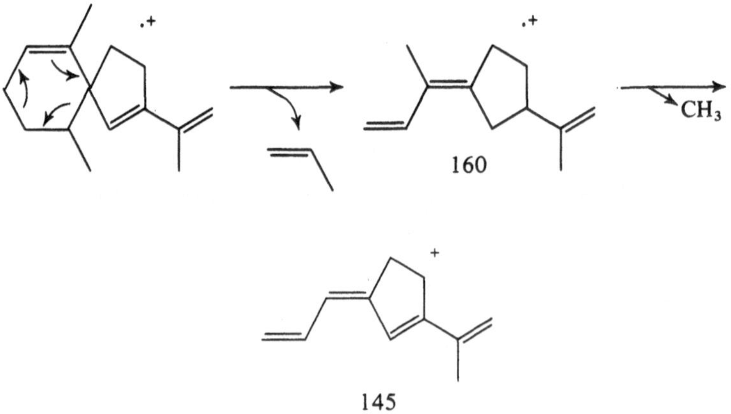
.+
.+
CH3
160
+
145

(*15*)

4. Synthesis

The acorane sesquiterpenes pose some interesting synthesis problems which have only recently attracted attention. Acorone itself has been synthesized by the sequence outlined in *Chart 10*. The final alkylation step caused problems however and a reproducible process could not be developed (67). An interesting aspect of this synthesis is the internal Michael addition (59)→(60) employed for construction of the spiro-[4.5]decane skeleton. This step is especially noteworthy from a stereochemical point of view in that the relationship between the spiro carbon atom and the cyclopentane methyl grouping is thereby established. The two remaining asymmetric centers in the final product (61), being epimerizable and having the more stable arrangement, present no stereochemical difficulties.

As mentioned earlier, the cation (2) first came to light as a proposed intermediate in the biosynthesis of cedrene (87). Ingenious laboratory syntheses of cedrene utilizing this concept have been completed by two research groups working independently (Chart 11) (33, 34). Both employ phenolic bromo esters, (62) (R=Me) and (63) (R=Et) respectively, as key intermediates which afford the spiro[4.5]decane derivatives (64) (R = Me or Et) upon treatment with potassium t-butoxide. This route to spiro[4.5]decadienones was conceived by BAIRD and WINSTEIN who suggested the term "Ar₁,₅-participation" to describe the alkylation process (20). The initial cyclization product of bromo esters (62) and (63), a mixture of cis- and trans-isomers, gave predominantly the indicated racemic trans-isomer (64) upon exposure

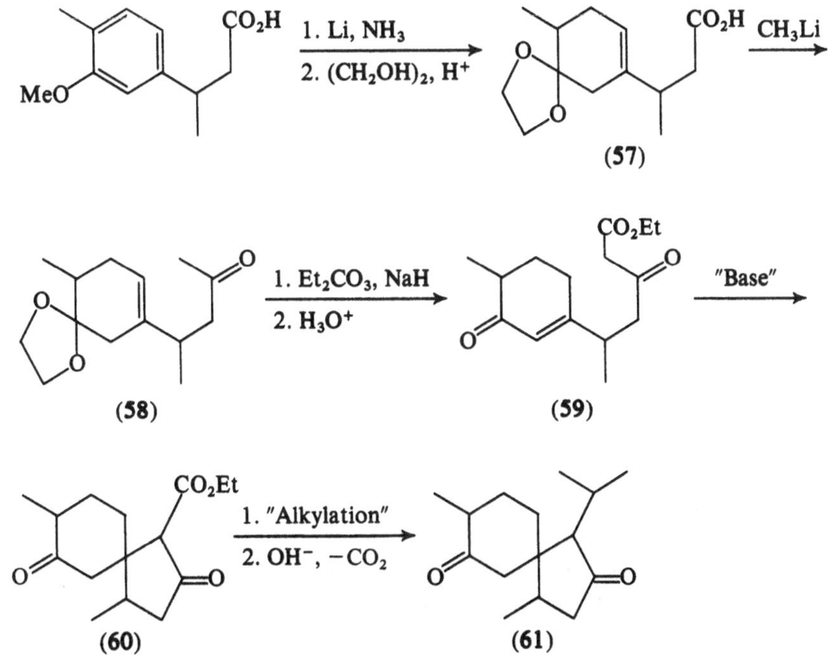

Chart 10. The synthesis of acorone

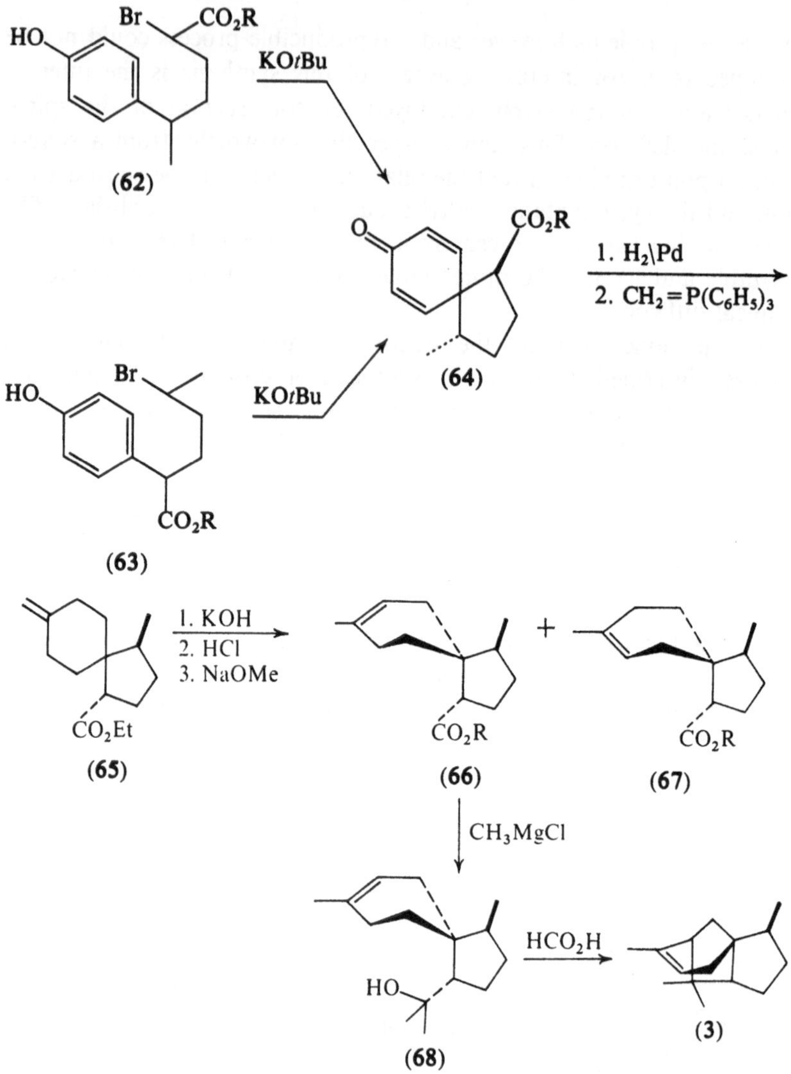

Chart 11. The synthesis of (±)-cedrene via spiro[4.5]decane cyclization

to base. Crandall and Lawton converted dienone (64) to a mixture of the unsaturated esters (66) (R = CH₃) and (67) (R = CH₃), mainly the former, by means of the indicated sequence. The apparent preference for the double bond isomer (66) may result from the ability of the carboxylate substituent of the chloro acid precursor to function as an internal base in the elimination reaction, thus directing the double bond toward the carboxylic grouping. Subsequent treatment with methylmagnesium chloride afforded the acorane-related alcohol (68), which was smoothly converted to cedrene (3) upon dissolution in formic acid.

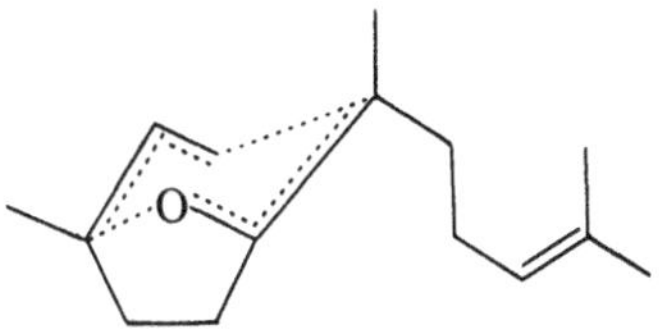
O

Additional syntheses of α-cedrene and its diastereomers by way of spiro[4.5]decane intermediates have recently appeared. The route developed by Demole et al. *(Chart 12)* is remarkable in that it proceeds through four distinct sesquiterpene skeletons starting from a *cis, trans*-mixture of nerolidol (**69**), which afforded the tetrahydrofuran (**70**) upon

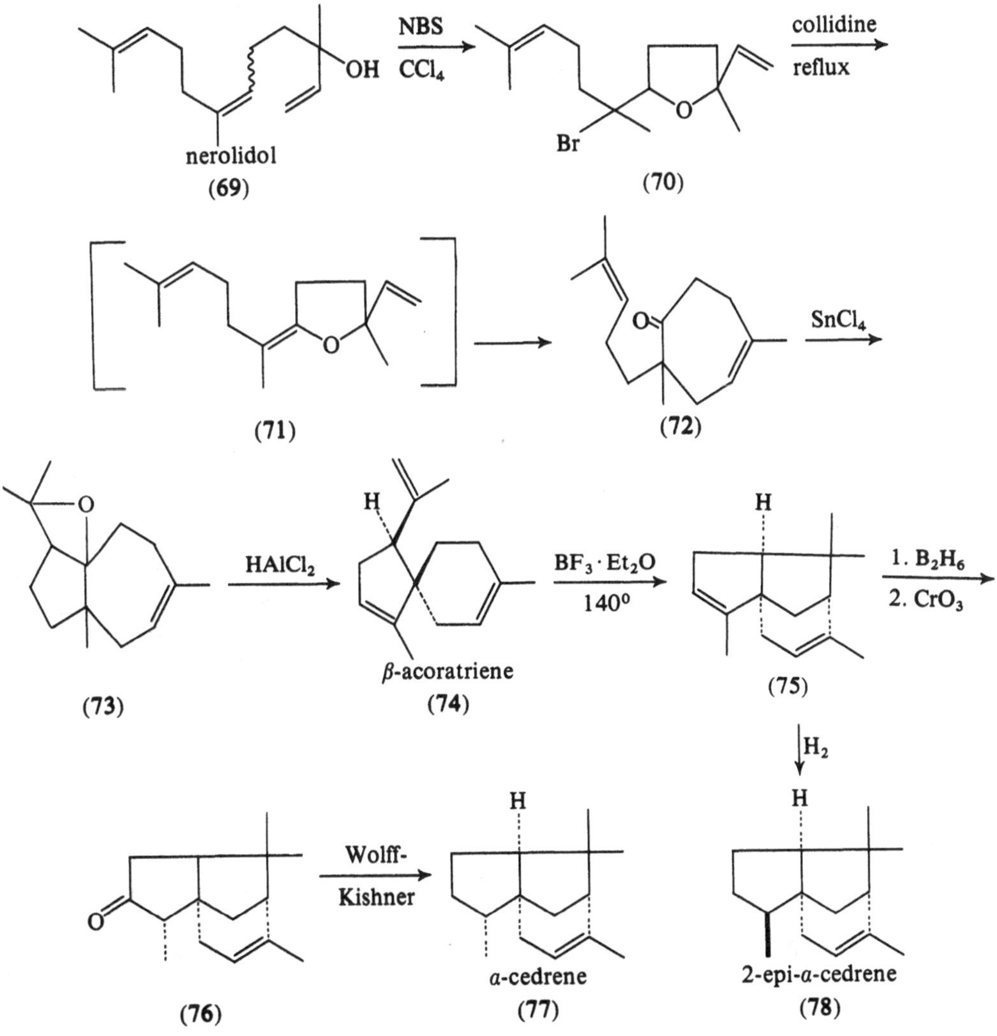
OH
NBS
CCl_4
Br
O
collidine
reflux
nerolidol
(69)
(70)
O
(71)
O
$SnCl_4$
(72)
O
$HAlCl_2$
H
β-acoratriene
(74)
(73)
$BF_3 \cdot Et_2O$
140°
H
(75)
1. B_2H_6
2. CrO_3
H_2
O
Wolff-
Kishner
H
α-cedrene
(77)
(76)
H
2-epi-α-cedrene
(78)

Chart 12. Synthetic conversions of nerolidol to acoratriene

hypobromination (35). Dehydrobromination yielded enol ether (71)
which then gave cycloheptenone (72) in 66% over-all yield from
nerolidol. The rearrangement of enol ether (71) can be viewed as a
[3.3]-sigmatropic process (Claisen rearrangement). From here the
sequence employs three consecutive Lewis acid catalyzed rearrangements.
A [2+2] addition afforded an ether of carotol (73), a daucane type
sesquiterpene which then underwent ring contraction to give β-acoratriene (74). None of the α-isomer, which would yield cedrenes directly
without double bond isomerization, could be detected. On treatment
with BF$_3$ · Et$_2$O in diglyme at 140° β-acoratriene cyclized to the cedra-
diene (75) presumably after double bond isomerization. Hydrogenation
of diene (75) afforded only the 2-epimer of cedrene. However. the ketone
(76) obtained by hydroboration-oxidation of diene (75) gave a mixture of
α-cedrene and its 2-epimer upon Wolff-Kishner reduction.

NAEGELI and KAISER reported another synthesis of β-acoratriene
(Chart 13) (71). Pyrolysis of dehydrolinalool afforded the allylic alcohol
(79) as the starting point for this synthesis. Conversion to the isopropenyl ether followed by Claisen rearrangement led to the ketone (80).

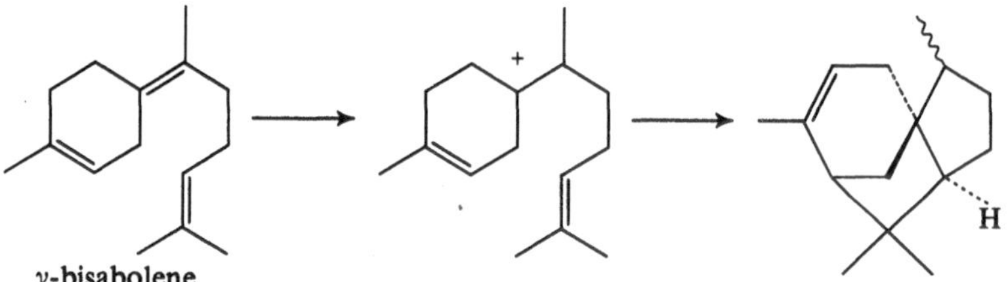

γ-bisabolene

cedrenes

Vinylmagnesium bromide addition gave the alcohol (81) which cyclized
upon treatment with stannic chloride. The stereoselectivity of this
cyclization is remarkable in that only the β-isomer of acoratriene (74)
could be detected. This contrasts with the results of the biogenetically
inspired studies of the cyclization of γ-bisabolene. In this case only
α-isomers were formed (14). An explanation for this divergence may
involve the intermediacy of a daucane structure in the cyclization

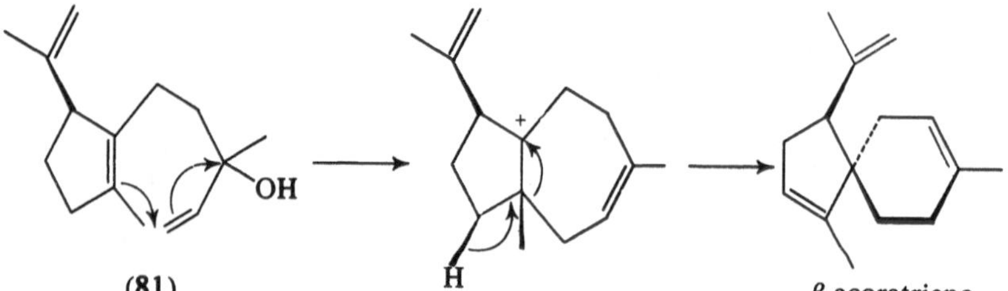

(81)

β-acoratriene
(74)

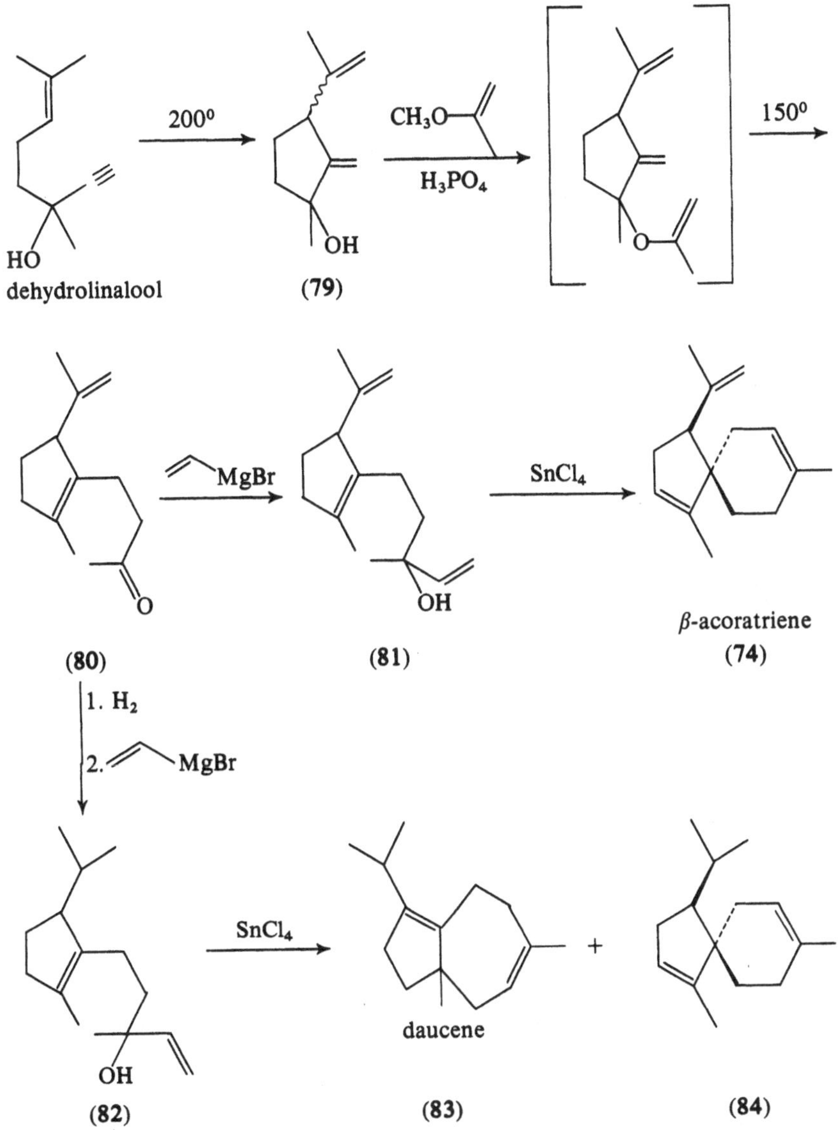

Chart 13. Synthetic conversions of dehydrolinalool to acorenes

of (81). In the dihydro series, daucene (83) is in fact the major cyclization product. The minor product, acoradiene (84), was identified as one
of the acoranes found in vetiver oil (48).

YAMASAKI has reported a synthesis of (−)-daucene proceeding
along similar lines starting with limonene (Chart 14) (111d). In this case
the reported optical activity is consistent with the stereoselectivity observed in the cyclization yielding β-accoratriene (Chart 13), (81)→(74).

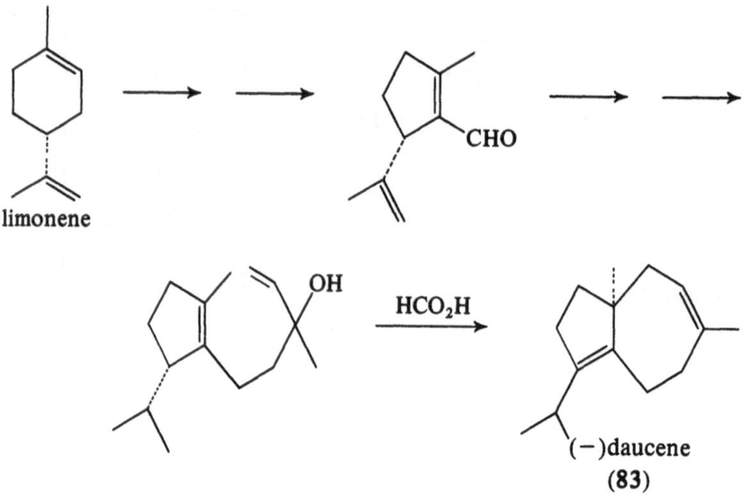

Chart 14. Synthetic conversion of limonene to daucene

A more direct route from nerolidol to cedrene which bears an even closer relationship to the postulated biosynthetic pathway has been reported by ANDERSEN and SYRDAL (Chart 15) (14). The dehydration of (+)-nerolidol (or various nerolidyl derivatives) proceeds largely with participation of the 6,7-olefinic bond giving monocyclic products (up to 86% bisabolenes) with significant induction of asymmetry. In fact, under certain conditions β-bisabolene (85) can be obtained in 80% yield. More vigorous acid treatment gives γ-bisabolene (87) and eventually yields a gross mixture of bicyclic sesquiterpenes. However the two phase system, pentane/CF$_3$CO$_2$H, affords a simpler mixture: 30% α-cedrene, 25% epi-α-cedrene, and 20% dihydro-α-curcumene. Even though optically active β-bisabolene was employed, these products were racemic thus implying that ion (88) was formed from γ-bisabolene. Detailed study of the reactions of the curcumenes, the alaskenes, and various acoradienes under the same reaction conditions clarified the course of the reaction. In CF$_3$CO$_2$H the ion (88) cyclizes readily, giving only spiranes of the α-series (ions 90 and 91) which cyclize directly to the cedrenes without the intermediacy of acoradienes. The high stereoselectivity of the spiro-cyclization suggests that the two stages may occur in concert. At least the transient ions (90) and (91) appear to be stabilized (relative to β-isomers) by the interaction of the π-bond and the cationic center. In a study patterned along similar lines, OHTA and HIROSE isolated α-cedrene and 2-epi-α-cedrene from the mixture obtained by treating farnesol with boron trifluoride etherate (78).

A potentially general synthetic approach to spiro[4.5]decane derivatives which should be applicable to the acoranes has been devised by CONIA (Chart 16) (30, 36). Enones of the type (94), available via Cope rearrangement of appropriately substituted cyclohexanones (e.g. **92** and **95**) and subsequent thermal deconjugation of the resulting conjugated ketones (**93**) and (**96**), undergo thermal cyclization at temperatures above 300° to give the spiro[4.5]decanones (**98**) and (**99**). In the case where R = Me, the former isomer predominates slightly [60% (**98**) and 40% (**99**)] indicating a slight preference for the mode of cyclization depicted by (**97a**)

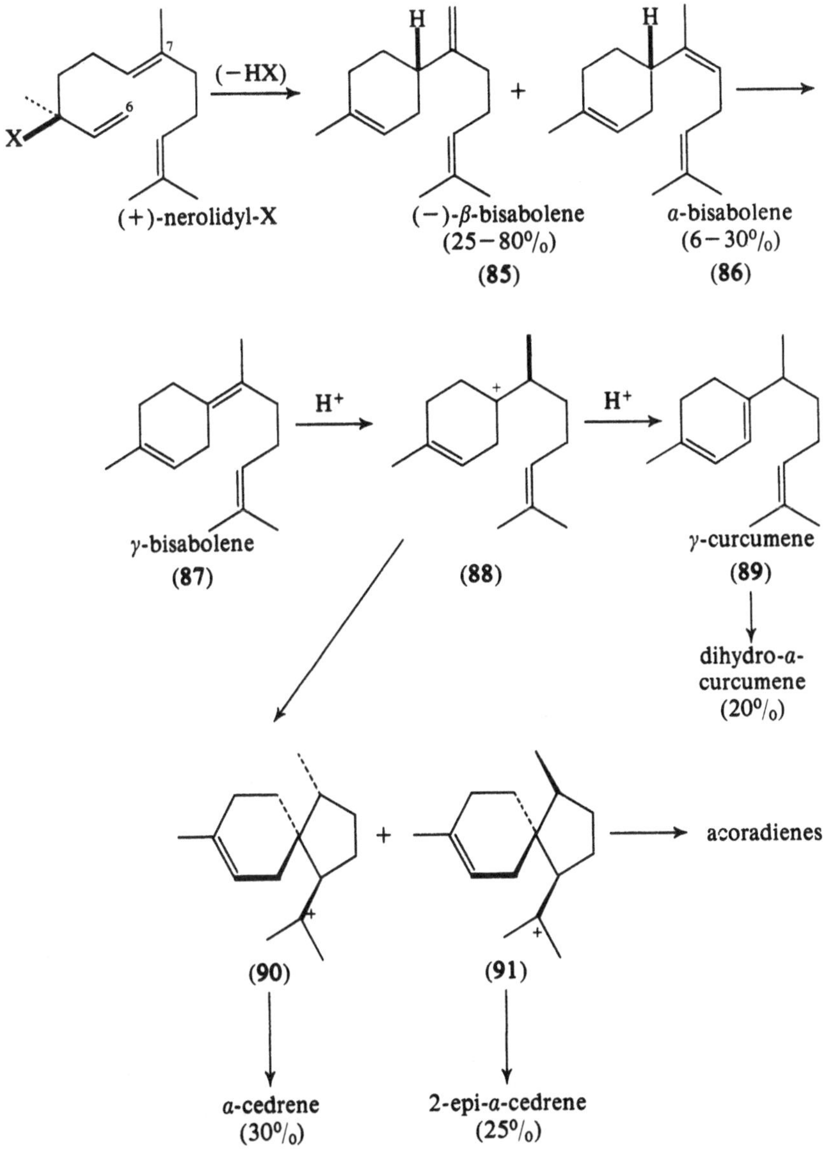

(+)-nerolidyl-X

(−)-β-bisabolene
(25−80%)
(**85**)

α-bisabolene
(6−30%)
(**86**)

γ-bisabolene
(**87**)

(**88**)

γ-curcumene
(**89**)

dihydro-α-
curcumene
(20%)

acoradienes

(**90**)

(**91**)

α-cedrene
(30%)

2-epi-α-cedrene
(25%)

Chart 15. Conversion of nerolidol to cedrenes

vs. (97b). *Chart 17* outlines a recent synthesis of acorone along these lines starting with the hydroxymethylene derivative of (+)-3-methylcyclohexanone (100) which afforded the enone (101) upon addition of 3-butenylmagnesium bromide (31, 32). Treatment with isopropenylmagnesium bromide then gave the dienone (102) which yielded a mixture of spiro[4.5]-decanones (103) (38% yield, 4:1 endocyclic/exocyclic double bond) and (104) (42% yield, 1:1 endocyclic/exocyclic double bond) upon thermo-

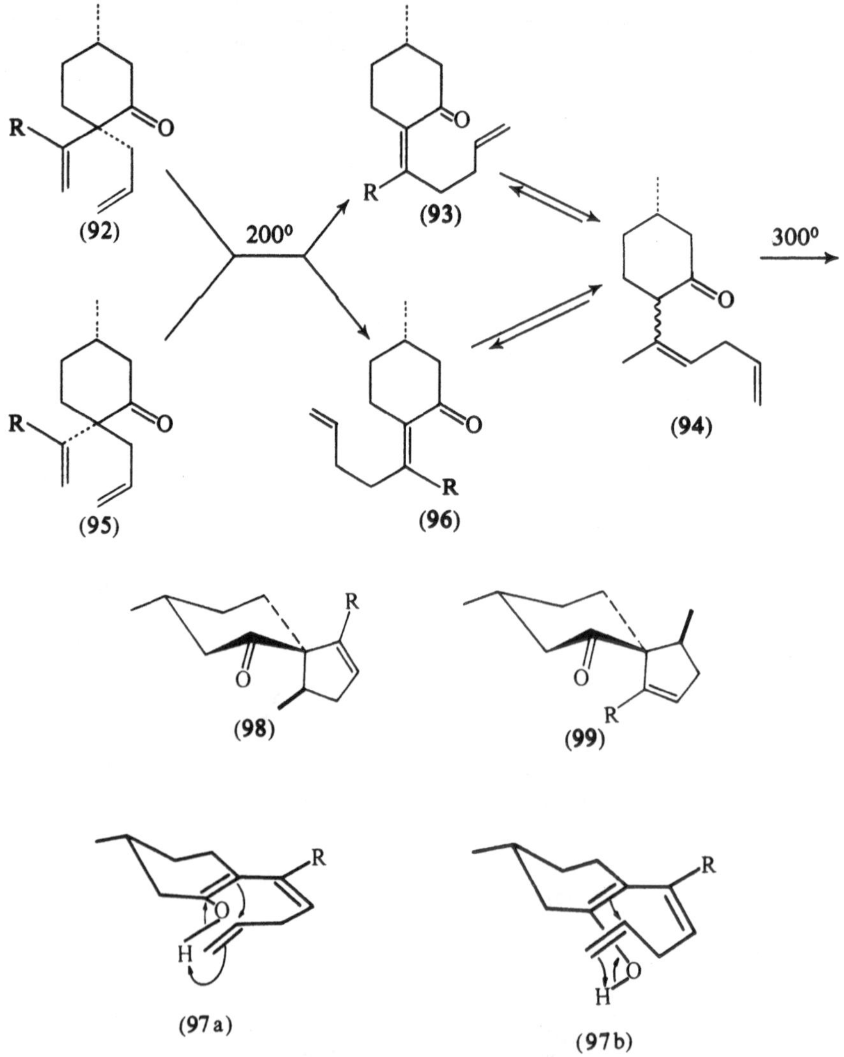

Chart 16. Thermal cyclizations leading to spiro[4.5]decanes

lysis at 220° for 36 hours. Interestingly, the desmethyl analog of dienone (102) gave the desmethyl analog of enone (103) containing only the exocyclic (isopropylidene) double bond isomer. Hydroboration-oxidation of the endocyclic olefins (103) and (104), secured via acid treatment of the related mixtures, yielded a mixture of acorones.

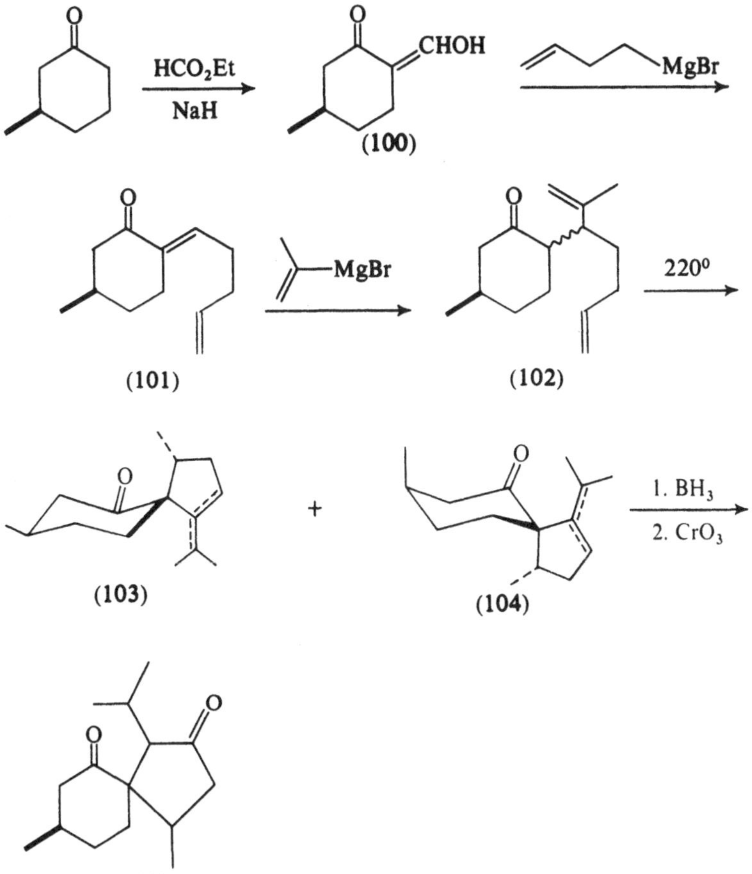

Chart 17. Synthesis of acorones via thermal cyclization

LAWTON and DUNHAM have reported an extension of their αα'-annelation scheme to the synthesis of spiro[4.5]decanes (Chart 18) (38). Addition of methyl α-(1-bromomethyl)acrylate to the enamine of 1-acetyl-2-methylcyclopentane (105) followed by treatment with triethylamine and hydrolysis led to the spiro[4.5]decanonecarboxylic ester (106), presumably by the indicated pathway. The steric course of the internal Michael addition appears to be controlled by the neighboring cyclopentyl methyl grouping. Keto ester (106) afforded a 70:30 mixture of the epimeric esters

20*

(107) and (106) upon treatment with sodium methoxide. The cyclization must therefore give rise to the kinetic product (106) which reverts to the thermodynamic product (107) when exposed to strong base.

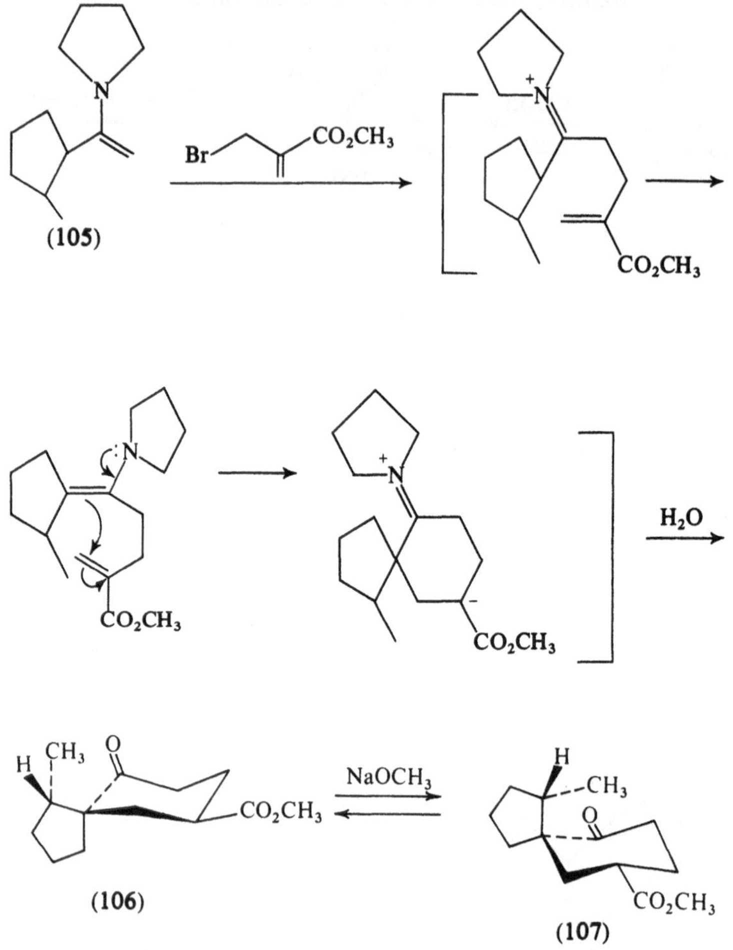

Chart 18. αα'-Annelation route to spiro[4.5]decanes

Another potential route to acorane-alaskane spiro[4.5]decanes is shown in *Chart 19*. In this case acid treatment of the carbinol (108) leads to five-membered ring formation via a cation-initiated olefin cyclization (50). Unfortunately the stereochemical aspects of such cyclizations have not yet been examined.

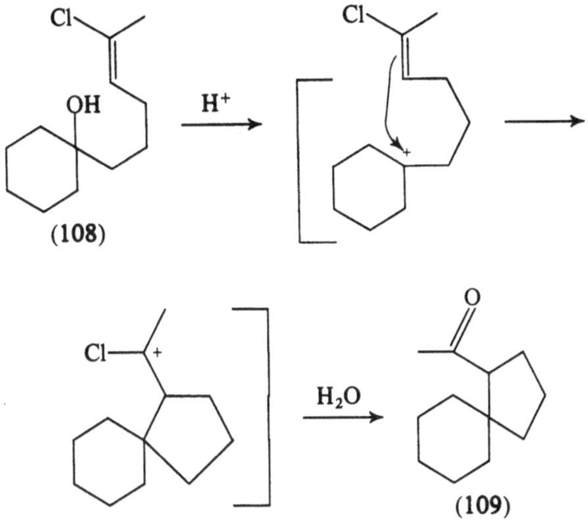

(108)

(109)

Chart 19. Chloro olefin annelation route to spiro[4.5]decanes

C. The Spirovetivanes

1. Structure

a) Agarospirol

A second type of spiro[4.5]decane sesquiterpene, which differs from the acoranes in the placement of substituents, was first recognized by BHATTACHARYYA and co-workers who isolated the substance agarospirol (110) from the oil of fungus-infected agarwood (*Aquilaria agollocha* Roxb.) (*104*). The chemical and physical properties of agarospirol showed it to be a bicyclic unsaturated tertiary alcohol. Pyrolysis of the benzoate derivative (111) afforded a non-conjugated diene (112) *(Chart 20)*. Dehydrogenation of this diene, or the corresponding monoene (116) secured via partial hydrogenation, afforded eudalene (117), along with trace quantities of unidentified azulenes. However, complete hydrogenation of the diene (112) led not to selinane (120), but gave instead a saturated bicyclic hydrocarbon whose "infrared spectrum is strikingly different from those of selinane or any other saturated skeletal hydrocarbon reported in the literature". Since the diene (112) could not be converted into a conjugated isomer by acidic or basic treatment, and since the nmr spectrum revealed no quaternary methyl groups which might block such an isomerization, a spiro ring system seemed a likely possibility for agarospirol. In that event, the aforementioned dehydrogenation

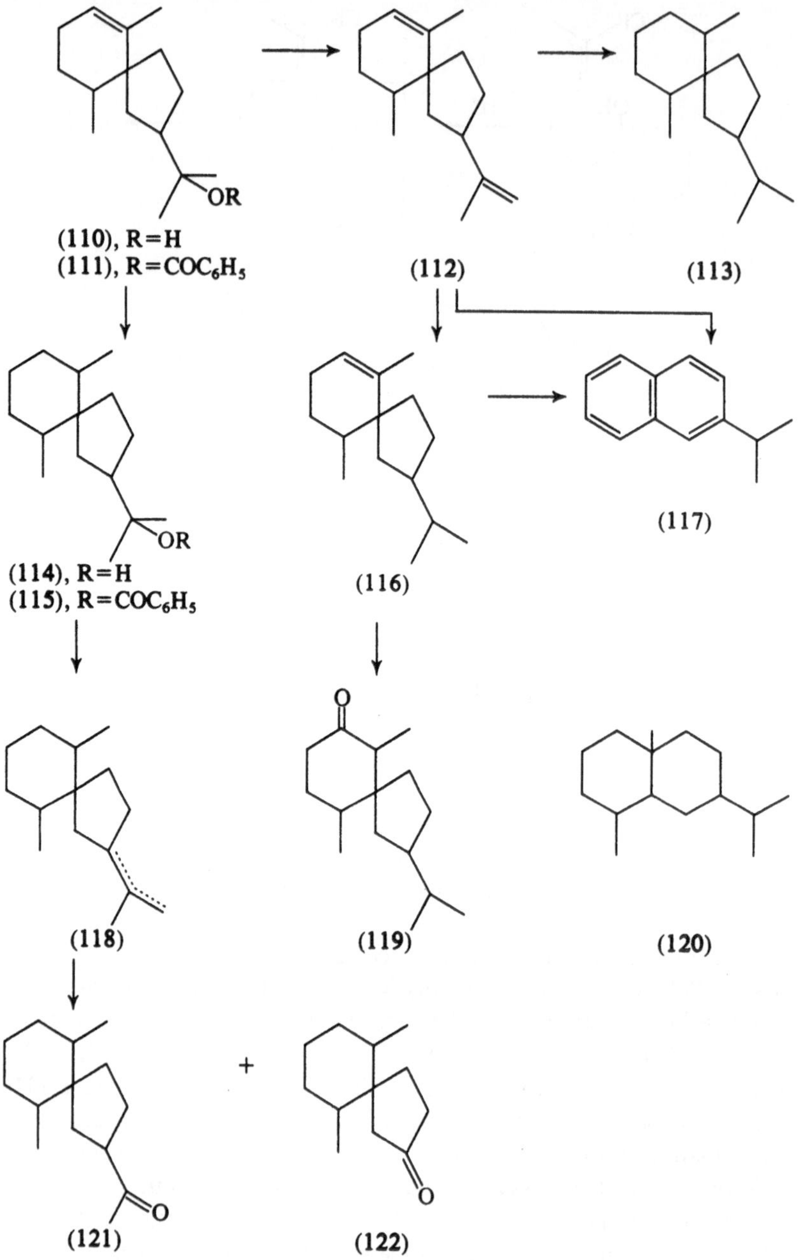

(110), R=H
(111), R=COC₆H₅

(112)

(113)

(114), R=H
(115), R=COC₆H₅

(116)

(117)

(118)

(119)

(120)

(121)

+

(122)

Chart 20. Degradation of agarospirol

reactions leading to eudalene would have proceeded with rearrangement of the carbon skeleton, as had previously been observed in the acorane structure work (*98*).

The nature of the two rings of agarospirol was revealed by oxidative conversions of the derived olefins (116) and (118) *(Chart 20)*. The former yielded the ketone (119) upon hydroboration-oxidation. The infrared spectrum of this ketone, a 70:30 mixture according to the gas chromatogram, indicated the presence of a $-CH_2CO-$ grouping in a six-membered ring. This mixture was converted to "an almost pure epimer" by treatment with strong base. These findings suggested that the cyclohexanone ring also possessed an epimerizable methyl-substituted α-position, a postulate supported by the nmr spectrum.

Pyrolysis of the benzoate derivative (115) of dihydroagarospirol (114) afforded a mixture of olefins (118). It should be noted at this point that the hydrogenation of agarospirol (110) leads to a 45:55 mixture of stereoisomers and therefore formula (118) actually represents a mixture of four isomers. Ozonolysis of this olefin mixture yielded the ketones (121) and (122) along with formaldehyde and acetone. Ketone (122), a 45:55 mixture of epimers, displayed a cyclopentanone split carbonyl band in the infrared spectrum thus showing that the second ring of agarospirol must be five-membered. The evidence at this point suggests a spiro[4.5]-decane carbon skeleton for agarospirol. The afore-men-

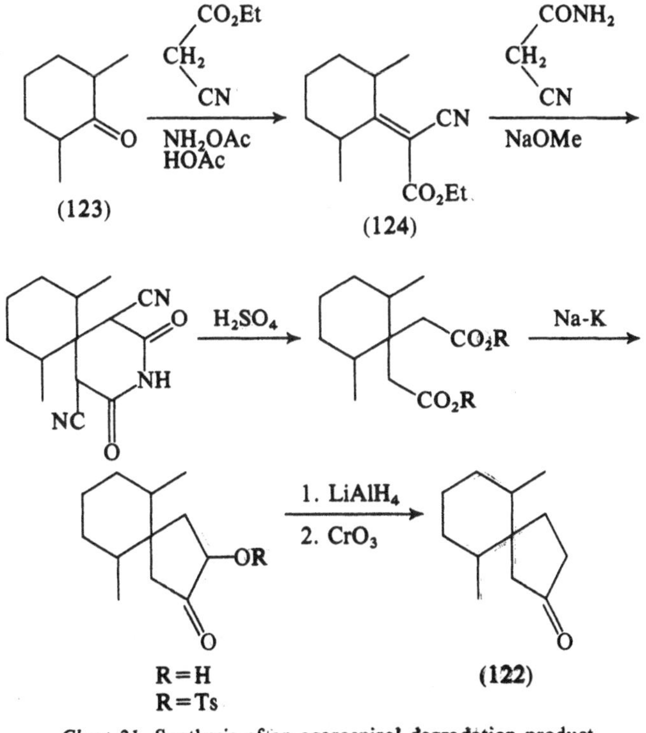

Chart 21. Synthesis of an agarospirol degradation product

tioned dehydration-ozonolysis sequence (114→118→121+122) also establishes the presence of an isopropylol side chain on the cyclopentane ring. The location of this side chain and further support for the gross structure of agarospirol was secured through a synthesis of the ketone (122) as outlined above *(Chart 21)*.

Condensation of 2,6-dimethylcyclohexanone (123) with ethyl cyano-acetate, under forcing conditions, afforded the cyanoacetate derivative (124). Michael addition of cyanoacetamide to this substance then gave the glutarimide which underwent hydrolysis and decarboxylation nearly quantitatively upon treatment with aqueous sulfuric acid, to yield the diacid. Condensation of the corresponding diester with 1:1 sodium-potassium alloy in refluxing xylene afforded the acyloin. Reduction of the corresponding *p*-toluenesulfonate derivative, followed by oxidation of the resulting alcohol gave the desired ketone (122). The infrared and nmr spectra of this substance, while not superimposable with the corresponding spectra of ketone (122), secured *via* degradation of agarospirol *(Chart 20)*, nonetheless showed striking similarities. In view of the fact that this latter sample actually consists of a 45:55 mixture of epimers, the lack of complete correspondence between the synthetic and degradative ketones might be expected.

Four stereoisomeric forms are possible for ketone (122), two *meso* (geometric) isomers (122a) and (122b) and one *dl* (enantiomeric) pair (122c) and (122d). The ketone derived via degradation of natural agarospirol (110) must contain one of the meso isomers (122a) or (122b) and one of the active isomers (122c) or (122d). The synthetic ketone, on the other hand, appears to consist mainly of one isomer. Under equilibrating conditions, steric factors should favor the *dl* (124a) (only one antipode is shown) over the meso cyclohexylidenecyanoacetate isomer

(122a)   (122b)   (122c)

(122d)   (124a)   (124b)

(124b) prepared by BHATTACHARYYA in his aforementioned synthesis (47). Accordingly, the synthetic ketone (122) should possess the *dl* configuration (122c, 122d).

The foregoing degradative experiments provide no clue to the stereochemistry of agarospirol. However, the nmr spectrum of agarospirol acetate (125) contains five closely spaced signals corresponding in area to three protons. BHATTACHARYYA attributed this feature of the spectrum to the tertiary methyl grouping of agarospirol acetate and suggested that "the explanation for this abnormal splitting may be sought in some steric factor which leads to congestion on passing from the alcohol to the acetate" (*104*). The corresponding methyl grouping of agarospirol itself gives rise to a normal doublet in the nmr spectrum. On this basis, BHATTACHARYYA and co-workers postulated a *cis*-relationship between the isopropyl side chain and the tertiary cyclohexyl carbon atom, as depicted in formula (126).

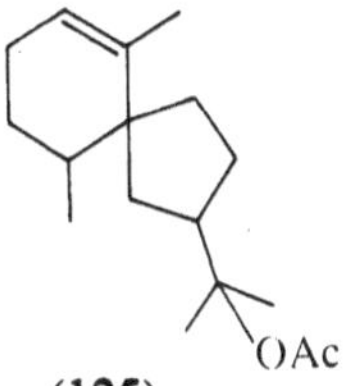

(125)

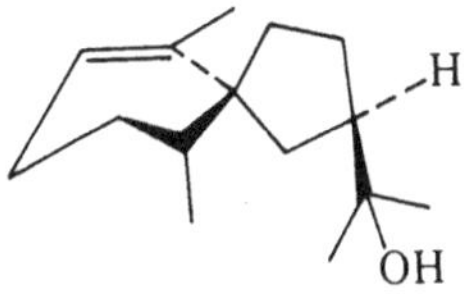

(126)

Unfortunately, this structure seems incompatible with that required by the biogenetic scheme proposed by the same authors in their 1965 paper. That scheme relates dihydroagarofuran, a constituent of agarwood

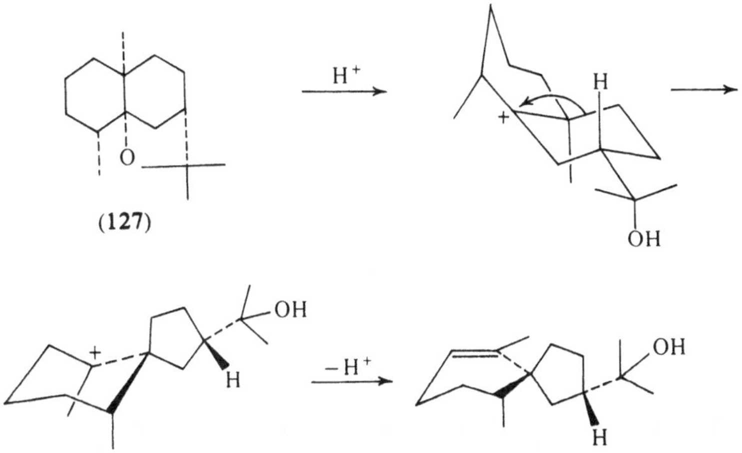

(127)

(128)

Chart 22. Proposed relationship between agarofuran and agarospirol oil whose structure was postulated as (127) by Bhattacharyya and co-workers, to agarospirol according to the indicated pathway *(Chart 22)*. Clearly, the derived structure for agarospirol (128) would require a *trans*-relationship between the isopropylol side chain and the tertiary cyclohexyl carbon atom and not a *cis*-relationship as suggested by the nmr results.

This disparity was partially resolved by Barrett and Büchi who revised the structure of agarofuran and, by implication, dihydroagarofuran. The correctness of this revision received support from three independent syntheses of agarofuran (*19, 21, 63*). The revised structure (129) for dihydroagarofuran was also confirmed by synthetic work (*19*). This revised structure provides support for the stereochemistry originally proposed for agarospirol (126) on the basis of the postulated biogenetic pathway. It should be noted, however, that this pathway now requires a *cis*-relationship between the migrating methylene group and the

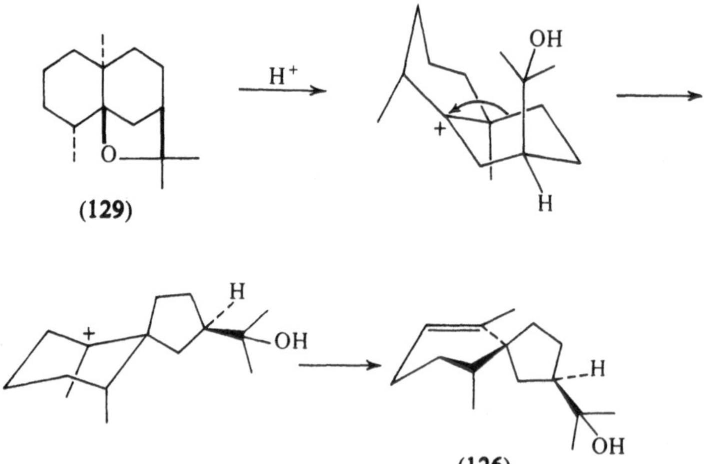

Chart 23. Revised relationship between agarofuran and agarospirol

departing ether bridge. Thus the postulated rearrangement could not proceed in a concerted fashion. Further speculation on the stereochemistry of agarospirol will be presented at a later point in this chapter.

b) β-Vetivone and Related Compounds

Whereas agarospirol (110) was the first spiro[4.5]decane of its kind to be recognized, another sesquiterpene, β-vetivone (130), was the first naturally occurring member of this class to receive extensive chemical study (*76, 83, 84*). However, owing to an undetected skeletal rearrangement which occurred in the degradative studies, the structure was incorrectly assigned as a hydroazulene derivative (82). Nonetheless, this assignment although erroneous for the vetivones led to the currently accepted formula for azulene thereby solving one of the most puzzling structure problems of the early 1900's. The key conversions underlying the original structure proposal are set forth in *Chart 24*.

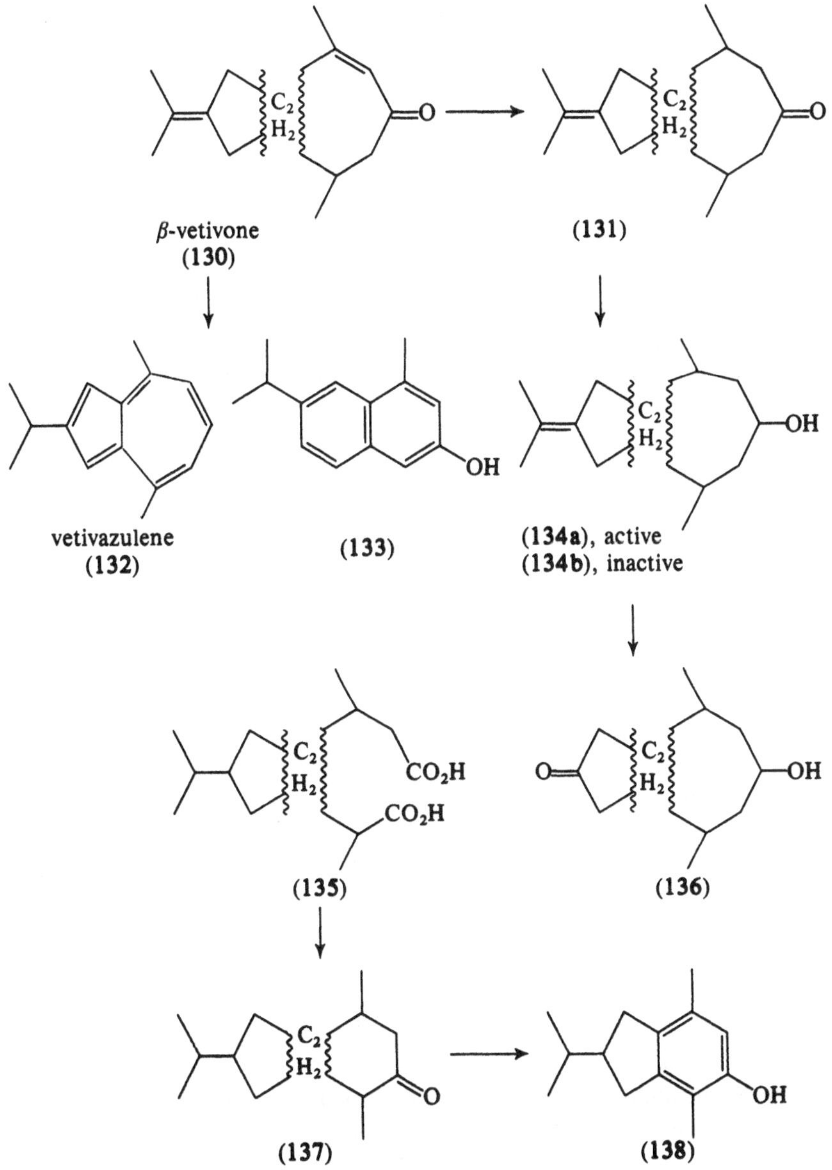

β-vetivone
(130)

(131)

vetivazulene
(132)

(133)

(134a), active
(134b), inactive

(135)

(136)

(137)

(138)

Chart 24. Degradation of β-vetivone

Upon treatment with sodium in alcohol, β-vetivone (**130**) yielded a mixture (stereoisomers) of dihydro-β-vetivols (**134a**) wherein the ketone grouping and a carbon-carbon double bond had been reduced (*84*). Ozonolysis of this alcohol mixture afforded acetone and the hydroxy ketone (**136**). These findings revealed the presence of an α,β-unsaturated ketone and an isopropylidene grouping. Furthermore, the compositional analysis required these derivatives to be bicyclic. Partial hydrogenation of β-vetivone ([α]$_D^{20}$ − 24°) gave a saturated ketone (**131**) and the corresponding alcohol (**134b**), both of which were devoid of optical activity. This finding constituted a key point in the structure analysis, as it suggested that the hydrogenation products contained a symmetry plane.

Dehydrogenation of β-vetivone with selenium afforded vetivazulene (**132**), along with a mixture of C_{14} and C_{15} naphthols, presumably largely (**133**). Interestingly and, as can now be appreciated, significantly, only traces of vetivazulene were secured upon dehydrogenation of β-vetivone with sulfur or palladium-on-carbon. In these cases the reactions led almost exclusively to the aforementioned naphthols.

As noted earlier, the isolation of vetivazulene (**132**) as a dehydrogenation product offered no immediate clue to the carbon skeleton of β-vetivone since, at that time, the characteristic bicyclo[5.3.0]decane carbon framework of azulene had not yet been recognized. The experimental basis for that structure assignment was obtained as follows.

Hydrogenation of dihydro-β-vetivol (**134**), followed by vigorous oxidation with chromic acid, afforded a dicarboxylic acid (**135**). This diacid, upon treatment with acetic anhydride at elevated temperature, yielded a cyclic ketone (**137**) with loss of carbon dioxide. Dehydrogenation of this ketone then gave 4,7-dimethyl-2-isopropyl-5-indanol (**138**) identified by comparison with an independently synthesized sample. Accordingly, PFAU and PLATTNER surmised that ketone (**137**) was a cyclopentane-fused cyclohexanone. This conclusion defined diacid (**135**) as a pimelic acid derivative and meant that the alcohol precursor (**134**), the related ketone (**131**), and β-vetivone (**130**) itself, must contain a seven-membered carbocyclic ring. Structure (**139**) could therefore be advanced for β-vetivone, and vetivazulene could be formulated as (**132**). The correctness of this latter assignment was quickly ascertained through synthesis (*82*). However, a similar test of the β-vetivone structure did not come forth for nearly thirty years (*61*).

Meanwhile, further studies on the oil of vetiver and other essential oils uncovered a number of sesquiterpenes whose structures were deduced through correlations with β-vetivone. For example, structures (**140**) and (**141**) were proposed for hinesol on the basis of chemical evidence and conversion to vetivazulene (*69, 113, 116*). Two other vetiver sesquiterpenes, α- and β-isovetivenene, were thought to be related to structure (141) by virtue of their spectral properties and their conversion to "isovetivane", the saturated hydrocarbon obtained upon total reduction of β-vetivone (86). Several additional unsaturated hydrocarbons, regarded as double bond isomers of the two isovetivenenes, also came to light in the course of these studies.

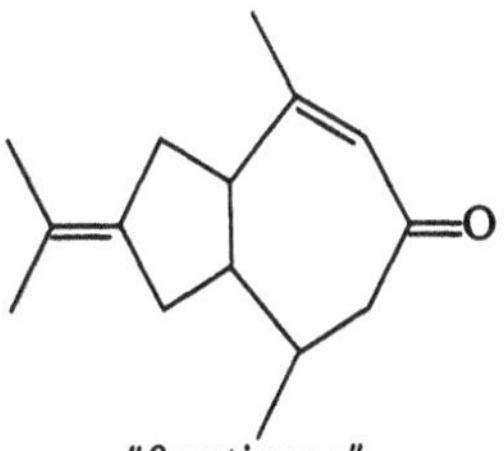

"β-vetivone"
"α-vetivone"
(139)

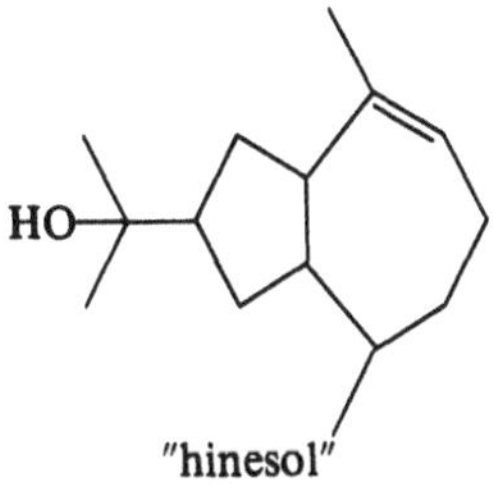

"hinesol"
(140)

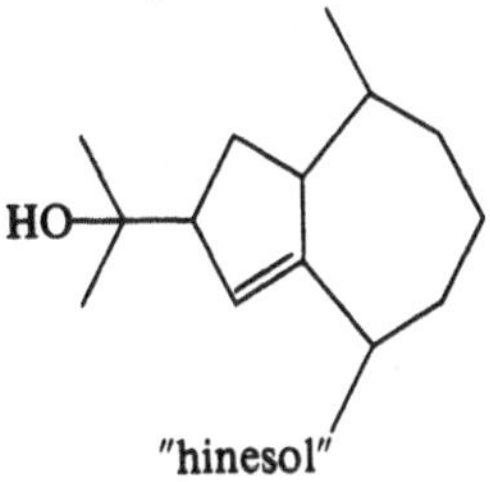

"hinesol"
(141)

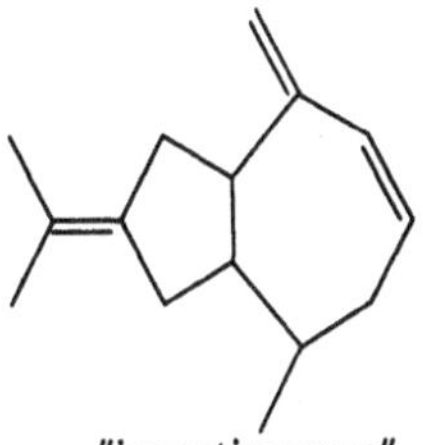

"isovetivenene"
(142)

One rather interesting aspect of ROMANUK and HEROUT's work was the finding that α- and β-vetivone afforded saturated hydrocarbons, vetivane and isovetivane respectively, whose infrared spectra were "significantly different" from each other (86). This result was somewhat surprising in view of the epimeric relationship postulated for the vetivones (79, 83). More surprising was the fact that the hydrocarbon spectra were fundamentally different from the spectrum of decahydrovetivazulene, the hydrogenation product of vetivazulene (132). These differences were attributed to stereoisomerism. However, as will be revealed later, the three "vetivanes" actually belong to three distinct skeletal classes.

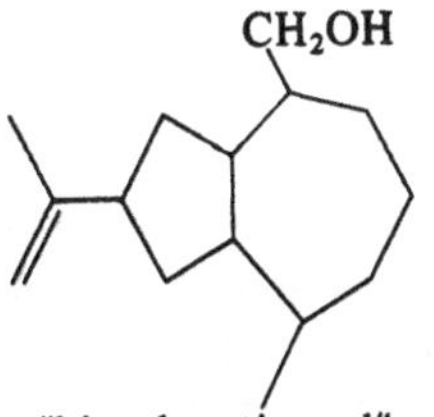

CH$_2$OH

"bicyclovetivenol"
(143)

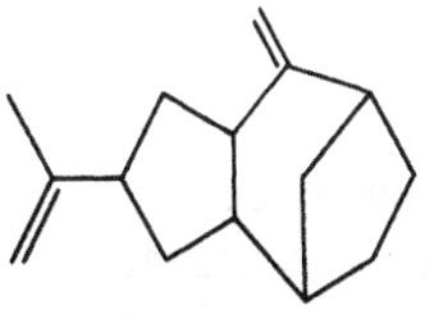

"tricyclovetivene"
(144)

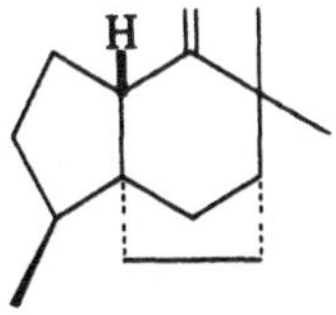

zizaene
(145)

Bicyclovetivenol, a primary alcohol isolated from oil of vetiver, was accorded structure (143) on the basis of its spectral properties and conversion to "α-vetivane" upon acetate pyrolysis followed by hydrogenation (26). A later report described the properties of "*tert*-bicyclovetivenol" and noted the absence of any primary alcohols in at least one variety of vetiver oil (44, 45). Tricyclovetivene, a hydrocarbon initially isolated from vetiver oil of Belgian Congo origin, was originally accorded structure (144) (27) but this assignment was later revised to (145) (72, 88).

The known chemistry of vetivone-related natural products at this point was still consistent with a hydroazulenic skeleton. The degradative studies associated with hinesol deserve some comment as they contain the first definite contraindications. Two isomeric structures were suggested by the Czech (140) (69) and Japanese (141) (116) groups. The key compound was the crystalline dihydro alcohol (147). ŠORM and co-workers converted this substance to ketone (153) which was essentially devoid of optical rotation, suggesting the indicated *cis*-methyl arrangement (28, 69). Dehydration of dihydrohinesol (147) afforded a mixture containing olefins (149) and (150). Hydrogenation of this mixture afforded a hydrocarbon identical to "isovetivane" according to infrared spectral comparison (86). The "isovetivane" sample was derived from "meso"-dihydro-β-vetivone, suggesting a *cis*-hydroazulene ring fusion and *cis*-relationship between the methyl groups. The correlation is even firmer in unpublished work of the Japanese group (113). Olefin (150) and ketone (153) were reported to be completely devoid of optical activity. In fact, ketone (153) afforded a benzylidene derivative identical to that obtained from "meso"-dihydro-β-vetivone. However dihydrohinesol (147) was clearly optically active. Furthermore hinesol dehydrates partially to olefin (151) on chromatography over basic alumina. Olefin (151) proved identical to isovetivene a congener of hinesol. Hydrogenation of isovetivene afforded optically active isovetivane. The optical activity of these two materials was clearly inconsistent with the hydroazulene formulation which had been accorded the vetivane sesquiterpenes.

ŠORM's evidence for the position of the double bond came from the nmr spectrum and further reactions of the ozonolysis product (148). In their early studies, YOSIOKA et al. obtained an ozonolysis product of hinesol that did not appear to be a methyl ketone (116). Additional evidence for their assignment came from the formic acid dehydration of hinesol which afforded a conjugated diene (λ_{max} 240 nm, log ε 4.07). Later, YOSIOKA and KIMURA confirmed ŠORM's assignment by allylic oxidation of hinesol, followed by dehydration, which afforded the enantiomer of β-vetivone (114). This correlation proved to be the

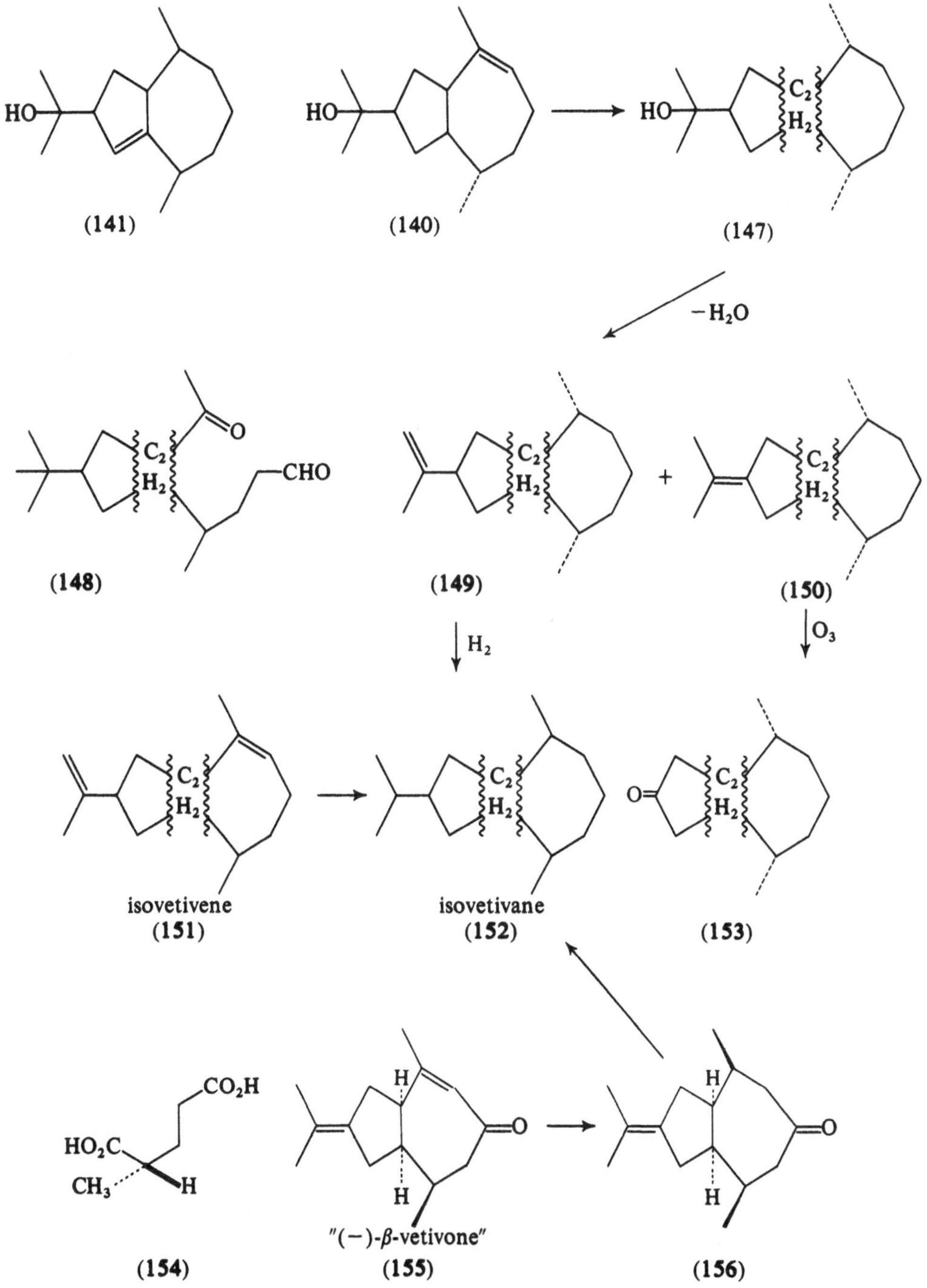
HO
(141)
HO
(140)
HO
C2
H2
(147)
−H2O
O
CHO
C2
H2
(148)
C2
H2
+
C2
H2
(149)
(150)
H2
O3
C2
H2
isovetivene
(151)
C2
H2
isovetivane
(152)
O
C2
H2
(153)
CO2H
HO2C
CH3
H
(154)
H
H
O
"(−)-β-vetivone"
(155)
H
H
O
(156)

Chart 25. Degradation of hinesol

keystone of the absolute stereochemistry of this class of sesquiterpenes, since vigorous oxidation of hinesol afforded (+)-α-methylglutaric acid (154).

The discovery that many of the vetivane sesquiterpenes actually belong to the spiro[4.5]decane family came about as a result of studies aimed at the synthesis of the postulated hydroazulene structures (61, 64). In one phase of these studies, the unsaturated aldehydes (159) and (160) were prepared via fragmentation of the respective oximes (157) and (158) followed by reduction of the resulting nitriles. Upon treatment with stannic chloride in benzene, these aldehydes cyclized to give the corresponding hydroazulenols (161) and (162). The former alcohol yielded a mixture of the meso and dl ketones (163) and (164) (only one enantiomer is shown) after hydrogenation and subsequent oxidation. The latter was likewise converted to the dl-ketone (164) and the second meso isomer (165) (Chart 26).

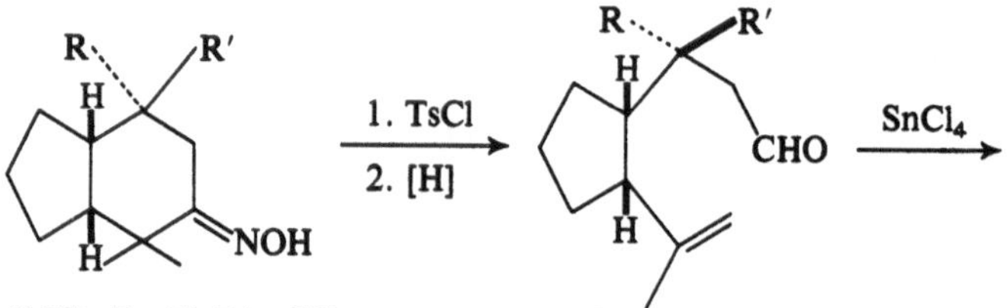

(157): R=H, R'=CH₃
(158): R=CH₃, R'=H

(159): R=H, R'=CH₃
(160): R=CH₃, R'=H

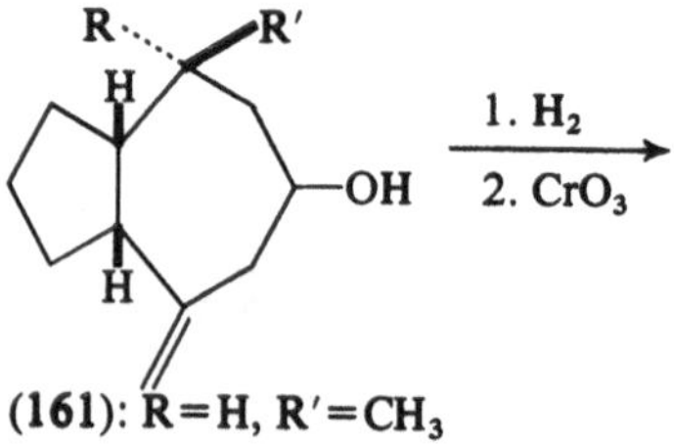

(161): R=H, R'=CH₃
(162): R=CH₃, R'=H

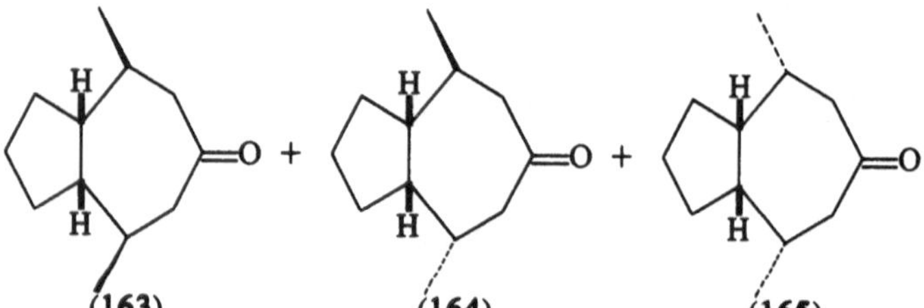

(163)   (164)   (165)

Chart 26. Synthesis of comparison dimethylhydroazulenones

In the second phase of these studies, β-vetivone was degraded (*Chart 27*) in order to obtain a comparison sample of what should have been either the *meso* ketone (163) or the isomer (165). Since the synthetic route to these ketones unambiguously defined their stereochemistry, the correspondence of one of them with the vetivone-derived ketone would likewise define the relative stereochemistry of β-vetivone. To this end, the aforementioned *meso* hydroxy ketone (136) was converted to the thioketal derivative and treated with Raney nickel to give the alcohol (166). Oxidation then afforded the desired ketone (167). However, the infrared spectrum of this ketone bore no resemblance to those of ketones (163)—(165). In fact, the CO stretching band of ketone (167) (5.81 μ) suggested that the carbonyl grouping was part of a six-membered ring. The synthetic ketones (163)—(165) all gave rise to an infrared band at 5.88 μ consistent with their assigned cycloheptanone structures.

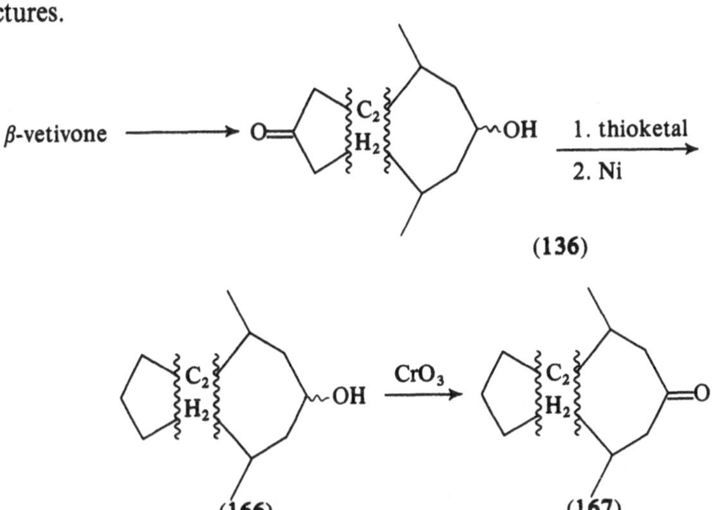

(136)

(166) **(167)**

Chart 27. Degradation of β-vetivone

ROMANUK and HEROUT had previously noted the anomalous carbonyl absorption maximum (5.82 μ) in the infrared spectrum of dihydro-β-vetivone, but could find no basis in this fact alone for rejecting the proposed structure (86). However, the lack of similarity between the synthetic ketones (163)—(165) and the degradation ketone (167) now clearly demanded a reevaluation of the hydroazulene structure. Since the spectral properties of ketone (167) conformed to those of a cyclohexanone derivative and the hydroxy ketone (136) displayed a cyclopentanone carbonyl band in its infrared spectrum, either a bicyclo-[4.3.0]-nonane or a spiro[4.5]decane skeleton could be considered for β-vetivone. The chemical facts strongly supported the latter possibility.

Accordingly, a simple revision could be postulated in terms of the previously proposed structure by a formalism (139)→(168), in which a ring fusion hydrogen and a cycloheptane ring bond merely exchanged places.

(139)   (168)

On this basis, PFAU and PLATTNER's degradation scheme could be represented as shown in *Chart 28*. It should be noted that the geometry of ketone (169) and alcohol (170) still permits a symmetry plane and these compounds can therefore exist as *meso* forms. Thus the postulated structure revision accommodates this crucial feature of the original structure proof. The dicarboxylic acid, initially regarded as a pimelic acid derivative (135, $C_2H_2 = CHCH$), would accordingly be formulated as the adipic acid derivative (171) (135, $C_2H_2 = CCH_2$). Cyclization of this diacid would result in the cyclopentanone (173), rather than the initially postulated cyclohexanone (137, $C_2H_2 = CHCH$). In the revised scheme, the formation of the indanol (138) would require that a skeletal rearrangement take place during dehydrogenation. Similar rearrangements would be required in the conversion of β-vetivone to vetivazulene (132) and the naphthol (133). These latter examples are reminiscent of those encountered in the dehydrogenation of the acorenes (98). The spirodecane formulation also explains the previously noted optical activity of hinesol degradation products which retain the isopropyl grouping.

Firm support for the postulated spiro[4.5]decane structure of β-vetivone was obtained from the synthetic and degradative studies outlined in *Chart 29* (64).

The cyclopropyl ketone (174) of known stereochemistry (49) yielded the dienone (175) upon treatment with sulfuric acid in acetic acid-acetic anhydride. Hydrogenation then gave the *meso* ketone (176) along with a lesser amount of the corresponding *dl* isomer. Removal of the ketonic grouping *via* desulfurization of the thioketal derivative afforded the hydrocarbon (179). The same hydrocarbon was secured from *meso*-dihydro-β-vetivone (169) by way of the olefin (177) (thioketal formation, desulfurization) and the ketone (178) (ozonolysis, thioketal formation, desulfurization). The ketones (176) and (178) exhibited similar, but clearly distinct spectral and chromatographic properties. The *cis*-arrangement of the methyl groups in both isomers follows from the *meso* nature of ketone (169) [and therefore ketone (178)] and their conversion to the same hydrocarbon. Since ketone (176) is stereochemically defined by the synthetic scheme, the relative stereochemistry of the β-vetivone-derived isomer must be epimeric at the spiro carbon atom, as shown in formula (178). The relative stereochemistry of β-vetivone is likewise defined.

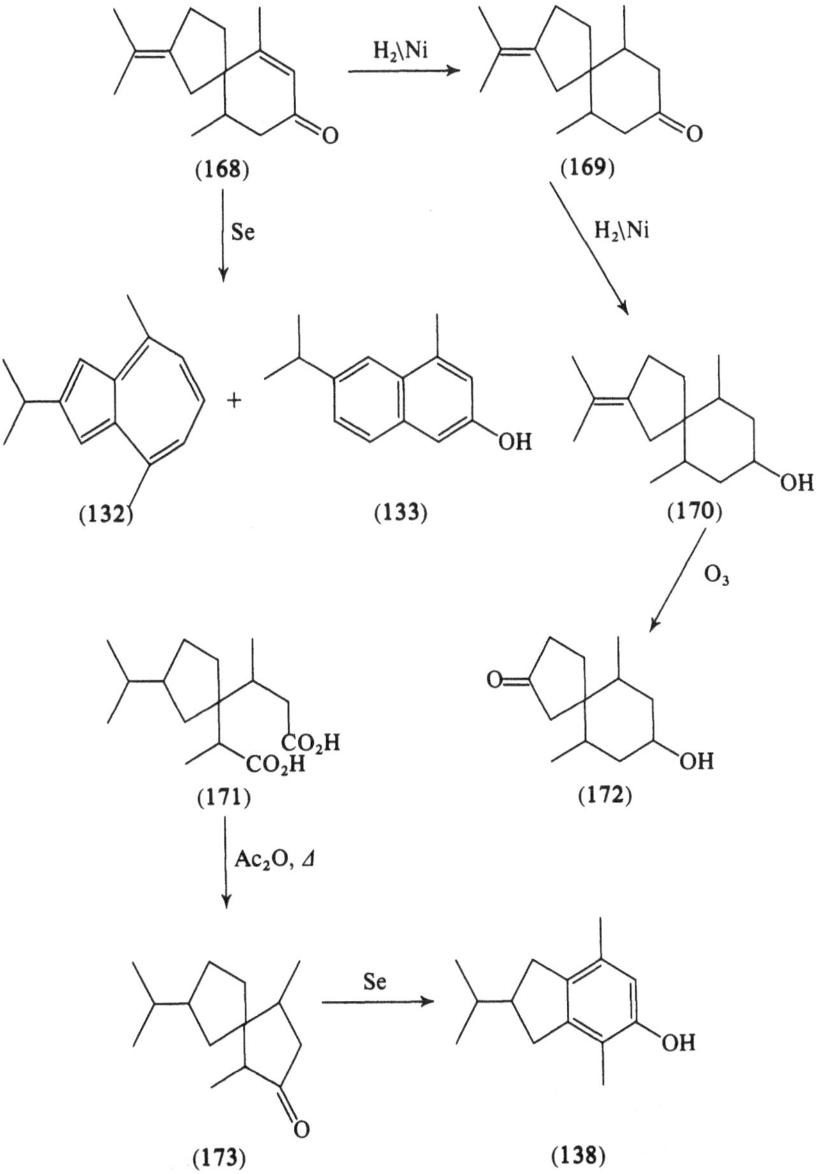

Chart 28. Revised interpretation of β-vetivone degradation

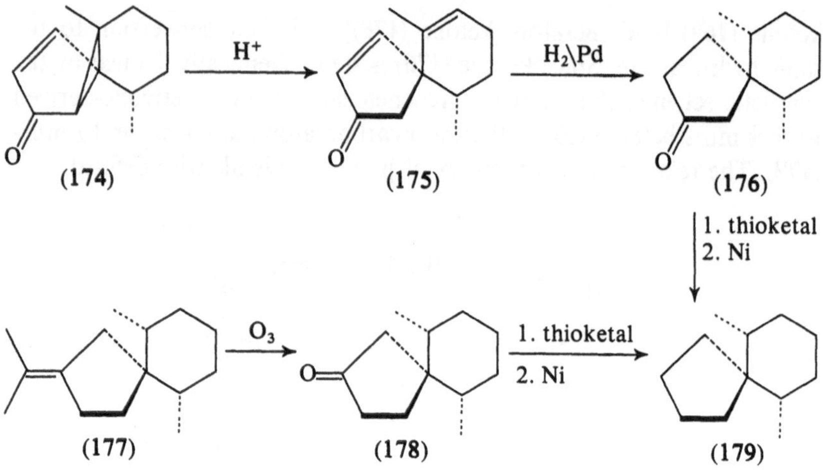

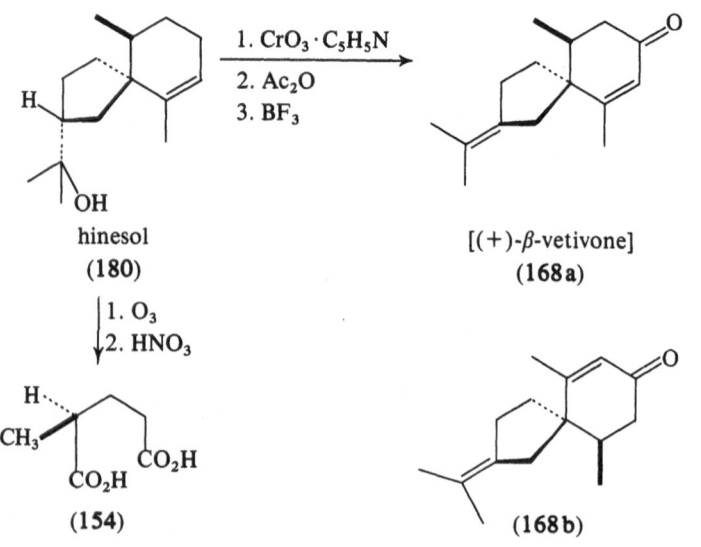

Chart 29. Stereochemistry of β-vetivone

The absolute stereochemistry of β-vetivone (**168b**) was deduced on the basis of its chemical correlation with hinesol (**180**). As noted earlier, hinesol afforded the enantiomer (**168a**) of natural (−)-β-vetivone upon treatment with chromic anhydride in pyridine followed by dehydration *(Chart 30)*. Hinesol must therefore not only possess the same spiro[4.5]decane carbon skeleton as β-vetivone, but the two compounds must also be enantiomeric at the spiro carbon and the adjacent tertiary center. More extensive oxidation of hinesol (**180**) led to (+)-α-methylglutaric acid (**154**) of known absolute configuration (*28*). This trans-

Chart 30. Absolute stereochemistry of β-vetivone

formation revealed the absolute stereochemistry of the tertiary methyl center in hinesol and, in view of the aforementioned correlation, β-vetivone as well. The orientation of the isopropylol side chain of hinesol could not be assigned on the basis of these studies. The stereochemistry shown in formula (180) was determined through an unambiguous total synthesis of racemic hinesol (62). Further support was provided by the degradative studies of YOSIAKA and KIMURA (114a).

The seemingly anomalous dehydration of hinesol to a conjugated diene (λ_{max}240 nm) can now be explained. The spirane formulation would allow for two possible rearrangements (Chart 31). Ion (181) could yield either (182) or (183). However ion (183) would yield (+)-δ-selinene (λ_{max}248 nm) (7), not a diene of the stated uv spectrum. Dienes (184) or (less likely) (185) thus represent possible structures for this dehydration product.

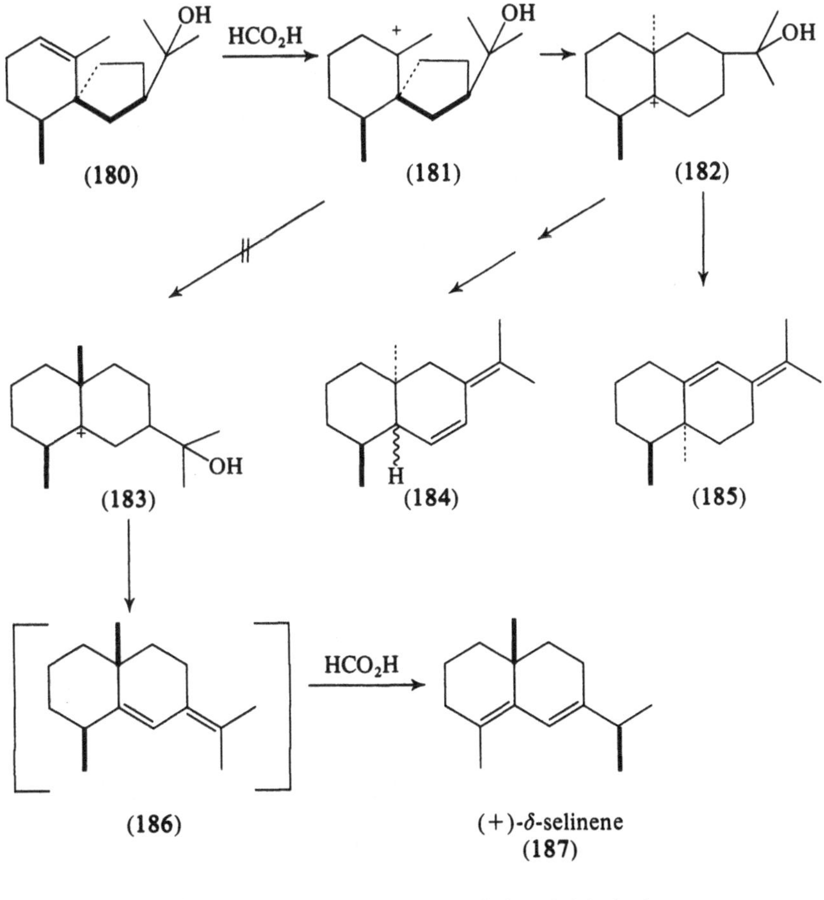

Chart 31. Revised interpretation of hinesol dehydration

Hinesol can now be recognized as a close relative of agarospirol. In view of the proposed biogenesis of agarospirol, the two compounds should differ in the relative orientation of their respective isopropylol chains and would be enantiomeric at the spiro carbon atom and the adjacent tertiary center, as depicted below. Support for this relationship was recently secured through total synthesis (*70*).

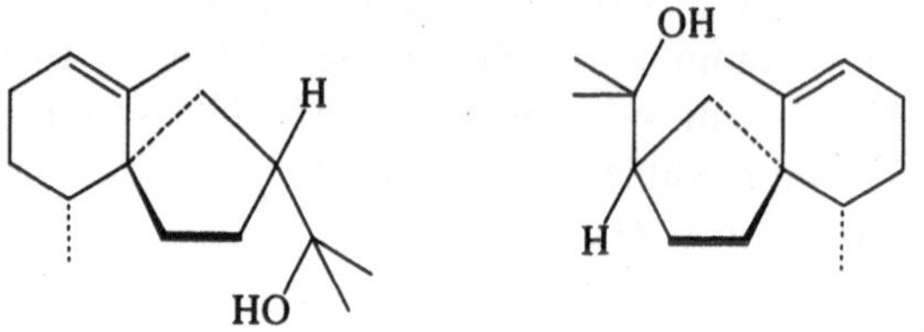

(**126**)
agarospirol

(**180**)
hinesol

Of direct relevance to this problem is the fact that the 1:1 mixture of the racemic forms of alcohols (**126**) and (**180**) obtained in the course of a total synthesis of racemic β-vetivone gave rise to an nmr spectrum which corresponded almost exactly with that of authentic hinesol (**180**) (*65*). The infrared spectra were also identical. Thus, spectral comparisons may not constitute sufficient criteria of identity for spiro compounds which may be epimeric at one or more centers unless the spectra of all possible isomers can be examined.

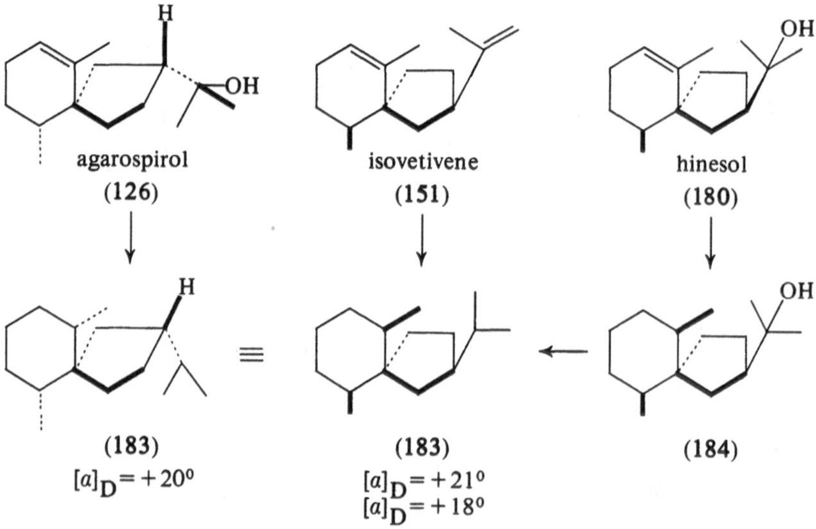

agarospirol
(**126**)

isovetivene
(**151**)

hinesol
(**180**)

(**183**)
$[a]_D = +20^\circ$

(**183**)
$[a]_D = +21^\circ$
$[a]_D = +18^\circ$

(**184**)

Chart 32. Absolute stereochemistry of agarospirol

The assignment of structure (126) to agarospirol is supported by the finding that hinesol, agarospirol, and isovetivene can be converted to hydrocarbon (183). The spectral properties and rotations of material derived from these three spirovetivanes are in good agreement *(Chart 32)*.

With the structure of β-vetivone established, it is of interest to scrutinize other vetiver-derived sesquiterpenes whose structure assignments are based on correlations with β-vetivone. Prior to the revision of the β-vetivone structure, independent studies on α-vetivone occasioned a reappraisal of the previously postulated epimeric relationship between these two similar compounds *(39, 60)*. These studies showed α-vetivone (185) to be a hydronaphthalene derivative and a close relative of nootkatone (186) *(57)*.

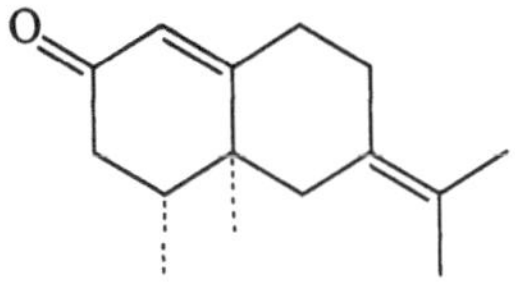

(185) 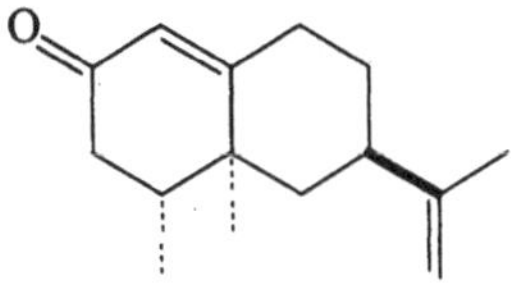   (186)

The striking differences observed for the "vetivanes" derived from the isomeric vetivones and vetivazulene can thus be understood on the following basis *(Chart 33)*.

α-vetivone —[H]→ (187) ("vetivane")
nootkatane

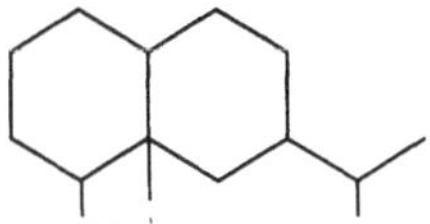

β-vetivone —[H]→ (188) ("isovetivane")
spirovetivane

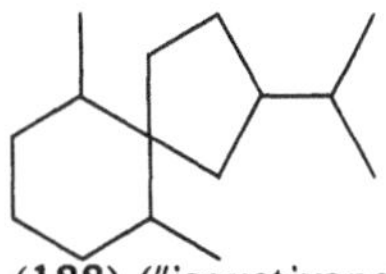

vetivazulene —[H]→ (189) (vetivane)

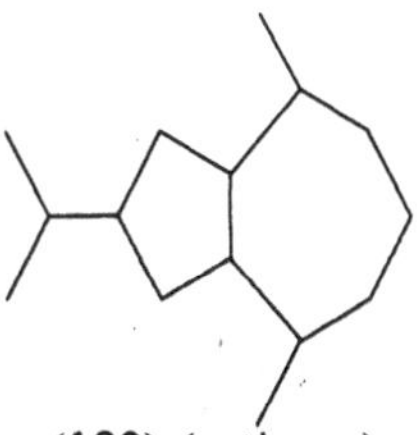

Chart 33. The "vetivanes"

Accordingly, the three "vetivanes" actually represent three different structural types with "vetivane" (187) corresponding to the hydrocarbon nootkatane (57) and "isovetivane" (188) corresponding to agarospirane (104). The name "spirovetivane" has been suggested for hydrocarbon (188) to point out its relationship to β-vetivone and vetivazulene, while distinguishing it from the hydrogenation product (189) of vetivazulene (62)*. This latter hydrocarbon, which should properly bear the name "vetivane", is the skeletal parent of a presently memberless family of sesquiterpenes. In all probability, authentic vetivanes will eventually be discovered in Nature.

ANDERSEN has recently isolated two hydrocarbons, β- and α-vetispirene, from oil of vetiver and assigned structures (190) and (191) to these compounds on the basis of spectral data and their correlation with spirovetivane (188) (7). The oil employed in these studies appears to be devoid of material corresponding to the α-isovetivenene reported by ROMANUK and HEROUT (86). Finally, since bicyclovetivenol affords the same hydrocarbon (187) as that derived from α-vetivone, the structure revision from (143) to (192) would appear reasonable.

β-vetispirene
(190)

α-vetispirene
(191)

CH₂OH
bicyclovetivenol
(192)

2. Synthesis

The first total synthesis of a spirovetivane sesquiterpene (Chart 34) employed a photochemical 2,5-cyclohexadienone rearrangement to generate stereoselectively an appropriately functionalized spiro[4.5]-decane derivative (65). Thus irradiation of the dienone (193) led to the cyclopropyl ketone (194) (49) which underwent cleavage in strong acid to give the isomeric dienone (195) having the requisite carbon skeleton. Selective hydrogenation of the conjugated double bond afforded the enone

---

* Some olefinic spiranes from vetiver oil have been called vetispirenes (7). In the interest of reducing confusion we suggest that the spirovetivane designation be used for the class of naturally occurring spiro[4.5]decanes related to β-vetivone. The individual compounds will undoubtedly retain their more common names, i.e., β-vetivone, isovetivene, hinesol and α-vetispirene.

(196). Condensation with ethyl formate took place at the less hindered alpha position of this ketone to give a hydroxymethylene derivative which yielded the thioether (197) upon subsequent treatment with butanethiol and boron trifluoride etherate.

The sequence of reactions from ketone (196) to thioether (197) set the stage for the introduction of the isopropylidene group of β-vetivone at the proper position on the cyclopentane ring. Further steps in this direction were achieved by sodium borohydride reaction of ketone (197), acidic hydrolysis and dehydration of the resulting alcohol, and subsequent addition of methyllithium to the aldehyde (198) thereby obtained. Oxidation of the alcohol mixture (diastereomers) thus secured with 2,3-dicyano-5,6-dichloro-1,4-benzoquinone (DDQ) then gave the dienone (199). Selective reduction of the conjugated double bond was accomplished using lithium in ammonia-ethanol, followed by oxidation of the saturated alcohol diastereomers to afford a nearly 1:1 mixture of two isomeric methyl ketones, differing with respect to the orientation

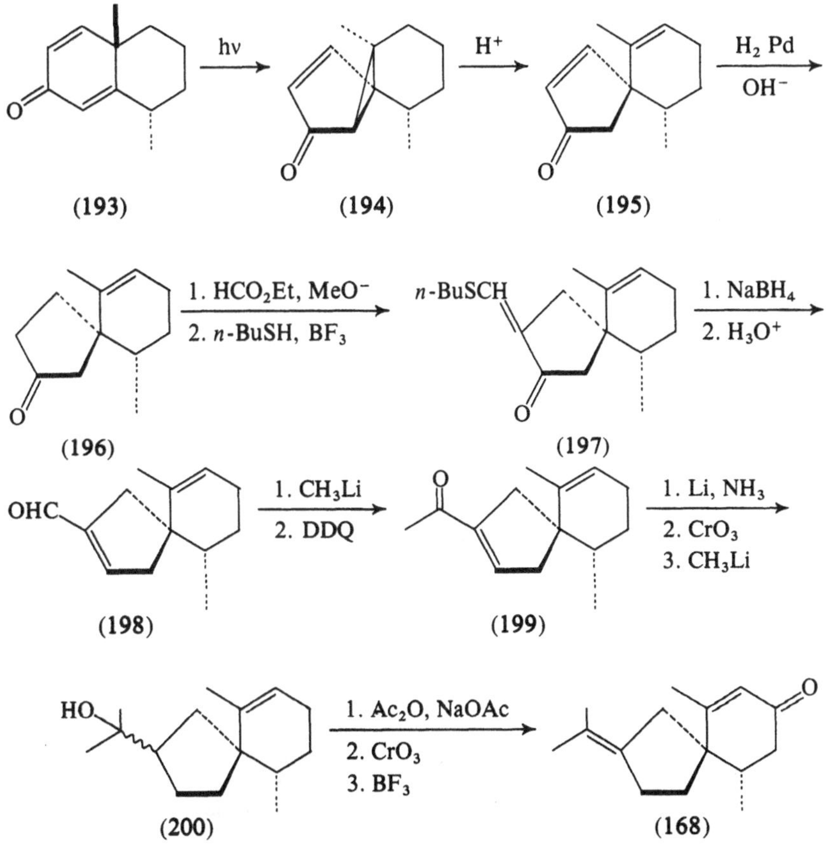

Chart 34. Synthesis of β-vetivone

of their acetyl substituents. Addition of methyllithium afforded the corresponding mixture of alcohols (200). As mentioned earlier, one of these isomers corresponds to racemic hinesol and the other to racemic agarospirol.

The conversion of the isomeric alcohols (200) to racemic β-vetivone (168) was achieved along the lines employed for the transformation of natural hinesol to (+)-β-vetivone (114).

The second total synthesis of racemic β-vetivone followed a completely different pathway from the first (46). In this case, the spiro[4.5]-decane system was elaborated in the final step by means of a stereoselective internal alkylation reaction (208)→(168). The requisite enone, (207), was neatly prepared using the anisole (201) as the starting material. Birch

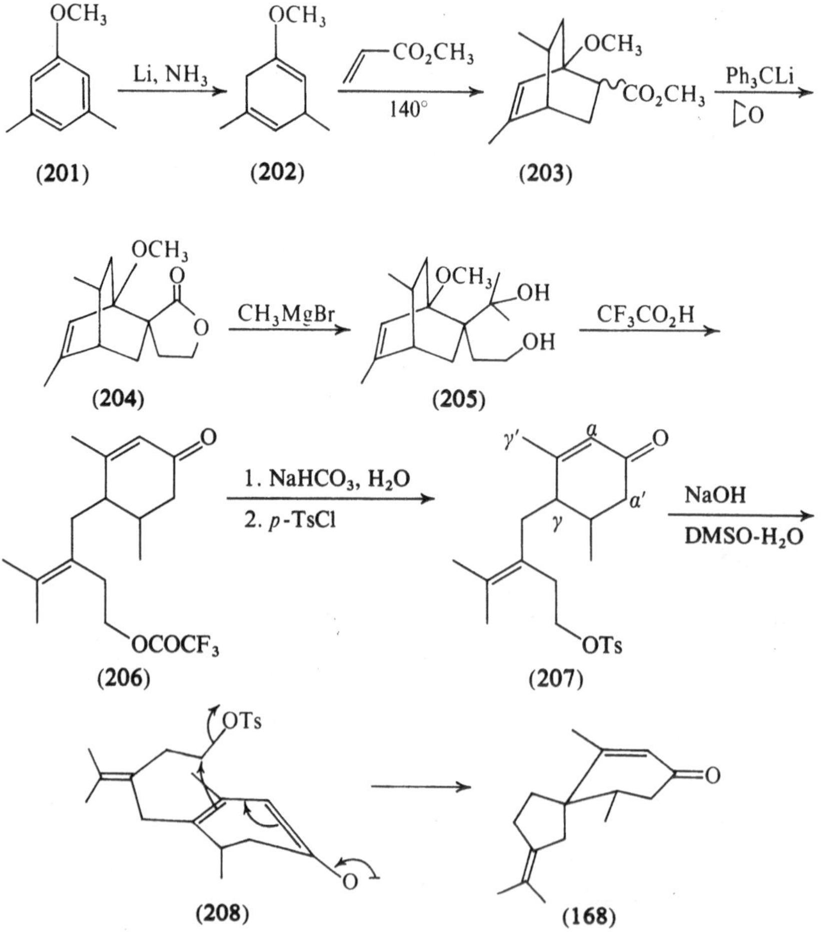

(201)   (202)   (203)

(204)   (205)

(206)   (207)

(208)   (168)

Chart 35. Synthesis of β-vetivone

reduction of the aromatic ring afforded the 1,4-diene (**202**). Treatment of this diene at 140° with methyl acrylate yielded the Diels-Alder adduct (**203**) as a 2.5 : 1 mixture of stereoisomers. This conversion requires prior thermal isomerization of diene (**202**) to the appropriate 1,3-isomer. Alkylation of this ester mixture with ethylene oxide afforded a 9 : 1 mixture of epimeric lactones, of which (**204**) must be the major isomer since it results from attack at the less hindered face of the intermediate ester enolate. Addition of methylmagnesium bromide to this lactone mixture gave rise to the corresponding diol mixture (mainly **205**). This diol mixture underwent an interesting fragmentation reaction upon treatment with trifluoroacetic acid to give the cyclohexanone (**206**) (*22*). Hydrolysis of the trifluoroacetic grouping and esterification of the resulting alcohol with *p*-toluenesulfonyl chloride then gave the desired enone (**207**).

It should be noted that enone (**207**) can give rise to three different enolates which could, in turn, yield six different internal alkylation products [one from α'-alkylation, two from γ-alkylation, two from γ'-alkylation and one from α-alkylation (double bond at γ')]. The successful outcome of this reaction requires not only the preferential formation of the proper (γ) enolate but the stereoselective alkylation of this enolate as well. After extensive experimentation, it was found that both requirements could be met through the use of sodium hydroxide in aqueous dimethyl sulfoxide, whereupon ketone (**207**) was efficiently transformed to racemic β-vetivone (**168**). Presumably this particular base-solvent allows rapid equilibration of all possible enolates with the one depicted by (**208**) being most prone to internal alkylation. The stereochemistry of this step may be controlled by the cyclohexyl methyl grouping, insofar as steric factors would favor the indicated arrangement.

A third synthetic approach to the spirovetivane sesquiterpenes was developed by DESLONGCHAMPS and his collaborators who employed the keto ester (**209**) as their starting material (*70*). The corresponding ketal (**210**), upon reduction with lithium aluminium hydride, gave the alcohol (**211**), which then underwent hydrolysis and dehydration in aqueous acid, leading to the dienone (**212**). Michael addition of diethyl malonate yielded the adduct (**213**) which gave the acid (R = H) directly in refluxing aqueous acid. The corresponding methyl ester (**214**, R = CH₃) was converted to the ketal (**215**), which was transformed via the acid chloride to the diazo ketone (**216**). This diazo ketone gave rise to a 90 : 10 mixture of two isomeric cyclopropyl ketones (mainly **218**) in high yield, upon treatment with copper in benzene. Thus the cyclization reaction (**216**)→(**218**) seems to bear some stereochemical similarity to the enolate cyclization reaction (**208**)→(**168**). The stereochemistry of both processes may be controlled by the cyclohexane methyl grouping with attack on the adjacent double bond being preferred *trans* to this substituent.

The keto ketal (218) was converted to the diol (221) *via* carbomethoxylation followed by selective ketone reduction and addition of methylmagnesium iodide to the resulting hydroxy ester (220). Treatment with aqueous acid then afforded the dienone (222) resulting from ketal hydrolysis and subsequent regioselective cyclopropane cleavage and dehydration.

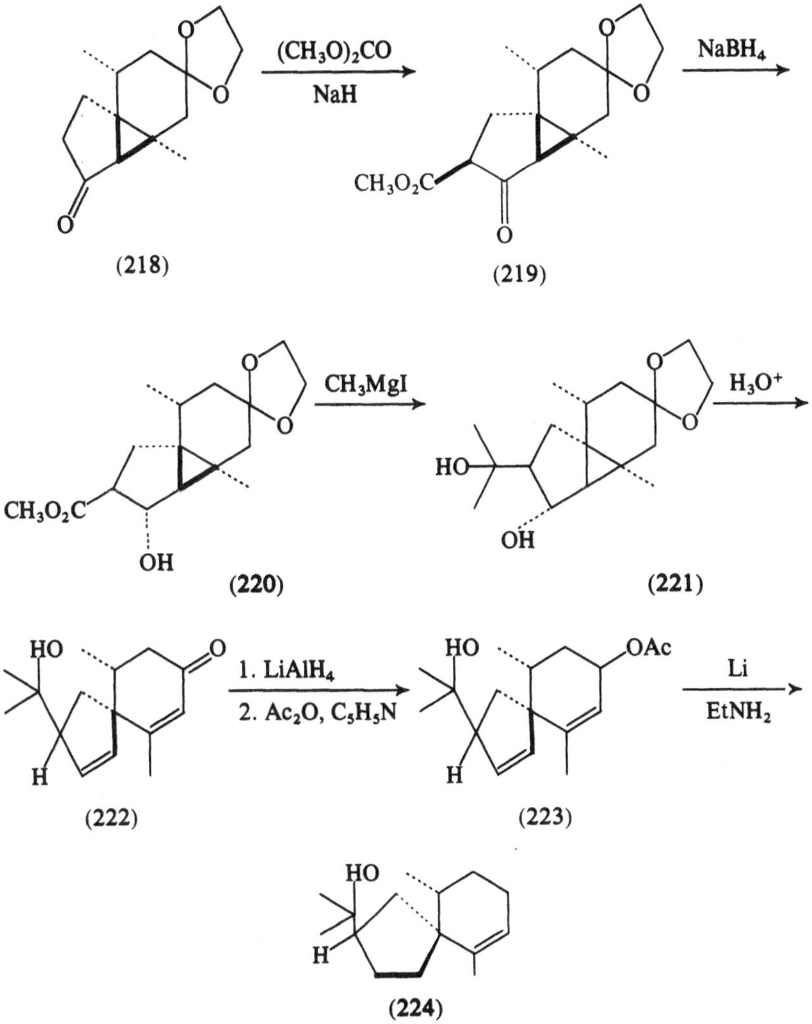

Chart 36. Synthesis of agarospirol

Reduction of the enone (222) gave an allylic alcohol whose acetate (223) was reduced with lithium in ethylamine to give a substance with spectral properties identical with those of epihinesol. Furthermore, the derived crystalline epoxide was identical with agarospirol epoxide according to the nmr spectra. Unfortunately an authentic sample of agarospirol could not be found, thus making a direct comparison impossible. However, the identity of the epoxide derivatives indicates that the synthetic alcohol (224) is very likely (±)-agarospirol.

A closely related scheme was independently conceived and executed by STORK and McCURRY who converted the 9:1 mixture of cyclopropyl ketones (218) and (217) to the unsaturated keto acid (225) via Reformatsky addition of ethyl α-bromoisobutyrate followed by acidic cleavage of the ketal and cyclopropane moieties and saponification of the ester grouping (53, 54). Numerous attempts to effect decarboxylation of this β,γ-unsaturated acid were to no avail.

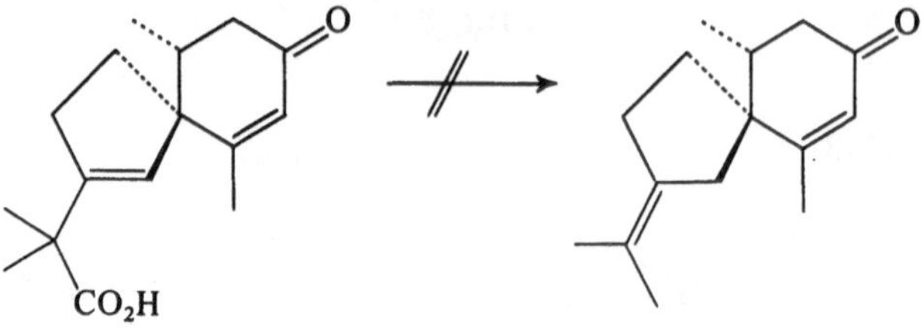

CO_2H

(225)   (±)-epi-β-vetivone

In a second approach, Stork and McCurry examined the sequence outlined in *Chart 37* as a possible route to β-vetivone. The enamine (226) afforded the unsaturated aldehyde (228) via Claisen-type rearrangement of the ammonium intermediate (227) upon alkylation with crotyl bromide followed by pyrolysis and subsequent hydrolysis. One might expect the stereochemistry of aldehyde (228) to be controlled by the orientation of the pyrollidine moiety as depicted below.

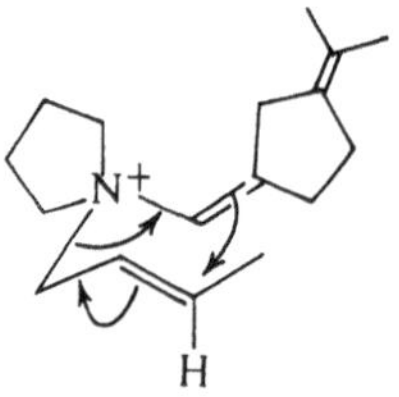

(227a) 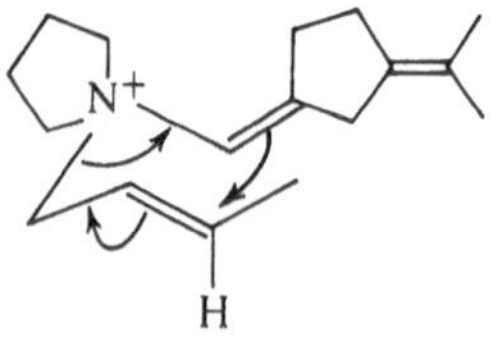   (227b)

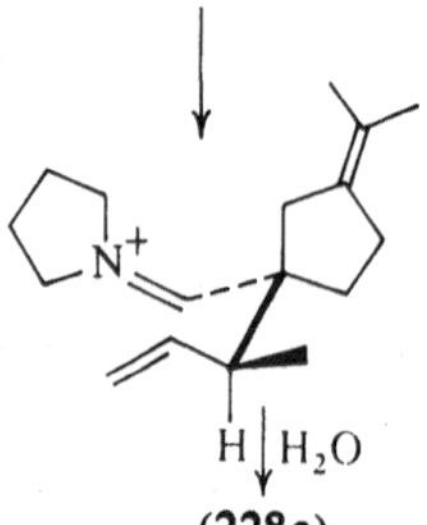

(228a) 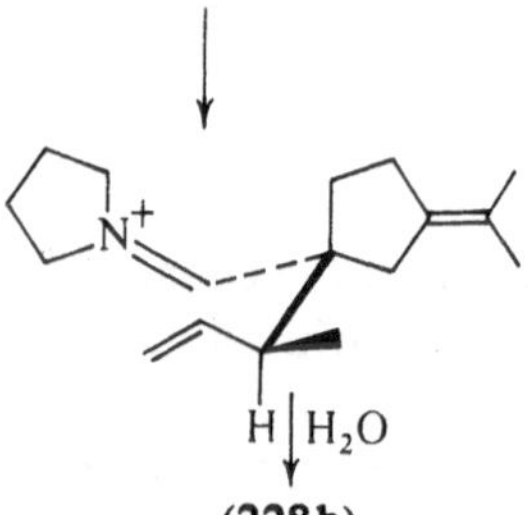   (228b)

Aldehyde (228) appeared homogeneous by gas chromatography but the possibility that the stereoisomers could not be separated appears likely. In any case subsequent transformations involving addition of methyllithium followed by terminal oxidation of the olefinic grouping with disiamylborane-hydrogen peroxide yielded the diol (229). Selective acetylation of the primary alcohol, dehydration of the tertiary alcohol with thionyl chloride in pyridine, saponification of the resulting unsaturated acetate, and oxidation of the derived primary alcohol gave the unsaturated aldehyde (230). This aldehyde cyclized to a mixture of epimeric methylene cyclohexanols upon treatment with silica gel (61). Oxidation with pyridine-sulfur trioxide in dimethylsulfoxide afforded a crystalline conjugated ketone in 20% yield. The spectral properties of this substance were similar to those of β-vetivone but its melting point

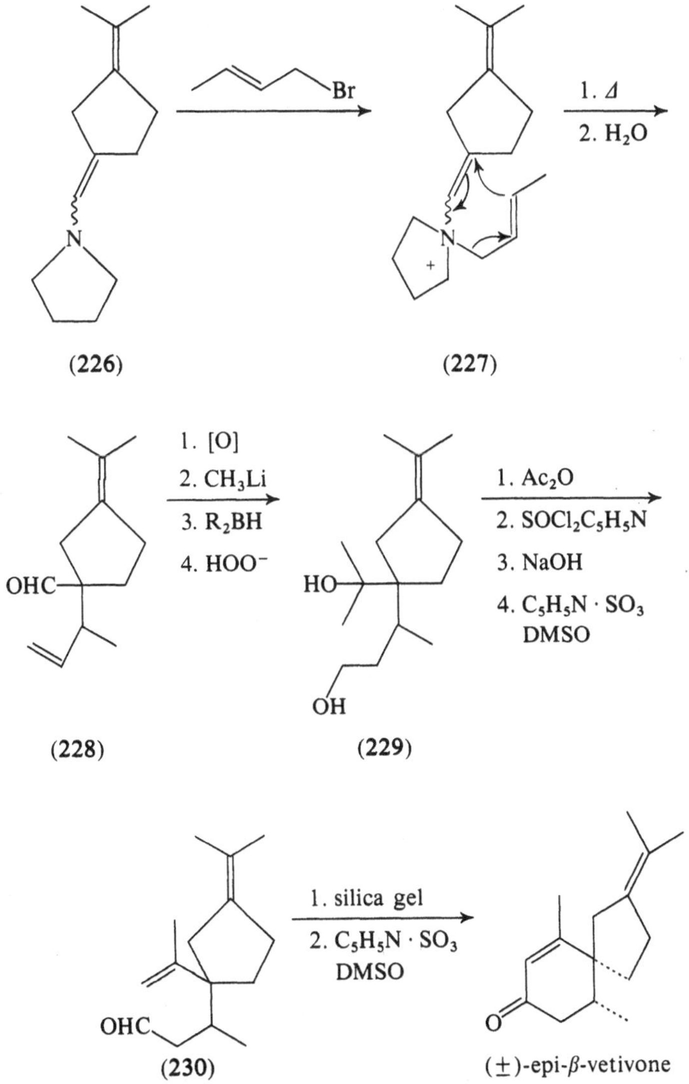

(226)   (227)

(228)   (229)

(230)   (±)-epi-β-vetivone

Chart 37. Synthesis of epi-β-vetivone

$(53-54.5°)$ was somewhat higher than that of $(\pm)$-β-vetivone $(45-46°)$. It was therefore concluded that the synthetic product was $(\pm)$-epi-β-vetivone not $(\pm)$-β-vetivone. The latter material may also have been formed in this sequence but its failure to crystallize apparently precluded isolation.

The structure of hinesol (**180**) was first deduced on the basis of its correlation with β-vetivone (*114*) and its degradation to $(+)$-α-methylglutaric acid (*28*). However, no information regarding the relative stereochemistry of the isopropylol grouping could be gleaned from these studies. The final basis for this assignment came from a total synthesis of racemic hinesol *(Chart 38)* (*62*). *A priori* there was no way of knowing which of the two possible isopropylol epimers corresponding to structure (**180**) was actually hinesol. An initial choice in favor of the isomer shown was made on the basis of biogenetic speculation which will be presented later. The *cis*-arrangement of the isopropylol side chain and C-6* in this isomer enabled synthetic approaches based on the tricyclic dienone (**231**) to be considered (*66*). One obvious advantage of this particular approach is the fact that the readily available starting dienone already contains the spiro[4.5]decane system (C-9,1,2,3,4,10,5,6,12,11) and additional carbons (C-7 and 8) to serve as the basis for the isopropylol chain and the vinylic methyl grouping of hinesol, after cleavage of the 7-8 or 8-9 carbon-carbon bond of an appropriate intermediate. Moreover, the *cis*-relationship of C-7 and C-9 in (**231**) (and thus C-11 and C-6 in **180**)* is ensured by virtue of the rigid geometry of the tricyclic structure. Finally, dienone (**231**) provides suitable activation for introduction of the C-10 methyl grouping of structure (**180**).

Introduction of this methyl group was achieved by treatment of dienone (**231**) with lithium dimethylcopper, whereupon a 1:3 mixture of two stereoisomers, (**232a**) and (**232b**), was produced. The stereochemical outcome of this reaction seems to be controlled by steric factors in the dienone (**231**), with approach from the bottom face of C-4 being hindered mainly by the C-5 methylene group. The indicated stereochemical assignments were initially made on the basis of subsequent chemical transformations of the epimeric ketones (**232a**) and (**232b**). Final support for these assignments came from a single crystal X-ray structure analysis of the *p*-bromophenylurethane derivative (**234b**) $(R = p\text{-}BrC_6H_4NHCO)$.

Alcohol (**234a**) $(R = H)$ was obtained *via* enol acetylation of enone (**232a**) and reduction of the resulting acetate (**233a**) with sodium borohydride. Oxidation of the unsaturated acetate (**234a**) $(R = Ac)$ with

---

* The numbering system follows the IUPAC convention for spiro compounds. Cf. J. Amer. Chem. Soc. **82**, 5545 (1960).

chromic acid afforded the acetoxy enone (235a) which readily lost acetic acid upon treatment with ethanolic hydrochloric acid to give the dienone (236a). Partial hydrogenation of this dienone over palladium on carbon yielded the enone (237a). Alternatively, both carbon-carbon double bonds of dienone (236a) could be hydrogenated and the resulting saturated ketone subjected to bromination followed by dehydrobromination to introduce the conjugated double bond of enone (237a).

The foregoing transformations of enone (232a) set the stage for cleavage of the 8, 9 bond of the tricyclic nucleus and the resulting stereoselective generation of the substituted spiro[4.5]decane system. To this end, enone (237a) was reduced with lithium tri-*t*-butoxyalumino-hydride, and the resulting alcohol was epoxidized with *m*-chloro-peroxybenzoic acid to give the epoxy alcohol (238a). Reduction of this epoxide with lithium aluminum hydride cleanly yielded the secondary, tertiary diol which was selectively esterified with methanesulfonyl chloride to give the hydroxy mesylate (239a). This intermediate afforded the spiro[4.5]decanone (240a) upon treatment with potassium *t*-butoxide in *t*-butyl alcohol.

Enone (240a) contains appropriate structural features which are properly oriented for the straightforward attainment of alcohol (180). Thus, addition of methyllithium followed by acetylation of the resulting tertiary alcohol and epoxidation of the vinylic grouping led to the epoxy acetate (241a), a mixture of diastereoisomers. Treatment with lithium aluminum hydride effected cleavage of the epoxide and acetoxy groupings, and selective acetylation of the resulting secondary, tertiary diol gave the hydroxy acetate (242a). Direct epoxidation of the alcohol (245a), derived *via* addition of methyllithium to ketone (240a), led not to the corresponding epoxide, but to the hydroxy ether (245), instead.

This undesired side reaction was easily circumvented by prior acetylation of the tertiary alcohol function.

(245)

Dehydration of the tertiary alcohol (242a) with phosphorus oxychloride yielded a mixture of olefins, predominantly the endocyclic isomer (243a). The mixture was not separated at this stage. Cleavage of the acetate grouping with lithium aluminum hydride followed by oxidation of the secondary alcohol function with Jones' reagent then gave the ketone (244a), contaminated with the exocyclic double bond isomer.

Addition of methyllithium to this material yielded a mixture of (±)-hinesol (180a) (R = H) and some of the exocyclic olefin isomer. These two substances were readily separated by conversion to the corresponding acetates (mainly 180a, R = Ac) followed by chromatography on

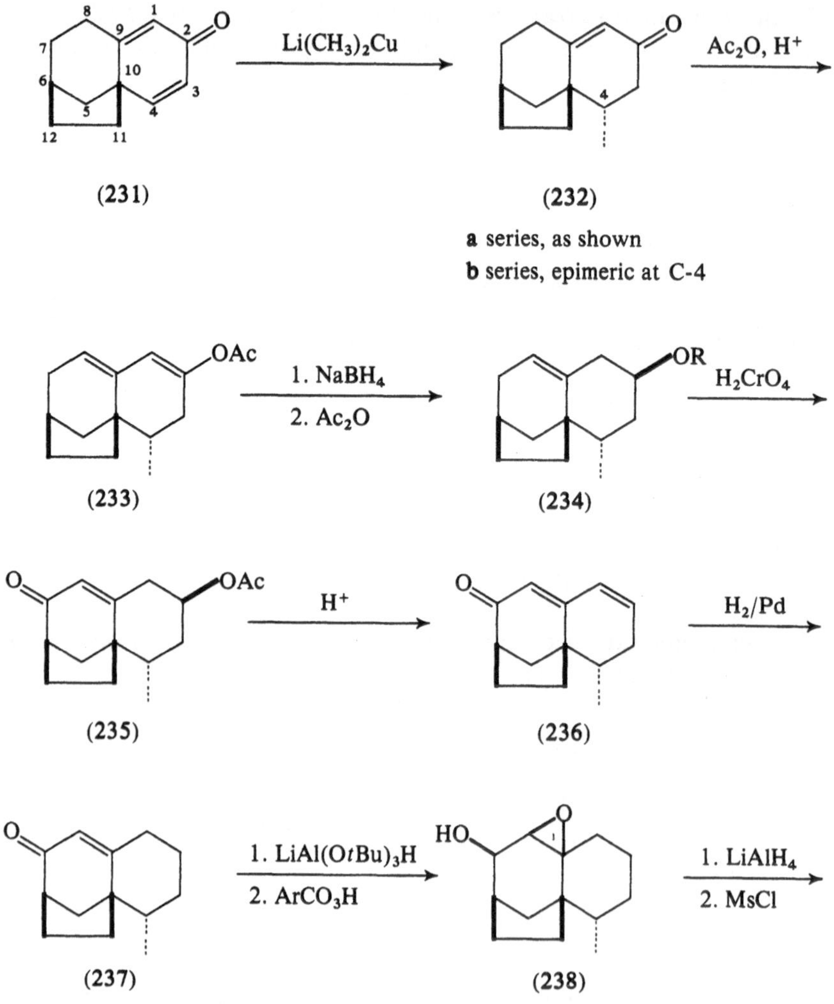

(231) $\xrightarrow{\text{Li(CH}_3)_2\text{Cu}}$ (232) $\xrightarrow{\text{Ac}_2\text{O, H}^+}$

a series, as shown
b series, epimeric at C-4

(233) $\xrightarrow[\text{2. Ac}_2\text{O}]{\text{1. NaBH}_4}$ (234) $\xrightarrow{\text{H}_2\text{CrO}_4}$

(235) $\xrightarrow{\text{H}^+}$ (236) $\xrightarrow{\text{H}_2/\text{Pd}}$

(237) $\xrightarrow[\text{2. ArCO}_3\text{H}]{\text{1. LiAl(O}t\text{Bu})_3\text{H}}$ (238) $\xrightarrow[\text{2. MsCl}]{\text{1. LiAlH}_4}$

Chart 38. Synthesis of hinesol (continued)

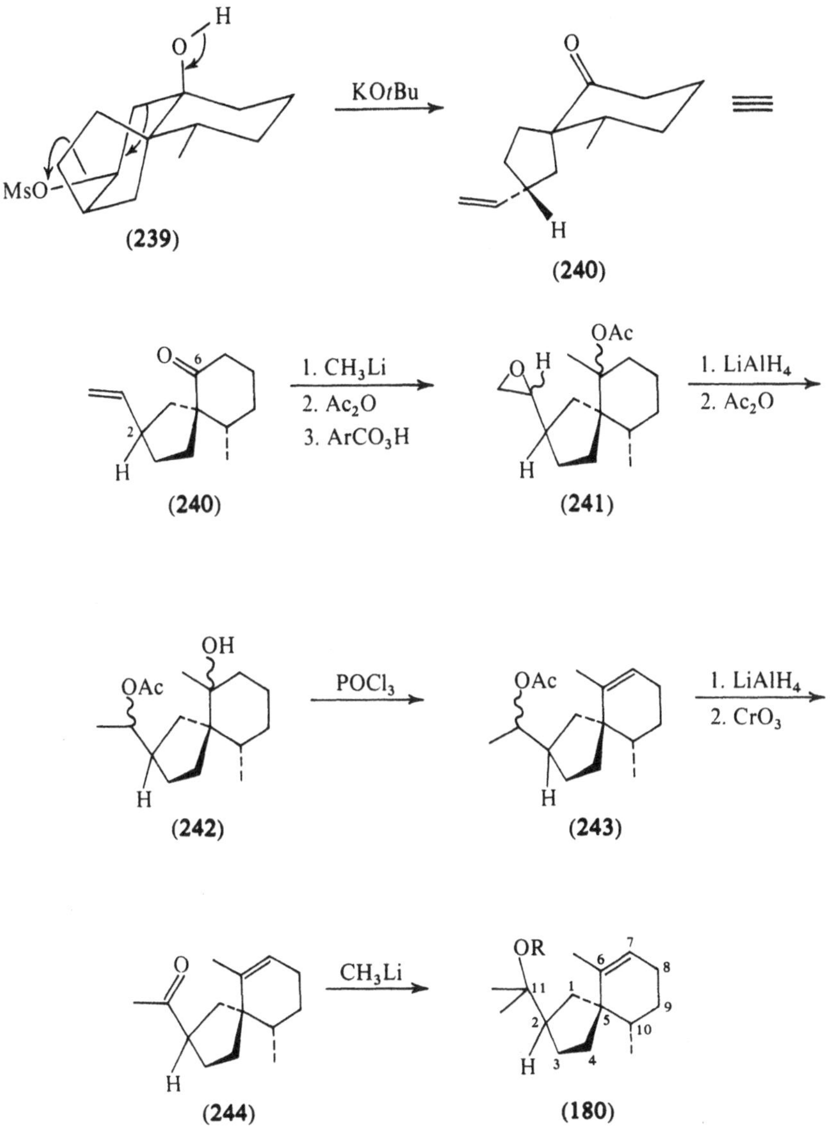

Chart 38. Synthesis of hinesol

silica gel impregnated with silver nitrate. The acetate (180a) (R=Ac) secured in this fashion was shown to be free of its C-2 epimer by gas chromatographic comparison with the acetate sample prepared previously from the epimeric alcohol mixture (200). Moreover, an authentic specimen of hinesol afforded an acetate having gas chromatographic retention time identical with that of the synthetic sample (180a) (R = Ac).

22*

Finally, treatment of this synthetic material with lithium aluminum hydride yielded racemic hinesol (**180a**) (R = H), thus completing the synthesis and resolving the final structural uncertainty of this spirovetivane sesquiterpene.

Repetition of the above described reaction sequence in the *b* series, starting from enone (**232b**), led to an alcohol, (**180b**) (R = H), whose spectral properties and gas chromatographic behavior were virtually indistinguishable from those of hinesol itself. However, the acetate derivative of this isomer showed a markedly different gas chromatographic retention time from that of hinesol acetate.

An exceedingly elegant synthesis of β-vetivone was recently completed by Stork and co-workers (*94*). The key transformation involved alkylation of the kinetic enolate of the enol ether of 5-methylcyclohexane-1,3-dione with 2-isopropylidene-1,4-dichlorobutane. The presumed intermediate chloro ketone (**246**) evidently undergoes a highly stereoselective cyclization reaction upon treatment with strong base; none of the epimeric spiro[4.5]decane product was formed. Addition of methylmagnesium iodide to the cyclized product afforded the alcohol (**248**) which yielded racemic β-vetivone (**168**) upon exposure to dilute acid.

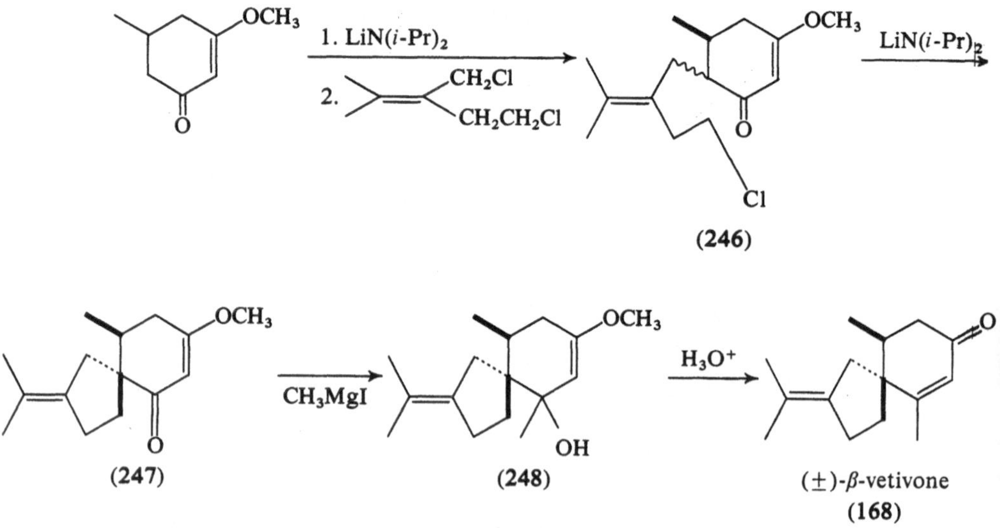

Chart 39. Synthesis of β-vetivone

A total synthesis of (±)-α-vetispirene has recently been completed by Caine, Dawson and Ingwalson starting with the methoxydienone (**249**) *(Chart 40)* (*24*). Photolysis of this material led to the spiro enone (**250**) in 90% yield (*25*). Reduction and acetylation followed by hydrogenolysis with lithium in ethylamine and acid hydrolysis of the enol ether afforded the hydroxy ketone (**252**). Dehydration of this intermediate with thionyl chloride in pyridine gave the enone (**253**) which yielded a 3:2 mixture of α-vetispirene (**254a**) and the double bond isomer (**254b**) upon treatment with isopropenylmagnesium bromide and thionyl chloride-pyridine.

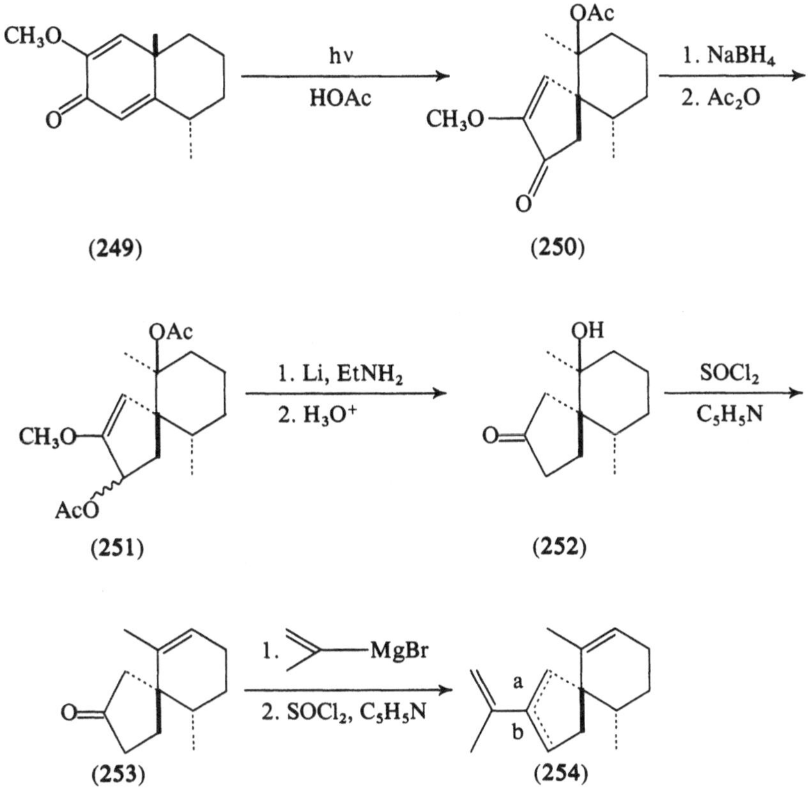

Chart 40. Synthesis of α-vetispirene

Another photochemical approach to the spirovetivane skeleton was described in the report of PIERS and WORSTER on the reductive cleavage of cyclopropyl ketones (*85*). Accordingly ketone (**255**), obtained *via* irradiation of dienone (**193**) followed by reduction of enone (**194**) (*49*), afforded a mixture of spiro[4.5]decanones (**256**) (76−94%) and (**257**) (6−24%) upon dissolving metal reduction. The composition of this mixture varied with temperature and proton source.

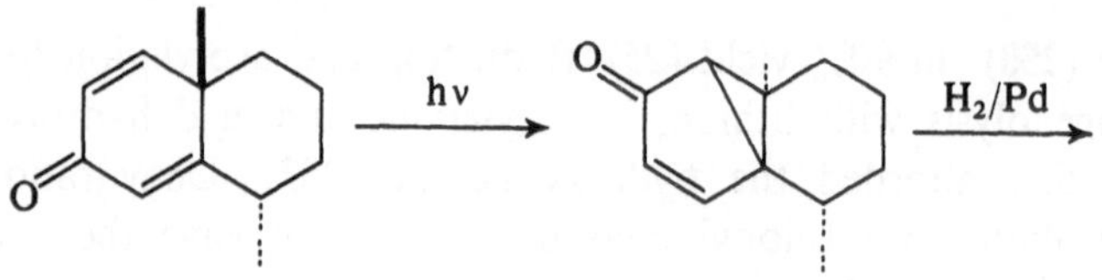

(193)   (194)

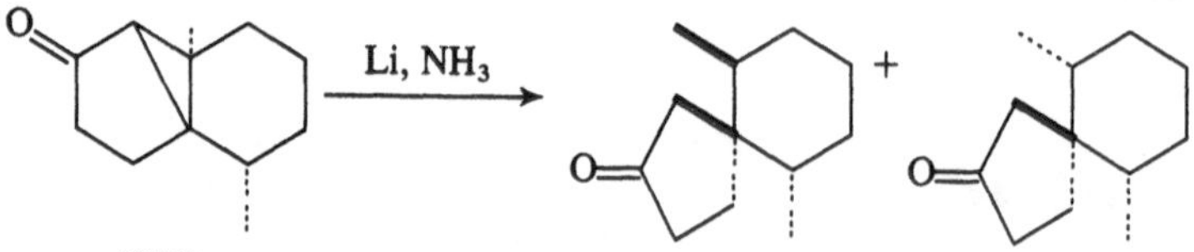

(255)   (256)   (257)

Chart 41. Synthesis of dimethylspiro[4.5]decanones

D. Biogenetic Considerations

As mentioned in the introduction, RUZICKA foresaw the possibility of spiro[4.5]decane intermediates when he proposed cation (2), an acorane skeletal type, as a possible intermediate in the biogenesis of cedrene (87). It now appears that the acorane-alaskane group plays an important part in the biogenesis of a number of tricyclic skeletons. The spirovetivanes most likely arise *via* conformationally and stereo-electronically controlled transformations of farnesol-derived inter-mediates. A self-consistent theory regarding the biosynthesis of such sesquiterpene types can be formulated on the basis of a detailed structure analysis of vetiver oil (2). This complex oil contains two spirane types, three tricyclic skeletons derived from them, and five classes of related hydronaphthalenic sesquiterpenes.

1. The Relationship between Spirovetivanes and Hydronaphthalenic Sesquiterpenes

When the hydroazulenone formulations of α- and β-vetivone were replaced by structures (185) and (168), the relationship of both to a 4,10-epieudesmane ion (258) was suggested (17). The vetivones were viewed as alternative rearrangement products of this ion (migration of either the C-10 methyl or C-9 ring methylene) with conjugation providing the driving force for each process. At the time, 10-epieudesmanes had been encountered only rarely in nature. The proposed α-vetivone pathway was supported by the reported structures for valencene (259) and

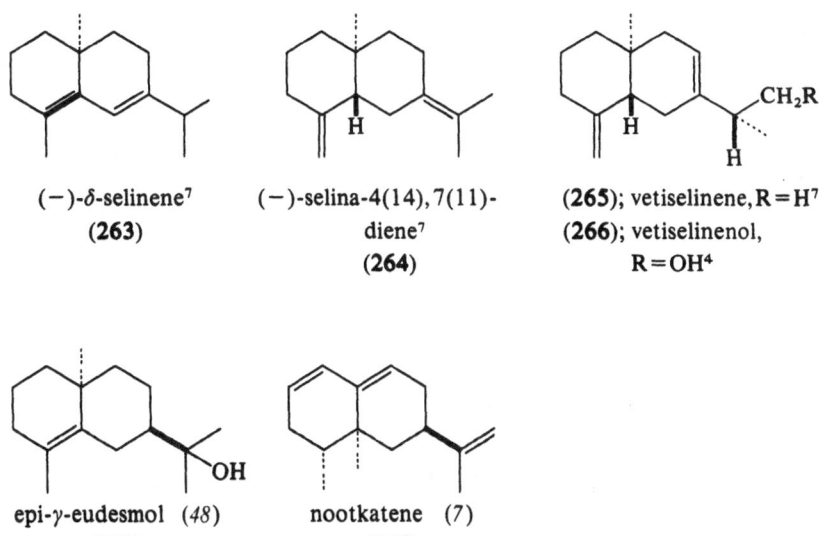

α-vetivone
(185)

(258)

β-vetivone
(168b)

nootkatone (**260**) (*57*) and the previous suggestion of ROBINSON (*81*) regarding the biogenesis of eremophilone (**262**) from alcohol (**261**) or some equivalent structure.

(**259**), valencene R=H$_2$
(**260**), nootkatone R=O

(**261**)

eremophilone
(**262**)

Further support for this suggestion came from the discovery of a number of 10-epi-eudesmanes in vetiver oil *(Chart 42)*. In addition, valencene was found in the oil. The four major hydrocarbons of vetiver

(−)-δ-selinene[7]
(**263**)

(−)-selina-4(14),7(11)-
diene[7]
(**264**)

(**265**); vetiselinene,R=H[7]
(**266**); vetiselinenol,
R=OH[4]

epi-γ-eudesmol (*48*)
(**267**)

nootkatene (*7*)
(**268**)

Chart 42. 10-*epi*-eudesmanes of vetiver oil

oil fall into pairs that can be viewed as Wagner-Meerwein rearrangement products of 4,10-*epi*-eudesmane intermediates, **(270)** and **(272)** *(Chart 43)*. This relationship lends considerable support to the proposed biogenetic scheme since it is difficult to accept the increasing co-occurrence of such closely related structures as coincidence.

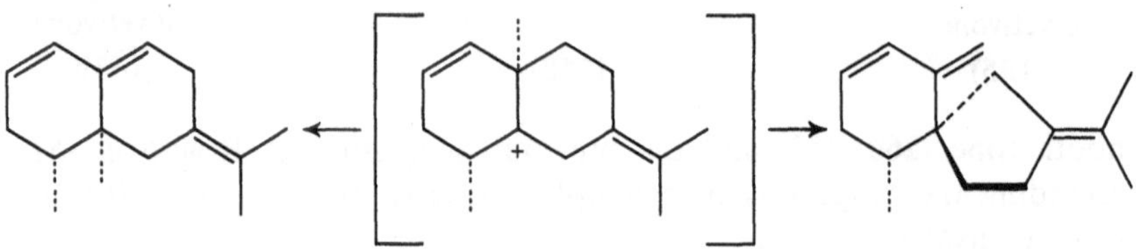

β-vetivenene **(269)** (**270**) β-vetispirene **(190)**

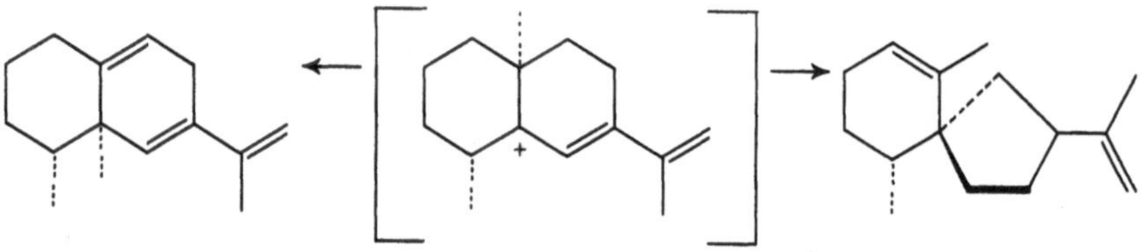

γ-vetivenene **(271)** **(272)** α-vetispirene **(191)**

Chart 43. Relationship of vetiver oil hydrocarbons

Similar biogenetic conjecture was the basis for a tentative assignment of the relative stereochemistry of the isopropylol side chain in hinesol (*7, 62*). A correlation with β-eudesmol, a congener of hinesol (*116*),

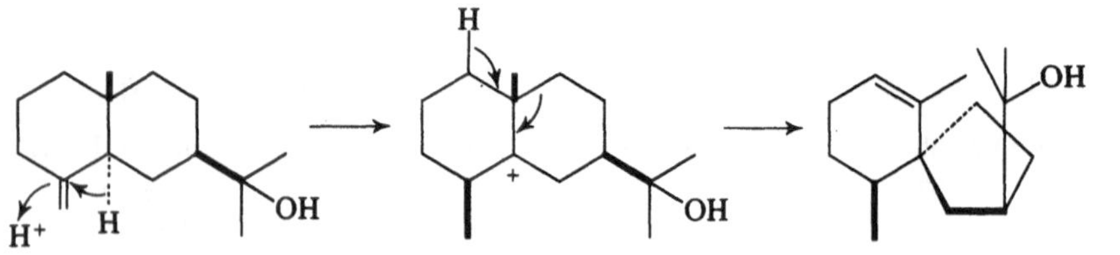
H
OH
H
H⁺
OH
OH

β-eudesmol **(273)** **(274)** hinesol **(180)**

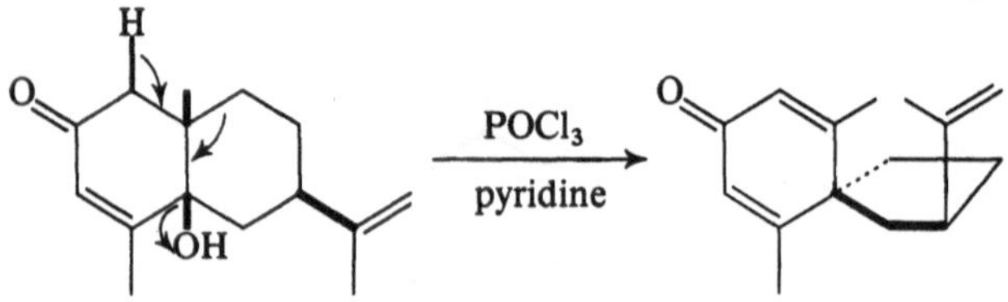
H
O
OH
POCl$_3$
pyridine
O

β-rotunol **(275)** **(276)**

References, pp. 371—376

correctly predicted the structure subsequently confirmed by a stereo-rational synthesis of hinesol (62). The degradation of β-rotunol (275) to the spirodienone (276) affords the first example of a chemically induced rearrangement of this type (41).

The literature on constituents of eudalene-yielding essential oils reveals two noteworthy features: 1) 10-epieudesmanes are more prone to rearrangement and 2) rearrangement always occurs from a eudesmane with *cis*-related methyl groupings *(Chart 44)*.

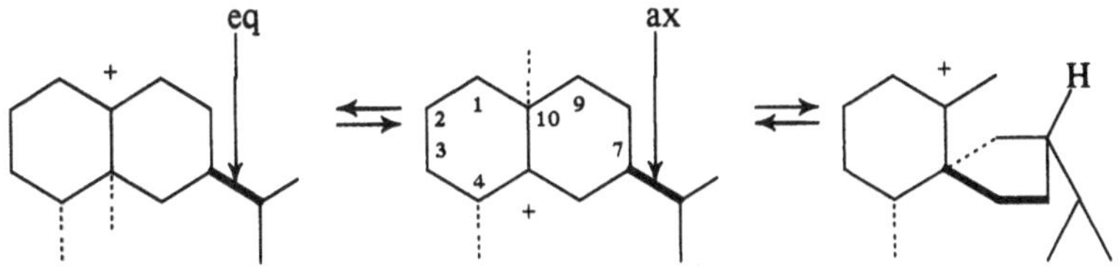

valencane (nootkatane) (277)	4,10-epieudesmane (278)	spirovetivane (279)
eremophilane (280)	eudesmane (281)	hinesane (282)

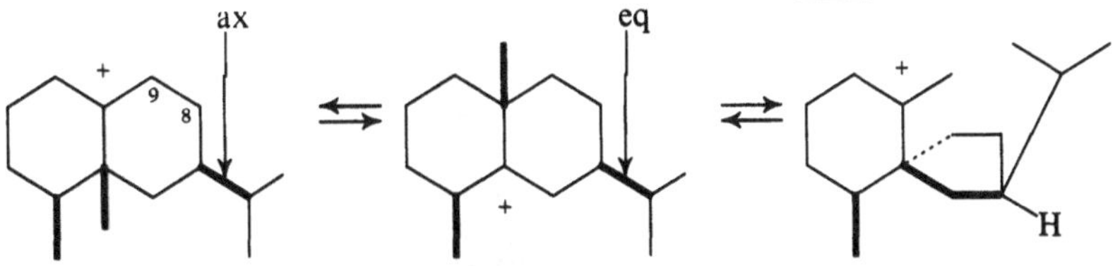

Chart 44. Relationship of eudesmane-derived skeletal types

These observations may be explained by postulating that relief of the strain associated with an axial isopropyl grouping and a 1,3-diaxial methyl interaction provides substantial driving force for the rearrangements. The functionalization of these sesquiterpenes may also be directed by strain relief. Thus, most natural eremophilanes have sp²-centers at C-9 and/or C-8 thereby reducing steric interactions involving the axial methyl or isopropyl groups. In the 10-epi series, the eudesmanes are generally unsaturated at C-7. Rearrangements leading to 5,10-dimethyl systems such as valeranone may represent another route for stabilizing a 4,10-*epi*-eudesmane (283)→(284) (23).

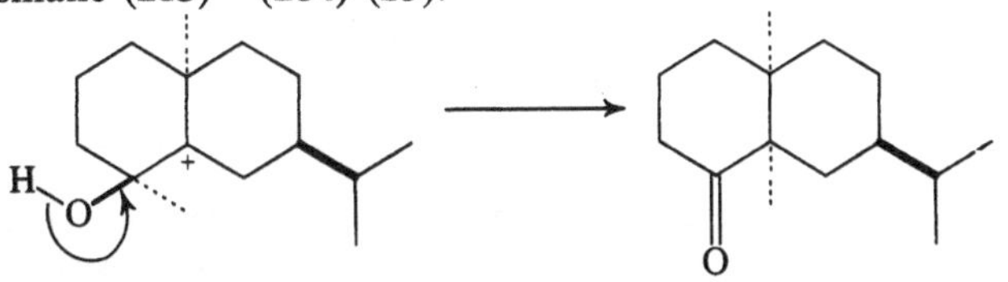

(283)   valeranone (284)

The spirovetivanes have also played a part in biogenetic proposals relating to khusimol *(Chart 45)*, a member of the tricyclovetivane (or zizaane) class which is a major constituent of vetiver oil *(58)*.

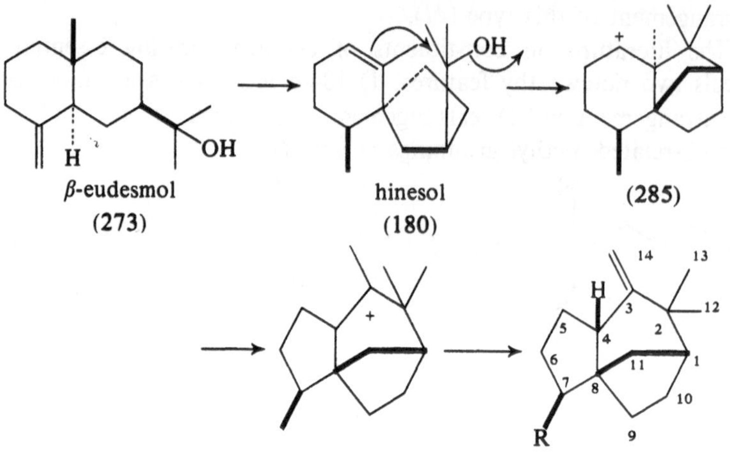

β-eudesmol
(273)

hinesol
(180)

(285)

(182)   (286), R=CH₂OH khusimol
(287), R=CH₃ zizaene
(288), R=CO₂H zizanoic
acid

Chart 45. Proposed biosynthesis of zizaenes

The indicated scheme is stereochemically consistent with the established structures of khusimol (286)*, zizaene (287) (4), and zizanoic acid (288) (58). The scheme can be faulted, however, on its failure to correlate the stereochemical features of the vetispirenes and β-vetivone which are found in the same oil (5, 7). Furthermore, although β-eudesmol can be detected in vetiver oil (4), the 4,10-*epi*-eudesmanes predominate.

Reconsideration of this problem led ANDERSEN to propose that the biosynthesis of tricyclovetivanes involves not the spirovetivane skeleton but rather the other spiro[4.5]decane sesquiterpene skeleton, acorane (13).

2. Alaskane-Acorane Spiranes as Precursors of Tricyclic Sesquiterpenes

That the alaskane skeleton, particularly ion (2), represents a most reasonable penultimate intermediate (biogenetic and synthetic) for cedrene was convincingly demonstrated by the synthesis work described in Part I. Further support comes from co-occurrence patterns; alaskadienes

---

* The relative stereochemistry has been confirmed by the X-ray crystal structure determination of khusimyl *p*-bromobenzoate (29). The absolute stereochemistry was suggested by ORD data for the degradation products of zizanoic acid and confirmed by CD comparison of tricyclovetivan-6-one and its 7-epimer (4).

and (−)-α-cedrene are found in *Juniperus* species (*101*) while acoradienes and (+)-α-cedrene are found in vetiver oil (*48*)*.

With the proposed genesis of zizaene via hinesol-related spiro-vetivanes (*58*) judged unlikely owing to the absence of this stereochemical series of spiranes in vetiver oil, a correlation with (+)-γ-curcumene and β-alaskene was suggested (*5*) *(Chart 46)*.

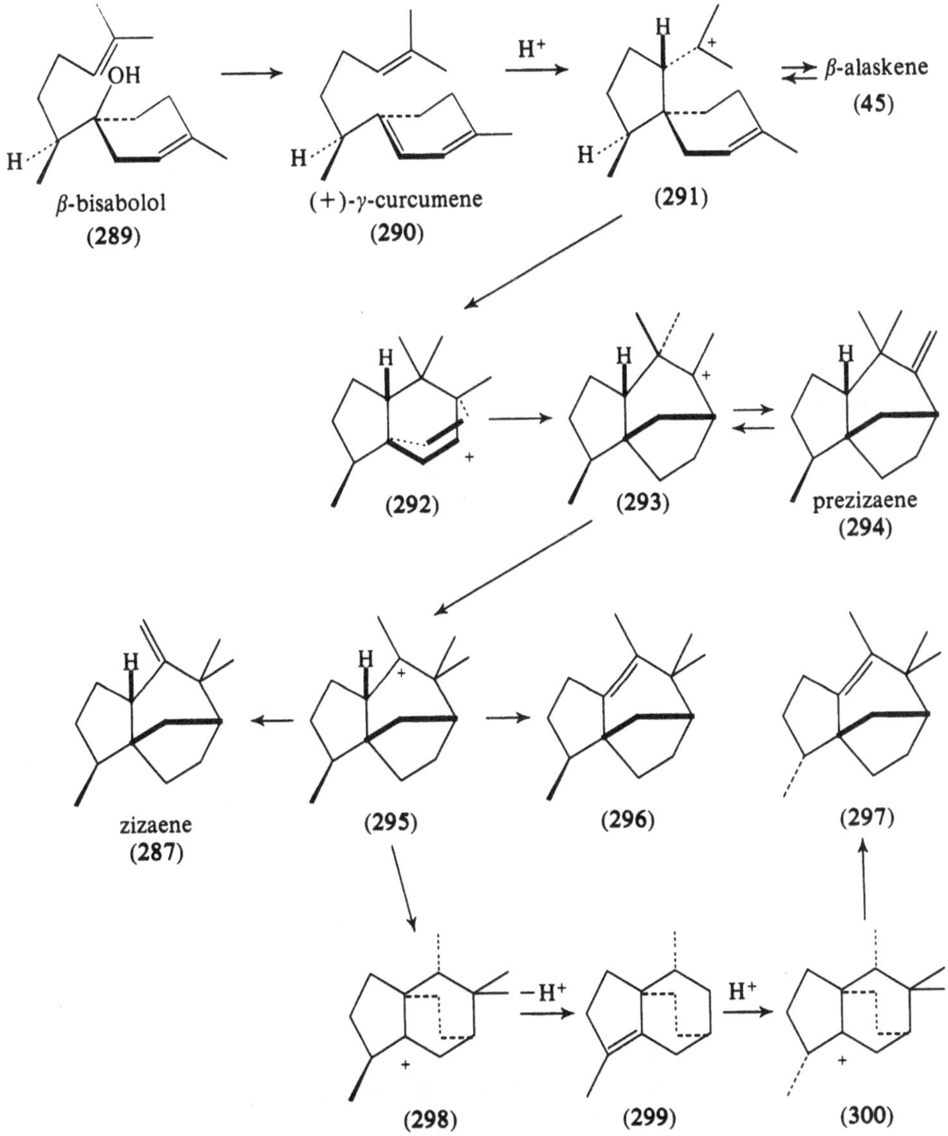

β-bisabolol (289)   (+)-γ-curcumene (290)   (291)   β-alaskene (45)

(292)   (293)   prezizaene (294)

zizaene (287)   (295)   (296)   (297)

(298)   (299)   (300)

Chart 46. Correlation of zizaene with β-alaskene

---

* This is the first recorded observation of the natural occurrence of ent-α-cedrene.

(2)   (−)-α-cedrene

The reported cyclization of "β-acoradiene" to "allocedrol" supports the proposed cyclization step leading to the initial tricyclic system (292) (*103*). Since the ion (293), although well disposed for an additional Wagner-Meerwein shift, must be a true intermediate in this scheme [models suggest that (292)→(293)→(295) cannot be concerted] the related exocyclic olefin (294) seemed a likely natural product. This reasoning led to a reexamination of the previously unidentified exocyclic olefins of vetiver oil. The major of these proved to be olefin (294) (named prezizaene), the first representative of a new sesquiterpene skeleton (*5*). The structure proof for prezizaene included, in addition to the usual spectroscopic data, its acid-catalyzed rearrangements which are identical to those of zizaene. The structures of the final products, olefins (296) and (297), were not established at that time, however the methyl shift (293)→(295) was confirmed by isolation of small amounts of zizaene formed during the course of the reaction (*8, 89*). The recent report of acoradienes and β-bisabolol (289) in vetiver oil is also consistent with this proposal (*48*).

To complete the picture, the two enantiomeric spiranes can be viewed as cyclization products of the curcumenes *(Chart 47)*. Unlike typical sesquiterpenes, which occur exclusively in one enantiomeric form in higher plants (*2*), the curcumenes occur in both forms (*12*). Among other features, this scheme offers an explanation for the observation that 2-epi-α-cedrene does not appear to be a common natural product. Since the α- and β-series of spiranes arise separately it is not necessary to postulate biogenetic pathways contrary to the known α-selectivity of the chemical spirane formation in the bisabolane series. The above sequence also provides an explanation for the co-occurrence of α- and β-alaskene. The choice of (−)-*trans*-nerolidol as the sesquiterpene precursor, rather than the naturally occurring dextrorotatory (often weakly so) form, is based upon the chemically observed asymmetric induction (*14*).

The unusual and obviously interrelated sesquiterpenes Δ^2-cedrene (313) and Δ^9-acorene (314) found in vetiver oil, appear to require a modified scheme in light of their functionalization pattern (*48*) *(Chart 48)*. One pathway proceeds from zizaene (309), a major hydrocarbon of vetiver (*3*), along lines suggested above for the murolene → cedrene scheme *(Chart 47)*. Zizanene has also been considered as a biogenetic relative of levojunenol (312) (*3*). If this is so the stereo-

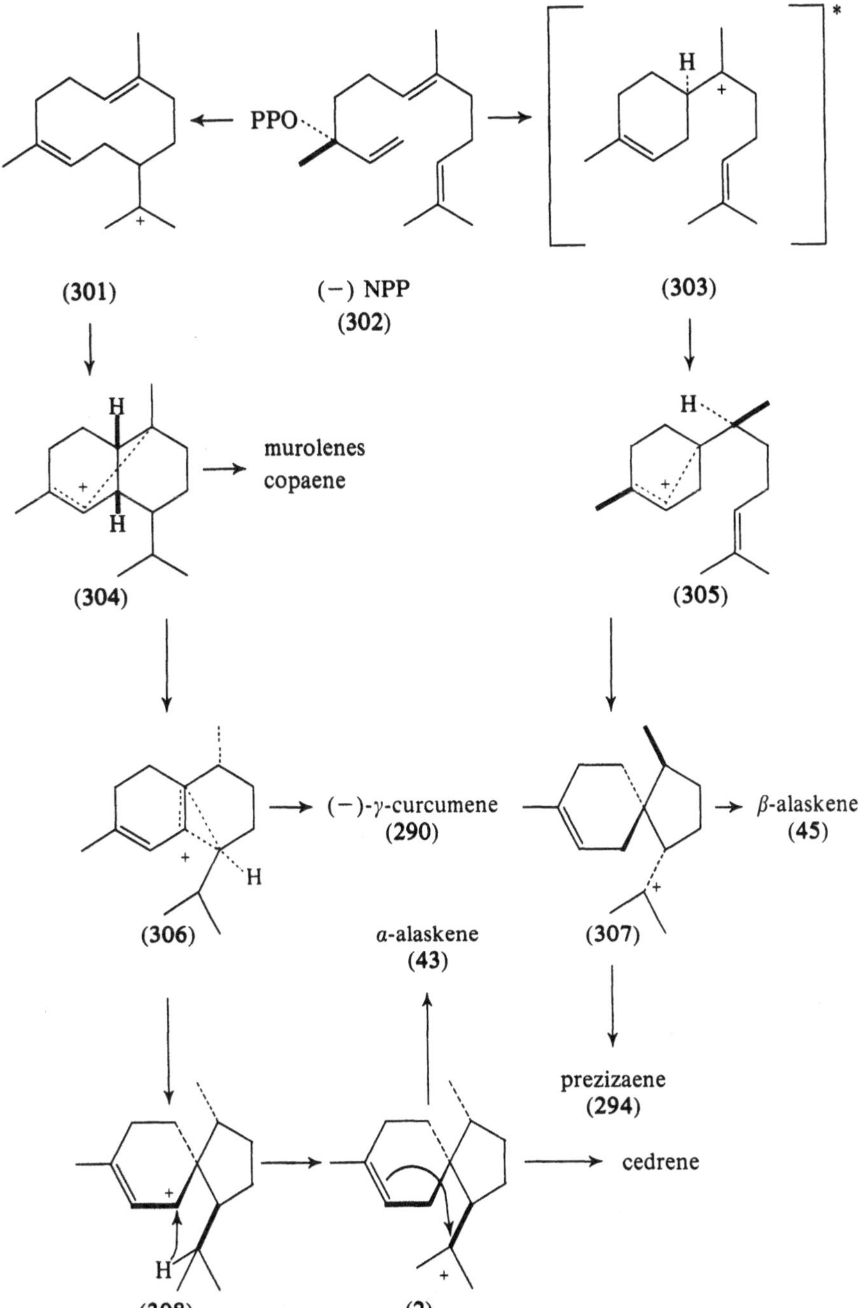
PPO
H
*
(301)
(−) NPP
(302)
(303)
H
H
murolenes
copaene
H
(304)
(305)
(−)-γ-curcumene
(290)
β-alaskene
(45)
H
(306)
α-alaskene
(43)
(307)
prezizaene
(294)
cedrene
H
(308)
(2)

Chart 47. Spiro[4.5]decanes involved in the biogenesis of tricyclic sesquiterpenes

---

* Ion **303** represents an unsymmetrically solvated ion derived from nerolidol. An alternative source of this ion would be the appropriate α- or β-bisabolol.

chemistry would be contrary to that tentatively assigned by KAISER and NAEGELI (48). Unfortunately, the stereochemistry of these recently discovered compounds has not yet been determined.

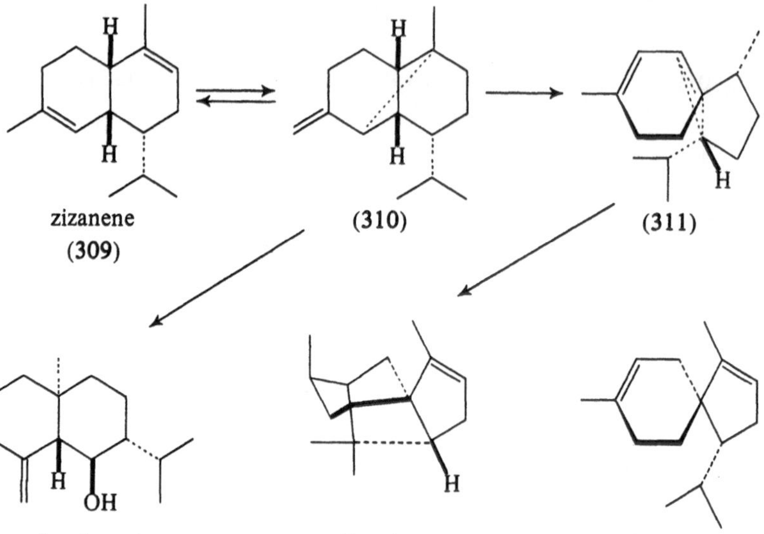

zizanene
(309)

(310)

(311)

levojuneol
(312)

Δ²-cedrene
(313)

Δ⁹-acorene
(314)

Chart 48. Correlation of Δ²-cedrene and Δ⁹-acorene

An alternative route to C-10 *epi*-eudesmanes involving spiro[4.5]decanes derived *via* migration of the C-1 ring methylene has been proposed [e. g. (315)→(317) (37)]. This suggestion, although consistent with observed chemically induced rearrangements of the same kind, appears unnecessary since 10-*epi*-eudesmanes are well established natural products which may be formed directly from farnesol-type precursors.

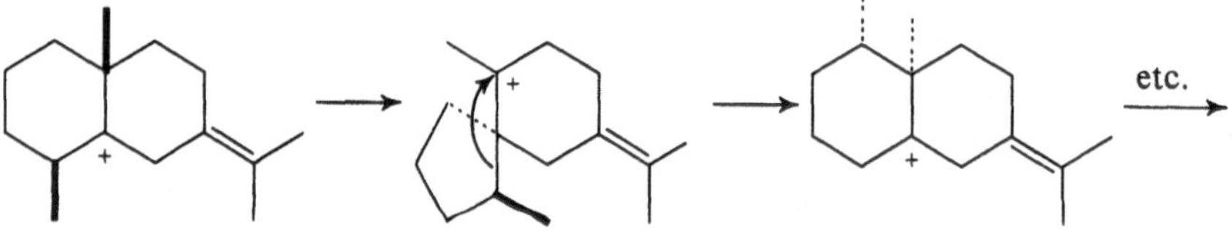

(315)

(316)

(317)

etc.

3. Chemical Simulation of Biogenetic Pathways involving Spiro [4.5]-Decanes

Total syntheses of natural products through sequences inspired by biogenetic conjecture have become part of present-day organic synthesis. The history of synthetic efforts aimed at cedrene is particularly illustrative of the simplification that frequently results from such strategy. The reported two-step, one-pot conversion of nerolidol to cedrene comes as close to the biogenetic hypothesis as any example in the sesquiterpene field (14). The second step undoubtedly proceeds through spirane intermediates although this could not be demonstrated by the isolation of

nerolidol   (318)   cedrene
(69)

spirodienes. Such intermediates evidently cyclize rapidly under the reaction conditions. On the other hand, spirane intermediates can be isolated from diene (319) (100). Upon treatment with trifluoroacetic acid, spirane (321) was isolated in 15% yield. This product was easily identified as a hydrogenation product of α-alaskene (43) (15). Other diastereomeric and isomeric spiranes were probably formed but these have not yet been characterized.

There have been a number of attempts to duplicate the Wagner-Meerwein shifts of eudesmanes and 10-*epi*-eudesmanes. In the eudesmane series, acid-catalyzed rearrangements of selinenes have been studied

(319)   (320)   (321)

extensively (*10*). The various paths leading to the stable product δ-selinene (**187**) have been defined and in excess of 97% of the material can be accounted for at all stages. No eremophilanes or spirovetivanes were encountered as intermediates. The 10-*epi* series [e. g. (**323**)] would

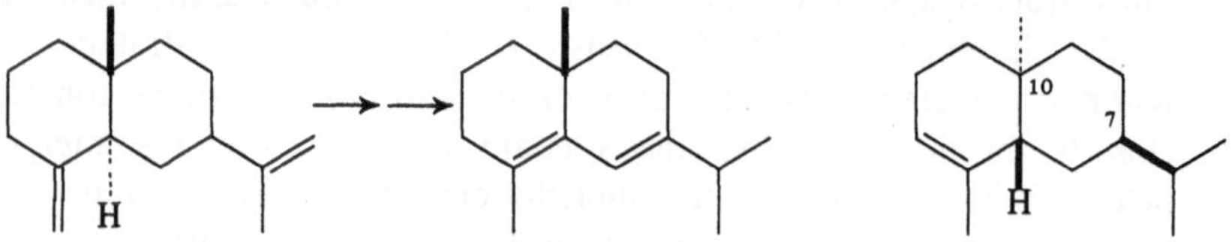

β-selinene
(**322**)

δ-selinene
(**187**)

(**323**)

presumably rearrange more readily as a consequence of the axial isopropenyl group. However, in acid, epimerization at C-7 proceeds and no rearrangement is observed (*10*). Ketones (**324**) and (**325**) gave no spirovetivanes or nootkatanes under various dehydration conditions (*42*). In the light of the close structural relationship to the presumed vetivone precursor, these findings were surprising. Thus far, the only chemical analogies for the proposed biogenetic conversion of the

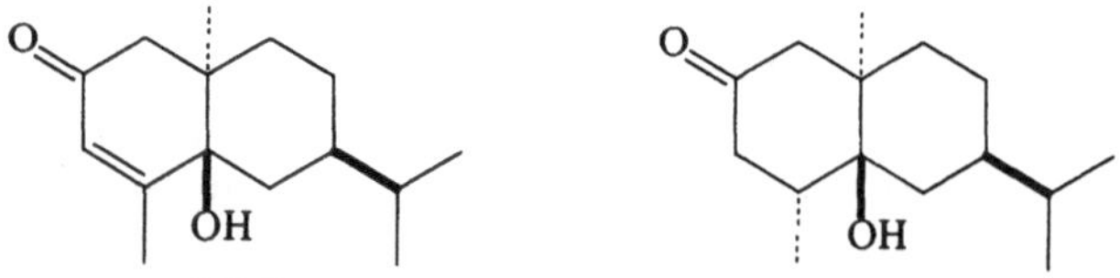

(**324**)

(**325**)

eudesmane to spirovetivane or nootkatane frameworks are the dehydration of β-rotunol (**275**) (*41*) and the observation that paradisiol (**326**) has the same mass spectrum as valencene (**259**) (*95*).

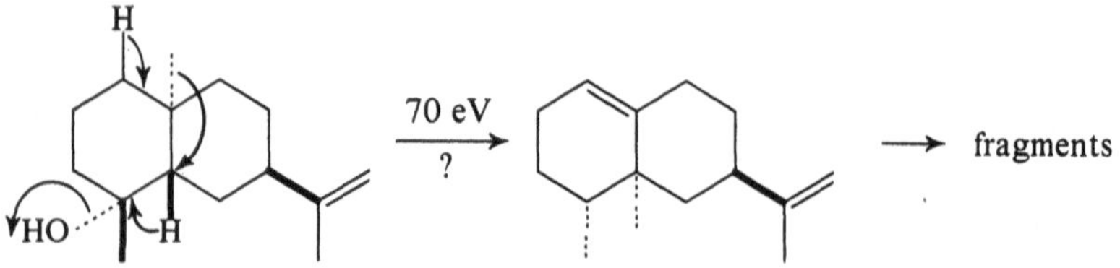

paradisiol
(**326**)

70 eV
?

valencene
(**259**)

fragments

E. Tables of Naturally Occurring Spiro[4.5]Decanes

The representation of relative and absolute stereochemistry in spiranes is particularly confusing. For this reason the tables of compounds to follow include full structural presentations with the ring bearing the greater number of asymmetric centers shown in the plane of the page. The stereo-designating nomenclature suggested by SYRDAL has been used for members of the acorane-alaskane group (*100*). The

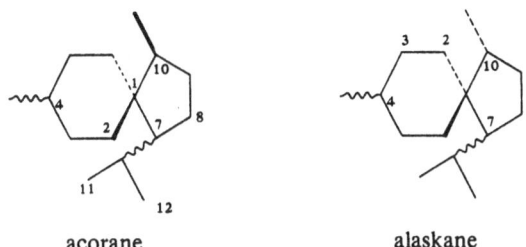

acorane   alaskane

numbering systems for the enantiomeric skeletons are as shown. The stereochemistry at C-4, C-7, and other centers, is indicated by specifying the relationship (*c* = *cis*, *t* = *trans*) between the hydrogen at that center and the lowest numbered carbon of the other ring. C-1/C-10 stereochemistry is indicated by the parent names acorane and alaskane.

Table 1. *Spiro[4.5]Decane Sesquiterpenes (Acorane-Alaskane Group)*

Formula	Name and Structure (References)	Physical Properties (References)	Sources (References)
$C_{15}H_{22}$	β-acoratriene	racemic	synthetic (35, 71)
$C_{15}H_{24}$	acoradiene acoradiene-IV acora-3,7-diene	$[\alpha]_D = +41°$ $\Delta\varepsilon_{202} = 13.2$ (11, 100) $[\alpha]_D = +34°$ (48)	from isoacorone *Vetiveria zizanoides*
	acoradiene-III	(70a) $[\alpha]_D = -4°$ (48)	*V. zizanoides* synthesis (71)
	β-alaskene	(11) $[\alpha]_D = -18°$ $\Delta\varepsilon_{207} = -3.2$	*Chamaecyparis nootkatensis* (12, 13)

Structure	Name	$[\alpha]_D$ / CD data	Source	Ref.
	4cH, 7cH-acora-2,8-diene	$[\alpha]_D = +171°$ $\Delta\varepsilon_{214} = -0.25$ $\Delta\varepsilon_{203} = 10.8$	from isoacorone	(11, 100)
	α-alaskene γ-acoradiene	$[\alpha]_D = -88°$ $\Delta\varepsilon_{212} = 3.5$ $\Delta\varepsilon_{207.5} = -4.5$ $[\alpha]_D = -66°$ $\Delta\varepsilon_{212} = -2.9$ $\Delta\varepsilon_{192} = 8$	Ch. nootkatensis Juniperus rigida Barbilophozia barbata	(12, 13) (11) (103) (15)
	δ-acoradiene ent-β-alaskene	$[\alpha]_D = +15.5°$	J. rigida	(103)
	α-acoradiene ("acorene")	$[\alpha]_D = +37°$	J. rigida, J. virginiana	(102, 110) (101)

Table 1 (continued)

Formula Name and Structure (References)	Physical Properties (References)	Sources (References)
β-acoradiene	$[\alpha]_D = +23°$	*J. rigida* (103)
4cH-acora-2,7-diene	$[\alpha]_D = -62°$ $\Delta\varepsilon_{202} = -22$	from isoacorone (16, 100)
7cH-acora-3,8-diene	$[\alpha]_D = +180°$ $\Delta\varepsilon_{198} = 8.5$	from isoacorone (16, 100)
4tH-acora-2,7-diene	$[\alpha]_D = -40°$ $\Delta\varepsilon_{207} = -33.6$	from acorone (16, 100)

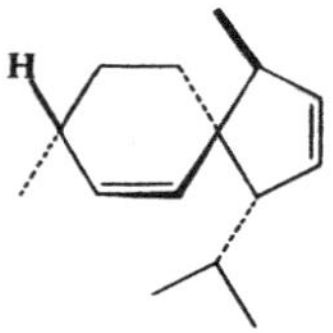

from acorone   *(100)*

$[\alpha]_D = +230°$ $\Delta\varepsilon_{203} = -8.3$

4tH, 7cH-acora-2,8-diene

$C_{15}H_{26}$

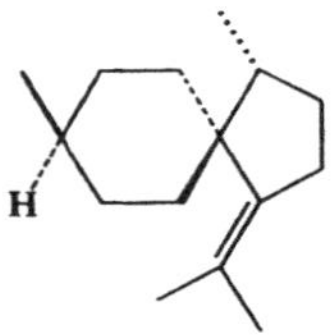

H₂/α-alaskene   *(15)*

$[\alpha]_{300} = +150°$ $\Delta\varepsilon_{204} = 11.4$

dihydro-α-alaskene

H₂/β-alaskene
H₂/4c, 7cH-acora-2,8-diene   *(11)*

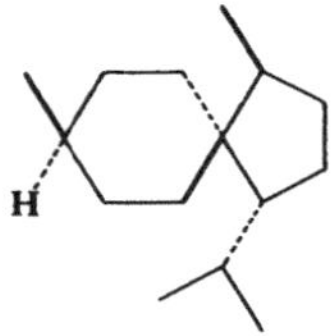

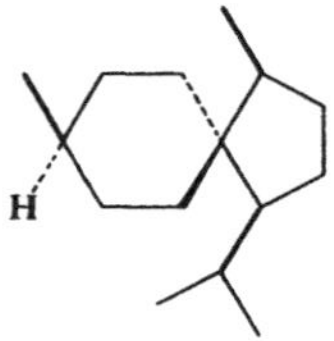

$[\alpha]_D = +43°$; $[\alpha]_{300} = +1110°$
$[\alpha]_D = +49°$; $[\alpha]_{200} = +1190°$

4cH, 7cH-acorane

$C_{15}H_{28}$

(6, 12)

(15)

$[\alpha]_{300} = -20°$

4cH, 7cH-acorane
ent-alaskane-II

Table 1 (continued)

Formula	Name and Structure (References)	Physical Properties (References)	Sources (References)
	4cH, 7cH-alaskane alaskane-I	$(11, 100)$ $[\alpha]_D = -46°$; $[\alpha]_{200} = -1250°$ (12)	H₂/α-alaskene $(6, 11, 12)$
	acoronene	(68)	
C₁₅H₂₄O	acorenone-B		Bothriochloa intermedia (51)
	acorenone	$n_D = 1.5039$, $[\alpha]_D = -22°$	Acorus calamus $(106, 107)$

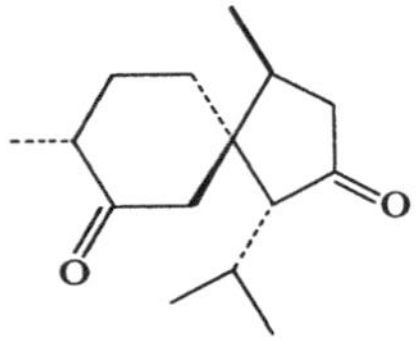

acorone
4tH, 7cH-acora-3,8-dione

$C_{15}H_{24}O_2$

(56, 110) mp 101°, $[\alpha]_D = +144°$
ORD-CD: a = +100
$\Delta\varepsilon_{308} = 2.77$

(98)
(99, 110)
(9)

A. calamus

(91)

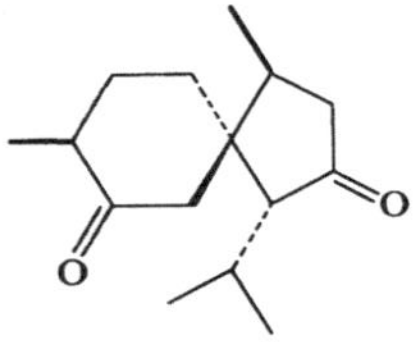

isoacorone

(110) mp. 98°, $[\alpha]_D = -91°$
ORD-CD: a = -160
$\Delta\varepsilon_{305} = -3.68$; $\Delta\varepsilon_{196} = 5.3$

(98)
(99, 110)
(9)

A. calamus

(91)

α-acorenol

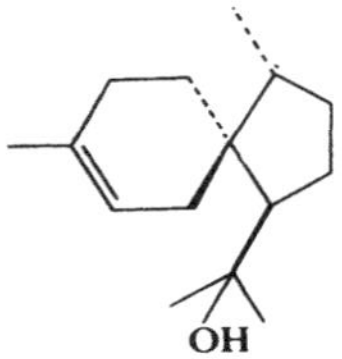

$C_{15}H_{26}O$

$[\alpha]_D = -36°$

J. rigida

(101)

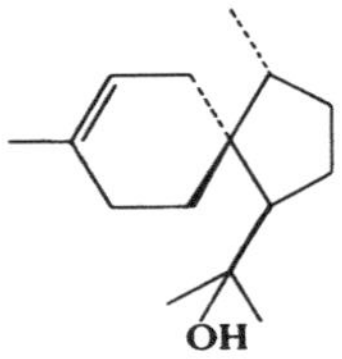

β-acorenol

$[\alpha]_D = 0°$

J. rigida

(103)

Table 1 (continued)

Formula	Name and Structure (References)	Physical Properties (References)	Sources (References)
	cryptoacorone 4cH, 7tH-acora-2,8-dione (110)	mp. 108°, $[\alpha]_D = +98°$ ORD-CD: a = 120 (108) (110)	A. calamus (108)
	neoacorone (110)	mp. 85°, $[\alpha]_D = +126°$	A. calamus (98)

Table 2. *Spiro[4.5]Decane Sesquiterpenes (Spirovetivane Group)*

Formula	Name and Structure	Physical Properties (References)	Sources (References)
$C_{15}H_{22}$	α-vetispirene	$[\alpha]_D = +220°$, $\varepsilon_{293.5} = 20,500$ (16) $\Delta\varepsilon_{239} = 24$	*Vetiveria zizanioides* (7)
	β-isovetivenene β-vetispirene	$[\alpha]_D = -68°$, $\varepsilon_{232} = 12,300$ $[\alpha]_D = -90°$, $\varepsilon_{232} = 13,500$ (7)	*V. zizanioides* (7, 86)
	α-isovetivenene	$n_D = 1.5222$, $[\alpha]_D = -120°$, $\varepsilon_{234} = 8,500$	*V. zizanioides* (86)
$C_{15}H_{24}$	α-isovetivene	$n_D = 1.5040$, $[\alpha]_D = -46°$	*Atractylodes lancea* (28)

Table 2 (continued)

Formula	Name and Structure	Physical Properties (References)	Sources (References)
$C_{15}H_{28}$	(+)-isovetivane hinesane racemate = isovetivane (spirovetivane)	$[\alpha]_D = +21°$, $n_D = 1.4746$ $[\alpha]_D = 0°$, $n_D = 1.4720$	H_2, α-isovetivene (28) "meso-dihydro-β-vetivone" degradation (86)
$C_{15}H_{20}O$	"anhydro"-β-rotunol		dehydration of β-rotunol (Cyperus rotundus) (41)
$C_{15}H_{22}O$	(−)-β-vetivone (64)	mp. 44° $[\alpha]_D = -24°$ (83, 84) mp. 44.5°, $[\alpha]_D = -39°$ (75, 76) $\varepsilon_{242} = 15{,}600$, $\varepsilon_{335} = 33$ (75) mp. 36—43°, $[\alpha]_D = -32°$ (1)	V. zizanioides (83, 76, 73) synthesis (65)
$C_{15}H_{24}O$	"meso-dihydro-β-vetivone"	bis-benzylidene, mp. 131.5° (84) 2,4-DNP, mp. 158° (1) semicarbazone, mp. 193° (64)	from β-vetivone

$C_{15}H_{26}O$

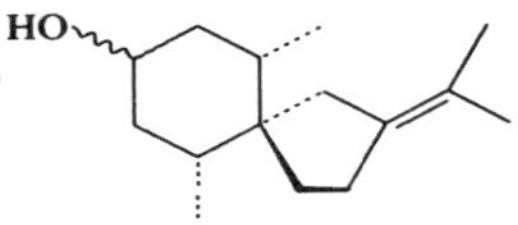

"meso-dihydrovetivol"
β-OH = isomer A
α-OH = isomer B
65 : 32 mixture

mp. 106°, 107°, 109°
mp. 107.5°
mp. 68 – 74°

(1, 84, 76) from β-vetivone
(1)
(61)

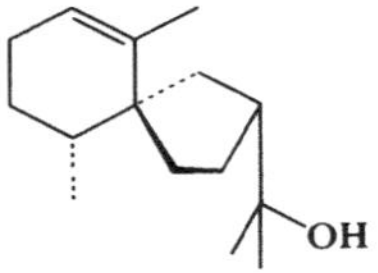

agarospirol

$n_D = 1.5080$, $[\alpha]_D = -6°$

(104)

Aquilaria agallocha
synthesis   (70)

$C_{15}H_{26}O$

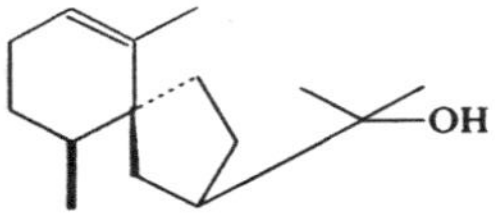

hinesol

mp. 58°, $[\alpha]_D = -48°$
mp. 59 – 60°, $[\alpha]_D = -40°$

(28, 69)
(116)

A. lancea
synthesis   (62)

$C_{15}H_{28}O$

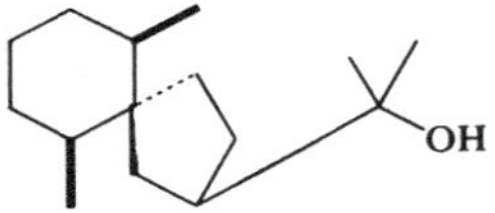

dihydrohinesol

mp. 43°, $[\alpha]_D = +21.6°$

(112)

from hinesol

Addendum

McCURRY and SINGH (*54 b*) have reexamined the enamine-Claisen route to *epi*-β-vetivone *(Chart 37)* and found that cyclization of the unsaturated aldehyde (**230**) with stannic chloride affords a separable mixture of alcohols (**327**) and (**328**) which differ in configuration at the spiro carbon atom.

(327)   (328)

The former epimer gave (±)-β-vetivone upon oxidation with pyridine-SO_3 complex in dimethyl sulfoxide. The latter epimer yielded (±)-10-epi-β-vetivone. The mesylate derivative (**329**) of alcohol (**327**) was converted to (±)-β-vetispirene in refluxing pyridine.

(329)   (±)-β-vetispirene

A cation cyclization route *(Chart 49)* which is formally related to the enolate cycloalkylation approach of Johnson *(Chart 35)* was examined by McCURRY and co-workers and was found to give (±)-10-epi-β-vetivone (*54 a*). The key cyclization steps of both syntheses appear to be controlled by stereochemical interactions which direct the incoming carbon of the forming cyclopentane ring *trans* to the adjacent methyl group [compare (**337**)→(**338**) with *Chart 35,* (**208**)→(**168**)]. Preparation of the penultimate intermediate (**336**) was effected through a convergent synthetic scheme involving alkylation of the lithio derivative (**334**) with the indicated chloro ketal. Reduction of the resulting product (**335**) with lithium in ethylamine effected hydrogenolysis of both the allylic sulfide and benzyl group to give alcohol (**336**).

References, pp. 371—376

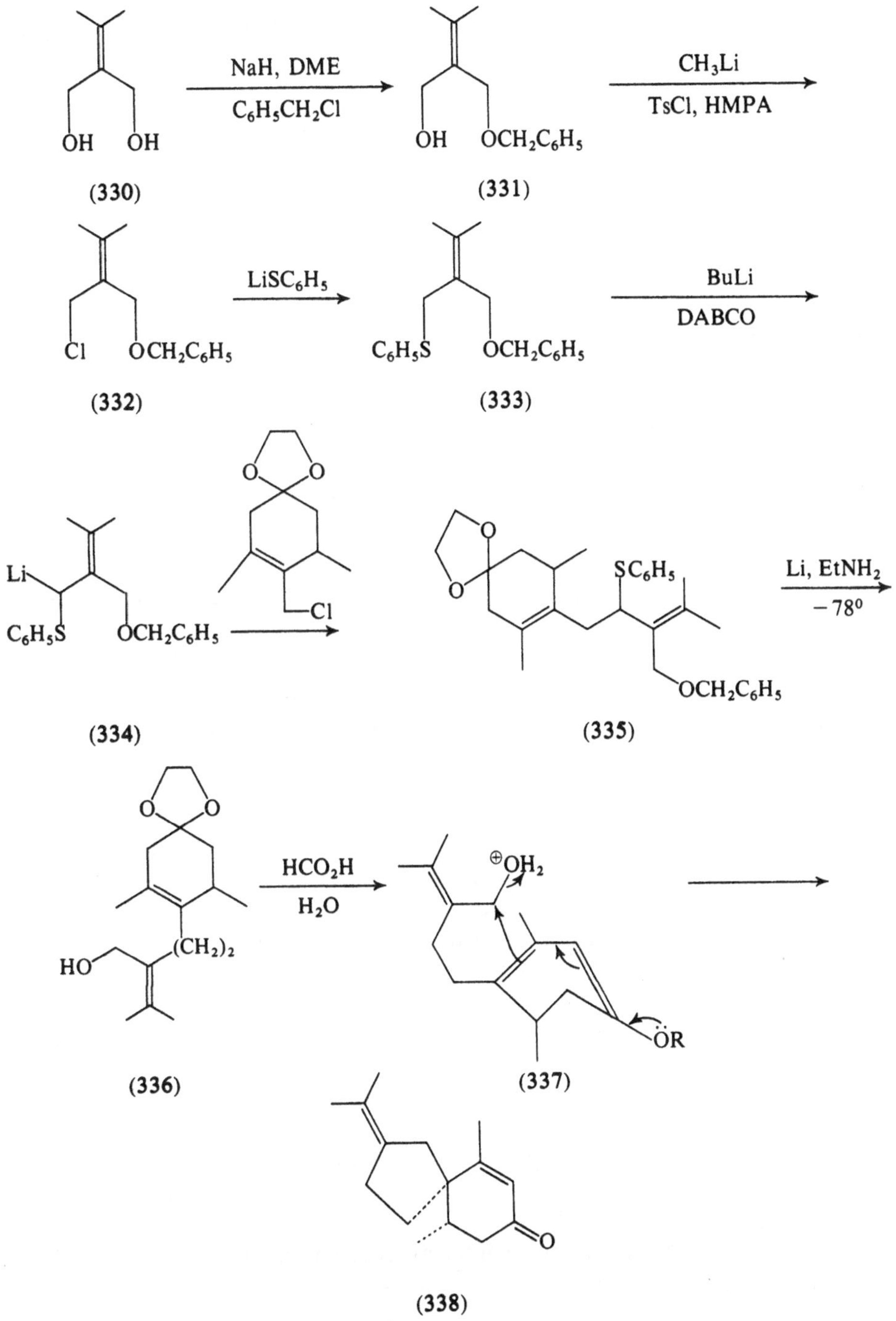

Chart 49. Synthesis of (±)-10-epi-β-Vetivone

Recently Yamada *(111c)* has completed total syntheses of $(\pm)$-β-vetivone, $(\pm)$-hinesol, and $(\pm)$-β-isovetivenene utilizing a cation cyclization approach *(Chart 50)*. In this case, the incoming carbon of the forming cyclopentanol ring must attach the cyclohexadienol ring *cis* to the adjacent carboxylic acid grouping during the course of the cyclization reaction (**340**)→(**341**).

The carboxylic acid grouping thus appears to exert a *cis* directing effect while the comparable methyl substituent exerts a *trans* directing effect on the cyclization reaction [*Chart 49*, (**337**)→(**338**)]. Of course, the cyclization of unsaturated keto acetal (**340**) may be a reversible process whereas cyclization of the unsaturated ketal alcohol (**336**) is undoubtedly irreversible. In that event, lactonization may provide the driving force for the formation of the stereoisomer represented by structure (**341**). Differentiation of the three oxygen functions of keto lactone (**341**) was accomplished by ketalization with ethylene glycol followed by reduction with lithium aluminum hydride to give a ketal diol which cyclized to the hydroxy keto ether (**342**) upon aqueous hydrolysis. The ketone was again ketalized with ethylene glycol after acetylation of the cyclopentanol (to prevent cyclic ether formation between the cyclopentanol and the CH_2OR function). Saponification afforded the ketal alcohol (**343**) which was oxidized and carbomethoxylated to give keto ester (**344**). Reduction of the cyclopentanone with sodium borohydride and basic elimination of the derived methanesulfonic ester afforded the unsaturated ester (**345**). Hydrogenation of the cyclopentene double bond expectedly took place from the less hindered face to give the ester (**346**). Introduction of the isopropylidene grouping was effected through addition of methyllithium followed by dehydration. Hydrolysis then afforded the keto ether (**347**). Ether cleavage with the triphenylphoshine bromine complex yielded the bromo ketone (**348**). Hydrogenolysis of the related iodo ketone with zinc in acetic acid led to $(\pm)$-β-vetivone.

The favorable steric course of the hydrogenation leading to ester (**346**) enabled the use of this intermediate in the synthesis of $(\pm)$-hinesol. This was achieved through hydrolysis of the ketal followed by ether cleavage (methyltriphenoxyphosphonium iodide) and hydrogenolysis of the resulting iodide with zinc in acetic acid. Desulfurization of the thioketal derivative of keto ester (**349**) followed by addition of methyllithium completed the synthesis of ($\pm$)-hinesol.

. The enol acetate derivative (**350**) of keto ester (**349**) was converted to ($\pm$)-β-isovetivenene through deacetoxylation with iron pentacarbonyl followed by addition of methyllithium to the resulting dienic ester and dehydration of the tertiary alcohol thus obtained.

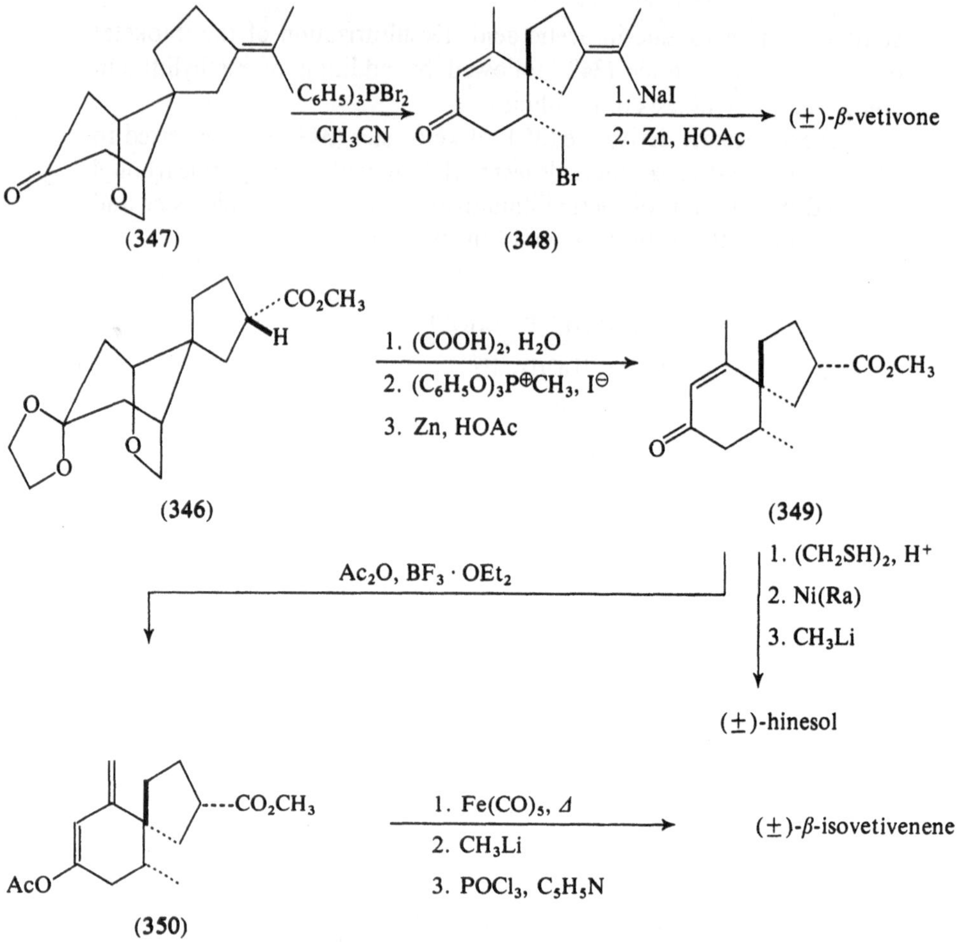

$(C_6H_5)_3PBr_2$
CH_3CN
(347)
(348)
Br
1. NaI
2. Zn, HOAc
(±)-β-vetivone
CO_2CH_3
H
1. $(COOH)_2$, H_2O
2. $(C_6H_5O)_3P^{\oplus}CH_3$, $I^{\ominus}$
3. Zn, HOAc
(346)
(349)
Ac_2O, $BF_3 \cdot OEt_2$
1. $(CH_2SH)_2$, H^+
2. Ni(Ra)
3. CH_3Li
(±)-hinesol
AcO
1. $Fe(CO)_5$, Δ
2. CH_3Li
3. $POCl_3$, C_5H_5N
(±)-β-isovetivenene
(350)

Chart 50. Synthesis of (±)-β-vetivone, (±)-hinesol and (±)-β-isovetivenene

A new approach to the stereoselective synthesis of spiro[4.5]decanes related to spirovetivanes has been devised by TROST *(Chart 51)* (*103a*). Spiroannelation of 2,6-dimethylcyclohexenone through addition of lithio cyclopropyl phenyl sulfide followed by acid catalyzed rearrangement of the resulting cyclopropyl vinyl carbinol (**352**) afforded the spiro[3.5]non-enone (**353**) stereoselectively. The dimethylaminomethylene derivative (**354**) was converted to the keto dithiane (**355**) upon treatment with 1,3-propanedithiol ditosylate. Addition of methyllithium followed by basic cleavage of the resulting methyl carbinol gave the keto thioacetal (**356**). Hydrolysis via the sulfonium derivative led to the keto aldehyde which underwent basic aldol cyclization to the spiro[4.5]decadienone (**358**), a potential intermediate for the synthesis of spirovetivanes.

References, pp. 371—376

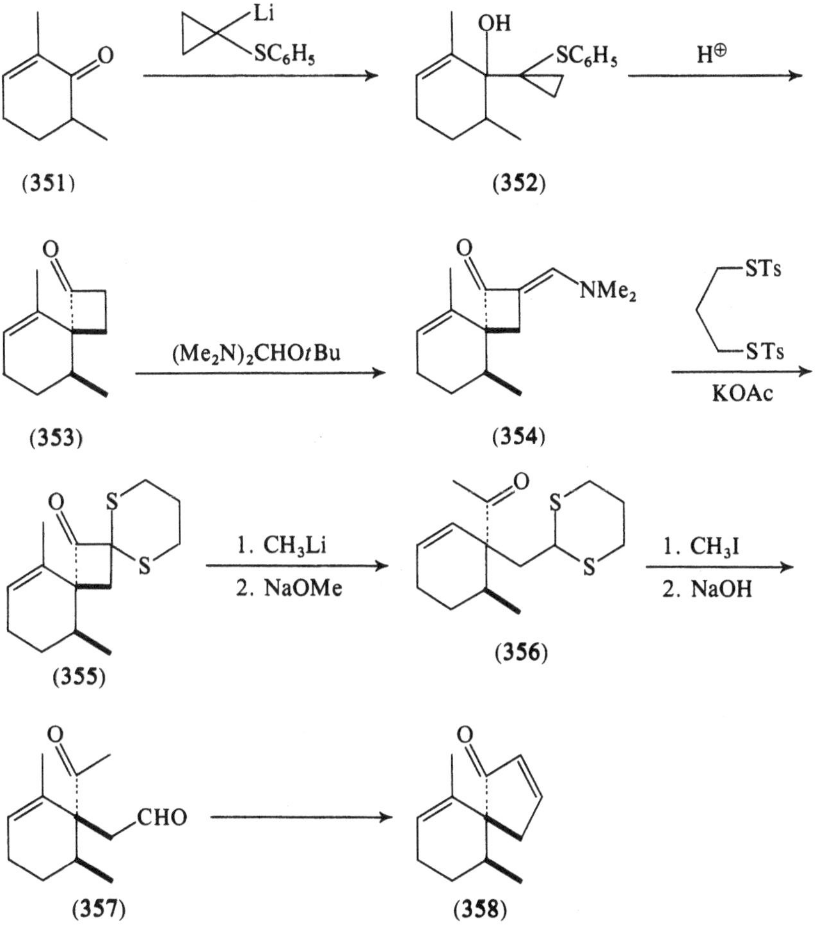

(351)　　　　　　　　　　　　(352)

(353)　　　　　　　　　　　　(354)

(355)　　　　　　　　　　　　(356)

(357)　　　　　　　　　　　　(358)

Chart 51. Spiroannelation approach to spiro[4.5]decanes

WILLIAMS and SARKISIAN (*111b*) have described two photochemical routes to spiro[4.5]decanones. The first involves photolysis of the bicyclo[4.5]undecenone (359) to give the enone (360). The second utilizes the α, β-epoxy ketone (361) derived from cyclopentylidenecyclopentanone which rearranges to the spirodione (362) upon irradiation.

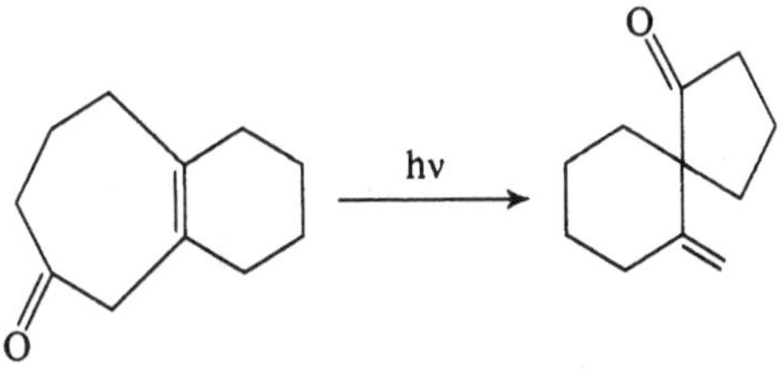

(359)　　　　　　　　　　　　(360)

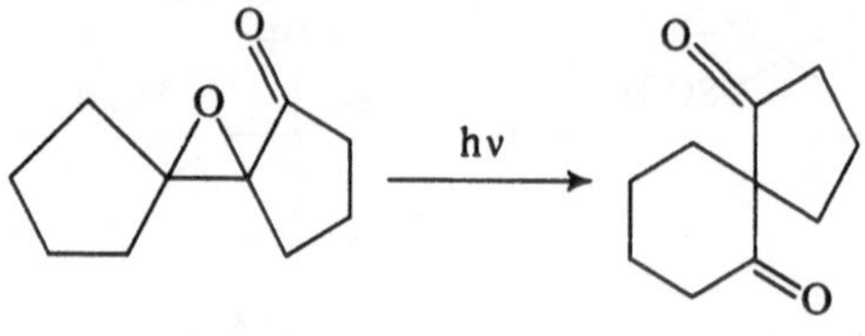

(361)   (362)

Several cases of angular methyl migration of the type noted in connection with the proposed eudesmane-nootkatane biogenetic conversion have recently been reported.

In a detailed study on solvolytically induced multiple alkyl shifts, Whitlock and Overman (*111a*) found that ethanolysis of the tosylate (**363**) gave, among other products, the spirodecene (**364**) (~ 20%) and the phenanthrene (**365**) (~ 10%).

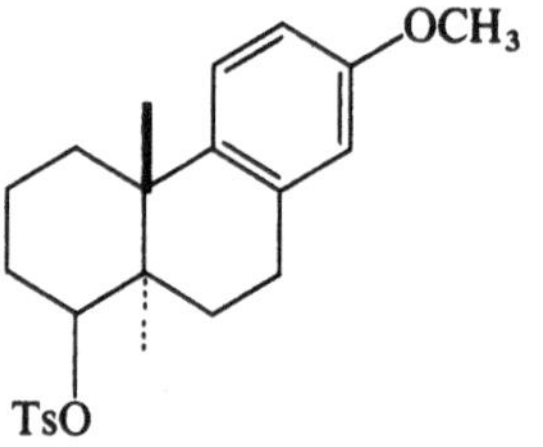

(363) 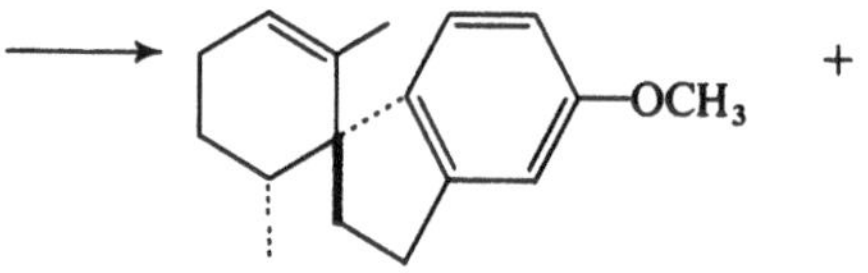   (364)

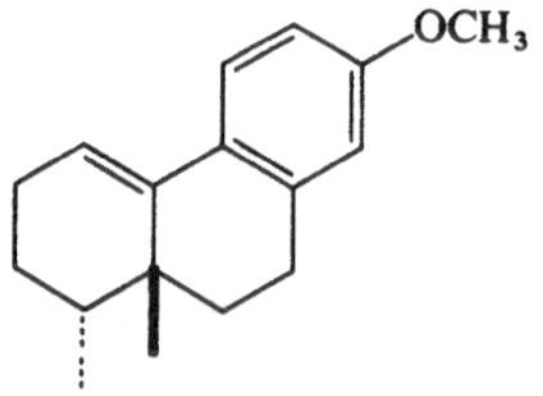

(365)

Kitagawa and co-workers prepared the epoxy lactone (**366**) from dihydroalantolactone and studied its solvolysis in formic acid-acetone (*47a*). Among the products isolated were the rearranged alcohol (**367**) (34%) and the corresponding formate (**368**) (10%).

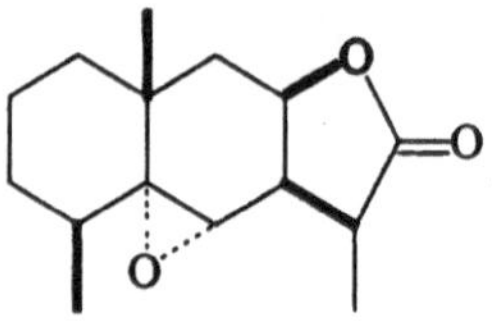

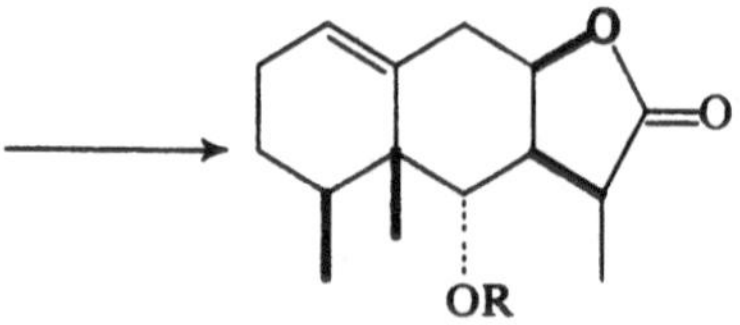

(366)   (367), R = H
           (368), R = CHO

HOCHSTETLER (*41a*) has shown that the rearrangement of olefin (**369**) to olefins (**370**) in 25% sulfuric acid-acetic acid involves the sequential alkyl shifts depicted below. This scheme is of special interest with regard to biosynthetic proposals since it involves both a methyl migration and a spiro[4.5]decyl cation.

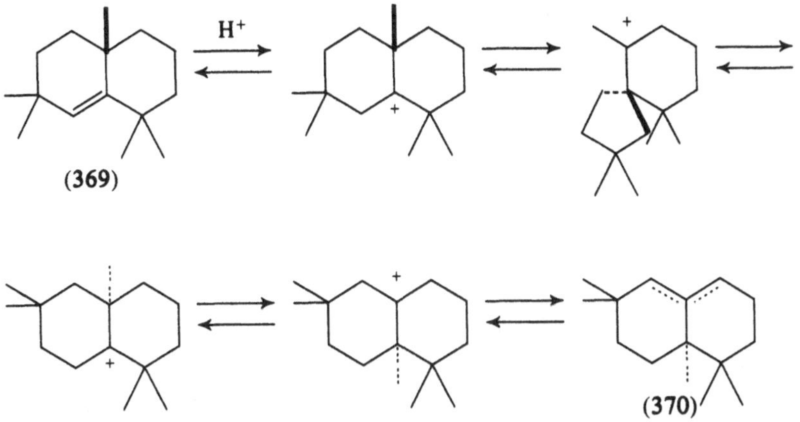

References

1. ANDERSEN, N. H.: Unpublished work.
2. — Biogenetic Implications of the Antipodal Sesquiterpenes of Vetiver Oil. Phytochemistry **9**, 145 (1970).
3. — On the Co-occurrence of Levojunenol and Zizanene [(+)-α-Amorphene]. Tetrahedron Letters **1970**, 4651.
4. — The Structures of Zizanol and Vetiselinenol. Tetrahedron Letters **1970**, 1755.
5. ANDERSEN, N. H., and M. S. FALCONE: Prezizaene and the Biogenesis of Zizaene. Chem. and Ind. **1971**, 62.
6. — — The Identification of Sesquiterpene Hydrocarbons from Gas-Liquid Chromatography Retention Data. J. Chromatogr. **44**, 52 (1969).
7. ANDERSEN, N. H., M. S. FALCONE, and D. D. SYRDAL: Structures of Vetivenenes and Vetispirenes. Tetrahedron Letters **1970**, 1759.
8. ANDERSEN, N. H., S. E. SMITH, and Y. OHTA: unpublished work.
9. ANDERSEN, N. H., S. E. SMITH, and D. D. SYRDAL: unpublished work.
10. ANDERSEN, N. H., and D. P. SVEDBERG: unpublished work.
11. ANDERSEN, N. H., and D. D. SYRDAL: The Absolute Stereochemistry of the Alaskenes and Acorone-Related Sesquiterpenes. Tetrahedron Letters **1972**, 899.
12. — Terpenes and Sesquiterpenes of *Chamaecyparis nootkatensis* Leaf Oil. Phytochem. **9**, 1325 (1970).
13. — — The Alaskenes — Precursors of Tricyclic Sesquiterpenes. Tetrahedron Letters **1970**, 2277.
14. — — Chemical Simulation of the Biogenesis of Cedrene. Tetrahedron Letters **1972**, 2455.
15. — — unpublished work.
16. ANDERSEN, N. H., D. D. SYRDAL, C. R. COSTIN, and D. P. SVEDBERG: The Optical Activity Associated with Isolated Olefinic Bonds. I. The Allylic Bond Polarization Model and its Extension to Homoconjugated Systems. J. Amer. Chem. Soc. **95**, 2049 (1973).

17. Andersen, N. H., D. D. Syrdal, and M. S. Falcone: Sesquiterpenes of Vetiver and Alaska Cedar, 158th National Meeting of the ACS. Abstracts of Papers, Abstract AGFD27 (Sept. 1969).

18. Andersen, N. H., D. D. Syrdal, and C. Graham: Aromatization of Cycloalkenes with Trifluoroacetic Acid. Tetrahedron Letters 1972, 903.

19. Asselin, A., M. Mongrain, and P. Deslongchamps: Syntheses of α-Agarofuran and Isodihydroagarofuran. Canad. J. Chem. 46, 2817 (1968).

20. Baird, R., and S. Winstein: Neighboring Carbon and Hydrogen. XLVI. Spiro[4.5]-deca-1,4-diene-3-one from $Ar_{1\ominus 5}$ Participation. J. Amer. Chem. Soc. 84, 788 (1962).

21. Barrett, H. C., and G. Büchi: Stereochemistry and Synthesis of α-Agarofuran. J. Amer. Chem. Soc. 89, 5665 (1967).

22. Birch, A. J., and J. S. Hill: Reactions of Cyclohexadienes. Part V. A New Synthesis of 4-Substituted Cyclohexenones. J. Chem. Soc. (C) 1966, 419.

23. Bundy, G.: Ph. D. Thesis, Part I. Synthesis of Cyclodecadienes from Decalylborane Derivatives. Part II: The Synthesis of Valeranone, Northwestern University (1968).

24. Caine, D.: Private Communication to J. A. Marshall.

25. Caine, D., and J. B. Dawson: The Synthesis and Acid-catalysed Rearrangement of a Spiro[4.5]dec-6-en-2-one. Chem. Commun. 1970, 1232.

26. Chiurdoglu, G., et J. Decot: Contribution à l'étude des composes sesquiterpeniques. II. Étude de la structure du bicyclovetivenol et du tricyclovetivenol. Alcols primaires de l'essence de vetiver. Tetrahedron 4, 1 (1958).

27. Chiurdoglu, G., et P. Tullen: Contribution à l'étude des composés sesquiterpeniques. I. Étude structurale du tricyclovetivene de l'essence de vetiver du Congo Belge. Bull. Soc. Chim. Belges 66, 169 (1957).

28. Chow, W. Z., O. Motl, and F. Šorm: On Terpenes CXL. Composition of the Oil from *Atractylodes lancea* Thunb. The Structure of Hinesol. Collect. Czech. Chem. Commun. 27, 1914 (1962).

29. Coates, R. M., R. F. Farney, S. M. Johnson, and I. C. Paul: The Crystal Structure of Khusimol p-Bromobenzoate. Chem. Commun. 1969, 999..

30. Conia, J.: Thermocyclisation de composés carbonylés non saturés. Bull. Soc. Chim. France 1968, 3057.

31. Conia, J., J. Gore, and J. Drouet: unpublished work.

32. Conia, J. M., J. P. Drouet, et J. Gore: Thermolyse et photolyse des cetones non saturées. 19ème Memoire. Sur l'obtention et la stéreochimie de quelques cetones de la famille de l'acorane formées dans la thermocyclisation de la (+)(isopropenyl-1-pentene-4-yl)-2-methyl-5-cyclohexanone. Tetrahedron 27, 2481 (1971).

33. Corey, E. J., N. N. Girotra, and C. T. Mathew: Total Synthesis of dl-Cedrene and dl-Cedrol. J. Amer. Chem. Soc. 91, 1557 (1969).

34. Crandall, T. G., and R. G. Lawton: A Biogenetic-Type Synthesis of Cedrene. J. Amer. Chem. Soc. 91, 2127 (1969).

35. Demole, E., P. Enggist, and (in part) C. Borer: Applications synthetiques de la cyclisation d'alcools tertiaires γ-ethyleniques en α-bromotetrahydrofurannes sous l'action du N-bromosuccinimide. II. Cyclisation du ±-nerolidol en dimethyl-2,5-(methyl-4-pentene-3-yl)-2-cycloheptene-4-one, tetramethyl-3,3,7,10-oxa-2-tricyclo-[5.5.0.0.^{1,4}]-dodecene-9, β-acoratriene, cedradiéne-2,8, epi-2-α-cédrene et α-cédrene. Helv. Chim. Acta 54, 1845 (1971).

36. Drouet, J.: Sur une voie d'accés au système acoranique. Ph. D. thesis, University of Caen, July (1969).

37. Dunham, D. J., and R. G. Lawton: Spiro Intermediates in Sesquiterpene Rearrangements and Synthesis. J. Amer. Chem. Soc. 93, 2075 (1971).

38. DUNHAM, D. J., and R. G. LAWTON: The Synthesis of Spiro Systems by the αα'-Annelation Process. J. Amer. Chem. Soc. **93**, 2074 (1971).

39. ENDO, K., and P. DE MAYO: α-Vetivone. Chem. Commun. **1967**, 89.

40. ERDTMAN, H., and T. NORIN: The Chemistry of the Order Cupressales. Fortschr. Chem. Organ. Naturstoffe **24**, 206 (1966).

41. HIKINO, H., K. AOTA, D. KUWANO, and T. TAKEMOTO: Structure of α-Rotunol and β-Rotunol. Tetrahedron Letters **1969**, 2741.

41a. HOCHSTETLER, A. R.: Acid Catalyzed Angular Methyl Migration in a Substituted Octalin 166th ACS National Meeting, Chicago, August 27—31, 1973. Abstracts, ORGN, Paper No. 128.

42. HUFFMAN, J. W.: Attempted Duplications of the Methyl Shift in Eremophilane Biosynthesis. J. Org. Chem. **37**, 2736 (1972).

43. JACQUES, J., M. HARISPE, D. MEA et A. HOREAU: Études stéréochimiques dans la série terpénique. II. Structure de l'acide obtenu par réarrangement de l'acide pinolique en milieu acid. Bull. Soc. chim. France **1963**, 472.

44. JENTSCH, J., and W. TREIBS: Constitution of Vetiver Oil Components Especially of the Tertiary Vetivenols. I. Isolation and Dehydrogenation of Tert-bicyclovetivenol and Tert-tricyclovetivenol. Parfüm und Kosmetik **49**, 29 (1968).

45. — — Constitution of Vetiver Oil Components, Especially Tertiary Vetivenols. II. Hydrogenation of Bi- and Tricyclovetivenol and Ozonization of Tricyclovetivenol. Parfüm und Kosmetik **49**, 143 (1968).

46. JOHNSON, A. P.: International Symposium on Synthetic Methods and Rearrangements in Alicyclic Chemistry, Oxford, July 22 – 24, 1969, Abstracts, p. 13.

47. JOHNSON, F.: Allylic Strain in Six-Membered Rings. Chem. Rev. **68**, 375 (1968).

47a. KITAGAWA, I., Y. YAMAZOE, R. TADEKA, and I. YOSIOKA: Conversion of Dihydroalantolactone to Eremophilane-Type Derivatives: A Biogenetic-Type Transformation. Tetrahedron Letters **1972**, 4843.

48. KAISER, R., and P. NAEGELI: Biogenetically Significant Components in Vetiver Oil. Tetrahedron Letters **1972**, 2009.

49. KROPP, P. J.: The Acid Catalyzed Cleavage of Cyclopropyl Ketones Related to Lumisantonin. J. Amer. Chem. Soc. **87**, 3914 (1965).

50. LANSBURY, P. T.: Chloro Olefin Annelation. Accounts Chem. Res. **5**, 311 (1972).

51. MC CLURE, R. J., K. S. SCHORNO, J. A. BERTRAND, and L. H. ZALKOW: The Structure and Stereochemistry of a New Sesquiterpene of the Acorane Type. Chem. Commun. **1968**, 1135.

52. MC CURRY, P. M. JR.: Private Communication to J. A. Marshall.

53. — Ph. D. Thesis, Columbia University (1970).

54. — Stereochemical Factors in the Intramolecular Keto-Carbene Addition to Substituted Cyclohexenes. Tetrahedron Letters **1971**, 1845.

54a. MCCURRY, P. M. Jr., R. K. SINGH, and S. LINK: Stereochemical Factors in a Cationic Cyclization Reaction. The Synthesis of 10-epi-β-Vetivone. Tetrahedron Letters **1973**, 1155.

54b. MCCURRY, P. M. Jr., and R. K. SINGH: Synthesis of Spirosesquiterpenes in the Vetivane Series Via an Aza Claisen Rearrangement. Tetrahedron Letters, in press.

55. MC EACHAN, C. E., A. T. MC PHAIL, and G. A. SIM: Sesquiterpenoids Part V. The Stereochemistry of Acorone: X-Ray Analysis of Acorone p-Bromophenylsulphonylhydrazone. J. Chem. Soc. (C) **1966**, 579.

56. — — — The Stereochemistry of Acorone. Chem. Commun. **1965**, 276.

57. MACLEOD, W. D. JR.: The Constitution of Nootkatone, Nootketone, and Valencene. Tetrahedron Letters **1965**, 4779.

58. MACSWEENEY, D. F., R. RAMAGE, and A. SATTER: Biogenetic Relationships of the Vetiver Sesquiterpenes. Tetrahedron Letters **1970**, 557.

59. Maheshwari, M. L., T. C. Jain, R. B. Bates, and S. C. Bhattacharyya: Terpenoids. XLI Structure and Absolute Configuration of α-Agarofuran, β-Agarofuran and Dihydroagarofuran. Tetrahedron 19, 1079 (1963).

60. Marshall, J. A., and N. H. Andersen: The Structure of α-Vetivone (Isonootkatone). Tetrahedron Letters 1967, 1611.

61. Marshall, J. A., N. H. Andersen, and P. C. Johnson: A Stereoselective Synthesis of Hydroazulenes. Grounds for Structure Revision of the Vetivane Sesquiterpenes. J. Amer. Chem. Soc. 89, 2748 (1967).

62. Marshall, J. A., and S. F. Brady: Stereochemical Relationships in Spirovetivane Sesquiterpenes: The Total Synthesis of (±)-Hinesol. Tetrahedron Letters 1969, 1387; J. Org. Chem. 35, 4068 (1970).

63. Marshall, J. A., and M. T. Pike: A Stereoselective Synthesis of α- and β-Agarofuran. J. Org. Chem. 33, 435 (1968).

64. Marshall, J. A., and P. C. Johnson: The Structure of β-Vetivone and Related Vetivane Sesquiterpenes. J. Amer. Chem. Soc. 89, 2750 (1967).

65. — — The Total Synthesis of (±)-β-Vetivone. Chem. Commun. 1968, 391.

66. Masamune, S.: Synthesis of 4a,6-Ethano-5,6,7,8-tetrahydro-2(4a)-naphthalenone. J. Amer. Chem. Soc. 83, 1009 (1961).

67. Mellor, J. M., and S. Munavalli: Synthesis of Sesquiterpenes. Quart. Rev. 18, 270 (1964).

68. Minato, H., R. Fujioka, and K. Takeda: Components of the Root of Acorus Calamus L. Chem. Pharm. Bull. 19, 638 (1971).

69. Motl, O., W. Z. Chow, and F. Šorm: Structure of the Sesquiterpenic Alcohol Hinesol. Chem. and Ind. 1961, 207.

70. Mongrain, M., J. Lafontaine, A. Belanger, and P. Deslongchamps: Stereoselective Synthesis of (±)-Epihinesol (Agarospirol). Canad. J. Chem. 48, 3273 (1970).

70a. Naegeli, P.: Private communication to N. H. Andersen.

71. Naegeli, P., and R. Kaiser: A New Synthetic Approach to the Acorane, Daucane, and Cedrane Skeleton. Tetrahedron Letters 1972, 2013

72. Nigam, I. C., H. Komae, G. A. Neville, C. Radecka, and S. K. Paknikar: Structural Relationships Between Tricyclic Sesquiterpenes in Oils of Vetiver. Tetrahedron Letters 1968, 2497.

73. Nigam, I. C., and L. Levi: Gas Liquid Partition Chromatography of Sesquiterpene Compounds. Canad. J. Chem. 40, 2083 (1962).

74. — — Preparation and Isolation of Isomeric Ketones by the Girard Reaction. Analyt. Chemistry 35, 1087 (1963).

75. Naves, Y. R.: Etudes sur les matières végétales volatiles (CXII). Sur l'absorption des vétivones α et β et de leurs derivés dan l'ultraviolet moyen. Bull. Soc. chim. France 1951, 369.

76. Naves, Y. R., et E. Perrottet: Études sur les matières végétales volatiles (XIII). Sur les α- et β-vetivones. Helv. Chim. Acta 24, 3 (1941).

77. Ogiso, A., M. Kurabayashi, H. Nagahori, and H. Mishima: Preparation of 6,10-Dimethylspiro[4.5]dec-6-en-2-one. Experiments directed towards the Synthesis of Spirovetivane Sesquiterpenes. Chem. Pharm. Bull. 18, 1283 (1970).

78. Ohta, Y., and Y. Hirose: Electrophile Induced Cyclization of Farnesol. Chemistry Letters 1972, 263.

79. — — New Sesquiterpenoids from Schisandra Chinensis. Tetrahedron Letters 1968, 2483.

80. Parker, W., J. S. Roberts, and R. Ramage: Sesquiterpene Biogenesis. Quart. Rev. 21, 331 (1967).

81. Penfold, A. R., and J. L. Simonsen: The Constitutions of Eremophilone, Hydroxyeremophilone, and Hydroxydihydroeremophilone. Part III. J. Chem. Soc. 1939, 87.

82. PFAU, A. ST., und PL. A. PLATTNER: Zur Kenntnis der flüchtigen Pflanzenstoffe (VIII). Synthese des Vetivazulens. Helv. Chim. Acta **22**, 202 (1939).

83. — — Études sur les matières végétales volatiles (X). Sur les vétivones, constituants odorants des essences de vetiver. Helv. Chim. Acta **22**, 640 (1939).

84. — — Études sur les matières vegetales volatiles (XI). Sur la constitution de la β-vétivone. Helv. Chim. Acta **23**, 768 (1940).

85. PIERS, E., and P. M. WORSTER: Stereochemistry of the Lithium Ammonia Reduction of cis-1,7-Dimethyltricyclo[4.4.0.0.^{2,6}]decan-3-one and Related Compounds. J. Amer. Chem. Soc. **94**, 2895 (1972).

86. ROMANUK, M., and V. HEROUT: On Terpenes. CXIV. On Stereoisomeric Vetivanes and Sesquiterpenic Hydrocarbons of Vetiver Oil. Collect Czech. Chem. Commun. **25**, 2540 (1960).

87. RUZICKA, L.: The Isoprene Rule and the Biogenesis of Terpenic Compounds. Experientia **9**, 357 (1953).

88. SAKUMA, R., and A. YOSHIKOSHI: Tricyclovetivene. Chem. Commun. **1968**, 41.

89. SMITH, S. E.: The Crystal and Molecular Structures of Azetidine-3-carboxylic Acid, the ortho-Chlorophenyl Urethan Derivative of 2-Methylene-4αH-(1βH)-bicyclo-[5.4.0]undecanol, and endo-1,2-dihydroxyzizaane. The Structure Elucidation of Postzizaene C, Ph. D. Thesis, University of Washington (1972).

90. ŠORM, F.: Medium-ring Terpenes. Fortschr. Chem. organ. Naturstoffe **19**, 2 (1961).

91. ŠORM, F., and V. HEROUT: On Terpenes (V). On the Constitution of the Essential Oil of Sweet Flag (*Acorus Calamus* L.). Collect. Czech. Chem. Commun. **13**, 177 (1948).

92. — — On Terpenes (XV). On the Constitution of Acorone and Isoacorone I. Collect. Czech. Chem. Commun. **14**, 723 (1949).

93. STORK, G.: Private communication to J. A. Marshall.

94. STORK, G., R. L. DANHEISER, and B. GANEM: Spiroannelation of Enol Ethers of Cyclic 1,3-Diketones. A Simple Stereospecific Synthesis of β-Vetivone. J. Amer. Chem. Soc. **95**, 3414 (1973).

95. SULSER, H., J. R. SCHERER, and K. L. STEVENS: The Structure of Paradisiol, a New Sesquiterpene Alcohol from Grapefruit Oil. J. Org. Chem. **36**, 2422 (1971).

96. SUZUKI, M., E. KUROSAWA, and T. IRIE: Spirolaurenone, A New Sesquiterpenoid containing Bromine from *Laurencia glandulifera* Kutzing. Tetrahedron Letters **1970**, 4995.

97. SÝKORA, V., V. HEROUT, J. PLÍVA, and F. ŠORM: Constitution of Acorone. Chem. and Ind. **1956**, 1231.

98. — — — — Über Terpene LXXXII. Die Konstitution von Acoron. Collect. Czech. Chem. Commun. **23**, 1072 (1958).

99. SÝKORA, V., V. HEROUT, A. REISER, and F. ŠORM: On Terpenes. XCVI. The Stereochemistry of Acorone and its Stereoisomers. Collect. Czech. Chem. Commun. **24**, 1306 (1959).

100. SYRDAL, D. D.: Sesquiterpenes of *Chamaecyparis nootkatensis*. I. Isolation and Structure Determination; II. Absolute Stereochemistry; III. Chemical Simulation of Biogenesis. Ph. D. Thesis, University of Washington (1971).

101. TOMITA, B., and Y. HIROSE: Terpenoids. XXVI. Acoradiene and Acorenol, Key Intermediates of Cedrane Group Sesquiterpenoids and their Transformation into (−)-α-Cedrene. Tetrahedron Letters **1970**, 143.

102. TOMITA, B., Y. HIROSE, and T. NAKATSUKA: Terpenoids. XXIII. Chemotaxonomy of Cuppressaceae. Sesquiterpenes in *Biota-(Thuja-)orientalis* wood. Mokuzai Gakkaishi **15**, 48 (1970).

103. TOMITA, B., T. ISONO, and Y. HIROSE: Terpenoids. XXVIII. Acorane Type Sesquiterpenoids from *Juniperus rigida* and Hypothesis for the Formation of New Tricarbocyclic Sesquiterpenoids. Tetrahedron Letters **1970**, 1371.

103a. TROST, B. M.: New Alkylation Reactions. Lecture to U.S.-Japan Seminar on Natural Product Synthesis. Tokyo. July 16—20, 1973.

104. VARMA, K. R., M. L. MAHESHWARI. and S. C. BHATTACHARYYA: Terpenoids-LXII. The Constitution of Agarospirol, a Sesquiterpenoid with a New Skeleton. Tetrahedron **21**, 115 (1965).

105. VENKATARAMANI, P. S., J. E. KAROGLAN, and W. REUSCH: Transformations of Cyclopropanol Intermediates. I. Synthesis of Angularly Substituted Perhydroindan Systems via a Stereospecific Cyclopropanol Rearrangement. J. Amer. Chem. Soc. **93**, 269 (1971).

106. VRKOČ, J., V. HEROUT, and F. ŠORM: On Terpenes. CXXII. Composition of Sesquiterpenic Ketonic Fraction of Sweet Flag Oil. Collect. Czech. Chem. Commun. **26**, 1021 (1961).

107. — — — On Terpenes. CXXXIII. Structure of Acorenone, a Sesquiterpenic Ketone from Sweet Flag Oil (*Acorus calamus L.*). Collect. Czech. Chem. Commun. **26**, 3183(1961).

108. — — — On Terpenes. CXLIII. Cryptoacorone, a New Stereoisomer of Acorone. Collect. Czech. Chem. Commun. **27**, 2709 (1962).

109. — — — On Terpenes. CXLIX. The Proof of Position of Cyclopentanone Carbonyl Group in Acorone and its Stereoisomers. Collect. Czech. Chem. Commun. **28**, 1084(1963).

110. VRKOČ, J., J. JONÁŠ, V. HEROUT, and F. ŠORM: On Terpenes. CLVII. Steric Structure of Acorone, Isoacorone and Cryptoacorone. Collect. Czech. Chem. Commun. **29**, 539 (1964).

111. WENNINGER, J. A., R. L. YATES, and M. DOLINSKY: High Resolution Infrared Spectra of Some Naturally Occurring Sesquiterpene Hydrocarbons. J. Assoc. Off. Agric. Chem. **50**, 1304 (1967).

111a. WHITLOCK, H. W. Jr., and L. E. OVERMAN: Solvolytic Rearrangements Accompanied by Multiple Alkyl Shifts. J. Amer. Chem. Soc. **93**, 2247 (1971).

111b. WILLIAMS, J. R., and G. M. SARKISIAN: [1,3]-vs, [1,2]-Sigmatropic Photorearrangements in Cyclic βγ-Unsaturated Ketones. Conversion of Bicyclo[5.4.0]undec-1(7)-en-3-one into 6-Methylenespiro[4.5]decan-1-one. Chem. Commun., 1564 (1971). Photochemical Syntheses of Spiro[4.5]decane-1,6-dione. J. Org. Chem. **37**, 4463 (1972).

111c. YAMADA, K.: Synthetic Studies on Spirovetivanes-Stereospecific Synthesis of (±)-β-Vetivone and (±)-Hinesol. Lecture to U.S.-Japan Seminar on Natural Product Synthesis, Tokyo, July 16—20, 1973.

111d. YAMASAKI, M.: Total Synthesis of the Sesquiterpene (−)-Daucene. Chem. Commun. **1972**, 606.

112. YOSIOKA, I., H. HIKINO, and Y. SASAKI: Studies on the Constituents of *Atractylodes*. IV. The Structure of Hinesol. (1). The Skeleton. Chem. Pharm. Bull. (Japan) **7**, 817 (1959).

113. — — — Abstracts Chem. Soc. Japan Symposium Nat. Prod., Kyoto (1960).

114. YOSIOKA, I., and T. KIMURA: Studies on the Constituents of Atractylodes. X. Correlation of Hinesol and β-Vetivone. Chem. Pharm. Bull. (Japan) **13**, 1430 (1961).

114a. — — The Structure and Absolute Configuration of Hinesol. Chem. Pharm. Bull. (Japan) **17**, 856 (1969).

115. YOSIOKA, I., Y. SASAKI, and H. HIKINO: Structure of Hinesol. Chem. Pharm. Bull. (Japan) **9**, 84 (1961).

116. YOSIOKA, I., S. TAKAHASHI, H. HIKINO, and Y. SASAKI: Studies on the Constituents of Atractylodes. III. Separation of Atraetylol into Eudesmol and Hinesol. Chem. Pharm. Bull. (Japan) **7**, 319 (1959).

117. ZALKOW, L. H., F. X. MARKEY, and C. DJERASSI: Terpenoids. XLVIII. The Absolute Configuration of Eremophilone and Related Sesquiterpenes. J. Amer. Chem. Soc. **82**, 6354 (1960).

(Received January 18, 1973)

Phorbolesters —
the Irritants and Cocarcinogens of Croton Tiglium L.*

By E. Hecker and R. Schmidt, Heidelberg

With 50 Figures

Contents

* Dedicated by E. H. to his distinguished teacher in natural product chemistry, Professor Dr. Dr. h. c. mult. Adolf Butenandt, Honorary President of the Max-Planck-Gesellschaft, on occasion of his 70th birthday in gratitude and admiration.

1. Introduction

Croton tiglium L. (family: Euphorbiaceae) is a leafy shrub native to Southeast Asia and cultivated in Ceylon, Malabar, Amboina, the Philippines, Mauritius and in China. Its fruits are three-celled capsules, each cell containing a single seed. They are toxic: 8−10 seeds may kill a dog and 15 seeds a horse; for human beings 4 seeds may be lethal (*123*). The seeds contain up to 20% of protein and 30−50% of lipids (*38, 50*). Isoguanine-D-riboside (crotonoside) and saccharose were isolated from the seeds (*35, 36, 37, 50*).

From the proteins of croton seeds a phytotoxin called "crotin" was isolated. For rabbits the lethal dose 50 (LD_{50}) is 0.05—0.08 mg/kg. In the blood of a number of species crotin causes hemolysis or agglutination, in others, including man, the erythrocytes are merely deformed (*113*). Crotin induces antibody formation protecting against the lethal dose (*24*). Also, the proteins contain various enzymes including trypsin and lipase (*151, 184*).

The lipids of the seeds may be obtained either by extraction or by expression and are known as croton oil. It is toxic to bacteria, insects, amphibia, fish and other vertebrates, including mammals. To skin the oil is an irritant and vesicant, in the eyes it causes severe conjunctivitis. Taken internally it is a drastic cathartic (*24, 29, 111, 160*).

On human skin croton oil causes severe inflammation with formation of erythema, oedema und pustules. The latter heal without visible lesions (*50, 137*). Subcutaneously injected the oil causes serosangineous inbibition and phlegmonous inflammation with bacterial free suppuration (*137*). In the oral cavity and throat the oil induces a long lasting burning sensation and stimulates salivary flush. Taken internally it causes tinnitus, vomiting, abdominal pain, sanguineous diarrhoea, gastroenteritis with tumefaction, irritation and degeneration of mucous membranes of the gut, cyanosis, delirium and attacks of suffocation (*123*).

Croton oil acts so powerfully that as such it is deemed unsafe for use in human medicine (*170*). Sporadically it served for denaturing of alcohol (*50*). Diluted with a suitable vehicle, it was used as a counterirritant (Malefizöl) and in combination with acupuncture (*50*). In homoeopathy it is also taken as a counterirritant and in certain cases of diarrhoea (*151*). The oil is also used in veterinary medicine. Croton seeds and/or croton oil are or were on record in the pharmacopoeae of a number of countries.

The irritant properties of croton oil led BERENBLUM (*15, 16*) to detect the augmentational or "cocarcinogenic" effect (for definition see *157*) of croton oil in tumorigenesis of mouse skin induced by carcinogenic aromatic hydrocarbons. The potency of the active principles of croton oil is of a much higher order than that of all other cocarcinogens known (*145*).

After an important modification of BERENBLUM's original experiment by MOTTRAM (*130*), BERENBLUM and SHUBIK (*21, 22*) devised an experiment of treating the back skin of mice with one single subthreshold dose of a carcinogenic aromatic hydrocarbon followed by repeated applications of croton oil ("BERENBLUM experiment", *77*). From the results of this and similar other experiments, the "two stage hypothesis" of skin carcinogenesis was derived (*17*). This hypothesis became an important although controversial approach to the analysis of the biological mechanism of chemical carcinogenesis. The interpretation of the first or "initiation" stage of BERENBLUM-experiments as the result of an essentially irreversible biological event (*23*) was readily accepted. However, the interpretation of the second or "promotion" stage remained controversial especially after a weak oncogenic activity of croton oil itself was found to exist (for reviews, see *17, 18, 19, 30, 47, 56, 57, 131, 144, 145, 146, 177*).

Numerous efforts have been made to isolate and chemically cha-
racterize the biologically active principles of croton oil.

Towards the end of the nineteenth century several investigators attributed the
vesicant properties of croton oil to an acid called "crotonol" or "crotonol acid"*.
It was believed to be related to oleic acid, although it was not obtained as a well
defined compound. Subsequently from the ethanol-soluble fraction of croton oil,
DUNSTAN and BOOLE (48) obtained a neutral and resinous mass which they called "croton
resin". It was strongly vesicant and was claimed to be more toxic to fish than rotenone
(160). DUNSTAN and BOOLE speculated that some sort of a lactone was the active principle
of the resin. BOEHM (25, 26) reported a modified procedure for preparation of a croton
resin, and investigated its acute toxic effects in frogs and rabbits. From one of these
preparations, BOEHM (27) obtained a crystalline, labile compound which he called
"phorbol" because of its origin from a species of the plant family Euphorbiaceae and its
alcohol nature. Later on, FLASCHENTRÄGER and coworkers investigated the fatty acids
of croton oil and developed a procedure for the preparation of phorbol, aiming at elucidation
of the structure of phorbol (28, 50, 51, 52, 53). In the same time period CHERBULIEZ et al. (37,
38), SPIES (161) and DRAKE and SPIES (46) were concerned with chemical investigations of
croton resin. Investigations on the chemistry of phorbol were again taken up by THOMAS
and MARXER (168) and KAUFFMANN et al. (114, 115, 116). However, the problem of the
chemical structure of phorbol remained unsettled.

In all experiments up to 1961, successful resolution of the active
principles of croton oil was not accomplished. Moreover, experimentation
with chemically poorly defined fractions obtained from the oil added
to the confusion existing already with regard to the interrelationships
of its biological activities.

CHERBULIEZ et. al. (37, 38) attempted to separate the cathartic from the vesicant
activity. BOEHM et al. (29) obtained by acetylation of phorbol what they called
"Acetylphorbol" and found it to be toxic to frogs and rabbits, whereas phorbol as
such was revealed to be non-toxic. Using the techniques described by CHERBULIEZ
et al. (37, 38), BERENBLUM (15, 16) separated croton resin from the remainder of the oil
and found quantitative differences in the cocarcinogenic activities of these two portions.
Using paper chromatography, DANNEEL and WEISSENFELS (44) claimed to have separated
the irritant from the carcinogenic activity of the oil. By column chromatography,
GWYNN (64) separated a number of fractions all of which produced gross epidermal
hyperplasia in mice, but only two of them appeared to have cocarcinogenic activity.
SICÉ (158) denied any parallelism between the cocarcinogenic activity of the oil and epidermal inflammation or hyperplasia and claimed to have found that the cocarcinogenic
activity of croton oil does not result from its vesicant activity; LIJINSKY (124, 125)
concluded from assays of different fractions obtained by high vacuum distillation techniques that croton oil may contain several cocarcinogens.

Thus, for a meaningful evaluation of the biological activities
and, in particular, for conclusive biological investigations of its

---

* For a review considering investigations of croton oil before 1895, see MEYER-
BERTENRATH (129).

cocarcinogenic activity, the isolation and chemical characterization of the active principles of croton oil remained of special importance. To achieve this goal, a systematic fractionation employing mild and efficient methods of separation was required, combined with careful follow-up of the physical and biological properties of all fractions produced (*71, 72*).

2. Fractionation of Croton Oil

In the fractionation of Croton oil, multiplicative liquid-liquid distribution procedures proved to be the separation tools of choice, complemented — at a certain stage — with a mild column chromatographic procedure (*71, 72*).

The sequence of fractionation steps is summarized in Charts 1—4 (p. 385, 387, 399, 400 respectively) and in Chart 6 (p. 445). Vertical arrows indicate the direction of increasing optical and/or biological activity; horizontal arrows the elimination of less active side fractions. Percentages given at individual fractionation steps refer to the weight of the starting material; optical activities are given as specific rotation $[\alpha]_D$ (in 1% dioxan solution at 24° to 28° C). Relative migration rates (R_f) refer to thin-layer chromatography on silica gel.

2.1. General Analytical Chemical Procedures and Biological Assays

2.1.1. Methods of Separation and Criteria of Purity

In multiplicative liquid-liquid distribution (*67, 70, 73, 76*) two principal types of flow scheme exist: depending on the amount of material to be fractionated and on the degree of separation sought, procedures with either steady state or non-steady state distribution may be used.

As a *preparative* separation procedure, the O'Keeffe distribution — with a steady state distribution scheme — employs repetitive feeding in the center of the distribution battery (*67, 70, 73, 76*). By this method of distribution a mixture of substances is resolved into three portions, according to the partition numbers (G) of its components: one portion is removed from the battery in the upper phase, the other in the lower phase; eventually a third portion remains on the battery.

O'Keeffe distributions may be performed according to either one of two different modes, called procedure 1 or 2 (for details see *73, 76*): If z is the number of elements used to perform procedure 1, a certain separation efficiency is obtained. In procedure 2, two times z elements are required to obtain that same separation efficiency. However, procedure 2 is advantageous with respect to the amount of material which can be separated, since the mixture can be fed in twice during each distribution cycle N.

In the fractionation of croton oil, O'Keeffe distribution procedure 1, with one feeding per cycle, was used together with *hand-operated* batteries* (*69*). Thus, the O'Keeffe distributions mentioned in Charts 1, 3 and 6 are performed in this manner providing, for example, separation of one liter of croton oil into its hydrophobic and hydrophilic portions within 6 to 8 hours (*71, 76*). Procedure 2, with two feedings per cycle, is preferentially used with automatic batteries (*1***, *140****, *182*).

In *analytical* separations (*70, 73*) using the different procedures of the Craig distribution — with non steady state distribution schemes — one single feed of the mixture at the beginning of the battery is employed (*42, 67*).

Depending on the size of battery available and the degree of separation required, the fundamental procedure may be extended by either recycling or by the single or double withdrawal procedure (*42, 67*). In the resulting distribution diagrams the components of the mixture appear as more or less separate bands, provided that the components exhibit large enough separation factors β in the solvent system chosen. According to standard procedures the shape of such bands may be calculated from the partition number G corresponding to r_{max} (i.e. the number r of the fraction containing the maximum of the band) and the total number n of transfers. The "fit" of the calculated band with the band obtained experimentally may be used to judge the purity of the material in the individual fractions r (*42, 67*).

For the Craig distributions employed in the fractionation of croton oil automatic batteries were used. For example, to further resolve bands A and B from the silica gel chromatography (see Fig. 1) a battery with $z = 200$ elements, each holding 25 ml of stationary phase (*181*) along with batteries of $z = 1020$ elements (*41*) one holding 3 ml and another holding 10 ml of stationary phase*** were used.

Table 1. *Specification of Craig-Distributions Used for Further Resolution of the Chromatographic Bands A and B (40, 91, 98, 104)*

Craig-distribution of fractions	Battery type	Volume ratio of phases (ml)	No. of transfers n	No. of complete shaking movements	Settling period min.	No. of tubes receiving sample	Example for distribution
Band A	Craig	12/10	1600	20	3.0	13	Fig. 3
Band B	v. Metzsch	25/25	450	20	3.0	1	Fig. 4
Factor group B_y	Craig	5/3	2500	20	1.5	6	Fig. 5
Factor group B_x	Craig	3/3	3020	20	1.5	3	Fig. 6

* Manufacturer: E. Bühler, Scientific Instruments, Reutlingerstraße 6, D-7400 Tübingen, Germany.

** Manufacturer: Quickfit Laborglas G.m.b.H., Hüttenstraße 8, D-6200 Wiesbaden-Schierstein, Germany.

*** Manufacturer: Spectrum Medical Industries Inc., 430 Middle Village Station, NY 11379, U.S.A. Formerly: H. O. Post, same address.

References, pp. 458—467

The material to be separated is introduced into as few elements as possible (Table 1). Usually at the beginning of distributions, longer settling periods than those recorded in Table 1 are necessary to achieve complete separation of phases. In the Craig distributions used for separation of factor groups B_x and B_y, it was convenient to finish the distributions by the "double phase withdrawal procedure" to remove most of the high-boiling di-n-butyl ether from the more important fractions.

Distribution curves are generally obtained by the weighed-sample method (*42, 67*). The material from a number of neighboring fractions was combined and the average fraction weights, thus obtained, plotted against the number of corresponding intermediate fractions (see for example Fig. 3).

As a criterion of purity of the material in the individual fractions the "fit" of calculated distribution curves cannot be used, since it was found, that the partition isotherms of the croton oil factors are non-linear in the systems and concentration ranges used (*40, 91, 98*). Also thin-layer chromatography was not always effective as a criterion of purity, since it does not separate the individual factors within each croton oil factor group (see below). Therefore, as a criterion of purity of individual croton oil factors contained in the fractions, gas-liquid chromatography of the methyl esters of their corresponding fatty acids was used. They are obtained by rigorous hydrolysis of the material from the fractions and subsequent methylation (*98*).

For the column chromatography stage silica gel with 13% water is used.

2.1.2. Monitoring of Fractionation Steps

To follow up the purification of the active principles during fractionation, measurements of the optical rotation and — with the reservation mentioned above — thin layer chromatography* proved useful, combined with standardized biological assays (*71, 72, 74, 75*) for acute toxicity in frogs, for irritant activity on the ear of mice and for cocarcinogenic activity on the back skin of mice (see also *87*).

(*a*) *Estimation of the acute toxicity in frogs.* For estimation of the acute toxicity, a weighed sample of the material to be assayed is dissolved in polyethylene glycol 400 and injected into the lymph sac of *Rana esculenta* L. The results of the assay are read 6—12 hours after injection and expressed as lethal dose (LD_{50}) in weight units per 50 gm frog (see Charts 1 and 2).

(*b*) *Estimation of the irritant activity on the mouse ear.* To assay irritant activity, the inner surface of the outer ear of mice is treated with an acetone solution of the sample to be investigated. The degree of ear redness as developed 24 hours after application is evaluated. A dose causing an ear redness of the degree + + is defined as "irritant

---

* Detection of spots by UV light of 254 nm and/or spraying with vanillin-sulfuric acid and heating at 110° C (see for example *149*).

unit" (IU) and expressed in micrograms per ear (see for example Table 2 and Chart 2). The IU is usually reproducible within a factor of 2 to 3 (96). It is a semiquantitative measure of irritant activity. — Prior to the standard procedure an irritant unit IU[+] neglecting possible sex differences was estimated in a preliminary standard procedure (75; see IU[+] in Charts 1 and 2).

Table 2. *Standard Procedure for Determination of the Irritant Unit (IU) (96)*

Croton oil GP 6 1958 dose/ear [µg]	Males	Females
0.6	−	+
1.2	+	+
2.4	+ +	+ +
4.8	+ + +	+ +
9.6	+ + +	+ + +

In a later stage of the work on croton oil factors a more precise measure of irritancy on the mouse ear — the "irritant dose 50" (ID_{50}) — was introduced (96). The ID_{50} may be fully evaluated by the usual statistical methods and is therefore more suited than the IU to quantitative comparison of irritant potencies (see Tables 4, 7 and 18). For any particular compound, the ID_{50} is normally lower than the IU. However, these two measures of irritant activity are not directly comparable.

(c) Estimation of the cocarcinogenic activity in Berenblum experiments on the back skin of mice. In the standard procedure with 14/14 NMRI mice, i = 25.6 µg, i.e., 0.1 µM of the carcinogen 7,12-dimethylbenz[a]anthracene (DMBA) dissolved in 0.1 ml of acetone are used as subthreshold (77) initiating dose. After one week a dose p of the material to be assayed for cocarcinogenic activity, dissolved in 0.1 ml of acetone, is administered twice weekly for at least 12 weeks. All tumors 1 mm in diameter or more are recorded. At the end of the twelfth week of treatment, i.e. after 24 applications, the cocarcinogenic activity is expressed as tumor rate (i.e. number of surviving mice carrying at least one skin tumor/number of survivors) in per cent and as average tumor yield (i.e. total number of skin tumors on survivors/number of survivors) (see Tables 3, 4, 7 and 18). — In the initial stages of the fractionation procedure 6/6 SIM mice and 300 µg of DMBA were used in a standardized manner (75) similar to that described above (see Table 3).

In the fractionation of croton oil the acute toxicity in frogs was used primarily to relate toxic data given by earlier investigators (e. g. 29) to irritant and cocarcinogenic activities. The estimation of the irritant activity was considered a rapid, albeit preliminary, measure for possible cocarcinogenic activity of the material, until the latter could be established separately in the more time-consuming assay for cocarcinogenic activity.

2.2. Preparation of the Hydrophilic and the Hydrophobic Portions

The fractionation procedure is started (Chart 1) with a sample of croton oil (GP 6, 1958)* exhibiting a LD_{50} of 5 mg/50 gm frog, an IU^+ of 2.4 µg/ear, and a specific optical rotation of $[\alpha]_D = +7°$ (*71. 72. 74, 93*). In the test for cocarcinogenic activity with a single dose p = 500 µg, the tumor rate is 100% and the average tumor yield 9.6 tumors/mouse (see Table 3).

In the first stage of the fractionation procedure croton oil is separated into hydrophobic and hydrophilic portions by O'Keeffe distribution employing a petroleum ether, methanol, water solvent system (Chart 1). Although the hydrophilic portion represents only 5% of the original oil, its IU^+ amounts to only one fifteenth of that of the original oil, indicating an approximately fifteen-fold increase in irritant activity. Also the optical (Chart 1) and the cocarcinogenic activities (Table 3) of the hydrophilic portion are considerably increased. The hydrophobic portion, representing 95% of the material, exhibits considerably reduced biological but nearly equal optical activity as compared with the original oil. Consequently, by the O'Keeffe distribution, practically the entire biological activity of croton oil is transferred to the hydrophilic portion.

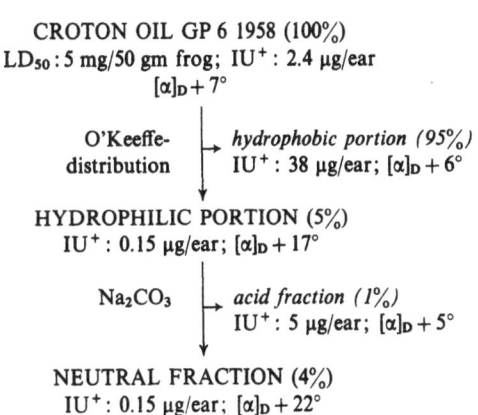

CROTON OIL GP 6 1958 (100%)
LD_{50} : 5 mg/50 gm frog; IU^+ : 2.4 µg/ear
$[\alpha]_D + 7°$

O'Keeffe-distribution → *hydrophobic portion (95%)*
IU^+ : 38 µg/ear; $[\alpha]_D + 6°$

HYDROPHILIC PORTION (5%)
IU^+ : 0.15 µg/ear; $[\alpha]_D + 17°$

Na_2CO_3 → *acid fraction (1%)*
IU^+ : 5 µg/ear; $[\alpha]_D + 5°$

NEUTRAL FRACTION (4%)
IU^+ : 0.15 µg/ear; $[\alpha]_D + 22°$

Irritation units (IU^+) estimated in the preliminary standard procedure (section 2.1.2, *75*).

Chart 1. Resolution of croton oil into hydrophobic and hydrophilic portions and initial stage of fractionation of the hydrophilic portion (*93, 97*)

---

* GP indicates that the oil satisfies the requirements specified in the German Pharmacopoeia, 6th ed.; the year indicates when this oil was purchased.

2.3. Phorbol Diesters from the Hydrophilic Portion

2.3.1. Isolation and Resolution of Croton Oil Factor Groups A and B

Treatment of the hydrophilic portion with sodium carbonate yields a small fraction of acids of relatively low biological and optical activities. The residual neutral fraction retains 4% by weight of the original oil and practically all of its biological activity (Chart 1, Table 3).

Table 3. *Cocarcinogenic Activities at Various Stages of the Fractionation of Croton Oil and of the Hydrophilic Portion of Croton Oil (97)*

Numbers in brackets refer to data obtained for the less active side fractions (Hydrophobic portion, acid fraction, Chart 1).

Fraction	single dose p [µg/appl.]	Cocarcinogenic activity[a] tumor rate[b] [%]	average tumor yield[b] [tumors/survivor]
Croton oil GP 6 1958	500	100	9.6
Hydrophilic portion	50 (500)	83 (18)	4.0 (0.5)
Neutral fraction	50 (500)	100 (60)	9.6 (0.9)
Factors group A[c]	5	100	10.2
Factors group B[c]	5	100	11.2

[a] Estimated in the standard procedure on SIM-mice (75); Initiator: i = 300 µg DMBA.
[b] After 12 weeks = 24 applications.
[c] Head and tail fraction of chromatography not assayed.

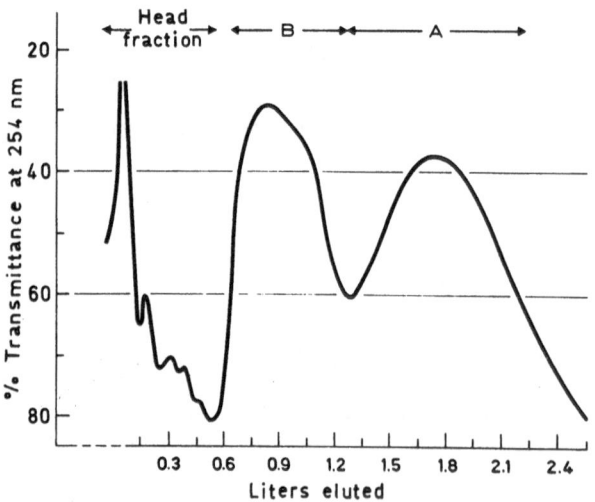

Fig. 1. Chromatography of the neutral fraction from croton oil on deactivated silica gel. Eluent: carbon tetrachloride-ether; elution curve recorded at 254 nm (from *104*)

In the next stage of the fractionation procedure, the neutral fraction is subjected to a column chromatography on deactivated silica gel using carbon tetrachloride-ether mixtures as eluents (Chart 2; Fig. 1). After a head fraction, containing relatively little irritant, but high optical activity (Chart 2), a broad band B is eluted from

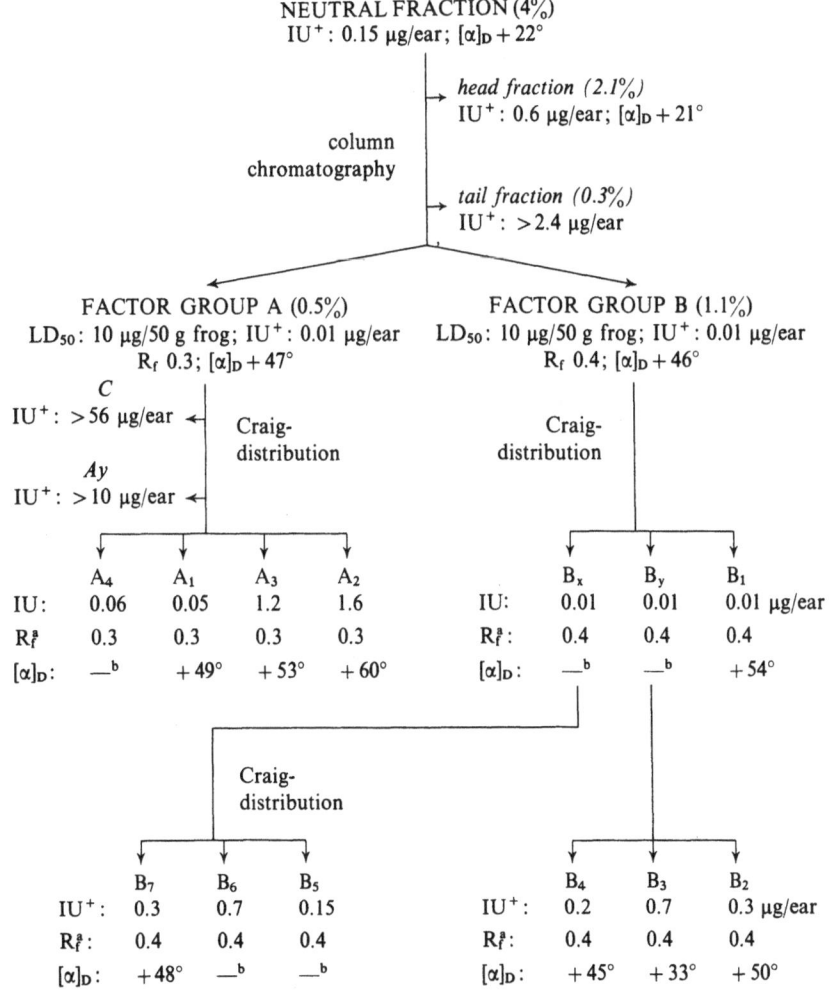

ᵃ System: methylene chloride/acetone = 3/1, silica gel Merck 254 HF.
ᵇ Not determined because of scarcity of compound.

Irritation units (IU⁺) estimated in the preliminary standard procedure (section 2.1.2, *75*), IU in the standard procedure (*96*).

Chart 2. Fractionation of the hydrophilic portion of croton oil, continued
(*40, 91, 93, 97, 98, 104*)

the column (Fig. 1). This is well separated from the subsequent broad band A. Finally, with acetone as eluent, a small and biologically relatively inactive tail fraction is obtained (Chart 2).

In thin-layer chromatography, each of the bands A and B are shown to exhibit single spots with R_f values of 0.3 and 0.4 respectively (see Chart 2). The spots can be seen in UV light (254 nm) and both develop a dark rusty brown color after spraying with vanillin-sulfuric acid followed by heating at 110° C. The material of both bands exhibits practically identical LD_{50}, IU^+, and optical (Chart 2) as well as cocarcinogenic activities (Table 3). Within the error of the assays used, the increase in the biological activities from the starting material to that of the bands A and B corresponds to the decrease in the weight of the material: the material contained in both bands together represents practically the entire toxic, irritant and cocarcinogenic activities of croton oil. In particular, this result indicates that the toxic principle is identical with the irritant and cocarcinogenic principles (*72, 74, 93, 94, 99*).

After the first successful trials, the fractionation procedure of the hydrophilic portion from croton oil was gradually improved by reduction of the number of stages employed between O'Keeffe distribution and column chromatography. In its most advanced version the hydrophilic portion is extracted with potassium bicarbonate instead of sodium carbonate, 40 g samples of the resulting neutral fraction being chromatographed on a column of 3.5 kg of deactivated silica gel (*104*). Similar modifications were reported also by Meyer-Bertenrath (*128*). Also, separations of 4 to 5 g portions of the alcoholic extract of croton oil by dry column chromatography have been described (*132*).

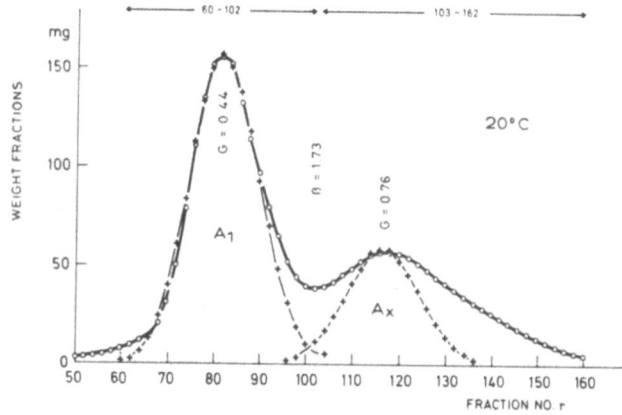

Fig. 2. Craig distribution of croton oil factor group A. Single withdrawal procedure: z = 200; n = 270; V = 25/25; system: carbon tetrachloride (2), methanol (1), water (0.15). (—o—) experimental; (--x--) calculated (from *91*)

Although in thin-layer chromatography the material contained in the chromatographic bands A and B appears to be uniform, further resolution of both bands is achieved by Craig distribution procedures (Chart 2).

In a preliminary Craig distribution (94), the material from the chromatographic band A is resolved to give two bands, A_1 and A_x (Fig. 2). Using the same solvent system as in this distribution (Fig. 2), resolution of band A could be improved employing $n = 1600$ transfers (Chart 2; Fig. 3).

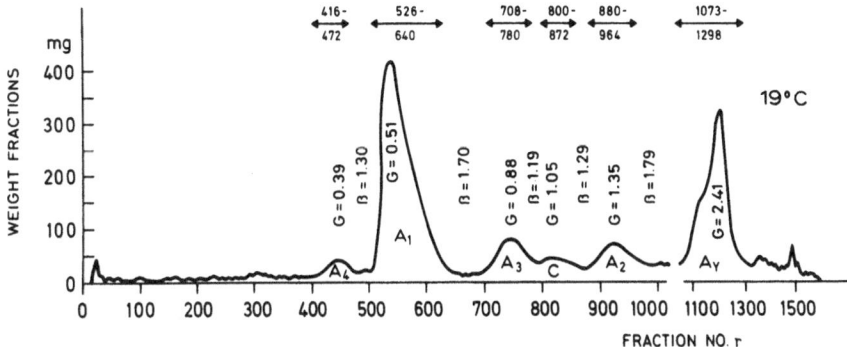

Fig. 3. Craig distribution of croton oil factor group A. Single withdrawal procedure: $z = 1020$; $n = 1600$; $V = 12/10$; system: carbon tetrachloride (2), methanol (1), water (0.15). Ordinate represents average weight of four neighboring fractions r to r-3, plotted against r-2; for fractions $r = 1021$ to 1600 average weight of nine neighboring fractions r to r-8, plotted against r-4 (from *104*)

In thin-layer chromatography, the materials contained in the fractions r of the bands with $G = 0.39$, 0.51, 0.88, 1.35 and 2.41 as they are collected from the battery, contain impurities of higher R_f values. These are formed from the main compounds by autoxidation during the distribution. Furthermore, the fractions of the bands $G = 0.88$ and 1.35 are contaminated with small amounts of the compound in band $C(G = 1.05)$. On thin-layer plates, contrary to the other bands the material of band C is not visible under UV light. In addition, band C develops pink colouration on spraying with vanillin-sulphuric acid, as opposed to the brown colour developed by the other bands. By thick-layer chromatography the materials of all bands except $G = 2.41$ are obtained in a pure state, as estimated by gas-liquid chromatography (see section 2.1.1).

The chromatographic band A thus consists of at least six different compounds $A_1 - A_4$, A_y and C, A_1 representing the main compound. A_y and C are not irritant in the doses assayed (Chart 2). Since A_y appeared to represent a mixture of inactive compounds, it was not further fractionated (see, however, section 2.4.2). Compound C was crystallized and identified as l-monopalmitine. The irritant factors $A_1 - A_4$ exhibit identical R_f values and similar optical activities (Chart 2). Also they show remarkable low ID_{50} values and different cocarcinogenic activities (Chart 2, Table 4). These substances will be collectively called croton oil factor group A.

Table 4. *Irritant and Cocarcinogenic Activities of the Croton Oil Factors Isolated from the Hydrophilic Portion (40, 91, 98, 104)*

Factor		Irritation[a] ID_{50} [mμM/ear]	Cocarcinogenic activity[b] tumor rate[c] [%]	average tumor yield[c] [tumors/survivor]
Group A	A_1	0.016[d]	82	3.6
	A_2	0.030	8	0.3
	A_3	0.020	29	0.6
	A_4	0.003	64	2.6
Group B	B_1	—	71	3.1
	B_2	—	86	6.2
	B_3	—	61	3.0
	B_4	0.024[e]	64	2.3
	B_5	—	32	2.0
	B_6	—	29	1.0
	B_7	0.089[e]	57	2.4

[a] Standard deviation σ: 1.3; significance level $\alpha = 0.05$.

[b] Estimated in the standard procedure on NMRI-mice (*91*); Initiator $i = 0.1$ μM DMBA; cocarcinogens group A: $p = 0.02$ μM/application; cocarcinogens group B: $p = 10$ μg/application.

[c] After 12 weeks = 24 applications.

[d] *96*.

[e] HECKER and SCHAIRER, unpublished.

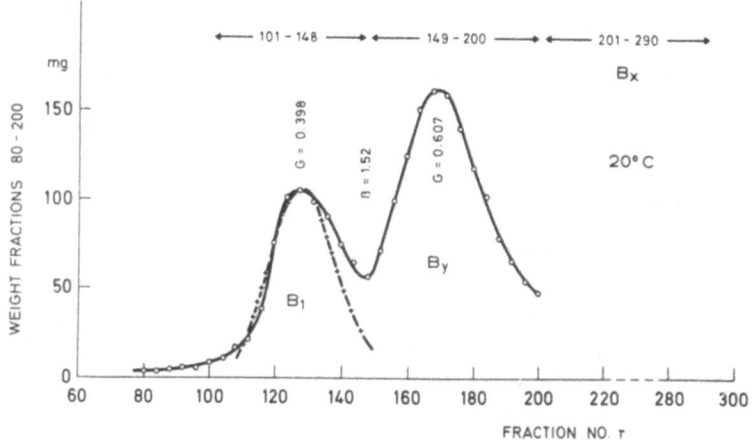

Fig. 4. Craig distribution of croton oil factor group B. Single withdrawal procedure: $z = 200$; $n = 450$; $V = 25/25$; system: carbon tetrachloride (2), methanol (1), water (0.15). (—o—) experimental; (--x--) calculated. Withdrawn fractions $r = 201 - 290$ pooled to give fraction B_x (from *98*)

In a similar manner, by Craig distribution (Chart 2; Fig. 4), the material from the chromatographic band B is separated into two bands $G = 0.398$ and $G = 0.607$ (*99*). Band $G = 0.398$ contains an essentially pure compound B_1. The material from band $G = 0.607$ (B_y) and that from the fractions $r > 200$ (B_x) is further fractionated, recombining similar fractions where appropriate (*39, 40*). However, only the more important stages of the entire procedure will be discussed.

B_y is fractionated further by Craig distribution (Chart 2) using a non-aqueous solvent system and $n = 2500$ transfers. From the fractions $r = 961 - 1080$ (Fig. 5) partially withdrawn from the battery, a further compound B_2 is obtained. Bands $G = 0.56$ and $G = 0.47$ remaining on the battery, contain additional compounds B_3 and B_4. Similarly B_x is resolved in a Craig distribution (Chart 2) using a different nonaqueous solvent system and $n = 3020$ transfers yielding the compounds B_5, B_6 and B_7 (Fig. 6). Again, as observed in the A group, the materials in the bands of all distributions contained autoxidation products. They were removed by thick-layer chromatography, to obtain the gas-liquid chromatographically pure compounds $B_1 - B_7$.

Compounds $B_1 - B_7$ all exhibit practically the same R_f values in thin-layer chromatography and, so far as determined, similar optical rotations (see Chart 2). B_1 and B_2 represent the main constituents. All compounds of the B-group exhibit irritant (Chart 2, Table 4) and cocarcinogenic activities (Table 4) of various degrees. They will be referred to as croton oil factor group B.

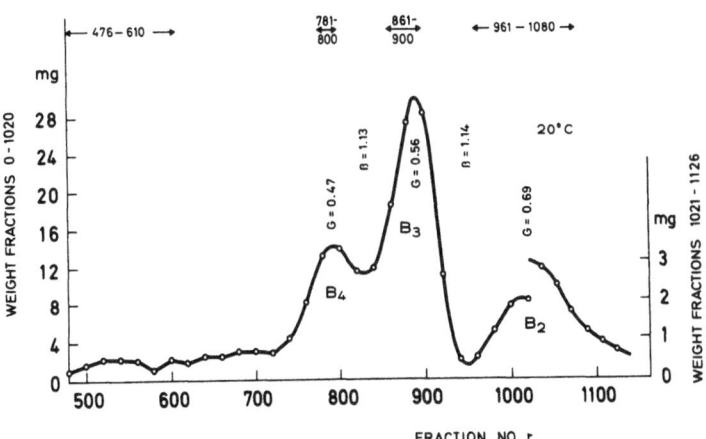

Fig. 5. Craig distribution of croton oil factor group B_y. Single withdrawal procedure: $z = 1020$; $n = 2500$; $V = 5/3$; system: cyclohexane (15), di-n-butyl ether (5), nitromethane (16). Ordinate represents average weight of two neighboring fractions r and r-1, plotted against r-1; for fractions $1021 - 1126$ average weight of three neighboring fractions r to r-2, plotted against r-2 (from *40*)

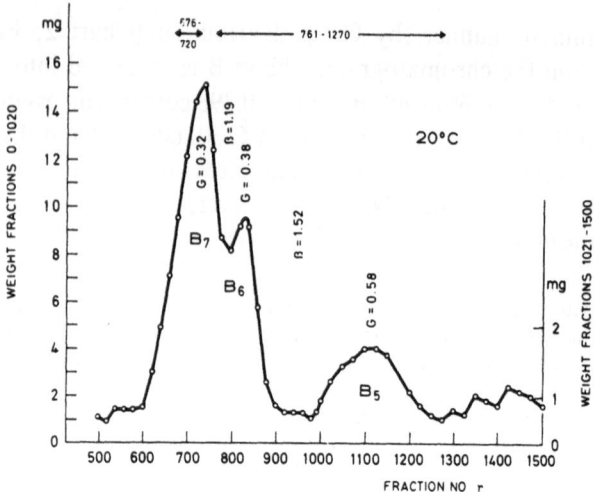

Fig. 6. Craig distribution of croton oil factor group B_x. Single withdrawal procedure for $n = 2000$; double withdrawal procedure for further $n = 1020$ transfers; $z = 1020$; $V = 3/3$; system: cyclohexane (13), di-n-butyl ether (7), nitromethane (20). Ordinate represents average weight of three neighbouring fractions r to r-2, plotted against r-1; for fractions $r = 1021$—1500 average weight of five neighbouring fractions r to r-4, plotted against r-2 (from 40)

It is interesting to see that the eleven pure croton oil factors of both groups exhibit similar irritant and cocarcinogenic activities. Furthermore it may be noticed that for all factors, within the limits of accuracy of the biological assays used, the increase in irritant — and also in toxic — potency is paralleled by a similar increase in cocarcinogenic potency (see Charts 1, 2 and Tables 3, 4). This indicates that the individual components of the croton oil factor groups A and B taken together represent practically all irritant and cocarcinogenic compounds of the oil.

In alternative fractionation trials employing combinations of column chromatography, Craig distribution and thick-layer chromatography, the biologically active "amorphous materials A and C" (171, 172, 173, 174, 175, 176) and a compound C-3 (4, 5) have been isolated from croton oil. Isolation of pure phorbolesters has also been claimed by MEYER-BERTHENRATH (127, 128). However, during all of these trials, quantitative measurement of biological activities was not carried out.

2.3.2. Chemical Characterization of the Croton Oil Factors from Groups A and B as Diesters of Phorbol

Ten out of the eleven pure croton oil factors $A_1 - A_4$ and $B_1 - B_7$ isolated from the hydrophilic portion were obtained as resinous, amorphous materials. Upon careful purification only croton oil factor A_1 crystallized.

All croton oil factors are sensitive to acid, alkali and oxygen. They are practically insoluble in water, but easily soluble in most organic solvents. From solutions in volatile organic solvents such as diethyl ether or methylene chloride, they may be obtained as brittle foams which can be broken to give light amorphous powders softening below 100° C. These powders and also the crystals of A_1 retain all kinds or organic solvents very strongly. Thus, the croton oil factors as such are not suitable for combustion analysis.

However, reaction of the factors with the acid chloride of 4-(4-nitrophenylazo-)benzoic acid (NPAB) (*68*) yields crystalline NPAB monoesters (NPABE) of which correct elemental analyses (Table 5) may be obtained (*93, 94, 99*). Also, by oxidation with manganese dioxide (*93, 94, 99*) from the croton oil factors amorphous aldehydes may be prepared, the 2,4-dinitrophenylhydrazones of which again yield correct elemental analyses (Table 5). Because of the high molecular weights of these derivatives and also because of their tendency to form solvates with all kinds of solvents, extrapolations from their equivalent molecular formulae as obtained by combustion analyses to that of the factors themselves must be considered somewhat uncertain. For some phorbol-derivatives the content of solvated molecules has been measured by using marker solvents for crystallization (*43, 116*).

Table 5. *Purities and Chemical Data Characterizing the Croton Oil Factors Isolated from the Hydrophilic Portion (40, 91, 98, 104)*

Croton oil factor	Purity by g.l.c.[a] [%]	Rf[b]	Crystalline derivative	m.p. °C	Molecular formula[d]
A_1	99	0.3	NPABE	86– 87	$C_{49}H_{63}N_3O_{11}$
			2,4-DNP	106–108	$C_{42}H_{58}N_4O_{11}$
A_2	97	0.3	–[c]	–	–
A_3	92	0.3	–[c]	–	–
A_4	92	0.3	–[c]	–	–
B_1	97	0.4	2,4-DNP	109–110	$C_{43}H_{60}N_4O_{11}$
B_2	97	0.4	2,4-DNP	120–121	$C_{41}H_{56}N_4O_{11}$
B_3	98	0.4	NPABE	90– 91	$C_{48}H_{59}N_3O_{11}$
B_4	96	0.4	NPABE	150–152	$C_{47}H_{59}N_3O_{11}$
B_5	94	0.4	–[c]	–	–
B_6	95	0.4	NPABE	94– 98	$C_{46}H_{55}N_3O_{11}$
B_7	98	0.4	NPABE	148–150	$C_{45}H_{55}N_3O_{11}$

[a] See section 2.1.1
[b] Methylene chloride/acetone = 3/1, silica gel "Merck HF 254".
[c] No derivative prepared from factors as isolated.
[d] By micro- or ultramicro combustion analyses.

NPABE = 4-(4-nitrophenylazo)benzoic acid ester, 2,4-DNP = 2,4-dinitrophenylhydrazone from corresponding aldehyde.

More reliable evidence for their molecular formulae is obtained by mass spectrometry of the croton oil factors. Here, molecules of solvent volatilize well before the molecules of the factors themselves. On the other hand, care must be taken to avoid thermal degradation of the labile substances prior to their ionization.

At first. the molecular ions could only be obtained with reasonable intensity using the comparatively mild *electron addition* mass spectrometry development by von Ardenne *et al.* (*179, 180*) (croton oil factors A_1, B_3—B_7; *39, 40, 93, 94, 99*). For croton oil factors B_1 and B_2 (*99*), A_2—A_4 (*104*), and later on for all others, satisfactory mass spectra have been obtained by conventional high-resolution *electron impact* mass spectrometry.

As a typical example, the mass spectrum of croton oil factor A_1 is reproduced in Fig. 7. With the careful techniques employed, especially during the heating procedure, in the spectra of all factors the molecular ions appear with reasonable intensity. Fragment ions arise from the molecular ion by loss of water and of fatty acid moieties and appear as more or less intense lines. Subsequently, the fragmentation of all factors channels into the same pattern, starting with the fragment ion $m/e = 328$ (Fig. 8) assigned to an entity, "parent alcohol minus 2 H_2O". The fragmentation patterns of all croton oil factors from groups A and B indicate the presence of one and the same diterpene parent alcohol $C_{20}H_{28}O_6$.

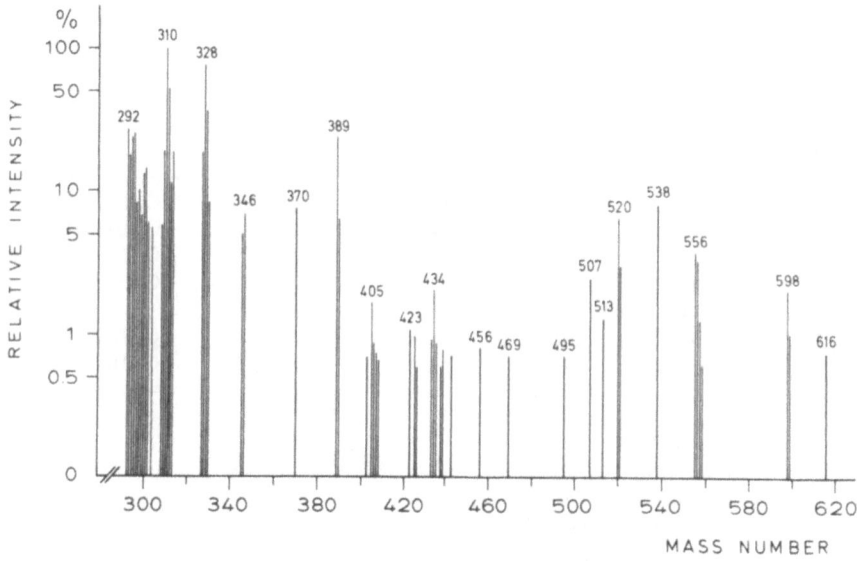

Fig. 7. Mass-spectrum of croton oil factor A_1, ordinate in logarithmic scale

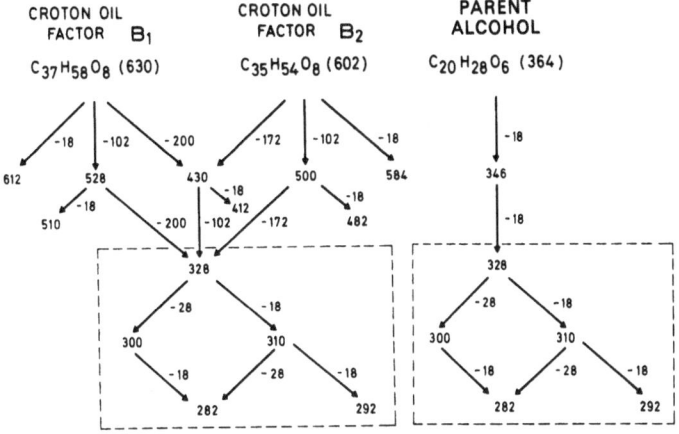

18 = H₂O,   28 = CO or C₂H₄ ,   102 = methyl butyric acid,   172 = decanoic acid ,   200 = dodecanoic acid

Fig. 8. Fragmentation of the croton oil factors B₁ and B₂ and of their parent alcohol by conventional mass spectrometry (from *98*)

In Table 6 the results of mass spectrometric characterization of the croton oil factors are summarized. In all factors, the structure of the long-chain fatty acid as determined by mass spectrometry is identical with that identified by the gas-liquid chromatographic method (see Section 2.1.1.). Certain common structural elements can be seen for all croton oil factors, independent of the factor group to which they belong. These include two ester functions, one with a short and the other with a long chain fatty acid. Most interestingly, two pairs

Table 6. *Mass Spectrometric Characterization of the Factors Isolated from the Hydrophilic Portion of Croton Oil (40, 91, 98, 104)*

Croton oil factor	R_f^a	Molecular formula	Acyl residues identified as	Parent-alcohol
A₁	0.3	$C_{36}H_{56}O_8$	acetic, tetradecanoic	
A₂[b]		$C_{32}H_{48}O_8$	*acetic, decanoic*	
A₃[b]		$C_{34}H_{52}O_8$	*acetic, dodecanoic*	
A₄		$C_{38}H_{60}O_8$	acetic, hexadecanoic	
B₁	0.4	$C_{37}H_{58}O_8$	(+)-S-2-methylbutyric, dodecanoic	
B₂		$C_{35}H_{54}O_8$	(+)-S-2-methylbutyric, decanoic	$C_{20}H_{28}O_6$
B₃		$C_{35}H_{52}O_8$	tiglic, decanoic	
B₄[b]		$C_{34}H_{52}O_8$	*acetic, dodecanoic*	
B₅		$C_{33}H_{50}O_8$	(+)-S-2-methylbutyric, octanoic	
B₆		$C_{33}H_{48}O_8$	tiglic, octanoic	
B₇[b]		$C_{32}H_{48}O_8$	*acetic, decanoic*	

[a] Methylene chloride/acetone = 3/1, silica gel "Merck HF 254".

[b] Italics: pairs of isomers.

of the croton oil factors, namely A_2 and B_7 and A_3 and B_4, exhibit identical molecular formulae and contain, apart from the same diterpene parent alcohol $C_{20}H_{28}O_6$, also the same acid moieties (see Table 6, italics). The individual members of each of these isomeric pairs differ, however, in their R_f values.

The ultraviolet (UV), infrared (IR), and nuclear magnetic resonance (NMR) spectra of the croton oil factors are very similar in type and rather uncharacteristic (*72, 93, 94, 99*). As typical examples the spectra of croton oil factor A_1 are recorded in Figs. 9 to 11.

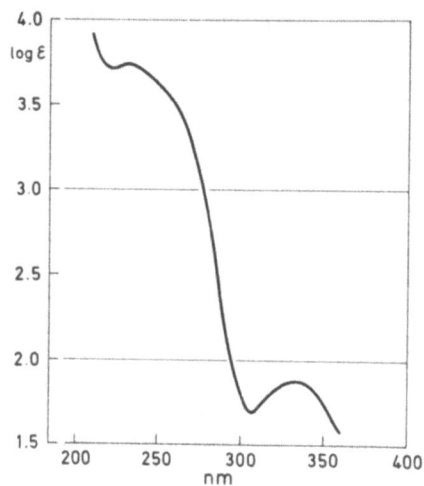

Fig. 9. UV-spectrum of croton oil factor A_1 in ethanol, $\lambda_{max}=232,\ 333$ nm, $\varepsilon_{max}=5400$, 73 (from *91*)

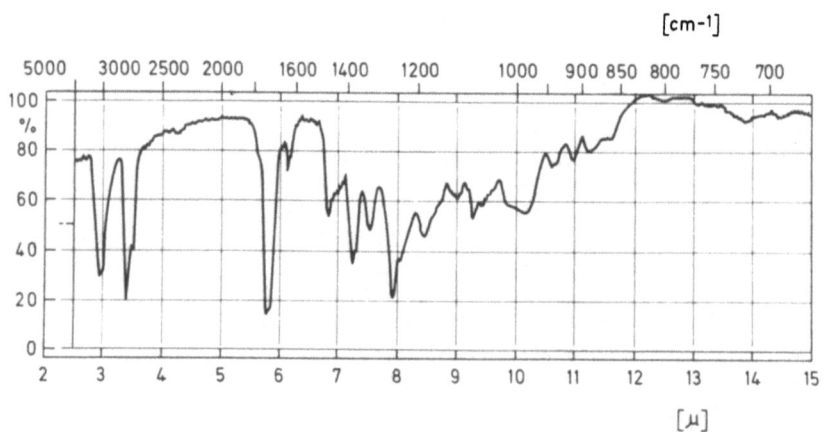

Fig. 10. IR-spectrum of croton oil factor A_1 in potassium bromide (from *91*)

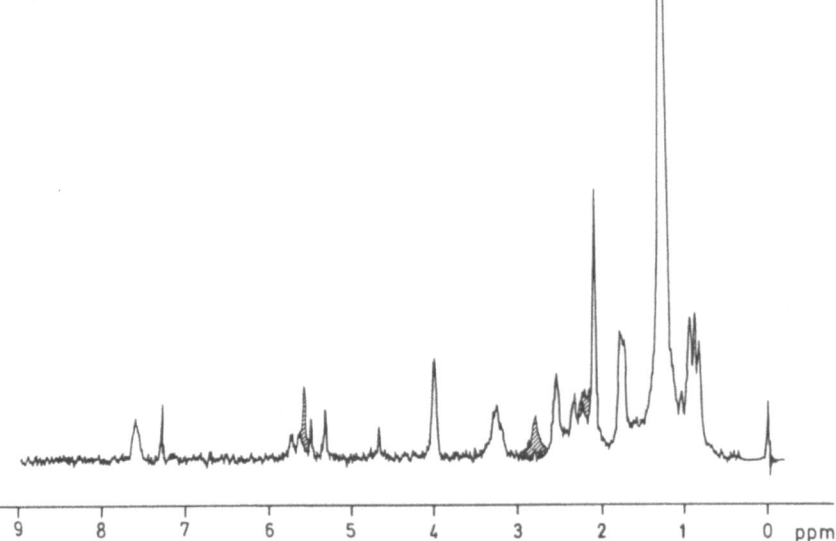

Fig. 11. NMR-spectrum (60 MHz, TMS 0.00 ppm) of croton oil factor A_1 in $CDCl_3$. Shaded
signals: protons exchanging in deuterium oxide

The UV spectra (see Fig. 9) exhibit absorption maxima, or in some cases an inflection
at about 230 nm with a further maximum at about 330 nm, indicating the presence of an
α,β-unsaturated carbonyl group. The IR spectra (see Fig. 10) show bands in the region
below 3 μ typical for hydroxyl groups, a broad or even split band between 5.75 and 5.85 μ
characteristic for carbonyls of ester- and keto-groups and a band between 6.1 and 6.15 μ
indicating an α,β-unsaturated carbonyl group.

In the NMR spectra (see Fig. 11), intense signals appear around δ = 1.3 ppm due to
the long-chain aliphatic acyl residues. In the NMR spectra of factors A_1—A_4, B_4 and
B_7, a singlet at δ = 2.10 ppm indicates the presence of an acetyl group. After proton
exchange in deuterium oxide, in the NMR spectra of all factors signals disappear partially
or completely at about δ = 2.2, 2.8 and 5.58 ppm, indicating the presence of three hydroxyl
groups.

As a first step in the chemical elucidation of the structures of
the croton oil factors, the ester residues are removed by mild reduction with lithium aluminium hydride (*40, 93, 94, 99*). This leads to a
mixture of three alcohols. In ether/water, two of them partition into the
ether phase and one into the water phase.

Treatment of the alcohols partitioning into the ether phase with
3,5-dinitrobenzoic acid chloride yields a mixture of 3,5-dinitrobenzoates,
which is separated in order to identify the individual alcohol moieties.
In this way, the chemical nature of the acid moieties in the two ester
groups in each of the factors $B_1 - B_7$ and A_1, as determined by mass
spectrometry, are confirmed (*39, 40, 93, 94, 98, 99*). In addition, from the
mixture of alcohols partitioning into the ether phase the long-chain

fatty alcohols may be characterized as the corresponding NPAB esters, as, for example, in croton oil factor A_1 (*91, 93, 94*).

For all croton oil factors, the alcohol partitioning into the water phase is identical. It exhibits the molecular formula $C_{20}H_{30}O_6$, m.p. $220° - 223°$ C (decomp.) and is extremely unstable. It may be characterized as tetraacetate, $C_{28}H_{38}O_{10}$, obtained by acetylation with acetic anhydride/pyridine. In the sensitive alcohol, as well as in its tetraacetate, the α,β-unsaturated carbonyl group, contained in all croton oil factors is no longer present. It is therefore the dihydro derivative of a parent alcohol $C_{20}H_{28}O_6$, which must contain an α,β-unsaturated carbonyl group (*39, 93, 94, 99*).

The molecular formula of this diterpene parent is reminiscent of that of phorbol as obtained by base catalyzed *trans*esterification directly from croton oil ($C_{20}H_{28-30}O_6$; *50*). Indeed, phorbol prepared essentially according to the procedure of Flaschenträger, as outlined below, exhibits a mass spectrometric fragmentation pattern (*98*) related to that of the croton oil factors (see Fig. 8). Below the fragmention "parent alcohol minus 2 H_2O" $m/e = 328$, the fragmentations of phorbol and of the croton oil factors are identical (*93, 94, 99*).

Further, "Acetylphorbol" (*29*), as obtained from phorbol by treatment with acetic anhydride/pyridine and characterized as phorbol-triacetate by mass spectrometry (*93, 94, 99*), reacts with lithium aluminium hydride to give the water-soluble and extremely unstable alcohol $C_{20}H_{30}O_6$ (*93, 94, 99*), identical in all respects with that obtained by reduction of the croton oil factors.

Thus, it is established unequivocally that the toxic, irritant and cocarcinogenic principles from croton oil are diesters of phorbol, each factor containing one short-chain and one long-chain fatty acid (see Table 6).

2.4. Higher Phorbol Esters from the Hydrophobic Portion

2.4.1. Isolation and Resolution of Croton Oil Factor Groups A' and B'

As shown in Section 2.2., the neutral fraction from croton oil exhibits considerable biological and optical activity whereas the hydrophobic portion, as well as the acidic fraction, show much lower biological activities with the optical activity remaining almost the same as that of the original oil (see Chart 1 and Table 3). In spite of its low biological activity, the hydrophobic portion contains an even higher amount of phorbol than the neutral fraction (*80, 81, 152*). Indeed, by carefully controlled acid-catalyzed *trans*esterification of the hydrophobic portion in methanol an "activated hydrophobic portion" may be obtained (Chart 3), which exhibits about 10 times the irritant activity of the hydrophobic portion. Thus, as an irritant, its activity is comparable with that of the original oil (*80, 81, 152*). It is obvious, that certain biologically relatively inactive phorbol derivatives are contained in the hydrophobic portion which, on acid catalyzed *trans*esterification, are converted to biologically active phorbol derivatives (*152*).

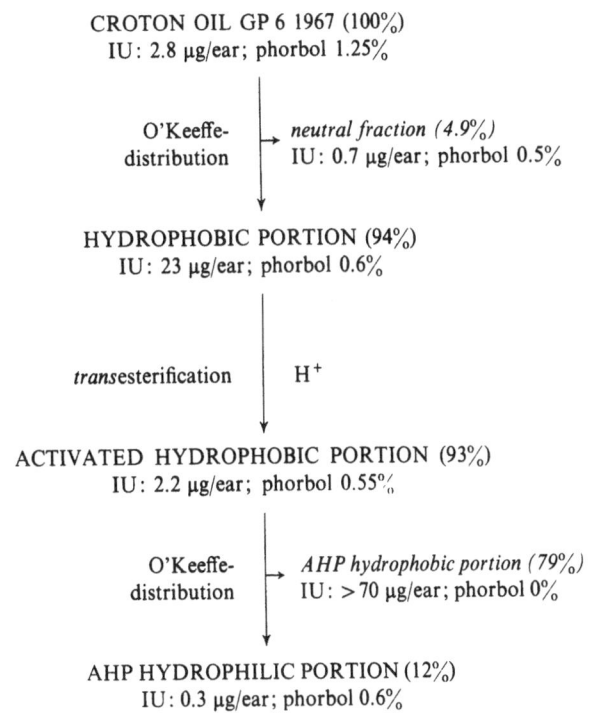

CROTON OIL GP 6 1967 (100%)
IU: 2.8 µg/ear; phorbol 1.25%

O'Keeffe-distribution → *neutral fraction (4.9%)*
IU: 0.7 µg/ear; phorbol 0.5%

HYDROPHOBIC PORTION (94%)
IU: 23 µg/ear; phorbol 0.6%

*trans*esterification   H⁺

ACTIVATED HYDROPHOBIC PORTION (93%)
IU: 2.2 µg/ear; phorbol 0.55%

O'Keeffe-distribution → *AHP hydrophobic portion (79%)*
IU: >70 µg/ear; phorbol 0%

AHP HYDROPHILIC PORTION (12%)
IU: 0.3 µg/ear; phorbol 0.6%

Irritation units (IU) estimated in the standard procedure (*96*).

Chart 3. Activation of the hydrophobic portion of croton oil and initial stages of fractionation of the "activated hydrophobic portion" (*80, 81, 83, 84, 85, 86, 87, 152*)

Consequently the activated hydrophobic portion is subjected to a fractionation procedure analogous to that of croton oil (Chart 3).

By O'Keeffe distribution of the activated hydrophobic portion the AHP hydrophilic portion is obtained*. It contains the entire irritant activity and phorbol content whereas

---

* To distinguish the fractionation steps for the original croton oil from those used in the fractionation of its "activated hydrophobic portion" the fractionation steps of the latter carry the initials of "*A*ctivated *H*ydrophobic *P*ortion". The resulting factor groups are labelled with a dash.

the corresponding AHP hydrophobic portion is practically inactive as an irritant and
does not contain any phorbol (Chart 3). In the next stage, by column chromatography
of the AHP hydrophilic fraction (Chart 4), practically inactive AHP head and tail
fractions as well as two highly irritant bands A′ and B′* are obtained. Together, the material
contained in these bands represents essentially all the irritant activity of the activated
hydrophobic portion (Chart 4).

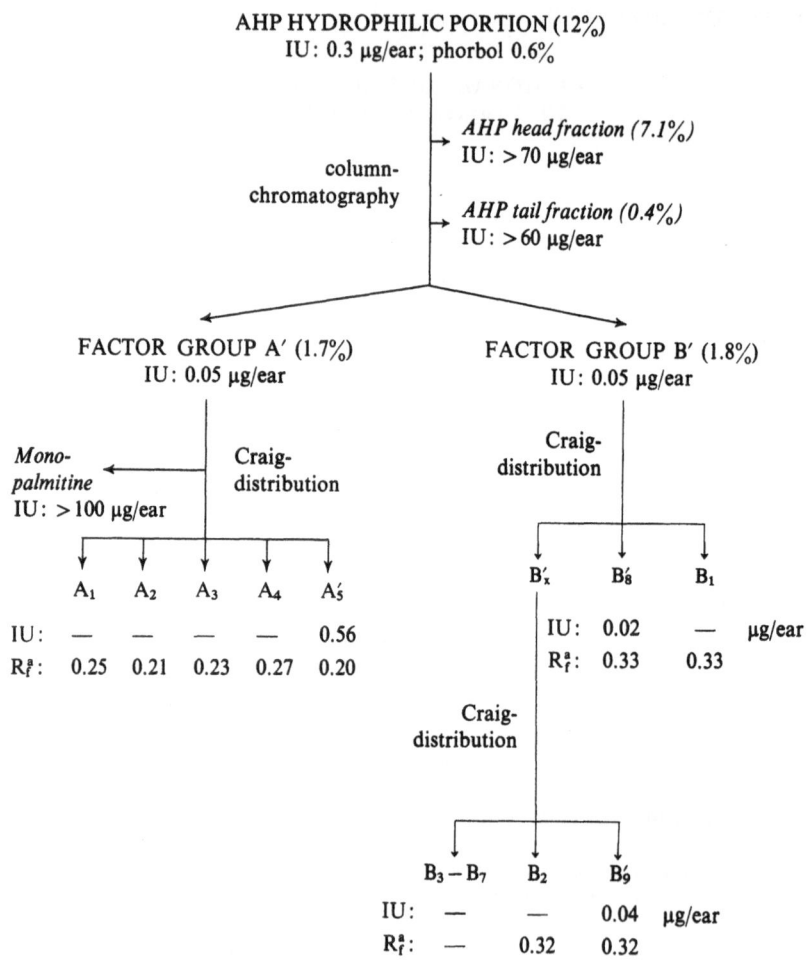

ᵃ System: chloroform/ethyl acetate = 2/3, silica gel Merck HF 254.
Irritation units (IU) estimated in the standard procedure (96).

Chart 4. Fractionation of the "activated hydrophobic portion" of croton oil, continued
(*80, 81, 83, 84, 85, 86, 87, 152*)

---

* See Footnote on p. 399.

References, pp. 458—467

2.4.2. Separation and Chemical Characterization of the Croton Oil Factors from Groups A' and B'

By Craig distribution of band A', employing the solvent system and machinery already used for resolution of band A of the hydrophilic portion (Fig. 3), a distribution diagram (Fig. 12) is obtained which is similar to that of factor group A. Using the gas chromatographic and mass spectrometric techniques as described above, the material in the bands with $G = 0.33$, 0.46, 0.67 and 1.12 (Fig. 12) were identified as croton oil factors $A_1 - A_4$, as isolated previously from band A. Also 1-monopalmitine, corresponding to compound C of Fig. 3, is present. In addition, the material in the fractions of the band $G = 2.06$ corresponding to A_y of Fig. 3 is identified as the new croton oil factor A_5' of relatively weak irritant and cocarcinogenic activity (Chart 4, Table 7, *152*).

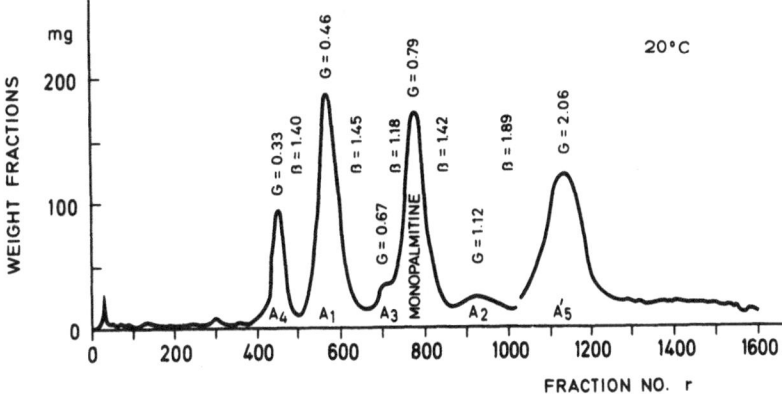

Fig. 12. Craig distribution of croton oil factor group A'. Single withdrawal procedure; $z = 1020$, $n = 1600$; $V = 12/10$; system: carbon tetrachloride (2), methanol (1), water (0.15). Ordinate represents average weight of four neighboring fractions r to r-3, plotted against r-2, for fractions $r = 1021$ to 1600 average weight of nine neighboring fractions r to r-8, plotted against r-4 (from *80, 81, 152*)

Table 7. *Irritant and Cocarcinogenic Activities of the Croton Oil Factors Isolated from the Activated Hydrophobic Portion compared to Croton Oil Factor A_1 (83, 84, 85, 86, 87, 152)*

Factor	Irritation[a] ID_{50} [mμM/ear]	single dose p [μM/appl.]	Cocarcinogenic activity[b] tumor rate[c] [%]	average tumor yield[c] [tumors/survivor]
A_1	0.016	0.02	82	3.6
A_5'	1.03	0.06	14	0.3
B_8'	0.002	0.02	42	1.0
B_9'	0.005	0.02	32	0.8

[a] Standard deviation σ: 1.3, significance level $\alpha = 0.05$.
[b] Assayed in the standard procedure on NMRI-mice (*91*); initiator $i = 0.1$ μM DMBA.
[c] After 12 weeks = 24 applications.

The similarity of band A_y (Fig. 3) — which was not investigated further — and the band containing A'_5 (Fig. 12) suggests that A_y may also contain one or more phorbol-12,13-diesters of relatively weak irritant activity (see also Chart 2).

Band B' is also resolved further by Craig distribution (Chart 4). In the first distribution (Fig. 13), essentially two bands are obtained. From band $G = 0.34$, by column chromatography on silica gel columns impregnated with silver nitrate, the croton oil factor B_1, previously obtained from band B of the hydrophilic portion, is isolated and subsequently identified using gas chromatographic and mass spectrometric techniques. In addition, a new croton oil factor B'_8 is obtained in small amounts. It exhibits remarkable irritant and cocarcinogenic activities (Chart 4, Table 7, *152*).

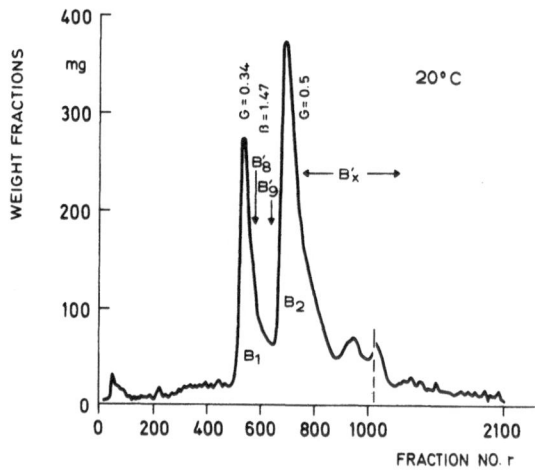

Fig. 13. Craig distribution of croton oil factor group B'. Single withdrawal procedure; $z = 1020$; $n = 2100$; $V = 10/10$; system: carbon tetrachloride (2), methanol (1), water (0.15). Ordinate represents average weight of four neighboring fractions r to r-3 plotted against r-2; for fractions $r = 1021 - 1600$ average weight of nine neighboring fractions r to r-8, plotted against r-4 (from *85, 152*)

The material above fraction $r = 600$ (B'_x, Fig. 13), is further resolved by redistribution in a different solvent system using $n = 6000$ transfers (Fig. 14). This leads to the isolation and identification of the factors $B_2 - B_7$, already obtained from band B of the hydrophilic portion. A third new croton oil factor B'_9 is obtained from this distribution, also showing considerable irritant and cocarcinogenic activities (Chart 4, Table 7, *152*).

According to its R_f value and by application of the techniques described in Section 2.3.2., the new croton oil factor A'_5 can be characterized as a tiglyl-phorbol-butyrate. Similarly, the factors B'_8 and B'_9 are recognized as tiglyl- and butyryl-phorbol-dodecanoates respectively

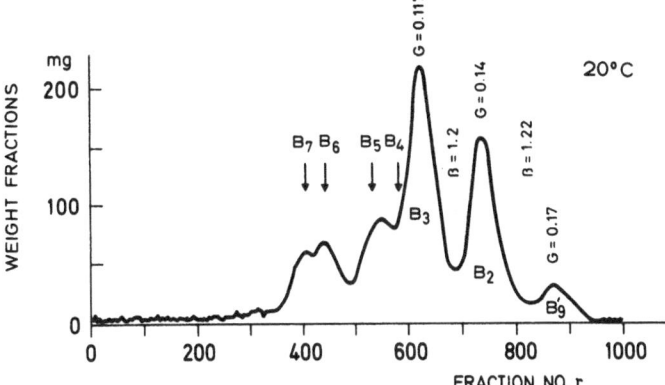

Fig. 14. Craig distribution of croton oil fraction B'ₓ. Single withdrawal procedure; z = 1020; n = 6000, V = 12/10; system: petroleum ether (2), carbon tetrachloride (0.4), methanol (1.75), water (0.1). Ordinate represents average weight of four neighboring fractions r to r-3 plotted against r-2 (from *85, 152*)

(Table 8). These croton oil factors therefore conform to the general pattern in that they are phorbol-diesters with one short- and one long-chain fatty acid.

Table 8. *Mass Spectrometric Characterization of the Croton Oil Factors Isolated from the Activated Hydrophobic Portion (83, 84, 85, 86, 87, 152)*

Croton oil factor	R_f^a	Molecular formula	Acyl residues identified as	Parent alcohol
A'₅	0.3	$C_{29}H_{40}O_8$	butyric, tiglic	
B'₈	0.4	$C_{37}H_{56}O_8$	dodecanoic, tiglic	$C_{20}H_{28}O_6$
B'₉	0.4	$C_{36}H_{56}O_8$	dodecanoic, butyric	

ᵃ Methylene chloride/acetone = 3/1, silica gel "Merck HF₂₅₄".

The successful activation of the hydrophobic portion indicates that in croton oil, besides the biologically active phorbol-diesters of the hydrophilic portion, accounting for approximately 50% of the phorbol content, another biologically relatively inactive source of phorbol-diesters is present, accounting for the rest of the phorbol content of the oil. The behaviour of the less active source of phorbol-diesters in liquid-liquid distribution and in *trans*esterification suggests that it represents higher esters of phorbol. They may be activated by partial hydrolysis of some of their ester groups to yield phorbol-diesters.

3. Chemistry of Phorbol and of the Croton Oil Factors

By base-catalyzed *trans*esterification, phorbol may be obtained directly from croton oil as colorless crystals of the solvate "alcohol phorbol" (*50*), i.e. phorbol·C_2H_5OH (*108*), in an average yield of 1%. It crystallizes from a viscous alcohol solution, which mainly consists of glycerol, the parent alcohol of the mass of the lipids of croton oil, and of phorbol·C_2H_5OH. Preparations of phorbol as obtained by crystallization of phorbol·C_2H_5OH from water are free of alcohol and are — contrary to phorbol·C_2H_5OH — relatively stable (*50, 105*).

To prepare phorbol (*105*) croton oil (500 gm) is shaken under nitrogen with a solution of 55 gm of $Ba(OH)_2$ · 8 H_2O in 2.25 liters of methanol for 10 to 12 hours. Subsequently the solution is filtered to remove the precipitate of barium soaps and rotated under vacuum until no more methanol volatilizes (temperature of bath, 40° C). The oily residue is taken up in 2 liters of water and extracted twice with 500 ml of ether. The aqueous phase contains phorbol and is adjusted to pH 5 with 2 N sulfuric acid. After addition of 40 ml of saturated sodium sulfate solution, the preparation is kept at 4° C. After 12 hours it is filtered from the precipitated barium sulfate, adjusted to pH 7 with 2 N sodium hydroxide, and extracted with 500 ml ethyl acetate followed by 500 ml of ether. The aqueous phase is rotated under vacuum until no more water volatilizes (temperature of bath, 45° C). The viscous residue is "digested" with 100 ml of ethanol and separated from sodium sulfate by suction through a filter plate. The sodium sulfate remaining on the filter is extracted with warm ethanol until a sample boiled with several milliliters of hydrochloric acid gives no or only a very slight red color ("phorbol reaction", according to *50*). From the filtrate, the ethanol is partly removed until 50 to 60 ml of an oily residue remain. This preparation is stored at 4° C. Crystallization of phorbol usually starts spontaneously. If not, the preparation may be too viscous and has to be diluted with a little ethanol. Also, seeding with phorbol may induce crystallization. After 4 weeks the crystals of phorbol · C_2H_5OH are separated from the mother liquor by suction filtration. The resulting mass is spread on a clay dish and stored under nitrogen at 4° C. Thus, 5.5 to 6.1 gm of pure white phorbol · C_2H_5OH are obtained. Since this solvate is not stable, it is converted to phorbol by dissolving e.g. 5.8 gm of phorbol · C_2H_5OH in 100 ml of water at 60° C. From this solution water is removed in the rotatory evaporator (temperature of bath, 60° C), until crystals begin to separate. This preparation is kept at 4° C for 1 week, and the crystals are collected by suction (e.g. 4.4 gm of phorbol mp. 250 to 51° C, decomposition). From the filtrate, additional phorbol may be obtained by further removal of water in the rotatory evaporator and precipitation of the remaining solution by addition of acetone (about 25% by volume). — Similar methods of preparation of phorbol are described by Kauffmann and Neumann (*116*) and by Crombie et al. (*43*).

As was observed for the croton oil factors, the parent alcohol phorbol is sensitive to oxygen, acid and alkali. It reduces Tollens's and Fehling's reagents. Also it exhibits a remarkable tendency to solvate with all kinds of organic solvents (*43, 93, 97, 105, 108*).

3.1. Structure and Stereochemistry of Phorbol

Phorbol (**1a**) is a tetracyclic diterpene with a five- (A), a seven- (B), a six- (C) and a three- (D) membered ring; the latter two rings represent a bicyclo[4.1.0]heptane system (Fig. 15). The six oxygen functions are located in the five-membered ring A as an α,β-unsaturated tertiary ketol grouping, in the seven-membered ring B, including C-20 as a primary allylic alcohol group, as a tertiary hydroxyl group at C-9, at the junction of ring B with the six-membered ring C, and as a glycol group at C-12 and C-13 of the bicyclo[4.1.0]heptane system. The molecule exhibits eight centers of asymmetry.

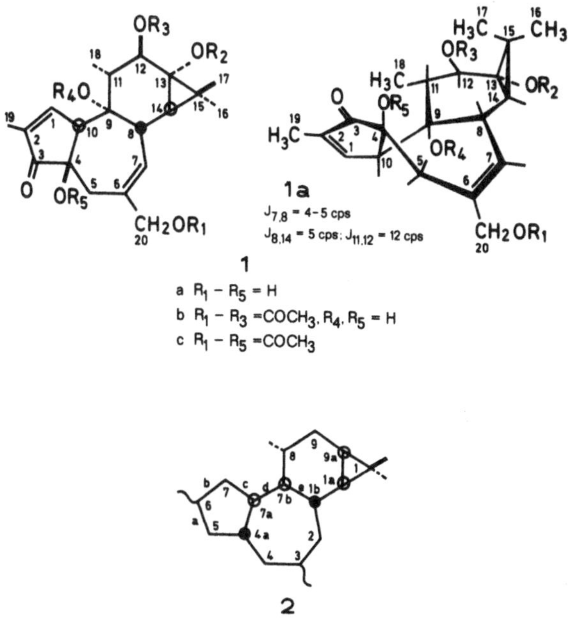

Fig. 15. Structure and stereochemistry of phorbol (**1a**), phorbol-12,13,20-triacetate (**1b**), phorbol-pentaacetate (**1c**) and the saturated parent hydrocarbon tigliane (**2**). Relevant coupling constants noted below the spatial formula of **1a** are taken from the NMR spectrum in d_5-pyridine/D_2O, see also Fig. 17 (from *60, 61, 79, 89*)

The functionality and structural elements of phorbol were derived from its mass spectrometric (Fig. 16), NMR (Fig. 17), UV (Fig. 18 and Table 9), IR and CD (Fig. 19) data, supplemented by evidence from chemical reactions and from spectrometric evaluation of the various products (*93, 94, 99, 102*). The partial structures thus obtained (*8, 77,*

105), were interlinked to yield the final chemical structure of phorbol (78, 79, 89). Relative configurations of asymmetric centres were established (excepting that at C-10) by NMR- and circular dichroism measurements, and by certain intramolecular chemical reactions (9, 10, 14, 58, 59, 60, 61, 62, 89, 118, 119, 120).

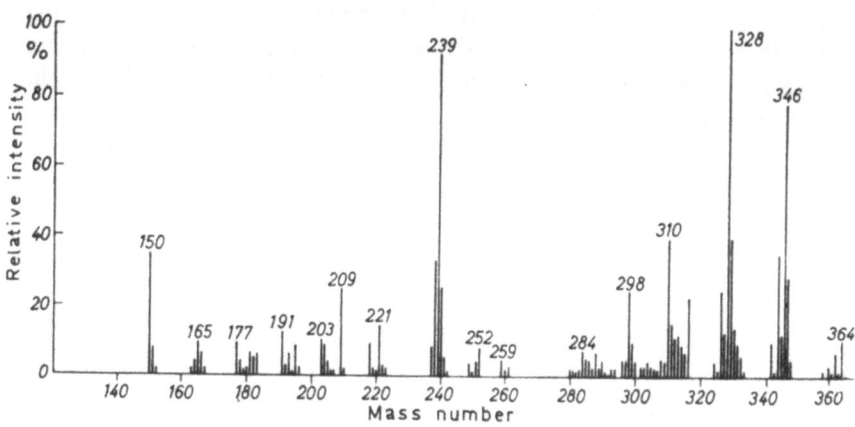

Fig. 16. Mass-spectrum (von Ardenne type) of phorbol, temperature of evaporator 130° C (from 105)

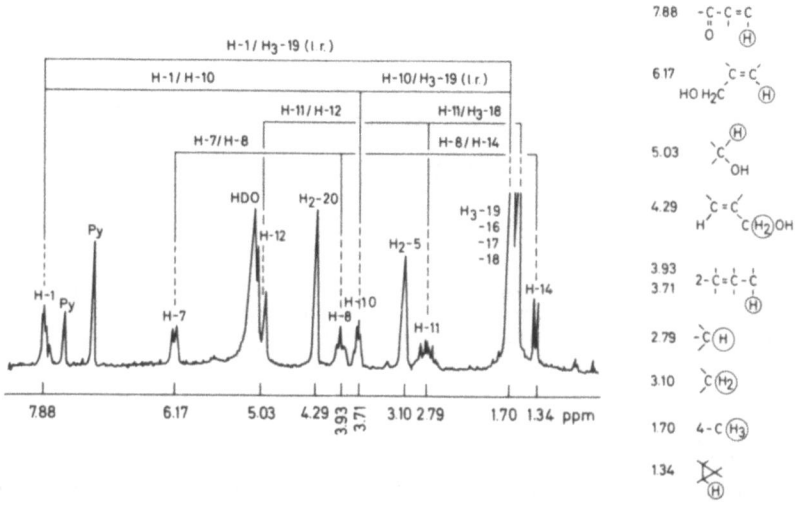

Fig. 17. NMR-spectrum (100 MHz, TMS 0.00 ppm) of phorbol (1a) in d₅-pyridine/D₂O (from 105)

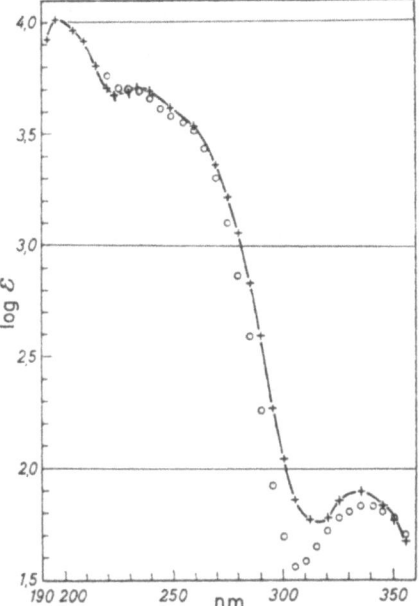

Fig. 18. UV-spectra of phorbol (**1a**) in ethanol, λ_{max} 196, 234, 332 nm, ε_{max} 10 600, 5060, 72 (x—x); and in dioxan λ_{max} 232, 338 nm, ε_{max} 5000, 70 (oooo) (from *105*)

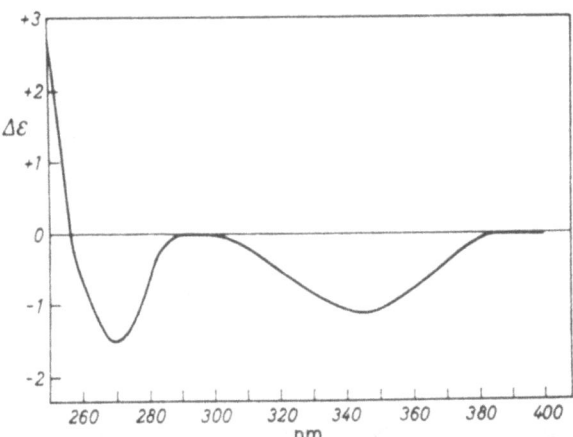

Fig. 19. CD-spectrum of phorbol (**1a**) in dioxan, 390 ($\Delta\varepsilon \pm 0$), 346 ($\Delta\varepsilon - 1.11$), 300 − 290 ($\Delta\varepsilon \pm 0$), 270 ($\Delta\varepsilon - 1.51$), 250 ($\Delta\varepsilon + 2.8$) (from *105*)

Table 9. *Comparison of the UV-Absorption of Bicyclo[10.1.0]tridecanyl-(13)-acetate with the Bicyclo[4.1.0]heptane System of some Derivatives of Phorbol in Methanol (14)*

Compound	Far UV		n-π*-band	
	λ_{max} [nm]	ε_{max}	λ_{max} [nm]	ε_{max}
Bicyclo[10.1.0]tridecanyl-13-acetate	185	2650	—	—
20-deoxy-tetrahydrophorbol-12,13-diacetate	195	2400	305	55
20-deoxy-tetrahydrophorbol-13-acetate	194	2300	307	44
12,20-dideoxy-12-oxo-tetrahydrophorbol-13-acetate	196	3700	302	72

The elucidation of the structure of phorbol was complicated by a number of unusual difficulties. Thus, because of the sensitivity of the molecule and its derivatives and because of their tendency to solvate organic solvents the problem of the molecular formula of phorbol and in particular of its functionality and chemical structure remained unsettled for a long time (*50, 114, 115, 116, 168*). Based upon the results of quantitative hydrogenation of phorboltriacetate (**1b**, *102*), its diterpene nature was recognized and a first proposal of a tetracyclic hydroaromatic structure made in 1964 (*93, 94, 102*). This structure was supported by NMR data (*100, 102*) and by the generation of azulenes on dehydrogenation and dehydration of phorbol and 3-deoxo-3β-hydroxyphorbol (*66, 102*). In agreement with the general chemical experience at that time, it was implicated in this first proposal, that the reducing properties of phorbol towards Fehling's and Tollens's reagents were due to a secondary 1,2-ketol group (*50, 102, 116*). Also, in our first proposal, the production of equivalent amounts of acetone by reaction of phorbol with lead tetraacetate as claimed by Kauffmann et al. (*114*) was accounted for by an appropriate ditertiary 1,2-glycol group involving the cyclopropane hydroxyl (*102*). At the same time, based upon results of dehydrogenation of phorbol with a palladium catalyst, Arroyo and Holcomb (*4*) proposed a tricyclic hydroaromatic structure for phorbol. Later they accepted the tetracyclic ring system as proposed by Hecker et al. (*102*), but translocated two of its structural elements (*5*). However, this translocation was not in agreement with NMR spectroscopic data (*100*).

Subsequent to our first structural proposal, we found that the production of acetone in the reaction with lead tetraacetate as claimed by Kaufmann et al. (*114*) was not reproducible. Also we were able to demonstrate that the reducing properties of phorbol are due to a tertiary cyclopropanol group rather than to a secondary α-ketol group. Up to that time cyclopropanol groups were unknown as entities capable of reducing Tollens's and Fehling's reagents. In addition it was found, that phorbol does contain a (nonreducing) tertiary 1,2-ketol group (*8, 77, 105, 119, 120*).

With this new chemical information at hand and together with the results of measurements of the circular dichroism, the interlinking of the structural elements and partial structures of phorbol as proposed at first (*102*) was revised to yield the correct chemical structure of phorbol (*78, 79*). By application of the "inverse octant rule" (*45, 148*) in connection with the CD of the cyclopropyl ketone group of neophorbol-13,20-diacetate (**54**, Fig. 40), S-chirality for the reference atom H-14 was suggested (*89*). However, the true configuration of phorbol (Fig. 15, **1a**), as subsequently determined by X-ray diffraction analysis (*107*) is the mirror image of the first proposal (*89*) for its absolute configuration.

The structure of phorbol, thus derived, was confirmed — and completed with respect to the relative configuration at C-10 — by X-ray diffraction analysis (*90, 107, 108, 138, 139*). Further, the absolute configuration of the molecule (Fig. 15), was established by X-ray diffraction analysis of heavy atom derivatives of phorbol (*107, 138, 139*).

As may be seen from Dreiding models, phorbol has a rigid, non flexible structure (Fig. 15). According to the model and to the rather low UV-extinction at about 230 nm ring A is not planar due to the strain imposed on it by its *trans*-connection with the unsaturated ring B. The latter is fixed in an "envelope" conformation with the fold between C-4 and C-8 (Fig. 15). Ring C is fixed in the half-chair conformation primarily by the planar *cis*-linked ring D. As suggested by the interpretation of the NMR coupling constants of relevant H-atoms (Fig. 15), according to the Karplus equation, the conformation of phorbol in solution is identical with the conformation exhibited by phorbol-20-[5-bromofuroate] (*138, 139*) and phorbol·C_2H_5OH (*108*) in X-ray diffraction analyses of single crystals.

The perhydrogenated parent hydrocarbon of phorbol is a derivative of a perhydrocyclopropabenzazulene called tigliane (**2**, Fig. 15). By definition, the configurations of the six asymmetric centres at the ring interlinkages are identical with the configurations of the corresponding centres in phorbol (*89, 102*). Thus, according to the IUPAC rules (*109a*) tigliane (**2**) is 1,1aα,1bβ,2,3,4,4aβ,6,7,7aα,7bα,8,9,9aα-tetradecahydro-1,1,3ξ,6ξ,8α-pentamethyl-5H-cyclopropa[3.4]benz[1.2–e]azulene and the systematic name of phorbol (**1a**) would be 1,1aα,1bβ,4,4a,7aα,7b,8,9,9a-decahydro - 4aβ,7bα,9β,9aα - tetrahydroxy - 3 - (hydroxymethyl) - 1,1,6,8α-tetramethyl-5H-cyclopropa[3.4]benz[1.2–e]azulen-5-one. A nomenclature system of phorbol and derivatives based upon tigliane in general provides more simple designations than the IUPAC system; accordingly phorbol (**1a**) is 4,9, 12β,13,20-pentahydroxy-1,6-tigliadien-3-one (*61, 89*).

By convention the structural formulae of tigliane and of phorbol are drawn with the A-ring to the left of the chirality center C-14 (see **2**, Fig. 15). To simplify the structural formula, C-atoms including those of methyl groups (if not substituted) and hydrogen atoms are omitted; to describe their configurations the following general rule is used:

The configuration of *substituents other than hydrogen atoms** in any position of the ring system is indicated by a thickened (β-position) or by a broken line (α-position). The configuration of hydrogen atoms at *ring junctions** is indicated by a filled (β-position) or by an open circle

---

* To describe the configuration of hydrogen atoms *and* substituents in all positions of carbo- and heterocyclic ring systems, the use of thickened and broken lines is generally accepted (*109a, 109b*). However, a generally accepted, consistent and logical rule for the use of thickened and broken lines together with filled and open circles seems not to

(α-position). For unequivocal description of configurations at certain C-atoms it may be necessary to label a dihedric pair of bonds by using thickened and/or broken lines or by a combination of either one with a filled or an open circle [e. g. hydroxyphorboisobutanone-9.13-semi-ketal-20-acetate (**58**, Fig. 40) at C-12 (two broken lines) or at C-14 (combination of broken line and open circle)]. Since it was shown recently that the cyclopropane ring of bicyclo[4.1.0]heptane can exist in *trans*-configuration (*135*) it is necessary also to include three-membered rings (including heterocyclics e. g. oxiranes) in the proposals above (see phorbol [**1a**] and phorbol-6.7-epoxides, e. g. **13a**, Fig. 24).

To our knowledge, phorbol was the first diterpene exhibiting a perhydro-benzazulene structure.

In the following sections 3.2, 3.3 and 3.4 dealing with the reactivity of phorbol also those reactions and reaction products which were used to elucidate its structure and stereochemistry will be described.

3.2. Reactions Altering the Functional Groups of Phorbol

3.2.1. Functional Derivatives of the Carbonyl Group

Reaction with Carbonyl Reagents. In phorbol and its esters the presence of a carbonyl group cannot be demonstrated with carbonyl reagents under usual reaction conditions. Under forcing conditions however a 2,4-dinitrophenylhydrazone may be obtained in low yield from phorbol-12,13,20-triacetate (*105*).

Reduction of the Carbonyl Group. As pointed out in Section 2.3.2., in the Croton oil factors and in phorbol-12,13,20-triacetate (**1b**) but also in phorbol-12,13,20-tribenzoate (*39, 43, 93, 94, 99, 116*) the carbonyl group is reduced by lithium aluminium hydride. The reaction yields — with concomitant loss of the ester groups — the extremely sensitive 3-deoxo-3β-hydroxyphorbol (FÜRSTENBERGER and HECKER, unpublished). The 3β-configuration of the newly introduced secondary allylic hydroxyl group follows from the rapid reaction of the 3,4-glycol group with lead tetraacetate (*116*) or with sodium periodate (*58*).

---

exist: LINSTEAD introduced the filled circle to indicate the β-configuration of hydrogen atoms (*126*). RUZICKA et al. (*143*) modified this proposal by using a filled circle to indicate β-configuration for both, hydrogen atoms *and* substituents. Later, VELLUZ et al. (*178*) introduced the open circle and used it together with the filled circle to differentiate α- and β-positions of hydrogen atoms *and* substituents at a ring junction only. BUCOURT et al. (*33a*) and other authors (e. g. *112, 133*) restricted the use of filled and open circles to hydrogen atoms at ring junctions. This method was extended to heterocyclics with the exception of the oxirane ring (e. g. *2*). Many uncertainties in the use of these different systems may be found in the literature. To avoid such confusion the general rule as stated above is proposed, and used consequently in the following.

Reduction of the carbonyl group in phorbol or its acetates may be accomplished also with sodium borohydride, with retention of the ester groups (see Table 10). In this reaction, in addition to the 3β-epimer as the main product, the 3α-hydroxy isomer is formed (FÜRSTEN-BERGER and HECKER, unpublished). Contrary to expectation, reduction of phorbol-pentaacetate (**1c**) does not provide the product with a free 3β-hydroxyl, but leads, by intramolecular acetyl migration, to the 3-deoxo-3β-hydroxyphorbol-3,9,12,13,20-pentaacetate (*149*). It was observed, that the ester groups of long chain fatty acids or benzoic acid with phorbol are more stable than acetyl groups towards sodium borohydride.

Table 10. *Selective Reduction of the Carbonyl Group in Phorbol and Phorbol Acetates and Oxidation of the 3β-Hydroxy Group (149)*

Phorbol	$\xrightarrow{\text{NaBH}_4{}^a}$	3-deoxo-3β-hydroxyphorbol
Phorbol-12,13,20-triacetate (**1b**)	$\xrightleftharpoons[\text{CrO}_3/\text{Py}]{\text{NaBH}_4{}^b}$	3-deoxo-3β-hydroxyphorbol-12,13,20-triacetate (**18**, Fig. 29)
Phorbol-9,12,13,20-tetra-acetate	$\xrightarrow{\text{NaBH}_4{}^b}$	3-deoxo-3β-hydroxyphorbol-9,12,13,20-tetraacetate
Phorbolpentaacetate (**1c**)	$\xrightarrow{\text{NaBH}_4{}^c}$	3-deoxo-3β-hydroxyphorbol-3,9,12,13,20-pentaacetate

  ª In ethanol.   ᵇ In isopropyl alcohol.   ᶜ In tetrahydrofuran/isopropyl alcohol.

The secondary allylic hydroxyl group of 3-deoxo-3β-hydroxy-derivatives of phorbol may be reoxidized by chromium trioxide/pyridine (see Table 10, *10, 149*) or by manganese dioxide (SCHMIDT and HECKER, unpublished).

3-Deoxo-3β-hydroxyphorbol-12,13,20-triacetate (**18**, Fig. 29) may be acetylated with acetic anhydride/pyridine to yield 3-deoxo-3β-hydroxyphorbol-3,12,13,20-tetraacetate (see Section 3.2.2). On the other hand it may be esterified with acylchlorides/pyridine (*149*); thus, for example, 3-deoxo-3β-[p-bromobenzoyloxy]phorbol-12,13,20-triacetate was obtained, forming crystals amenable to X-ray structural analysis (*90*).

3.2.2. Functional Derivatives of the Hydroxyl Groups

Phorbol Esters. By reaction of phorbol with acetic anhydride/pyridine at room temperature phorbol-12,13,20-triacetate (**1b**, Fig. 15), the "Acetylphorbol" of BOEHM *et al.* (*29*) is obtained (Table 11). Reaction of **1b** with acetic anhydride/p-toluenesulfonic acid yields phorbolpenta-acetate (**1c**, Fig. 15).

Table 11. *Preparation of Phorbol Acetates*

Py = pyridine, Ac₂O = acetic anhydride, DMF = dimethylformamide, pTsOH = p-toluenesulfonic acid

Reaction products	Reactants		temp. [°C]	react. time [hours]	yield [% of th.]	lit.
phorbol-12,13,20-triacetate	phorbol	Ac₂O/Py	20	19	90–98	(43, 105, 116)
phorbol-12-acetate	phorbol-12,13,20-triacetate	NaOCH₃/CH₃OH	4	45	68	(183)
phorbol-13-acetate	phorbol	Ac₂O/DMF	20	14	50	(183)
12-O-tetradecanoylphorbol-20-acetate	12-O-tetradecanoylphorbol	Ac₂O/Py	−20	24	52	(121)
phorbol-12,13-diacetate	phorbol-12,13,20-triacetate	HClO₄/CH₃OH	20	26	80	(183)
phorbol-12,20-diacetate	phorbol-12-acetate	Ac₂O/Py	−20	40	50	(183)
phorbol-13,20-diacetate	phorbol	Ac₂O/DMF/CaCO₃	20	96	80	(183)
phorbol-pentaacetate	phorbol-12,13,20-triacetate	Ac₂O/pTsOH	20	48	80	(105)
phorbol-9,12,13-triacetate	phorbol-pentaacetate	CH₃OH/NaOCH₃	4	20	90	(149)
phorbol-9,12,13,20-tetraacetate	phorbol-9,12,13-triacetate	Ac₂O/Py	20	8	90	(149)
phorbol-9,12,13,20-tetraacetate	phorbol-12,13,20-triacetate	Ac₂O/pTsOH	20	27	62	(149)
phorbol-4,12,13,20-tetraacetate	phorbol-12,13,20-triacetate	Ac₂O/NaOAc	115	26	48	(154)

Under appropriate reaction conditions, involving partial acetylation of phorbol or partial acid or base catalyzed *trans*esterification of phorbol-12,13,20-triacetate, the 12-, 13- and 20-monoacetates and the corresponding 12,13-, 12,20- and 13,20-diacetates are obtained (Table 11). From phorbol-pentaacetate, the 9,12,13-triacetate and from phorbol-12,13,20-triacetate the 4,12,13,20- and the 9,12,13,20-tetraacetates may be prepared (Table 11). By reaction of phorbol with p-bromobenzoyl- or 5-bromofuroyl chloride/pyridine at 0° C the corresponding 20-esters may be obtained (*43*). At room temperature, treatment of phorbol with benzoyl chloride/pyridine or substituted benzoyl chlorides(p-nitro-, p-chloro-)/pyridine yields the corresponding 12,13,20-triacylates (*43, 105, 116*). Phorbol-12,13,20-tribenzoate may be converted to phorbol-12,13-di-benzoate by base catalyzed *trans*esterification in methanol (*149*).

In esterifications and *trans*esterifications of the three hydroxyl functions at positions 12, 13 and 20 of phorbol OH-20 is in general the most reactive, the least reactive being OH-12. In particular, as compared to simple tertiary hydroxyl or ester groups the free or esterified tertiary cyclopropanol exhibits remarkable reactivity in both esterification or *trans*esterification reactions (Table 11). For example, in acetic anhydride/DMF, OH-13 of phorbol is acetylated even more rapidly than OH-20. Thus, this hydroxyl group was considered as "quasi enolic" or "quasi phenolic" (*183*). Further-more it is of interest that *trans*esterification of phorbol 12,13,20-triacetate to yield phorbol-12-acetate, can be accomplished using alkali concentrations well below those necessary to obtain phorbol-9,12,13-triacetate from phorbol-pentaacetate. It is remarkable that in phorbol-pentaacetate, the acetoxyl group at C-4 as part of an α-acetoxy-ketone is more easily *trans*esterified than the secondary acetoxyl at C-12, the tertiary acetoxyl at C-9 or even the con-siderably reactive cyclopropane acetoxyl at C-13. One observes that the acetoxyl group at C-13 even appears to be stable towards alkali catalyzed *trans*esterification if the hydroxyl in position 9 is acetylated. A hydrogen bridge from 9-OH to the carbonyl group of ester functions at C-13 (KREIBICH and HECKER, unpublished) seems to be responsible for these effects (for a review of OH-group participation in ester hydrolysis see *34*).

The variable reactivity of the hydroxyl groups in phorbol was used in the partial synthesis of some of the croton oil factors (see Section 3.3.). C-20 derivatives of the croton oil factors and of phorbol-12,13-diacetate were prepared by reaction with substituted benzoyl chlorides/pyridine. In this way, mixed 12,13,20-triesters of phorbol, some of which were labelled with heavy atoms, were obtained for characterization (see Table 5) and for X-ray analysis (*43, 90, 107*).

Phorbol Ethers. For alkylation of the hydroxyl groups of phorbol, comparatively mild methods have to be chosen. The hydroxyl groups at the 4-, 12-, 13- and 20-positions of phorbol or its derivatives may be alkylated using the specific conditions shown in Table 12.

By treatment of phorbol with trityl chloride/pyridine 20-OH is selectively tritylated in high yield (Table 12). Using diazomethane/ Al-i-propylate, the 13,20-dimethylether is obtained (Table 12). Similarly, by treatment of phorbol-20-tritylether with diazoethane the corresponding 12- and 13-O-ethyl-derivatives are obtained. To avoid loss of acetyl groups in the alkylation of the primary hydroxyl group of suitable phorbolacetates, diazoalkane/Al-i-propylate and methyl iodide/silver oxide/ethylacetate may be used. If, in reactions with silver oxide, 13-OH is free in the reactant, it will be oxidized to yield from e. g. phorbol-12,20-diacetate, bisdehydrophorbol-12,20-diacetate* (Table 12, see also below and Section 3.4.2.). With methyl iodide/silver oxide, DMF and with methyl iodide/barium oxide/barium hydroxide/DMSO, alkylation of 4-OH takes place in accordance with the reactivity of this ketol hydroxyl group in esterification reactions (see above). Under more basic conditions (see Table 12) concomitant exchange of acetyl with alkyl groups may take place. However, epimerization of 4-OH (see Section 4.1.) was not observed. Under the most drastic conditions, i.e. in the last system recorded in Table 12, phorbol-12,13,20-triacetate furnishes, together with 4,13,20-tri-O-methylphorbol-12-acetate, traces of 4,9,13,20-tetra-O-methylphorbol-12-acetate.

In the silver oxide/DMF reaction, it is important to use a large excess of methyl iodide and catalytic amounts of DMF. One of the more interesting side products which arise if DMF is present in excess is the 20-OH elimination product 3 (Fig. 21). Furthermore it is interesting to see (Table 12) that in phorbol-12,13,20-triacetate — besides methylation of 4-OH — during prolonged exposure to methyl iodide/silver oxide/DMF the acetyl group at 13-OH is exchanged but not that at 20-OH as in the barium oxide/ barium hydroxide system.

The ether group in tritylethers of phorbol and derivatives may be removed selectively under mild acidic conditions (*121*). This group has proved of particular preparative value in the chemical modification of the phorbol structure, including labelling with tritium (e. g. *12, 31, 121*). It was further observed that 20-O-methylphorbol-9,12,13-triacetate, on

---

* Bisdehydrophorbol is a trivial name and thus the position of dehydrogenation as regards the numbering system of phorbol is not given. In general, in the following presentation of phorbol chemistry the trivial names introduced already in the literature will be used, if they are shorter than the corresponding systematic names referring to the parent hydrocarbon tigliane.

References, pp. 458—467

Table 12. *Preparation of Phorbol Ethers*

Py = pyridine, DMF = dimethyl formamide, DMSO = dimethylsulfoxide

Reaction products	Reactants		temp. [°C]	react. time [hours]	yield [% of th.]	lit.
phorbol-20-tritylether	phorbol	tritylchloride/Py	20	20	77	(12, 120)
phorbol-13,20-dimethylether	phorbol	diazoalkane/	20	6	42	(120)
phorbol-12-ethyl-20-tritylether	phorbol-20-tritylether	Al-i-propylate,	20	6	52	(120)
phorbol-13-ethyl-20-tritylether		div. ethers		6	19	(120)
20-O-methylphorbol-12,13-diacetate	phorbol-12,13-diacetate		20	6	70	(120)
20-O-methylphorbol-9,12,13-triacetate	phorbol-9,12,13-triacetate		20	6	78	(120)
4,20-di-O-methylphorbol-12,13-diacetate	4-O-methylphorbol-12,13-diacetate		20	2	71	(120)
20-O-methylphorbol-12,13-diacetate	phorbol-12,13-diacetate	CH_3J/Ag_2O, ethyl-acetate	20	100	50	(120)
bisdehydrophorbol-12,20-diacetate	phorbol-12,20-diacetate		20	13	76	(120)
4-O-methylbisdehydrophorbol-12,20-diacetate	4-O-methylphorbol-12,20-diacetate		20	72	78	(60)
4-O-methylphorbol-12,13,20-triacetate	phorbol-12,13,20-triacetate	CH_3J/Ag_2O, DMF[a]	20	44	30[c]	(120)[d] / d
4,13-di-O-methylphorbol-12,20-diacetate					28	d
4-O-methylphorbol-13,20-diacetate	phorbol-13,20-diacetate	CH_3J/Ag_2O, DMF[b]	20	15	45[c]	(120) / d
4-O-methylphorbol-12,13,20-triacetate	phorbol-12,13,20-triacetate		20	6	82	d
4-O-methylphorbol-12,13,20-tridecanoate	phorbol-12,13,20-tridecanoate		20	7	80	d
4,20-di-O-methylphorbol-12,13-diacetate	phorbol-12,13,20-triacetate	$CH_3J/BaO/Ba(OH)_2$ DMSO	20	8	42	d
4,13,20-tri-O-methylphorbol-12-acetate					12	d

[a] DMF as solvent.   [b] DMF in catalytic amounts.   [c] product plus many side products.   [d] JACOBI and HECKER, unpublished.

treatment with acetic anhydride/p-toluenesulfonic acid, was converted to phorbol pentaacetate (*120*).

Oxidation of Single Hydroxyl Groups. The primary allylic hydroxyl of phorbol-12,13-diesters may be oxidized with manganese dioxide or chromium trioxide/pyridine to yield the corresponding 20-aldehydes (*98, 121, 149, 154*, Borchert and Hecker, unpublished). Such aldehydes are formed also by autoxidation of the croton oil factors during their isolation.

2,4-dinitrophenylhydrazones were prepared from 20-aldehydes of the croton oil factors for characterization (see Table 5). 20-Deoxy-20-oxophorbol-12,13-diacetate was used to obtain a 2,4-dinitro-5-iodophenylhydrazone for X-ray structural analysis (*90*) and it was converted to the corresponding nitrile *via* its oxime (*149*). Furthermore, 20-aldehydes may be used to introduce tritium into the phorbol molecule by reduction with sodium borohydride-^{3}H (*121*).

By treatment with selenium dioxide/hydrogen peroxide in t-butanol, 20-deoxy-20-oxo-phorbol-12,13-diacetate may be converted to the corresponding acid (*154*).

Oxidation of the secondary 12-OH in phorbol-13,20-diacetate by chromium trioxide/pyridine yields 12-deoxy-12-oxophorbol-13,20-diacetate (*14*). Even under drastic conditions, the latter forms only a mono-2,4-dinitrophenylhydrazone involving the carbonyl group at C-3. Thus the newly introduced carbonyl at C-12 exhibits comparatively low reactivity as do the corresponding 12-hydroxyl- or acyl-groups. This allows selective reduction of the carbonyl at C-3 in 12-deoxy-12-oxo-phorbol-13,20-diacetate with sodium borohydride: the 13,20-diacetate of a positional isomer of phorbol with a 3β-hydroxy and a 12-keto-group is formed; it is called "neophorbol" (*14*).

In neophorbol-13,20-diacetate, 3β-OH may be esterified to yield 3-O-[p-bromobenzoyl] neophorbol-13,20-diacetate, which was the key compound for X-ray structural analysis of phorbol (*90, 107*). Attempts to react the carbonyl group at C-12 in neophorbol-13,20-diacetate with carbonyl reagents failed (Borchert and Hecker, unpublished).

Simple tertiary hydroxyls and also tertiary 1,2-ketols are quite resistant to oxidation under moderate conditions. That tertiary cyclopropanols are easily oxidized became apparent (Table 13) in comparative studies of the reduction of Tollens's and Fehling's reagents by phorbol, 3-deoxo-3β-hydroxyphorbol and suitable phorbol-ethers (*8, 77, 119, 120*). At the same time, the facile oxidation of simple tertiary cyclopropanols was independently detected by Schaafsma, Steinberg and DeBoor (*147*).

As may be seen from Table 13 the tertiary cyclopropane hydroxyl is the structural unit which causes reduction of the reagents by phorbol and derivatives with free 13-OH. Similarly, in attempts to methylate hydroxyl groups in phorbol-12,20-diacetate and 4-O-methylphorbol-12,20-diacetate, using methyl iodide/silver oxide, ethylacetate as methylating agent (Table 12) oxidation at 13-OH takes place in the same manner as is observed on treatment of phorbol-12,20-diacetate with lead dioxide or lead tetraacetate (see Section 3.4.2). Phorbol-12,13,20-triacetate remains essentially unchanged with lead dioxide or lead tetraacetate (*61, 119, 120*).

Table 13. *Reduction of Tollens's and Fehling's Reagent by Phorbol and Derivatives (119, 120)*

Compound	Tollen's reagent	Fehling's reagent
phorbol	+	+
3-deoxo-3β-hydroxyphorbol	+	+
phorbol-20-tritylether	+	+
phorbol-13,20-dimethylether	0	0
13-O-ethylphorbol-20-tritylether	0	0
12-O-ethylphorbol-20-tritylether	+	+

Oxidation of cyclopropanols may be started by abstraction of the proton from the hydroxyl group generating a cation (Fig. 20) or a radical which can be stabilized as a carbonyl group with displacement of charge over to one of the neighbouring C-atoms, in phorbol preferably to the tertiary center at C-15. This tertiary cation or radical can follow several possible stabilization pathways, as will be discussed in Sections 3.4.2 and 3.4.3.

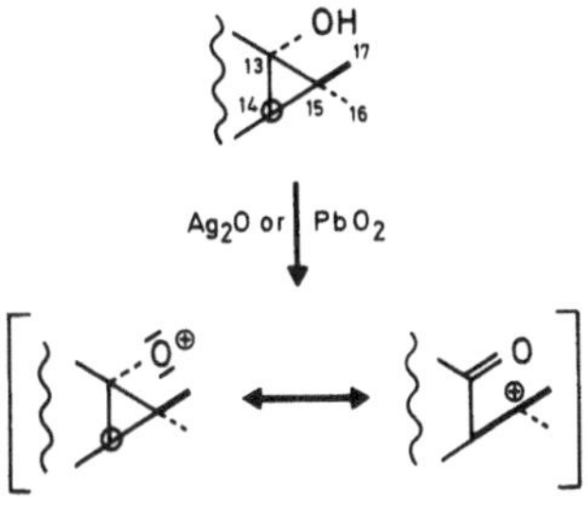

Fig. 20. Oxidation of the tertiary cyclopropanol group as cationic mechanism

Substitution and Elimination of Hydroxyl Groups. In phorbol-12,13-diacetate and other phorbol derivatives treatment with mesyl-chloride/pyridine results in the *substitution* of OH-20 by chlorine (*154, 165*).

In attempts to substitute OH-12 in phorbol-13,20-diacetate by halogen even with mildly active reagents such as methyl iodide/triphenylphosphite or thionylchloride/magnesium oxide (Borchert and Hecker, unpublished), rearrangements of the cyclopropane ring were induced (see also Section 3.4.3). All trials to replace the secondary 12-OH in phorbol derivatives by hydrogen failed (Borchert and Hecker, unpublished). However, esters of 12-deoxyphorbol have been isolated from natural sources (63).

In phorbol-12,13,20-triacetate, phorbol-12,13-didecanoate-20-acetate and in neophorbol-13,20-diacetate-3-benzoate, substitution of the primary allylic hydroxyl or acetoxyl group by hydrogen can be accomplished by Raney Nickel/ethanol. The reaction conditions here must be carefully controlled in order to avoid concomitant saturation of at least the 6,7-double bond. Minor amounts of the 20-deoxy-6,7-dihydro products formed may be removed on columns of silica gel impregnated with silvernitrate. Under similar conditions phorbol-12,13,20-tridecanoate does not react (Borchert and Hecker, unpublished). The reduction of 20-chloro compounds to yield 20-deoxy-phorbol derivatives is less sensitive with respect to the reaction conditions (154).

In phorbol-12,13,20-triacetate, the 3,4-ketol group undergoes the typical acyloin reduction with zinc/acetic acid to yield — not the expected 4-deoxyphorbol-12,13,20-triacetate — but rather 4-deoxy-4α-phorbol-12,13,20-triacetate (58). A mechanism for this reaction will be discussed in Section 4.2.

Using strongly basic conditions, the 20-OAc group in phorbol-12,13, 20-triacetate may be *eliminated* with concomitant isomerisation of $\Delta^{6,7}$ (Fig. 21) to yield 12β,13-diacetoxy-4,9-dihydroxy-1,6(20),7-tigliatrien-3-one (3) (Jacobi and Hecker, unpublished).

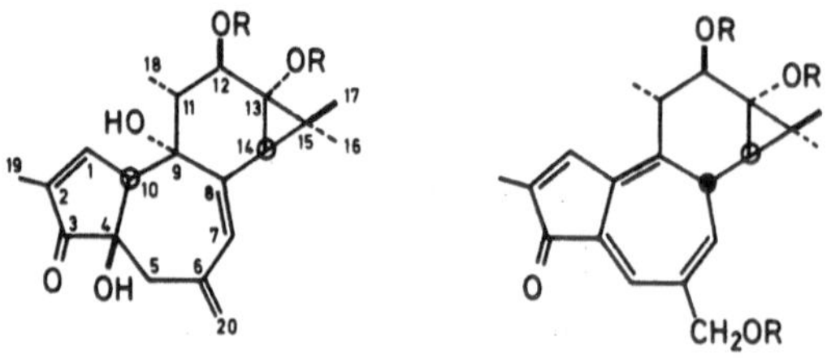

3
R = acetyl

4
R = decanoyl

Fig. 21. Base catalyzed elimination of 20-OAc from phorbol-12,13,20-triacetate (1b) to yield 3 and dehydration of both, 9-OH and 4-OH, from phorbol-12,13,20-tridecanoate to yield 4 (from 154, Jacobi and Hecker, unpublished)

Attempts to introduce a 11,12-double bond into the phorbol molecule by elimination of 12-OH induced homoallyl rearrangement of the cyclopropane ring as will be described in Section 3.4.3. In principle elimination of 12-OH to form a 11,12-double bond is possible, if the neighboring cyclopropane ring is missing (see Section 3.4.2.).

From 4-O-acetylphorbol-12,13,20-tridecanoate, 9-OH is eliminated by thionylchloride/pyridine to yield 4-O-acetyl-9,10-anhydrophorbol-12,13,20-tridecanoate. Under similar conditions, phorbol-12,13,20-tridecanoate provides 4,5:9,10-dianhydrophorbol-12,13,20-tridecanoate (**4**) (Fig. 21), exhibiting a cycloheptatrienyl-system (*154*).

3.2.3. Functional Derivatives Involving the C=C-Bonds

Catalytic Hydrogenation. If phorbol-12,13,20-triacetate is subjected to catalytic hydrogenation with palladium on charcoal, uptake of three moles of hydrogen is observed (*102, 105*). Also, phorbol consumes three moles of hydrogen with platinum as catalyst (*116*). In addition to the two moles of hydrogen used for saturation of $\Delta^{1,2}$ and $\Delta^{6,7}$ a third mole is consumed by hydrogenolysis of the allyl acetoxyl or allyl hydroxyl group to yield 20-deoxy-1,2ξ,6ξ,7-tetrahydrophorbol-12,13-diacetate (**5**, Fig. 22) or 20-deoxy-1,2ξ,6ξ,7-tetrahydrophorbol respectively.

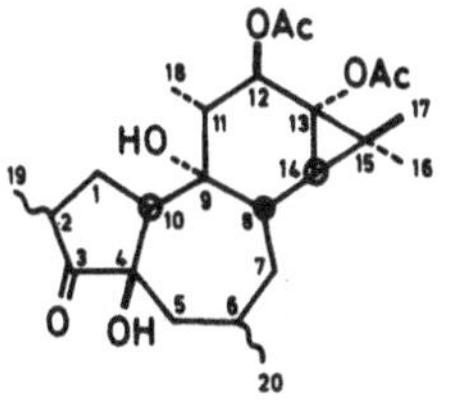

5

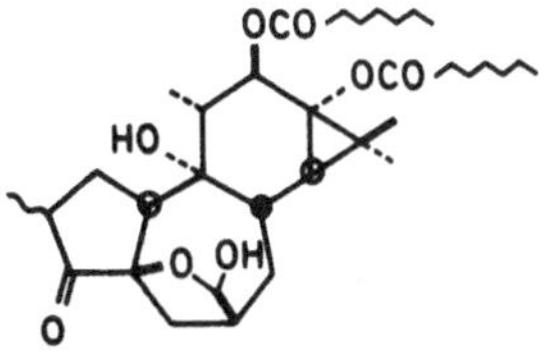

6

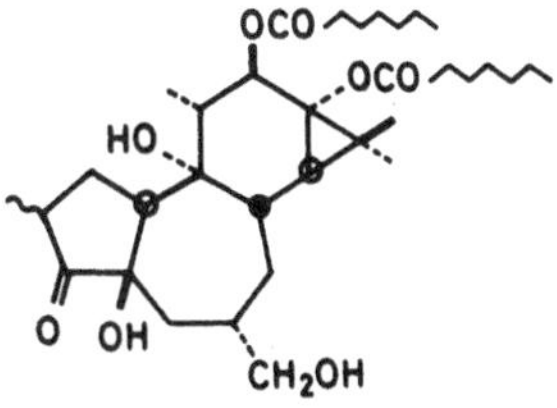

7

Fig. 22. Catalytic hydrogenation of phorbol-12,13,20-triacetate (**1b**) to yield **5** and of 20-deoxy-20-oxo-phorbol-12,13-didecanoate to yield **6** and **7**, respectively (from *105, 154*)

Under milder hydrogenation conditions (Borchert and Hecker, unpublished), but also using Adams catalyst (43), uptake of only two moles of hydrogen was observed to yield derivatives of 20-deoxy-6ξ,7-dihydrophorbol. By hydrogenation of phorbol with Pd/charcoal and subsequent acetylation the 20-deoxy compound 5 and in addition 1,2ξ,6ξ,7-tetrahydro-phorbol-12,13,20-triacetate are obtained (Schmidt and Hecker, unpublished). — Some 20-deoxy-tetrahydro-derivatives of phorbol have been used to demonstrate the UV-absorption of its cyclopropane ring below 200 nm (see Table 9).

By catalytic hydrogenation of 20-deoxy-20-oxophorbol-12,13-dide-canoate with palladium/charcoal, both double bonds may be saturated to yield a mixture of the semiacetal 6 and compound 7 (Fig. 22). The spontaneous formation of 6 during hydrogenation indicates that the configuration of the formyl group involved is β. Alternatively, the formyl-group of the 6α-epimer is not protected and thus is reduced to yield 1,2ξ,6β,7-tetrahydrophorbol-12,13-didecanoate (7) (154).

Paal and Roth (134) have reported that croton resin loses its sharp taste and toxicity on hydrogenation with colloidal platinum. The formation of a phenolic compound, mp. 96° C, by catalytic hydrogenation of an unidentified mixture of cocarcinogenic compounds from croton oil (175) remains unexplained.

Bromination and Hydrobromination. Phorbol (**1a**) consumes bromine and iodine vigorously, but derivatives were not obtained (50, 116).

By treatment of phorbolpentaacetate (**1c**, Fig. 23) with bromine in carbon tetrachloride at room temperature, 10-H is substituted with reten-

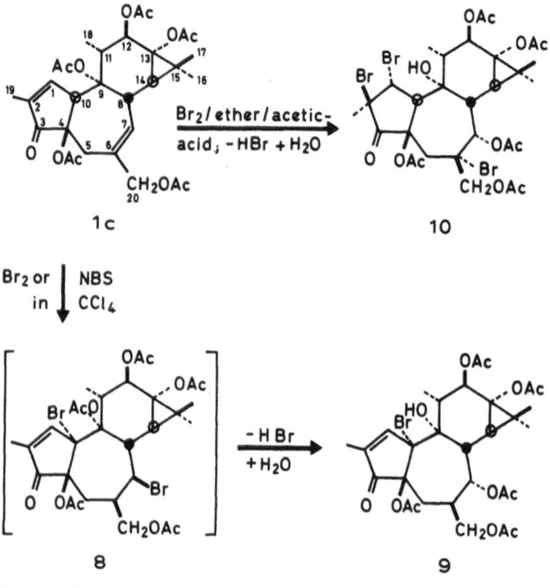

Fig. 23. Bromination of phorbol pentaacetate (**1c**) under various conditions (from 167)

tion of the configuration. Concomitantly, hydrogen bromide formed in the reaction saturates $\Delta^{6,7}$ to yield, as primary product of the reaction, the labile 4,9,12β,13,20-pentaacetoxy-7β,10-dibromo-6αH-1-tiglien-3-one (8). Most probably, during chromatographic purification, 8 undergoes intramolecular acetylmigration 9→7 with hydrolysis, yielding as the final main product 4,7α,12β,13,20-pentaacetoxy-10-bromo-9-hydroxy-6αH-1-tiglien-3-one (9). — Compound 9 can also be obtained by reaction of 1c with N-bromosuccinimide in carbon tetrachloride and chromatographic purification (*167*).

Treatment of 1c with bromine in ether/glacial acetic acid yields as primary product the labile 1,2,6,7-tetrabromo derivative. Again, during chromatographic purification this compound is stabilized by intramolecular acetyl migration 9→7 with hydrolysis, to yield as final main product, 4,7α,12β,13,20-pentaacetoxy-1α,2,6-tribromo-9-hydroxy-2βH,6αH-tiglian-3-one (10) (*167*).

Oxidation. By reaction of phorbol-12,13,20-triacetate (1b) (Fig. 24) with osmium tetroxide, the corresponding 6β,7β-dihydroxy-12,13,20-triacetate 11a is formed. In addition as a side product of this reaction some 6β,7β-dihydroxy-12,13-diacetate 11b is obtained. Similarly, in the reaction of phorbol-pentaacetate (1c) with osmium

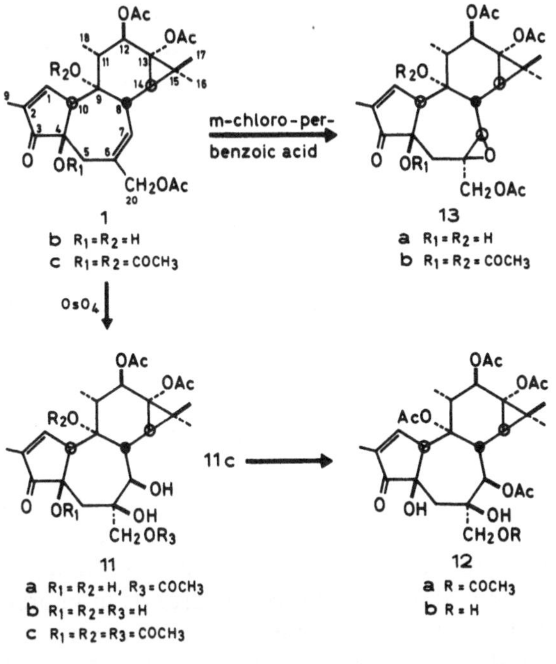

Fig. 24. Oxidation of phorbol-12,13,20-triacetate (1b) and phorbolpentaacetate (1c) with osmium-tetroxide and with peracids (from *166*, SCHMIDT, THIELMANN and HECKER, unpublished)

tetroxide formation of the 6β,7β-dihydroxy-4,9,12,13,20-pentaacetate (11c) is expected, but — obviously under the weakly basic working up conditions — 11c is not stable: By an acetyl migration 4β-OAc→7β-OH, 11c yields the corresponding 6β-hydroxy-7β,9,12β,13,20-pentaacetate 12a besides some 7β,9,12β,13-tetraacetate 12b (*166*). This acetylmigration is reminiscent of the migration 4β-OAc→3β-OH following reduction of phorbol-pentaacetate with sodium borohydride (Table 10). Also, it accords with the easy base-catalyzed *trans*esterification of 4-OAc and 20-OAc in phorbol-pentaacetate to yield phorbol-9,12,13-triacetate (see Section 3.2.2, Table 11).

The 6,7-double bond in phorbol (1a), phorbol-12,13,20-triacetate (1b) and phorbol pentaacetate (1c) may be epoxidized by peracids (*43, 115, 149, 155*). In all cases the 6β,7β-epoxides are formed preferentially. From oxidations of phorbol with perbenzoic acid besides the β-epoxide the 6α,7α-isomer was isolated in low yield (SCHMIDT and HECKER, unpublished).

3.3. Structure of the Croton Oil Factors

After clarification of the structure of phorbol and of the reactivity of its functional groups, three problems remain regarding the croton oil factors isolated from the hydrophilic and the activated hydrophobic portions of croton oil:

(a) which two of the hydroxyl groups are esterified in the croton oil factors,

(b) what kind of isomerism is involved in the two pairs of croton oil factors, A_2/B_7 and A_3/B_4, with different R_f values, but, nevertheless, with the same acid moieties and identical molecular formulae (Table 6),

(c) what is the nature of the higher phorbol esters contained in in the hydrophobic portion of croton oil.

3.3.1. Phorbol Diesters from the Hydrophilic Portion

As revealed by spectroscopic and chemical methods (see Sections 2.3.2 and 2.4.2), in all of the 11 croton oil factors $A_1 - A_4$ and $B_1 - B_7$ which have been isolated from the hydrophilic portion, the primary hydroxyl in position 20 and two tertiary hydroxyls in phorbol are free. Further, of the two acid residues present in each of the croton oil factors, according to NMR-data, one can be localized at the secondary hydroxyl group at C-12 of phorbol. Thus one of the three tertiary hydroxyls in phorbol must carry the second acyl residue. Unlike the tertiary hydroxyl groups at C-4 and C-9, the tertiary cyclopropane hydroxyl is easily acetylated with acetic anhydride/pyridine at room temperature (*102, 183*). This remarkable reactivity suggests that the second esterified hydroxyl in the croton oil factors is the cyclopropane hydroxyl (13-OH). Since they contain a long and a short chain fatty acid each, esterification of the 12,13-glycol group in **1a** would provide two types of phorbol-12,13-diesters: type a = phorbol-12-long-chain-13-short-chain-diesters and type b = phorbol-12-short-chain-13-long-chain-diesters. This proposal was checked by partial synthesis of phorbol-12,13-diesters with acetic acid as the short chain and various straight chain, saturated fatty acids as the long chain moieties.

The synthetic program for the preparation of isomeric phorbol-12,13-acetate-acylates is summarized schematically in Fig. 25. Esterification of phorbol (**1a**) with either long-chain fatty acid chlorides or with acetic anhydride in pyridine yields phorbol-12,13,20-triacylates (**14**) or phorbol-12,13,20-triacetate (**1b**), respectively. In the next step, by base catalyzed *trans*esterification the phorbol-12-monoesters (**15a**) and (**15b**) are prepared. Acetylation of **15a** with acetic anhydride and of **15b** with long chain fatty acid chlorides, yields the corresponding mixed 12,13,20-triesters **16a** and **16b**, respectively. Finally, in **16a** and **16b** the ester groups at C-20 are removed selectively by acid catalyzed *trans*esterification. By these reaction sequences, the mixed phorbol-12,13-diesters **17a**, i. e., 12-O-acylphorbol-13-acetates (type a) and the corresponding positional isomers **17b**, i. e., 12-O-acetyl-phorbol-13-acylates (type b) are obtained. In their physical and chemical properties the synthetic phorbol-12,13-acetate-acylates of types a and b exhibit characteristic differences (*8, 31, 101*).

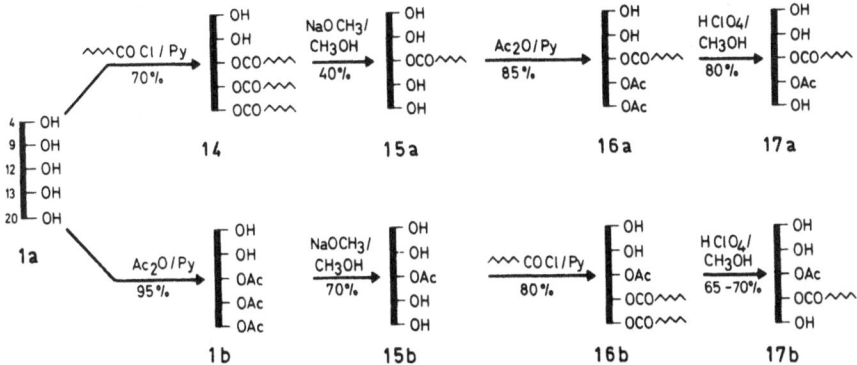

Fig. 25. Reaction sequences for partial synthesis of isomeric phorbol-12,13-acetate-acylates from phorbol (**1a**, schematically) Py = pyridine (from *31*)

As summarized in Table 14, diesters of type a show practically the same R_f value. This is smaller than the R_f values of the diesters of type b, which is again practically identical in all esters of this type. Also a characteristic difference between the two types of positional isomers with respect to the melting points of their NPAB-20-esters is observed. As a rule, type a NPAB-20-esters melt below 100° C, whereas type b NPAB-20-esters melt above 140° C (see Table 14). Further, in mass spectrometric fragmentation, the long chain acyloxy residue attached to C-12 of phorbol — as in type a esters — leaves the molecule as a radical. However, the long chain acyloxy residue attached to C-13 — as in type b esters — leaves the molecule as the acid (see Table 14, Fig. 26). Also, the IR spectra of 12-O-acylphorbol-13-acetates (type a) exhibit a rather broad band with a maximum around 1700 cm^{-1}, but those of 12-O-acetyl-phorbol-13-acylates (type b) show two clearly separated maxima (Table 14, Fig. 27). This difference in IR spectra is also apparent, if the spectra are measured in potassium bromide instead of in methylene chloride.

Table 14. *Physical Properties of Isomeric Phorbol-12,13-Acetate-Acylates Type a and b* (*31, 104*)

Properties	Type a	Type b
R_f value[a]	0.3	0.4
NPAB-20-ester[b]	m.p.< 100° C	m.p. >140° C
Mass spectrum[c]	M − $\wedge\!\wedge$COO˙	M − CH₃COO˙
	M − CH₃COOH	M − $\wedge\!\wedge$COOH
IR-spectrum	1710 cm^{-1}	1705 and 1736 cm^{-1}

[a] Methylene chloride/acetone = 3/1, silica gel "Merck HF 254".
[b] NPAB = 4-(4-nitrophenylazo-)benzoic acid.
[c] M = molecular ion, $\wedge\!\wedge$COOH long chain fatty acid, $\wedge\!\wedge$COO˙ corresponding radical.

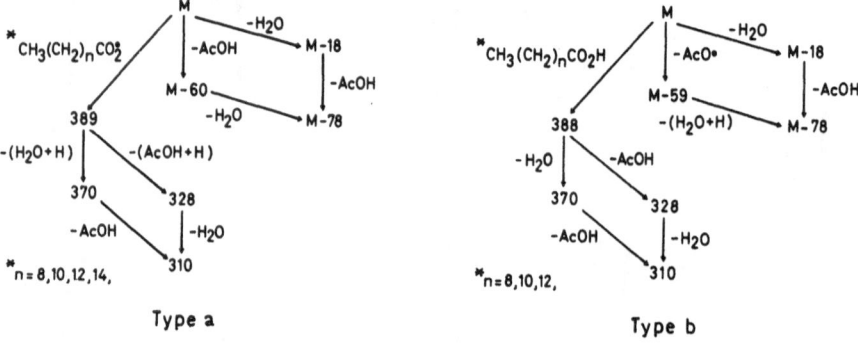

Fig. 26. General fragmentation pattern of isomeric phorbol-12,13-acetate-acylates types a and b in conventional high-resolution mass spectrometry (from *31, 104*)

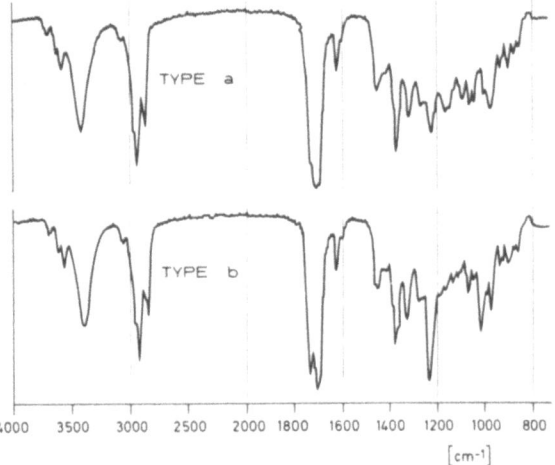

Fig. 27. IR-spectra of isomeric phorbol-12,13-acetate-acylates types a and b, e. g., 12-O-decanoylphorbol-13-acetate (type a) and 12-O-acetylphorbol-13-decanoate (type b) in methylene chloride (from *31, 104*)

Using these criteria, of the croton oil factors isolated from the hydrophilic portion all those carrying an acetyl group as short-chain acid (see Table 6) were identified with corresponding phorbol-12,13-acetate-acylates prepared synthetically (*31, 77, 79, 101*). Further, based upon the results of mass spectrometric investigation, several of the isolated croton oil factors of the hydrophilic portion with short chain acyl residues other than acetyl, were classified in a similar manner (*31*). The following generalization can be made: in factor group A, the 12,13-diesters carry their long-chain fatty acid residue in position 12 and their short-chain fatty acid residue in position 13 of phorbol (see Fig. 28). In factor group B the long- and short-chain fatty acid residues are located in 13- and 12-position, respectively.

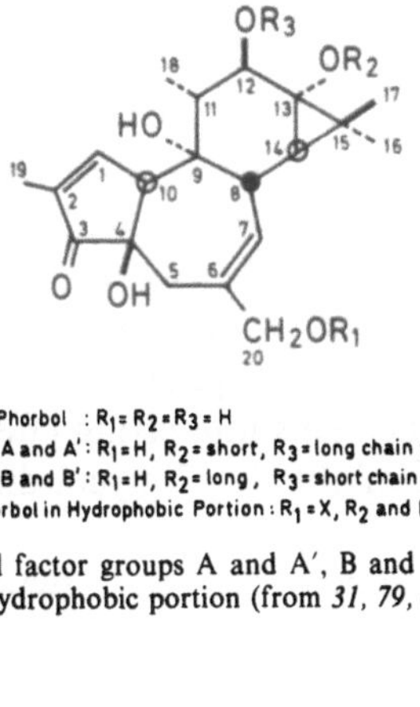

Phorbol : $R_1 = R_2 = R_3 = H$
Factor Group A and A' : $R_1 = H$, $R_2 =$ short, $R_3 =$ long chain fatty acid
Factor Group B and B' : $R_1 = H$, $R_2 =$ long, $R_3 =$ short chain fatty acid
Source of phorbol in Hydrophobic Portion : $R_1 = X$, R_2 and R_3 as above

Fig. 28. Structures of the croton oil factor groups A and A', B and B' and of the source of phorbol present in the hydrophobic portion (from *31, 79, 80, 101, 152*)

As a consequence, the isomeric pairs of factors A_2/B_7 and A_3/B_4 (Table 6) were recognized as positional isomers, as indicated in Table 15. Croton oil factor A_2 is 12-O-decanoylphorbol-13-acetate, whereas croton oil factor B_7 is the corresponding positional isomer. Similarly the factor A_3 carries its dodecanoyl moiety at the 12-position and its acetyl moiety at the 13-position of phorbol. In factor B_4, the position of both acid residues is inverted (see Fig. 28).

Table 15. *Identification of the Pairs of Isomeric Croton Oil Factors A_2/B_7 and A_3/B_4 as Positional Isomers (8, 31, 79, 80, 81)*

Croton oil factor isolated	R_f[a]	Identical with synthetic phorbol-12,13-diester carrying acyloxy residues at	
		C-12	C-13
A_2	0.3	decanoic	acetic
B_7	0.4	acetic	decanoic
A_3	0.3	dodecanoic	acetic
B_4	0.4	acetic	dodecanoic

[a] Methylene chloride/acetone = 3/1, silica gel "Merck HF 254".

By direct comparison of thin-layer chromatograms and mass spectra of authentic samples, the "amorphous material C" isolated by VAN DUUREN and co-workers (see Section 2.3.1) was found to be essentially pure and identical to croton oil factor A_1 (80, 81). However, mass spectrometry of the "amorphous material A" of these authors (see Section 2.3.1) was shown to be a mixture consisting mainly of croton oil factors B_1, B_2, B_4 and B_7 together with small amounts of croton oil factor A_1 (80, 81). Partial separation of the "amorphous materials A and C" as the corresponding NPAB-esters has been described (173). MEYER-BERTENRATH (127, 128) reported on the characterization of the four phorbol esters A_1, A_4^*, B_1 and B_2. — According to the physical and chemical data reported for compound C-3 (4, 5), it is identical with croton oil factor A_1. Yet it seems somewhat puzzling, that in a simple one-step reaction followed by thin-layer chromatographic purification (4, 5) a pure sample of croton oil factor A_1, *alias* C-3, could have been obtained.

Thus, the occurrence of the two factor groups A and B in the hydrophilic portion of croton oil is fully understood. It is surprising, that within one and the same factor group, variation in the length of the long-chain acyl residues of individual factors does not cause a significant variation in their R_f values. However, the position of the long-chain acyl residue in the phorbol moiety exerts a remarkable influence on the R_f value. This interesting effect may be connected with the *trans*-configuration of the 12,13-glycol-diester group (Fig. 28).

---

* This ester was originally called "A_3" (127). However it is identical with A_4.

3.3.2. Phorbol Triesters from the Hydrophobic Portion

The structural principles established for the phorbol-12,13-diesters of the hydrophilic portion are also valid for the 14 croton oil factors isolated from the activated hydrophobic portion as revealed by identification of eleven of them with the phorbol-12,13-diesters $A_1 - A_4$ and $B_1 - B_7$ occurring in the hydrophilic portion (see Section 2.4.2). Moreover, by the mass spectrometric and the IR criteria (Table 14), A_5' as well as B_8' and B_9' were seen to belong to the A and B groups as suggested by their R_f-values (Table 8). Thus they were identified as 12-O-tiglyl-phorbol-13-butyroate (A_5'), 12-O-tiglyl-(B_8') and 12-O-butyryl-(B_9')-phorbol-13-dodecanoates.

The phorbol-12,13-diesters isolated from the activated hydrophobic portion occur in the hydrophobic portion of croton oil as biologically relatively inactive entities. From the acid catalyzed *trans*esterifications in phorbol-12,13,20-triesters as well as in phorbol-pentaacetate, by which esters at C-20 can be selectively removed (see Table 11, and Fig. 25), it may be concluded, that the phorbol derivatives contained in the hydrophobic portion are phorbol-12,13,20-triesters, carrying an unknown acyl-residue X at the 20-position (Fig. 28). Indeed, acetylation (KUBINYI and HECKER, unpublished) or acylation of 20-OH in phorbol-12,13-diesters causes a decrease in irritant activity (*163*).

The greater difficulty of selective *trans*esterification of the 20-ester group in phorbol-12,13,20-tribenzoate and in phorbol-12,13,20-triacylates with long chain aliphatic fatty acids at the 20-position (see Section 3.2.2) suggests that X may be a relatively short chain aliphatic acyl group, perhaps an acetyl residue.

It may be noted here that the 14 different phorbol diesters isolated contain seven of the 12 fatty acids shown to be present in croton oil (*53*). Phorbol esters with formic, stearic, oleic, linoleic or arachidonic acid were not detected. In particular, none of the phorbol diesters isolated contains crotonic acid, nor was this acid found in croton oil (*53*). Thus, as early as 1872, Kekulé rightly stated (*117*) that of the postulated existence of this acid in croton oil the only connection was in the name. 100 years after this statement, it is amusing to see several text books of chemistry and biochemistry still recording crotonic acid as a constituent of croton oil.

The finding of phorbol-12,13,20-triesters in the hydrophobic portion of croton oil clarifies the previously inexplicable observation by BUCHHEIM (*32, 33*), that mild treatment of the ethanol insoluble (i. e. hydrophobic) portion of croton oil with alkali restored its sharp taste and irritant properties. Under the conditions employed this was obviously caused by partial hydrolysis of biologically relatively inactive phorbol-12,13,20-triesters to give the active phorbol-12,13-diesters. Since selective *trans*esterification is needed to release the active phorbol-12,13-diesters, a reaction possibly caused also *in skin* by esterases, phorbol-12,13,20-triesters comprise a sort of "cryptic" biological activity.

3.4. Reactions Altering the Tigliane Skeleton of Phorbol and of Neophorbol

3.4.1. Dehydrogenation of Phorbol and 3-Deoxo-3 β-hydroxyphorbol

By dehydrogenation of phorbol (**1a**) or of 3-deoxo-3β-hydroxyphorbol with selenium or with palladium/charcoal, mixtures of blue and violet azulenes are obtained in low yield. It was demonstrated that they consist of five to six components, some of them still containing oxygen. Mixtures of the same azulenes are obtained in slightly better yields by thermal decomposition of 3-deoxo-3β-hydroxyphorbol. Under these conditions phorbol does not yield azulenes. As compared with the UV-spectrum of guaiazulene the main bands of the UV-spectra of the azulenes are shifted towards higher wave length; they are reminiscent of the spectra of benz-azulenes (*66*).

Phorbobutanone (see Section 3.4.3) on treatment with palladium/charcoal or selenium also yields a mixture of azulenes. Three of these are identical with the azulenes obtained under similar conditions from phorbol. Hence phorbobutanone may be considered as a precursor of some of the azulenes formed from phorbol (*12*).

Due to the low yields of the azulenes, and hence due to their restricted value for the elucidation of the structure of the phorbol skeleton, more detailed investigations of the structures of these azulenes were not undertaken (*66*).

3.4.2. Oxidative Ring Opening of Phorbol and Derivatives

Ring A. In phorbol-12,13,20-triacetate the tertiary 3,4-ketol group is relatively stable towards sodium periodate and lead tetraacetate (see Section 3.2.2). However, the corresponding *cis*-3β,4-glycol group in 3-deoxo-3β-hydroxyphorbol-12,13,20-triacetate (**18**) is rapidly split with sodium periodate in dioxan/water to yield as a crystalline precipitate from the reaction mixture the 3,4-*seco*-compound (**19**) (Fig. 29) i.e. 12β,13,20-triacetoxy-9-hydroxy-4-oxo-3,4-seco-1,6-tigliadien-3-al. As a derivative of glutacondialdehyde **19** is very sensitive. A similar chromophore is exhibited by $\triangle^5$-cholestene-3,7-dione (**20**).

19 may be dissolved in methanol to yield — as the result of eno-lization — a yellow solution, the colour of which disappears on acidification (Table 16).

Both carbonyl groups in 19 may be reduced by sodium borohydride to yield the more stable tetrahydro derivative 21 (58).

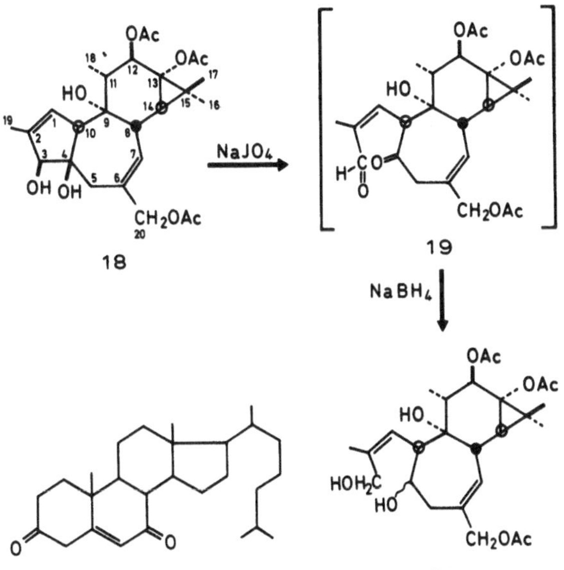

Fig. 29. Scission of the 3β,4-glycol group in 3-deoxo-3β-hydroxyphorbol-12,13,20-triacetate (18) by sodium periodate to yield the glutacondialdehyde derivative 19 and stabilization of 19 by reduction with sodium borohydride to yield 21; comparison of the enone system of 19 with that in 20, see text (from 58)

Ring B. The seven membered ring in the 6β,7β-dihydroxy-compound 11 a, obtained from phorbol-12,13,20-triacetate (1 b) by treatment with osmium tetroxide (see Section 3.2.3.), is cleaved by lead tetraacetate or sodium periodate (Fig. 30) to yield, *via* the 6,7-*seco*-compound, the 4,7-semiacetal 22 i. e. 12β,13,20-triacetoxy-7β,9-dihydroxy-4,7-epoxy-6,7-*seco*-1-tigliene-3,6-dione (*166*).

In the 6β,7β-diol-12,13-diacetate 11b all hydroxyl functions of the glycerol-like partial structure at C-20, C-6 and C-7 are accessible to oxidation. Accordingly 11b may be oxidized with lead tetraacetate or sodium periodate to yield the corresponding 20-*nor*-6,7β-ketol-12,13-diacetate 23. Reduction of 23 with sodium borohydride yields the corresponding 3β,6ξ-diol-12,13-diacetate 24 and, on subsequent ace-

Table 16. *UV-Absorption of Glutacondialdehyde and Derivatives (58)*

Compound	$\lambda_{max}, \varepsilon_{max}$ [nm]	$\lambda_{max}, \varepsilon_{max}$ [nm]	$\lambda_{max}, \varepsilon_{max}$ [nm]
glutacondialdehyde, Na-salt[a]	228,5300	365,6700*	—
Seco-compound 19	225,18700	323,7600**	—
Seco-compound, Na-salt	225,2000	295,11000	408,44000**
3-Hydroxy-$\Delta^{3,5}$-cholestadien-7-one[b]	—	320,24300***	393,62200***
3-Hydroxy-$\Delta^{3,5}$-cholestadien-7-one[b], Na-salt	—	—	

* In water. ** In acidified or alcaline methanol (~0.1 n). *** In ethanol or alcaline ethanol (~0.1 n).
[a] (150). [b] (156).

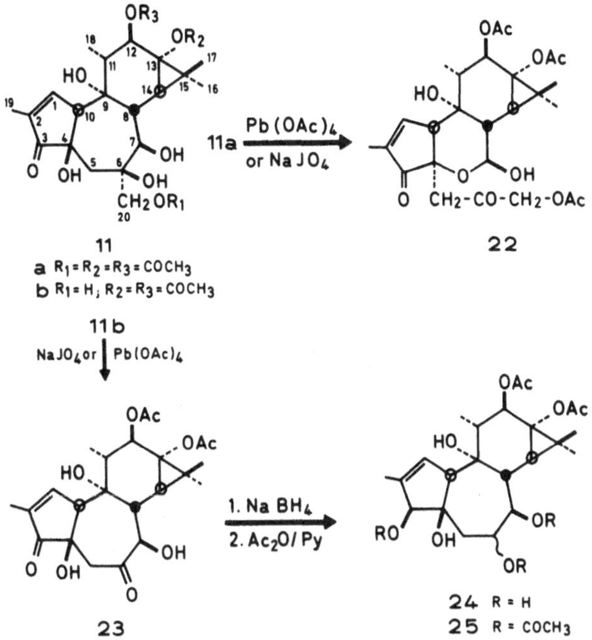

Fig. 30. Oxidative ring scission at C-6/C-7 of phorbol-derivatives to yield the 6,7-*seco*-compound **22** and at C-6/C-20 to yield the 20-nor-compound **23** (from *166*)

tylation, the pentaacetate **25**. These derivatives proved to be key compounds for the location of CH_2-5 in the phorbol skeleton by NMR (*149, 166*).

Rings C and D (Bicyclo[4.1.0]heptane System). As pointed out in Section 3.2.2 the 13-cyclopropanol group of phorbol is easily oxidized with Tollens' and Fehling's reagents as well as silver oxide and lead dioxide, resulting in oxidative scission of ring D. The products of oxidation with lead tetraacetate played an important role in the elucidation of the structure and stereochemistry of phorbol.

With one mole of lead tetraacetate in dioxan, phorbol-12,20-di-acetate yields (see Figs. 20, 31) bisdehydrophorbol-12,20-diacetate (**26a**): the intermediate C-15 cation (see Fig. 20) is stabilized by loss of a proton from C-16 or C-17 (*61*). Also phorbol · C_2H_5OH, dissolved in water, readily consumes one mole of lead tetraacetate (*114*) to yield essentially an equimolar mixture of bisdehydrophorbol (**26b**) and tiglophorbol (**27**) (*61*). The latter precipitates from the reaction mixture. In the formation of **27**, the intermediate C-15 cation (Fig. 20) is intramolecularly trapped by 12β-OH with concomitant formation of the 9→13 semiketal. Oxidation of phorbol · C_2H_5OH with one mole of sodium periodate in water (Fig. 31) also yields mainly bisdehydrophorbol (**26b**) and tiglophorbol (**27**) in approximately equal amounts but, in addition,

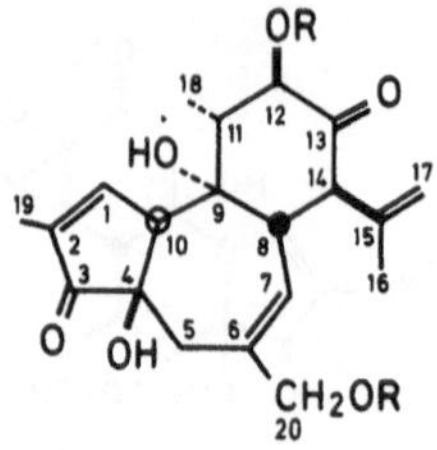

Fig. 31. Products of oxidative scission of ring D of phorbol (1a) with one mole of lead tetraacetate or sodium periodate: bisdehydrophorbol (26b), tiglophorbol (27), hydroxybisdehydrophorbol-semiketal (28a) (from *61, 62*)

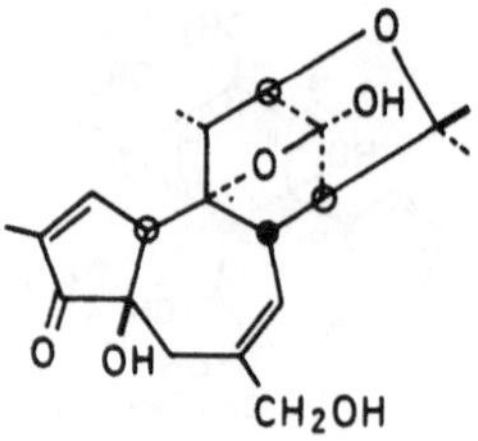

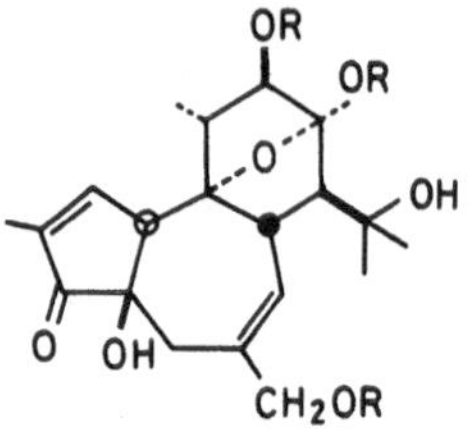

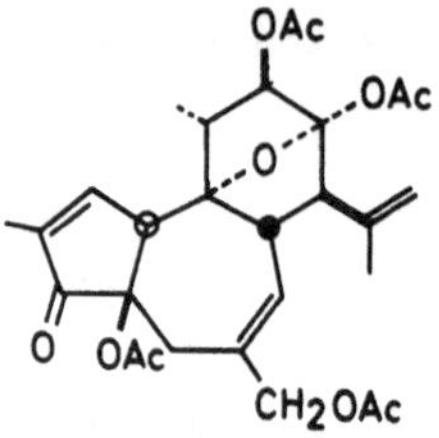

also hydroxybisdehydrophorbol-9,13-semiketal (28a) (*62*). In the latter case the C-15 cation reacted with water from the solvent. Again, as in 27, concomitant formation of the 9→13 semiketal is observed. 26b and 28a are further cleaved at C-12, 13 by either lead tetraacetate or sodium periodate (see below and Fig. 32). Thus, it is not surprising that even in reactions with one mole of oxidant also traces of the corresponding 12,13-seco-compounds are found (*61, 62*).

Cleavage of the 13,15-bond as the result of the oxidation of phorbol with lead tetraacetate and sodium periodate indicates that the selective oxidation of the tertiary cyclopropanol group is much faster than the possible scission of the *trans*-12,13-glycol group. If the latter would compete for oxidant, a primary reaction product would be the 12,13-*seco*-compound carrying an aldehyde group at C-12 and a cyclopropanecarbonyl at C-13.

While during formation of 26b a free carbonyl group at C-13 is maintained, this carbonyl group forms a semiketal with 9-OH in 27 and 28a. Obviously, the formation of an ether bridge by the cation at C-15 induces a boat conformation of Ring C in 27 allowing for semiketal-bridging (see also Fig. 33, *62*). In contrast to 26b, which yields 26a on treatment with acetic anhydride/pyridine, under similar conditions from

28a the 12,13,20-triacetate **28b** is obtained. This indicates that the formation of the 9→13-semiketal is not induced by the conditions employed during the acetylation, but rather occurs during the oxidation reaction itself (*62*). Thus, in case of **28a** and **b** it may be concluded, that the oxygen function present at C-15 in these molecules may stabilize a boat form of ring C (see also **28b** in Fig. 33), e. g. by hydrogen bonding between 15-OH and 12-OH.

From **26a**, by lithium aluminium hydride reduction and subsequent reaction with acetic anhydride/pyridine, a mixture of the epimers 3,13-dideoxo-3β,13α- and 3,13-dideoxo-3β,13β-dihydroxybisdehydrophorbol-3,12,20-triacetate is obtained (*61*, see Section 3.2.1). By treatment of **26a** with acetic anhydride/p-toluenesulfonic acid, the 4,9,12,20-tetraacetate is obtained. As to be expected because of its secondary 12,13-ketol group, **26b** reduces Fehling's and Tollens's reagents — the latter at room temperature (*61*).

From **27** a 3-[2,4-dinitrophenylhydrazone] may be obtained. Reduction of **27** with lithium aluminium hydride and subsequent acetylation with acetic anhydride/pyridine yields 3-deoxo-3β-hydroxytiglophorbol-3,13,20-triacetate (*61*). With acetic anhydride/pyridine **27** yields tiglophorbol-13,20-diacetate. Treatment of the latter with acetic anhydride/p-toluenesulfonic acid yields — besides the 4,13,20-triacetate — by acetolysis of the ether bridge the 4,12,13,20-tetraacetate **29** (Fig. 31). The latter may also be obtained by treatment of the 4,13,20-triacetate with acetic anhydride/p-toluenesulfonic acid. **27** reduces Tollens's solution only at elevated temperatures (*114*). Furthermore, oxidation of **27** by lead tetraacetate in acetic acid only takes place on heating at 60° to 70° C (*114*).

Oxidation of phorbol · C_2H_5OH with two moles of sodium periodate in water yields mainly phorbolactone-semiacetal (**32**), together with about equal amounts of tiglophorbol (**27**) and hydroxyphorbolactone-semiacetal (**33**, Fig. 32). While **27** is essentially stable under the oxidative conditions used, compounds **26b** and **28a**, resulting from oxidation by one equivalent of periodate, react further with cleavage of ring C: reaction at the secondary 12,13-ketol in **26b** as well as the glycol group in **28a** furnishes the intermediates **30** and **31** (Fig. 32). In a secondary reaction the aldehyde groups at C-12 in **30** and **31** trap 4β-OH intramolecularly to yield the semiacetal groups in **32** and **33**. While in **30** the carboxyl group at C-13 reacts with 9-OH to form the lactone group of **32**, the lactone group in **33** most probably is formed directly by oxidation of the semiketal hydroxyl in **28a** (Fig. 32, *62*).

CROMBIE et al. (*43*) also investigated the oxidation of phorbol with lead tetraacetate, sodium periodate and sodium bismuthate under various conditions. Apart from **26b**, **27** and **32** they isolated in minor amounts what they called tiglophorbol B. From their spectral data they suggested for this compound the structure of the 12-oxo-13-hydroxy isomer of **26b**.

1a
NaJO4
tiglophorbol (27)
26b
28a
NaJO4
NaJO4
30
31
32
33

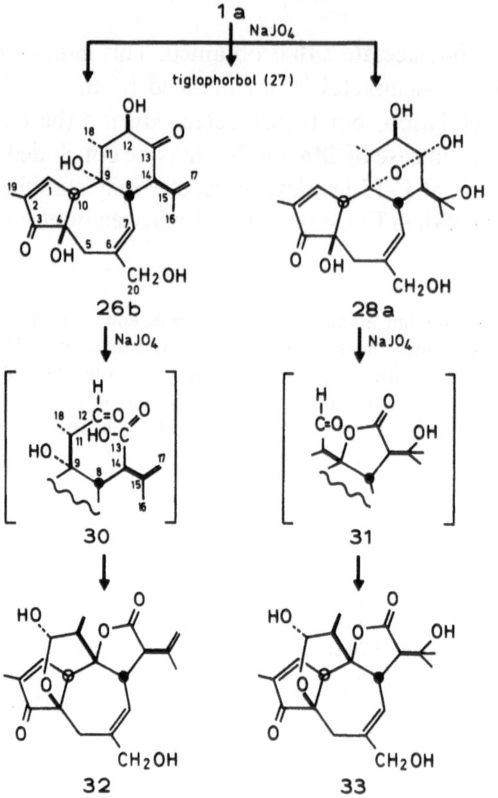

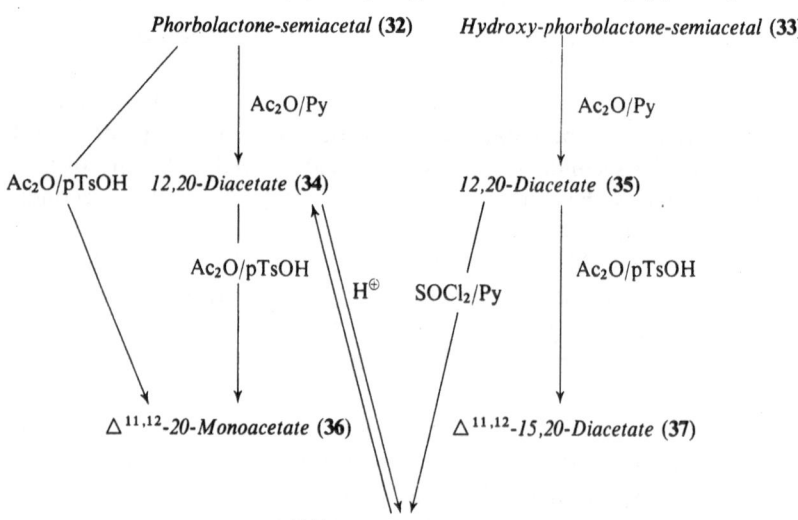

Fig. 32. Oxidation of phorbol (**1a**) with two moles of sodium periodate to yield tigliophorbol (**27**), phorbolactone-semiacetal (**32**) and hydroxyphorbolactone-semiacetal (**33**) (from *62*)

Phorbolactone-semiacetal (**32**)
Hydroxy-phorbolactone-semiacetal (**33**)
Ac2O/Py
Ac2O/Py
Ac2O/pTsOH
12,20-Diacetate (**34**)
12,20-Diacetate (**35**)
Ac2O/pTsOH
H⊕
SOCl2/Py
Ac2O/pTsOH
Δ 11,12-*20-Monoacetate* (**36**)
Δ 11,12-*15,20-Diacetate* (**37**)
Δ 14,15-*iso-12,20-diacetate* (**38**)

Chart 5. Some reactions of phorbolactone-semiacetal and hydroxy-phorbolactone-semiacetal (*62*)

If treated with acetic anhydride/pyridine, both **32** and **33** yield the corresponding 12,20-diacetates **34** and **35** (see Chart 5). Reaction with acetic anhydride/p-toluenesulfonic acid results in the elimination of 12-OAc from **34** and **35** to give the $\Delta^{11,12}$-20-monoacetate **36** and the $\Delta^{11,12}$-15,20-diacetate **37**, respectively. **36** may also be obtained by direct reaction of **32** with acetic anhydride/p-toluenesulfonic acid. Elimination of OH-15 is accomplished in **35** by treatment with thionyl chloride/pyridine to yield — not the $\Delta^{15,16}$ — but, instead, the $\Delta^{14,15}$-iso-12,20-diacetate **38**. By acid catalyzed isomerization of **34**, an equilibrium mixture of **34** with the $\Delta^{14,15}$-iso-compound (**38**) is obtained (*62*).

In Fig. 33, the stereochemistry of typical products of the oxidative scission of the bicyclo[4.1.0]heptane system in phorbol is given together with relevant NMR coupling constants.

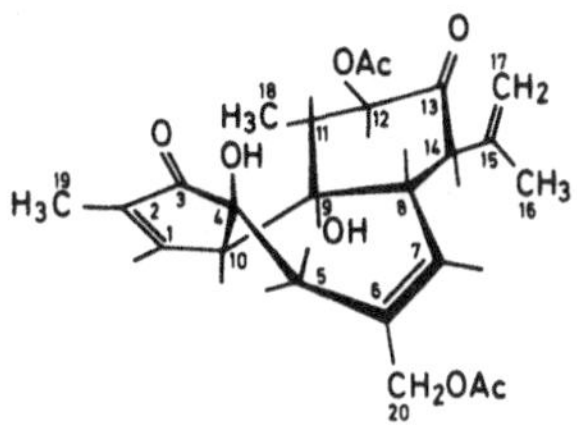

26a

$J_{7,8}$ = 4-5 cps; $J_{8,14}$ = 13 cps
$J_{11,12}$ = 12 cps

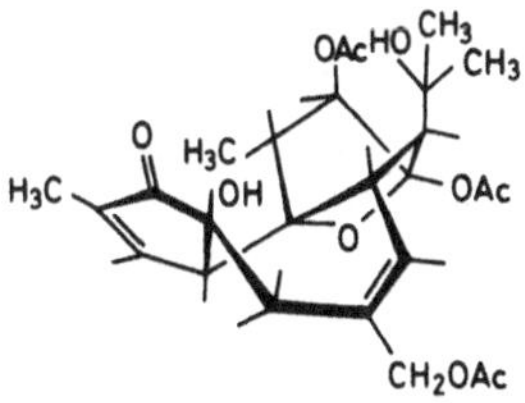

28b

$J_{7,8}$ —; $J_{8,14}$ = 7 cps
$J_{11,12}$ = 7 cps

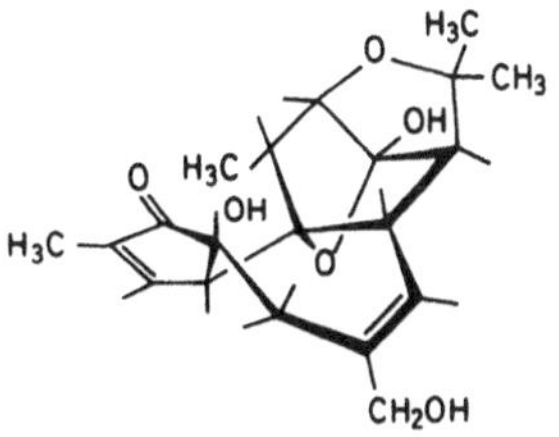

27

$J_{7,8}$ = 2-3 cps; $J_{8,14}$ = 3-4 cps
$J_{11,12}$ < 1 cps

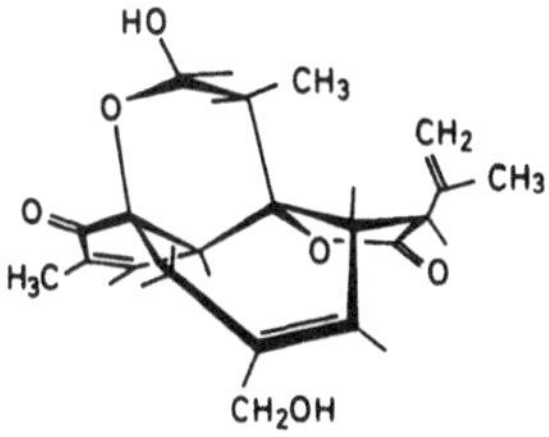

32

$J_{7,8}$ = ca. 4 cps; $J_{8,14}$ —
$J_{11,12}$ = 8,5 cps

Fig. 33. Stereochemistry of bisdehydrophorbol-12,20-diacetate (**26a**), hydroxybisdehydrophorbol-semiketal-12,13,20-triacetate (**28b**), tiglophorbol (**27**) and phorbolactone-semiacetal (**32**); relevant coupling constants from NMR-spectra of **26a** in $CDCl_3 + d_6$-DMSO + D_2O, of **28b** in $CDCl_3$ and of **27** and **32** in d_5-pyridine (from *61, 62*)

In the derivatives **26a, 27** and **32** the conformation of the perhydroazulene-system (rings A and B) is essentially identical with that of phorbol (**1a**, see Fig. 15). In **27** $J_{7,8}$ is smaller than in **1a, 26a** and **32**: obviously the angle between H-7 and H-8 is smaller in the presence of the ether- and semiketal-bridges. In **28b** $J_{7,8}$ is not measurable. In **26a**, ring C exists in a chair conformation as indicated by $J_{8,14}$ and $J_{11,12}$. In **27** and **28b** both of these coupling constants are compatible with a boat conformation of ring C. In **27** the angle between H-11 and H-12 seems to be close to 90°. Also it is interesting to see that $J_{8,14}$ decreases gradually from **26a** to **28b** to **27**. The same is true of $J_{11,12}$, which is in satisfactory agreement with the structures proposed for these compounds. In **32**, $J_{8,14}$ cannot be determined, but $J_{11,12}$ is compatible with a chair conformation.

3.4.3. Rearrangements in Phorbol and Neophorbol involving the Bicyclo-[4.1.0]heptane System

Crotophorbolone-enol-13,20-diacetate and Acetoxycrotophorbolone-20-acetate. With mesyl chloride/pyridine at low temperatures, phorbol-13,20-diacetate yields mainly crotophorbolone-enol-13,20-diacetate (**39**, Fig. 34). As a minor product of the reaction, acetoxycrotophorbolone-20-acetate (**41**) is obtained. By base catalyzed *trans*esterification, **39** may be converted to crotophorbolone-20-acetate (**40a**) and to crotophorbolone (**40b**) (*11*).

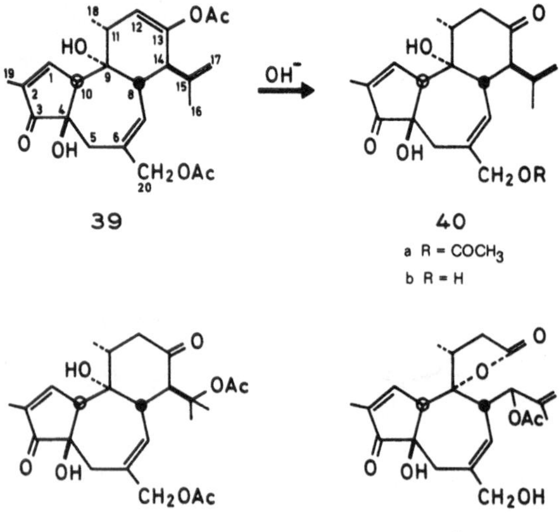

Fig. 34. Products of homoallyl rearrangement of phorbol-13,20-diacetate to yield crotophorbolone-enol-13,20-diacetate (**39**) and acetoxycrotophorbolone (**41**) and product of oxidation of crotophorbolone (**40b**) with lead tetraacetate to yield **46** (from *11, 43, 89, 165*)

Under similar conditions phorbol-12,13,20-triacetate (**1b**) does not react with mesylchloride/pyridine (*11*), but on refluxing with lithium aluminium hydride/aluminium chloride followed by acetylation, **39** was isolated (*43*). Treatment of **1b** with acetic acid/boron trifluoride at room temperature was reported to yield **41** (Fig. 34, *43*).

The products of these reactions are the result of a homoallyl rearrangement of the [α-acetoxycyclopropyl]-carbinol group in **42** (Fig. 35): after the formation of a carbonium ion at C-12 (**43**) opening of the cyclopropane ring is induced, producing a cation at C-15 (**44**). **44** may be stabilized either by formation of the isopropenyl group as in **39** or by intramolecular migration of the acetyl group at position 13 with concomitant hydrolysis to yield the corresponding 15-acetoxyisopropyl-ketone as in **41**. A similar acetyl migration with hydrolysis was observed in the bromination of phorbol-pentaacetate (see Section 3.2.3).

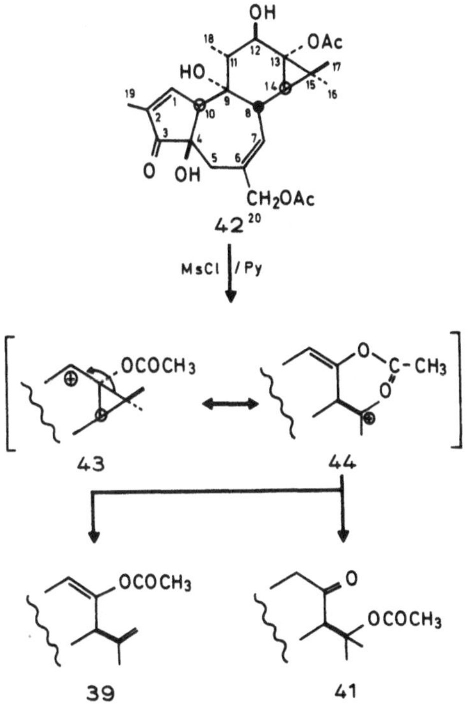

Fig. 35. Mechanism of homoallyl rearrangement of phorbol-13,20-diacetate (**42**): Crotophorbolone-enol-13,20-diacetate (**39**) and acetoxycrotophorbolone (**41**) (from *11*)

In **40b,** only the carbonyl group at C-13 may be detected by carbonyl reagents (*49*). Treatment of **40b** with acetic anhydride/pyridine gives rise to the 20-acetate **40a,** which may be peracetylated with acetic anhydride/p-toluenesulfonic acid. Similarly, reaction with benzoylchloride/pyridine, diazomethane/Al-i-propylate or mesylchloride/pyridine provides the 20-benzoate, 20-methyl ether or the 20-deoxy-20-chloro derivative respectively (*165*). By catalytic hydrogenation of **40b** formation of a 1,2ξ,6ξ,7,15,16-hexahydro-20-deoxy compound was reported (*114*). Treatment of **41** with acetic anhydride/p-toluene-sulfonic acid yields the corresponding peracetylated product (*11*).

Lead tetraacetate oxidizes **40b,** however, much more slowly than phorbol (*114*). According to Crombie et al. (*43*) the main product of this oxidation is the 13,14-*seco* compound **46** (Fig. 34). As mechanism of the reaction, intermediary formation of a lead tetraacetate adduct with the 9→13 semiketal of **40b** with subsequent cleavage of the 13,14-bond was proposed (*43*).

Phorbobutanone and Phorboisobutanone. Dehydration of phorbol-20-acetate (**45a**), -20-benzoate (**45b**) or -20-tritylether (**45c**) with phosphorus oxychloride/pyridine at 0° C induces pinacol rearrangement of the [α-hydroxycyclopropyl]-carbinol group to yield phorbobutanone-20-acetate (**48a**), -20-benzoate (**48b**) or -20-tritylether (**48c**), respectively (Fig. 36). They may be converted to phorbobutanone (**48d**). Under similar conditions phorbol-12,13,20-tri- and -12,20-diacetate do not react with phosphorous oxychloride/pyridine (*12*).

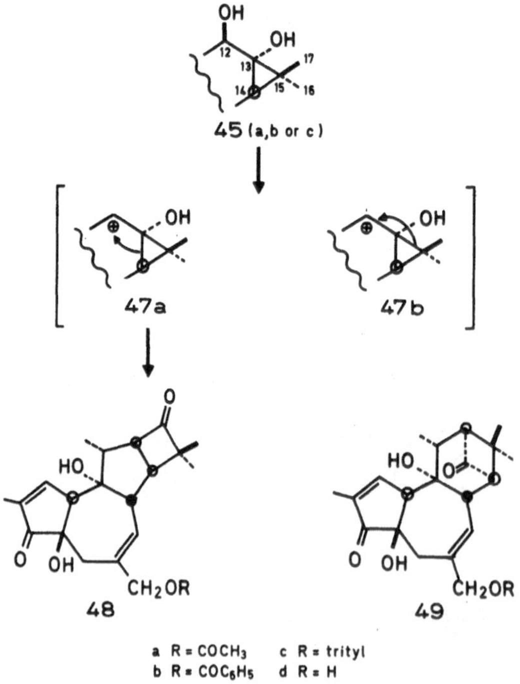

Fig. 36. Mechanism of rearrangement of phorbol-20-acetate (**45a**), -20-benzoate (**45b**) and -20-tritylether (**45c**): corresponding derivatives of phorbobutanone (**48d**) and phorboisobutanone (**49d**) (from *12*)

Because of the particular structure of the [α-hydroxycyclopropyl]-carbinol group, in addition to migration of the C-14 anion (**47a**), yielding the bicyclo[3.2.0]heptane system of phorbobutanone (**48d**) migration of the C-15 anion (**47b**) may also be expected (Fig. 36) to give, alternatively, the bicyclo[3.1.1]heptane system of phorboisobutanone (**49d**). However, the main product of the reaction, as seen by NMR-double resonance measurements, clearly contains the sequence C-7,8,14,12,11 of the phorbobutanone type **48d**. The isomer phorboisobutanone (**49d**) is not found in this reaction (see, however, below, under "Flaschenträger Reaction"). — The CD of **48d** — $\Delta\varepsilon$ negative between 390 and 260 nm — suggests $12\alpha H, 14\alpha H$-configuration of the cyclobutanone ring (Fig. 37).

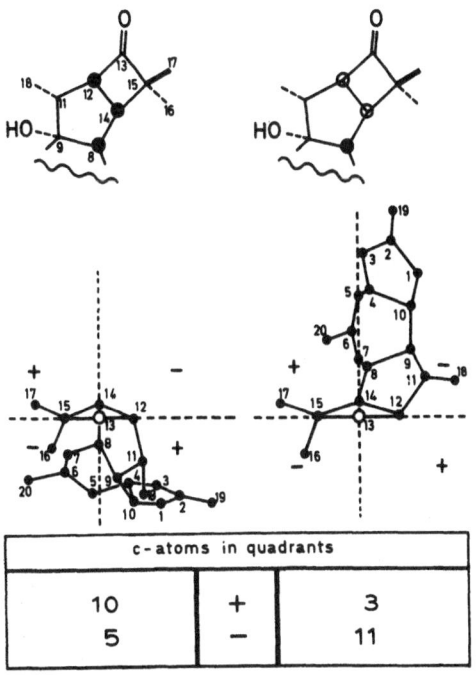

c-atoms in quadrants		
10	+	3
5	−	11

Fig. 37. Octant projections of the possible 12,14-*cis*-phorbobutanones, $\Delta\varepsilon_{max}^{307}$ found: − 3.44 (from *12*)

Fig. 37 shows the octant projections of both possible 12,14-*cis*phorbobutanones and a balance of their contributions to the sign of $\Delta\varepsilon$ in the CD. The $12\alpha H, 14\alpha H$-configuration clearly exhibits the majority of C-atoms in the negative quadrants. Hence this configuration is compatible with the CD of the cyclobutanone carbonyl as determined in **48d** (*12*).

Sodium borohydride reduction of **48d** may be carried out partially (at C-3) or fully (at C-3 and C-13) depending on the particular conditions chosen. If treated with acetic anhydride/pyridine or benzoylchloride/pyridine, **48d** yields the corresponding 20-monoesters **48a** and **48b**. Catalytic hydrogenation of **48b** with palladium/charcoal yields 20-deoxy-1,2ξ,6ξ,7-tetrahydrophorbobutanone (*12*).

The „Flaschenträger Reaction". This complex reaction was detected and investigated already by FLASCHENTRÄGER and v. FALKENHAUSEN (*50, 51*) in their trials aiming at the determination of the structure of phorbol.

In their attempts to find a suitable procedure for degradation of phorbol FLASCHEN-TRÄGER and v. FALKENHAUSEN (*51*) treated phorbol · C₂H₅OH with hot dilute sulfuric acid, and obtained small amounts of a crystalline and slightly levorotatory anhydro derivative of phorbol, which they called "crotophorbolone". Further, they obtained from this reaction traces of: acetone, of a light, non-crystalline resin and of carbon dioxide. This reaction was reinvestigated by KAUFFMANN *et al.* (*114*), HECKER *et al.* (*89*), CROMBIE *et al.* (*43*) and is called "Flaschenträger reaction" (*165*).

As stable products of the Flaschenträger reaction (Fig. 38), both phorbobutanone (**48d**) and crotophorbolone (**40b**) are isolated (*43, 89, 165*) together with a trinor compound, 4,9,20-trihydroxy-15,16,17-trinor-1,6-tigliadiene-3,13-dione (**52**) and acetone (*165*). If, as a variant of the Flaschenträger reaction, phorbol·C₂H₅OH is treated with hydrochloric acid in dry ether (SCHMIDT and HECKER, unpublished), again **48d** and **40b** are obtained, and in addition the $\Delta^{14,\,15}$ isomer of **40b**. Also, in the fraction containing **48d**, by careful thin layer chromatography, traces of phorboiso-

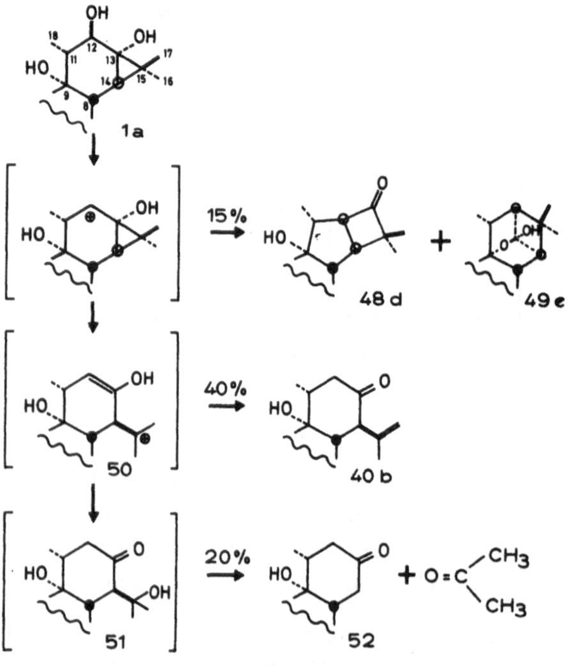

Fig. 38. Mechanism of the Flaschenträger reaction of phorbol (**1a**) yielding: phorbobutanone (**48d**) and phorboisobutanone-semiketal (**49e**), crotophorbolone (**40b**) and the trinor compound (**52**) plus acetone (from *165*, SCHMIDT and HECKER, unpublished)

butanone-9,13-semiketal (**49e,** Fig. 38) are to be detected; **49e** is also detectable as a product from the original version of the Flaschenträger reaction.

The Flaschenträger reaction can be interpreted as the consequence of acid catalyzed elimination of OH-12 from the 12,13-glycol group in the bicyclo[4.1.0]heptane system (*165*). The carbonium ion at C-12 (Fig. 38) is stabilized in three ways: partly by pinacol rearrangement to yield phorbobutanone (**48 d**) and some **49e,** partly by homoallyl rearrangement to yield a cation at C-15 (**50**). The latter collapses preferably to yield crotophorbolone (**40b**) and its $\Delta^{14,15}$-isomer. To a lesser extent the C-15 cation reacts with water to form the β-hydroxyketone **51** which is unstable and undergoes retroaldol cleavage, to yield as stable end products the trinorcompound **52** and acetone.

By reduction of the 20-acetate of **52** with sodium borohydride, separation of the reaction mixture and treatment of the main product with acetic anhydride/pyridine, the 3β,13α,20-triacetate can be isolated. Treatment of the trinor compound **52** with acetic

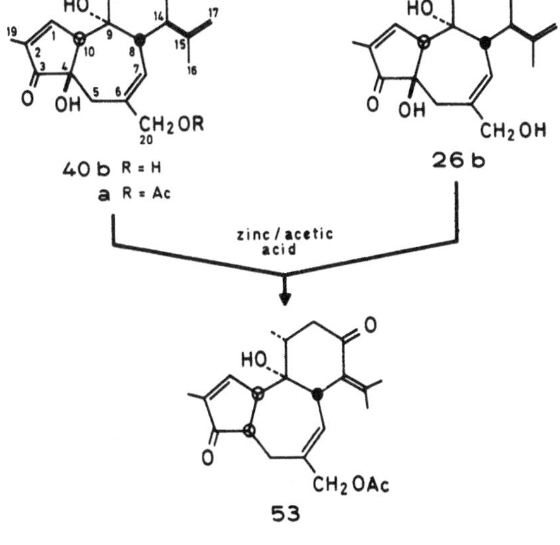

Fig. 39. Correlation of structures of crotophorbolone as 20-acetate (**40a**) and bisdehydrophorbol (**26b**): treatment of both compounds with zin/acetic acid yields **53** (from *165*)

anhydride/pyridine yields the 20-monoacetate, which may be converted to the 4,9,20-tri-acetate with acetic anhydride/p-toluenesulfonic acid (*165*).

The compounds obtained by oxidative ring opening of the bicyclo-[4.1.0]heptane system (Section 3.4.2.) and by the homoallyl rearrangement of this system (Section 3.4.3.) may be correlated by reduction of both, the tertiary 3,4-ketol-group in crotophorbolone-20-acetate (**40a**) (Fig. 39), and the tertiary 3,4-ketol as well as the secondary 12,13-ketol groups in bisdehydrophorbol (**26b**), to yield the common deoxy-compound   20-acetoxy-9-hydroxy-13,15-*seco*-4αH-1,6,14-tiglia-triene-3,13-dione (**53**) (*165*). It may be noted, that, besides reduction of the 1,2-ketol groups and epimerization at C-4 (see also Section 4.1), in both these reactions isomerization of the isopropenyl group to an isopropylidene group takes place.

A similar isopropenyl-isopropylidene isomerization was observed as consequence of the treatment of crotophorbolone (**40b**) with 0.01 n base (*43*). Also, under similar conditions, this kind of isomerization was observed with phorbolactone-semiacetal-12,20-diacetate (**34**) and the corresponding 15-hydroxy derivative 35 (see Chart 5).

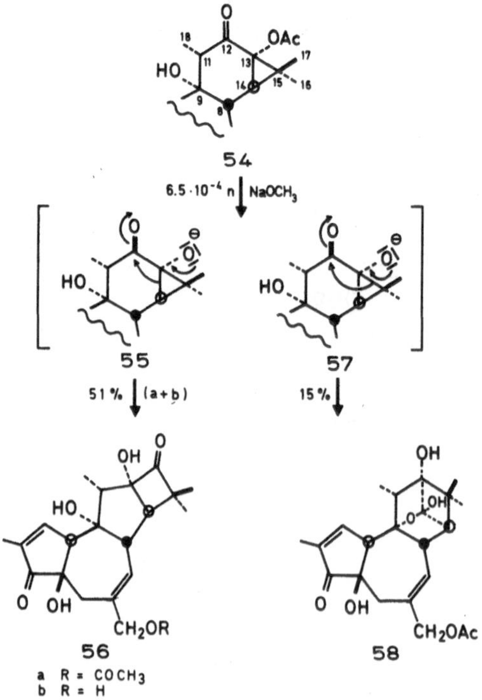

Fig. 40. 12,13-ketol rearrangement of neophorbol-13,20-diacetate (**54**) yielding hydroxyphorbobutanone (**56b**), -20-acetate (**56a**) and hydroxyphorboisobutanone-9,13-semiketal-20-acetate (**58**) (from *13*)

12,13-Ketol Rearrangement in Neophorbol. Neophorbol-13,20-diacetate (**54**, Fig. 40), contains two tertiary 1,2-ketol groups: one in the free 1,2-unsaturated-3,4-ketol group of the A-ring and another in the acetylated 12,13-ketol group of the bicyclo[4,1,0]heptane system (see also p. 416). On reaction with $6.5 \cdot 10^{-4}$ n sodium methoxide in methanol the 13-acetyl group is readily and selectively *trans*esterified to form neophorbol-20-acetate as intermediate. Even under these mild basic conditions, it undergoes an acyloin shift to yield, resulting from migration of the C-13/C-14 bond (**55**), as main products a mixture of 12-hydroxy-phorbobutanone-20-acetate (**56a**) and the corresponding compound with free 20-OH (**56b**). As a result of the alternative migration of the C-13/C-15 bond (**57**, Fig. 40) small amounts of the 9→13-semiketal of 12-hydroxyphorboisobutanone-20-acetate (**58**) are obtained. It is not known whether semi-ketal formation in **58** occurs during the reaction or as a consequence of the working up procedure (*13*). — At the low alkali concentration chosen epimerization of 4β-OH is excluded in both **56** and **58** (see Section 4.1).

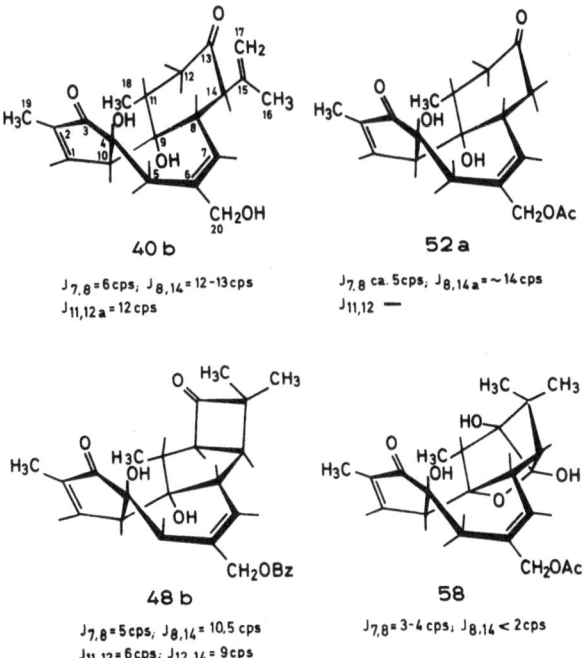

Fig. 41. Stereochemistry of the products of the Flaschenträger reaction: crotophorbolone (**40b**), trinor-compound-20-acetate (**52a**), phorbobutanone-20-benzoate (**48b**) and of hydroxyphorboisobuta-none-9,13-semiketal-20-acetate (**58**); relevant coupling constants from NMR-spectra of **40b** in d$_5$-pyri-dine, of **48b**, **52a** and **58** in CDCl$_3$ (from *12, 13, 43, 165*)

In Fig. 41 the stereochemistry of some of the products typical for the rearrangements of the bicyclo[4.1.0]heptane system is depicted, together with relevant coupling constants.

As suggested by their mode of formation and supported by their coupling constants $J_{7,8}$ compared to $J_{7,8}$ of phorbol (**1a**, Fig. 15) the conformation of the perhydroazulene system of rings A and B is retained in all four compounds. It is interesting, however, that $J_{7,8}$ in **58** is smaller than in **1a**, **40b**, **52a** and in **48b**. This may result from compression of the angle between 7-H and 8-H by formation of the $9 \rightarrow 13$ semiketal bridge and was similarly found in tiglophorbol (**27**, Fig. 33). Also in **40b** and **52a**, $J_{8,14}$ indicates an almost ideal *trans*-diaxial relationship for 8-H and 14-H. The exception in **58** again is reminiscent of the conformational situation of tiglophorbol (Fig. 33). In **40b**, both $J_{8,14}$ and $J_{11,12}$ indicate a *trans*-diaxial relationship for 11-H and 12-H and thus a chair conformation of ring C. In **52a**, $J_{11,12}$ is not measurable, but the chair conformation for the cyclohexane ring can be assumed from the size of $J_{8,14}$ and by analogy with **40b**. In **48b**, as compared to **40b**, $J_{8,14}$ and $J_{11,12}$ indicate deviations from the ideal *trans*-diaxial conformation. Similarly $J_{12,14}$ also suggests departure from the ideal *cis*-configuration and indicates a twisted cyclobutanone ring in **48b**. Thus, **48b** seems to be rather strained as compared to **40b** and **52a**. As indicated by $J_{8,14}$ in **58**, such strain seems to be even more pronounced in this molecule and is similar again to the situation in tiglophorbol (**27**, Fig. 33).

4. Further Diterpenes and Diterpene Esters from Croton Oil

Phorbol was shown to be parent alcohol of the toxic, irritant and cocarcinogenic principles and of its "cryptic" counterparts occuring in croton oil. However further diterpenes might have escaped detection during the fractionation procedures (Charts 1–4) because of their scarcity and/or biological inactivity.

Additional diterpenes would be expected to be concentrated in the filtrate of phorbol · C_2H_5OH obtained on base catalyzed *trans*esterification of croton oil, if they partition in a manner similar to phorbol or glycerol (see Section 3 and Chart 6). Indeed, in thin-layer chromatograms of such filtrates, in addition to phorbol and glycerol, several other spots may be observed under UV light and on staining with vanillin/sulfuric acid. The most intense of these spots has an R_f value slightly smaller than that of phorbol, whereas three less intense spots exhibit larger R_f values (*105, 110*).

To obtain definite evidence for the presence of further diterpenes in the filtrate of phorbol · C_2H_5OH preparations, the glycerol is removed by O'Keeffe distribution using a n-butanol/water system. The glycerol goes into the water-rich phases, whereas from the butanol-rich phases resinous material is obtained (Chart 6). The latter is acetylated and subjected to Craig distribution in a carbon tetrachloride, methanol, water solvent system (Chart 6, Fig. 42). In Fig. 42 the bands with distribution numbers $G = 3.56$ and 2.12 represent 4α-phorbol-(**59**)

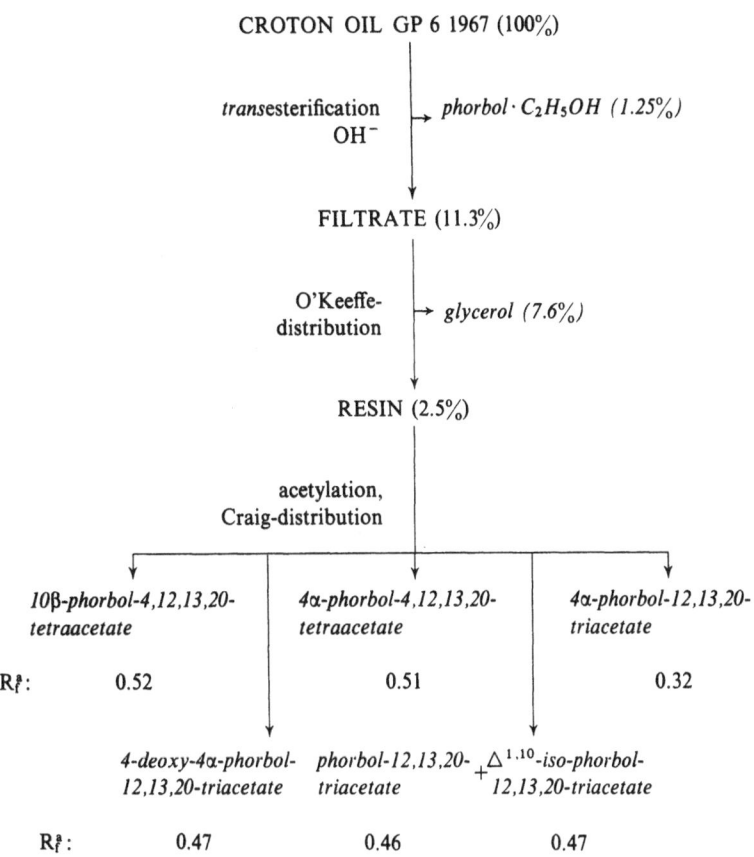

CROTON OIL GP 6 1967 (100%)

*trans*esterification
OH⁻ → *phorbol·C₂H₅OH (1.25%)*

FILTRATE (11.3%)

O'Keeffe-
distribution → *glycerol (7.6%)*

RESIN (2.5%)

acetylation,
Craig-distribution

10β-phorbol-4,12,13,20-tetraacetate	*4α-phorbol-4,12,13,20-tetraacetate*	*4α-phorbol-12,13,20-triacetate*
R_f^a: 0.52	0.51	0.32

4-deoxy-4α-phorbol-12,13,20-triacetate	*phorbol-12,13,20-triacetate* + $\Delta^{1,10}$-*iso-phorbol-12,13,20-triacetate*	
R_f^a: 0.47	0.46	0.47

ª System: chloroform/ethylacetate = 2/3, silica gel Merck HF 254.

Chart 6. Preparation of phorbol and other diterpenes directly from croton oil (*85, 153*)

(see Fig. 43) and phorbol-12,13,20-triacetate (**1b**), respectively. The band of the latter contains small amounts of $\Delta^{1,10}$-isophorbol-12,13,20-triacetate (**62b**) (see Fig. 44). Of the remaining three bands, that with G = 1.15 is 4α-phorbol-4,12,13,20-tetraacetate. The small band with G = 0.75 represents 10β-phorbol (**61**)-4,12,13,20-tetraacetate (see Fig. 43) and the material with G = 0.95 is 4-deoxy-4α-phorbol-12,13,20-triacetate (**63**) (see Fig. 44) (*85, 87, 153*).

In Fig. 43 the four different possibilities to interlink rings A and B of the tigliane skeleton are shown. Of the two possible *trans*-isomers, the 4β,10α-configuration is that of phorbol (**1a**). The *trans*-isomer with the 4α,10β-configuration **60**, an enantiomer of **1a**, which may be called allophorbol, is unknown as yet. Of the two possible *cis*-isomers, that with the 4α,10α-configuration is called 4α-phorbol (**59**), and its *cis*-enantiomer **61**, with 4β,10β-configuration, is called 10β-phorbol. Whereas in Dreiding models both 4,10-

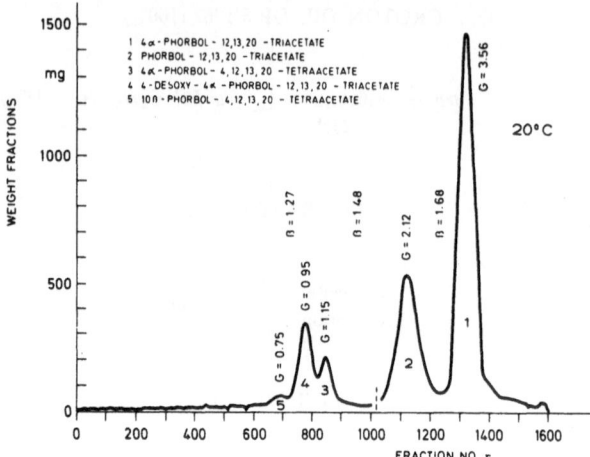

Fig. 42. Craig distribution of the "acetylated resin" obtained from the filtrate of phorbol preparations. Single withdrawal procedure: z = 1020, n = 1600; V = 10/10; system: carbon tetrachloride (2), methanol (1), water (0.13). Ordinate represents average weight of four neighboring fractions r to r-3 plotted against r-2; for fractions r = 1021 − 1600 average weight of nine neighboring fractions r to r-8, plotted against r-4 (from *85, 153*)

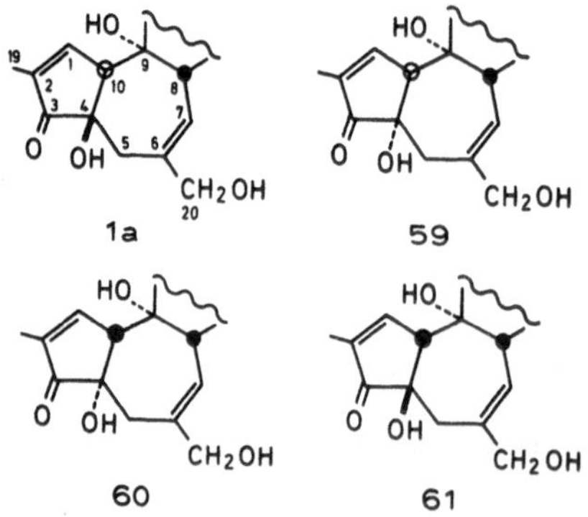

Fig. 43. The four possibilities of interlinking rings A and B in phorbol: **1a** phorbol, **59** 4α-phorbol, **60** allophorbol, **61** 10β-phorbol

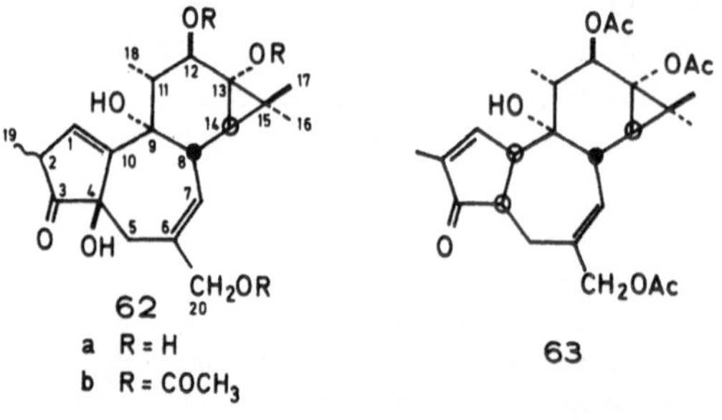

Fig. 44. $\Delta^{1,10}$-isophorbol-12,13,20-triacetate (**62b**) and 4-deoxy-4α-phorbol-12,13,20-triacetate (**63**) (from *153*)

References, pp. 458—467

trans-isomers **1a** and **60** appear conformationally rigid and remarkably strained, the molecules of both *cis*-isomers **59** and **61** are more flexible. This is essentially due to the increased conformational mobility of ring B in these isomers.

4.1. Chemistry of 4α-Phorbol

4α-phorbol (**59**, *87, 95, 110*), as well as 10β-phorbol (**61**) and $\Delta^{1,10}$-isophorbol (**62a**) (*87, 153, 154*) may be prepared by treatment of phorbol with sodium methoxide in methanol. Use of certain methoxide concentrations and reaction times is essential and special care has to be taken to exclude oxygen from the system. The diterpenes formed are isolated as the corresponding 12,13,20-triacetates (*110, 153, 154*).

In $6.5 \cdot 10^{-4}$ m sodium methoxide the tertiary 3,4-ketol group does not epimerize (see Section 3.4.3.). However, in 10^{-2} m sodium methoxide, epimerization of the 3,4-ketol group takes place (*87, 110*). In this reaction, in addition to 4α-phorbol (**59**), 10β- (**61**) and $\Delta^{1,10}$-isophorbol (**62a**) are obtained in small amounts, if short term exposure to sodium methoxide is used (SCHMIDT and HECKER, unpublished). Similar isomerizations and epimerizations of a cyclopentenone *trans*-fused to a seven membered ring, although under acidic conditions, are observed in compounds of the pseudoguaianolide type (e. g. *106*).

The major product of this reaction, 4α-phorbol (**59**), was investigated in some detail (*95, 110*). The pure resinous material, which has not yet been obtained in crystalline form, exhibits a mass spectrometric fragmentation pattern almost identical with that of phorbol. In the UV spectrum the maximum of the π-π^*-band of **59** is more pronounced than in case of phorbol (Fig. 45). Also, the NMR spectrum of **59** is similar to that of phorbol; the differences which do exist, however, are highly characteristic for the stereochemistry of the molecule (see below, Fig. 47).

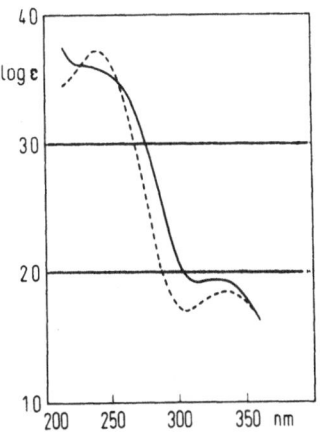

Fig. 45. UV-spectra of phorbol (**1a**) (———) and 4α-phorbol (**59**) (----) in MeOH. λ_{max} 241, 336 nm, ε_{max} 5300, 71 (from *110*)

The reactivity of the functional groups in 4α-phorbol is similar to that in phorbol, except for OH-4 (see below).

Reduction of the 12,13,20-triacetate of **59** with sodium borohydride provides 3-deoxo-3β-hydroxy-4α-phorbol-12,13,20-triacetate as main product together with some 3-deoxo-3β-hydroxy-1,2ξ-dihydro-4α-phorbol-12,13,20-triacetate. With acetic anhydride/pyridine **59** yields the corresponding 12,13,20-triacetate **59a** (see Fig. 47). However, in contrast to phorbol-12,13,20-triacetate, **59a** may be acylated further either with acetic anhydride/pyridine (*110, 152, 153*) or with benzoylchloride/pyridine (*110*) to yield the corresponding 4-O-acyl-12,13,20-triacetates. The tetraacetate obtained by ARROYO and HOLCOMB (*5*) and described as a phorbol tetraacetate in fact is not identical with either authentic phorbol-9,12,13,20-tetraacetate (*149*) or authentic phorbol-4,12,13,20-tetraacetate (*154*) but with 4α-phorbol-4,12,13,20-tetraacetate. On treatment with acetic anhydride/p-toluenesulfonic acid, **59a** provides 4α-phorbol pentaacetate. As **1a**, also **59** yields the corresponding 20-tritylether on treatment with trityl chloride/pyridine. Reaction of **59a** with methyliodide/silver oxide/DMF gives, as the main product, 4-O-methyl-4α-phorbol-12,13,20-triacetate, together with some 4-O-methyl-4α-phorbol-12,20-diacetate and 4,13-di-O-methyl-4α-phorbol-12,20-diacetate. Thus, as in the methylation of **1b** (Table 12), OH-13 is selectively set free and methylated under the conditions of this reaction (*110*).

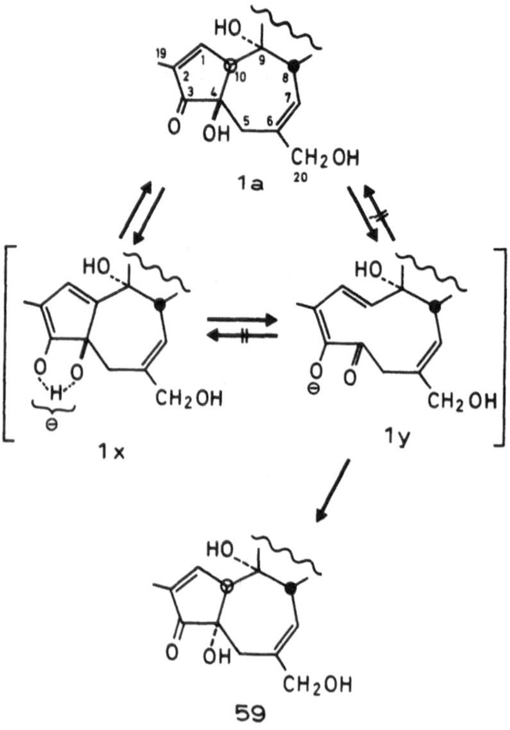

Fig. 46. Phorbol (**1a**) as a δ-hydroxy-α,β-enone and its vinylogous retroaldol scission and recondensation to yield 4α-phorbol (**59**) (from JACOBI and HECKER, unpublished)

In zinc/acetic acid, **59a** undergoes the typical acyloin reduction even more readily than phorbol-12,13,20-triacetate (see Section 3.2.2) to yield 4-deoxy-4α-phorbol-12,13,20-triacetate (**63**, Fig. 44).

The epimerization of phorbol (**1a**) to 4α-phorbol (**59**) is irreversible (*110*). It may be understood, if the functional groups of ring A in phorbol are considered as a δ-hydroxy-α,β-enone, i.e. a vinylogous 1,3-ketol (see Fig. 46). In a manner analogous to that of simple 1,3-ketols, such as, for example, the β-hydroxyketone **51** (see Fig. 38), **1a** may undergo vinylogous retro-aldol cleavage between C-4 and C-10. Thus, with sodium methoxide, **1a** may form not only the regular enolate **1x**, but also the enolate **1y** (Fig. 46). In **1y** the bond C-4/C-10 may be formed again and irreversibly with OH-4 in the thermodynamically more stable 4α(*cis*)-configuration (JACOBI and HECKER, unpublished).

The stereochemistry of 4α-phorbol-12,13,20-triacetate (**59a**) is shown in Fig. 47, together with relevant coupling constants. From the coupling constants it can be seen that the conformations of rings C and D in **1b** (Fig. 15) and in **59a** are essentially identical. The main stereochemical differences are located in ring B. In **1b** this ring is an envelope with the fold through C-4 and C-8 (see Fig. 15). It exhibits considerable rigidity, whereas in **59a**, with the fold through C-5 and C-9 (see Fig. 47), ring B is more flexible. The greater flexibility of ring B obviously imposes less conformational strain on ring A, than it does in **1b**, as can be seen for example from the UV-data of **59** (Fig. 45).

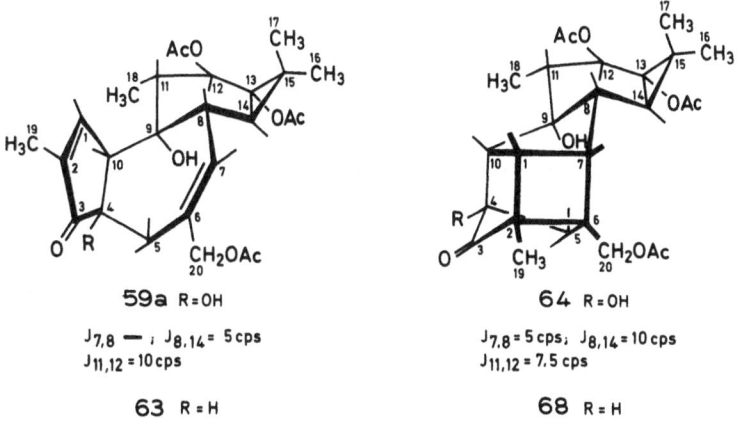

59a R=OH

$J_{7,8} = \;$; $J_{8,14} = 5$ cps
$J_{11,12} = 10$ cps

63 R=H

64 R=OH

$J_{7,8} = 5$ cps; $J_{8,14} = 10$ cps
$J_{11,12} = 7.5$ cps

68 R=H

Fig. 47. Stereochemistry of 4α- (**59a**), 4-deoxy-4α- (**63**) phorbol-12,13,20-triacetate, of lumi- (**64**) and 4-deoxylumi- (**68**) phorbol-12,13,20-triacetate; relevant coupling constants from NMR-spectra of **59a** and **64** in CDCl₃, see also Fig. 48 (from *65*)

As may be seen from Dreiding models, ring B in **59a** and **b** allows for two con-formations A and B. In conformation A, shown in Fig. 47, with the plane of ring A and the plane through C-5, 6, 7, 8 and 9 of ring B β-*cis*-oriented with respect to the plane through C-5, 4, 10 and 9 of ring B, CH₂-5 is in an *endo*-position with respect to

9-OH. Therefore, in contrast to **1b**, in **59a** the signal of 5-H$_2$ in the NMR-spectrum is split to an AB-system because of the vicinity of 9-OH. In the alternative conformation B, 5-H$_2$ would be in an *exo*-position with respect to 9-OH, precluding any interaction between 9-OH and 5-H$_2$.

By irradiation of **59a** with UV-light of wave length 254 nm, the isomeric lumiphorbol-12,13,20-triacetate (**64**) is obtained (Fig. 47). It is a cage compound of the cyclobutane type as deduced from NMR data (Fig. 48) and confirmed by X-ray diffraction analysis (*65, 95*). The cyclobutane ring is formed by intramolecular cycloaddition of the double bonds of **59a** (see Fig. 47). The 4,12,13,20-tetraacetate, the -tetrabenzoate and the pentaacetate of 4α-phorbol also undergo this light catalyzed cycloaddition. The determination of the structure of **64** contributed significantly to the clarification of the structure of 4α-phorbol (*65, 95, 110*).

The stereochemistry of **64** is depicted in Fig. 47, together with relevant coupling constants.

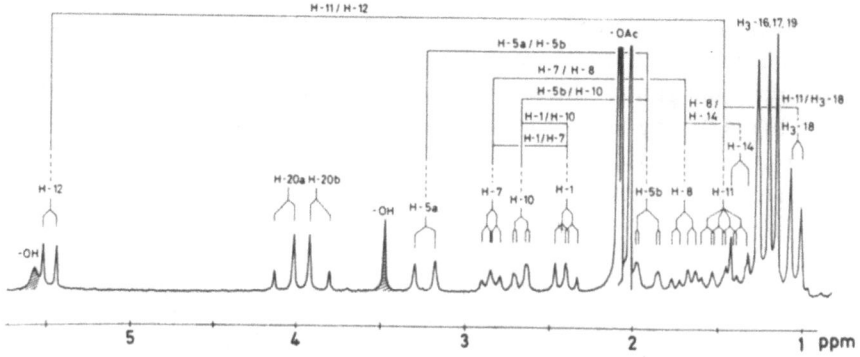

Fig. 48. NMR-spectrum (100 MHz, TMS 0.00 ppm) of lumiphorbol-12,13,20-triacetate (**64**) in CDCl$_3$ (from *65*)

4.2. Chemistry of 4-Deoxy-4α-phorbol and 4-Deoxyphorbol

4-Deoxy-4α-phorbol-12,13,20-triacetate (**63**) may be obtained by reduction of either phorbol- or 4α-phorbol-12,13,20-triacetate with zinc/ acetic acid (see Sections 3.2.2 and 4.1). The following mechanism permits one to understand the formation of one and the same reduction product from both of the 4-epimeric triacetates **1b** and **59a** (Fig. 49): From **1b** or **59a**, with one electron from zinc, the radical anion **65** is formed. 4-OH leaves **65** as an anion, forming the enol-radical **66**. The latter captures one further electron to yield the enolate **67**, which is stabilized by formation of the thermodynamically preferred 4α-configuration (*110*).

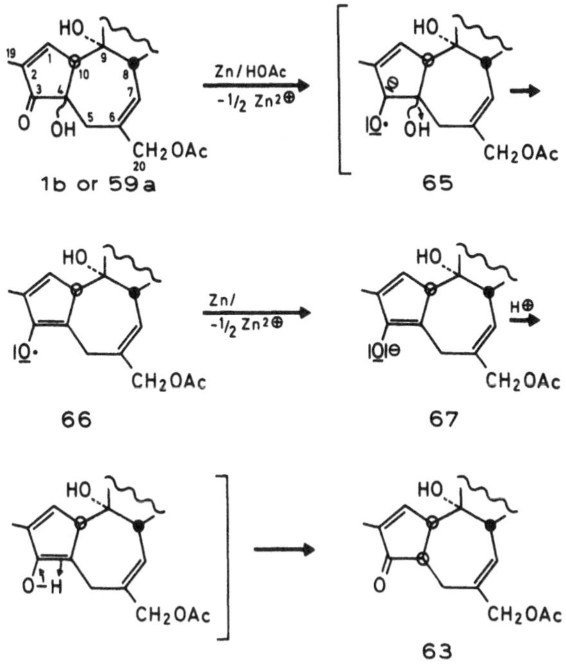

Fig. 49. Mechanism of formation of 4-deoxy-4α-phorbol-12,13,20-triacetate (**63**) from both phorbol-
(**1b**) and 4α-phorbol- (**59a**) -12,13,20-triacetate by zinc/acetic acid (from *110*)

This consideration suggests that 4-deoxyphorbol with its A/B-*trans*-configuration will be rather labile under both acidic and alkaline conditions, because of the possible formation of the enolate **67**. However, it was demonstrated recently, that under carefully controlled conditions it is possible to isolate 4-deoxyphorbol-esters from natural sources (*54, 55*).

Reduction of **63** with lithium aluminium hydride with subsequent acetylation yields a mixture of 3-deoxo-3β- and 3-deoxo-3α-hydroxy-4-deoxy-4α-phorbol-3,12,13,20-tetra-acetate (SCHMIDT and HECKER, unpublished).

Irradiation of **63** also induces the lumireaction to yield 4-deoxy-lumiphorbol-12,13,20-triacetate **68** (Fig. 47) (*65*).

4.3. Compound Groups D, E, D' and E' from Croton Oil

The occurrence of 4α-, 10β-, Δ^{1,10}-iso- and 4-deoxy-4α-phorbol in the filtrates of phorbol·C$_2$H$_5$OH preparations raises the question whether these compounds really occur in croton oil as parent alcohols of natural diterpene esters or if they are merely artificial side products formed during the preparation of phorbol from croton oil.

As demonstrated by the sodium methoxide catalyzed epimerization of phorbol and phorbol-12,13,20-triacetate, the epi- and isomers of phorbol may be produced during the base catalyzed *trans*esterification of the phorbol esters contained in croton oil. However, such conditions would not produce 4-deoxy-4α-phorbol, if this diterpene were not originally present in croton oil. Therefore, either 4-deoxyphorbol or 4-deoxy-4α-phorbol or both can be considered as native companion diterpenes of phorbol in croton oil.

To test this proposition, both the hydrophilic and the hydrophobic portions of croton oil (see Chart 1) were subjected to base catalyzed *trans*-esterification in methanol. In both preparations, thin layer chromatography demonstrated the presence of phorbol, $\Delta^{1,10}$-isophorbol, 4α- and 10β-phorbol, together with 4-deoxy-4α-phorbol. Therefore the hydrophilic and the hydrophobic portions both definitely contain esters of 4-deoxy- or 4-deoxy-4α-phorbol in addition to the esters of phorbol. Also, it may be expected that the 4-deoxyphorbol-12,13-diesters in the hydrophilic portion exist as two groups of positional isomers (called D and E), by analogy with the groups A and B observed with phorbol-12,13-diesters.

In the fractionation of the hydrophilic portion of croton oil such diesters may have been accumulated mainly in the "head fraction (2.1%)" separated from the hydrophilic portion by chromatography (see Chart 2). This suggestion is supported by the relatively high amount and optical activity of this fraction and, in particular, by analogy with the results obtained in a more detailed investigation of the activated hydrophobic portion as described below.

The occurrence of esters of a 4-deoxy-derivative of phorbol was investigated in more detail in the hydrophobic portion of the oil. If the "AHP head fraction (7.1%)" obtained by column chromatography of the activated hydrophobic portion (see Chart 4) is rechromatographed two bands D' and E' are obtained. According to their R_f values, these bands correspond to the factor groups A' and B' of phorbol-12,13-diesters, as demonstrated also by Craig distribution of the material in band E' to show presence of at least nine individual 12,13-diesters of 4-deoxy-4α-phorbol (Fig. 50). However, contrary to the factor groups A' and B' the materials of both of the bands D' and E' do not exhibit irritant activity in doses up to 100 μg/ear. Also, at single doses of p = 100 μg the material of band D' showed no observable cocarcinogenic activity (83,84,85,86,153). Moreover, also 4-deoxy-4α-phorbol-12, 13-didecanoate obtained by partial synthesis from phorbol proved to be inactive in the assay for irritant and for cocarcinogenic activity (154).

With regard to the structure of the congeners of phorbol in croton oil the problem of the configuration at C-4 in the naturally occurring 4-deoxyphorbol derivatives remains to be clarified.

References, pp. 458—467

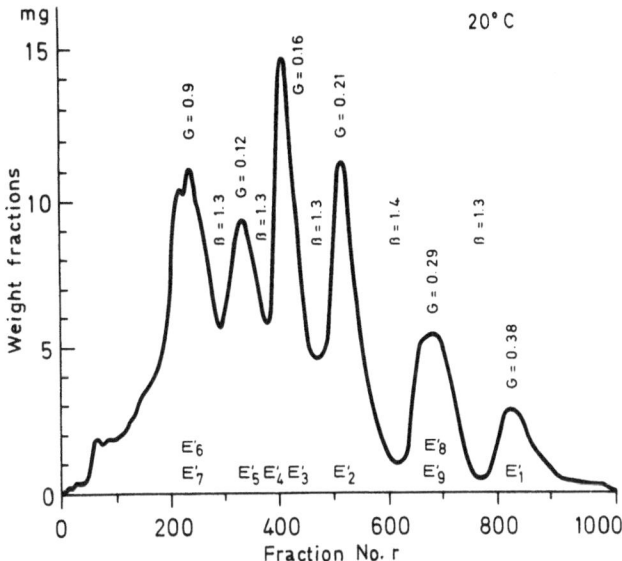

Fig. 50. Craig distribution of compound group E', single withdrawal procedure: z = 1020; n = 3000; V = 13/10; system: petroleum ether (15), methanol (10), water (1). Ordinate represents average weight of fractions r and r-1 plotted against r-3; r-2, r-3 and r-4 pooled (from *153*)

Esters of 4-deoxyphorbol occurring in croton oil might epimerize even under mild acidic or alkaline conditions (see Section 4.2). Such conditions may exist already in the seed oil itself or in its lipid fraction as suggested by the more or less high acid number of croton oil. At times the latter was used together with the optical activity as a measure for quality and age of the oil (*50, 75*). Further the experimental conditions used during the fractionation procedure of both the hydrophilic and the hydrophobic portions of croton oil (Charts 1 – 4) and during the preparation of phorbol directly from croton oil (Chart 6), may cause epimerization of 4-deoxyphorbol to 4-deoxy-4α-phorbol.

In an attempt to solve the problem the lipid fraction from freshly harvested croton seeds was extracted at room temperature. It exhibited an acid number of about 12 and was subjected to an O'Keeffe distribution in the usual (acid free) system (see Chart 1) to obtain the hydrophilic and hydrophobic portions. Both these portions were reduced separately with lithium aluminium hydride and each of the product mixtures treated with acetic anhydride/pyridine. The two fractions of acetylated 3-deoxo-3β-hydroxyditerpenes thus obtained were separated by thin layer chromatography. In both cases, besides the 3,12,13,20-tetraacetate of 3-deoxo-3β-hydroxyphorbol, only the tetraacetate of 3-deoxo-3β-hydroxy-4-deoxy-4α-phorbol and of its 3α-hydroxy epimer were detected. As a reference for the corresponding reduction products from 4-deoxyphorbol an authentic sample prepared from a natural source of 4-deoxyphorbol (*54*) was used. This result suggests that the 4-deoxy derivative of phorbol occurring in the seeds is in fact 4-deoxy-4α-phorbol and not 4-deoxyphorbol (Schmidt and Hecker, unpublished). Nevertheless, the possibility cannot be excluded that the native esters of 4-deoxyphorbol may be epimerized within the seeds themselves.

4.4. Composition of Croton Oil with Regard to Diterpenes and their Esters

Based upon quantitative data obtained from the fractionation schemes employed in the isolation of the diterpenes and their esters a balance of the composition of croton oil may be drawn up (87, Schmidt and Hecker, unpublished).

Table 17. *Estimation of the Amounts of Phorbol and 4-Deoxy-4α-Phorbol and their Esters Contained in Croton Oil*

Compound	Diterpenes isolated directly from croton oil [%]	Diterpene diesters isolated [%]	Diterpene diesters calculated [%]
Phorbol	3.4[a]	5.5[c]	5.6[e]
4-deoxy-4α-phorbol	0.29[b]	0.27[d]	0.52[e]

[a] Total of phorbol, 4α-, 10β-phorbol and $\Delta^{1,10}$-isophorbol obtained directly from croton oil.

[b] Total of 4-deoxy-4α-phorbol obtained directly from croton oil.

[c] Total of 12,13-diesters isolated from the hydrophilic and the activated hydrophobic portions.

[d] Total of esters isolated from the activated hydrophobic portion only.

[e] Calculated from data in first column, assuming av. mol. weight of 600 per diester.

According to Table 17 phorbol accounts for 3.4% and 4-deoxy-4α-phorbol for 0.29% of the weight of croton oil as determined experimentally (Chart 6). The 25 phorbol-12,13-diesters isolated from the hydrophilic (Charts 1, 2) and the hydrophobic (Charts 3, 4) portions altogether amount to 5.5%. Of the 4-deoxy-4α-phorbol-diesters, only those isolated from the activated hydrophobic portion have been determined quantitatively and amount to 0.27% of the oil. If it is assumed that — as in case of phorbol-diesters — the hydrophilic portion contains approximately the same amount of diesters of 4-deoxy-4α-phorbol, a total of 0.54% would result. Assuming an average molecular weight of 600 per diester from the data of isolated phorbol and 4-deoxy-4α-phorbol, respectively, a content of 5.6% and 0.52% of diesters, respectively, may be calculated. These figures are in satisfactory agreement with those obtained from isolations of diesters and indicate that satisfactory yields were obtained in the isolation procedures used.

References, pp. 458—467

5. On the Biological Activities of the Isolated Diterpenes and their Esters

Even at comparatively high dose levels the esters of 4-deoxy-4α-phorbol from croton oil are inactive, either as irritants or as cocarcinogens (*154*).

Of the 25 phorbol-12,13-diesters isolated from croton oil, those 14 ($A_1 - A_4$, A'_5, $B_1 - B_7$, B'_8 and B'_9) obtained from the hydrophobic portion actually represent phorbol-12,13,20-triesters of croton oil which are *per se* of little irritant and cocarcinogenic activity. However, their presence in croton oil has contributed a great deal to the confusion to be found in earlier investigations regarding the distribution of biological activities to the different fractions of croton oil (see Section 3.3.2.). Due to the ease of partial hydrolysis of their ester groups in 20-position to yield highly active phorbol-12,13-diesters, native phorbol-12,13,20-triesters may be considered as "cryptic" irritants and cocarcinogens.

The biological activities of croton oil are essentially due to the 11 phorbol-12,13-diesters isolated from the hydrophilic portion of croton oil. All of them are more or less toxic to frogs and are irritants to mouse skin. A_1, A_4, B_1 and B_2 are biologically the most active, whereas A'_5 is by far the least active of the factors isolated. The cocarcinogenic doses of some of them, e. g. of A_1 (Table 18), range in the order of magnitude of hormone doses, such as, for example, estrogenic hormones (*77, 80, 81*). They represent the most active cocarcinogens to mouse skin known so far (*80, 81*).

A comparison of the biological activities of the 14 different phorbol-12,13-diesters (Tables 4 and 7) indicates that these differences in activity may be related to the overall number of C-atoms in the acid moieties of the factors. Also it is interesting to note, that in the pair of isomers A_2/B_7 and A_3/B_4 (see Table 4) the factors of the B-group seem to be more active as cocarcinogens than the corresponding factors of the A-group, whereas the pairs of isomers exhibit practically identical irritant activities.

In contrast to croton oil and to the phorbol-12,13-diesters, phorbol and phorbol-12,13,20-triacetate exhibit little if any biological activity. Phorbol is relatively nontoxic to frogs (Table 18) as found already by BOEHM et al. (*29*). Also, in comparatively high doses, phorbol is not irritant and not cocarcinogenic to the skin of NMRI-mice (Table 18). Yet, recently, BAIRD and BOUTWELL (*5a*) reported very weak cocarcinogenic activity of phorbol in the STS strain of mice, especially bred for high sensitivity toward skin carcinogenesis. Phorbol-12,13,20-triacetate which was reported to be toxic to frogs already by BOEHM et al. (*29*) exhibits some activity in all of the three assys used (Table 18). To frogs it is more toxic than croton oil, but less toxic than croton oil factor A_1. However, it is less irritant and cocarcinogenic than both croton oil and croton oil factor A_1.

Table 18. *Toxic, Irritant and Cocarcinogenic Activities of Croton Oil, Phorbol-12.13,20-triacetate and Croton Oil Factor A_1*

Compound	LD_{50} [mg/50 g frog]	ID_{50} [µg/ear]	Cocarcinogenic activity[c] single dose [µg/appl.]	tumor rate[d] [%]	average tumor yield[d] [tumors/survivor]
Croton oil (GP 1958)	6[e]	0.5[h]	500[f]	64	4.0
Phorbol	>5[g]	>100[h]	183[i,k]	0	0
Phorbol-12,13,20-triacetate	0.15[g]	3.8[h]	490[i]	4	0.04
Croton oil factor A_1 (TPA)	0.01[g]	0.01[h]	10[k]	93	5.6

[a] Estimated in the standard procedure (75).
[b] Estimated in the standard procedure (96).
[c] Cocarcinogenic activities estimated in the standard procedure on NMRI-mice (91). Initiator: i = 0.1µ M DMBA.
[d] At the end of 12 weeks = 24 applications. [e] (75). [f] (40). [g] (93). [h] (96). [i] Hecker, unpublished. [k] (77).

Slight differences in the cocarcinogenic activities of croton oil as recorded in Tables 3 and 18 and of croton oil factor A_1 as recorded in Tables 4, 7 and 18 reflect the reproducibility of the biological assays in the standard procedures used (see Section 2.1.2).

With pure and highly active croton oil factors at hand the important question if their effects in BERENBLUM experiments are due to *co*- or to *pluri*carcinogenesis (*88*), i. e. if they represent cocarcinogens or merely new carcinogens, was reinvestigated. In BERENBLUM experiments using pure phorbol-12,13-diesters it was found that the augmentational effects of these substances are due to essentially reversible responses of the skin (*77, 80, 81, 82*). Thus in BERENBLUM experiments phorbol-12,13-diesters truly represent that type of "syncarcinogenic action" which is called "cocarcinogenesis" (*88*). Therefore, the BERENBLUM experiment can now be considered as the most advanced biological model for investigations of the biochemical mechanism of chemical carcinogenesis (*77, 80, 81, 82, 88*).

6. Conclusions and Perspectives

Phorbol is a polyfunctional tetracyclic diterpene of perhydro-cyclopropabenzazulene structure. Its 14 different 12,13-diesters, isolated from croton oil are the most active cocarcinogens known so far. They extend the number of known carcinogenic factors of plant origin considerably (*88*). In addition, a hitherto unknown principle of "cryptic" carcinogenic factors of plant origin has been detected with the fraction of phorbol-12,13,20-triesters. Thus, with respect to problems of environmental hygiene, it will be of interest to learn more about the distribution and chemical nature of free and cryptic cocarcinogens in the plant kingdom as well as about their biogenesis (see, for example, *83, 84, 85, 86, 88*).

Further, studies of the interrelationships between the chemical structures and the biological activities of irritants and cocarcinogens of the diterpene ester type deserve increasing interest. Such investigations shed new lights on basic problems of the biological mechanism of carcinogenesis: Already several investigations in this area indicate quite clearly that the extent of the toxicity of phorbol-12,13-diesters to frogs and of the irritant and cocarcinogenic activity to mouse skin depends on both, the number of C-atoms in the ester groups and the particular structure of the diterpene moiety (*54, 80, 81, 154, 164*). Thus, the biological activity of phorbol-12,13-diesters is associated with the specific structure of the entire molecule. According to general pharmacological experience, the ester groups may provide merely enough lipophilicity for an appropriate diterpene moiety to reach the receptor site or sites on or in the target cell (*164*). Indeed, cocarcinogenic activity of phorbol in tissues other than skin was recently reported by BERENBLUM and LONAI (*20*) (leukemia in Swiss mice) and by ARMUTH and BERENBLUM (*3*) (liver and lung cancer of AKR mice).

Finally and most importantly the availability and use of molecularly uniform phorbol esters as tumor promotors has stimulated a renaissance of biochemical investigations employing the two-stage initiation-promotion system of mouse skin *in vivo* and *in vitro* as a model system to gain deeper insights into the mechanism of chemical carcinogenesis at the molecular level (e. g. *6, 7, 92, 103, 121, 122, 136, 141, 142, 159, 162, 169, 185*).

The isolation and chemical as well as biological characterization of the active principles of croton oil has opened new avenues of investigation interlinking both natural product chemistry and experimental cancer research *.

Acknowledgement. Many thanks are due to Dr. A. BALMAIN for his kindness in revising the English and for critical reading of this survey.

References

1. ALDERWEIRELDT, F. C.: Steady-State Distribution. New Instrument for Continued Batchwise Separation by Extraction. Analyt. Chem. **33**, 1920 (1961).

2. ARIGONI, D., W. VON DAEHNE, W. O. GODTFREDSEN, A. MELERA, and S. VANGEDAL: The Stereochemistry of Fusidic Acid. Experientia **20**, 344 (1964).

3. ARMUTH, V., and I. BERENBLUM: Systemic Promoting Action of Phorbol in Liver and Lung Carcinogenesis in AKR Mice. Cancer Res. **32**, 2259 (1972).

4. ARROYO, E. R., and J. HOLCOMB: Isolation and Structure Elucidation of a Highly Active Principle from Croton Oil. Chem. and Ind. **1965**, 350.

5. — — Structural Studies of an Active Principle from *Croton tiglium* L. J. Med. Chem. **8**, 672 (1965).

5a. BAIRD, W. M., and R. K. BOUTWELL: Tumor-promoting Activity of Phorbol and Four Diesters of Phorbol in Mouse Skin. Cancer Res. **31**, 1074 (1971).

6. BAIRD, W. M., J. A. SEDGWICK, and R. K. BOUTWELL: Effects of Phorbol and Four Diesters of Phorbol on the Incorporation of Tritiated Precursors into DNA, RNA and Protein in Mouse Epidermis. Cancer Res. **31**, 1434 (1971).

7. BALMAIN, A., and E. HECKER: On the Biochemical Mechanism of Tumorigenesis in Mouse Skin. VI. Early Effects of Growth Stimulating Phorbol Esters on Phosphate Transport and Phospholipid Synthesis in Mouse Epidermis. In press.

8. BARTSCH, H., H. BRESCH, M. GSCHWENDT, E. HÄRLE, G. KREIBICH, H. KUBINYI, H. U. SCHAIRER, CH. VON SZCZEPANSKI, H. W. THIELMANN und E. HECKER: Kombination wirksamer Trennverfahren mit modernen analytischen Methoden in der Naturstoffchemie, Isolierung und Strukturaufklärung der biologisch aktiven Substanzen aus Crotonöl. Z. Analyt. Chem. **221**, 424 (1966).

9. BARTSCH, H., und E. HECKER: Circulardichroismus bei Phorbolderivaten. Angew. Chem. **79**, 994 (1967); Angew. Chem. Intern. Ed. Engl. **6**, 974 (1967).

---

* 12-O-Tetradecanoyl-phorbol-13-acetate (TPA = croton oil factor A_1) and related compounds are commercially available from: Consolidated Midland Corporation, 195 East Main Street, Brewster, NY 10509, U.S.A.

10. BARTSCH, H., und E. HECKER: Circulardichroismus und Röntgenstrukturanalyse des Phorbols. In: Aktuelle Probleme aus dem Gebiet der Cancerologie (H. LETTRÉ und G. WAGNER eds.). Vol. II, p. 162. Berlin-Heidelberg-New York: Springer. 1968.

11. — —Zur Chemie des Phorbols, XI. Crotophorbolon-enolacetat und Acetoxy-crotophorbolon. Z. Naturforsch. **24b**, 91 (1969).

12. — — Zur Chemie des Phorbols, XII. Phorbobutanon. Z. Naturforsch. **24b**, 99 (1969).

13. — — Zur Chemie des Phorbols XIII. Über eine Acyloin-Umlagerung des 12-Desoxy-12-oxophorbol-13,20-diacetats. Liebigs Ann. Chem. **725**, 142 (1969).

14. BARTSCH, H., G. SNATZKE und E. HECKER: Zur Chemie des Phorbols, VI. Funktionelle Derivate des Phorbol-12-ons und deren Circulardichroismus. Z. Naturforsch. **23b**, 1453 (1968).

15. BERENBLUM, I.: The Cocarcinogenic Action of Croton Resin. Cancer Res. **1**, 44 (1941).

16. — The Mechanism of Carcinogenesis. A Study of the Significance of Cocarcinogenic Action and Related Phenomena. Cancer Res. **1**, 807 (1941).

17. — Cocarcinogenesis. Brit. Med. Bull. **4**, 343 (1947).

18. — Carcinogenesis and Tumor Pathogenesis. Advan. Cancer Res. **2**, 129 (1954).

19. — The Two-Stage Mechanism of Carcinogenesis as an Analytical Tool. In: Cellular Control Mechanisms and Cancer (P. EMMELOT and O. MÜHLBOCK eds.), p. 259. Amsterdam: Elsevier Publishing Corp. 1964.

20. BERENBLUM, I., and V. LONAI: The Leukemogenic Action of Phorbol. Cancer Res. **30**, 2744 (1970).

21. BERENBLUM, I., and P. SHUBIK: The Role of Croton Oil Applications, Associated with a Single Painting of a Carcinogen, in Tumour Induction of the Mouse's Skin. Brit. J. Cancer **1**, 379 (1947).

22. — — A New, Quantitative Approach to the Study of the Stages of Chemical Carcinogenesis in the Mouse's Skin. Brit. J. Cancer **1**, 383 (1947).

23. — — The Persistence of Latent Tumor Cells Induced in the Mouse's Skin by a Single Application of 9:10-Dimethyl-1:2-Benzanthracene. Brit. J. Cancer **3**, 384 (1949).

24. BERGER, F.: Handbuch der Drogenkunde, Vol. VI, p. 278. Wien: W. Maudrich. 1964.

25. BOEHM, R.: Über Krotonharz (mit Anhang über Euphorbiumharz). Arch. Exptl. Pathol. Pharmakol. **79**, 138 (1915).

26. — Über Krotonharz. Arch. Pharmaz. **253**, 574 (1915).

27. — (1923), quoted according to: *28, 29* and *50.*

28. BOEHM, R., und B. FLASCHENTRÄGER: Über den Giftstoff im Krotonöl. Arch. Exptl. Pathol. Pharmakol. **157**, 115 (1930).

29. BOEHM, R., B. FLASCHENTRÄGER und L. LENDLE: Über die Wirksamkeit von Sub-stanzen aus dem Krotonöl. Arch. Exptl. Pathol. Pharmakol. **177**, 212 (1935).

30. BOUTWELL, R. K.: Some Biological Aspects of Skin Carcinogenesis. Progr. Exptl. Tumor Res. **4**, 207 (1964).

31. BRESCH, H., G. KREIBICH, H. KUBINYI, H. U. SCHAIRER, H. W. THIELMANN und E. HECKER: Über die Wirkstoffe des Crotonöls, IX. Partialsynthese von Wirkstoffen des Crotonöls. Z. Naturforsch. **23b**, 538 (1968).

32. BUCHHEIM, R.: Über die pharmakologische Gruppe des Crotonöls. Virchow's Arch. path. Anat. Physiol. klin. Med. **12**, 1 (1857).

33. — Über die „scharfen" Stoffe. Wagner's Archiv der Heilkunde **14**, 1 (1873).

33a. BUCOURT, R., M. LEGRAND, M. VIGNAU, J. TESSIER et V. DELAROFF: Structure et stéréochimie de l'acide fusidique, antibiotique d'apparentement Stéroide. C. R. Acad. Sci. (France) **257**, 2679 (1963).

34. CAPON, B.: Neighbouring group participation. Quart. Rev. **18**, 45 (1964).

35. Cherbuliez, E., et K. Bernhard: Recherches sur la graine de croton, I. Sur le crotonoside (2-oxy-6-amino-purine-d-riboside). Helv. Chim. Acta 15, 464 (1932).
36. — — Remarques sur le crotonoside. Helv. Chim. Acta 15, 978 (1932).
37. Cherbuliez, E., K. Bernhard et E. Ehninger: Recherches sur la graine de croton, III. Sur l'improbabilité de l'existence de l'alcaloide de Tuson, et sur la présence de saccharose, et de glucosides du glucose, dans la graine de croton. Helv. Chim. Acta 15, 855 (1932).
38. Cherbuliez, E., E. Ehninger et K. Bernhard: Recherches sur la graine de croton, II. Le principe vésicant. Helv. Chim. Acta 15, 658 (1932).
39. Clarke, E., and E. Hecker: Isolation, Purification and Chemical Characterization of further Irritant and Cocarcinogenic Compounds from Croton Oil. Naturwiss. 52, 446 (1965).
40. — — On the Active Principles of Croton Oil, V. Purification and Characterization of Further Irritant and Cocarcinogenic Compounds of the B-group. Z. Krebsforsch. 67, 192 (1965).
41. Craig, L. C., and T. P. King: Design and Use of a 1000-tube counter-current distribution apparatus. Federation Proc. 17, 1126 (1958).
42. Craig, L. C., T. P. King, and E. G. Scheibel: Laboratory Extraction and Countercurrent Distribution. Techn. Org. Chem. (A. Weissberger ed.) 2nd ed., Vol. 3, Part 1, p. 149. New York: Interscience Publ. 1956.
43. Crombie, L., M. L. Games, and D. J. Pointer: Chemistry and Structure of Phorbol, the Diterpene Parent of the Cocarcinogens of Croton Oil. J. Chem. Soc. (C) (London) 1968, 1347.
44. Danneel, R., und N. Weissenfels: Über die tumorrealisierende Wirkung verschiedener Crotonölfraktionen. Naturwiss. 42, 128 (1955).
45. Djerassi, C., W. Klyne, T. Norin, G. Ohloff, and E. Klein: Rotatory Dispersion Curves of Cyclopropyl-Ketones and Epoxy-Ketones. Tetrahedron 21, 163 (1965).
46. Drake, N. L., and J. R. Spies: Croton Resin, III. The Combined Acids. J. Amer. Chem. Soc. 57, 184 (1935).
47. Druckrey, H.: Fortpflanzung und Wachstum. In: Physiologische Chemie (B. Flaschenträger and E. Lehnartz, eds.). Vol. II, Part 2c, p. 1 (see p. 232). Berlin-Göttingen-Heidelberg: Springer. 1959.
48. Dunstan, W. R., and L. E. Boole: Croton Oil "An Enquiry into the Nature of its Vesicating Constituent". Pharmaz. J. 55, 5 (1895); Vesicating constituent of Croton Oil. Proc. Roy. Soc. 58, 238 (1895).
49. Falkenhausen, F. Frh. von: Mikrobestimmung von Carbonylgruppen. Z. Analyt. Chem. 99, 241 (1934).
50. Flaschenträger, B.: Über den Giftstoff im Krotonöl. „Festschrift Heinrich Zangger", Vol. II, p. 857. Zürich-Leipzig-Stuttgart: Rascher. 1935.
51. Flaschenträger, B., und F. Frh. von Falkenhausen: Über den Giftstoff im Krotonöl, II. Zur Konstitution von Krotophorbolon. Liebigs Ann. Chem. 514, 252 (1934).
52. Flaschenträger, B., und G. Wigner: Über den Giftstoff des Krotonöls, V.: Die Gewinnung von Krotonharz, Dünnem Öl und Phorbol aus dem Crotonöl durch Alkoholyse. Helv. Chim. Acta 25, 569 (1942).
53. Flaschenträger, B., und R. v. Wolffersdorff: Über den Giftstoff des Crotonöls, I. Die Säuren des Crotonöls, Helv. Chim. Acta 17, 1444 (1934).
54. Fürstenberger, G., und E. Hecker: Zum Wirkungsmechanismus cocarcinogener Pflanzeninhaltsstoffe. Planta Medica. 22, 241 (1972).
55. Fürstenberger, G., E. Henseleit und E. Hecker: Über den Zusammenhang zwischen entzündlicher und cocarcinogener Wirkung von Phorbolderivaten. 11. wissenschaftliche Tagung der Deutschen Krebsgesellschaft, Hannover 1971. Abstracts p. 78. Berlin-Heidelberg-New York: Springer. 1971.

56. GRAFFI, A.: Experimente und Betrachtungen zur Natur und Ursache des Krebses. Sitzber. Dtsch. Akad. Wiss. Berlin, Kl. Med. Nr. **2**, 1 (1964).

57. GRAFFI, A., und H. BIELKA: Probleme der experimentellen Krebsforschung. Leipzig: Akad. Verlagsges. 1959.

58. GSCHWENDT, M., E. HÄRLE und E. HECKER: Zur Chemie des Phorbols, VII. Die α,β-ungesättigte tertiäre 1,2-Ketolgruppierung im Phorbol. Z. Naturforsch. **23b**, 1579 (1968).

59. GSCHWENDT, M., und E. HECKER: Ermittlung von Partialstrukturen des Phorbols durch Perjodat- und Bleitetraacetatspaltung. Angew. Chem. **79**, 994 (1967); Angew. Chem. Intern. Ed. Engl. **6**, 974 (1967).

60. — — Ermittlung von Partialstrukturen des Phorbols durch Perjodat- und Bleitetraacetatspaltung. In: Aktuelle Probleme aus dem Gebiet der Cancerologie (H. LETTRÉ und G. WAGNER eds.), Vol. II, p. 170. Berlin-Heidelberg-New York: Springer. 1968.

61. — — Zur Chemie des Phorbols, VIII. Die Oxydation von Phorbol mit Bleitetraacetat. Z. Naturforsch. **23b**, 1584 (1968).

62. — — Zur Chemie des Phorbols, IX. Oxydation von Phorbol mit Natriumperjodat. Z. Naturforsch. **24b**, 80 (1969).

63. — — Tumor Promoting Compounds from Euphorbia Triangularis: Mono- and Diesters of 12-Desoxy-Phorbol. Tetrahedron Letters **1969**, 3509.

64. GWYNN, R. H.: Tumour-Promoting Action of Croton Oil Fractions. Brit. J. Cancer **9**, 445 (1955).

65. HÄRLE, E., und E. HECKER: Zur Chemie des Phorbols, XVII. Lumiphorbol und einige seiner Derivate. Liebigs Ann. Chem. **748**, 134 (1971).

66. HÄRLE, E., W. KOCH und E. HECKER: Zur Chemie des Phorbols, II. Über das Kohlenstoffskelett des Phorbols. Chem. Ber. **100**, 795 (1967).

67. HECKER, E.: Verteilungsverfahren im Laboratorium, Monographie Nr. 67 zu Angew. Chem. und Chem. Ing. Techn. Weinheim: Verlag Chemie. 1955.

68. — 4'-Nitro-azobenzol-carbonsäure-(4)-chlorid als Reagens auf Alkohole. Chem. Ber. **88**, 1666 (1955).

69. — Kontinuierliche Verteilungsverfahren im Laboratoriumsmaßstab. Chem. Ing. Techn. **29**, 23 (1957).

70. — Neuere Entwicklungen und Ergebnisse der Substanztrennung durch multiplikative Verteilung. Z. Analyt. Chem. **181**, 284 (1961).

71. — Kontinuierliche Verteilungsverfahren im Laboratorium. Chemiker Ztg. **86**, 272 (1962).

72. — Über das toxische, entzündliche und carcinogene Prinzip des Crotonöls. Angew. Chem. **74**, 722 (1962); Angew. Chem. internat. ed. Engl. **1**, 602 (1962).

73. — Der Mechanismus multiplikativer Verteilungsverfahren. Naturwiss. **50**, 165 and 290 (1963).

74. — Biochemische Aspekte des Krebsproblems. Mitt. Max-Planck-Ges. Nos. 1 – 2, **1963**, 41.

75. — Über die Wirkstoffe des Crotonöls, I. Biologische Teste zur quantitativen Messung der entzündlichen, cocarcinogenen und toxischen Wirkung. Z. Krebsforsch. **65**, 325 (1963).

76. — Multiplikative Verteilung. In: Handbuch der Lebensmittelchemie (I. SCHORMÜLLER, ed.), Vol. II, Part 1, p. 713. Berlin-Heidelberg-New York: Springer. 1965.

77. — Die cocarcinogene Wirkung der Phorbolester. In: Molekulare Biologie des malignen Wachstums (H. HOLZER und A. W. HOLLDORF, eds.), p. 105. Berlin-Heidelberg-New York: Springer. 1966.

78. — Quality and Quantity of Cocarcinogenic Action in Mouse-Skin of Phorbol Esters from Croton Oil. Proc. 9th Intern. Cancer Congr., Tokyo, 1966, Abstracts, p. 17.

462   E. HECKER and R. SCHMIDT:

79. HECKER, E.: Phorbol Esters from Croton Oil, Chemical Nature and Biological Activities. Naturwiss. **54**, 282 (1967).
80. — Cocarcinogenic Principles from the Seed Oil of Croton tiglium and from Other Euphorbiaceae. Cancer Res. **28**, 2338 (1968).
81. — Cocarcinogene Wirkstoffe aus Euphorbiaceen. Planta Medica **1968** (Supplement), 119.
82. — Biochemische und molekularbiologische Probleme der Tumorgenese. Arzneimittelforschung (Drug Res.) **18**, 978 (1968).
83. — Carcinogens from Euphorbiaceae. Symp. Naturally Occuring Carcinogens, Prague, Abstracts, p. 2 (1970).
84. — New Phorbol Esters and Related Cocarcinogens, in: Oncology, Proc. 10th Intern. Cancer Congr., Houston, Year book Medical Publishers, Inc., Chicago, Vol. V, p. 213, 1971.
85. — Cocarcinogens from Euphorbiaceae and Thymelaeaceae. In: Pharmacognosy and Phytochemistry, Proc. 1st Intern. Symp. Munich (H. WAGNER and L. HÖRHAMMER, eds.), p. 14. Berlin-Heidelberg-New York: Springer. 1971.
86. — Phorbol-esters from Euphorbiaceae — a New Class of Highly Potent Cocarcinogens of Plant Origin. 1st. Congr. European Assoc. Cancer Res., Brussels, 1970, Abstracts, p. 5.
87. — Isolation and Characterization of the Cocarcinogenic Principles from Croton Oil. Methods in Cancer Research (H. BUSCH, ed.), Vol. VI, p. 439, New York-London: Academic Press, Inc. 1971.
88. — Aktuelle Probleme der Krebsentstehung. 11. wissenschaftliche Tagung der Deutschen Krebsgesellschaft, Hannover 1971. See also: Z. Krebsforsch. **78**, 99 (1972).
89. HECKER, E., H. BARTSCH, H. BRESCH, M. GSCHWENDT, E. HÄRLE, G. KREIBICH, H. KUBINYI, H. U. SCHAIRER, CH. VON SZCZEPANSKI, and H. W. THIELMANN: Structure and Stereochemistry of the Tetracyclic Diterpene Phorbol from *Croton tiglium* L. Tetrahedron Letters **1967**, 3165.
90. HECKER, E., H. BARTSCH, G. KREIBICH und CH. VON SZCZEPANSKI: Zur Chemie des Phorbols, X. Schweratom-Derivate zur Röntgenstrukturanalyse des Phorbols. L. Tetrahedron Letters **1967**, 3165.
91. HECKER, E., und H. BRESCH: Über die Wirkstoffe des Crotonöls, III. Reindarstellung und Charakterisierung eines toxisch, entzündlich und cocarcinogen hochaktiven Wirkstoffes. Z. Naturforsch. **20b**, 216 (1965).
92. — — Incorporation of Thymidine, Uridine and Leucine in the Skin of Mice after Treatment with Croton Oil Factor A_1 (TPA). Proc. Amer. Ass. Cancer Res. **1969**, 37.
93. HECKER, E., H. BRESCH und J. G. MEYER: Über cocarcinogene Wirkstoffe des Crotonöls. Abstr. Papers, 1st World Fat Congr. Hamburg, 1964, p. 176; see also: Fette, Seifen, Anstrichmittel **67**, 78 (1965).
94. HECKER, E., H. BRESCH und CH. VON SZCZEPANSKI: Cocarcinogen A_1 — der erste reine hochaktive Wirkstoff aus Crotonöl. Angew. Chem. **76**, 225 (1964); Angew. Chem. Intern. Ed. Engl. **3**, 227 (1964).
95. HECKER, E., E. HÄRLE, H. U. SCHAIRER, P. JACOBI, W. HOPPE, J. GASSMANN, M. RÖHRL und H. ABEL: Lumiphorboltriacetat — ein käfigartiges Derivat des Diterpens 4α-Phorbol. Angew. Chem. **80**, 913 (1968); Angew. Chem. Intern. Ed. Engl. **7**, 890 (1968).
96. HECKER, E., H. IMMICH, H. BRESCH und H. U. SCHAIRER: Über die Wirkstoffe des Crotonöls, VI. Entzündungsteste am Mäuseohr. Z. Krebsforsch. **68**, 366 (1966).
97. HECKER, E., H. JARCZYK, J. G. MEYER, H. BRESCH und I. BRACHMANN: Über die Wirkstoffe des Crotonöls, II. Eine systematische Fraktionierung des Crotonöls. Z. Krebsforsch. **66**, 478 (1965).

98. HECKER, E., und H. KUBINYI: Über die Wirkstoffe des Crotonöls, IV. Reindarstellung und Charakterisierung der entzündlichen und cocarcinogenen Wirkstoffe B₁ und B₂. Z. Krebsforsch. **67,** 176 (1965).

99. HECKER, E., H. KUBINYI und H. BRESCH: Eine neue Gruppe von Cocarcinogenen aus Crotonöl. Angew. Chem. **76,** 889 (1964); Angew. Chem. Intern. Ed. Engl. **3,** 747 (1964).

100. HECKER, E., H. KUBINYI, H. BRESCH, and CH. VON SZCZEPANSKI: The Chemical Structure of a Cocarcinogen and of Phorbol Isolated from Croton Oil. J. Med. Chem. **9,** 246 (1966).

101. HECKER, E., H. KUBINYI, H. U. SCHAIRER, CH. VON SZCZEPANSKI und H. BRESCH: Partialsynthese einiger cocarcinogener Wirkstoffe aus Crotonöl. Angew. Chem. **77,** 1076 (1965); Angew. Chem. Intern. Ed. Engl. **4,** 1072 (1965).

102. HECKER, E., H. KUBINYI, CH. VON SZCZEPANSKI, E. HÄRLE und H. BRESCH: Phorbol — ein neues tetracyclisches Diterpen aus Crotonöl. Tetrahedron Letters **1965,** 1837.

103. HECKER, E., und D. PAUL: Zum biochemischen Mechanismus der Tumorgenese der Mäusehaut, I. Verteilung und Stoffwechsel intragastral verfütterten 9,10-Dimethyl-(1,2)-benzanthracens in der Maus. Z. Krebsforsch. **71,** 153 (1968).

104. HECKER, E., und H. U. SCHAIRER: Über die Wirkstoffe des Crotonöls, VIII. Verbessertes Isolierungsverfahren für die Wirkstoffgruppen A und B sowie Isolierung und Charakterisierung weiterer Wirkstoffe der Gruppe A. Z. Krebsforsch. **70,** 1 (1967).

105. HECKER, E., CH. VON SZCZEPANSKI, H. KUBINYI, H. BRESCH, E. HÄRLE, H. U. SCHAIRER und H. BARTSCH: Über die Wirkstoffe des Crotonöls VII. Phorbol. Z. Naturforsch. **21b,** 1204 (1966).

106. HERZ, W., M. V. LAKSHMIKANTHAM, and R. N. MIRRINGTON: Constituents of Helenium Species — XVIII. 1-Epiisotenulin and Its Transformations. Tetrahedron **22,** 1709 (1966).

107. HOPPE, W., F. BRANDL, I. STRELL, M. RÖHRL, J. GASSMANN, E. HECKER, H. BARTSCH, G. KREIBICH und CH. VON SZCZEPANSKI: Röntgenstrukturanalyse des Neophorbols. Angew. Chem. **79,** 824 (1967); Angew. Chem. Intern. Ed. Engl. **6,** 809 (1967).

108. HOPPE, W., K. ZECHMEISTER, M. RÖHRL, F. BRANDL, E. HECKER, G. KREIBICH, and H. BARTSCH: Structure Determination of a Solvate of Phorbol — The Diterpene Parent of the Tumor Promotors from Croton Oil. Tetrahedron Letters **1969,** 667.

109a. IUPAC, Nomenclature of Organic Chemistry, Sections A, B, C. London: Butterworths. 1969.

109b. IUPAC, 1968 Tentative Rules, Section E. Fundamental Stereochemistry. See e.g.: Europ. J. Biochem. **18,** 151 (1971) and J. organ. Chem. (USA) **35,** 2849 (1970).

110. JACOBI, P., E. HÄRLE, H. U. SCHAIRER und E. HECKER: Zur Chemie des Phorbols, XVI. 4α-Phorbol. Liebigs Ann. Chem. **741,** 13 (1970).

111. JACOBSON, M.: Insecticides from Plants. A review of the literature 1941—1953, Agriculture Handbook No. 154, p. 99, United States Department of Agriculture, U.S. Government Printing Office, Washington D.C., 1958.

112. JACQUESY, J. C., R. JACQUESY et G. JOLY: Milieux hyperacides — V — reaction d'isomerisation a longue distance. Tetrahedron Letters **1972,** 4739.

113. KARRER, P., F. WEBER und J. VAN SLOOTEN: Über Toxine, II. Zur Kenntnis des Crotins. Helv. Chim. Acta **8,** 384 (1925).

114. KAUFFMANN, T., A. EISINGER, W. JASCHING und K. LENHARDT: Zur Konstitution des Phorbols, II. Über die α-Glykolgruppe des Phorbols. Chem. Ber. **92,** 1727 (1959).

115. KAUFFMANN, T., W. JASCHING und J. SCHULZ: Zur Konstitution des Phorbols, III. Epoxydierung von Phorbol und Phorbolderivaten. Chem. Ber. **99,** 1569 (1966).

116. KAUFFMANN, T., und H. NEUMANN: Zur Konstitution des Phorbols. I. Über die reduzierende Gruppe des Phorbols. Chem. Ber. **92,** 1715 (1959).

117. Kekulé, A.: Über einige Condensationsprodukte des Aldehyds. Liebigs Ann. Chem. **162,** 77 (1872), see p. 123.

118. Kreibich, G., und E. Hecker: Über die Verätherung des Phorbols: Angew. Chem. **79.** 993 (1967); Angew. Chem. Intern. Ed. Engl. **6,** 973 (1967).

119. — — Über die Verätherung des Phorbols. In: Aktuelle Probleme aus dem Gebiet der Cancerologie. (H. Lettré und G. Wagner, eds.), Vol. II, p. 155. Berlin-Heidelberg-New York: Springer. 1968.

120. — — Zur Chemie des Phorbols, V. Über einige Äther des Phorbols. Z. Naturforsch. **23b,** 1444 (1968).

121. — — On the Active Principles of Croton Oil, X. Preparation of Tritium Labelled Croton Oil Factor A_1 and Other Tritium Labelled Phorbol Derivates. Z. Krebsforsch. **74,** 448 (1970).

122. Kreibich, G., I. Witte, and E. Hecker: On the Biochemical Mechanism of Tumorigenesis in Mouse Skin. IV. Methods for Determination of Fate and Distribution of Phorbolester TPA. Z. Krebsforsch. **76,** 113 (1971).

123. Lewin, L.: Gifte und Vergiftungen, 4. Auflage des Lehrbuchs der Toxikologie. p. 861. Berlin: Verlag von Georg Stilke. 1929.

124. Lijinsky, W.: A Tumor Promoting Principle from Croton Oil. Biochem. J. **70,** 5P (1958).

125. — A Tumor-Promoting Agent from Croton Oil. Commun. 5th Intern. Congr. Biochem. Moscow, 1961, Abstr. p. 9.

126. Linstead, R. P.: Chem. and Ind. **15,** 510 (1937).

127. Meyer, J. G.: Eine verbesserte Methode zur Isolierung von Cocarcinogenen aus Crotonöl und Charakterisierung des neuen Phorbolesters A_3. Experientia **22,** 482 (1966).

128. Meyer-Bertenrath, J. G.: A Simplified Isolation Procedure of Four Cocarcinogenic Phorbolesters from Croton Oil. Experientia **24,** 1295 (1968).

129. — 150 years of croton oil research. Experientia **25,** 1 (1969).

130. Mottram, J. C.: A Developing Factor in Experimental Blastogenesis. J. Pathol. Bacteriol. **56,** 181 (1944).

131. Nakahara, W.: Critique of Carcinogenic Mechanism. Progr. Exptl. Tumor Res. **2,** 158 (1961).

132. Ocken, P. R.: Dry-column Chromatographic Isolation of Fatty Acid Esters of Phorbol from Croton Oil. J. Lipid Res. **10,** 460 (1969).

133. Oxley, P.: Cephalosporin P_1 and Helvolic acid. Chem. Commun. **1966,** 729.

134. Paal, C., und K. Roth: Über katalytische Wirkungen kolloidaler Metalle der Platingruppe. V. Die Reduktion der Fette. Ber. dtsch. Chem. Ges. **42,** 1541 (1909).

135. Paukstelis, J. V., and J. Kao: Synthesis of *trans*-Bicyclo[4.1.0]heptanes. J. Amer. Chem. Soc. **94,** 4783 (1972).

136. Paul, D., and E. Hecker: On the Biochemical Mechanism of Tumorigenesis in Mouse Skin, II. Early Effects on the Biosynthesis of Nucleic Acids Induced by Initiating Doses of DMBA and by Promoting Doses of Phorbol-12,13-diester TPA. Z. Krebsforsch. **73,** 149 (1969).

137. Petri, E.: Pathologische Anatomie und Histologie der Vergiftungen. In: Handbuch der spez. pathol. Anatomie und Histologie (Henke-Lubarsch), p. 434. Berlin: Springer. 1930.

138. Pettersen, R. C., G. Ferguson, L. Crombie, M. L. Games, and D. L. Pointer: The Structure and Stereochemistry of Phorbol, Diterpene Parent of Co-carcinogens of Croton Oil. Chem. Commun. **1967,** 716.

139. Pettersen, R. C., G. I. Birnbaum, G. Ferguson, K. M. Islam, and I. G. Sime: X-Ray Investigation of Several Phorbol Derivates. The Crystal and Molecular Structure of Phorbol Bromofuroate-Chloroform Solvate at $-160°$. J. Chem. Soc. (B) (London) **1968,** 980.

140. POST, O., and L. CRAIG: A New Type of Stepwise Countercurrent Distribution Train. Analyt. Chem. **35**, 641 (1963).

141. RAICK, A. N., K. THUMM, and B. R. CHIVERS: Early Effects of 12-O-Tetradecanoyl-phorbol-13-acetate on the Incorporation of Tritiated Precursor into DNA and the Thickness of the Interfollicular Epidermis, and Their Relation to Tumor Promotion in Mouse Skin. Cancer Res. **32**, 1562 (1972).

142. ROHRSCHNEIDER, L. R., D. H. O'BRIEN, and R. K. BOUTWELL: The Stimulation of Phospholipid Metabolism in Mouse Skin following Phorbol Ester Treatment. Biochim. Biophys. Acta **280**, 57 (1972).

143. RUZICKA, L., M. FURTER und M. W. GOLDBERG: Über Steroide und Sexualhormone (42. Mitteilung). Zur Stereochemie epimerer steroider Alkohole mit einem Hydroxyl in den Stellungen 3 oder 17. Helv. Chim. Acta **21**, 498 (1938).

144. SAFFIOTTI, U., and P. SHUBIK: Studies on Promoting Action in Skin Carcinogenesis. Natl. Cancer Inst. Monograph. **10**, 489 (1963).

145. SALAMAN, M. H.: Cocarcinogenesis. Brit. Med. Bull. **14**, 116 (1958).

146. SALAMAN, M. H., and F. J. C. ROE: Cocarcinogenesis. Brit. Med. Bull. **20**, 139 (1964).

147. SCHAAFSMA, S. E., H. STEINBERG, and TH. J. DE BOER: Isomerisation and Oxidative Dimerisation of 1-Substituted Cyclopropanols (Preliminary communication). Recueil Trav. Chim. Pays-Bas **85**, 73 (1966).

148. SCHAFFNER, K., und G. SNATZKE: Circulardichroitische Messungen an gesättigten und α,β-ungesättigten Cyclopropyl-ketonen. Helv. Chim. Acta **48**, 347 (1965).

149. SCHAIRER, H. U., H. W. THIELMANN, M. GSCHWENDT, G. KREIBICH, R. SCHMIDT und E. HECKER: Zur Chemie des Phorbols, IV. Polybenzoate und -acetate des Phorbols und Phorbol-3-ols und funktionelle Derivate der Allylgruppierung des Phorbols. Z. Naturforsch. **23b**, 1430 (1968).

150. SCHEIBE, G., D. BRÜCK und F. DÖRR: Über die Ähnlichkeit des Absorptions-spektrums einfacher organischer Moleküle und Farbstoffe mit dem Spektrum des Wasserstoffatoms. Chem. Ber. **85**, 867 (1952).

151. SCHINDLER, H.: Die Inhaltsstoffe von Heilpflanzen und Prüfungsmethoden für pflanzliche Tinkturen. Arzneimittelforschung (Drug Res.) **3**, 313 (1953); see also H. SCHINDLER: Inhaltsstoffe und Prüfungsmethoden homöopathisch verwendeter Heilpflanzen. Aulendorf i. Württ.: Editio Cantor. 1955.

152. SCHMIDT, R., und E. HECKER: Neue Phorbolester aus Crotonöl. Fette, Seifen, Anstrichmittel **70**, 851 (1968).

153. — — Ein neues Diterpen und seine Fettsäureester aus Crotonöl. Fette, Seifen, Anstrichmittel — Die Ernährungsindustrie **73**, 676 (1971); SCHMIDT, R., und E. HECKER, unpublished.

154. — — Untersuchungen über die Beziehungen zwischen Struktur und Wirkung von Phorbolestern. In: Aktuelle Probleme aus dem Gebiet der Cancerologie (H. LETTRÉ und G. WAGNER, eds.), Vol. III, p. 98. Berlin-Heidelberg-New York: Springer. 1971.

155. SCHMIDT, R., H. W. THIELMANN, and E. HECKER, in preparation.

156. SCOTT, A. I.: Interpretation of the Ultraviolet Spectra of Natural Products, p. 407. Oxford: Pergamon Press. 1964.

157. SHEAR, M. J.: Studies in Carcinogenesis, V. Methyl Derivates of 1:2-Benzanthracene. Amer. J. Cancer **33**, 499 (1938).

158. SICÉ, I.: Tumor Promoting Principles in Seeds of *Croton tiglium* L. Arch. Intern. Pharmacodyn. **115**, 408 (1958).

159. SIVAK, A., B. T. MOSSMAN, and B. L. VAN DUUREN: Activation of Cell Membrane Enzymes in the Stimulation of Cell Division. Biochem. Biophys. Res. Comm. **46**, 605 (1972).

160. SPIES, J. R.: Croton Resin. I. Toxicity Studies Using Goldfish. J. Amer. Chem. Soc. **57**, 180 (1935).

161. — Croton Resin, II. The Toxic and Vesicant Action of Certain of its Derivatives. J. Amer. Chem. Soc. **57**, 182 (1935).

162. SÜSS, R., and V. KINZEL: Phorbolesters as a tool in Cell Research. Europ. J. Cancer **8**, 299 (1972).

163. THIELMANN, H. W., und E. HECKER: Beziehungen zwischen der Struktur von Phorbolestern und ihren entzündlichen sowie tumorpromovierenden Eigenschaften. Hoppe-Seyler's Z. physiol. Chem. **349**, 17 (1968).

164. — — Beziehungen zwischen der Struktur von Phorbolderivaten und ihren entzündlichen und tumorpromovierenden Eigenschaften. In: Fortschritte der Krebsforschung (C. G. SCHMIDT und O. WETTER, eds.), p. 171. Stuttgart-New York: Schattauer. 1969.

165. — — Zur Chemie des Phorbols, XIV. Die Flaschenträger-Reaktion. Liebigs Ann. Chem. **728**, 158 (1969).

166. — — Zur Chemie des Phorbols, XV. Oxydation der Δ^6-Doppelbindung des Phorbols mit Osmiumtetroxid und Spaltung einiger 6β,7β-Diole. Liebigs Ann. Chem. **735**, 113 (1970).

167. THIELMANN, W., P. JACOBI und E. HECKER: Zur Chemie des Phorbols, XVIII. Bromierung des Phorbol-pentaacetats. Liebigs Ann. Chem. **765**, 171 (1972).

168. THOMAS, A. F., and A. MARXER: Crystalline Acetates from "Croton Resin". Experientia **14**, 320 (1958).

169. TRAUT, M., G. KREIBICH und E. HECKER: Über die Proteinbindung carcinogener Kohlenwasserstoffe und cocarcinogener Phorbolester. In: Aktuelle Probleme aus dem Gebiet der Cancerologie (H. LETTRÉ und G. WAGNER, eds.), Vol. III, p. 91. Berlin-Heidelberg-New York: Springer. 1971.

170. TSCHIRCH, A.: Handbuch der Pharmacognosie, Vol. II, Abteilung 1., p. 579, Leipzig, 1912.

171. VAN DUUREN, B. L.: Tumor-promoting Agents in Two-Stage Carcinogenesis. Progr. Exptl. Tumor Res. **11**, 31 (1969).

172. VAN DUUREN, B. L., E. R. ARROYO, and L. ORRIS: The Tumor-enhancing and Irritant Principles from *Croton tiglium* L. J. Med. Chem. **6**, 616 (1963).

173. VAN DUUREN, B. L., and L. ORRIS: The Tumor-enhancing Principles of *Croton tiglium* L. Cancer Res. **25**, 1871 (1965).

174. VAN DUUREN, B. L., L. ORRIS, and E. R. ARROYO: The Tumor-enhancing Principle from Croton Oil and Croton-Resin. Proc. Am. Assoc. Cancer Res. p. 69 (1963).

175. — — Tumor-enhancing Activity of the Active Principles of *Croton tiglium* L. Nature **200**, 1115 (1963).

176. VAN DUUREN, B. L., A. SIVAK, A. SEGAL, L. ORRIS, and L. LANGSETH: The Tumor-promoting Agents of Tobacco Leaf and Tobacco Smoke Condensate. J. Natl. Cancer Inst. **37**, 519 (1966).

177. VAN POTTER, R.: Biochemical Perspectives in Cancer Research. Cancer Res. **24**, 1085 (1964).

178. VELLUZ, L., G. NOMINÉ und G. MATHIEU: Neuere Ergebnisse bei der Totalsynthese von Steroiden. Angew. Chem. **72**, 725 (1960).

179. VON ARDENNE, M., K. STEINFELDER und R. TÜMMLER: Elektronenanlagerungs-Massenspektrogramme kondensierter aromatischer Kohlenwasserstoffe. Angew. Chem. **73**, 136 (1961).

180. VON ARDENNE, M., K. STEINFELDER, R. TÜMMLER und K. SCHREIBER: Molekül-Massenspektrographie von Naturstoffen. Steroide. Experientia **19**, 178 (1963).

181. Von Metzsch, F. A.: 200-stufige, vollautomatische Apparatur zur fraktionierten Gegenstromverteilung. Chem. Ing. Techn. **25,** 66 (1953).

182. — Vollautomatische Laboratoriumsapparatur für Verteilungsverfahren mit schubweise bewegten Phasen. Chem. Ing. Techn. **31,** 262 (1959).

183. Von Szczepanski, Ch., H. U. Schairer, M. Gschwendt und E. Hecker: Zur Chemie des Phorbols, III. Mono- und Diacetate des Phorbols. Liebigs Ann. Chem. **705,** 199 (1967).

184. Wehmer, C.: Die Pflanzenstoffe, p. 425. Jena: G. Fischer. 1911.

185. Weissmann, G., W. Troll, B. L. Van Duuren, and G. Sessa: Studies on Lysosomes, X. Effects of Tumor-Promoting Agents upon Biological and Artificial Membrane Systems. Biochem. Pharmacol. **17,** 2421 (1968).

(Received February 19, 1973)

Stereoselektive Totalsynthese von Indolalkaloiden

Von E. Winterfeldt, Hannover

Die synthetischen Arbeiten in der Reihe der polycyclischen Indolalkaloide haben in den letzten 20 Jahren, beginnend mit der eindrucks-
vollen, eine neue Epoche synthetischer Strategie und Planung ein-
leitenden Totalsynthese des Alkaloid-Veteranen Strychnin durch
R. B. Woodward und Mitarbeiter (21), erhebliche Fortschritte gemacht.
Da in den letzten Jahren aber außerdem auch die Biogenese dieser
Alkaloide durch die Arbeiten von Arigoni, Battersby und Scott
weitgehend geklärt werden konnte, wird in dem folgenden Überblick
über stereoselektive totalsynthetische Vorhaben auf diesem inzwischen
etwa 800 Alkaloide umfassenden Gebiet eine Einteilung gemäß der
kürzlich von I. Kompis, M. Hesse und H. Schmid (32) vorgeschlagenen
biogenetischen Klassifikation der Indolalkaloide vorgenommen. Als
Kriterium für diese Einteilung der Indolalkaloide in Klassen ist die
Anzahl der Bindungen zwischen den beiden wichtigen biogenetischen
Bausteinen, nämlich dem Tryptophan (1) und dem Loganin (2) bzw.
Secologanin (3), anzusehen.

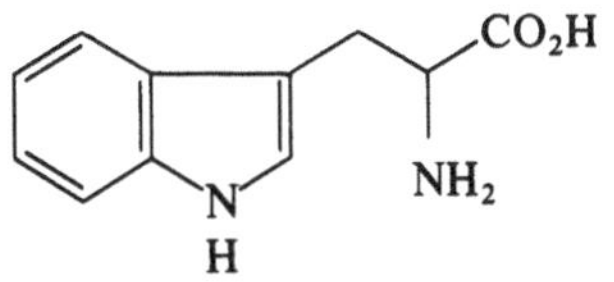

(1) 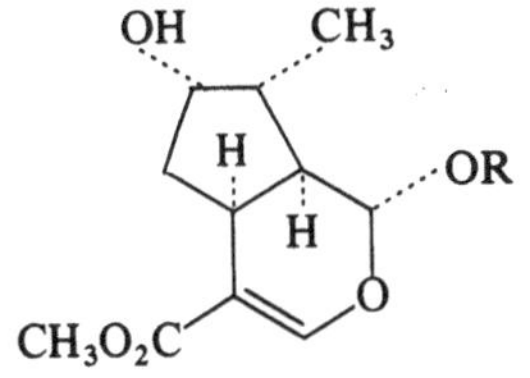   (2) 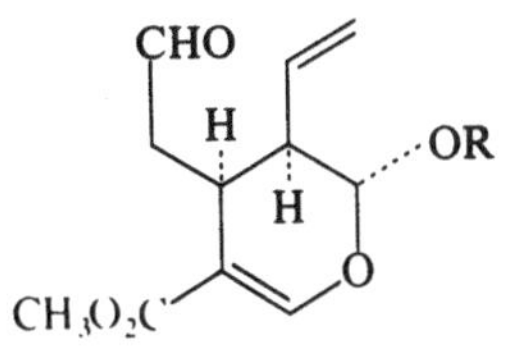   (3)

Danach fallen in die Klasse I alle Alkaloide mit unverändertem Loganin-
Grundgerüst. Sie ist mit Abstand die größte Klasse, da hier neben
dem Corynanthein-Typ auch so große Gruppen wie der Sarpagin-Typ,
der Ajmalin-Typ, eine große Zahl von Oxindol-Alkaloiden, sowie die

Condylocarpin-Gruppe, die Akuammicin-Gruppe und schließlich der in vielen Variationen auftretende Strychnos-Typ anzutreffen sind. Es nimmt daher nicht wunder, wenn in dieser Gruppe auch ein Schwerpunkt totalsynthetischer Vorhaben liegt. Ein Alkaloid, das noch sehr gut den Loganin-Baustein erkennen läßt, ist z. B. das Corynanthein (4), das von mehreren Arbeitskreisen synthetisiert wurde.

(4)

Bei der ersten von VAN TAMELEN und WRIGHT (43) publizierten Synthese ist die Stereoselektivität beim Aufbau der entscheidenden Zentren nicht sehr groß (s. u.), aber das Ausgangsmaterial (5), das aus Tryptamin, Formaldehyd und dem aus Glutaconester und Acetessigester gut darstellbaren Keto-triester (6) gewonnen werden kann, hat sich als ein sehr vielseitiges Zwischenprodukt erwiesen.

(5)   (6)

Durch Cyclisierung und Reduktion sowie anschließende Verseifung und Decarboxylierung erhält man das Gemisch der an C-15 isomeren Keto-ester (7) und (8) im Verhältnis 60:40, die getrennt werden müssen.

(7)   (8)

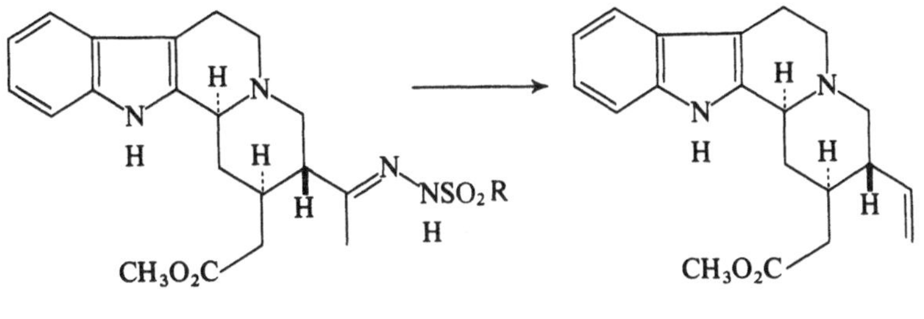

(9)   (10)

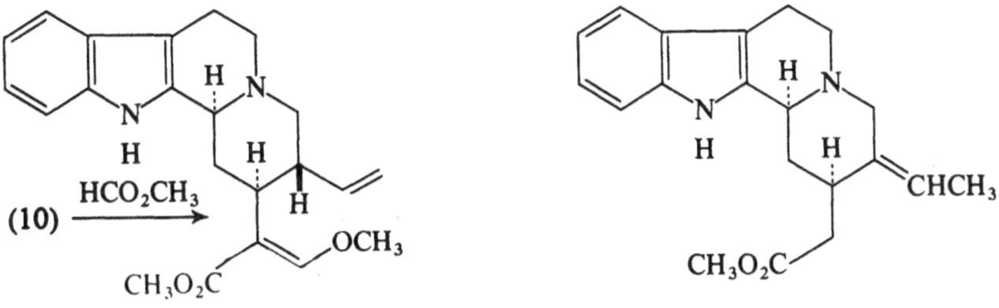

(10)
HCO₂CH₃
→

(4)   (11)

Die Konfiguration an diesen beiden Zentren 15 und 20 stellt in der Tat das zentrale Problem bei der Totalsynthese von Verbindungen dieses Typs dar, da man außer der Überführung in Verbindungen bekannter Konfiguration keine überzeugenden Kriterien für die Konfigurationszuweisung zur Hand hat, so daß gerade hier Syntheseschritte, die mit hoher Stereoselektivität nach Möglichkeit nur ein Produkt erzeugen, sehr erwünscht sind.

Natürlich erfolgt auch die Einführung der Doppelbindung, die über das Hydrazon (9) bzw. das entsprechende Carben verläuft, weder regioselektiv noch stereoselektiv. Die nach Trennung der Olefinkomponenten isolierbare Verbindung (10) läßt sich durch Kondensation mit Ameisensäureester in Corynanthein (4) überführen. Die ebenfalls anfallende Verbindung (11) mit exocyclischer Doppelbindung war seinerzeit als sterisch einheitliches Produkt einer stereoselektiven Carben-Eliminierung mit E-konfigurierter, „unnatürlicher" Doppelbindung formuliert worden, weil die anschließende Kondensation mit Ameisensäureester nicht das entsprechende Alkaloid mit exocyclischer Z-konfigurierter Doppelbindung — das Geissoschizin — hervorbrachte.

Einen wichtigen Ansatzpunkt für den stereoselektiven Aufbau der Zentren 15 und 20 liefert die Totalsynthese des Corynantheidins (4a) (12) mit cis-ständigen α-Wasserstoffen in diesen Positionen.

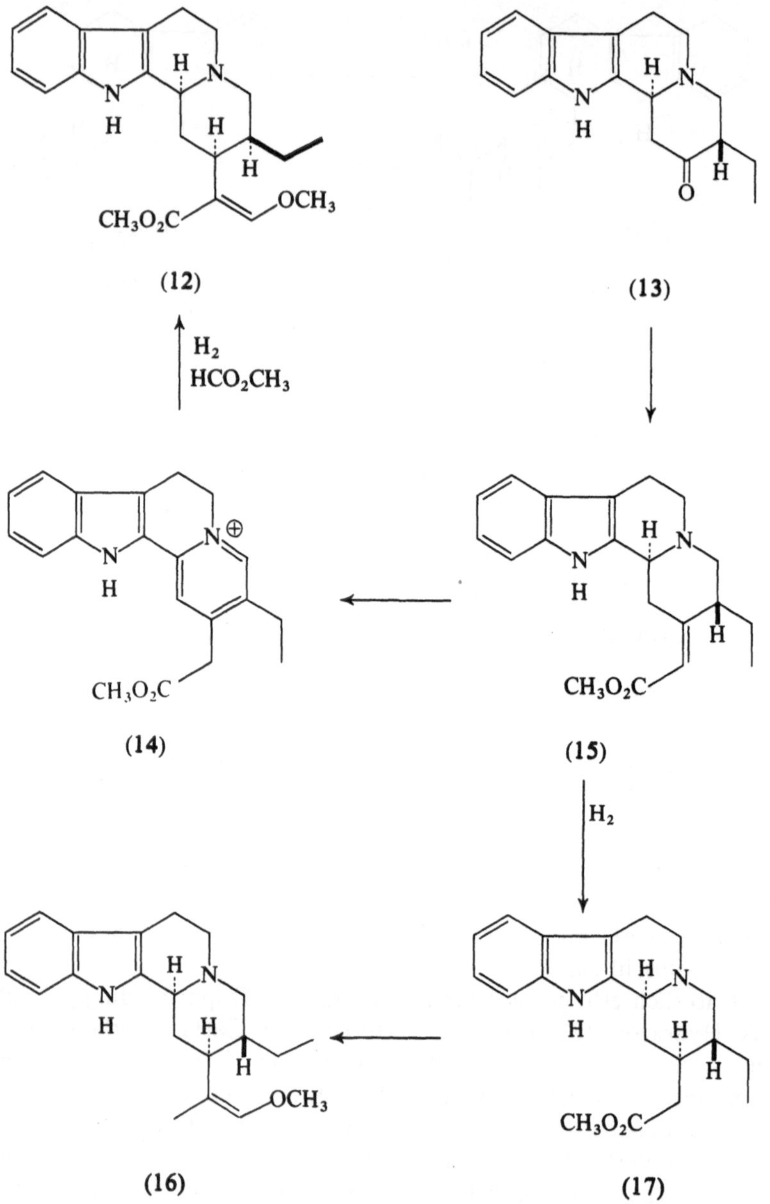

(12)   (13)

H_2
HCO_2CH_3

(14)   (15)

H_2

(16)   (17)

Ausgehend vom Keton (13) von definierter und thermodynamisch stabiler Konfiguration wurde nach der Phosphonat-Methode der ungesättigte Ester (15) präpariert. Direkte Hydrierung dieser Substanz liefert die 15α, 20β-Verbindung (17), die sich durch Formylierung in Dihydrocorynanthein überführen läßt.

Literaturverzeichnis: SS. 518—520

Nach Dehydrierung zum Pyridiniumsalz (14) erzeugt die anschließende Hydrierung und Formylierung das 15α, 20α konfigurierte Corynantheidin. Es führt also hier ein zentrales und gut zugängliches Ausgangsmaterial zu beiden relativen Konfigurationen an diesen Zentren.

Das gleiche Keton wird von SZANTAY und BARCZAI-BEKE (8) als Ausgangsmaterial für eine stereoselektive Totalsynthese des (—)-Corynantheidins verwendet. Bei der Kondensation mit Cyanessigester tritt erwartungsgemäß Inversion an C_{20} auf, und man erhält (18).

(18)   (19)

In dieser Arbeit wird auch eine elegante Überführung der Malonatgruppe in den Methoxyacrylester angegeben. Reduktion mit Lithiumaluminiumhydrid bei $-50°$ ergibt die Formylverbindung, die als Aluminiumkomplex ausfällt. Anschließende Methylierung mit Dimethylsulfat in Benzol liefert dann in fast quantitativer Ausbeute das racemische Corynantheidin, das anschließend mit Dibenzoyl-D-Weinsäure in die Antipoden gespalten wurde. Ebenfalls über ein Pyridiniumsalz läuft die Synthese von WENKERT und Mitarbeitern (23).

Das durch Quartärnisierung des Pyridinesters (20) darstellbare Pyridiniumsalz (21) geht bei der Hydrierung am Palladium-Katalysator stereoselektiv in das cis-konfigurierte Enamin (23) über. Wird (23) alkalisch verseift, so cyclisiert es im sauren Medium unter Decarboxylierung zu (22). Dehydriert man diese 3β-konfigurierte Verbindung, so erzeugt die anschließende Boranat-Reduktion den Ester (24), der nach bereits angegebenen Methoden in Corynantheidin überführbar ist. Wird zunächst sauer cyclisiert und anschließend verseift, so decarboxyliert die Säure bei der anschließenden Dehydrierung, die (25) liefert. Auch aus dieser Verbindung geht erwartungsgemäß bei der Boranat-Reduktion das wichtige Zwischenprodukt (24) hervor.

Auch Oxindolverbindungen dieser Serie wurden stereoselektiv gewonnen. So sicherte z. B. BAN (7) die Konfiguration des Rhynchophyllins

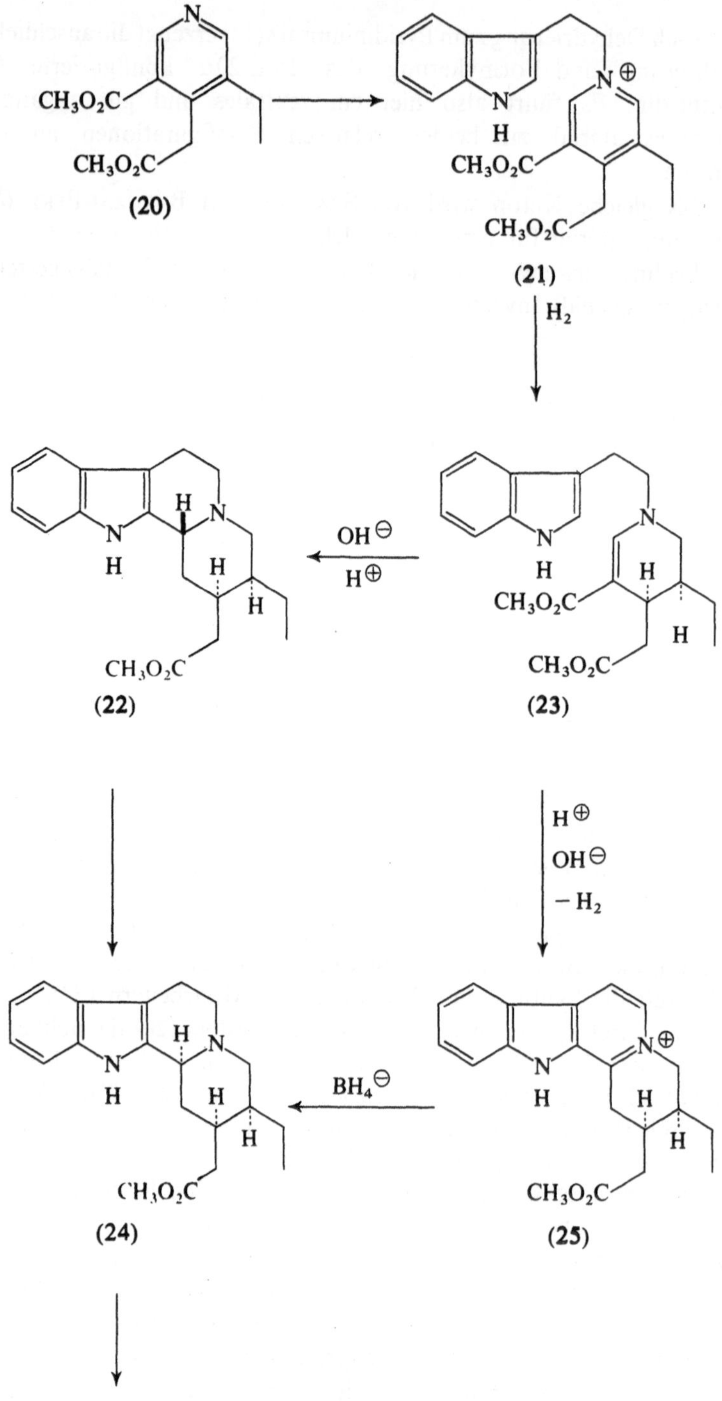

Literaturverzeichnis. SS. 518—520

(26) durch eine stereoselektive Synthese des Rhynchophyllans (27), das durch Hydrolyse des Naturstoffes zum Aldehyd und anschließende WOLFF-KISHNER-Reduktion gewonnen wurde. Ausgangspunkt waren die von VAN TAMELEN bereiteten Lactone (28a) und (28b) (3), aus denen nach Überführung in Methylanilide und Chlorierung die Chloramide (29a) und (29b) zugänglich werden.

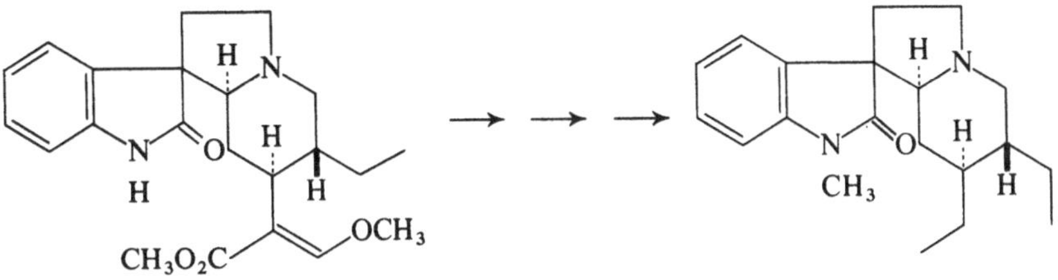

(26)   (27)

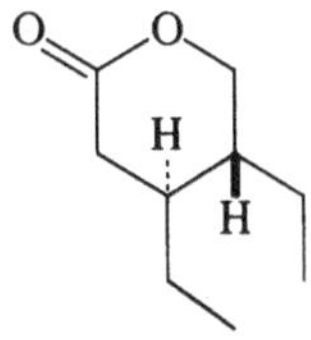

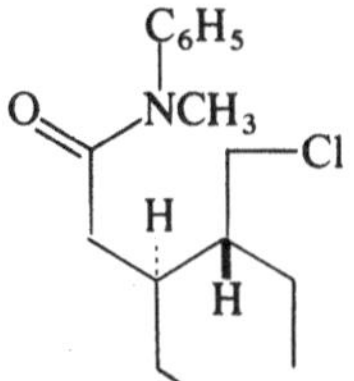

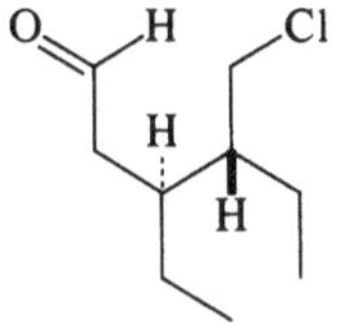

(28a)   (29a)

(28b) (H cis)   (29b) (H cis)

(30)

Alanatreduktion des *trans*-Amids (29a) liefert den Aldehyd (30), der dann bei der Kondensation mit N-Methyl-oxytryptamin in die beiden Spiro-Epimeren N-Methyl-Rhynchophyllan und Epi-N-Methyl-Rhynchophyllan übergeht, womit die *trans*-Anordnung der Reste in C_{15} und C_{20} bewiesen ist.

Sehr gut erkennbar ist der terpenoide Baustein auch in den pentacyclischen Enoläthern vom Typ des Ajmalicins (31).

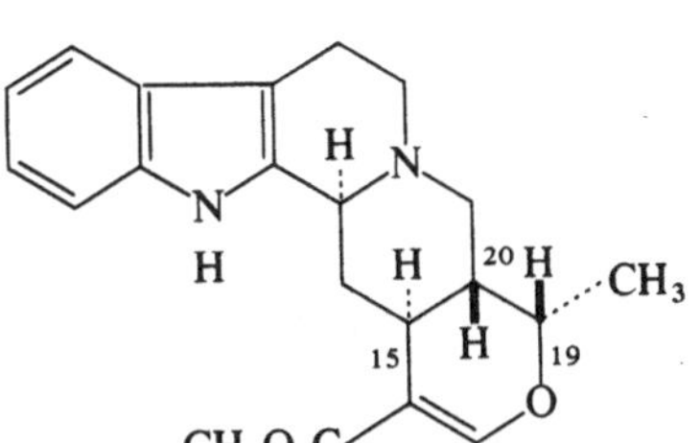

(31) 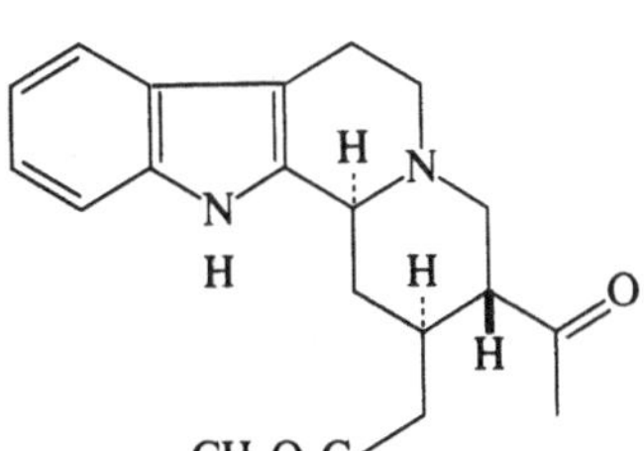   (7)

In dieser Verbindungsklasse stellt der Aufbau der Konfiguration
am C_{19} ein spezielles Problem dar. Für beide Möglichkeiten, nämlich
die axiale, α-ständige Methylgruppe (s. 31) wie auch die äquatoriale
β-ständige, sind stereoselektive Methoden erarbeitet worden. Van
Tamelen (38) wählte den bereits beschriebenen Ketoester (7) als
Ausgangsmaterial, der dann in einer stereoselektiven Boranatreduktion
das Lacton (32) lieferte.

Die Überführung in Ajmalicin wird durch Kondensation mit Ameisensäureester vorgenommen, die sich bildende β-Dicarbonylverbindung (33)
geht bei der sich sofort anschließenden Korte-Umlagerung in Ajmalicin

(38)   (39)

(40)   (41)

(42)   (43)

über. Ein großer Nachteil besteht unter den angegebenen Bedingungen darin, daß sich Addition des als Lösungsmittel verwendeten Methanols an die Enoläther-Doppelbindung nicht vermeiden läßt, so daß in allen Fällen ein beträchtlicher Teil des entsprechenden cyclischen Acetals gebildet wird (25).

Der totalsynthetische, stereoselektive Aufbau praktisch aller wichtigen Konfigurationskombinationen dieser Naturstoffgruppe wird von Chemikern der Firma Hoffmann-La Roche beschrieben (27, 28, 29). Aus dem Pyridinderivat (34) bereiteten sie über Hydrierung und Chlorierung die N-Chlor-Verbindung (36), während das bekannte Lactam (35) über den Iminoäther, Boranatreduktion und Chlorierung in die entsprechende *trans*-Verbindung (37) überführt wurde.

In sehr eleganter Weise wird dann in einer Photoreaktion das Halogenatom in die Äthylgruppe eingeführt und nach Benzoylierung mit Kaliumtert.butylat eliminiert, so daß die beiden Vinylverbindungen (40) und (41) resultieren. Amidacetal-Kondensation bringt daraus die β-Dicarbonylverbindungen (42) und (43) hervor, die bereits alle wichtigen Strukturmerkmale der Corynanthein- und der Ajmalicin-Serie erkennen

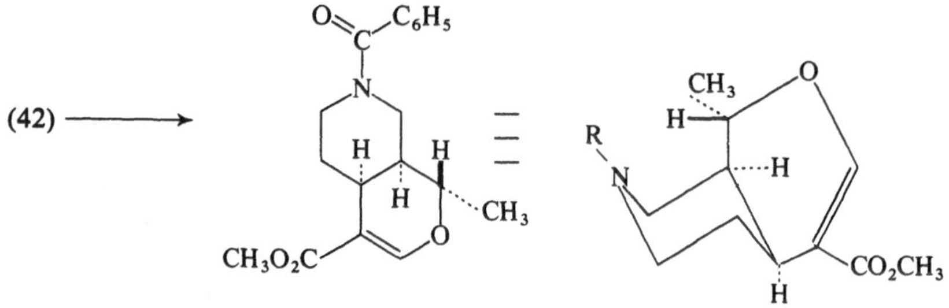

(44)   (44a)

lassen. Die anschließende Oxymercurierung verläuft bei (42) stereoselektiv unter Bildung des cyclischen Äthers (44) mit äquatorialer Methylgruppe, eine Erscheinung, die offenbar typisch ist für die

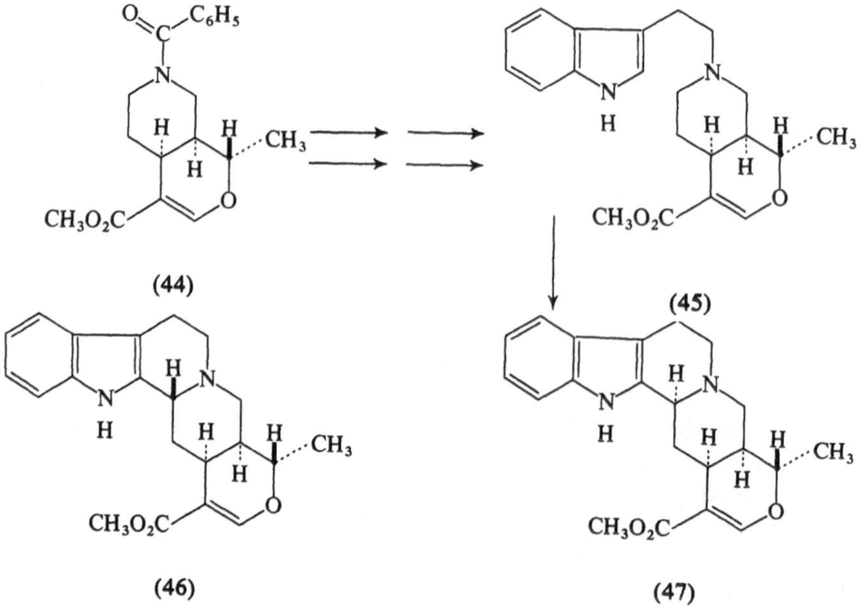

(44)   (45)

(46)   (47)

15,20-*cis*-Serie. Eine ähnliche Beobachtung wurde bei der Reduktion eines entsprechenden Ketons gemacht (s. u.), und die Erklärung ist sicher in einer speziellen Hinderung für den zur axialen β-ständigen Methylgruppe führenden Übergangszustand zu finden (s. **44a**).

Partielle Reduktion und Hydrolyse führt zur debenzoylierten Zwischenstufe, die dann mit β-Indolyl-äthylbromid in das Indolderivat (**45**) übergeht.

Obwohl bei der anschließenden Dehydrierungs-Cyclisierungssequenz das Imoniumsalz sich prinzipiell in zwei verschiedenen Positionen ausbilden und somit zu zwei verschiedenen Cyclisierungsprodukten

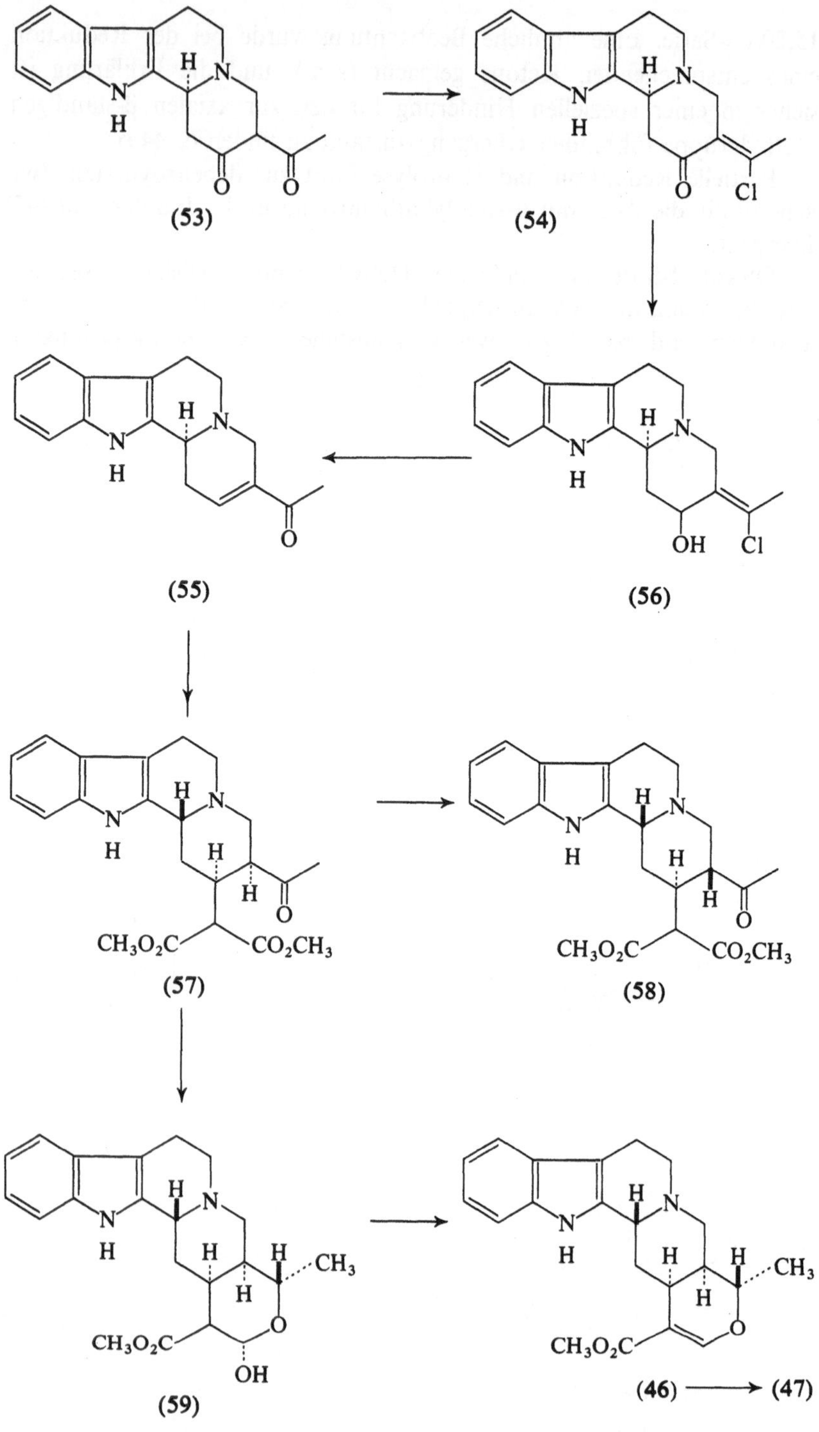

(53)

(54)

(55)

(56)

(57)

(58)

(59)

(46) ⟶ (47)

Anlaß geben kann, verläuft diese Reaktion offenbar nur in der gewünschten Weise und bringt das Tetrahydroalstonin (47) bzw. die 3β-H-Verbindung Akuammicin (46) hervor. Weniger gerichtet verläuft die Oxymercurierung bei der *trans*-Verbindung (43). Hier werden die beiden Epimeren (48) und (49) gewonnen, die dann mit β-Indolyl-äthylbromid in (50) und (51) übergehen.

Wiederum verläuft die letzte Cyclisierungsreaktion nur in der gewünschten Weise und liefert Ajmalicin (31) sowie die 19-epi-Verbindung (52) mit äquatorialer Methylgruppe. Als ein universelles Ausgangsmaterial in dieser Serie hat sich das ungesättigte Keton (55) erwiesen, das aus dem 1,3-Diketon (53) über Chlorierung zum vinylogen Säurechlorid (54) und anschließende Boranat-Reduktion (56) und Hydrolyse in einfacher Weise gewonnen werden kann (36).

Die Michaeladdition von Malonester liefert bei kinetischem Reaktionsabschluß unter gleichzeitiger Chinolizidin-Inversion die all-*cis*-Verbindung (57) mit axialer Ketomethylgruppe. Basenkatalyse epimerisiert zur thermodynamisch stabilen Verbindung (58).

Die Boranatreduktion von (57) erzeugt nun in hoher Ausbeute
stereoselektiv sofort das Cyclohalbacetal (59), ohne daß das intermediär
auftretende Lacton isolierbar ist. Auch hier bildet sich also, wie
bereits erwähnt, nur die Verbindung mit α-ständiger, äquatorialer
Methylgruppe. Anschließende Wasserabspaltung führt dann in glatter
Reaktion zum Akuammicin (46), das bei C_3-Epimerisierung über
Bleitetraacetat-Oxydation und Boranatreduktion in Tetrahydroalstonin
(47) übergeht.

Beide Konfigurationen am C_{19} können indessen aus dem *trans*-Keto-
ester (58) hervorgebracht werden (25). Während Hydrierung das Lacton (60)
mit äquatorialer Methylgruppe erzeugt, erhält man bei der Boranat-
reduktion bei tiefer Temperatur stereoselektiv das Lacton (61) mit
axialer Methylgruppe. Die Konfiguration am C_{19} kann durch Spin-
entkopplung des entsprechenden Protons klar zugeordnet werden.
Weitere Boranatreduktion von (60) und anschließende Wasserabspaltung
führt zu (62), das durch C_3-Epimerisierung (s. o.) 19-Epi-Ajmalicin
und durch Hypochlorit-Oxydation zum Chlor-Indolenin und anschließende saure Hydrolyse Formosanin (63) liefert. Diese Sequenz von
Reaktionen läßt sich jedoch nicht auf das Lacton (61) übertragen, da
offenbar wegen der axialen Lage der Methylgruppe das Cyclohalbacetal
nicht hinreichend stabil ist und daher zu verschiedenen recht polaren
Substanzen weiterreduziert wird.

(61) $\xrightarrow{CN^{\ominus}}$ (64) $\longrightarrow$ (65) $\longrightarrow$ (31)

Es wurde daher die Estergruppe mit Cyanidionen verdrängt und das
resultierende Lacton durch Ameisensäureester-Kondensation und Korte-
Umlagerung in 3-Iso-Ajmalicin (65) überführt. Anschließende C_3-
Epimerisierung ergibt Ajmalicin (31).

Das ungesättigte Keton (55) war auch die Basis für eine stereoselektive Totalsynthese des Roxburghins D (66) (39), das Hauptalkaloid
einer Gruppe stereoisomerer Verbindungen mit biogenetisch interessan-
tem Baumuster.

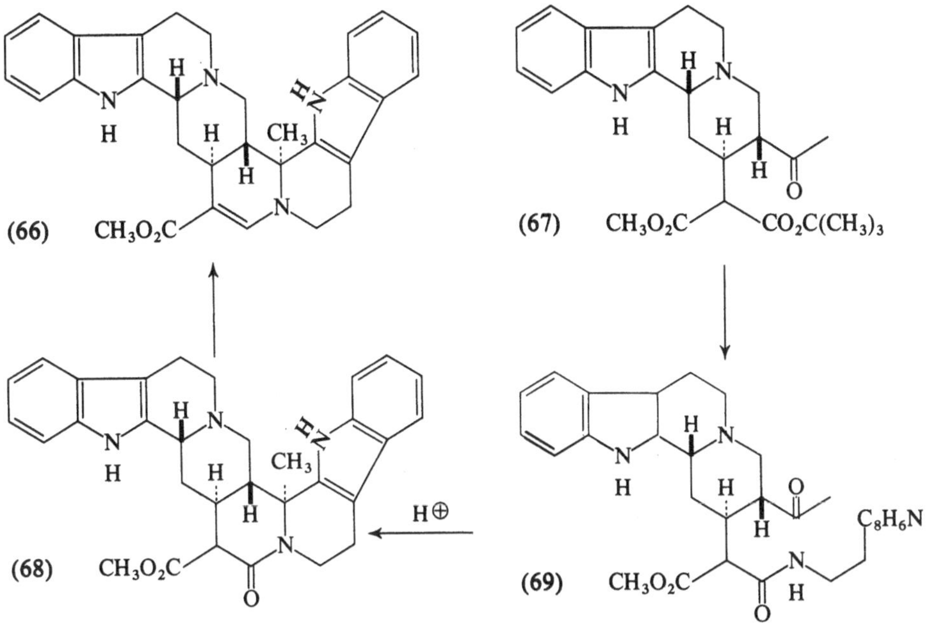

(66) CH₃O₂C

(67) CH₃O₂C CO₂C(CH₃)₃

(68) CH₃O₂C

(69) CH₃O₂C

Das thermodynamisch gelenkte Addukt von Mono-tert. butyl-malonester (67) ließ sich mit kalter Trifluoressigsäure selektiv spalten und mit Tryptamin und DCCI in das Amid (69) überführen. Protonenkatalysierte Cyclisierung liefert stereoselektiv das Lactam (68), dessen selektive Reduktion nach Säurebehandlung das octacyclische Enamin mit α-ständiger Methylgruppe hervorbringt.

Ein prominenter Vertreter der pentacyclischen Reihe mit carbocyclischem Ring E ist das Yohimbin (70), dessen stereoselektive Totalsynthese von zwei Arbeitsgruppen erfolgreich betrieben wurde. Die erste Synthese wurde von VAN TAMELEN (4) publiziert, der das Dien-Addukt (71) als Ausgangsmaterial wählte.

Reduktion und Glycidester-Kondensation führt unter Epimerisierung neben der verbleibenden Carbonylgruppe zu (72), das nach Spaltung und Hydrolyse des dabei resultierenden Acetals den Aldehyd (73) als thermodynamisch stabiles Produkt liefert. Versuche, das primäre Monotosylat, welches bei Reduktion und selektiver Tosylierung von (73) erhalten wird, in einer nucleophilen Verdrängungsreaktion mit Tryptamin zu kondensieren, schlugen fehl. Man oxydierte daher zur Säure und setzte nach Überführung in das entsprechende Säurechlorid mit Tryptamin um. Das dabei resultierende Amid (74) wurde reduziert und hydroxyliert zum Triol (76), dessen Perjodsäurespaltung einen Dialdehyd liefert, der durch Wechselwirkung mit der Amidgruppe und der OH-Gruppe jedoch vollständig in der Cyclohalbacetal-Form vorliegt. Daher erfolgt glatte Cyclisierung zu (75). Wasserabspaltung zum

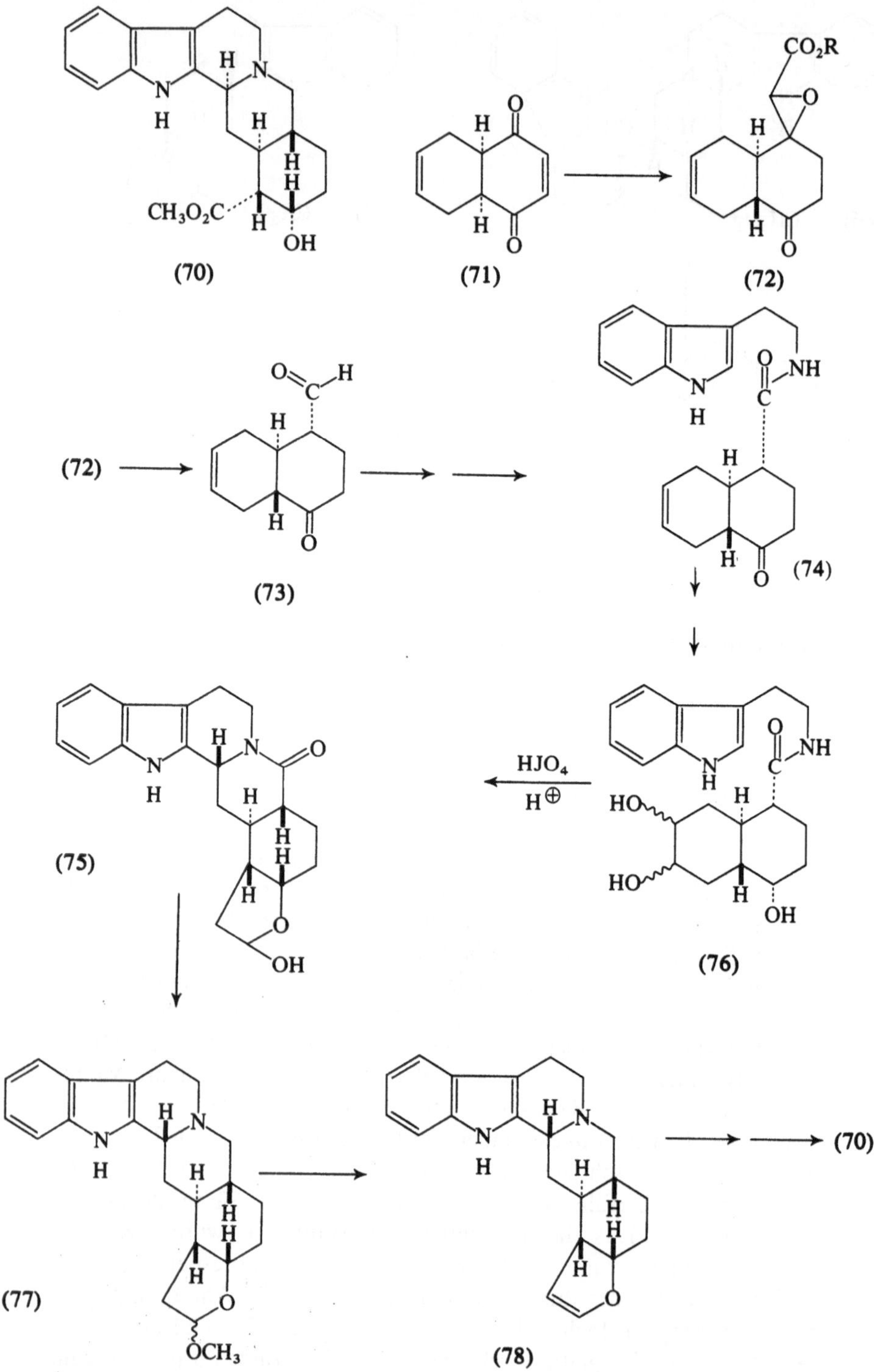

(70)   (71)   (72)

(72) →

(73)

(74)

(75)

$\xleftarrow{\text{HJO}_4 \atop \text{H}^\oplus}$

(76)

(77)

(78)

→ → (70)

Dihydrofuran gelingt auf dieser Stufe nicht, es wird daher zunächst das Vollacetal gebildet und dann mit Lithium-Alanat reduziert (**77**). Wiederum jedoch macht die Eliminierung Schwierigkeiten, und es mußte

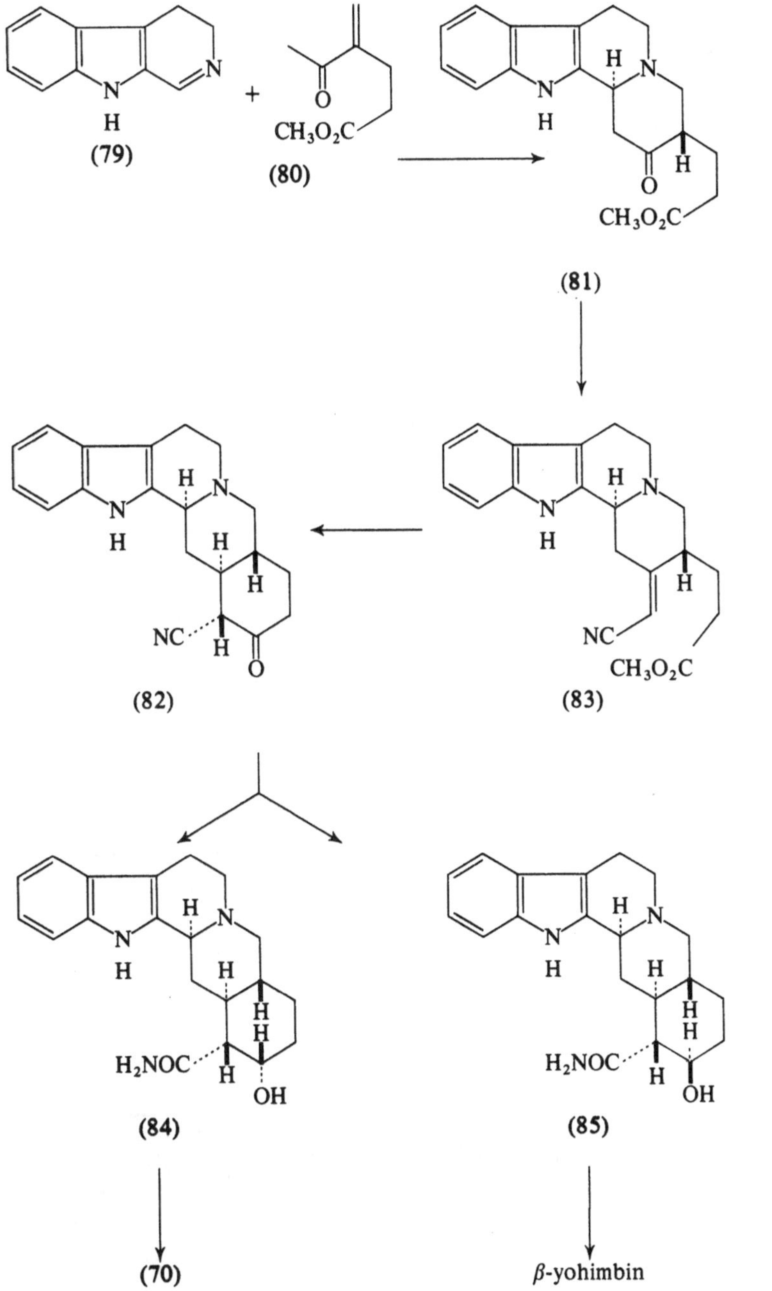

(79)   (80)   (81)

(82)   (83)

(84)   (85)

(70)   β-yohimbin

erst der Äther in das Acetat überführt werden, das dann in einer Pyrolyse-Reaktion unter recht drastischen Bedingungen zum Enoläther (78) führt. Mit dieser Substanz sind die sterischen Voraussetzungen für eine Yohimbin-Synthese geschaffen. Oxydative Spaltung liefert 3-Epi-Yohimbin, das nach Epimerisierung am C_3 in (70) übergeht.

Sehr viel einfacher gestaltet sich die Totalsynthese von Szantay (9), der als Ausgangsmaterial (79) und das ungesättigte Keton (80) verwendet.

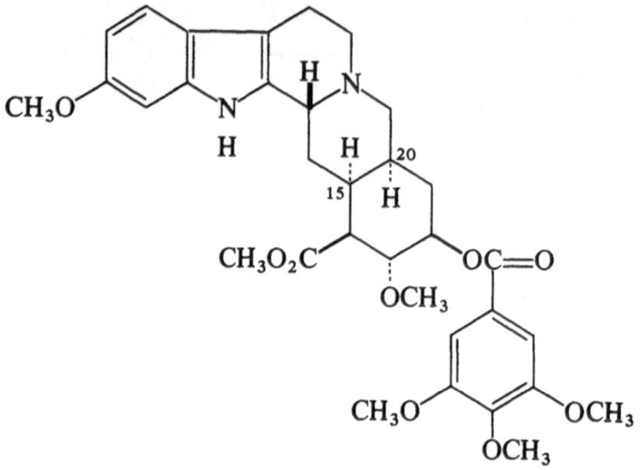

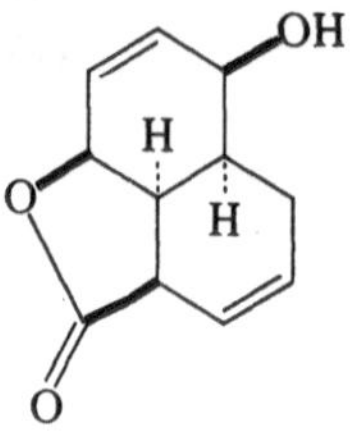

(87)

(86)

Thermodynamische Reaktionslenkung führt zu (81), das nach der Phosphonat-Methode in (83) übergeht. Im Gegensatz zum entsprechenden Diester kondensiert der nach Hydrierung anfallende Nitrilester

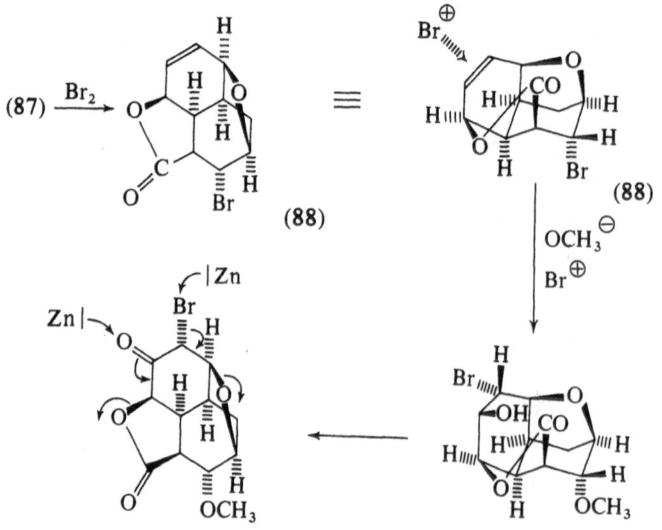

(88)

(88)

(89)

(90)

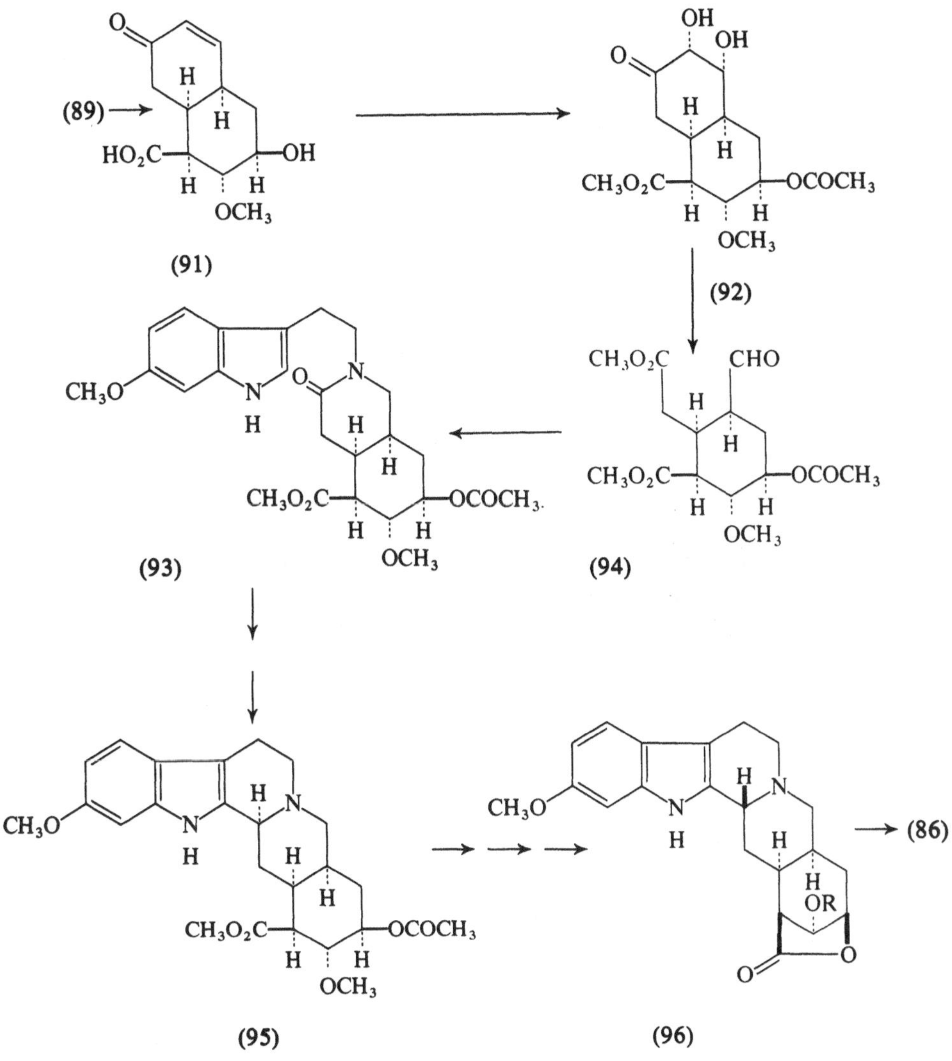

(91)

(92)

(93)

(94)

(95)

(96)

→ (86)

ausschließlich zum gewünschten Ketonitril (82), das nach Boranat-reduktion und alkalischer Hydrolyse als wichtige Produkte (84) und (85) erzeugt, die dann in Yohimbin bzw. β-Yohimbin überführt werden. Anmerkung bei der Korrektur: Im Rahmen dieser Untersuchun-gen hat sich gezeigt, daß die Konfiguration des Alloyohimbins einer Revision bedarf (Ester- wie OH-Gruppe α-ständig) C. S. SZANTAY Privat-mitteilung. (Lit. J. org. Chem. 1973, 2501, 2497.)

Die Synthese des Reserpins (86) stellt wegen seiner speziellen Stereo-chemie und seiner interessanten physiologischen Wirksamkeit ein beson-ders reizvolles Problem dar. R. B. WOODWARD und Mitarbeiter lösten diese Aufgabe bereits vor 15 Jahren (6).

Wegen der *cis*-Konfiguration der Zentren 15 und 20 ist auch hier ein Dien-Addukt als Ausgangsmaterial angezeigt. Reduktion des Adduktes von Chinon und Butadiencarbonester liefert das Lacton (**87**), das jetzt wegen seiner speziellen Faltung die konvexe Seite sehr viel offener angreifenden Reagentien darbietet. Somit führt die Halogenierung unter gleichzeitigem Äther-Ringschluß zum Halogenäther (**88**).

Methanolyse erfolgt unter Retention, und das im zweiten Reaktions-schritt eintretende $Br^{\oplus}$ nähert sich ebenfalls von der konvexen Seite unter Ausbildung des Bromhydrins (**90**). Oxydation zum Keton erzeugt dann eine Verbindung (**89**), die leicht reduktiv abgebaut werden kann, zum ungesättigten Keton (**91**), in dem bereits die wichtigsten chemi-schen und stereochemischen Probleme gelöst sind. Hydroxylierung zu (**92**) und Perjodsäurespaltung liefern eine Aldehydsäure, die nach Ver-esterung (**94**) in Form des Acetats mit Methoxytryptamin eine Schiffsche-Base bildet, die sofort mit Boranat reduziert wird und dabei zum Lactam (**93**) cyclisiert.

Cyclisierung und Boranatreduktion ergeben erwartungsgemäß die 3α-Verbindung (**95**), also ein Derivat des 3-Isoreserpins. In sehr elegan-ter Weise wird jetzt von der Tatsache Gebrauch gemacht, daß die Estergruppe und die γ-ständige Hydroxylgruppe nur über Ringinversion unter Umklappen des gesamten Ringsystems die für den Lacton-Ring-schluß notwendige diaxiale Lage einnehmen können.

(95A)   (95B)

Die aus der Verseifung des Esters (**95**) hervorgehende Hydroxy-Säure (**95A**) muß aus dieser sicher thermodynamisch stabilsten Konformation in die sehr viel ungünstigere Konformation (**95B**) übergehen, die dann jedoch durch Lactonringschluß verklammert wird. Die C-C-Bindung neben dem Stickstoff ist jetzt axial. Daher kann nun leicht saure Epi-merisierung diese Konfiguration ändern, unter Ausbildung des Lactons (**96**). Von hier aus führen unproblematische Schritte — nämlich Methanolyse und Veresterung mit Trimethoxybenzoylchlorid — zum Ziel.

Eine sehr interessante Variation dieses Strukturtyps stellt das Ulein (**97**) dar, das strukturell mit dem Condylocarpin-Typ verknüpft ist, über den noch zu berichten sein wird.

Literaturverzeichnis: SS. 518—520

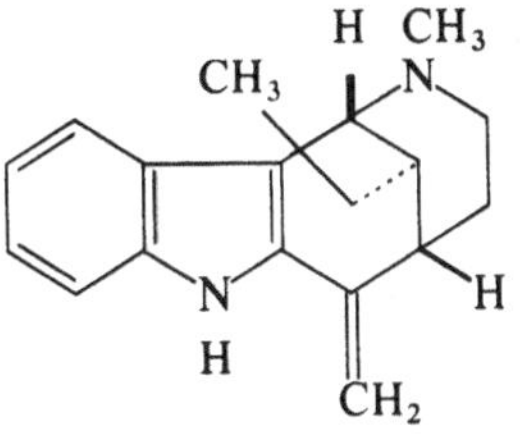

(97)

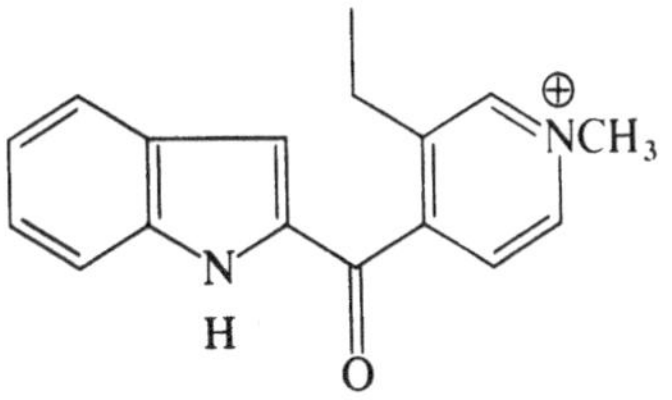

(98)

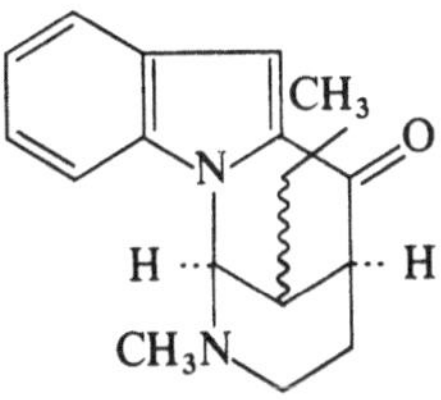

(99)

H⊕

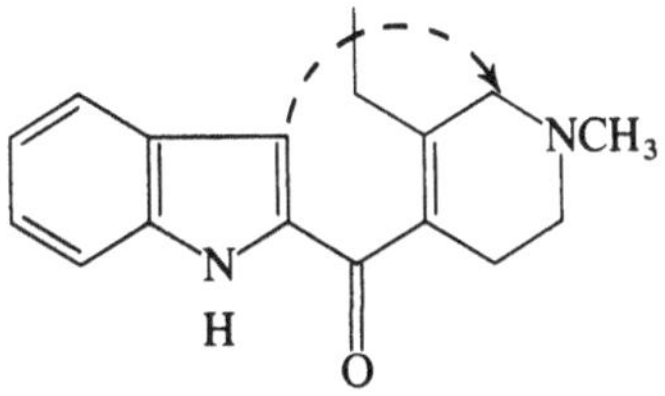

(100)

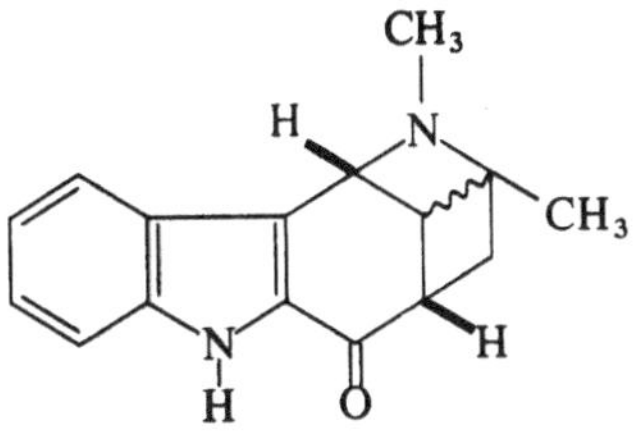

(101)

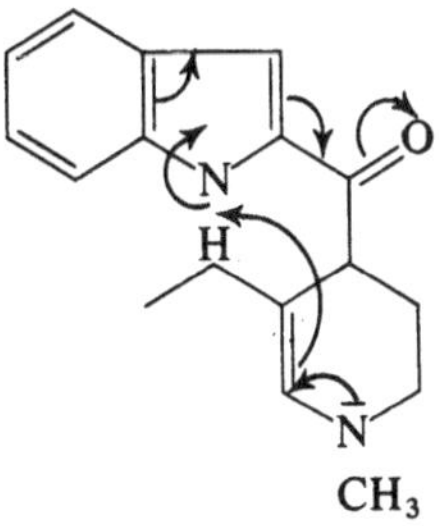

(102)

Die erste Synthese, obwohl nicht stereoselektiv, was das die Äthylgruppe tragende Zentrum anbelangt, wurde von Joule (26) mitgeteilt, sie sei hier vor allem wegen einiger überraschender und nicht vorhersehbarer Reaktionen erwähnt. Durch Quartärnisierung einer durch Fischer-Synthese darstellbaren Pyridinverbindung wird das Salz (98) erhalten, das bei Boranat-Reduktion des Tetrahydro-Pyridinderivat (100) liefert. Die Hoffnung, daß das Allylamin unter Einwirkung starker Protonenacceptoren in ein Enamin übergehen würde, das dann als das entsprechende Imoniumsalz mit der β-Indolposition cyclisieren würde (s. Pfeil in 100), erfüllte sich nicht. Vielmehr erhielt man die tetracyclische Verbindung (99). Offenbar wird das Enamin vom Indol-

NH protoniert (s. **102**), und das resultierende Acceptorsystem reagiert
dann mit dem Stickstoff des Indols. Anschließende Säurebehandlung .

(103)   (104)   (105)

(106)   (107)

erzeugt dann zwar das Keton (**101**), das in Ulein überführt werden kann,
jedoch gemeinsam mit der entsprechenden Epi-Verbindung, so daß
Ulein und Epi-Ulein anfallen.

(106) ⟶

(108)   (108 A)

(109)   (110)

Literaturverzeichnis: SS. 518—520

Eine stereoselektive Synthese beider Alkaloide haben BÜCHI und Mitarbeiter durchgeführt (*17*). Ausgangsmaterial ist das bei Kondensation des Aldehyds (**103**) und des Aminoketons (**104**) anfallende Piperidin-Keton (**105**).

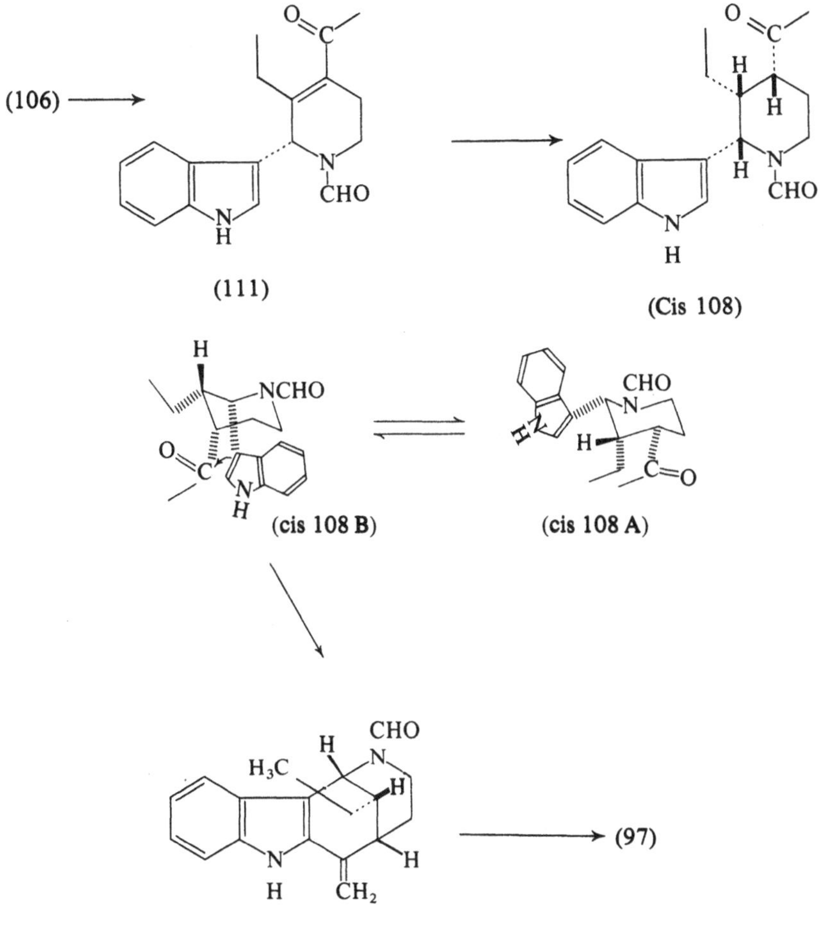

(**106**) ⟶

(**111**)

(Cis 108)

(cis 108 B)   (cis 108 A)

(Cis 110)   (**97**)

Die Äthinylierung der entsprechenden Formyl-Verbindung gelingt nur mit Kalium-acetylid in tert. Butanol-Tetrahydrofuran (**107**), mit Lithium-acetylid wird lediglich im Zuge einer β-Eliminierung der Piperidonring geöffnet. Hydratisierung der Dreifachbindung führt dann zum Hydro-oxyketon (**106**), das als gemeinsame Vorstufe sowohl zum Ulein als auch Epi-Ulein dient. Reduktion mit Lithium in Ammoniak liefert das Keton (**108**), in dem Methylketon und Indolrest *cis*-ständig sind, wobei jedoch die Äthylgruppe die Epi-Konfiguration einnimmt.

Aus der Konformation (**108A**) kann unter der Einwirkung von Bortrifluorid Cyclisierung erfolgen, die von einer Wasserabspaltung zum Olefin (**110**) begleitet wird, anschließende Alanat-Reduktion führt dann zum Epi-Ulein.

Da im Ulein die drei Wasserstoffatome im Piperidinring *cis*-ständig sein müssen, wird hier zunächst in einer Pyrolyse-Reaktion das Acetat des Hydroxy-Ketons (**106**) in das Olefin (**111**) überführt, das das Hauptprodukt dieser Reaktion darstellt, neben einer geringen Menge des isomeren trisubstituierten Olefins.

Die Hydrierung erfolgt nun von der dem Indolring abgewandten Seite und liefert die *cis*-Verbindung (**108**), die zwar vollständig in der stabileren Konformation (**cis-108A**) vorliegt, aber dennoch mit Bortrifluorid aus der Konformation (**cis-108B**) cyclisiert, unter Bildung des Olefins (**cis-110**), das dann bei der Alanatreduktion in Ulein (**97**) übergeht.

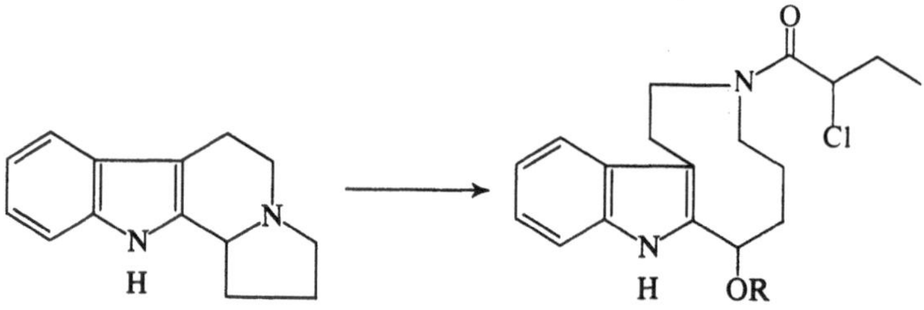

(112)

(113) R = CO – CH(Cl) – C_2H_5

(113a) R = H

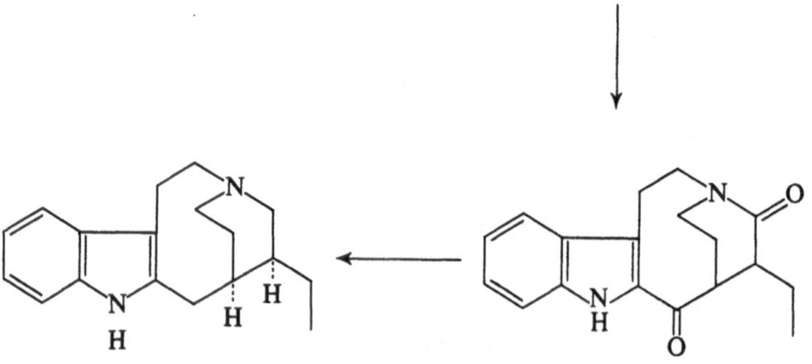

(114) (115)

Literaturverzeichnis: SS. 518—520

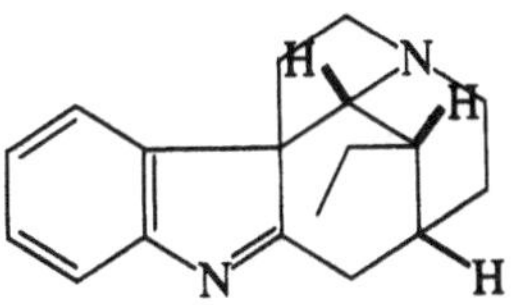

(116)

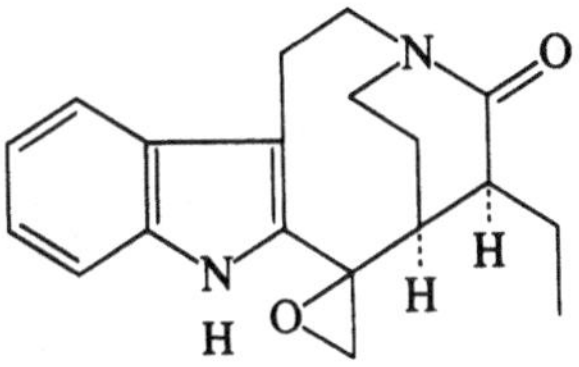

+

(117)

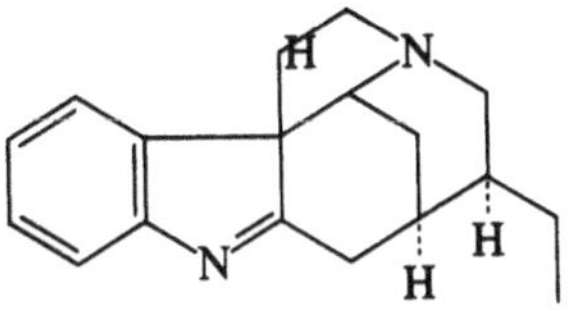

(118)

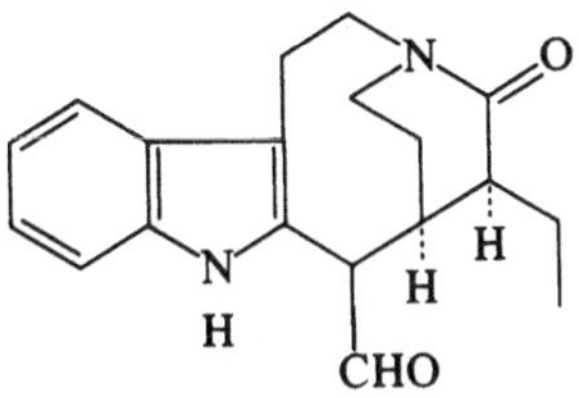

(119)

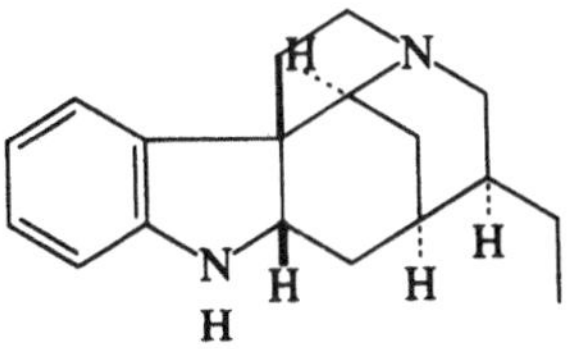

(120)

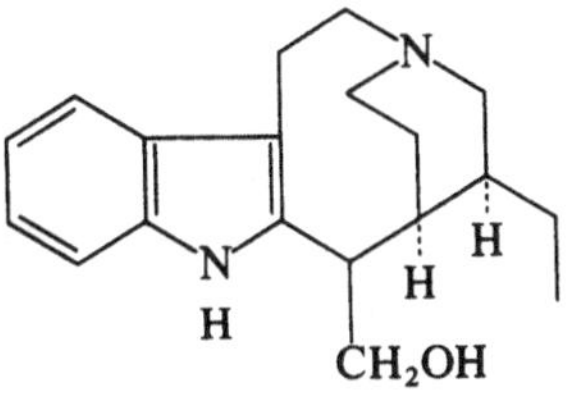

(121)

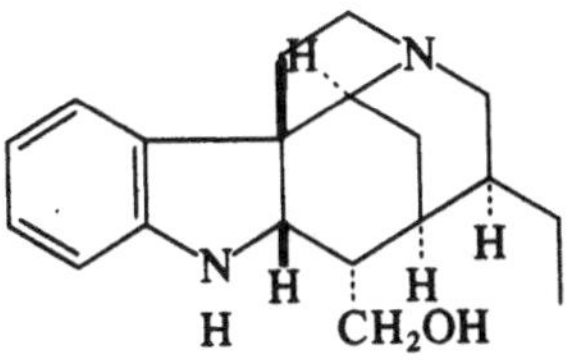

(122)

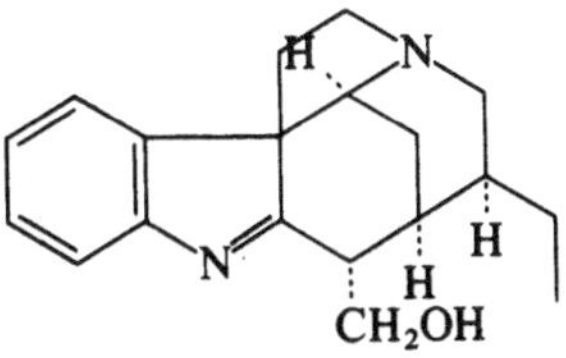

(123)

Eine Reihe sehr interessanter stereoselektiver Synthesen in dieser Serie von Alkaloiden hat Harley-Mason, ausgehend von der gut zugänglichen Verbindung (112), durchgeführt (22, 30). Hier ist der Angriff eines Acylrestes auf das freie Elektronenpaar am Stickstoff begleitet von einer nucleophilen Ringöffnung. So liefert α-Chlor-buttersäure-anhydrid das Amid (113), das bei milder Hydrolyse in das Carbinol (113a) übergeht.

Oxydation mit Braunstein und basenkatalysierte Cyclisierung bringen dann das Keto-Lactam (115) hervor, von dem aus verschiedene Alkaloide gut zugänglich sind. Nach Wolff-Kishner-Reduktion und Behandlung mit Lithiumalanat gewinnt man (114), das bei der Oxydation in einer *trans*-annularen Mannich-Kondensation in ein Gemisch des Alkaloids Condyfolin (116) und der Base (118) übergeht, die bei der anschließenden Hydrierung dann stereoselektiv in Tubofolidin (120) überführbar ist. Führt man in (115) über die Schwefel-Ylid-Kondensation ein weiteres C-Atom ein unter Bildung des Epoxids (117), so erhält man nach Isomerisierung zum Aldehyd (119) und anschließender Alanatreduktion (121), das ebenfalls cyclisiert werden kann, und dann nach Reduktion des Indolenins (123) stereoselektiv Geissoschizolin (122) erzeugt.

Schützt man in (119) die Aldehyd-Gruppe durch Acetalisierung und reduziert dann mit Alanat, so erhält man die Base (124), die ebenfalls oxydativ cyclisiert. Im sauren Medium isomerisiert das Cyclisierungsprodukt dann zum Dihydronorfluorocurarin (126).

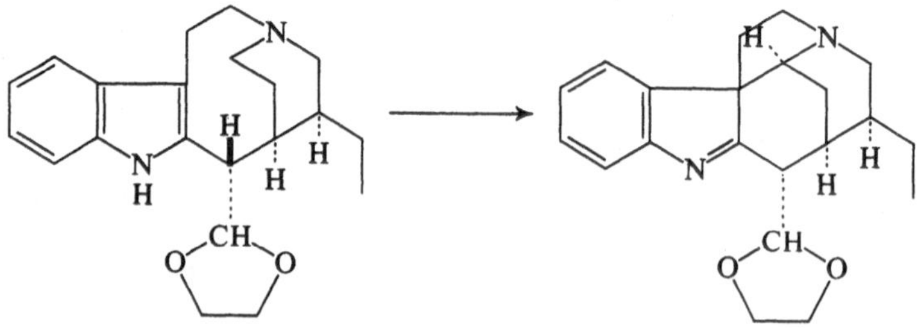

(124)   (125)

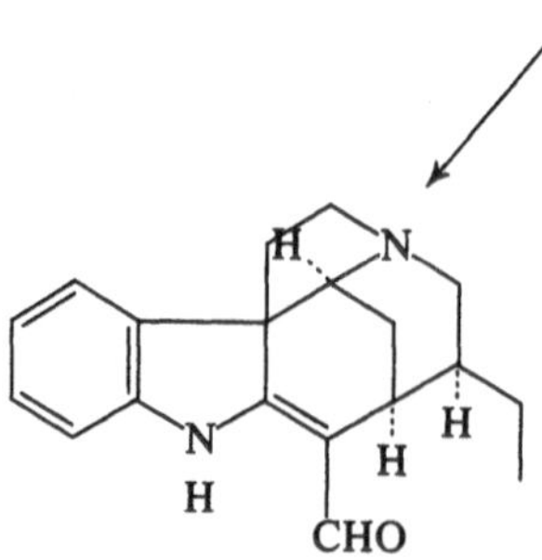

(126)

Literaturverzeichnis: SS. 518—520

Einen etwas komplizierteren Vertreter der Gruppe der Indol-Alkaloide mit Corynanthein-Grundgerüst stellt das Ajmalin (**127**) dar. VAN TAMELEN (*37*) bediente sich hier einmal mehr der bewährten Dialdehyd-Technik, indem der aus der Säure (**128**) darstellbare Aldehyd (**130**) mit N-Methyl-Tryptophan kondensiert und zu (**129**) hydriert wurde.

Diese ersten Schritte erfahren zwar keine sterische Lenkung, aber dennoch sei dieses Verfahren hier erwähnt, weil es eine sehr elegante und originelle Technik zum Aufbau des polycyclischen Systems verwendet. Alkalische Verseifung und Perjodsäure-Spaltung erzeugt (**131**), das bei der Behandlung mit DCCI unter Decarbonylierung (über **132**) ein Imoniumsalz hervorbringt, das dann in einer innermolekularen Mannich-Kondensation in ein Gemisch der Stereoisomeren der allgemeinen Struktur (**133**) übergeht.

Durch präparative Schichtchromatographie gewinnt man nunmehr in relativ geringer Menge den Aldehyd (**133A**), der mit dem Aldehyd (**133B**) ins Gleichgewicht gesetzt werden kann. Leider liegt dieses Gleichgewicht sehr stark auf der Seite der nicht cyclisierbaren Kon-

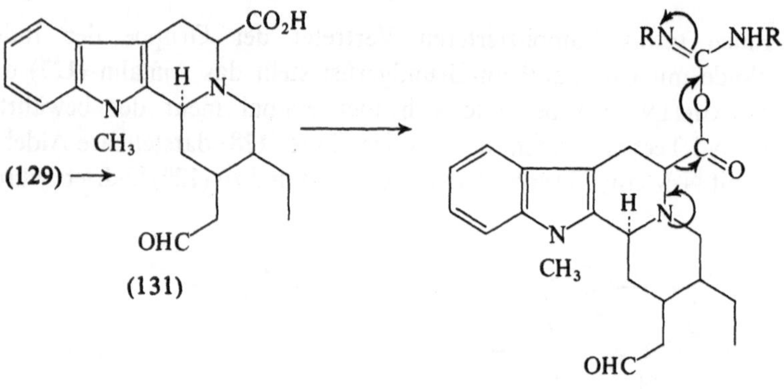

CO2H
H
N
N
CH3
(129) →
OHC
RN
NHR
C
O
C
O
H
N
N
CH3
OHC

(131)

(132)

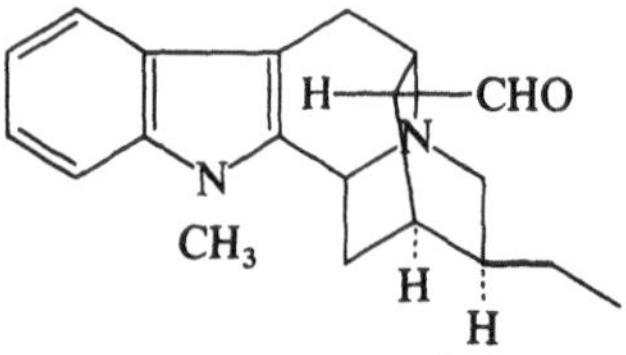

H
CHO
N
N
CH3
H
H

(133 A) (85%)

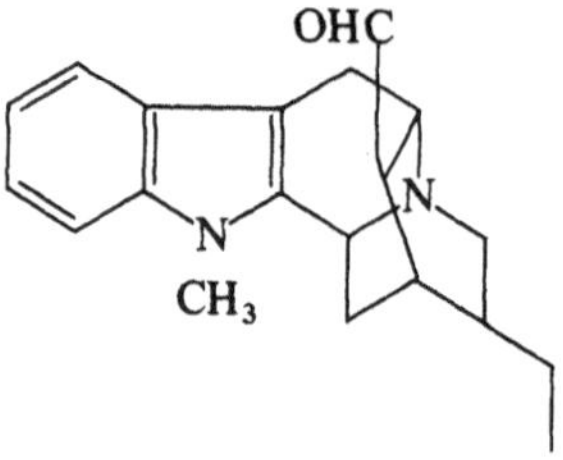

OHC
N
N
CH3

(133)

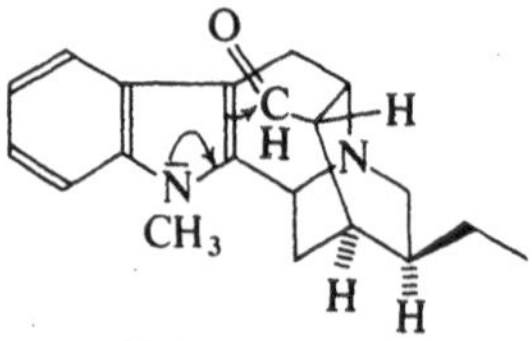

O
C
H
H
N
N
CH3
H
H

(133 B) (15%)

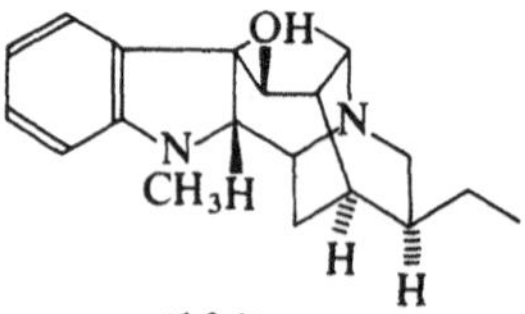

OH
N
N
CH3H
H
H

(134)

figuration (**133A**) (85%), während nur 15% der Konfiguration (**133B**) anfallen, die die Aldehydgruppierung in günstiger Lage zur β-Indolposition trägt. Säurebehandlung und Reduktion bringt aus dieser Konfiguration Desoxyajmalin (**134**) hervor, das dann durch Einwirkung von Phenoxykohlensäurechlorid und Lithiumjodid unter oxydierenden Bedingungen nach einem prinzipiell bekannten Verfahren in den Aldehydammoniak überführt wird.

Über eine oxydative Spaltung einer Cyclopentenverbindung zum entsprechenden Dialdehyd verläuft auch die Synthese von MASAMUNE (*5*). Der aus Indolylessigsäurechlorid und dem Magnesium-malonesterderivat (**135**) darstellbare Ketoester (**136**) wird nach Bildung des Oxims reduziert und das resultierende Aminocarbinol ins Bisbenzoat (**137**) überführt.

Literaturverzeichnis: SS. 518—520

(135)   (136)   (137)

(138)   R = COC₆H₅   R' = CHO   R'' = H   (139)
(138A) R = COC₆H₅   R' = CN   R'' = C₂H₅
(138B) R = H   R' = CN   R'' = C₂H₅

Bei der oxydativen Spaltung des 5-Rings bildet sich dann spontan der Aldehydammoniak (139), der unter Protoneneinwirkung zu (138) cyclisiert. Zur Einführung der Äthylgruppe wird der Aldehyd zunächst über das Oxim in das entsprechende Nitril überführt, das dann nach Deprotonierung mit Triphenyl-methylnatrium mit Äthyljodid alkyliert wird zum Dibenzoat (138A).

Partielle Verseifung dieses Benzoats mit Methylat führt zum Alkohol (138B), der nun seinerseits bereits durch Abbau von Ajmalin gewonnen werden kann. Oxydation zum Aldehyd (140) und anschließende saure Cyclisierung zu (142) sind natürlich ebenfalls vom Äquilibrierungs-gleichgewicht der Aldehyde abhängig, da auch hier ein Epimeren-gemisch erzeugt wird. Hydrierung und partielle Reduktion führen dann zum Nitril (143), das bei der nachfolgenden Reduktion zum Imin bzw. Aldehyd (144) spontan in Ajmalin übergeht.

Am Strychnin sind eine große Zahl von Umwandlungs- und Abbau-reaktionen durchgeführt worden (41). In diesem Zusammenhang sind auch Versuche zur Synthese dieses komplizierten Alkaloids unternom-men worden, und natürlich muß an dieser Stelle wenigstens in großen Zügen die bereits 1954 durchgeführte Totalsynthese von WOODWARD (20) beschrieben werden, bei der biogenetische Vorstellungen und stereo-selektive Reaktionen die Strategie bestimmt haben.

(138 A) R=COC$_6$H$_5$ R'=CN   R''=C$_2$H$_5$

(138 B) R=H   R'=CN   R''=C$_2$H$_5$ (140)

(141)   (142)

(143)   (144)   (127)

Ganz besonders bemerkenswert ist bereits die durch Toluolsulfochlorid ausgelöste Cyclisierung der Schiffschen Base aus der Verbindung (145). Der Veratryl-Rest in der α-Position des Indols verhindert einerseits, daß die sonst übliche Umlagerung von β nach α unter Rückbildung des Indols erfolgt und liefert außerdem wichtige Bauelemente für zwei weitere Ringe. Auch die Reduktion des Indolenins verläuft stereoselektiv, so daß der nach Acetylierung und Ozonisierung von (146) anfallende Diester (148) ein sterisch einheitliches Produkt darstellt. Der Bildung des Pyridonringes schließt sich die Spaltung des N-Tosylates an, wobei natürlich auch die Estergruppen verseift werden. Nach Rückveresterung kann der fünfte Ring über eine Diekmann-Cyclisierung gebildet werden, und die dabei sich bildende β-Dicarbonylverbindung wird als Enol-tosylat aktiviert. Verdrängung mit Benzyl-thiolat und

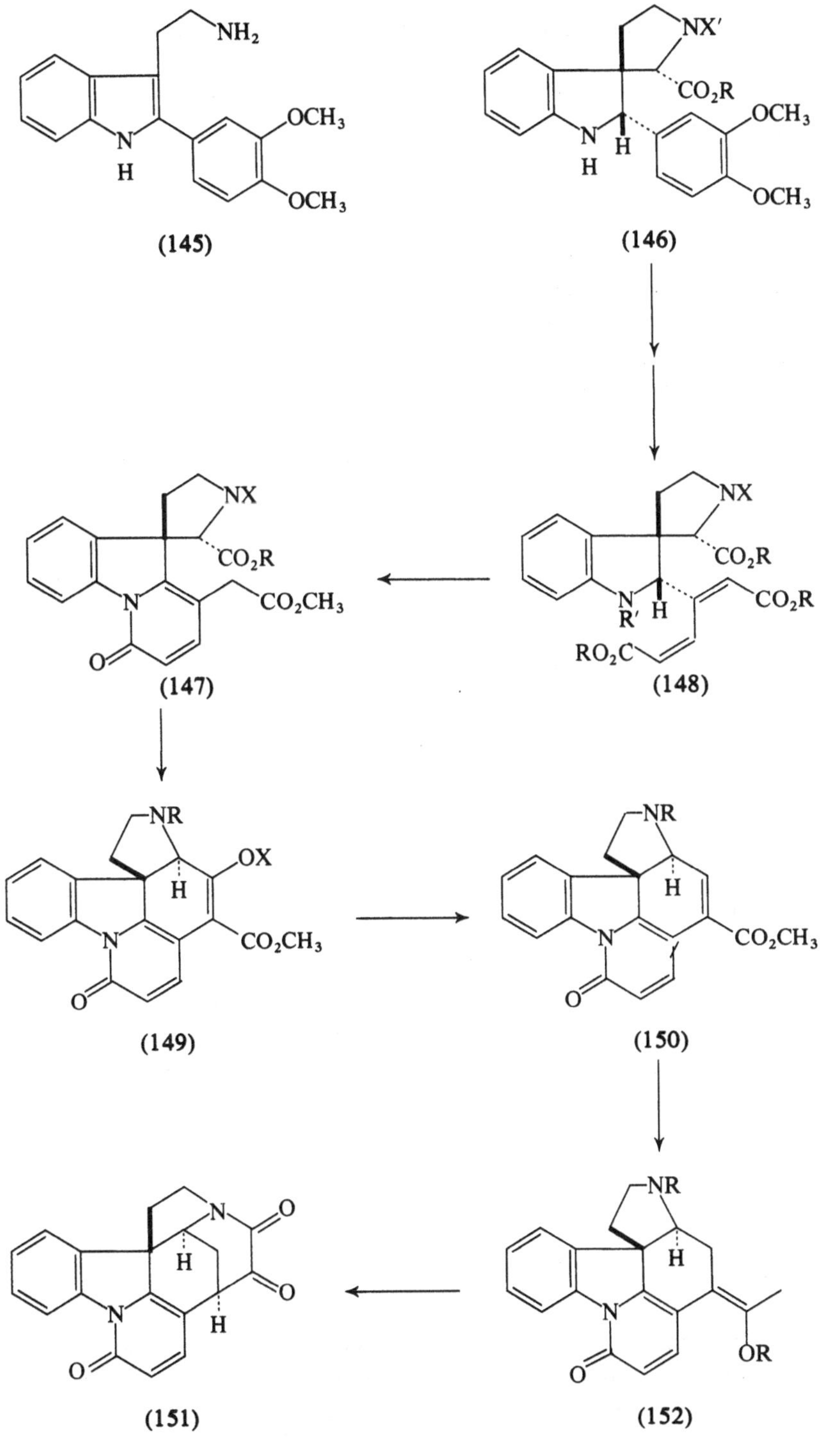

(145)

(146)

(147)

(148)

(149)

(150)

(151)

(152)

Raney-Nickel-Behandlung entfernt die Sauerstoffgruppierung vollständig. Der resultierende ungesättigte Ester wird durch Hydrierung in den
gesättigten und durch Verseifung in die Säure überführt. Die Umsetzung dieser Säure mit Acetanhydrid und Pyridin liefert in einer sehr
wichtigen Schlüsselreaktion das Enolacetat (152), dessen Bildung wahrscheinlich über das gemischte Anhydrid verstanden werden muß, wobei
die Tatsache, daß das betroffene Zentrum eine quasi-Benzylposition
darstellt, sicher ebenfalls von großer Bedeutung ist. Nach Verseifung
der Acetate führt die Selendioxid-Oxydation unter Angriff auf die
Ketomethylgruppe schließlich zum Keto-Lactam (151). Zum besseren
Verständnis der folgenden Reaktionen muß man sich die Architektur
dieses Moleküls vergegenwärtigen, weil hier wiederum ein schönes
Beispiel dafür vorliegt, wie die spezielle Form gefalteter, polycyclischer
Systeme einem angreifenden Agens nur wenig Spielraum läßt.

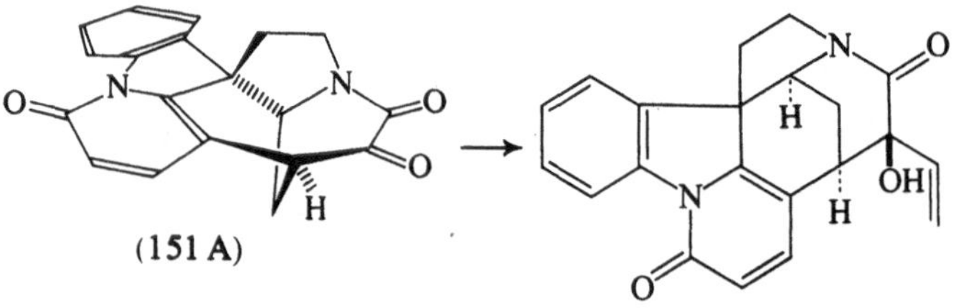

(151 A)   (153)

Es fällt angesichts der in (151A) wiedergegebenen Gestalt dieses
Moleküls nicht schwer, zu verstehen, daß der Angriff des Natrium-
acetylids nur von der α-Seite des Moleküls stattfindet. Die Alternative
wäre, daß das Nucleophil in das Innere einer sehr schwer zugänglichen
Höhle vordringen müßte. Mit dieser stereoselektiven Einführung des
Vinylrestes ist das stereochemische Problem praktisch gelöst. Partielle

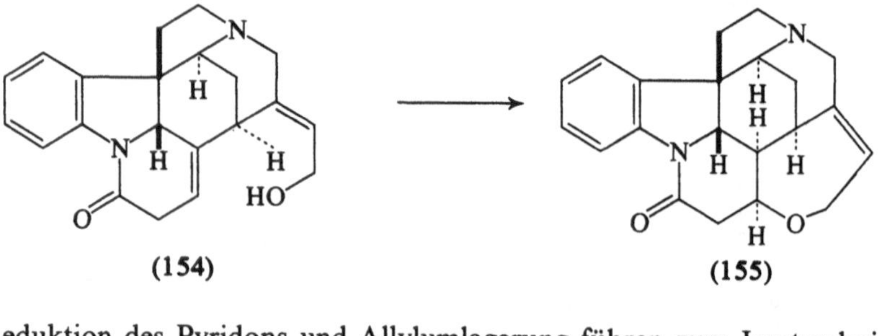

(154)   (155)

Reduktion des Pyridons und Allylumlagerung führen zum Isostrychnin
(154), das sich durch Protonenacceptoren zum Strychnin (155) cyclisieren läßt. Hingewiesen sei jedoch auf die Tatsache, daß die β-Konfigura-

tion des durch partielle Reduktion eingeführten Wasserstoffatoms mit hoher Wahrscheinlichkeit nur durch eine innere Übertragung über ein primär sich bildendes Aluminiumalkoholat unter Einbeziehung der allylischen Hydroxylgruppe überzeugend erklärt werden kann.

In der folgenden Gruppe werden nun Alkaloide mit umgelagertem Seco-Loganin Grundgerüst behandelt, und zwar Verbindungen vom Typ des Quebrachamins, Aspidospermins und Vincamins. Eine wichtige Verbindung in dieser Serie ist das Velbanamin (**156**), das als Baustein der dimeren Indolalkaloide Vinblastin und Vincristin angetroffen wird, die sich als Leukämie-Mittel einen Namen gemacht haben.

Ausgangsmaterial in der von Büchi durchgeführten Totalsynthese (*18*) ist das Amid-Keton (**157**), das durch Dien-Addition von Methyl-vinylketon an das Dihydropyridin-Derivat (**158**) gewonnen wird.

Bei der Hydroxylierung zum α-Hydroxyketon (**159**) setzt man mit Vorteil Triäthylphosphit zu, um eventuell gebildete N-Oxide wieder zu reduzieren. Nach Reduktion zum 1,2-Diol kann Perjodsäurespaltung erfolgen, jedoch wird das sich bildende Keton sofort anschließend wieder ketalisiert (**160**). Hofmann-Abbau führt dann zum Keton-Ketal (**162**), das nach Hydrogenolyse des Benzylrestes mit Indolylessigsäure in (**161**) überführt wird. Säurebehandlung bewirkt Cyclisierung und Ätherspaltung zu (**163**), das im Zuge einer Retroaldol-Spaltung zu (**164**) übergeht. Boranat reduziert beide Ketogruppen und die zum Indolrest α-ständige Hydroxylgruppe wird durch chemische Reduktion entfernt. Anschließende Rückoxydation mit Dimethylsulfoxid und DCCI erzeugt das Ketolactam (**165**), das in der angegebenen Konformation nur einen α-ständigen Substituenten aufnehmen kann. Daher liefert eine Grignard-reaktion mit Äthylmagnesiumbromid und anschließende Alanatreduktion sterisch einheitliches Velbanamin (**156**).

Auf einem völlig anderen Wege wurde Velbanamin von Kutney synthetisiert, jedoch müssen hier zum besseren Verständnis einige wichtige von dieser Gruppe erarbeitete Experimente (*1, 2, 11, 13, 34, 35*) vorausgeschickt werden, die in enger Beziehung stehen zu den Unter-

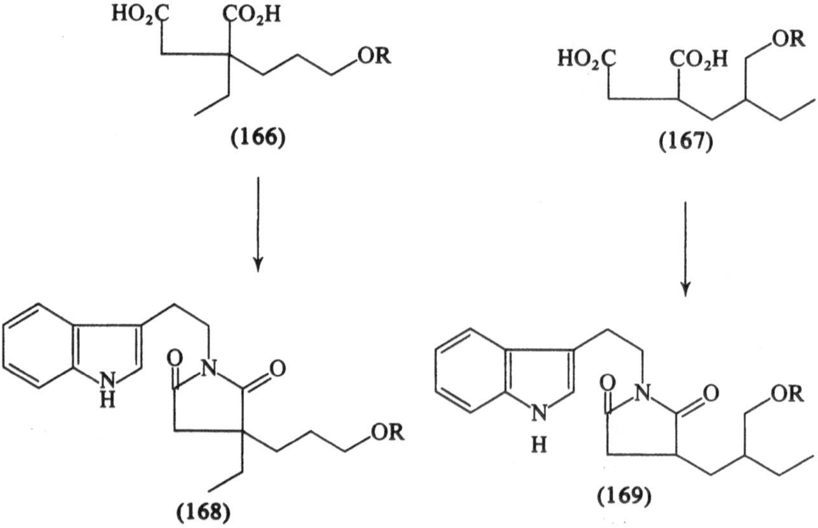

(**166**)   (**167**)

(**168**)   (**169**)

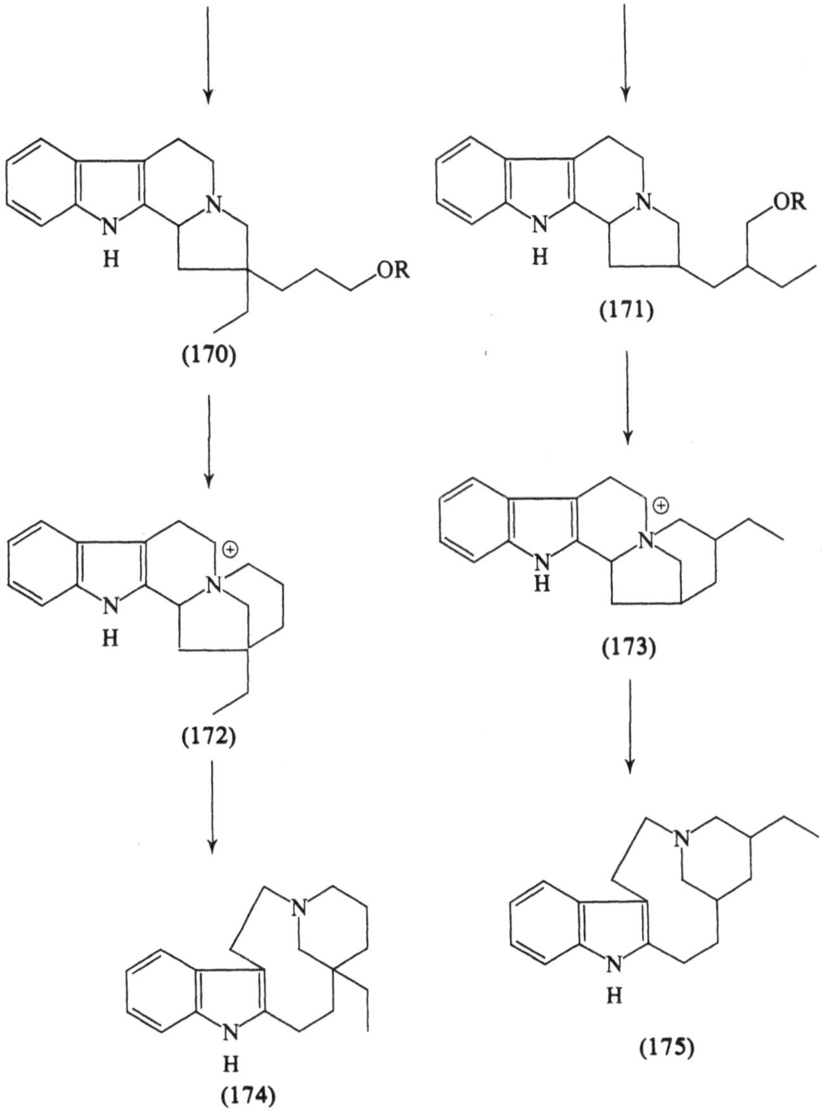

(170)

(171)

(172)

(173)

(174)

(175)

suchungen HARLEY-MASONS über die bereits berichtet wurde. Aus (166) und Tryptamin wurde (168) und aus (167) auf die gleiche Weise (169) gewonnen.

Alanatreduktion dieser Zwischenstufen führt zu Pyrrolidinderivaten, die bei der Dehydrierung mit Quecksilberacetat mit anschließender Boranatreduktion zur Entfernung der Imoniumgruppen in ein Gemisch aus mehreren Verbindungen übergehen, aus dem (170) und (171) jeweils chromatographisch abgetrennt werden. Nach Spaltung der Benzyläther erfolgt Quartärnisierung zu (172) bzw. (173). Anschließende reduktive Spaltung liefert dann die ringoffenen Basen (174) (Quebrachamin) und (175).

In (175) kann über Oxydation mit Hypochlorid (176) und Verdrängung mit Cyanid-Ionen eine Nitrilgruppe (177) bzw. eine Estergruppe eingeführt werden.

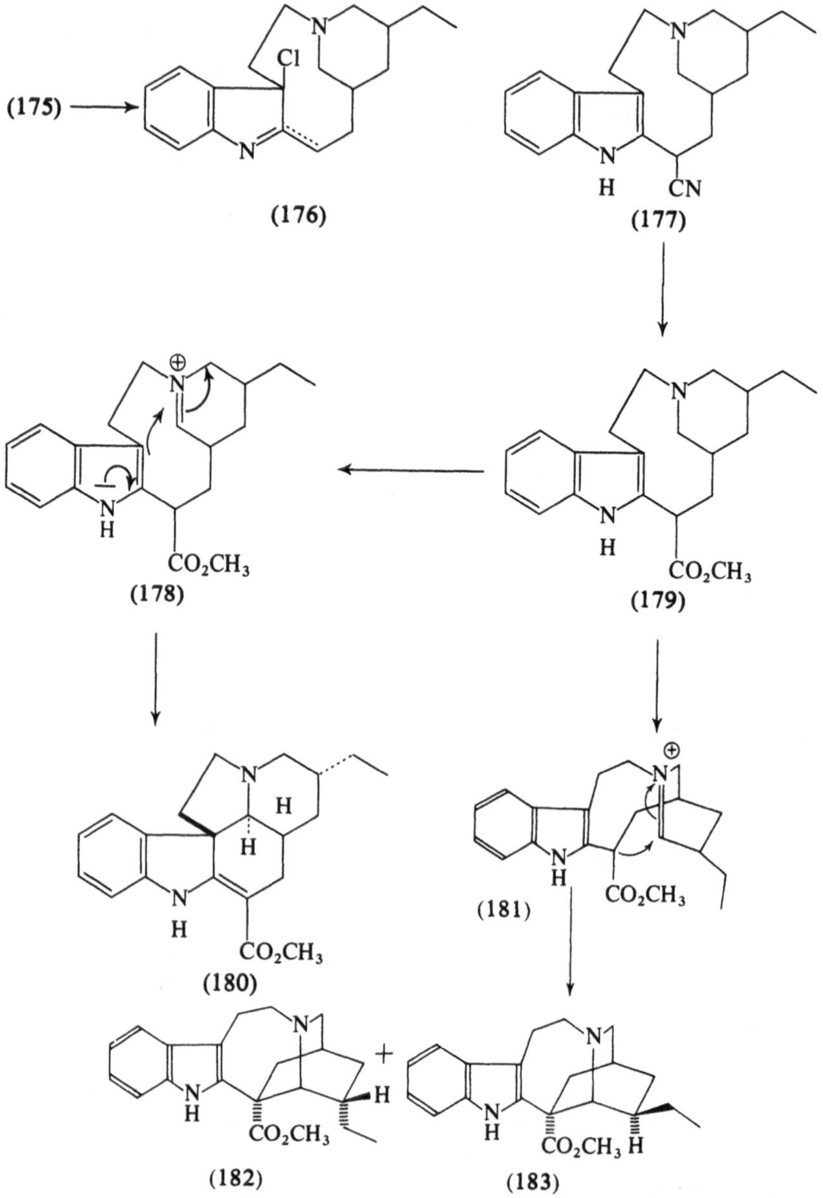

(175) ⟶

(176)

(177)

(178)

(179)

(180)

(181)

(182)

(183)

Der aus (177) hervorgehende Ester (179) hat nun, wie Verbindungen dieses Typs allgemein, zwei Reaktionsmöglichkeiten zur Verfügung. Während das Imoniumsalz (178) zum Pseudovincadifformin (180) cyclisiert, geht das isomere Immoniumsalz aus der Konformation (181) in die Verbindungen Coronaridin (182) und Dihydrocatharanthin (183) über. Dieses Dihydrocatharanthin ist nun das Ausgangsmaterial für KUTNEYs Velbanamin-Synthese (19).

Alanatreduktion und Behandlung mit Toluolsulfosäurechlorid liefert das Tosylat (184), das in Gegenwart von Triäthylamin fragmentiert und zum Enamin (185) isomerisiert.

Hydroxylierung dieser Verbindung und anschließende Boranatreduktion führt zum Triol (187) mit α-ständiger Hydroxylgruppe, wie es bei der Konformation von (185) (s. 165) zu erwarten war. Durch Perjodsäurespaltung wird sodann das Keton (186) bereitet, das bei der Alanatreduktion Isovelbanamin ergibt, dessen saure Epimerisierung auch Velbanamin (156) zugänglich macht.

Verknüpft man nun den Piperidinring auch mit der β-Position des Indols, so gelangt man zum Bauprinzip des Aspidospermins (188).

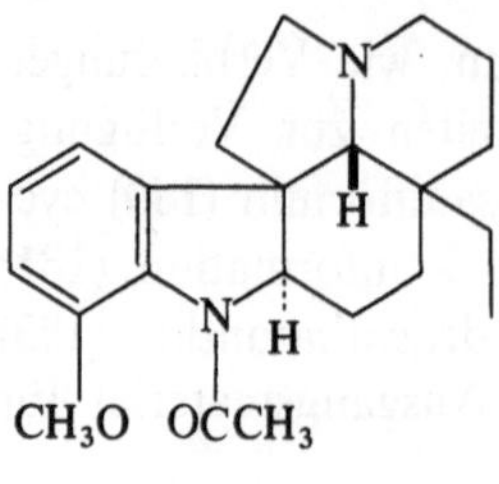

(188) 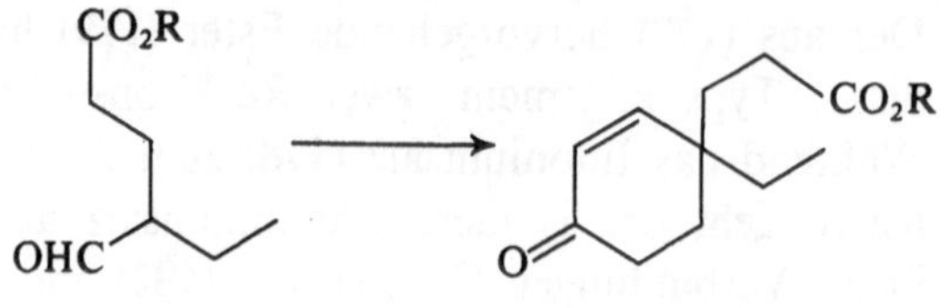   (189)   (190)

Die erste Synthese wurde von STORK (24) durchgeführt, wobei jedoch der Konfiguration der verschiedenen Zentren zunächst noch nicht allzu viel Aufmerksamkeit geschenkt wurde, da man hoffen konnte, daß im Zuge des Aufbaus des Moleküls jeweils die Möglichkeit der Äquilibrie-

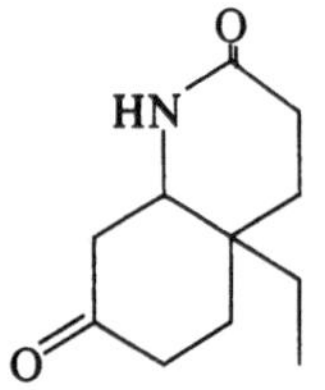

(191) 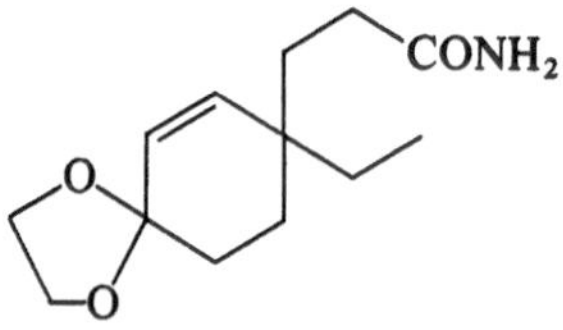   (192) 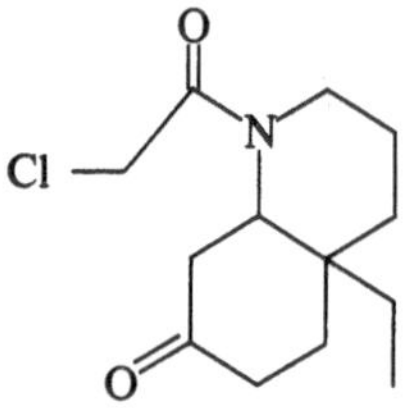   (193)

(193) ⟶   ⟶

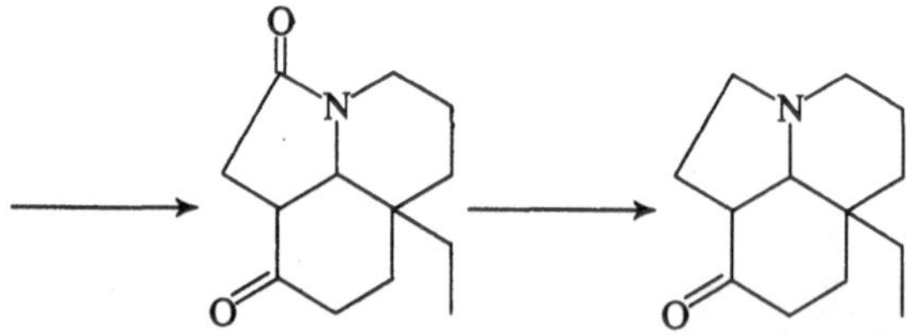

(194)   (195)

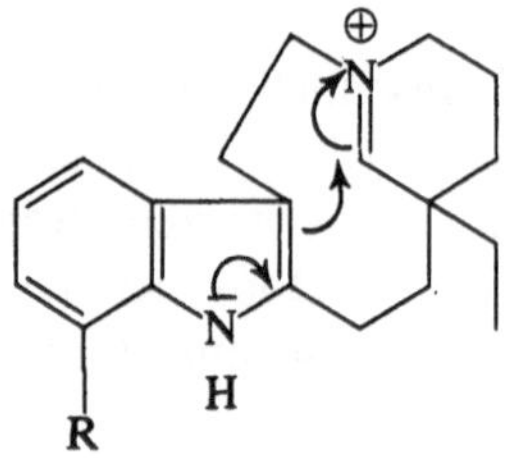

(196) 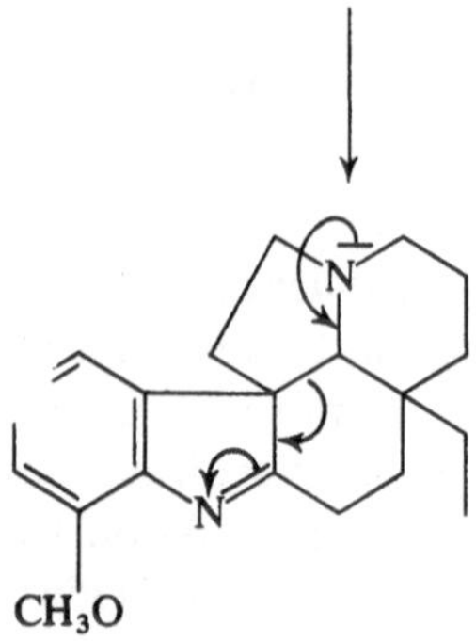   (197)

Literaturverzeichnis: SS. 518—520

rung besteht (s. **197**). Der über die Enamin-Technik zugängliche Aldehyd-ester (**189**) liefert in einer weiteren Enamin-Alkylierung mit Methyl-vinylketon und anschließender Cyclisierung das Cyclohexanonderivat (**190**). Aber weder das daraus leicht zugängliche Ketolactam (**191**) noch das über (**192**) und (**193**) präparierte tricyclische Ketolactam (**194**) cyclisieren in der gewünschten Weise beim Versuch, sie mit dem ent-sprechenden Phenylhydrazin einer Fischer-Indol-Synthese zu unterwerfen. Entfernt man jedoch nach Ketalisierung des Ketons die Amidgruppe durch Alanatreduktion, so gelingt die Indolsynthese. Das sich bildende Indolenin (**197**) wird durch Alanatreduktion und Acetylierung in Aspidospermin überführt. Daß die Äquilibrierung tatsächlich auf der Stufe (**197**) über Retro-Mannich-Kondensation erfolgt, unter Aus-bildung des Imoniumsalzes (**196**) (R = OCH₃), lehrt die Cyclisierung mit unsubstituiertem Phenylhydrazin. Hier kann man nämlich die ringoffene Form (**196**) (R = H) durch Boranatreduktion abfangen und erzeugt Quebrachamin (**174**).

Mit erstaunlich hoher Stereoselektivität, die aber sehr wahrscheinlich auch auf die Äquilibrierbarkeit von Zwischenstufen zurückzuführen ist, verläuft die Aspidospermidin-Synthese von HARLEY-MASON (*31*). Hier wird ebenfalls ein hochsubstituiertes Aldehyd-Derivat, und zwar der Acetal-Ester (**198**), mit Tryptamin kondensiert.

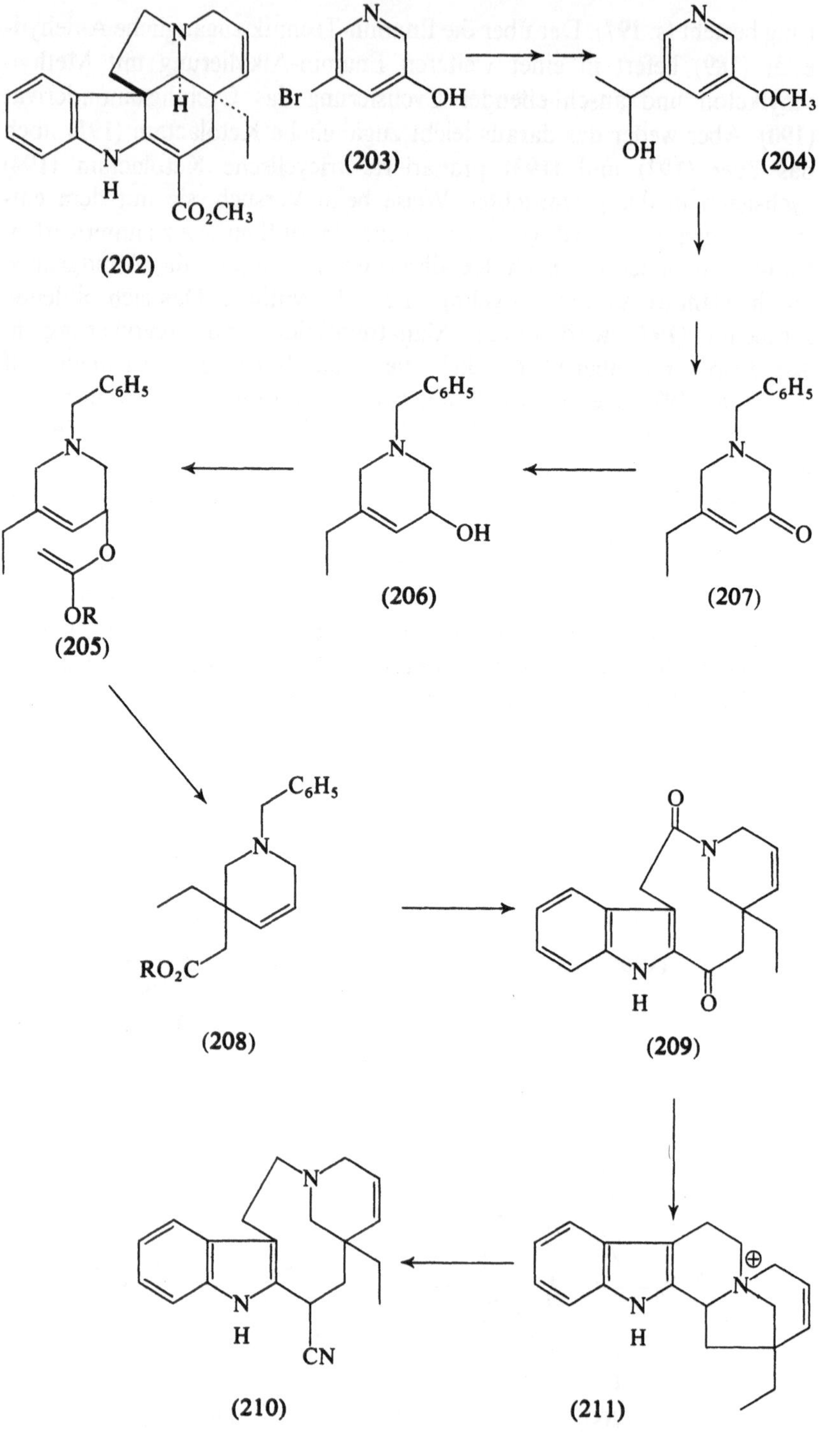

(202)

(203)

(204)

C_6H_5

(205)

C_6H_5

(206)

C_6H_5

(207)

C_6H_5

(208)

(209)

(210)

(211)

Man erhält das Lactam (**199**), das bei der Behandlung mit Bortrifluorid stereoselektiv in das Indolenin (**201**) übergeht, dessen Alanatreduktion dann Aspidospermidin (**200**) erzeugt.

Die Erkenntnisse von HARLEY-MASON und KUTNEY wurden von F. E. ZIEGLER (*12*) bei einer sehr einfachen Synthese des Tabersonins (**202**) zum Tragen gebracht.

Das wichtige Zwischenprodukt (**208**) erhält man aus Br-Nicotinamid durch Hofmann-Abbau und Verätherung der Hydroxylgruppe, anschließende Grignard-Reaktion mit Acetaldehyd zu (**204**), sowie Quartärnisierung und Alanatreduktion, die nach saurer Aufarbeitung unter Wasserabspaltung und Isomerisierung das ungesättigte Keton (**207**) erzeugt. Das daraus bei Boranat-Reduktion hervorgehende Carbinol (**206**) wird in einer [3,3]-sigmatropen Reaktion über (**205**) in (**208**) überführt. Nach Abspaltung der Benzylgruppe erfolgt Kondensation zum Lactam-Keton (**209**), das nach Alanat-Reduktion zum Carbinol mit Methansulfosäurechlorid über das entsprechende Mesylat in das Quartärsalz (**211**) übergeht. Nucleophile Ringöffnung mit Cyanid-Ionen liefert dann das Nitril (**210**), das in den entsprechenden Ester (**212**) umgewandelt wird. Dieser Ester (**212**) läßt sich nun in einer katalytischen Oxydation im Sinne der in der Formel angegebenen Pfeile zu Tabersonin cyclisieren.

(212)   ⟶ (202)   (213)

Auf einem völlig anderen Weg, der an biogenetischen Vorstellungen orientiert ist, synthetisierte ZIEGLER das Minovin (**213**), ein sehr ähnliches Alkaloid (*42*).

Der Acrylester (**215**), der aus dem α-Ketoester (**214**) in einer Wittig-Reaktion gewonnen wird, kondensiert in einer Enamin-Alkylierung zu (**216**), das dann über (**218**) in die beiden epimeren Ester vom Typ (**217**)

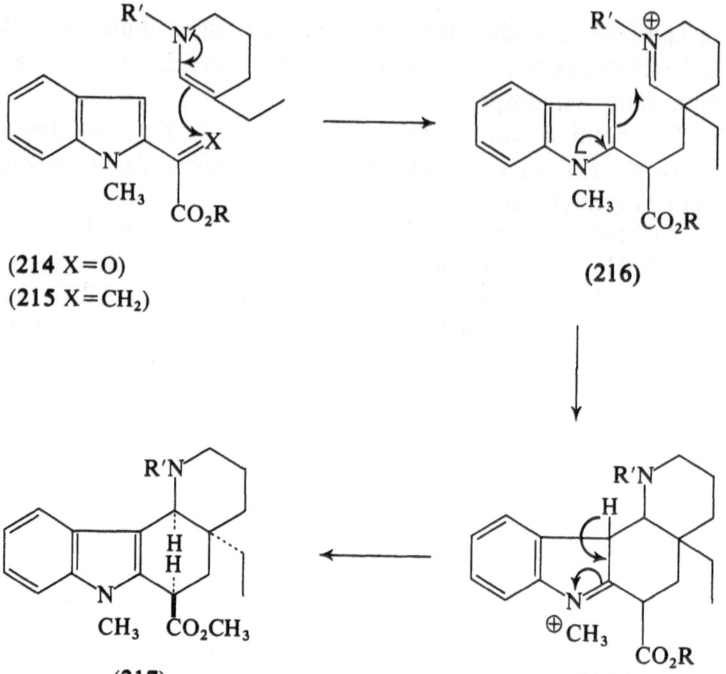

(214 X=O)
(215 X=CH₂)

(216)

(217)

(218)

übergeht. Durch Behandlung mit Alkoholat jedoch erzeugt man die thermodynamisch stabilere Konfiguration (217). Hydrogenolyse und Kondensation mit 1,2-Dibromäthan bringen daraus (213) hervor.

Eine durch ihre Eleganz und hohe Stereoselektivität besonders bestechende Synthese in dieser Serie ist die von Büchi (16) publizierte Totalsynthese des Vindorosins (219).

Das interessante Ausgangsmaterial (220) wird aus N-Methyltryptamin durch Kondensation mit Chlorbutenon und anschließende Acetylierung gewonnen. Bei der Cyclisierung mit Bortrifluorid bildet sich neben dem Tetrahydroharman-derivat, das aus der Kondensation mit der α-Position des Indols hervorgeht, als thermodynamisch stabilstes Produkt das tetracyclische Keton (222). Nach Verseifung wird es mit Acrolein zur pentacyclischen Verbindung (221) kondensiert. Somit ist bereits in wenigen Stufen das Aspidospermin-Grundgerüst aufgebaut, dessen spezielle Konfiguration und Konformation jetzt die weiteren Umwandlungsreaktionen sterisch lenkt. Alkylierung liefert (223), und die Einführung der Estergruppe mit Kohlensäuredimethylester führt zum Gemisch der epimeren Ester vom Typ (224). Obwohl eine β-Dicarbonylverbindung, erweist (224) sich als stabil gegen Kalium-tert.butylat und Sauerstoff, verwendet man jedoch H₂O₂, so entsteht der Hydroxyester (225), wahrscheinlich über das intermediär sich bildende Epoxid (226).

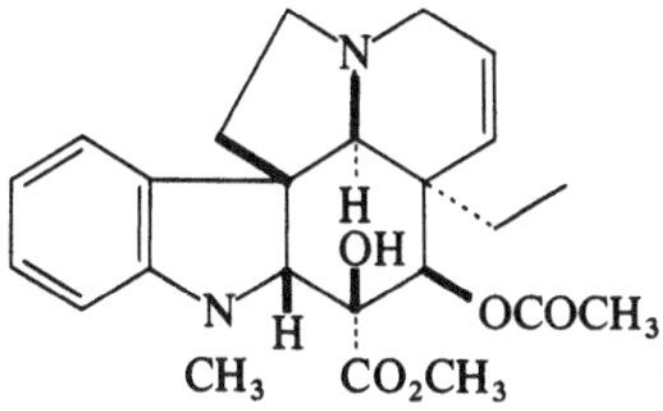

(219)

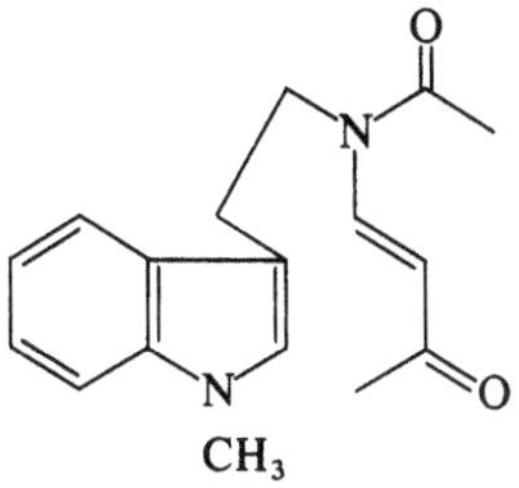

(220)

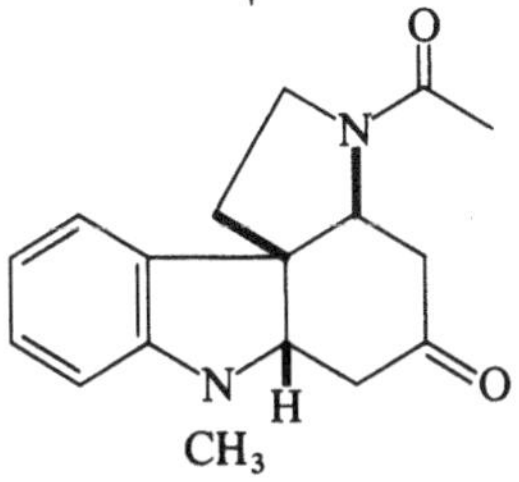

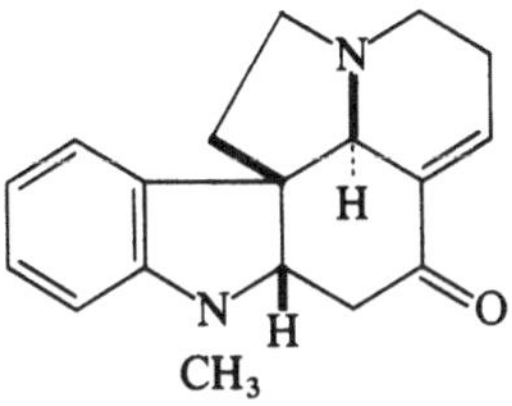

(221)

(222)

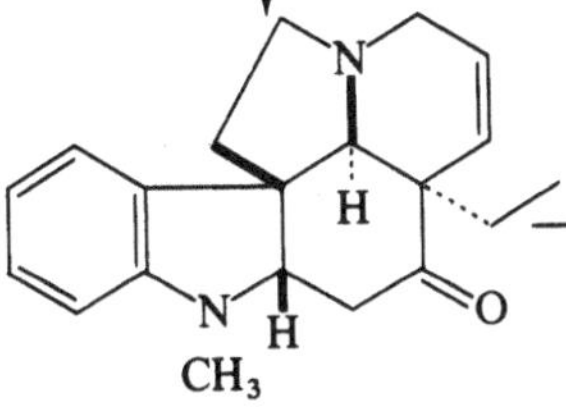

(223)

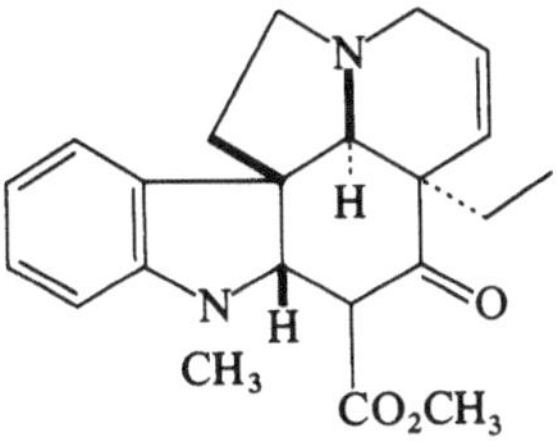

(224)

Die partielle Reduktion mit Lithiumalanat erzeugt dann eine β-ständige Hydroxylgruppe, so daß man nach Acetylierung (219) gewinnt.

(224) →

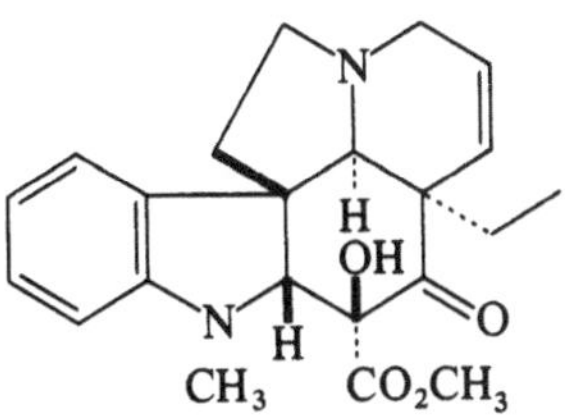

(225)

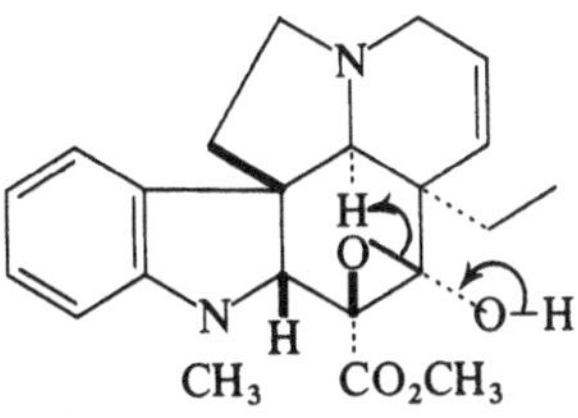

(226)

Auch in der Eburna-Reihe wurden stereoselektive Methoden erar-beitet. HARLEY-MASON *et al.* (*10*) verwendeten das Zwischenprodukt (**227**), um daraus das Lactam (**228**) zu bereiten. Hydroxylierung und Perjodsäurespaltung erzeugen einen Aldehyd, der jedoch sofort in das

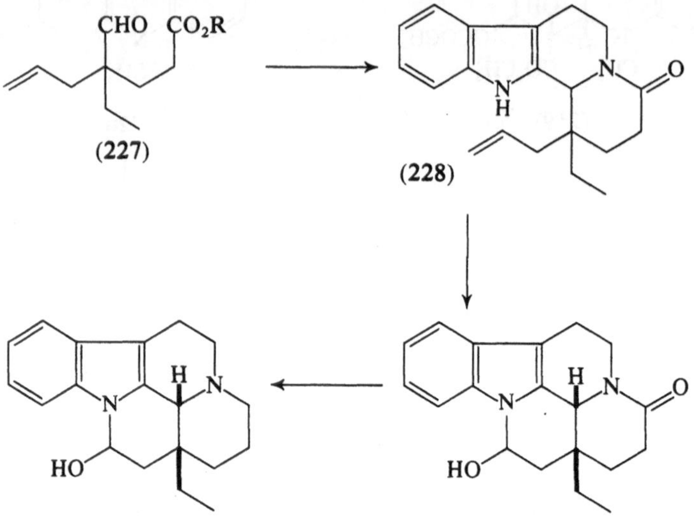

(**227**)

(**228**)

(**229**)

(**230**)

Cyclohalbacetal (**230**) übergeht. Reduktive Beseitigung der Lactam-gruppe führt dann interessanterweise nur zum Eburnamin (**229**).

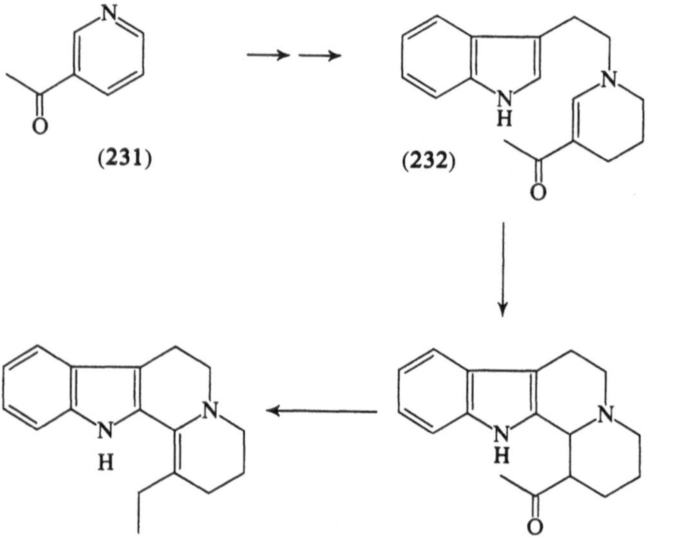

(**231**)

(**232**)

(**233**)

(**234**)

Literaturverzeichnis: SS. 518—520

Die *cis*- wie die *trans*-Verbindung dagegen erhielt WENKERT (*44*) bei einer vom Pyridinderivat (**231**) ausgehenden Synthese.

Das nach Quartärnisierung und Boranatreduktion anfallende Enamin (**232**) cyclisiert im sauren Medium (**234**) und wird durch Wolff-Kishner-Reduktion und anschließende Quecksilberacetat-Dehydrierung in das Enamin (**233**) überführt. Anmerkung bei der Korrektur: Auf der Basis dieses Enamins haben SZANTAY, SZABO und KALAUS (Tetrahed. Lett. 1973, 191) eine stereoselektive Totalsynthese des Vincamins durchgeführt.

(235)   (236)

(236)

(237)   (238)

Alkylierung von (**233**) mit Jodessigester führt über (**235**) zum penta-cyclischen Imoniumsalz (**236**), das bei der Boranatreduktion Epieburna-monin (**237**) und daneben DL-Eburnamonin (**238**) liefert. Bei der kata-lytischen Hydrierung von (**236**) jedoch überwiegt die gewünschte *cis*-Konfiguration des Eburnamonins.

Als letzte Alkaloidgruppe erscheinen umgelagerte Indolalkaloide vom Iboga-Typ. Hier sind vor allem das Ibogain (**239**, R = OCH₃) und das Ibogamin (**239**, R = H) bearbeitet worden.

Beide synthetisierte BÜCHI (*14, 15*), ausgehend vom bereits erwähnten Dien-Addukt (**157**). Beim nach der Boranatreduktion erfolgenden Hofmann-Abbau greift die Hydroxylgruppe zur Bildung des Azahalb-acetals (**241**) ein. Dieses wird sauer gespalten, die Hydroxylgruppe

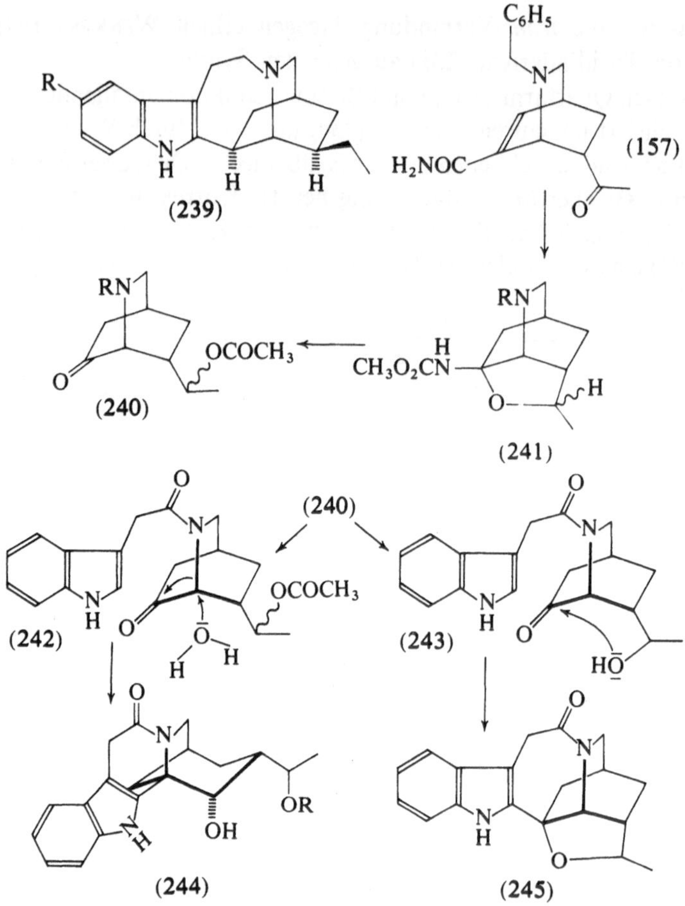

C_6H_5

(157)

H_2NOC

O

R

(239)

H   H

RN

O

(240)

OCOCH$_3$

CH_3O_2CN—

O—

(241)

H

H

(240)

(242)

O

OCOCH$_3$

O$^-$

H$^{\,\,}$O$^-$H

(243)

H$\bar{O}$

(244)

OR

OH

(245)

O

acetyliert und anschließend reduktiv debenzyliert zum wichtigen Zwischenprodukt (240). Mit Indolylessigsäurechlorid geht daraus direkt (242) hervor, verseift man jedoch vorher, so erhält man das Carbinol (243). Diese Verbindung cyclisiert zwar im sauren Medium zum Iboga-Grundgerüst (245), jedoch, wie ersichtlich, unter gleichzeitiger Ausbildung eines cyclischen Äthers, und obwohl vergleichbare Methyläther durchaus wieder spaltbar sind, erwies dieses Produkt nach Reduktion zum Amin sich als völlig stabil gegen Säure. Die Cyclisierung von (242) dagegen, bei der ein vergleichbarer Eingriff der Sauerstoffgruppierung nicht möglich ist (Acetat!), verläuft unter Umlagerung zu (244), und da selbst in Essigsäure als Solvens sich dieses Carbinol bildet, kann davon ausgegangen werden, daß die Hydroxygruppe im Zuge einer inneren Rückkehr aus dem Ionenpaar aufgenommen wird, das aus dem bei der Cyclisierung primär sich bildenden tertiären Carbinol hervorgeht.

Literaturverzeichnis: SS. 518—520

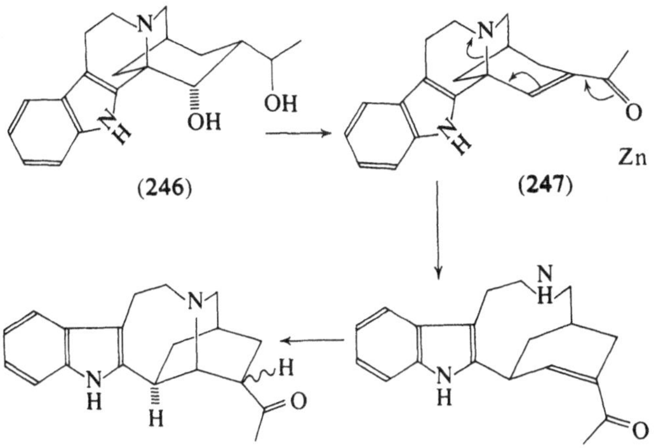

(246) (247) Zn

(248) (249)

In der Hoffnung, daß eine Rückisomerisierung zum Iboga-Gerüst möglich ist, wurde (244) mit Alanat reduziert und das resultierende Diol (246) anschließend mit Dimethylsulfoxid und DCCI selektiv zum Methylketon oxydiert, das dann als β-Hydroxyketon leicht dehydratisiert werden kann zu (247).

(250) (251)

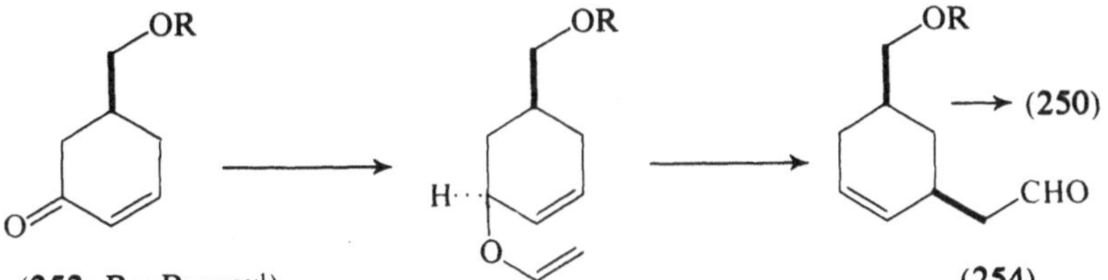

(252; R = Pyranyl) (253) (254) → (250)

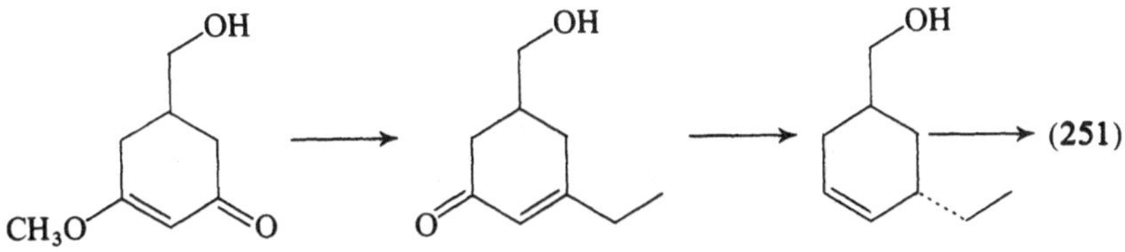

(255) (256) (257) → (251)

33*

Als vinylog α-funktionalisiertes Keton ist (**247**) mit Zink reduzierbar, unter Sprengung des unerwünschten Ringgerüstes, das intermediär auftretende ungesättigte Keton (**249**) cyclisiert dann sofort zum Keton (**248**). Diese Verbindung ist nicht einheitlich, sondern stellt ein Epimerengemisch dar, das sich als nicht äquilibrierbar erwies. Somit ergibt die anschließende Wolff-Kishner-Reduktion das Gemisch von Ibogamin und Epi-Ibogamin, das jedoch aufgetrennt werden kann. Ganz analog wird zur Synthese des Bogains vorgegangen.

Da die Konfiguration der Äthylgruppe der entscheidende Punkt in dieser Synthese ist, werden abschließend noch zwei Synthesen erwähnt, bei denen dieser Frage besondere Aufmerksamkeit geschenkt wird.

Eine japanische Arbeitsgruppe (*33*) synthetisierte stereoselektiv die beiden Amine (**250**) und (**251**) auf den folgenden Wegen.

Die Reduktion des ungesättigten Ketons (**252**) erfolgt stereoselektiv, so, daß die daran sich anschließende Claisen-Umlagerung des Vinyläthers (**253**) den *cis*-Aldehyd (**254**) erzeugt. Nach Huang-Minlon-Reduktion wird der Pyranyläther sauer gespalten und das Tosylat des resultierenden Carbinols durch Gabriel-Synthese ins Amin überführt. Die *trans*-Verbindung (**257**) erhält man aus dem vinylogen Ester (**255**) durch Grignard-Reaktion und anschließende Hydrolyse, die von einer Wasserabspaltung zu (**256**) begleitet wird. Wolff-Kishner-Reduktion erzeugt dann vorwiegend (**257**) neben 10 Prozent der *cis*-konfigurierten Substanz. Erneut führt Gabriel-Synthese mit dem Tosylat zum Amin. Die beiden Amine werden sodann durch Bleitetraacetat-Oxydation zu den Aziridinen (**258**) und (**259**) cyclisiert.

Literaturverzeichnis: SS. 518—520

Der weitere Gang der Synthese wird nun an (258) demonstriert, das mit dem Anhydrid der Indolylessigsäure unter Ringöffnung den Amidester (260) erzeugt. Verseifung und Oxydation liefern (261), das dann zu (262) cyclisiert werden kann. Diese Verbindung ist reduktiv in Ibogamin (239) überführbar. Verfährt man analog mit (259), so wird die Epi-Verbindung erhalten. Ausschließlich Ibogamin erhielt SALLAY (40), der das Diketon (263) als Ausgangsmaterial wählte.

Nachdem die besser zugängliche Ketogruppe durch selektive Ketalisierung geschützt war, wurde an der verbleibenden über das Oxim die Beckmann-Umlagerung zu (264) durchgeführt. Epoxidierung, reduktive Spaltung zum Carbinol und anschließende Oxydation erzeugen (265), das durch Wittig-Olefinierung und Hydroborierung in das primäre Carbinol (268) überführt wird. Nach der anschließenden Alanat-Reduktion wird die NH-Gruppe als Benzylurethan geschützt (267) und dann das primäre Tosylat (266) präpariert. Diese Verbindung geht bei Behandlung mit Bromwasserstoffsäure in das Keton (269) über. An der Konformation (269A) ist gut ersichtlich, daß nucleophile Verdrängung der Tosylatgruppe durch den basischen Stickstoff zum Keton (270) führt, das dann in einer Fischer-Indolsynthese Ibogamin ergibt.

Zum Abschluß sei bemerkt, daß der Autor sich darüber im klaren ist, keine lückenlose Darstellung der synthetischen Vorhaben in dieser Serie geliefert zu haben. Der Plan war vor allem, zu den prominentesten Typen von Indolalkaloiden führende, stereoselektive Techniken zusammenzustellen und auf die spezielle Problematik in den einzelnen Bereichen hinzuweisen. Wenn der Artikel dem Leser als Überblick und Sammlung des Erreichten sowie als Stimulans und Anregung für andere Gebiete dient, so ist das Ziel des Autors erreicht.

Literaturverzeichnis

1. ABDURAHMAN, N., J. P. KUTNEY, E. PIERS, P. LE QUESNE, and J. VLATTAS: New total synthesis of dl-Quebrachamine and dl-Aspidospermidine. J. Amer. chem. Soc. **88,** 3656 (1966).
2. ABDURAHMAN, N., C. GLETSOS, J. P. KUTNEY, E. PIERS, and J. VLATTAS: Total synthesis of indole and dihydroindole alkaloids. The total synthesis of dl-Quebrachamine and dl-Aspidospermidine. J. Amer. chem. Soc. **92,** 1727 (1970).
3. ALDRICH, P., T. J. KATZ, and E. E. VAN TAMELEN: Stereochemistry of Corynantheine and Corynantheidine. J. Amer. chem. Soc. **79,** 6426 (1957).
4. ALDRICH, P. E., A. W. BURGSTAHLER, M. SHAMMA, E. E. VAN TAMELEN, R. TAMM, and J. WOLINSKY: Total synthesis of Yohimbine. J. Amer. chem. Soc. **91,** 7315 (1969).
4a. ANDERSON, E. L., B. DOUGLAS, J. L. KIRKPATRIK, J. A. WEISBACH, K. K. WILLIAMS, and N. C. YIM: Stereospecific total synthesis of dl-Corynantheidine and dl-Dihydrocorynantheine. Tetrahedron Letters **11,** 1405 (1968).
5. ANG, S. K., CHR. EGLI, N. NAKATSUKA, S. K. SARKAR, Y. YASUNARI, and S. MASAMUNE: The synthesis of Ajmaline. J. Amer. chem. Soc. **89,** 2506 (1967).
6. BADER, F. E., H. BICKEL, A. J. FREY, R. W. KIERSTEAD, and R. B. WOODWARD: The total synthesis of Reserpine. Tetrahedron **2,** 1 (1958).
7. BAN, Y., and T. OISHI: Stereospecific synthesis of rac N-methylrhynchophyllane for Stereochemistry of Rhynchophylline. Tetrahedron Letters **22,** 791 (1961).
8. BARCZAI-BEKE, M., and Cs. SZANTAY: Synthesis of l-Corynantheidine. Tetrahedron Letters **11,** 1405 (1968).
9. — — Die stereospezifische Synthese des (—)-Corynantheidins. Chem. Ber. **102,** 3963 (1969).
9a. HONTY, K., L. TÖKE und Cs. SZANTAY: Die Totalsynthese des (+)-Yohimbins und β(—)-Yohimbins. Chem. Ber. **102,** 3248 (1969).
10. BARTON, J. E. D., and J. HARLEY-MASON: Total synthesis of Hunteria and Aspidosperma alkaloids from a common intermediate. Chem. Comm. **14,** 298 (1965), Chem. Comm. **10,** 197 (1965).
11. BECK, J., F. BYLSMA, W. J. CRETNEY, and J. P. KUTNEY: Studies on the synthesis of dimeric Vinca alkaloids. J. Amer. chem. Soc. **90,** 4504 (1968).
12. BENNET, G. B., and F. E. ZIEGLER: Total synthesis of (±)-Tabersonine. J. Amer. chem. Soc. **93,** 5930 (1971).
13. BROWN, R. T., J. R. HADFIELD, and J. P. KUTNEY: Total synthesis of indole and dihydroindole alkaloids. III. The transannular cyclization of Carbomethoxycleavamine

derivatives. An approach to Vinca and Iboga alkaloids. J. Amer. chem. Soc. **92**, 1708 (1970).

14. Büchi, G., D. L. Coffen, K. Kocsis, P. E. Sonnet, and F. E. Ziegler: The total synthesis of (±)-Ibogamine and of (±)-Epiibogamine. J. Amer. chem. Soc. **9**, 2073 (1965).

15. — — — — — The total synthesis of Iboga alkaloids. J. Amer. chem. Soc. **88**, 3099 (1966).

16. Büchi, G., K. E. Matsumoto, and H. Nishimura: Total synthesis of (±)-Vindorosine. J. Amer. chem. Soc. **93**, 3299 (1971).

17. Büchi, G., St. J. Gould, and F. Näf: Stereospecific synthesis of Uleine and Epiuleine. J. Amer. chem. Soc. **93**, 2492 (1971).

18. Büchi, G., P. Kulsa, K. Ogasawara, and R. Rosati: Synthesis of Velbanamine and Catharanthine. J. Amer. chem. Soc. **92**, 999 (1970).

19. Bylsma, F., and J. P. Kutney: Studies on the synthesis of monomeric and dimeric Vinca alkaloids. The total synthesis of Isovelbanamine, Velbanamine, Cleavamine, 18β-Carbomethoxycleavamine and Catharanthine. J. Amer. chem. Soc. **92**, 6090 (1970).

20. Cava, M. P., H. U. Daeniker, A. Hunger, W. D. Ollis, K. Schenker, and R. B. Woodward: The total synthesis of Strychnine. J. Amer. chem. Soc. **76**, 4749 (1954).

21. Cava, M. D., A. Hunger, W. P. Ollis, and R. B. Woodward: The total synthesis of Strychnine. Tetrahedron **19**, 247 (1963).

22. Dadson, B. A., G. H. Foster, and J. H. Mason: The synthesis of (±)-Tubifoline, (±)-Tubifolidine and (±)-Condyfoline. Chem. Comm. **20**, 1233 (1968).

23. Dave, K. D., R. G. Lewis, P. W. Strague, and E. Wenkert: General methods of synthesis of indole alkaloids VI. Synthesis of dl-Corynantheidine and a Camptothecin model. J. Amer. chem. Soc. **25**, 6741 (1967).

24. Dolfini, J. E., and G. Stork: The total synthesis of dl-Aspidospermidine and of dl-Quebrachamine. J. Amer. chem. Soc. **85**, 2872 (1963).

25. Gaskell, A. J., T. Korth, H. E. Radunz, M. Walkowiak, and E. Winterfeldt: Die stereoselektive Totalsynthese von DL-Ajmalicin, DL-19-Epiajmalicin, DL-Formosamin und Isoformosamin. Chem. Ber. **102**, 3558 (1969).

26. Gaskell, A. J., A. Jackson, J. A. Joule, and N. D. V. Wilson: The total synthesis of Dasycarpidone and 3-epi-Dasycarpidone. Chem. Comm. **7**, 364 (1968); The synthesis of (±)-Dasycarpidone, (±)-3-epi-Dasycarpidone, (±)-Uleine and (±)-3-epi-Uleine. J. chem. Soc. C **1969**, 2738.

27. Grethe, G., H. L. Lee, T. Mitt, and M. R. Uskokovic: Synthesis of Chinchona alkaloids via Quinuclidine precursors. J. Amer. chem. Soc. **93**, 5904 (1971).

28. Grethe, G., J. Gutzwiller, H. L. Lee, C. Reese, and M. R. Uskokovic. Epimeric-3-vinyl-4-piperidineacetic acids. Synthetic precursors of Chinchona and indole alkaloids. J. Amer. chem. Soc. **93**, 5902 (1971).

29. Gutzwiller, J., G. Pizzolato, and M. R. Uskokovic: A novel synthesis of racemic Ajmalicine, Tetrahydroalstonine and Akuammigine. J. Amer. chem. Soc. **93**, 5907 (1971).

30. Harley-Mason, J., and C. G. Taylor: Total synthesis of (±)-dihydronorfluoracuarine and an improved total synthesis of (±)-Geissoschizoline. Chem. Comm. **13**, 812 (1970).

31. Harley-Mason, J., and M. Kaplan: A simple total synthesis of dl-Aspidospermine and of dl-Quebrachamine. Chem. Comm. **18**, 915 (1967).

32. Hesse, M., J. Kompis, and H. Schmid: An approach to the biogenetic classification of Indole Alkaloids. Lloydia **34**, 269 (1971).

33. Hirai, S., K. Kawata, W. Nagata, and T. Ohumura: A stereo-chemically controlled total synthesis of dl-Ibogamine and dl-Epiibogamine. J. Amer. chem. Soc. **90**, 1650 (1968).

34. McKAGUE, B., W. J. KRETNEY, J. P. KUTNEY, E. PIERS, and P. LE QUESNE: The total synthesis of dl-Dihydrocleavamine, dl-Carbomethoxydihydrocleavamine. dl-Coronaridine and dl-Dihydrocatharanthine. J. Amer. chem. Soc. **88,** 4756 (1966).
35. McKAGUE, B.. W. J. KRETNEY. J. P. KUTNEY, P. LE QUESNE, and E. PIERS: Total synthesis of indole and dihydrocleavamine, dl-Carbomethoxydihydrocleavamine, dl-Coronaridine, dl-Dihydrocatharanthine and dl-Ibogamine. J. Amer. chem. Soc. **92,** 1712 (1970).
36. KORTH, T., H. RADUNZ und E. WINTERFELDT: Die partielle stereoselektive Total-synthese von DL-Akuammigin und DL-Tetrahydroalstonin. Chem. Ber. **101,** 3172 (1958).
37. OLIVER, L. K., and E. E. VAN TAMELEN: The biogenetic type total synthesis of Ajmaline. J. Amer. chem. Soc. **92,** 2136 (1970).
38. PLACEWAY, C., G. P. SCHIEMENZ, E. E. VAN TAMELEN, and L. G. WRIGHT: Total syn-thesis of dl-Ajmalicine and Emetine. J. Amer. chem. Soc. **91,** 7359 (1969).
39. RIESNER, H., and E. WINTERFELDT: Biogenetically patterned stereoselective total syn-thesis of the indole alkaloid Roxburghine D. Chem. Comm. 786 (1972).
40. SALLAY, S. J.: The total synthesis of dl-Ibogamine. J. Amer. chem. Soc. **89,** 6762 (1967).
41. SMITH, G. F.: In: The Alkaloids, Vol. VIII. Ed. R. H. F. MANSKE, p. 591. New York: Academic Press. 1965.
42. SPITZNER, E. B., and F. E. ZIEGLER: The biogenetically modeled total synthesis of (±)-Minovine. J. Amer. chem. Soc. **92,** 3492 (1970).
43. VAN TAMELEN, E. E., and J. G. WRIGHT: Total synthesis of dl-Corynantheine. Tetra-hedron Letters **6,** 295 (1964); J. Amer. chem. Soc. **91,** 7349 (1969).
44. WENKERT, E., and B. WICKBERG: General methods of synthesis of indole alkaloids. A synthesis of dl-Eburnamonine. J. Amer. chem. Soc. **87,** 1580 (1965).

(Eingelaufen am 19. September 1972)

Structure, Chemistry, and Biosynthesis of the Melanins

By G. A. Swan, Newcastle upon Tyne

Contents

I. Introduction

The term "melanin" (μέλας = black) is a purely descriptive one, which conveys no chemical information and merely denotes a black pigment of biological origin, although in fact some melanins are brown or even yellow. Different authors have accepted various definitions of exactly what constitutes a melanin; and melanins are sometimes loosely described as pigments of high molecular weight formed by the enzymic oxidation of phenols. The latter is not however a satisfactory definition. According to Thomson (*149*), who has written admirable reviews on the chemistry of melanins (*149, 150*), the term melanin appears to have been used first with some precision in 1902 by Fürth and Schneider, in respect of the black precipitate they obtained by the action *in vitro* of insect tyrosinase on tyrosine. The fact that the general properties and the carbon, hydrogen, and nitrogen analyses of this material were in approximate agreement with those reported for natural pigments from animal hair, melanoma, *Sepia* black, *etc.*, implied that these natural pigments also were products of the tyrosine-tyrosinase reaction. An excellent, comprehensive, and detailed account of these pigments has been given by Nicolaus (*107*) in his book, which includes references up to 1967, and in which he has classified the pigments into eumelanins, phaeomelanins, and allomelanins. In the present article, an attempt will be made to give a much briefer, although up-to-date review of the structure, chemistry, and biosynthesis of melanins. A number of other useful reviews on melanin, and on the chemistry and biochemistry of melanogenesis are also available (*42, 53, 85, 86, 94, 95, 106, 130, 137, 148*).

Tyrosine	Dopa	Dopamine
(1)	(2)	(3)

The eumelanins, which are usually black, and which occur especially in the animal kingdom, contain nitrogen, and are derived from tyrosine (**1**), dopa, *i. e.* 3,4-dihydroxyphenylalanine (**2**), and perhaps dopamine, *i. e.* 3,4-dihydroxyphenethylamine (**3**), tyramine, *etc.* Typical examples are the pigments of human skin, black hair, feathers, the malignant melanoma, and the ink sac of the squid. The phaeomelanins are responsible for the color of red hair and of chicken feathers, and contain nitrogen and sulphur; it is believed that both tyrosine and cysteine are involved in their formation. The allomelanins, which are usually black, occur particularly in the plant kingdom, e.g. in certain fungi, and in the seeds of some flowering plants, and are formed by oxidation of diphenols such as catechol. Certain species of bacteria also produce black pigments, e. g. *Microspira tyrosinatica* of ditch water, some *Actinomyces, Bacterium symbioticum*, and *Bacillus salmonicida*.

Without prejudice regarding NICOLAUS' classification, in this review the synthetic pigments are dealt with under the heading which seems most appropriate structurally, e.g. adrenochrome-melanin, although not definitely known to occur naturally, is treated under the general heading of eumelanins, because it is an indole-type of pigment.

More research has been carried out on the eumelanins than on the phaeomelanins and allomelanins. These eumelanins appear to be polymers of irregular structure, and often occur conjugated with protein. They are usually insoluble in almost all solvents, although sometimes soluble in alkali; they lack well defined spectral and other physical characteristics, and are difficult to separate from the other constituents of the organism in which they occur. Indeed, isolation of a melanin usually simply involves removing or destroying all other compounds present. Even when the pigment has been isolated, it is difficult to know whether or not it is pure, if indeed the term "pure" can rightly be applied to any melanin. Moreover, fruitful degradation of the pigment is difficult to achieve. Progress in the elucidation of the structures of melanins has therefore been slow. In fact, the history of melanin research has been unusual, in that, in contrast with most other natural products, the starting material and some of the intermediates in its biosynthesis were discovered before the structure of the melanin itself. A good deal of what is known about the structure of melanins has been gained by studying melanogenesis from known precursors. Oxidation under certain conditions *in vitro* of tyrosine, dopa, *etc.,* can give rise to black, polymeric materials, known as "synthetic melanins". However, although the structures and mechanisms of formation of these have been studied, proof is lacking that the natural and synthetic melanins are identical, although they do show great similarity to one another. There are no satisfactory tests for the identification of melanins, and the histochemical tests which have been used are not specific. A dark pigment, insoluble in most solvents, and decolorized by hydrogen peroxide, and which reduces silver nitrate has often been deemed to be a melanin. There is also no means of proving that two melanin samples are identical.

II. Enzymic Nature of Melanogenesis

1. Melanogenesis in Invertebrates and Plants

The enzymic nature of the melanin-producing reaction was first satisfactorily demonstrated in the plant kingdom. The enzyme, first recognized in 1895 in the fungus *Russula nigricans* by E. BOURQUELOT and G. BERTRAND, and in 1896 in dahlia tubers by BERTRAND, and named by him tyrosinase, acted on tyrosine *in vitro* to give successively red and black products. BERTRAND also isolated tyrosine from these plants. (For references, see *107, 148* and *149*.) At this point it is worth mentioning that the presence of other amino-acids in the solution alters the colors formed; and in fact, pigments formed in the presence of cysteine are probably related to phaeomelanins.

In his pioneering research RAPER (*131*) studied the oxidation of tyrosine by molecular oxygen in the presence of an enzyme obtained from mealworms (*Tenebrio molitor*). The advantage of this particular enzyme is that it can be precipitated by dilute acid, so that its action is stopped, and it can then be removed from the solution by filtration. However, broadly similar results were also obtained using tyrosinase from mushroom or potato. RAPER concluded that melanin-formation occurs in three stages:

(a) oxidation of tyrosine to give a red pigment,
(b) transformation of the red pigment into a colorless substance, and
(c) oxidation of this colorless substance to give melanin.

Often a period of induction was observed, before oxidation of tyrosine began; but the presence of a small amount of dopa was effective in shortening the lag period. Deamination apparently did not occur during the reaction, because the formation of ammonia was not detected, and the percentage of nitrogen in the melanin formed was slightly higher than in tyrosine.

At the end of stage (a) RAPER obtained a red solution at pH 6, which was relatively stable. However, when this solution was kept in a vacuum, or was treated with a little sulphurous acid, it became colorless; and in the resulting solution RAPER detected the presence of unchanged tyrosine, dopa (isolated as the insoluble lead salt), 5,6-dihydroxyindole (**8**), and 5,6-dihydroxyindole-2-carboxylic acid (**7**) (the two latter products were isolated as dimethyl ethers). Later, PIATTELLI, FATTORUSSO and MAGNO (*117*) isolated 5,6-dihydroxyindole by extracting with ether a solution of pH 6.8 obtained by enzymic oxidation of dopa.

It was supposed that the enzyme first catalyzed the hydroxylation of tyrosine (**1**) to give dopa (**2**), which was then oxidized (again enzymically)

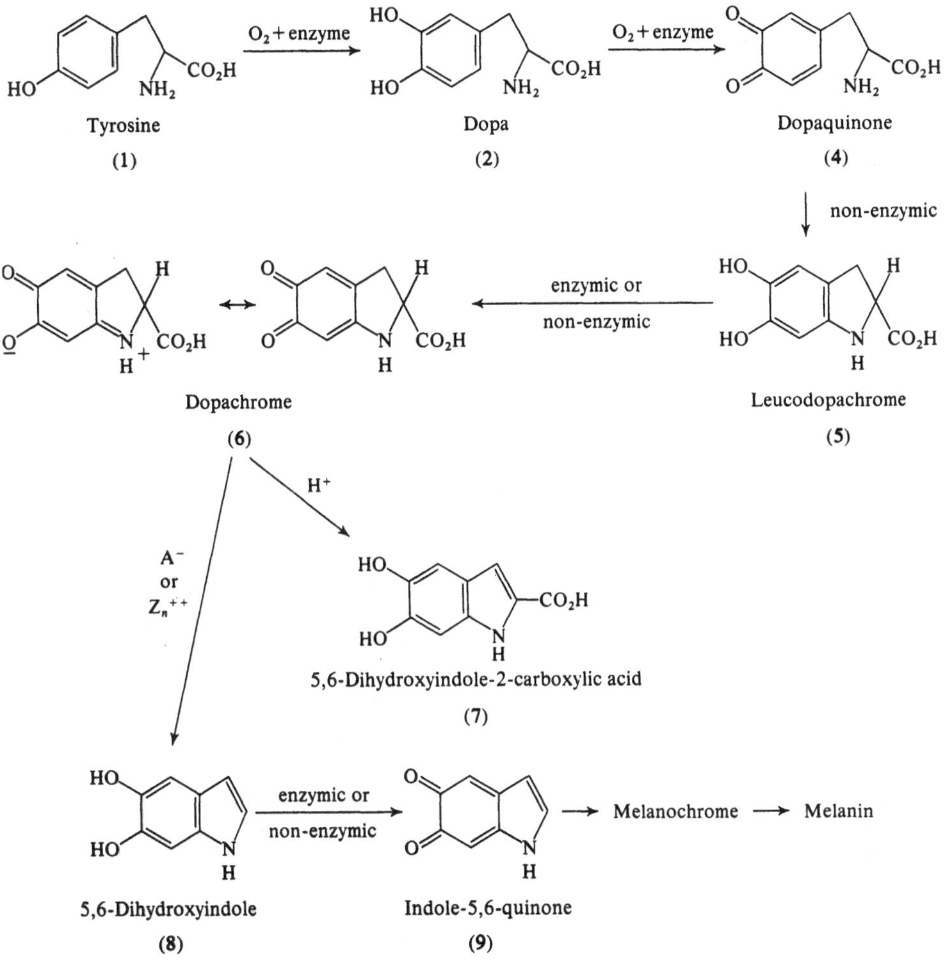

Chart 1. The Raper-Mason scheme of melanogenesis

to dopaquinone (**4**). It is probable that a high concentration of the latter does not build up in the solution (except under acidic conditions), as the side-chain amino-group rapidly adds on intramolecularly to the quinone system, giving leucodopachrome (**5**), which is then rapidly oxidized to dopachrome (**6**), i.e. RAPER'S red compound.

In the second stage, the dopachrome undergoes rearrangement and (to some extent) decarboxylation, being converted partly into 5,6-dihydroxyindole (**8**) and to some extent into 5,6-dihydroxyindole-2-carboxylic acid (**7**). The latter product predominated under acidic conditions. It has been shown (*66*) that zinc ions catalyze this rearrangement, base catalysis of which is a special feature of such *o*-quinonoid structures (aminochromes) (*68*); the corresponding *p*-quinones are stable to base, their oxidation-reduction potentials being too low for an internal hydrogen-shift to occur (136). It is noteworthy that in the choroid of the vertebrate eye (which contains melanin) there is a remarkably high concentration of zinc, and other metallic elements, greatly exceeding that in non-pigmented tissues (22, 23, 88).

In the third stage, further oxidation of 5,6-dihydroxyindole was supposed to yield melanin.

The enzyme was essential only for the first stage, and for the second stage up to dopaquinone (44), although the possibility remained that it could increase the rate of some subsequent reactions.

In a tyrosine-tyrosinase reaction which had proceeded for two to four hours, dopa could be isolated in yield of 10—20% of the actual tyrosine oxidized, in spite of the fact that the tyrosinase could oxidize dopa more rapidly than tyrosine. It seems likely that this is a consequence of the following oxidation-reduction potentials:

Dopa   ⇆ Dopaquinone $E_0 = +0.511$ V at pH 4.6
Leucodopachrome ⇆ Dopachrome $E_0 = +0.170$ V at pH 4.6

Dopaquinone is thus able to oxidize leucodopachrome to dopachrome, itself thereby being reduced back to dopa (44). Reducing agents such as ascorbic acid act as inhibitors of melanogenesis. In the presence of ascorbic acid melanin cannot be formed by the action of tyrosinase on tyrosine or dopa until all the ascorbic acid has been oxidized (86).

The oxidation of dopa by oxygen in the presence of an enzyme isolated from mushrooms was studied spectroscopically by Mason (92), who observed three chromophoric stages:

(a) with absorption maxima at 305 and 475 nm, (red),
(b) with absorption maxima at 300 and 540 nm, (purple), and
(c) with general absorption, considered to correspond to melanin.

The first stage is considered to correspond to Raper's dopachrome. A broad absorption maximum at 305—310 nm also appeared when dopa was oxidized in the presence of melanoma tyrosinase (91).

It was originally suggested that the second stage corresponded to indole-5,6-quinone (9), because enzymic oxidation of 5,6-dihydroxindole yielded a solution, the spectrum of which corresponded to (b); but it now seems clear that spectrum (b) in fact represents a low polymer, or a mixture of polymers, derived from indole-5,6-quinone. Thus Beer, Broadhurst and Robertson (6) pointed out that the light absorption of the purple pigment (melanochrome) was similar to that of the indolylbenzoquinones prepared by Bu'Lock and Harley-Mason (30). Then Cromartie and Harley-Mason (36) showed that autoxidation, or oxidation with silver oxide in methanol, or with aqueous ferricyanide

solution, of 2,3-dimethyl-5,6-dihydroxyindole gave a comparatively
stable, red product, with absorption maxima at 297 and 470 nm, which
was presumably 2,3-dimethylindole-5,6-quinone (10). Bu'Lock (28, 29)
concluded that in the spectrum of the purple solution formed by enzymic
oxidation of 5,6-dihydroxyindole at pH 6.8, the initial peak at 530 nm
may represent a dimer, e. g. (11), but this soon flattened out to give a maximum at 540 nm, perhaps corresponding to a mixture of oligomers.

Evidence for the transient existence of uncyclized dopaquinone-types
of intermediates, having a yellow color, and absorption maximum at
ca. 385 nm has been found by Kodja and Bouchilloux (82).

   Dulière and Raper (43) measured the uptake of oxygen during the
enzymic conversion of tyrosine or dopa into melanin and found that 5
or 4 atoms of oxygen (respectively) were required per molecule of
tyrosine or dopa. Later, Mason and Wright (102) studied the conversion
of dopa into dopachrome, and noticed that no evolution of carbon
dioxide occurred during this phase. In the subsequent phase, carbon
dioxide was evolved, but oxygen uptake diminished. They found that
oxygen consumption varied between 2.9 and 4.6 atoms/molecule of
dopa oxidized, while 0.6—1.0 molecules of carbon dioxide were evolved.
The exact values depended on numerous factors, such as the pH of the
solution, the temperature, the purity of the enzyme, and the concentration
of the enzyme and substrate.

   According to the Raper-Mason Scheme of Melanogenesis (Chart 1),
melanin is formed with the consumption (per molecule of tyrosine) of
5 atoms of oxygen, and evolution of 1 molecule of carbon dioxide
derived from the carboxy-group of tyrosine; and the melanin is a polymer
of indole-5,6-quinone, ($C_8H_3NO_2$) . We shall take up the story again
from this point later (IV.1); but meanwhile we shall turn our attention to
melanogenesis in vertebrates.

2. Melanogenesis in Vertebrates

For a long time it was thought that melanin was formed in mammalian tissue by a mechanism different from that operating in invertebrates. Skin pigmentation in man has protective, social, and cosmetic significance, while the presence of melanin in the (malignant) melanona causes the tumor to be black. For these and other reasons, melanogenesis has attracted the attention of workers in many different fields, including chemists, biochemists, cytologists, geneticists, anthropologists, and many clinicians, including dermatologists, pathologists, and surgeons. Although only a few of the very many papers touching on melanogenesis can be referred to here, attention should be drawn to the publication of papers presented at a series of conferences, initiated by, and continued in memory of (the late) MYRON GORDON — the International Pigment Cell Conferences (39, 56—58, 89a, 132, 133).

In his research on melanogenesis BLOCH, a Swiss dermatologist, (10) was guided by two clinical observations. First, melanin pigmentation is a prominent feature of Addison's disease, which results from hypofunction of adrenal glands; the increased pigmentation might perhaps be related (in some paradoxical way) to the metabolism of epinephrine-like compounds. Second, the urine of patients in cases of metastatic melanoma with melanuria contains significant amounts of catechol derivatives. These facts suggested to BLOCH a chemical similarity between the precursor of melanin and the compounds epinephrine and catechol, so he selected dopa as substrate for histochemical studies.

BLOCH immersed sections of pigmented human skin in dilute solutions of dopa at pH 7.3—7.4, and noted that after 24 hours at room temperature melanin granules were deposited in the cytoplasm of cells in the basal layer of the epidermis. The intensity of the response corresponded to the known capacity of the skin to form melanin. Thus melanin deposition did not occur in albino skin, nor in the skin of patients with vitiligo, a skin disease characterized by localized areas of complete loss of melanin pigmentation. BLOCH therefore concluded that the "dopa reaction" is a reliable indication of the capacity of cells to form pigment. He also found evidence that the catalytic effect of certain cells on the oxidation of dopa to melanin is due to the presence of an enzyme, which he called dopa-oxidase.

BLOCH's histochemical studies with dopa did not provide a complete explanation for the mechanism of melanogenesis since, at that time (i.e. 1917—1927) dopa had not been detected in mammalian tissue. However, as described above, in 1928 RAPER showed that dopa was an oxidation product of tyrosine; but BLOCH's dopa-oxidase reaction could not be obtained by treatment of skin sections with tyrosine instead of dopa. And histochemists questioned the applicability to vertebrates of a metabolic pathway that had been worked out in an extract or homogenate obtained from invertebrates. Then in 1942 HOGEBOOM and ADAMS (71) demonstrated the presence of tyrosinase in mammalian tissue (Harding-Passey mouse melanoma); but biologists were still unhappy because the enzyme had been obtained from pathological material, and because it seemed doubtful whether what occurred in an extract or homogenate would also occur in the melanocyte. Soon, however, the misconception that tyrosine was inactive as a substrate in histochemical experiments on mammalian skin was resolved. Tyrosine was shown by FITZPATRICK, BECKER, LERNER, and MONTGOMERY to be, in fact, a melanogenic substrate in human skin that had been irradiated in vivo with ultraviolet rays (46). Moreover, tyrosinase was found to be present in the hair bulbs of mice. Tyrosine has been shown by a number of workers to be an in vivo precursor of melanin in vertebrates (47, 48). Perhaps the most convincing evidence that tyrosine is a precursor of melanin in man is the demonstration by DUCHON and PECHAN (41) that oral administration of tyrosine to patients with melanoma with melanuria resulted in a marked increase in the excretion of urinary indole and phenol melanogens. Of a number of compounds isolated from the urine, two were recognized as being 5,6-dihydroxyindole-glucuronoside and -sulphate (40). As the darkening of the urine when exposed to air is presumably a result of autoxidation of 5,6-dihydroxyindole to give melanin, it is more correct to use the term "melanogenuria" rather than "melanuria" (41).

In man and other higher organisms eumelanins are synthesized in specialized cells, known as melanocytes, pigment formation requiring the presence of the enzyme (tyrosinase), the substrate (tyrosine or dopa), and molecular oxygen. Electron microscopy has revealed the presence within the melanocyte of a unique organelle, known as the melanosome, in which melanin synthesis occurs (13, 47, 48). In its early stages the melanosome may be considered to consist largely of a protein matrix, upon what the melanin polymer is destined to be formed. Part of the protein provides the necessary enzyme activity. After the synthesis of this matrix, the melanin precursors may then diffuse into the melanosome and become incorporated into polymer. As the polymer is formed, the protein matrix is eventually covered up until finally no catalytic sites are available for the conversion of further precursor, and melanization is complete. The fully melanized organelle thus constitutes an apparently uniformly dense and structureless particle, a micron or less in diameter, of melanin polymer admixed with protein, which has often been referred to as a melanin granule. However, a study (84) of human and animal hair, using the electron microscope, has shown that the granules do indeed have a composite structure, made up of a colorless matrix upon which a sheath or envelope of melanin is deposited. In this work the hair protein was removed by treatment with a hot mixture of thioglycollic acid and phenol.

In man an excessive production of melanocytes occurs in moles and melanoma, but melanotic tumors occur also in insects and fishes. Pituitary and other hormones have been shown to influence melanin pigmentation, but this work is outside the scope of the present review (*139*), as is the subject of adaptive coloration of axolotls and fish in response to light (*1a, 88*).

3. The Enzyme

Enzymes known as polyphenol-oxidases catalyze the aerobic oxidation of *o*-diphenols to *o*-quinones, and are widespread in the animal and vegetable kingdoms (*78, 86, 95*). They are abundant in some species of mushrooms, but occur also in moulds, bacteria, algae, and in higher plants, *e.g.* potato.

The confusing and misleading name tyrosinase has been applied to enzymes capable of transforming tyrosine or dopa into melanin. Such an enzyme has two distinct activities:

(a) the aerobic oxidation of tyrosine to dopa, and

(b) the arobic oxidation of dopa to yield ultimately a melanin.

However, so far as (b) is concerned, it seems only to have been established that the enzyme is concerned with the initial oxidation of dopa to dopaquinone, subsequent stages being possible in the absence of enzyme. Some authors (*107*) limit the use of the name tyrosinase to enzymes originating from the animal kingdom. Other names, such as phenolase (*94*), are also used to describe some of these enzymes; and BLOCH's dopa-oxidase has already been mentioned.

However, whether prepared from mammalian, insect, or plant tissues, these enzymes appear to be copper-protein complexes, and the function of the copper has been discussed (*26, 78*). It is possible to oxidize monohydric phenols to *o*-quinones simply by oxygen in the presence of a copper-secondary amine complex (*24, 25, 155*). Copper is essential for normal pigmentation in mammals (*86*).

Examples of enzymes isolated from different parts of the phylogenetic scale have been tabulated and classified (*159*). *In vitro* melanogenesis has been studied extensively using a polyphenol-oxidase isolated from commercial mushrooms, probably largely because of the relatively easy availability of this enzyme. However, it is of relatively low specificity, acting equally well on **L**- and **D**-dopa, and oxidizing a wide range of *o*-dihydroxy-compounds (although melanins are not formed in all cases).

In fact it oxidizes catechol much faster than dopa; and polyphenol-oxidase from potato oxidizes catechol 1000 times faster than dopa (76, 77). On the other hand, tyrosinase isolated from human melanoma is said to be more specific for L-dopa and L-tyrosine, and does not significantly oxidize catechol (86, 122).

Details are available of methods for the extraction and purification of the enzyme from mushrooms (79, 105), *Neurospora crassa* (75), and hamster melanoma (122, 123). The "cresolase activity" (representing the ability of the enzyme to hydroxylate a monohydric phenol such as tyrosine) may be determined by measuring the rate of oxygen uptake in the oxidation of *p*-cresol in the presence of the enzyme. The "catecho-lase activity" (representing the ability to oxidize *o*-diphenols) may be similarly determined using catechol in the place of *p*-cresol (105, 107). However, during recent years a number of other methods have become available (75, 105, 123). During the process of "purification" the catecho-lase: cresolase ratio is often increased, so that if due care is not taken, the final product may be essentially a polyphenol-oxidase.

Many substances inhibit tyrosinase activity *in vitro*, notably thiol compounds, which combine with copper (33) and a few of these have been shown to be active also *in vivo* (86). 2-Mercaptoethylamine and *N*-(2-mercaptoethyl)dimethylamine hydrochloride, when topically ap-plied to the skin, are very potent agents that cause depigmentation in the treated areas only (55). Inhibitors of this type have been found in human epidermis, and are probably mainly responsible for the inactivation of tyrosinase in normal unpigmented skin. FLESCH (49) has shown that the concentration of thiol groups in human skin diminishes on exposure to ultraviolet radiation; and it is known that tyrosine can be converted into dopa under the same conditions. These photochemical oxidations are important factors in the formation of melanin, which occurs during sun-tanning (86). Oxidation of thiol groups releases the bound copper by removal of the natural inhibitor, and the formation of dopa catalyzes the tyrosine-tyrosinase reaction. The tyrosinase system in human epidermal melanocytes is normally inhibited unless activated by radiation, whereas that in the melanocytes of pigmented human hair is active (83); this accounts for the coexistence of black hair and white skin.

By using $^{18}O_2$ it has been shown that in the enzymic hydroxylation of 3,4-dimethylphenol to dimethylcatechol, in the presence of ascorbic acid, the oxygen of the new hydroxy-group originates from the molecular oxygen (98). According to most authors the enzyme is first activated by reduction by an *o*-diphenol, and then introduces the oxygen atom into the monophenol molecule. However, KERTESZ and ZITO (78) believe that it is the *o*-quinone produced by the oxidation of the *o*-diphenol which is the active agent of the hydroxylation.

From histochemical studies, OKUN, EDELSTEIN, and their associates (112, 113) have recently concluded that mammalian peroxidase (rather than tyrosinase) is the enzyme responsible for the conversion of tyrosine into dopa, as well as the conversion of dopa into melanin. They have also carried out biochemical studies (115), which they interpret to show that either human myeloperoxidase or horseradish peroxidase in the presence of hydrogen peroxide catalyzes the conversion of tyrosine into dopa in the presence of dopa as cofactor. However, in our opinion (138) this interpretation is not entirely unambiguous. Moreover, BAYSE and MORRISON (5) have shown that although lactoperoxidase, horseradish peroxidase, and mushroom polyphenol-oxidase all catalyze in vitro the oxidation of dopa at reasonably comparable rates, the rate of oxidation of dopa by lactoperoxidase is greatly increased by the presence of phenolic compounds such as tyrosine. It therefore seems that tyrosine may act simply as a cofactor, and may not be converted into dopa. The formation of radical-coupling dimers from phenols and aromatic amines in the presence of hydrogen peroxide and peroxidase is well known; and GROSS and SIZER (62) have shown the formation from tyrosine, not only of (12), but also of brown polymeric products. It is possible that such polymers, if formed histochemically, could be mistaken for melanins. However, the in vivo situation clearly requires further investigation, and the possible participation of both tyrosinase and peroxidase cannot be excluded on present evidence. See Addendum, p. 582.

(1)   (12)

III. Allomelanins

1. Catechol-Melanin

When catechol (13) is oxidized in the presence of polyphenol-oxidase, a brown, insoluble pigment, known as catechol-melanin, is formed (96). Chemical (128), spectroscopic (93), and polarographic (107) investigations suggest that o-benzoquinone (14) is an intermediate in the reactions involved, although the nature of the subsequent stages is less certain. However, FORSYTH and QUESNEL (51, 52) examined by paper chromatography the products formed by enzymic oxidation of an aqueous solution of catechol. They found that at very low concentration of catechol, where the oxygen-consumption amounted to 2.5 atoms per mole, only one intermediate, a purple-red pigment of undetermined structure, could be detected. However, at concentrations of catechol higher than 10^{-3} M, when the oxygen uptake was less than 2.5 (approximately 2) atoms per mole, the presence of the following compounds was detected: 3,3',4,4'-tetrahydroxy-(15), 2,3,3',4'-tetrahydroxy-(16), and 2,2',3,3'-tetrahydroxy-(17)-biphenyl, as well as compound (18). Thus, both C—C and C—O—C linkages are formed, and the formation of the products could be explained in terms of radical-coupling reactions, i.e. from the semiquinone (19). However, THOMSON (150) pointed out that the same products could also be formed by nucleophilic attack of catechol on o-bezoquinone, as shown in Chart 2. Incidentally, spinach contains an enzyme which is specific to catechol, which it oxidizes to compound (18) (104). More recently WATERS (156) has suggested that biochemical oxidations of phenols which lead to

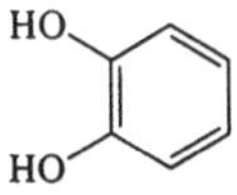

Catechol

(13)

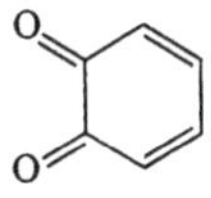

o-Benzoquinone

(14)

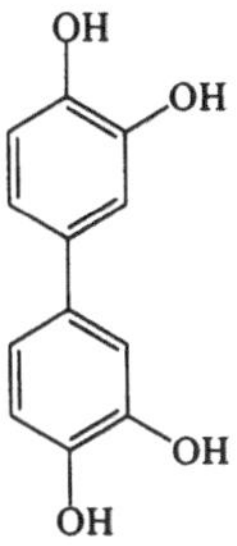

(15) 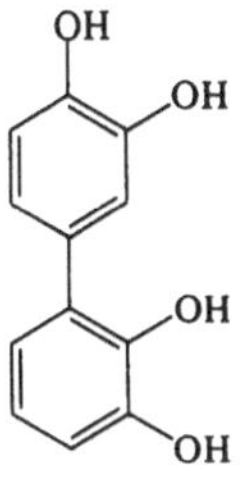   (16) 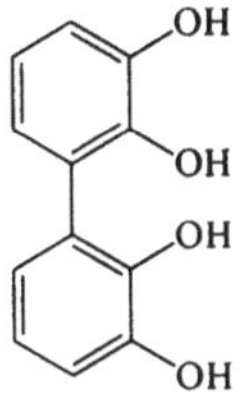   (17) 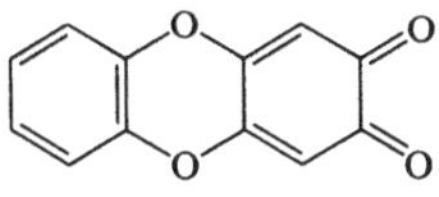   (18)

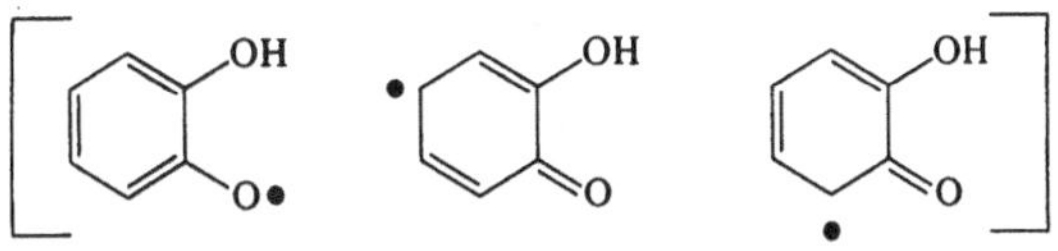

Semiquinone from catechol

(19)

C—O—C linkages may involve radicals, whereas those which result in C—C linkages may involve electrophilic attack on phenol molecules by aryloxy-cations.

MASON (93, 96) suggested that o-benzoquinone polymerizes directly to give phenolic polyphenyls susceptible to further oxidation so that catechol-melanin might be something like (20). However, DAWSON and TARPLEY (38) have produced considerable evidence that hydroxy-p-quinone (22) takes part in the polymerization process, despite MASON's (101) failure to detect spectroscopically the formation of this quinone. o-Benzoquinone is very unstable in the presence of water, particularly at high pH values, owing to nucleophilic attack leading to hydroxyquinol (21) (See Chart 3). Moreover, it was demonstrated spectroscopically that catechol is reformed when an aqueous solution of o-benzoquinone is allowed to stand, pre-sumably through the rapid oxidation of hydroxyquinol by o-benzoqui-none. DAWSON and TARPLEY suggested that the sequence of reactions shown in Chart 3 was relevant to the oxidation of very dilute solutions of catechol, where the enzyme could rapidly convert all of the original catechol into o-benzoquinone. However, in more concentrated solutions, it was suggested that a side-reaction, involving nucleophilic attack of catechol on o-benzoquinone could lead to the formation of the tetra-hydroxybiphenyls, reported by FORSYTH and QUESNEL. DAWSON and TARPLEY suggested that catechol-melanin might be a copolymer formed from hydroxyquinone and o-benzoquinone, an idea which would seem to be in keeping with analytical data, which correspond to the presence of approximately 2.5 oxygen atoms per 6 carbon atoms of the polymer. The

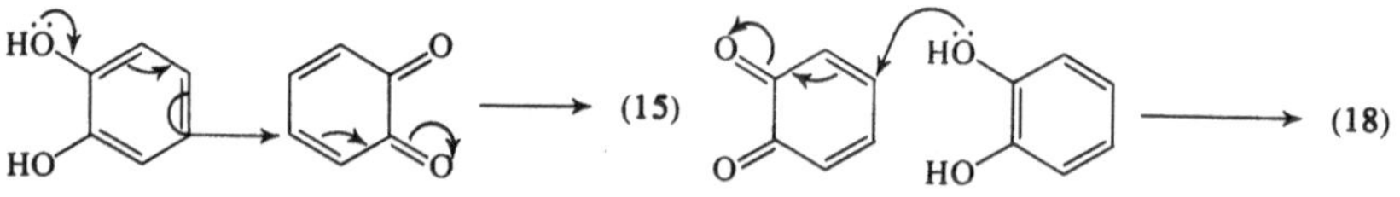

Chart 2. Nucleophilic attack of catechol on o-benzoquinone

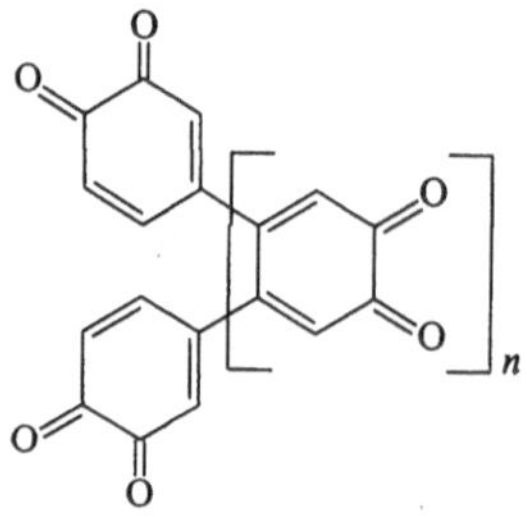

(20)

u. v. spectrum of a solution of catechol-melanin was found to be unchanged after addition of sodium borohydride.

However, PIATTELLI, FATTORUSSO, NICOLAUS, and MAGNO (*109, 120*) carried out degradative experiments which appear to suggest that the kind of coupling demonstrated by FORSYTH and QUESNEL does play some

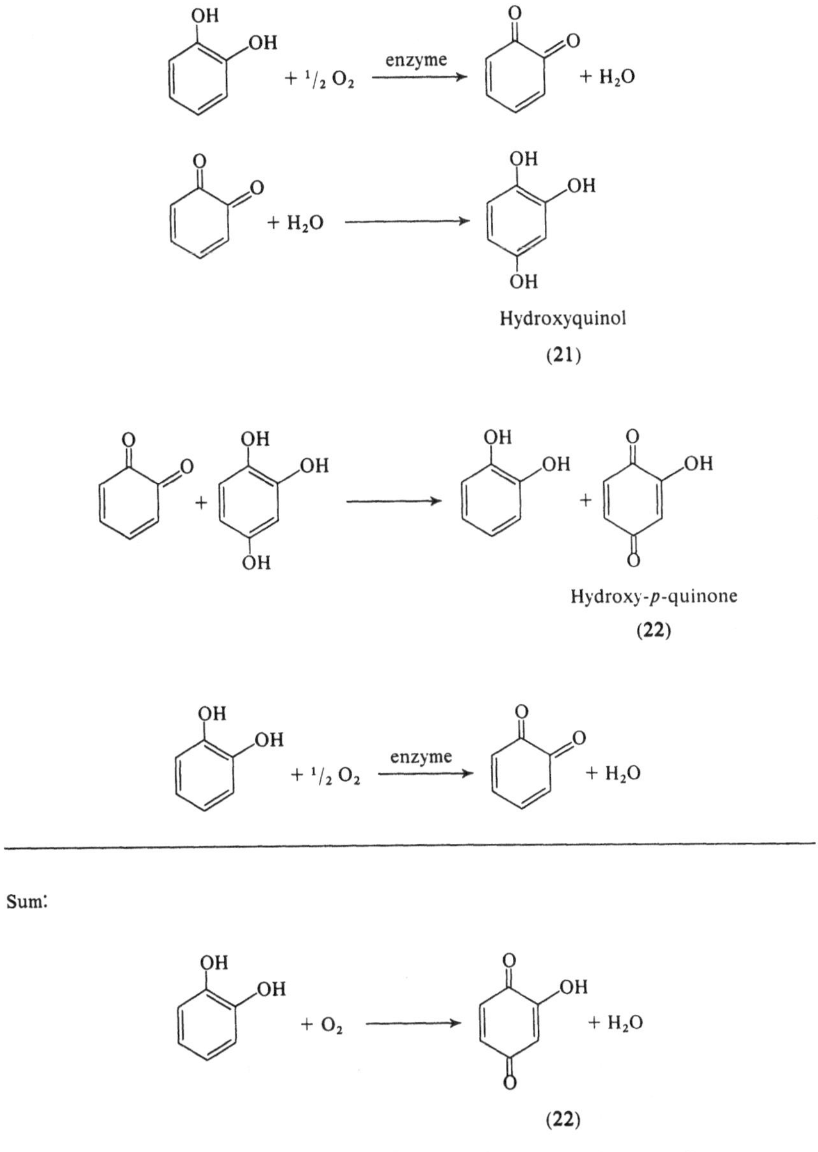
enzyme
Hydroxyquinol
(21)
Hydroxy-*p*-quinone
(22)
enzyme
Sum:
(22)

Chart 3. Enzymic oxidation of catechol (DAWSON and TARPLEY)

part in the polymerization process. Thus, when catechol-melanin was treated with hydrogen at 200° in the presence of a palladium catalyst it
yielded, in addition to catechol a small quantity of 3,3',4,4'-tetra-
hydroxybiphenyl (15). Moreover, fusion of catechol-melanin with alkali
afforded catechol, protocatechuic acid (23), and salicylic acid (24).
However, it is implicit in DAWSON and TARPLEY'S hypothesis that there
could be a series of different catechol-melanins, differing in the proportions of different units present in the copolymer, and depending upon the
conditions (e.g., concentration of the catechol and of enzyme, pH, etc.)
under which the melanin was formed.

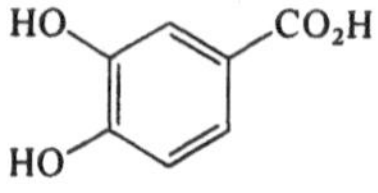

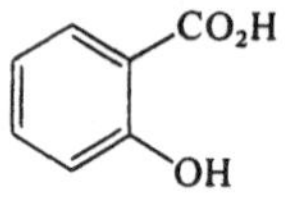

Protocatechuic acid   Salicylic acid

(23)   (24)

When catechol-melanin was treated with methanolic hydrogen chloride
or with diazomethane it afforded a product containing 3.2 or 23%,
respectively, of methoxy-group (120). Thus it probably contains a few
carboxy-groups and a considerable number of phenolic hydroxy-groups.
Carboxy-group might arise by oxidative fission of benzene rings of the
polymer by hydrogen peroxide. However, DAWSON and LUDWIG (37)
were unable to detect the formation of hydrogen peroxide in the enzymic
oxidation of catechol, although it is formed under conditions of autoxidation, with evolution of carbon dioxide.

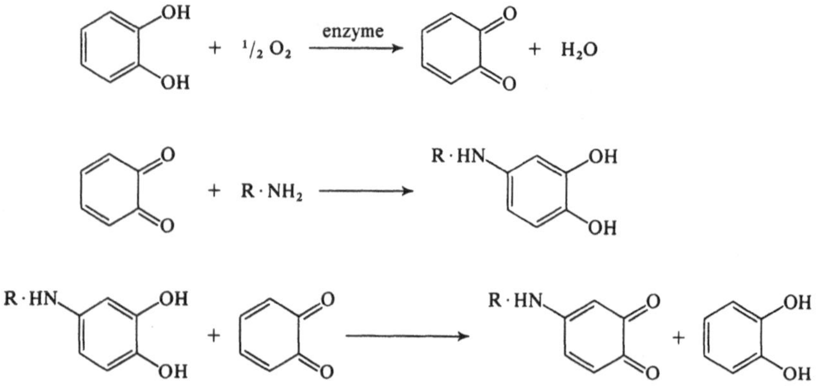

Chart 4. Enzymic oxidation of catechol in the presence of an amine

References, pp. 575—582

Catechol-melanin contains a little nitrogen (*ca.* 1.4%), believed to be derived from protein of the enzyme, perhaps through nucleophilic attack of an amino-group on the quinone system. The formation of catechol-melanoproteins has been studied by MASON and others (*96*). When an *o*-diphenol is oxidized enzymically in the presence of an amino- or sulphydryl-compound, this attacks the initially formed *o*-quinone to yield a product which, even though it may not act as a substrate for the enzyme, can nevertheless be oxidized by the *o*-quinone; see Chart 4. Thus, if catechol is oxidized enzymically in the presence of other protein, oxidation of the amino-catechol to the amino-quinone (and hence to melanoprotein) can occur only if the ratio of catechol to protein is high enough.

2. Natural Allomelanins

Whereas fusion with alkali of black pigments of animal origin usually gives rise to 5,6-dihydroxyindole, similar fusion of black polymers from the plant kingdom often yields nitrogen-free compounds, such as catechol, 1,8-dihydroxynapthalene (**27**), *etc.*

NICOLAUS and PIATTELLI (*109*) found that the black pigment, present in some seeds, could be isolated in a simple manner. The pigment is present in the outer cells; and when the seeds are simply stirred with water, the outer layer collapses, and the pigment can be isolated by centrifugation of the resulting suspension, the product being practically lignin-free. The pigments thus isolated from a number of species of plants were fused with alkali, and in all cases yielded catechol, so these pigments may be essentially catechol-melanins.

The spores of the corn parasite *Ustilago maydis* D. C. contain a black pigment, which has been studied by PIATTELLI, FATTORUSSO, NICOLAUS, and MAGNO (*109, 120*). The spores were isolated from the parasite mechanically, and the pigment was purified by treatment with acid, and extraction with solvents, after which it contained approximately 1% of nitrogen; and its composition (C, 62.7; H, 3.4; N, 1.0%) was fairly close to that of catechol-melanin prepared *in vitro* (C, 63.6; H, 3.5; N, 1.1%). Like the latter, when fused with alkali it yielded catechol, protocatechuic acid, and salicylic acid; and when treated with hydrogen in the presence of a palladium catalyst at 200° it afforded catechol and 3,3′,4,4′-tetrahydroxy-biphenyl (**15**). It was suggested that *Ustilago*-melanin is an irregular polymer made up of catechol units at different stage of oxidation, linked to each other by C—C and C—O—C linkages. However, there is a significant difference in methoxy-content (12.5 and 23.0% respectively) in the products obtained by treatment of *Ustilago*-melanin and catechol-

melanin with diazomethane, suggesting that the natural pigment is at a
higher state of oxidation than the synthetic melanin. The products obtained by treating these two melanins with methanolic hydrogen chloride
contained almost the same amount (3.3 and 3.2%, respectively) of methoxy-
group.

A black pigment, aspergillin, from the spores of the fungus *Aspergillus
niger* has been investigated by QUILICO and his collaborators (*107, 130*).
This was extracted from the fungus by dilute ammonium hydroxide, and
was precipitated when the extract was acidified. The material was then
extracted with ethanol in a Soxhlet apparatus, and the residue was further
purified by gel-filtration on Sephadex (*107*). The pigment contains carbon,
hydrogen, nitrogen, oxygen, and a little iron. It is acidic, and when
hydrolyzed yields sugars and amino-acids, although it is doubtful whether
the sugars are chemically bound to the pigment. However, about one
third of the nitrogen appears to be in a structure which is difficult to
hydrolyze. A nitrogen-free product has been obtained by LUND, ROBERT-
SON, and WHALLEY (*89*) by long boiling with 20% sulphuric acid; and
this gave the same oxidation and reduction products as did the aspergillin

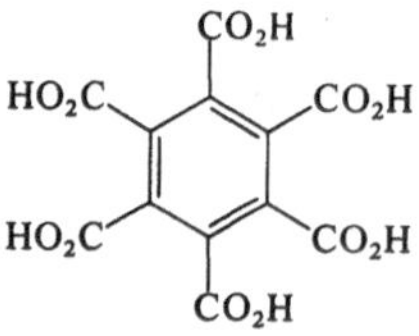

Mellitic acid
(25)

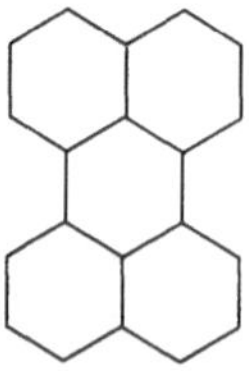

Perhydroperylene
(26)

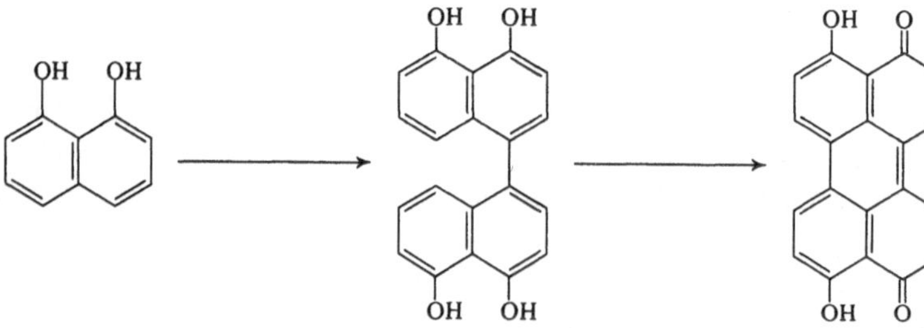

1,8-Dihydroxynaphthalene
(27)

(28)

obtained by the Italian workers. Oxidation of aspergillin with hydrogen peroxide yields mellitic acid (25). Reduction of aspergillin in alkaline solution in the presence of Raney nickel at 200° affords perhydroperylene (26). The pigment may therefore be a polymer derived from perylene units substituted with oxygen functions. Biosynthetically one could envisage the oxidative dimerization of 1,8-dihydroxynaphthalene (27), followed by further oxidation to give a product (28) with a perylenequinone structure (1).

A black pigment produced by the large Ascomycete *Daldinia concentrica* (Bolt.) Ces. and de Not., which is parasitic on ash trees, is possibly related chemically to aspergillin. When fused with alkali, this pigment gives catechol, and 1,8-dihydroxynaphthalene (1, 109).

The formation of dark brown or black pigments during the normal development of green plants is a well-known phenomenon, commonly observed as markings on leaves, petals, *etc.*, and in senescent leaves, seed pods, seedcoats, *etc.* (148). Also, many plant tissues darken on injury. Moreover, specific names such as *niger* or *nigricans* indicate that these pigments have not been overlooked by botanists. However, these dark plant pigments have been studied much less than the melanins of animal origin, perhaps because they seem to perform no obvious function, but probably also because in many cases it would be very difficult to isolate them in amounts sufficient for chemical investigation. Except for what is mentioned above, very little is therefore known about the chemistry of these, although it is likely that many of the pigments arise by enzymic oxidation of phenols to quinones, which are subsequently converted into complex products by polymerization and interaction with proteins. Both tyrosine and tyrosinase appear to be widely distributed throughout the plant kingdom, although the occurrence of melanins is more restricted. The occurrence of tyrosine and dopa together in the same plant [e. g. in broom, *Sarothamnus scoparius* (L.) Wimm. ex Koch or in *Vicia faba* L.] might be a hint that the dark colored products in these species could be eumelanins. The localization of pigmentation might perhaps be associated with the presence of inhibitors.

The formation of dark pigments from plant material accompanies some manufacturing processes, for example, the manufacture of black tea, and the Chinese or Japanese lacquers, or the extraction of liquorice; but little is known about the chemistry involved. Other dark pigments originate from glycosides or polyacetylenic compounds. For information regarding these the reader is referred to articles by THOMSON (149, 150).

The humic acids, which occur in soil and peat, are also polymeric and phenolic in nature and may be related to aspergillin. They present problems in structural determination similar to melanins. Current views on the structure of humic acids have been discussed by HAWORTH (67).

IV. Eumelanins

1. Chemical Investigations on Dopa-Melanin and 5,6-Dihydroxyindole-Melanin

1.1. Introduction

The present-day knowledge regarding the structures of the allomelanins has been briefly summarized in III. The eumelanins have been studied in much greater detail. In II.1 we had followed through the RAPER-MASON Scheme of Melanogenesis, and concluded that melanins might be polymers of indole-5,6-quinone (**9**).

A number of speculative suggestions regarding melanin structure have been made from time to time; but only structures for which some chemical evidence can be adduced will be discussed here. A structure based on 5,5',6,6'-tetrahydroxyindigo was disproved by HARLEY-MASON (*63*).

1.2. Studies on Model Compounds

It has been suggested that indole-5,6-quinone may function both as a quinone and an indole, the nucleophilic centre in the pyrrole ring of one molecule linking up with an electrophilic centre in the quinonoid ring of a second molecule, repetition of the process leading to a polymer (*30, 74*). BU'LOCK and HARLEY-MASON (*30*) showed in model experiments, that *o*-benzoquinone reacts with indole at room temperature. Reaction normally takes place in the 3-position of the indole nucleus to give (**29**), although a second condensation with another quinone molecule may occur at an unsubstituted 2-position. However, if the 3-position of the indole is blocked, the reaction takes a quite different course. The reactions are markedly subject to steric effects and proceed with difficulty, or not at all, if overlapping substituents are present, which prevent the formation of a coplanar molecule. Planar mesomeric structures make a considerable contribution to the light absorption of the indolylquinones, which are deeply colored.

(**29**)

Bu'LOCK and HARLEY-MASON proposed that melanin is formed by repeated oxidative condensations of indole-5,6-quinone (9), between position 3 of one molecule and position 4 or 7 of another. As the product obtained by 3—4 linkage could not be coplanar a 3—7 linkage was considered more likely, giving the coplanar structure (30). The extensive conjugation in this structure was supposed to account for the general light absorption of the polymer. To explain the extreme insolubility of melanin, it was further suggested that some cross-linking between position 2 with 4 or 7 of another molecule might occur, leading to a three-dimensional polymer. The repeating unit in structure (30) has the formula $C_8H_3NO_2$. To attain something approaching agreement between found and calculated analytical values, it was suggested that the cross-linked

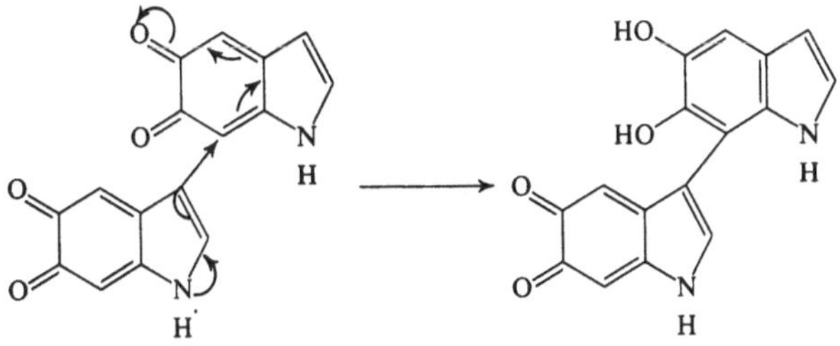

(9)   (11)

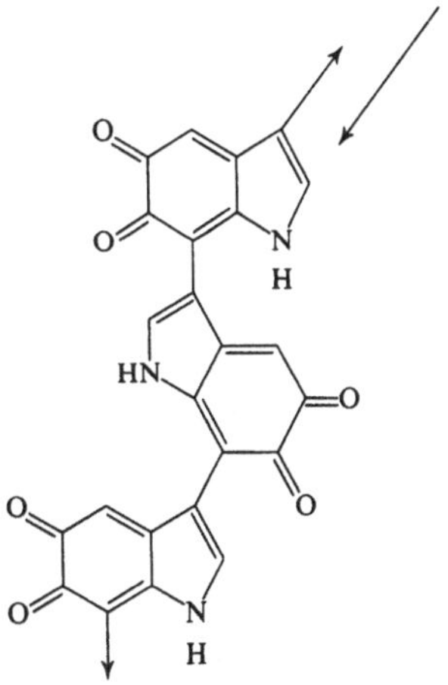

(30)

polymer retains water firmly, giving an empirical formula approximating to $(C_8H_3NO_2 \cdot H_2O)$. However, samples of melanins which have been dried are usually very hygroscopic and it is difficult to obtain reproducible elemental analyses on the same sample.

To test this hypothesis, CROMARTIE and HARLEY-MASON (36) studied the autoxidation at pH 6.85 of a series of 5,6-dihydroxyindole derivatives, in which different nuclear positions were blocked by methyl groups. It was shown by preliminary experiments that the effect of enzyme on the rate and on the visible phenomena of the reaction was negligible. 1-Methyl-, 2-methyl-, 4-methyl-, and 7-methyl-5,6-dihydroxyindole, like the unsubstituted 5,6-dihydroxyindole, gave black solids, insoluble in the common organic solvents, although slightly soluble in pyridine. On the other hand, 3-methyl-, and 4,7-dimethyl-5,6-dihydroxyindole gave highly colored solids, which were soluble in ethanol. These results were considered to demonstrate that for melanin-formation it was essential to have an unsubstituted 3-position, together with a free position at either C-4 or C-7; but it was recognized that positions 2, 3, 4, and 7 were probably all concerned to some extent. The precipitates formed from the 2-methyl-, 4-methyl-, and 7-methyl-compounds were more slowly deposited, less flocculent in appearance and more readily redispersed to a colloidal solution by washing with water than those from the unsubstituted 5,6-dihydroxyindole; this suggests that they were of lower molecular weight on account of inhibition of the normal cross-linking by the methyl groups. These results were in broad agreement with those of BEER, BROADHURST, and ROBERTSON (6), obtained by autoxidation of 5,6-dihydroxyindoles at pH 8, except in the case of the 2,3-dimethyl compound, where only the latter workers observed polymer formation.

However, the above turned out to be an oversimplification of the structure; and the structure of the eumelanins must now be discussed in greater detail.

1.3. Isotopic and Degradative Studies

What we (7) refer to as "enzymic dopa-melanin" is prepared by passing air through a dilute solution of dopa in phosphate buffer of pH 6.8 in the presence of enzyme (e. g. mushroom polyphenol-oxidase). During this aeration, the initially colorless solution rapidly becomes red, and finally black. Although the reaction is probably mainly complete after a few hours, as a routine in our experiments we allowed air to pass through the solution for 3 days at 20°. At the end of this period, the melanin was precipitated from the solution either by acidification, or by the addition of saturated sodium chloride solution. The melanin was then separated, and washed with water by centrifugation, and dried over phosphorus pentoxide in a vacuum for 3 days at room temperature. A melanin could also be similarly obtained at pH 8 in the absence of enzyme, and this is referred to as "autoxidative dopa-melanin". The autoxidation was much slower than the enzymic oxidation; but the reaction was allowed to proceed for the same length of time in both cases. In enzymic experiments it is important to use a sufficient amount of enzyme to ensure that autoxidation is not competitive; in many experiments described in the literature, the amount of enzyme is not recorded. From tyrosine, a melanin was obtained at pH 6.8 in the presence of tyrosinase. 5,6-Dihydroxyindole and dopamine yielded melanins under either enzymic or autoxidative conditions.

If the RAPER-MASON Scheme of Melanogenesis is correct, then one might expect to obtain an identical melanin from tyrosine, dopa, dopamine and 5,6-dihydroxyindole. However, although these melanins may look superficially alike, their properties are different (*134, 140, 141*). For example, whereas autoxidative dopamine-melanin and 5,6-dihydroxy-indole-melanin both separate from the solution at pH 8 in which they were formed, the former is readily redissolved by aqueous alkali, whereas the latter is insoluble in alkali. On the other hand, melanins from tyrosine or dopa remain in the solution in which they were formed until this is either acidified or treated with sodium chloride solution. Moreover, dopamine-melanin differs from the others in being basic; when treated with dilute hydrochloric acid it evidently forms a salt.

Evidence that melanogenesis from tyrosine and dopa is more complicated than as suggested by RAPER and MASON, and BU'LOCK and HARLEY-MASON came from experiments by CLEMO, DUXBURY, and SWAN (*34*), in which ($\pm$)-[carboxy-^{14}C]tyrosine and ($\pm$)-[carboxy-^{14}C] dopa were converted into melanins enzymically at pH 6.8 and by autoxidation at pH 8, respectively. The amount of carbon dioxide which arose from dopa was found to be at least equal to that which could theoretically arise from the total dopa, although the yield of melanin was only 60% of the weight of the dopa used. The specific activity of the evolved carbon dioxide showed that only about half of this carbon dioxide arose from the carboxy-group, the remainder presumably coming from one or more of the other eight carbon atoms of the amino-acid. Moreover, some radio-activity was retained in the melanin, indicating that approximately one in

(31)   (32)   (33)

every six polymer units contained a carboxy-group derived from the carboxy-group of the original amino-acid. Although these experiments were relatively crude, later more refined experiments (*134, 143*) confirmed their essential correctness.

It was also found that evolution of carbon dioxide occurred during the autoxidation of dopamine or 5,6-dihydroxyindole (*34*); and SWAN and WRIGHT (*144*) found that this evolution of carbon dioxide was diminished when catalase was present in the solution undergoing melanogenesis. Hydrogen peroxide is produced in solutions of *o*-diphenols, including 5,6-dihydroxyindole, which are undergoing autoxidation; in the case of melanogenesis, this attacks the benzene ring of some polymeric units, breaking open these rings, with evolution of carbon dioxide as shown in formulae (**31**), (**32**) and (**33**), (*6, 144*). Hydrogen peroxide has been said not to be formed in the enzymic oxidation of *o*-diphenols; and indeed DAWSON and TARPLEY (*38*) failed to detect its formation during the enzymic oxidation of catechol. If, however, it were produced, this would explain the formation of acid in solutions in which dopa is undergoing enzymic oxidation (*21*).

Moreover, the formation of hydrogen peroxide in the enzymic oxidation of 5,6-dihydroxyindole might be inferred from the work of FATTORUSSO, NICOLAUS, SUSSMANN, and KERTESZ (*107*), in which they followed the uptake of oxygen during the enzymic and autoxidative conversion of 5,6-dihydroxyindole into melanin at pH 6.8, each in the presence and also in the absence of catalase. The curves for the enzymic and autoxidative experiments in the absence of catalase were virtually identical. Those in the presence of catalase were also identical with one another, although lower than the first pair. The oxygen consumption per molecule of the indole was 2.4 and 1.4 atoms, in the absence and presence, respectively, of catalase. It was therefore suggested that 5,6-dihydroxyindole is oxidized with the formation of hydrogen peroxide, and that the resulting quinone or semiquinone polymerizes without

(34)

(35)

consumption of oxygen, although the polymer is subsequently oxidized to some extent with consumption of oxygen.

Hydrogen peroxide may also be able to attack amino-acid residues present in melanin, as the retention of the carboxy-group of [carboxy-^{14}C] dopa is slightly increased in the presence of catalase (143).

We (8, 80, 140, 141) later obtained other clear evidence that the BU'LOCK and HARLEY-MASON structure of melanin was oversimplified as follows. Melanins prepared either enzymically or by autoxidation from ($\pm$)-[β-^{2}H]dopa (34) or [β-^{2}H]dopamine (35) showed very considerable retentions of deuterium. If the RAPER-MASON-Scheme were followed to indole-5,6-quinone, which then yielded melanin as a 3—7-linked polymer according to the hypothesis of BU'LOCK and HARLEY-MASON, the melanin should not be enriched with respect to deuterium.

PIATTELLI, FATTORUSSO, MAGNO, and NICOLAUS (118) measured the amount of carbon dioxide evolved when melanins were heated at 190—200°; and found similar amounts (7.5, 9.3, and 9.6%, respectively) from enzymic dopa-melanin, autoxidative 5,6-dihydroxyindole-melanin, and enzymic 5,6-dihydroxyindole-melanin. However, the pigments prepared in the presence of tyrosinase and catalase together, when decarboxylated yielded only approximately half the amount of carbon dioxide, i.e. 5.2 and 4.7%, respectively, in the case of dopa-melanin and 5,6-dihydroxyindole-melanin. These results are in line with those of SWAN and WRIGHT (144), already mentioned, and are taken to imply the presence in the melanin of carboxy-groups formed by oxidative fission of benzene rings by hydrogen peroxide, e.g. (32).

All the melanins, when methylated with diazomethane, gave brown, amorphous, insoluble, infusible products, which contained similar percentages of methoxy-group (approximately 20%) (118). This means that the melanins are not fully quinonoid, as had been thought earlier, but are in fact at approximately half way between the quinonoid and diphenolic stages. On the contrary, HORAK and GILLETTE (72) apparently still believe melanins to exist predominantly in the oxidized state. Melanins prepared in the presence of catalase showed methoxy-values very little lower than for those prepared in its absence. This can still be explained if the oxidative fission occurs as shown in (31)→(32), remembering that the proportion of pyrrolecarboxylic acid units in the melanin is, in any case, small. Although some phenolic hydroxy-groups are lost as a result of the fission of benzene rings, some carboxy-groups are gained. The methylated melanins were said to give positive Labat test for methylenedioxy-group.

The Italian workers (118) also demonstrated by paper chromatography, that when any of these melanins were oxidized with potassium permanganate, the following were produced: pyrrole-2,3-dicarboxylic acid (36), pyrrole-2,3,5-tricarboxylic acid (37), and pyrrole-2,3,4,5-tetra-

(36)   (37)   (38)

(39)   (40)

carboxylic acid (38). BINNS and SWAN (9) also found pyrrole-2,3-dicarboxylic acid and pyrrole-2,3,5-tricarboxylic acid among the products of oxidation (by alkaline hydrogen peroxide) of melanins prepared from dopa, dopamine, tyrosine, and 5,6-dihydroxyindole. In these oxidation experiments, some polymeric material (known as "melanic acid") can be recovered; and further oxidation of this usually affords additional quantities of the same pyrrole acids. Moreover, fusion of 5,6-dihydroxy-indole-melanin with alkali yielded catechol, 5,6-dihydroxyindole, pyrrole-2-, and pyrrole-3-carboxylic acid, and pyrrole-2,4- (39), and pyrrole-2,5-dicarboxylic acid (40) (119).

The Italian workers (118) showed that when the decarboxylated mela-nin was oxidized, the same acids as before were obtained, but in addition pyrrole-2,4- (39) and pyrrole-2,5-dicarboxylic acid (40) were formed, the two latter being absent among the oxidation products of the undecarboxylated melanin. This supports the assumption that pyrrole-carboxylic acid units are present in the melanin.

(41)   (42)   (37)

In the cases of dopa-melanin and 5,6-dihydroxyindole-melanin, no difference in yield was observed of pyrrole-2,3-dicarboxylic acid and pyrrole-2,3,5-tricarboxylic acid produced by oxidation of the decarboxy-lated and undecarboxylated melanin. Moreover, when methylated dopa-melanin was oxidized, and the products were hydrolyzed, pyrrole-2,3,5-

tricarboxylic acid was not found. This was taken to show that there could not be carboxy-group in position-2 of the indole nucleus in the polymer. As we had shown the presence in the melanin of carboxy-groups derived from the amino-acid, this suggested the presence of uncyclized amino-acid units in the melanin. Moreover, the Italian workers (*118*) showed that trimethylamine is formed on oxidation of methylated dopamelanin, in confirmation of the presence of such units (**41**).

BINNS, CHAPMAN, ROBSON, SWAN, and WAGGOTT (*7*) stated that the formation of pyrrole-2,3,5-tricarboxylic acid by oxidation of melanin derived from precursors not themselves containing a carboxy-group suggested that this acid may arise by disruption of units such as (**42**). It could also be that part of the pyrrole-2,3,5-tricarboxylic acid arising from tyrosine-melanin or dopa-melanin is formed in this way. They found that when autoxidative dopa-melanin was boiled with dilute hydrochloric acid, it yielded a small amount of pyrrole-2,3,5-tricarboxylic acid. The radioactivity of melanin prepared from [$^{14}CO_2H$]dopa showed the melanin to contain approximately one carboxy-group of the original amino-acid in

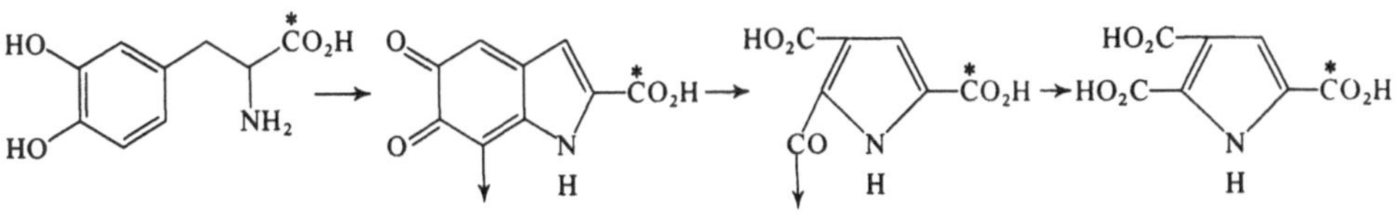

(**43**)

Chart 5. Oxidation of melanin derived from [$^{14}CO_2H$] dopa

five polymer units; and this radioactivity diminished when the melanin was boiled with dilute hydrochloric acid (*143*). If the pyrrole-2,3,5-tricarboxylic acid were derived from end-groups containing ^{14}C (**43**) in the polymer, as shown in Chart 5, their loss would result in lowering of the specific activity of the melanin.

SWAN and WAGGOTT (*143*) isolated, by preparative paper chromatography, pyrrole-2,3-dicarboxylic acid and pyrrole-2,3,5-tricarboxylic acid formed by oxidation of melanin prepared from [$^{14}CO_2H$]dopa. As expected, the dicarboxylic acid was not radioactive. The specific activity of the pyrrole-2,3,5-tricarboxylic acid was lower than would be expected if it was all derived as shown in Chart 5. It was therefore concluded that the radioactive acid, produced according to Chart 5, was being diluted in the ratio 1:1.3 with inactive acid formed by oxidation of indole units (**42**) linked through their 2-position to other units of the polymer. There was little difference in the dilution ratio observed between enzymic and autoxidative melanins which had been treated with hydrochloric acid under the same conditions.

(44) (45) (46) (47)

Melanin prepared from $[^{14}CO_2H]$dopa was decarboxylated by being heated at 200° in a vacuum, it being assumed that under these conditions decarboxylation occurred of all carboxy-groups in the melanin, except those in uncyclized amino-acid side-chains. The radioactivity of the decarboxylated melanin was therefore taken as an indication of the number of the latter type of units (41) in the melanin. The total amount of carbon dioxide evolved during decarboxylation was taken to indicate the number of carboxy-groups capable of undergoing decarboxylation, *i. e.* the number of units of types (44) and (45), together with pyrrolecarboxylic acid units such as (46), formed by attack of hydrogen peroxide on indole-5,6-quinone units. Radioactive carbon dioxide arising from units (44) and (45) would be diluted with inactive carbon dioxide from the pyrrolecarboxylic acid units, so that the specific activity of the evolved gas should provide information as to the relative proportions of the two different groups of units.

The amount of pyrrole-2,3,4,5-tetracarboxylic acid (38) formed on oxidation of dopa-melanin seems to be extremely small, so the proportion of units of type (47) is probably very small.

Both autoxidative and enzymic melanins prepared in the presence of catalase contain a considerably lower fraction of carboxylated pyrrole units than melanins similarly prepared in the absence of catalase. No significant difference was seen between autoxidative and enzymic melanins.

When dopa-melanin was treated with excess of diazomethane, it was assumed that all carboxy-groups and phenolic hydroxy-groups were methylated, and that Zeisel determinations gave values corresponding to the combined methoxy-content. However, from analytical results, it appeared that a substantial amount of CH_2 derived from diazomethane was introduced into the melanin in a form which did not yield methyl iodide under Zeisel conditions. Although it is possible that this could be in the form of methylenedioxy-group, as suggested by the work of PIATTELLI, FATTORUSSO, MAGNO, and NICOLAUS (*118*), it seems rather likely that at least part of it is in the form of *N*-methyl groups. In fact, when pyrrole-2,3,5-tricarboxylic acid is treated with excess of diazomethane, *N*-methylation occurs, besides esterification (*142*). It was assumed that all pyrrole and indole (but *not* indoline) NH-groups reacted with diazomethane.

SWAN and WAGGOTT (*143*) concluded that the various radioactivity and analytical (C,H,N,O,OMe) results on melanins, decarboxylated melanins, methylated melanins, *etc.*, could best be accommodated on the assumption that autoxidative dopa-melanin consists in the main of four types of units: 10% of uncyclized units, which are diphenolic (**48**), 10% of indolinecarboxylic acid type (**49**), 65% of indole type (**50**), and 15% of pyrrole type (**51**); and moreover, that of the units of types (**49**) and (**50**) taken together, one half are quinonoid, and the other half are diphenolic; 0.5 H_2O was included per average polymer unit (Chart 6).

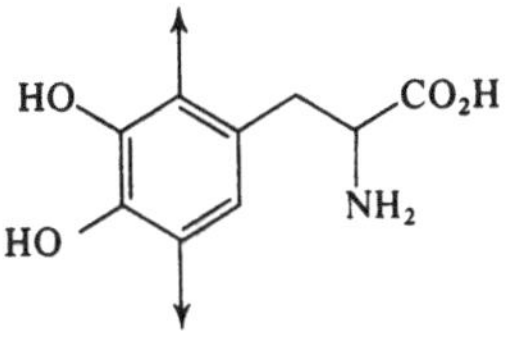

0.1

(**48**)

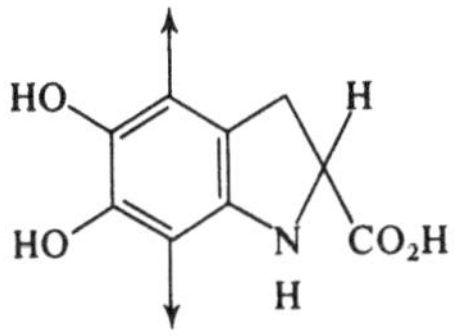

0.1

(**49**)

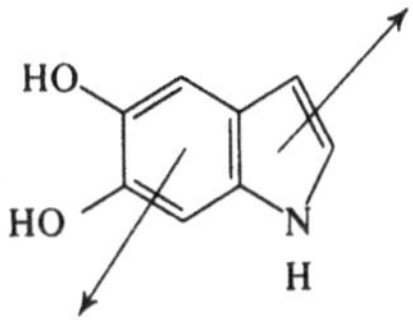

0.65

(**50**)

half quinonoid, half dihydroxy

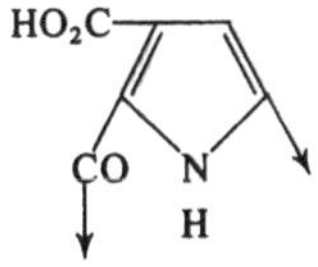

0.15

(**51**)

Chart 6. Autoxidative dopa-melanin

The proportion of units (**48**) received some confirmation from the specific activity of a sample of methylated melanin which had been treated with [^{14}COCl]benzoyl chloride, it being assumed that only these units were (*N*)-benzoylated. A check on the carboxy-content of the melanin was obtained from Zeisel methoxy-determination on melanin which had been treated with methanolic hydrogen chloride. The difference between the methoxy-values on melanin treated with diazomethane and melanin treated with methanolic hydrogen chloride was used to determine the proportion of phenolic hydroxy-group.

MASON (*97*) concluded from titrations that enzymic dopa-melanin contains six phenolic hydroxy-groups and fewer than one carboxy-group per ten units. These values are lower than those of SWAN and WAGGOTT (*143*) on autoxidative dopa-melanin. Under MASON's conditions of melanogenesis, up to 4.6 atoms of oxygen were consumed, and up to 1 molecule of carbon dioxide was evolved per molecule of substrate. MASON concluded: "The formation of melanin under these conditions precludes the participation of dopa or 5,6-dihydroxyindole-2-carboxylic acid in the polymerization to any important extent." However, this conclusion is difficult to accept, in view of the fact that the conversion of dopa into melanin was not quantitative.

To determine the relative numbers of linkages at different positions of the polymer units, and to check the proportion of indoline-type units in the melanin, KING, PERCIVAL, ROBSON and SWAN (*80*) converted samples of ($\pm$)-dopa, specifically deuterated at each of the α- and β-positions of the side-chain, and the 2-, 5-, and 6-positions of the benzene ring (separately) into melanins, and compared the deuterium enrichment of the precursor and melanin. As a kinetic isotope effect could obscure the interpretation of the results, two series of experiments were carried out; in one series the precursor contained only a tracer concentration of deuterium, while in the other the relevant position of the molecule was deuterated to as nearly 100% as possible. Further, each series was carried out both by autoxidation and enzymically (using an excess of enzyme). The enzymic melanins contained protein, the removal of which required long boiling with 2N-hydrochloric acid.

Melanins prepared from α- and β-deuterated ($\pm$)-dopa were boiled with 2N-hydrochloric acid for various periods, after which their deuterium contents were measured. The results suggested that two distinct exchange processes occurred, which could be explained on the basis of the above structure. It was assumed that at the end of the first (rapid) exchange process, deuterium remained only in units of type (**48**) and (**49**), and that the deuterium content thus provided an estimate of the combined amounts of these units present in the melanin. Deuterium retentions in enzymic melanin after long boiling with acid were not greatly different from the corresponding values on autoxidative melanins, which had been boiled with acid for similar periods. There was nothing in the results to suggest a fundamental difference in structure between autoxidative and enzymic dopa-melanins. There was little difference as between the tracer and 100%deuterium series for the ratio of deuterium enrichments in melanin and precursor, which suggests that an isotope effect cannot have any great influence. Assuming that units (**48**), (**49**), and (**51**) are linked as shown, it was calculated that the following fractions of the main units (**50**) are linked at the positions indicated:

Position	2	3	4	7
Fraction	0.36	0.37	0.34	0.28

Thus the linkages at these positions appear to be fairly equally shared. The results also imply that the majority of units are linked to two other units.

KIRBY and OGUNKOYA (*81*) also investigated the structure of dopa-melanin in essentially the same way, although the details of their work are different. They used tritium instead of deuterium, so a considerable isotope effect might occur. They prepared their melanins by oxidation of a mixture of the specifically tritiated ($\pm$)-dopa with ($\pm$)-[α-^{14}C]dopa, so that through measurement of ^{3}H and ^{14}C activity separately, the fraction of tritium incorporated into the melanin could be measured. This method has the advantage that the presence of protein in the melanin would not affect the results, the degree of hydration of the melanin would be of no consequence, and it is not necessary to know the composition of the melanin. KIRBY and OGUNKOYA's experiments were carried out in the presence of tyrosinase and catalase in unspecified concentration, and in the presence of oxygen (rather than air, used by KING, PERCIVAL, ROBSON, and SWAN). It is therefore impossible to relate one set of results exactly to the other. In the Newcastle work, few experiments were carried out in the presence of catalase, but in these the catalase appeared to have no great effect on the retention of deuterium in the melanin. Table 1 shows a comparison of KIRBY and OGUNKOYA's results with KING, PERCIVAL, ROBSON, and SWAN's results on autoxidative melanin, expressed as fractional retention of isotope.

Table 1. *Comparison of the Results obtained by* KIRBY *and* OGUNKOYA, *with those of* KING, PERCIVAL, ROBSON, *and* SWAN

Position of label in dopa	α	β	2	5	6
Fractional retention of ^{3}H (*K.* and *O.*).	0.52	0.54	0.37	0.42	0.13
Fractional retention of ^{2}H (*K., P., R.,* and *S.*).	0.49	0.41	0.31	0.37	0.1

The validity of the results of KING, PERCIVAL, ROBSON, and SWAN is dependent upon the absence of variation in degree of hydration between many different melanin samples, and the quantitive results of SWAN and WAGGOTT are to some extent dependent on this degree of hydration (although the percentage composition of the sample was not used directly in the calculations). The agreement shown in Table 1 is therefore very satisfactory.

1.4. Conclusion Regarding the Structure of Melanin

It seems impossible to escape the conclusion that autoxidative dopa-melanin, prepared in the way we have described, is an irregular polymer, containing a number of different types of units, linked fairly randomly in various ways (*141a*). It is probable that units other than those already represented are also present, to a lesser extent. It also seems likely that enzymic dopa-melanin, prepared *in vitro* in the way we have described is not greatly different structurally from the autoxidative melanin. However, MASON (*97*) has continued to defend the homopolymer theory.

As described earlier, enzymic oxidation of catechol gives rise to catechol-melanin, in which both C—C and C—O—C linkages may be involved. One might therefore guess that any such type of polymerization which could occur in the oxidation of catechol might also be able to occur to some extent in the oxidation of dopa or dopamine, although in these latter cases there would be competition with the processes involved in the RAPER-MASON Scheme. So now one could think of a melanin as being a copolymer containing indole-5,6-quinone units, dopachrome units, uncyclized dopa units, and pyrrolecarboxylic acid units, the latter being formed by oxidative fission of indole-5,6-quinone units already part of the polymer. The proportions in which these different units are present in the polymer will doubtless depend not only upon which precursor (tyrosine, dopa, dopamine, or 5,6-dihydroxyindole) was used, but also upon the chemical or biological conditions (pH, concentration, *etc.*) under which polymerization occurred. And moreover, one could then envisage the possibility of other "foreign" molecules being incorporated into melanins, if they had suitable chemical reactivity.

This leads to the concept that melanin might be a unique biopolymer, in which a range of variation is possible, and to which one cannot give an exact chemical structure, but only determine statistically the proportions of different units (*11, 13*). Moreover, if radical reactions were involved in the polymerization process, such radicals might be trapped in the polymer and in fact melanins have been shown to have radical properties, although the fraction of the total number of units which exist as radicals must be small. For this reason, discussion of the radical structure will be deferred until later (IV.3.3).

(52)

Although tyrosinase has been found to have little effect on the rate of oxidation of 5,5',6,6'-tetrahydroxybiphenyl-3,3'-ylenedialanine (**52**) (*114*), a melanin has been prepared by autoxidation of the latter; and oxidation of the melanin afforded pyrrole-2,3-dicarboxylic acid and pyrrole-2,3,5-tricarboxylic acid (*31*).

1.5. Studies Relevant to the Structure of the Melanoproteins

As melanins usually occur naturally conjugated with protein, there is a possibility that eumelanins may arise *in vivo* through the oxidation, catalyzed by tyrosinase, of tyrosyl residues in peptides (*97*). Bu'Lock and Harley-Mason (*30*) showed that oxidation of tyrosylglycine (**53**) in the presence of mushroom polyphenol-oxidase gave a red solution, the absorption spectrum of which was almost identical with that formed by a similar oxidation of tyrosine. This solution presumably therefore contained (**54**); but prolonged oxidation resulted in the formation of a brown solution, although no melanin was precipitated. When the red solution was kept under nitrogen, it became colorless, possibly owing to the formation of (**55**). The latter could presumably be oxidized to the corresponding quinone, which might then be able to react at the 7-position with 5,6-dihydroxyindole, eventually leading to a polymer having an end-group such as (**55**) in which the glycine residue was replaced by a protein chain. Since mammalian tyrosinase cannot catalyze the oxidation of *N*-acetyl- or *N*-formyltyrosine, it is unlikely that tyrosine which is linked to another amino-acid through its amino-group could be oxidized by this enzyme (*86*). The problem has also been studied by Yasunobu, Peterson, and Mason (*160*).

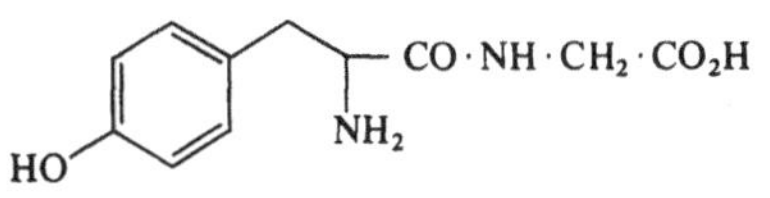

Tyrosylglycine

(53)

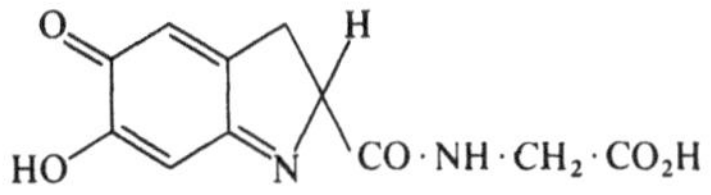

(54)

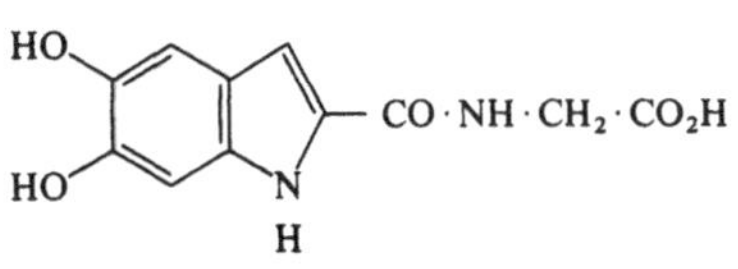

(55)

Mason and Peterson (*100*) investigated the spectral changes during the enzymic oxidation of dopa, in the presence of various amino-acids and amines. In most cases, no significant change of spectrum was brought about by the presence of the amino-compound. However, the presence of certain sulphydryl compounds (*e. g.* cysteine or glutathione) altered the spectroscopic course of the reaction, which it was suggested was as a result of the sulphydryl group attacking the indolequinone. Similar results have been reported by BOUCHILLOUX and KODJA (*20*). Thus if melanogenesis occurs in the presence of protein, a sulphur-linked melanoprotein may result.

A reaction between proteins and quinones derived from dopa may also be involved in the hardening process of the outer skin of insects and arthropods (*127*).

2. Chemical and Biochemical Investigations on Natural Eumelanins

2.1. Introduction

Turning now to the natural eumelanins, we shall see that there is much evidence to suggest that, like synthetic dopa-melanin and 5,6-dihydroxy-indole-melanin, these also contain carboxy-groups and phenolic-groups. Thus CHEN and CHAVIN (*32*) studied melanogenesis from [$^{14}CO_2H$] tyrosine by skin enzyme preparations, and found that incorporation of the carboxy-group occurred in 33 different vertebrates investigated. D-Tyrosine was also used by 9 vertebrates. It seems possible that although the enzyme might be specific for L-tyrosine so far as hydroxylation is concerned, nevertheless, on the basis of the present concept of melano-genesis, D-tyrosine could become incorporated into the melanin non-enzymically at a later stage of the process. Also, eumelanins readily bind metals (*27*) *e.g.* calcium and magnesium in cuttlefish, iron in black hair, and zinc and barium in choroid; and in fact, melanin granules act as cation-exchange resins, presumably owing to the presence of either carboxy- or *o*-diphenolic-groups. The melanin from squid (*Loligo opalescens*) has a higher capacity for cation-exchange than does melanin from the transplantable Harding-Passey tumor of mice (*157*).

2.2. Sepiomelanin

Sepiomelanin occurs in the ink sac of the cuttlefish (*Sepia officinalis*), which puts forth its ink as camouflage when frightened. By cutting open the sac and squeezing it gently, 1—2 g of raw ink can be obtained from a medium sized cuttlefish (300—400 g) (*106, 121*); this is the pigment used by artists. The melanin in the ink is in the form of small granules (0.2—0.3 μ diameter) suspended in a clear fluid. The ink gland contains an enzyme which is effective in the oxidation of tyrosine to melanin. Reduction of the melanin with sodium dithionite or ascorbic acid causes. the color to change to light tan, but the black color can be restored by oxidation with potassium ferricyanide (*86*).

NICOLAUS and his collaborators (107) carried out tracer experiments, injecting cuttlefish with radioactive precursors. From these, they concluded that melanin formation in cuttlefish is slow, the relatively large amount of pigment having accumulated over a period of time. The amount of radioactivity incorporated after injection of L-[$^{14}CO_2$H]tyrosine was approximately one tenth of that incorporated when L-[^{14}C-U]tyrosine was injected. Thus, during melanogenesis, a considerable part (but not all) of the tyrosine carboxy-group is eliminated. Some other compounds (e.g. phenylalanine and tryptophan) also showed some incorporation. This implies that besides the main units derived from tyrosine, other units (especially aromatic amino-acids), if present in the cell during melano-genesis, can be incorporated into sepiomelanin.

The pigment occurs as a protein-melanin complex (sepiomelanopro-tein) which is homogeneous on electrophoresis. In the hydrolyzate of this, 18 different amino-acids were identified; and from the amount of amino-acids recovered it was estimated that protein constituted 10% of the weight of the sepiomelanoprotein.

To obtain the protein-free pigment, PIATTELLI and NICOLAUS (121) used two different methods of purification, of which the first is the more drastic, and is accompanied by the loss of carbon dioxide and ammonia: (a) by washing the sepiomelanoprotein with 1% hydrochloric acid, followed by extraction with hot acetone, and finally prolonged boiling with concentrated hydrochloric acid, and (b) by dialysis of the contents of the ink sac against distilled water, followed by washing with 1% hydro-chloric acid, and finally digestion for 370 hours with concentrated hydro-chloric acid. The second method does not effect the complete hydrolysis of protein. It is stated that method (a) does not affect the fundamental structure of the pigment, although it is not clear as to how this is known. There is some difference in the percentage composition of the two preparations.

Like dopa-melanin, sepiomelanin contains less carbon and nitrogen, but more hydrogen and oxygen than does the theoretical indole-5,6-quinone polymer. To some extent this discrepancy can be explained by the presence of carboxy-groups in the melanin; these were determined by titration, by thermal decarboxylation, and by methoxy-determination on the methyl ester formed by treating the melanin with methanolic hydrogen chloride (121). Treatment of the melanin with diazomethane produced an insoluble pigment which gave a positive Labat test for methylenedioxy-group. The difference between the methoxy-values for specimens treated with methanolic hydrogen chloride (5.7%) and with diazomethane (18.8%) can be attributed to the presence of phenolic hydroxy-group. As in dopa-melanin, the number of hydroxy-groups was surprisingly high;

(56)   (57)

and free carboxy-groups are said to make up 9% by weight of the melanin (119).

Alkaline fusion of sepiomelanin at 300° yielded 5,6-dihydroxyindole (8), 5,6-dihydroxyindole-2-carboxylic acid (7), 4-methylcatechol (56) and 5,6-dihydroxyindole-4,7-dicarboxylic acid (57), identified by paper chromatography (119, 121). These products indicate the presence of indole units in sepiomelanin; and it is interesting to note that synthetic 5,6-dihydroxyindole-melanin did not yield 5,6-dihydroxyindole-2-carboxylic acid when fused with alkali. The formation of 5,6-dihydroxyindole-4,7-dicarboxylic acid indicates that some indole units are linked through positions 4 and 7. The formation of 4-methylcatechol from sepiomelanin, but not from 5,6-dihydroxyindole-melanin is ascribed to the presence of dopachrome units in the natural pigment (109). Reduction of sepiomelanin in ethanol at 150° with hydrogen in the presence of a palladium catalyst yielded 5,6-dihydroxyindole and unidentified products, one of which may be a dimer derived from the latter (109). When the pigment was boiled with 4% aqueous sodium hydroxide solution, pyrrole-2,3,5-tricarboxylic acid was liberated, which was considered to be formed by hydrolysis of terminal, carboxylated pyrrole units linked through a carbonyl group (58) (109).

(58)   (59)   (60)

Oxidation of sepiomelanin with potassium permanganate gave the same pyrrolecarboxylic acids as those similarly obtained from dopamelanin (107, 116), along with pyrrole-2,3,4-tricarboxylic acid (59) (116); and again, oxidation of sepiomelanin which had been decarboxylated yielded, in addition, pyrrole-2,4-dicarboxylic acid (39) and pyrrole-2,5-dicarboxylic acid (40) (118).

The yields (*107*) of acids isolated from 10 g of sepiomelanin were as follows:

Pyrrole-2,3-dicarboxylic acid (**36**) 0.0001 mg (estimated).
Pyrrole-2,3,4-tricarboxylic acid (**59**) 3.2 mg.
Pyrrole-2,3,5-tricarboxylic acid (**37**) 200 mg.
Pyrrole-2,3,4,5-tetracarboxylic acid (**38**) 61 mg.

PIATTELLI, FATTORUSSO, MAGNO, and NICOLAUS (*118*) have compared the physical and chemical properties of sepiomelanin with those of dopa-melanin, 5,6-dihydroxyindole-melanin, and pyrrole black (formed by oxidation of pyrrole). Although the three melanins resemble one another to a considerable extent, they are quite different from pyrrole black. Similar amounts of carbon dioxide are evolved when sepiomelanin, enzymic dopa-melanin, and enzymic 5,6-dihydroxyindole-melanin are heated.

The presence of carboxy-groups in position 2 of the indole nucleus of some units in sepiomelanin is shown by the fact that when decarboxylated sepiomelanin is oxidized, the yield of pyrrole-2,3-dicarboxylic acid is higher, and that of pyrrole-2,3,5-tricarboxylic acid is lower than the yields obtained from the undecarboxylated pigment. Methylated sepiomelanin, when oxidized, yielded 5-carbomethoxypyrrole-2,3-dicarboxylic acid (**60**), which was not found among the products similarly obtained from dopa-melanin, thus confirming the above conclusion (*118, 119*). On the other hand, methylated sepiomelanin does not give trimethylamine on oxi-dation, pointing to the absence of uncyclized amino-acid side-chains in sepiomelanin (*107, 118*).

$CH_2 \cdot SO_3H$	$CH_2 \cdot SO_3H$	$CH_2 \cdot CO_2H$	$CH_2 \cdot CO_2H$
$CH(NH_2)$	$CH_2 \cdot NH_2$	$CH(NH_2)$	NH_2
CO_2H		CO_2H	
Cysteic acid	Taurine	Aspartic acid	Glycine
(**61**)	(**62**)	(**63**)	(**64**)

NICOLAUS (*107*) concluded that sepiomelanin is a macromolecule, or mixture of macromolecules, in which the most important part is played by polymers of type:

$$\ldots\ldots[-(C_8H_5NO_2)_x - C_8H_4NO_2 \cdot CO_2H -]y \ldots\ldots$$

where x varies between 1 and 4, according to the method of purification used and, y is very high. The type of linkages and the sequence of the units were undetermined, but the main units were thought to be of the indole type, including 5,6-dihydroxyindole, dopachrome, leucodopachrome, and

5,6-dihydroxyindole-2-carboxylic acid units, some being in the quinonoid state. It was suggested that about one carboxylated-pyrrole end group [such as (58)] per 100 units is present, derived by an oxidative breakdown of the benzene ring of an indole unit *in vivo*. Results of elemental analyses are in reasonable agreement with the proposed structure. However, sepiomelanin contains 0.2—0.3% of sulphur, presumably derived from the amino-acid responsible for binding the pigment to the protein; and this is probably cysteine, bound by reaction of the −SH group. This was shown by oxidation with hydrogen peroxide in acetic acid of sepiomelanin which had been exhaustively treated with 6N-hydrochloric acid until no more amino-acids were released hydrolytically, when cysteic acid (61), taurine (62), aspartic acid (63), and glycine (64) were obtained. The two latter were also obtained by similar oxidation of 5,6-dihydroxyindole-2-carboxylic acid (7), which provides some confirmation of the presence of dopachrome units in the melanin (*119*).

2.3. Melanin from Melanoma

Melanin from rat melanoma has been studied similarly by NICOLAUS, PIATTELLI, and FATTORUSSO (*110*). As for sepiomelanin, purification involved treatment with 6N-hydrochloric acid. Although it was concluded that this material like sepiomelanin, may be derived from dopa, there were some differences. Thus, oxidation of the melanoma melanin yielded not only pyrrole-2,3-dicarboxylic acid, pyrrole-2,3,5-tricarboxylic acid, and pyrrole-2,3,4,5-tetracarboxylic acid, but also pyrrole-2,4- and pyrrole-2,5-dicarboxylic acid. In the case of sepiomelanin, the two latter acids were obtained only by oxidation of the decarboxylated pigment. Fusion with alkali of melanoma melanin afforded 5,6-dihydroxyindole and 5,6-dihydroxyindole-2-carboxylic acid. Oxidation of the melanin which had been treated with diazomethane yielded 5-carbomethoxy-pyrrole-2,3-dicarboxylic acid (60) and trimethylamine, the latter indicating the presence in the melanin of uncyclized amino-acid side-chains.

HEMPEL (*69, 70*) attempted to investigate the structure of melanoma melanin without isolating it. He injected mice bearing Harding-Passey melanoma simultaneously with $(\pm)$-[α-^{14}C]dopa and specifically tritiated $(\pm)$-dopa. In autoradiographs of the melanomas 2—10 days later, it was found that the silver grains lay over the melanin granules, and it was therefore assumed that all the radioactivity was contained in the melanin. The melanomas were homogenized, acid-soluble and lipid fractions were removed, and the ^{3}H and ^{14}C activities of the residual material were measured by scintillation counting. On the reasonable assumption that the α-carbon atom of dopa is incorporated quantitatively into the melanin, the ratio of ^{3}H to ^{14}C activity was taken as a measure of the incorporation of the particular hydrogen atom labelled by ^{3}H in the precursor. He found that at least 49% of the polymer units of the melanin must retain the hydrogen atom of position 6 of dopa (*i. e.* these units were uncyclized), and that 71% of the units retain the carboxy-group of the original amino-acid; and he concluded that although all positions of dopa were involved to some extent in the polymerization process, the melanin is a copolymer of dopaquinone (**4**), indole-5,6-quinone (**9**), and indole-5,6-quinone-2-carboxylic acid (**65**) in the approximate ratio 3:2:1.

(**65**)

HEMPEL'S method is elegant, but its value is based entirely on the assumption that all the measured radioactivity originates from the melanin. The validity of this assumption is open to question (*108*), and indeed there are indications (*48, 146*) that dopa can be released by hydrolysis of melanoma homogenates such as those used by HEMPEL, *i. e.* there is present dopa which is not part of the melanin polymer proper, but is at most linked to it as a peptide. In any case, the highly reactive 6-position of dopaquinone is unlikely to remain unsubstituted to such a high extent; and only about one tenth of the nitrogen of melanoma melanin has been found by BLOCH and SCHAAF to be in the amino-form (*107*). HEMPEL wrote all the units of his structure in the quinonoid form. However, his method provides no information on this aspect, and it is likely that the polymer is more highly reduced.

Work by BLOIS and his collaborators (*13, 17*) is relevant. Chart 7 shows the biochemical pathway leading from phenylalanine and tyrosine on to melanin, and shows the branch leading off to epinephrine (**66**) synthesis. Selected melanin precursors along this pathway were labelled with ^{14}C and administered intraperitoneally to a series of mice. At a fixed time after administration the animals were killed. Samples of different tissues were homogenized, and known amounts of these tissues were counted, so that the distribution of the administered radioactivity could be determined for each of the different tissues. Phenylalanine showed approximately equal incorporation into tumor, adrenal, spleen, liver, and kidney; and the same could be said of tyrosine. However [^{14}C]dopa showed a quite selective effect, the radioactivity being concentrated primarily in the tumor tissue and the adrenals, as would be expected from the above metabolic pathway. This suggests the possibility of radiotherapeutic use of labelled dopa for melanoma.

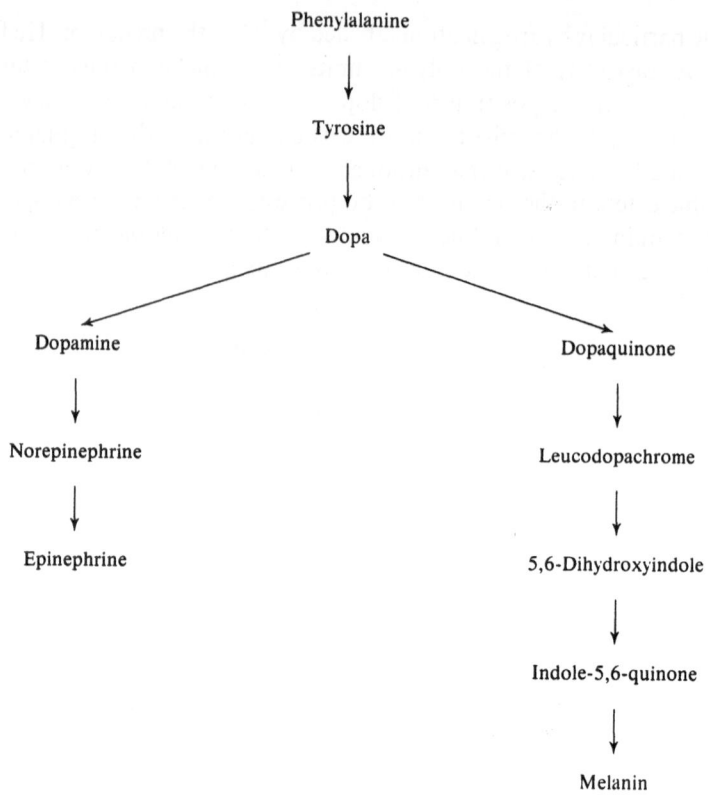

Phenylalanine

↓

Tyrosine

↓

Dopa

Dopamine   Dopaquinone

↓   ↓

Norepinephrine   Leucodopachrome

↓   ↓

Epinephrine   5,6-Dihydroxyindole

↓

Indole-5,6-quinone

↓

Melanin

Chart 7. The pathway of phenylalanine metabolism, leading to melanin and epinephrine

These results suggest that under these experimental conditions, dopa tends to be metabolized in the forward direction (*i. e.* onward to melanin or epinephrine, rather than backward to tyrosine). Using labelled dopamine, the main incorporation was into the adrenals, showing that dopamine too is metabolized primarily to epinephrine, rather than undergoing the reverse reaction to dopa and down the pathway to melanin.

The radioactivity of the melanoma tissues was shown to be largely in the melanin by extracting the pigment from the homogenate, washing it with acid, and showing it to contain most of the tumor activity.

The behaviour of dopa does not distinguish between enzymic and non-enzymic processes. However, administration of catechol or quinol (non-natural melanin precursors) showed some (although not specific) incorporation into melanins. This would not be expected if the synthesis were totally enzymic, and is in keeping with the conception of melanin put forward in IV. 1.4.

Evidence that in the melanoprotein of mouse melanoma the protein is conjugated to the pigment though sulphur bonds (*60*) is based on the observation that the melanin obtained by pancreatic digestion of the melanoprotein contains a higher proportion of sulphur than does the melanoprotein.

2.4. Melanin from Hair

It seems that hair bulbs can produce both eumelanins and phaeomelanins. NICOLAUS, PIATTELLI, and FATTORUSSO (110) found that black hair melanin, purified by boiling with 6 N-hydrochloric acid, on oxidation or on fusion with alkali yielded the same products as rat melanoma melanin similarly treated. The behaviour on methylation, followed by oxidation, was also quite similar to that of melanoma melanin. Other workers (84) claim a hot mixture of thioglycollic acid and phenol to be effective in removing the protein from hair pigment; but no chemical investigation of melanin so isolated is known to have been reported. The importance of tyrosine in pigmentation is seen in the disease phenylketonuria, in which the patient has light colored hair, because of an incapacity to convert phenylalanine into tyrosine.

2.5. Melanin from the Eye

The color of the eye is produced by scattering of the shorter wavelengths of white light by melanin granules, the size of which controls the color. However, there is evidence that melanins can complex with heavy metals, with consequent change in shade and intensity (22, 23, 88). The incorporation of radioactivity from $(\pm)$-[2-^{3}H]dopa into the melanosomes of chick embryo epitheleum has been demonstrated (73). NICOLAUS, PIATTELLI, and FATTORUSSO (110) found that ox-choroid melanin, when oxidized, yielded pyrrole-2,3,5-tricarboxylic acid, pyrrole-2,3,4,5-tetracarboxylic acid, and pyrrole-2,3-dicarboxylic acid; and when fused with alkali it afforded 5,6-dihydroxyindole and 5,6-dihydroxyindole-2-carboxylic acid.

2.6. Other Melanins

The feathers of many birds are black, owing to the presence of eumelanins. The color of red feathers is due to the presence of phaeomelanins, which are soluble in alkali. Integumentary melanins are indirectly responsible for many structural (as distinct from pigmentary) colors displayed by animals, such as the Tyndall blues seen in the skin of fish, reptiles, and mammals, and in the feathers of birds. Here the melanin provides a dark background (53). NICOLAUS, PIATTELLI, and FATTORUSSO (110) found that oxidation or fusion with alkali of bird eumelanins yielded the same products as did melanin from melanoma or hair. An interesting case of genetic control of melanogenesis is seen in the Silky fowl, with pure white plumage and purple-black skin, and with melanin elsewhere in the body.

Melanin from axolotl liver yields 5,6-dihydroxyindole when fused with alkali (*107*). No research on the structure of melanin from skin appears to have been published, presumably because of the difficulty of isolation in sufficient quantity.

In a study of the conversion of tyrosine into the propylhygric acid moiety of lincomycin, it was noted that *Streptomyces lincolnensis* produced a dark, melanin-like pigment in media containing tyrosine; and that in media that were sulphur limited, propyl- and ethyl-proline accumulated (*158*). The production of the two latter compounds was stimulated by L-tyrosine and L-dopa. It was suggested that these products arose through cyclization of tyrosine to an indole derivative, followed by opening of the benzene ring, although it is not very clear as to exactly how the propyl group would thus arise.

3. Investigations on Melanins by Spectroscopic and other Physical Methods

3.1. Ultraviolet and Infrared Spectra

Spectroscopic methods have been of little value in elucidating the structure of melanin. In the visible range of the spectrum melanin shows high absorption, but no definite bands; this absorption increases towards shorter wavelengths in the u.v. However, the earlier suggestion that this absorption is due to a highly conjugated system is now doubtful (*12, 14*).

The i.r. spectra of melanins show absorption bands expected of an aromatic substance, but line-broadening and overlapping precludes a detailed interpretation. The i.r. spectra of even chemically different melanins (*e.g.* synthetic dopa-melanin and catechol-melanin) are quite similar (*13*) although detailed differences have been claimed (*19*) between the spectra of natural melanins derived from different species. These similarities must reside in electronic properties which are shared by these materials, but which do not reflect closely the elemental composition or detailed chemical structure.

3.2. X-Ray Diffraction

Melanins are probably not truly soluble, and at the best form colloidal systems under biological conditions; they have not been obtained crystalline and are described as amorphous. Cross-linking is really only assumed from the polyfunctional nature of the intermediates in melanogenesis.

Recent work by THATHACHARI and BLOIS (14, 147) using the more refined methods of X-ray diffraction now available, has demonstrated the photograph from a powdered melanin sample of animal origin to show a single, diffuse halo, which corresponds to a Bragg spacing of approximately 3.4 Å. This spacing is interpreted as arising from indole or other aromatic units which are stacked upon one another to form a π-complex; and it is assumed that these units are members of different polymer chains or distant members of the same chain. In the case of plant melanins, the spacing between the stacked, planar rings was 4.2 A. It thus appears that the π-stacking is less compact in the plant melanins than in the eumelanins. It was concluded that the diffraction pattern of Ustilago-melanin and other plant melanins is consistent with a random structure, in which adjacent monomeric units of o-benzoquinone, etc., are in random mutual orientations that satisfy the steric requirements. Somewhat greater regularity seems to occur in the structure of the animal melanins and synthetic catechol-melanin. They seem to aggregate with their adjacent planar groups nearly parallel and as close together as possible. It has also been found that dopa-melanin is much denser than melanin from Ustilago (14).

Certain drugs (e.g. chloroquine, and chlorpromazine) are known to bind to melanin. This binding has been observed in the eye, and in melanoma, as well as in vitro (15, 153), and is of potential pharmacological importance. BLOIS believes that the mechanism of this binding may be similar to the π-coupling in the local, stacking region of melanin. Chlorpromazine also activates tyrosinase (153).

3.3. Electron Spin Resonance

Electron spin resonance (e.s.r.) signals were detected in natural melanin (the first time in living things) in 1954 by COMMONER, TOWNSEND, and PAKE (35), who proposed that the paramagnetism was due to free radicals trapped in the pigment. The trapping of free radicals by a growing polymer has indeed been demonstrated in the case of polymethacrylate (54).

BLOIS, in studying the autoxidation of metal-free phenolic compounds to give melanins, found that the e.s.r. of these polymers was generally characterized by a single, usually structureless, absorption line, a g-value near 2, and a paramagnetism that was permanent (14). There appeared to be a systematic correlation of g-values with the molecular structure of the radical, in general those molecules with a greater delocalization of the odd electron also having lower g-values (16). The natural melanins had g-values which implied electron delocalization over one, or at most, two aromatic rings. The rather generally held, earlier belief that melanins were extensively conjugated was thus not borne out, and moreover the results could not be explained by a melanin structure based on a single monomer type.

As the temperature was lowered, this paramagnetism increased, being inversely proportional to the temperature down to 4° K, *i.e.* the Curie Law was obeyed (*13, 18*). It had been proposed earlier (*87, 129*) that melanin might be a semiconductor. However, in the case of an intrinsic semiconductor, the number of unpaired (conducting) electrons is proportional to the temperature — the opposite of the experimental result. Moreover, melanin acts as an insulator in the e.s.r. cavity.

The free radicals in melanin are extremely stable. The e.s.r. spectrum of melanin heated in air or in a vacuum, then cooled, was found to be unchanged; only at temperatures of the order of 300° did the signal begin to diminish slightly. It seems that atmospheric oxygen is unable to diffuse into the radical-containing regions of the polymer. The free radical signal of melanin was also unaffected by treatment of the melanin with acid or alkali or ascorbic acid, although in the latter case a lightening of the color occurred (reversible reduction). However copper ions can quench the paramagnetism of the melanins, although the color of the product remains unchanged. Thus the color and the paramagnetism of melanin are independent.

In the alkaline autoxidation of dopa, when the reaction is initiated, the dominant paramagnetic entity present is found to be a semiquinone of dopa (*14*). As the reaction continues, other semiquinone radicals appear and the superposition of the hyperfine structure of these produces a complex spectrum. When the reaction has gone to completion and the insoluble melanin has been formed, all hyperfine structure is seen to have vanished, and the single, broad absorption of melanin remains. For these radicals to be stably trapped, it has been suggested that the unpaired electrons must be limited to localized regions of the polymer, otherwise they would leak out to the edges of the melanin "particles" ("molecules") and become paired off by interaction with the environment. On the other hand, a rather similar situation prevails in humic acid, where Haworth (*67*) has attributed the radical stability to the fact that "the free electron could be dispersed amidst and stabilized by a large number of resonance structures".

The e.s.r. results suggest that melanins usually contain approximately one free radical per 100—200 units of monomer according to Blois (*18*), or one per 800—1000, according to Mason (*97*). Such radical units thus make only a very small contribution to the overall structure of the melanin.

Tollin and Steelink (*151*) found that when humic acid or dopa-melanin was converted into a sodium salt, the free-radical content increased greatly. In view of Blois' finding of localized areas of π-stacking in melanins, one may speculate as to whether such stacking might represent something in the nature of a quinhydrone, which under alkaline conditions yields a semiquinone. Sepiomelanin failed to give a sodium salt.

Studies, which included hair from Egyptian mummies, showed correlation between e.s.r. signals and hair pigmentation. Treatment of black hair with hydrogen peroxide led to a reduction in the signal intensity, which could be increased again by u.v. irradiation, although the color was not simultaneously restored (*99, 135*).

Most e.s.r. studies on melanin have been carried out at X-band, where essentially characterless singlet signals were obtained. GRADY and BORG (*59*) studied the spectra of various melanins at Q-band, and obtained broader, asymmetrical spectra composed of two components with different pH-dependencies. Eumelanins (synthetic or of animal origin) gave spectra at pH 10—12 which were very similar to one another, but which were significantly different from the spectra of allomelanins. The pH dependency was explained in terms of dissociation of protons from radical centres in alkaline solution producing an anion of higher g value through alteration of the delocalization of the unpaired electron (less spin density on the oxygen atom of the phenoxy-group).

It has been suggested (*88, 99*) that in some organisms melanin can act as an electron exchange polymer capable of protecting a tissue against reducing or oxidizing conditions, or of trapping free radicals which could disrupt the metabolism of living cells; and it has been speculated that protection against solar ultraviolet radiation extends beyond its mere opaqueness. But the above studies show that the radicals in melanin are buried so deeply that only very tiny radicals could diffuse into the matrix and become paired off. Melanins catalyze the non-enzymic oxidation of reduced nicotinamide-adenine dinucleotide and p-phenylenediamine, and this could be related to the free-radical property of the melanin (*152*), although if so, it would raise the counter-argument. Chloroquine and chlorpromazine inhibit this effect. It has been shown that ultraviolet irradiation increases the spin concentration in melanins.

However, WASSERMANN (*155a*) has questioned the popular assumption that racial pigmentation developed in certain geographical areas as an adaptation to the environment; and has suggested that it may be the consequence of an adaptational adjustment to tropical disease. Moreover, he has suggested that melanin may influence cellular homeostasis, in the skin and elsewhere in the body, so acting like a hormone.

4. Adrenochrome-Melanin

Solutions of adrenaline (epinephrine) (**66**), on oxidation, give first a red color, then a brown or black precipitate of "adrenaline black", the formation of which proceeds *via* adrenochrome (**67**) (*64*). However, the pigment can also be obtained directly from adrenochrome by a non-oxidative process, by merely keeping an aqueous solution of it under nitrogen for 48 hours. It is an amorphous material, insoluble in organic solvents except pyridine, but soluble in sodium hydroxide solution, from which solution it can be reprecipitated by addition of acid. It can be reversibly reduced to a leuco-compound (64).

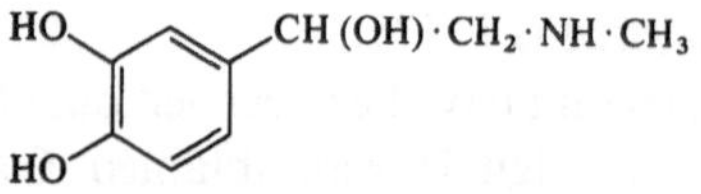

Epinephrine (Adrenaline)

(66)

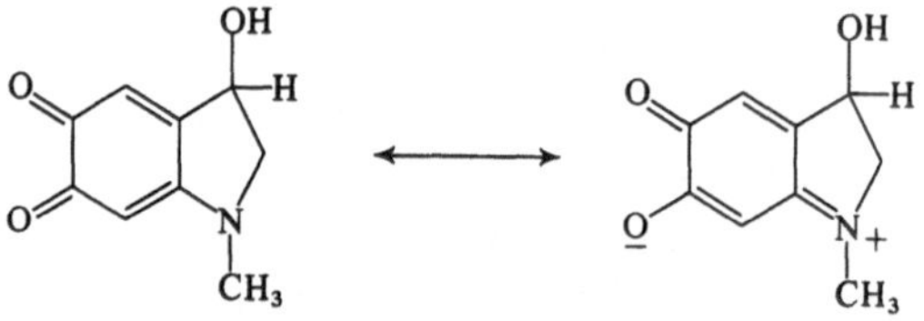

Adrenochrome

(67)   (68)

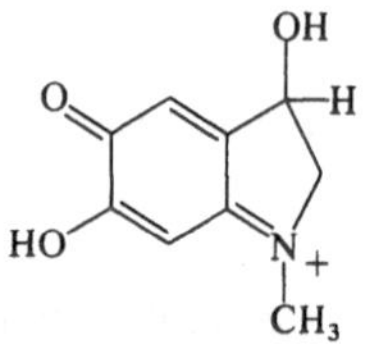

Bu'Lock (29) found that on addition of acid to adrenochrome, a yellow salt (68) was formed reversibly. If the solution is neutralized immediately, most of the adrenochrome is regenerated, although the salt is very unstable, especially in acid solution, giving rise to adrenochrome-melanin. It was suggested that it underwent an acid-catalyzed reaction to give 1-methyl-indole-5,6-quinone (69), which could be trapped in the presence of 1,4-naphthaquinone affording (70). The acid-catalyzed melanin formation can be considered to start with the self-combination of 1-methyl-indole-5,6-quinone to give a dimer such as (71), followed by a reaction between a dihydroxyindolyl group of the dimer or oligomer with a free or combined indolequinone group.

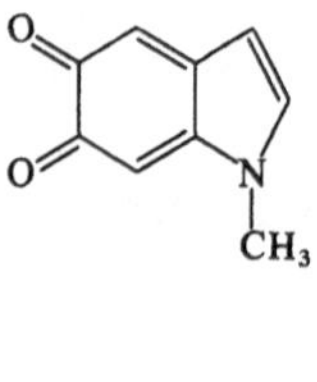

(69) 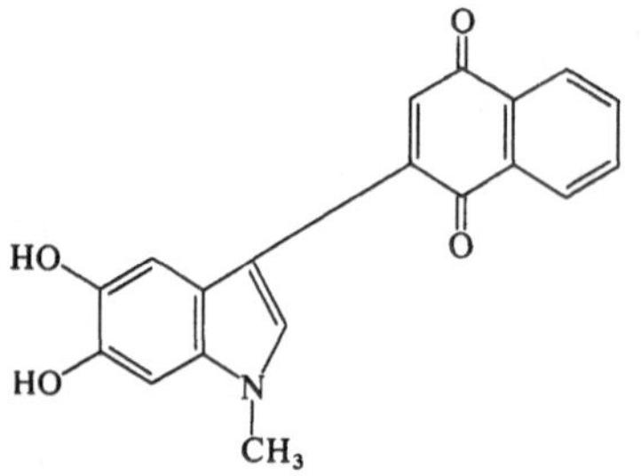   (70) 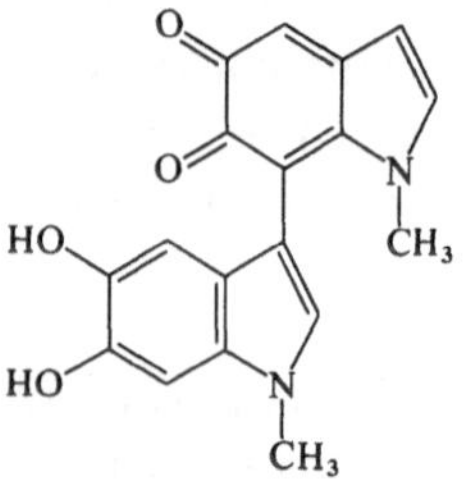   (71)

If under anaerobic conditions this were the only process, the final polymer should contain only one quinonoid group per molecule. However, since the oxidation potential of an indolequinone such as (71) should be lowered by the dihydroxyindolyl substituent, it follows that dihydroxyindolyl residues in the polymer can be oxidized by 1-methyl-indole-5,6-quinone, giving rise to a number of quinonoid groups in the polymer. But the progress of this reaction is limited by the fact substitution of a hydrogen from 1-methylindole-5,6-quinone by a quinonoid group raises its oxidation potential. Thus in a linear or random polymer, not more than half of the units will be readily oxidized by the free quinone.

5. Dopamine-Melanin

Surprisingly little research appears to have been done on dopamine-melanin. Dopamine has been isolated from banana (61), which contains a polyphenol-oxidase, so the blackening of the fruit may be the result of the formation of a melanin. Almost nothing is known about the structure of the melanin of *Substantia nigra* in the brain of man and primates, although it has been suggested that this may be formed by oxidative polymerization of epinephrine (66) or norepinephrine (45, 90, 90a, 154). Dopa and dopamine have been obtained by hydrolysis of a brain pigment (111). The presence of a dopa-oxidase appears not to have been demonstrated in *Substantia nigra*. However, it has been proposed that the neuromelanin granule, unlike the melanosome, is a combination of melanin and lipofuscin (3, 4).

As indicated above, dopamine-melanin differs from dopa-melanin and 5,6-dihydroxyindole-melanin markedly. Its structure has been investigated by SWAN and WRIGHT (144, 145) and by BINNS, KING, MISHRA, PERCIVAL, ROBSON, SWAN, and WAGGOTT (8).

Autoxidative dopamine-melanin was found to be soluble in dilute sodium hydroxide solution and when the resulting solution was shaken with benzoyl chloride, it gave a precipitate of "benzoylated melanin". By using [^{14}COCl] benzoyl chloride, the latter was shown to contain approximately one benzoyl group for every eight carbon atoms of the original melanin polymer. The i.r. spectrum of the benzoylated melanin showed bands at 2929 and 1441 cm^{-1} (CH$_2$) and strong peaks at 1746 (C=O) and 1265 (C−O) cm^{-1}, indicating that benzoylation had occurred predominantly at phenolic hydroxy-groups, although a weak band at 1646 cm^{-1} suggested that N-benzoylation had occurred to a lesser extent.

Autoxidative dopamine-melanin which had been washed with 2N-hydrochloric acid, and then dried, contained approximately 0.37 chlorine atoms (as Cl^-) per C_8-unit. It seems that this chlorine must represent uncyclized phenethylamine side-chains in the form of the hydrochloride. The weakness of the amide carbonyl band in the i. r. spectrum of the benzoylated melanin casts doubt as to the completeness of benzoylation.

By using dopamine specifically labelled with ^{14}C at each of the α-, -β-, 3-, 4-, and 5-positions (separately) SWAN and WRIGHT (144, 145) showed that in autoxidative dopamine-melanin which had been boiled with 2N-hydrochloric acid for 24 hours, one unit in five had lost carbon atoms 4, 5 and 6 of the indole ring (during melanogenesis by attack of hydrogen peroxide on the benzene ring) while retaining carbons 2 and 3. However, the oxidative process was only partly complete before the boiling with acid.

When autoxidative dopamine-melanin, which had been collected at pH 8, and washed with water, was heated at 200°, a negligible amount of carbon dioxide was evolved; but the same melanin which had been boiled for 24 hours with 2N-hydrochloric acid, when heated at 200°, evolved up to 2% by weight of carbon dioxide (8).

Dopamine-melanin was methylated by diazomethane, yielding a product which was sufficiently soluble in deuteriochloroform to obtain an n.m.r. spectrum, although the peaks were broad and ill-defined. It gave a signal at τ 6.4, attributed to the methoxy-protons, together with a peak at 8.8, with a shoulder at 9.1. The benzoylated melanin showed peaks presumably owing to aromatic protons of the benzoyl group, around τ 2.05, 2.55, and 2.8, in addition to signals at ca. 8.8 and 9.0. The $CH_2.CH_2$ groups of 3,4-dimethoxyphenethylamine and 5,6-dimethoxyindoline both give signals in the τ 7 region. The presence of CH_2 groups in dopamine-melanin is seen from the results of deuteriation experiments, as well as from the i.r. spectra of methylated and benzoylated melanins. If the peaks at τ 8.8 and 9.0—9.1 in the n.m.r. spectra of these two derivatives do indeed represent $CH_2.CH_2$ groups, their appearance at such high field might be the result of shielding, because in the polymer such a group might be held above the plane of an aromatic ring other than the one to which it is directly attached.

Samples of dopamine, specifically deuteriated at each of the α-, and β-positions of the side chain, and the 2-, 5- and 6-positions of the benzene ring (separately), were converted into melanins, and the incorporation of deuterium was determined (8). Two series of experiments were carried out; in one series the precursor contained only tracer concentration of deuterium, while in the other the relevant position of the molecule was deuteriated to as nearly 100% as possible. Again, the experiments of each series were carried out both by autoxidation and enzymically. When the melanins resulting from the α- and β-labelled dopamine were boiled with 2N-hydrochloric acid, two exchange processes appeared to occur —

a rapid loss of deuterium, followed by an extremely slow one. It was thought that at the end of the first (rapid) exchange, deuterium remained only in the side-chains of units which had not undergone cyclization; and isotopic analysis therefore indicated the proportion of these units present in the polymer. The absence of indoline-type units was inferred from the agreement of this value with that deduced from the percentage of chlorine in the acid-washed melanin.

In the enzymic experiments, the initially precipitated melanin contains protein, which can be removed by long boiling with acid. The deuterium retention of such a melanin was not greatly different from that of a corresponding autoxidative melanin boiled for an equal time with acid. There was nothing in the results to suggest a fundamental difference in structure between autoxidative and enzymic dopamine-melanins. The results also suggested that a kinetic isotope effect had little influence.

The ratios of deuterium-enrichment in the melanin to that in the precursor were, almost always, higher in the case of dopamine-melanin than in dopa-melanin (for the same position of labelling). If one attempts to fit these values to a model structure for dopamine-melanin, as was done for dopa-melanin, one has to conclude that there must be many polymeric linkages which do not involve any of the carbon atoms which had been labelled with attached deuterium (*i.e.* α-, β-, 2-, 5-, and 6- of dopamine); or in other words, their must be a considerable number of polymeric linkages at positions other than those throught to be involved in dopa-melanin, *e.g.* linkages through oxygen or nitrogen might be possible.

Dopaminequinone	2,4,5-Trihydroxyphenethylamine	
(72)	(73)	(74)

If in the formation of melanin, dopamine if first oxidized to dopaminequinone (72), there would be competition for reaction in the 6-position of the latter between, on the one hand, intramolecular cyclization, and on the other hand, attack by hydroxide ion to give 2,4,5-trihydroxyphenethylamine (73), or intermolecular reaction with the amino-group of another dopamine molecule. However, 2,4,5-trihydroxyphenethylamine appears not to be very important so far as melanogenesis is concerned (*31*). SENOH and WITKOP (*136*) have shown that it undergoes oxidative cyclization to give the aminochrome (74), although this was not isolated. On the other hand, HARLEY-MASON (65) was able to obtain 5,6-dihydroxyindole by oxidation of 2,4,5-trihydroxyphenethylamine, and BAIRD and BAXTER (2) have recently shown that 2,5-bis-(β-aminoethyl) benzoquinone undergoes cyclization through attack of the amino-group at the carbonyl group, rather than by conjugate-addition. Support for the possibility of C—N linkages occurring between units in dopamine-melanin was obtained by preparing a melanin from dopamine in the presence of $[\alpha\text{-}^{14}C]$phenethylamine, when some radioactivity was incorporated into the melanin. However, this appears to be contradictory to MASON and PETERSON's (100) failure to detect any ability of primary amines to affect the u.v. or visible spectra observed during the oxidation of dopa. A melanin was also obtained by autoxidation of 5,6-dihydroxyindoline (75), but was not found to contain basic groups. Tracer evidence by BINNS, CHAPMAN, ROBSON, SWAN, and WAGOTT (7) suggested that a very small proportion of units in dopamine-melanin might be 2—6 linked. No evidence of C—O linked polymer units is forthcoming.

5,6-Dihydroxyindoline

(75)

The results at present available have been explained in terms of a combination of units shown in Chart 8 for autoxidative dopamine-melanin, which has been washed with water. The state of oxidation of individual units in the polymer is unknown — at the most, the relative numbers of quinonoid and phenolic units as a whole can be deduced from methoxy-group values.

V. Phaeomelanins

Hair color of mammals varies from black through brown to yellow. It is generally believed that two different classes of pigments are responsible for the production of the wide range of hair color; the eumelanins giving dark colors, and the phaeomelanins responsible for the lighter ones. Both classes of pigments are found as granules in melanocytes within the hair follicle and are transferred with the cells which will constitute the hair. Eumelanins and phaeomelanins, together with

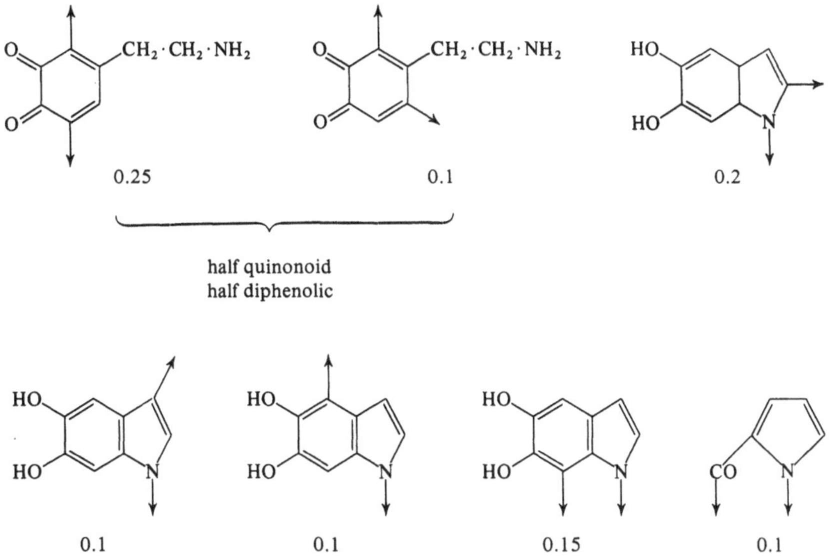

half quinonoid
half diphenolic

Chart 8. Autoxidative dopamine-melanin

carotenoids, are also responsible for the great variety of tints in bird feathers. Whereas eumelanins are widely distributed throughout the various tissues of mammals and birds, as well as in those of other animals, the occurrence of phaeomelanins appears to be restricted to hair and feathers, and perhaps freckles. Typical phaeomelanins are responsible for the red color of human hair, the color of New Hampshire or Rhode Island chicken feathers, and the fur of many animals.

Through the work of NICOLAUS, PROTA, and their collaborators (*124*), it has recently become fairly clear that phaeomelanins are formed in nature by a modification of the eumelanin pathway, involving interaction of cysteine with dopaquinone, produced by enzymic oxidation of tyrosine.

Earlier investigators used acid treatment for isolation of pigments from red hair or feathers, although this is now known to be unsuitable, giving only minor fractions of pigment, the so-called trichosiderins, and even these may be structurally modified. The Italian workers therefore developed an alkaline extraction procedure. Feathers of New Hampshire chickens were extracted with 0.1 N-sodium hydroxide at room temperature; and the extract was acidified, and a red-brown, amorphous precipitate was separated, from the yellow-orange supernatant.

The precipitate was dialyzed, and subjected to chromatography on Sephadex, giving four gallophaeomelanin fractions, which were polymeric, all containing sulphur, and some also containing protein. The yellow-orange supernatant was subjected to ion-exchange chromatography, followed by chromatography on Sephadex, yielding a number of pigments.

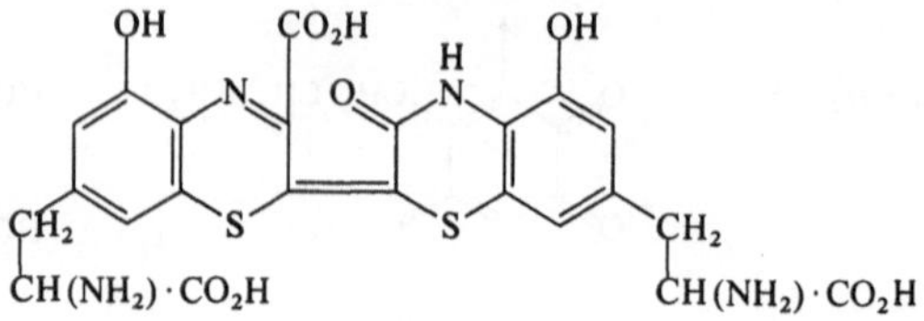

(76)

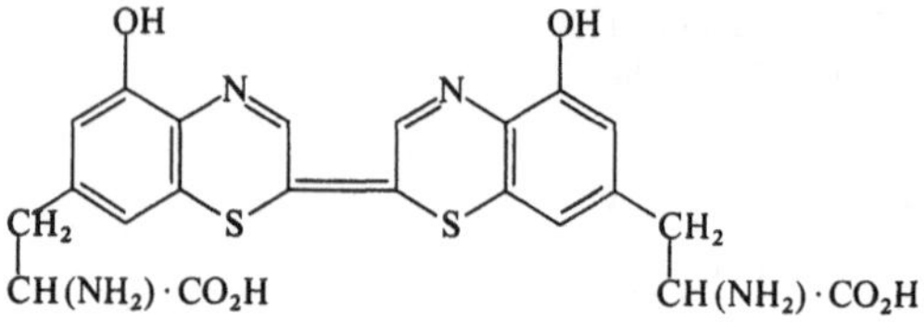

(77)

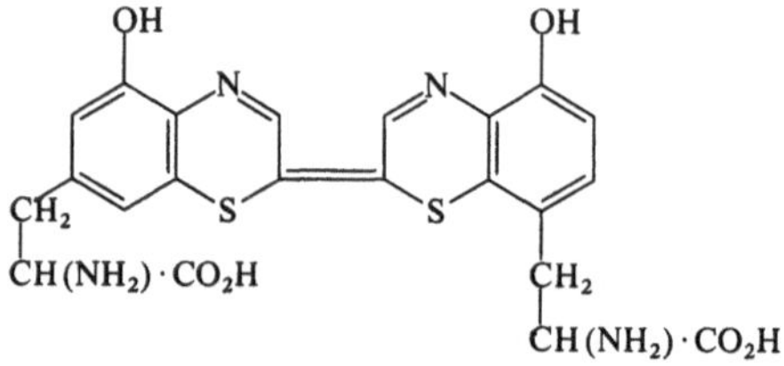

(78)

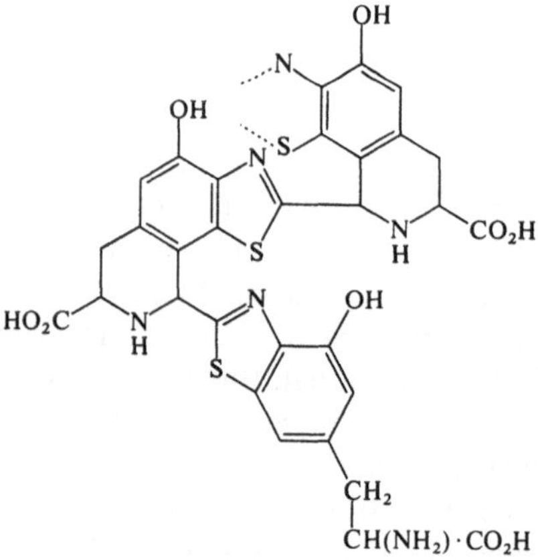

(79)

It was concluded that one group of these pigments, the acid-soluble fraction, formerly regarded as trichosiderins, are derivatives of $\Delta^{2\ 2'}$-bi-(2H-1,4-benzothiazine), a chromophore not previously found in natural products; and indeed these pigments do not contain iron, as suggested by Flesch (50). Structures (76), (77), and (78) have been proposed for three of these pigments (126). The structures of the gallophaeomelanins are still

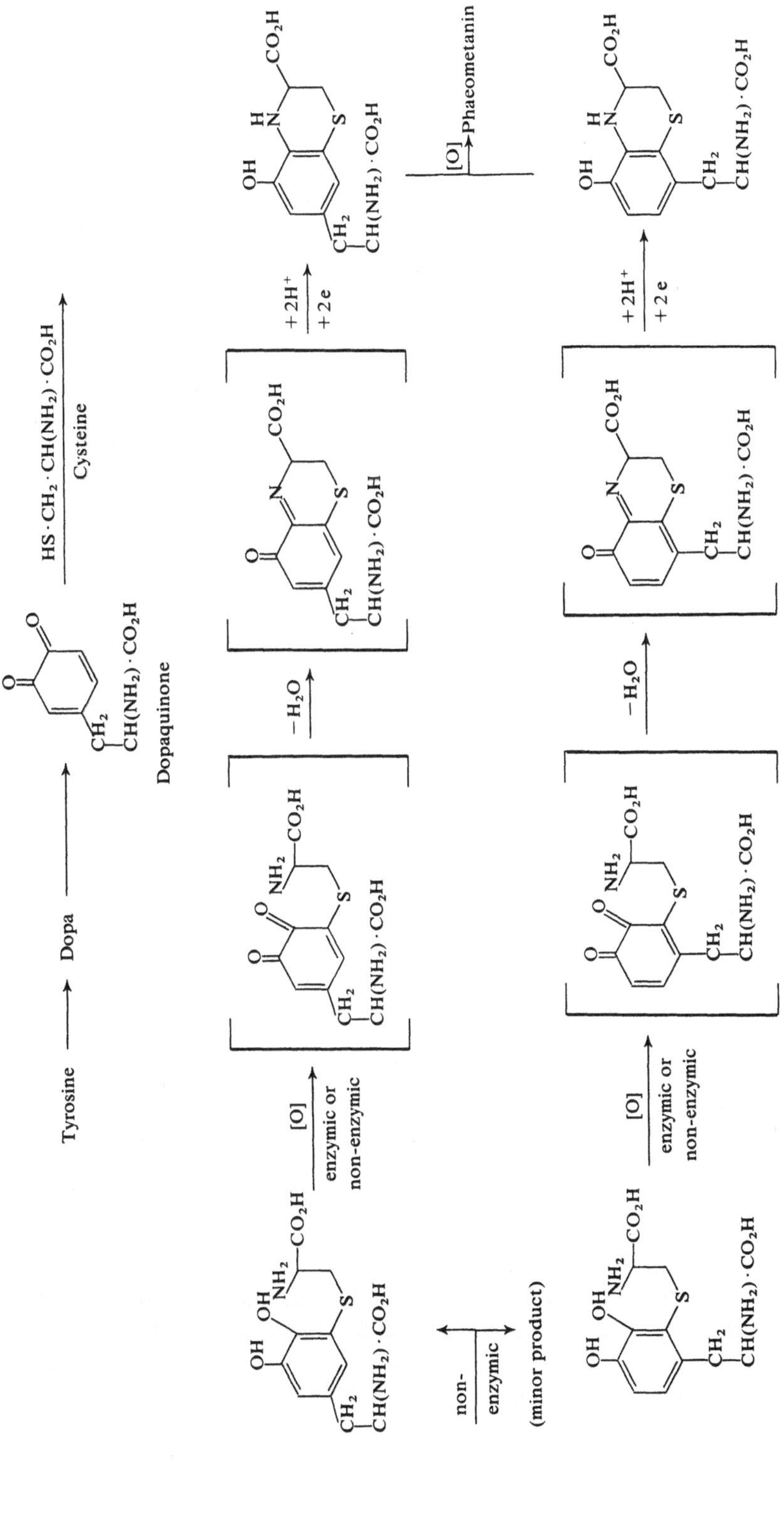

Chart 9. The initial stages of the biogenesis of phaeomelanin

unknown, although the main units which constitute these complex macromolecules are thought to be derived from benzothiazole, linked as shown in partial structure (**79**).

Chemical and biochemical evidence suggested that phaeomelanins might be formed *in vivo* by a deviation of the eumelanin pathway involving interaction of cysteine with dopaquinone produced by enzymic oxidation of tyrosine (*124, 125*). In fact, it was found that both [3-^{14}C]cysteine and [2-^{14}C]tyrosine were incorporated specifically into pigment cells of feather papillae of New Hampshire chick embryos; and that this incorporation was inhibited by a tyrosinase inhibitor (*103*).

The suggested biogenetic route (Chart 9) is supported by a study of the products produced by oxidation of dopa with mushroom polyphenol-oxidase in the presence of cysteine. The main product was a red-brown pigment, the elemental composition and properties of which were very similar to those of natural gallophaeomelanin. If the enzymic reaction was stopped at a very early stage, by acidification, a pale yellow solution was obtained, which became intensly violet within a few minutes. Paper chromatography showed the presence of two red-violet compounds, identical with two of the natural pigments.

VI. Conclusions

To summarize, the structures of eumelanins have been investigated by a number of research groups, each using different methods; but the conclusions drawn regarding the structure of natural melanins can in no case be accepted without question (*141a*). Thus, melanins isolated by methods involving strongly acidic conditions could have undergone structural change; and the yields of products obtained from degradative experiments have been very low. The interpretation of Hempel's results is doubtful in view of the possible occurrence of dopa as a peptide in melanoma. The uncertainty of the degree of hydration and elemental composition renders the quantitative interpretation of isotopic and other experiments on synthetic melanins open to question; and in the case of enzymic experiments, again treatment with acid is required to free the melanin from protein. Moreover, no means is known of deciding whether a synthetic and a natural melanin are identical.

Yet, in spite of all these and other doubts and the inherent difficulty of the research, the remarkable fact emerges that a number of groups of workers, using quite different approaches, have now come to very much the same conclusion, at any rate in a qualitative sense, regarding the structures of both natural and synthetic eumelanins. Moreover, it seems that the allomelanins, eumelanins, and phaeomelanins can be fitted into a common biogenetic scheme; but whether or not these pigments can participate in metabolism remains to be decided in the future.

References

1. ALLPORT, D. C., and J. D. BU'LOCK: The Pigmentation and Cell-wall Material of *Daldinia* Sp. J. Chem. Soc. (London), 4090 (1958).

1a. BAGNARA, J. T., and M. E. HADLEY: Chromatophores and Color Change. New Jersey: Prentice-Hall. 1973.

2. BAIRD, D. B., and I. BAXTER: Private communication.

3. BARDEN, H.: The Histochemical Relationship of Neuromelanin and Lipofuscin. J. Neuropathol. and Exptl. Neurology **28**, 419 (1969).

4. BARDEN, H., and E. MARTIN: Electron Probe Microanalysis of Neuromelanin and Lipofuscin. In: V. RILEY (Ed.): Pigmentation, its Genesis and Biological Control, p. 631. New York: Appleton-Century-Crofts. 1972.

5. BAYSE, G. S., and M. MORRISON: The Role of Peroxidase in Catalyzing Oxidation of Polyphenols. Biochim. Biophys. Acta **244**, 77 (1971).

6. BEER, R. J. S., T. BROADHURST, and A. ROBERTSON: The Chemistry of Melanins. Part V. The Autoxidation of 5:6-Dihydroxyindoles. J. Chem. Soc. (London), 1947 (1954).

7. BINNS, F., R. F. CHAPMAN, N. C. ROBSON, G. A. SWAN, and A. WAGGOTT: Studies Related to the Chemistry of Melanins. Part VIII. The Pyrrolecarboxylic Acids formed by Oxidation or Hydrolysis of Melanins derived from 3,4-Dihydroxyphenethylamine or ($\pm$)-3,4-Dihydroxyphenylalanine. J. Chem. Soc. (London) (C), 1128 (1970).

8. BINNS, F., J. A. G. KING, S. N. MISHRA, A. PERCIVAL, N. C. ROBSON, G. A. SWAN, and A. WAGGOTT: Studies Related to the Chemistry of Melanins. Part XIII. Studies on the Structure of Dopamine-Melanin. J. Chem. Soc. (London) (C), 2063 (1970).

9. BINNS, F., and G. A. SWAN: Oxidation of Some Synthetic Melanins. Chem. and Ind. 396 (1957).

10. BLOCH, B.: Das Pigment. In J. JADASSOHN: Handbuch der Haut- und Geschlechtskrankheiten. **1**, part 1, 434. Berlin: J. Springer. 1927.

11. BLOIS, M. S.: A Note on the Problem of Melanin Structure. In: W. MONTAGNA and F. HU (Eds.): Advances in Biology of Skin. Vol. VIII. The Pigmentary System, p. 319. Oxford: Pergamon Press. 1967.

12. — Recent Developments in the Physics and Chemistry of the Melanins. In: F. URBACH (Ed.): The Biological Effects of UV Radiation, p. 299. Oxford: Pergamon Press. 1969.

13. — Biological Free Radicals and the Melanins. In: S. J. WYARD (Ed.): Solid State Biophysics, p. 243. New York: McGraw-Hill. 1969.

14. BLOIS, M. S., Jr.: Physical Studies of the Melanins. In: T. KAWAMURA, T. B. FITZPATRICK, and M. SEIJI (Eds.): Biology of Normal and Abnormal Melanocytes, p. 125. University of Tokyo Press. 1971.

15. BLOIS, M. S.: The Binding Properties of Melanin: In Vivo and in Vitro. In: W. MONTAGNA, R. B. STOUGHTON, and E. J. VAN SCOTT (Eds.): Advances in Biology of Skin, Vol. XII, Pharmacology and the Skin, p. 65. New York: Appleton-Century-Crofts. 1972.

16. Blois, M. S., Jr., H. W. Brown, and J. E. Maling: Precision g-Value Measurements on Free Radicals of Biological Interest. In: M. S. Blois, Jr., H. W. Brown, R. M. Lemmon, R. O. Lindblom, and M. Weissbluth (Eds.): Free Radicals in Biological Systems, p. 117. New York and London: Academic Press. 1961.

17. Blois, M. S., Jr., and R. F. Kallman: The Incorporation of C^{14} from 3,4-Dihydroxyphenylalanine-2′-C^{14} into the Melanin of Mouse Melanomas. Cancer Res. **24**, 863 (1964).

18. Blois, M. S., A. B. Zahlan, and J. E. Maling: Electron Spin Resonance Studies on Melanin. Biophys. J. **4**, 471 (1964).

19. Bonner, T. G., and A. Duncan: Infra-Red Spectra of Some Melanins. Nature **194**, 1078 (1962).

20. Bouchilloux, S., and A. Kodja: Combinaison des Thiols avec les Quinones se Formant au Cours de la Mélanogénèse. Bull. soc. chim. biol (Paris) **42**, 1045 (1960).

21. Bouchilloux, S., S. Lissitzky, and D. Kertesz: Sur L'Existence d'un Phénomène d'Acidification au Cours de l'Oxydation de Divers Substrats par la Polyphénoloxydase. Bull. soc. chim. biol. (Paris) **39**, 1049 (1957).

22. Bowness, J. M., and R. A. Morton: The Association of Zinc and other Metals with Melanin and a Melanin-Protein Complex. Biochem. J. **53**, 620 (1953).

23. Bowness, J. M., R. A. Morton, M. H. Shakir, and A. L. Stubbs: Distribution of Copper and Zinc in Mammalian Eyes. Occurrence of Metals in Melanin Fractions from Eye Tissues. Biochem. J. **51**, 521 (1952).

24. Brackman, W., and E. Havinga: The Oxidation of Phenols with Copper-Amine Catalysts and its Relation to the Mode of Action of Tyrosinase. I. The Catalytic Oxidation of Monohydric Phenols to orthoquinone Derivatives. Rec. trav. chim. Pays-Bas **74**, 937 (1955).

25. — — The Oxidation of Phenols with Copper-Amine Catalysts and its Relation to the Mode of Action of Tyrosinase. V. Reaction Mechanisms. Rec. trav. chim. Pays-Bas **74**, 1107 (1955).

26. Bright, H. J., B. J. B. Wood, and L. I. Ingraham: Copper, Tyrosinase, and the Kinetic Stability of Oxygen. Ann. New York Acad. Sci. **100**, 965 (1963).

27. Bruenger, F. W., B. J. Stover, and D. R. Atherton: The Incorporation of Various Metal Ions into in Vivo- and in Vitro-Produced Melanin. Radiation Research **32**, 1 (1967).

28. Bu'Lock, J. D.: Intermediates in Melanin Formation. Arch. Biochem. Biophys. **91**, 189 (1960).

29. — The Formation of Melanin from Adrenochrome. J. Chem. Soc. 52 (1961).

30. Bu'Lock, J. D., and J. Harley-Mason: Melanin and its Precursors. Part II. Model Experiments on the Reactions between Quinones and Indoles, and Consideration of a Possible Structure for the Melanin Polymer. J. Chem. Soc. (London) 703 (1951).

31. Chapman, R. F., A. Percival, and G. A. Swan: Studies Related to the Chemistry of Melanins. Part XII. Some Spectroscopic Experiments regarding Intermediates in Melanogenesis. J. Chem. Soc. (London) (C) 1664 (1970).

32. Chen, Y. M., and W. Chavin: Incorporation of Tyrosine Carboxyl Groups and Utilization of D-Tyrosine in Melanogenesis. Analyt. Biochemistry **27**, 463 (1969).

33. — — Effects of Depigmentary Agents and Related Compounds upon in Vitro Tyrosinase Activity. In: V. Riley (Ed.): Pigmentation, its Genesis and Biological Control, p. 593. New York: Appleton-Century-Crofts. 1972.

34. Clemo, G. R., F. K. Duxbury, and G. A. Swan: Formation of Tyrosine Melanin. Part III. The Use of Carboxyl-labelled Tyrosine and Dihydroxyphenylalanine in Melanin Formation. J. Chem. Soc. (London) 3464 (1952).

35. Commoner, B., J. Townsend, and G. E. Pake: Free Radicals in Biological Materials. Nature **174**, 689 (1954).

36. CROMARTIE, R. I. T., and J. HARLEY-MASON: Melanin and its Precursors. 8. The Oxidation of Methylated 5:6-Dihydroxyindoles. Biochem. J. 66, 713 (1957).
37. DAWSON, C. R., and B. J. LUDWIG: On the Mechanism of the Catechol-Tyrosinase Reaction. II. The Hydrogen Peroxide Question. J. Amer. Chem. Soc. 60, 1617 (1938).
38. DAWSON, C. R., and W. B. TARPLEY: On the Pathway of the Catechol-Tyrosinase Reaction. Ann. New York Acad. Sci. 100, 937 (1963).
39. DELLA PORTA, G., and O. MÜHLBOCK (Eds.): Structure and Control of the Melanocyte. Berlin-Heidelberg-New York: Springer. 1966.
40. DUCHON, J., B. MATOUŠ, and Z. PECHAN: On the Chemical Nature of Urinary Melanogens. In: G. DELLA PORTA and O. MÜHLBOCK (Eds.): Structure and Control of the Melanocyte, p. 175. Berlin-Heidelberg-New York: Springer. 1966.
41. DUCHON, J., and Z. PECHAN: The Biochemical and Clinical Significance of Melanogenuria. Ann. New York Acad. Sci. 100, 1048 (1963).
42. DUCHON, J., and Z. PECHAN: Biochemie melaninu a melanogenese. Prague: Státní zdravotnické nakladatelství. 1964.
43. DULIÈRE, W. L., and H. S. RAPER: The Tyrosinase-Tyrosine Reaction. VII. The Action of Tyrosinase on Certain Substances Related to Tyrosine. Biochem. J. 24, 239 (1930).
44. EVANS, W. C., and H. S. RAPER: The Accumulation of 1 – 3:4-Dihydroxyphenylalanine in the Tyrosinase-Tyrosine Reaction. Biochem. J. 31, 2162 (1937).
45. FELLMAN, J. H.: Epinephrine Metabolites and Pigmentation in the Central Nervous System in a Case of Phenylpyruvic Oligophrenia. J. Neurol. Neurosurg. Psychiat. 21, 58 (1958).
46. FITZPATRICK, T. B., S. W. BECKER, JR., A. B. LERNER, and H. MONTGOMERY: Tyrosinase in Human Skin: Demonstration of its Presence and of its Role in Human Melanin Formation. Science 112, 223 (1950).
47. FITZPATRICK, T. B., M. MIYAMOTO, and K. ISHIKAWA: The Evolution of Concepts of Melanin Biology. In: W. MONTAGNA and F. HU (Eds.): Advances in Biology of Skin, Vol. VIII. The Pigmentary System. p. 1. Oxford: Pergamon Press. 1967.
48. — — — The Evolution of Concepts of Melanin Biology. Arch. Dermatol. 96, 305 (1967).
49. FLESCH, P.: Inhibitory Action of Extracts of Mammalian Skin on Pigment Formation. Proc. Soc. Exptl. Biol. Med. 70, 136 (1949).
50. — The Epidermal Iron Pigments of Red Species. J. Invest. Dermatol. 51, 337 (1968).
51. FORSYTH, W. G. C., and V. C. QUESNEL: Intermediates in the Enzymic Oxidation of Catechol. Biochim. Biophys. Acta 25, 155 (1957).
52. FORSYTH, W. G. C., V. C. QUESNEL, and J. B. ROBERTS: Diphenylenedioxide-2,3-quinone: an Intermediate in the Enzymic Oxidation of Catechol. Biochim. Biophys. Acta 37, 322 (1960).
53. FOX, D. L.: Animal Biochromes and Structural Colours. London: Cambridge University Press. 1953.
54. FRAENKEL, G. K., J. M. HIRSHON, and C. WALLING: Detection of Polymerization Radicals by Paramagnetic Resonance. J. Amer. Chem. Soc. 76, 3606 (1954).
55. FRENK, E., M. A. PATHAK, G. SZABÓ, and T. B. FITZPATRICK: Selective Action of Mercaptoethylamines on Melanocytes in Mammalian Skin. Arch. Dermatol. 97, 465 (1968).
56. GORDON, M. (Ed.): Biology of Melanomas. New York: Academy of Science. 1948.
57. — Pigment Cell Growth. New York: Academic Press. 1953.
58. — Pigment Cell Biology. New York: Academic Press. 1959.
59. GRADY, F. J., and D. C. BORG: Electron Paramagnetic Resonance Studies on Melanins I. The Effect of pH on Spectra at Q Band. J. Amer. Chem. Soc. 90, 2949 (1968).
60. GREENSTEIN, J. P., F. C. TURNER, and W. V. JENRETTE: Chemical Studies on the

Components of Normal and Neoplastic Tissues. IV. The Melanin-Containing Pseudolo-globulin of the Malignant Melanoma of Mice. J. Natl. Cancer Inst. **1**, 377 (1940).

61. GRIFFITHS, L. A.: Detection and Identification of the Polyphenoloxidase Substrate of the Banana. Nature **184**, 58 (1959).

62. GROSS, A. J., and I. W. SIZER: The Oxidation of Tyramine, Tyrosine, and Related Compounds by Peroxidase. J. Biol. Chem. **234**, 1611 (1959).

63. HARLEY-MASON, J.: Melanin and its Precursors. Part I. The Synthesis of 5:6:5′:6′-Tetrahydroxyindigo. J. Chem. Soc. (London) 1244 (1948).

64. — The Chemistry of Adrenochrome and its Derivatives. J. Chem. Soc. (London) 1276 (1950).

65. — Melanin and its Precursors. Part VI. Further Syntheses of 5:6-Dihydroxyindole and its Derivatives. J. Chem. Soc. (London) 200 (1953).

66. HARLEY-MASON, J., and J. D. BU'LOCK: Synthesis of 5:6-Dihydroxyindole Derivatives: An Oxido-reduction Rearrangement Catalysed by Zinc Ions. Nature **166**, 1036 (1950).

67. HAWORTH, R. D.: The Chemical Nature of Humic Acid. Soil Science **111**, 71 (1971).

68. HEACOCK, R. A.: The Aminochromes. In: A. R. KATRITZKY (Ed.): Advances in Heterocyclic Chemistry **5**, p. 205. New York and London: Academic Press. 1965.

69. HEMPEL, K.: Investigation on the Structure of Melanin in Malignant Melanoma with ^{3}H- and ^{14}C-Dopa Labelled at Different Positions. In: G. DELLA PORTA and O. MÜHLBOCK (Eds.): Structure and Control of the Melanocyte, p. 162. Berlin-Heidelberg-New York: Springer. 1966.

70. HEMPEL, K.: Über Biosynthese und Struktur des tierischen Melanins. Z. Naturforsch. **22 B**, 173 (1967).

71. HOGEBOOM, G. H., and M. H. ADAMS: Mammalian Tyrosinase and Dopa Oxidase. J. Biol. Chem. **145**, 273 (1942).

72. HORAK, V., and J. R. GILLETTE: A Study of the Oxidation-Reduction State of Synthetic 3,4-Dihydroxy-DL-phenylalanine Melanin. Molecular Pharmacology **7**, 429 (1971).

73. HORI, Y.: Ultrastructural Study of ^{3}H Incorporation from 3,4-Dopa-2,3-^{3}H into Retinal Pigment Epithelium of Chick Embryo. In: V. RILEY (Ed.): Pigmentation, its Genesis and Biological Control, p. 143. New York: Appleton-Century-Crofts. 1972.

74. HORNER, L., and W. SPIETSCHKA: Zur Kenntnis der o-Chinone VI. Das Verhalten von o-Chinonen gegenüber tert. Aminen als Beitrag zum Vorgang der Melaninbildung. Liebigs Ann. Chem. **591**, 1 (1955).

75. HOROWITZ, N. H., M. FLING, and G. HORN: Tyrosinase (*Neurospora crassa*). In: H. TABOR and C. W. TABOR (Eds.): Methods in Enzymology, XVII. A, p. 615. New York and London: Academic Press. 1970.

76. KERTÉSZ, D.: The Phenol-Oxidizing Enzyme System of Human Melanomas; Substrate Specificity and Relationship to Copper. J. Nat. Cancer Inst. **14**, 1081 (1954).

77. — The Relative Oxygen Affinity of Human and Vegetal Phenoloxidase. J. Nat. Cancer Inst. **14**, 1093 (1954).

78. KERTÉSZ, D., and R. ZITO: Phenolase. In: O. HAYAISHI (Ed.): Oxygenases, p. 307. New York and London: Academic Press. 1962.

79. — — Mushroom Polyphenol Oxidase, I. Purification and General Properties. Biochim. Biophys. Acta **96**, 447 (1965).

80. KING, J. A. G., A. PERCIVAL, N. C. ROBSON, and G. A. SWAN: Studies Related to the Chemistry of Melanins. Part XI. The Distribution of the Polymeric Linkages in Dopa-melanin. J. Chem. Soc. (London) (C) 1418 (1970).

81. KIRBY, G. W., and L. OGUNKOYA: Structure of Melanin derived from ($\pm$)-3,4-Dihydroxy-[^{14}C, ^{3}H]phenylalanine by Oxidation with Tyrosinase. Chem. Commun. 546 (1965).

82. KODJA, A., and S. BOUCHILLOUX: Sur la Caracterisation des Orthoquinones Mono-cycliques au Cours de l'Oxidation d'Acides Aminés et Amines Orthodiphenoliques. Biochim. Biophys. Acta **41**, 345 (1960).

83. KUKITA, A., and T. B. FITZPATRICK: Demonstration of Tyrosinase in Melanocytes of the Human Hair Matrix by Autoradiography. Science **121**, 893 (1955).

84. LAXER, G., J. SIKORSKI, C. S. WHEWELL, and H. J. WOODS: The Electron Microscopy of Melanin Granules Isolated from Pigmented Mammalian Fibres. Biochim. Biophys. Acta **15**, 174 (1954).

85. LERNER, A. B.: Metabolism of Phenylalanine and Tyrosine. Adv. Enzymology **14**, 73 (1953).

86. LERNER, A. B., and T. B. FITZPATRICK: Biochemistry of Melanin Formation. Physiol. Rev. **30**, 91 (1950).

87. LONGUET-HIGGINS, H. C.: On the Origin of the Free Radical Property of Melanins. Arch. Biochem. Biophys. **86**, 231 (1960).

88. LUKIEWICZ, S.: The Biological Role of Melanin. I. New Concepts and Methodical Approaches. Folia Histochemica et Cytochemica **10**, 93 (1972).

89. LUND, N. A., A. ROBERTSON, and W. B. WHALLEY: The Chemistry of Fungi. Part XXI. Asperxanthone and a Preliminary Examination of Aspergillin. J. Chem. Soc. (London) 2434 (1953).

89a. McGOVERN, V. J., and P. RUSSELL (Eds.): Pigment Cell, Vol. 1: Mechanisms in Pigmentation (Series Ed.: V. RILEY). Basel: Karger. 1973.

90. MARSDEN, C. D.: Brain Pigment and its Relation to Brain Catecholamines. Lancet 475 (1965, part 2).

90a. — Brain Melanin. In: M. WOLMAN (Ed.): Pigments in Pathology, p. 395. New York: Academic Press. 1969.

91. MASON, H. S.: The Chemistry of Melanin. II. The Oxidation of Dihydroxyphenyl-alanine by Mammalian Dopa Oxidase. J. Biol. Chem. **168**, 433 (1947).

92. — The Chemistry of Melanin. III. Mechanism of the Oxidation of Dihydroxyphenyl-alanine by Tyrosinase. J. Biol. Chem. **172**, 83 (1948).

93. — The Chemistry of Melanin. VI. Mechanism of the Oxidation of Catechol by Tyrosinase. J. Biol. Chem. **181**, 803 (1949).

94. — Comparative Biochemistry of the Phenolase Complex. Adv. Enzymology **16**, 105 (1955).

95. — Mechanisms of Oxygen Metabolism. Adv. Enzymology **19**, 79 (1957).

96. — Structure of Melanins. In: M. GORDON (Ed.): Pigment Cell Biology, p. 563. New York: Academic Press. 1959.

97. — The Structure of Melanin. In: W. MONTAGNA and F. HU (Eds.): Advances in Biology of Skin, Vol. VIII, p. 293. Oxford: Pergamon Press. 1967.

98. MASON, H. S., W. L. FOWLKS, and E. PETERSON: Oxygen Transfer and Electron Transport by the Phenolase Complex. J. Amer. Chem. Soc. **77**, 2914 (1955).

99. MASON, H. S., D. J. E. INGRAM, and B. ALLEN: The Free Radical Property of Melanins. Arch. Biochem. Biophys. **86**, 225 (1960).

100. MASON, H. S., and E. W. PETERSON: Melanoproteins I. Reactions Between Enzyme-Generated Quinones and Amino-Acids. Biochim. Biophys. Acta **111**, 134 (1965).

101. MASON, H. S., L. SCHWARTZ, and D. C. PETERSON: The Allergenic Principles of Poison Ivy. IV. On the Mechanism of the Enzymatic Oxidation of Catechols. J. Amer. Chem. Soc. **67**, 1233 (1945).

102. MASON, H. S., and C. I. WRIGHT: The Chemistry of Melanin. V. Oxidation of Dihydroxyphenylalanine by Tyrosinase. J. Biol. Chem. **180**, 235 (1949).

103. MISURACA, G., R. A. NICOLAUS, G. PROTA, and G. GHIARA: A Cytochemical Study of Phaeomelanin Formation in Feather Papillae of New Hampshire Chick Embryos. Experientia **25**, 920 (1969).

104. Nairn, P. M., and L. C. Vining: Enzymic Oxidation of Catechol to Diphenylene-dioxide-2,3-quinone. Arch. Biochem. Biophys. **106**, 422 (1964).

105. Nelson, R. M., and H. S. Mason: Tyrosinase (Mushroom). In: H. Tabor, and C. W. Tabor (Eds.): Methods in Enzymology, XVII A, p. 626. New York and London: Academic Press. 1970.

106. Nicolaus, R. A.: Biogenesis of Melanins. Rassegna di Medicina Sperimentale **9**, Suppl. 1 (1962).

107. — Melanins. Paris: Hermann. 1968.

108. Nicolaus, R. A., K. Hempel, and H. S. Mason: Comments on Howard S. Mason's Paper "The Structure of Melanin". In: W. Montagna and F. Hu (Eds.): Advances in Biology of Skin, Vol. VIII, p. 313. Oxford: Pergamon Press. 1967.

109. Nicolaus, R. A., and M. Piattelli: Progress in the ·Chemistry of Natural Black Pigments. Rend. Accad. Sci. fis. mat. (Napoli) [4] **32**, 1 (1965).

110. Nicolaus, R. A., M. Piattelli, and E. Fattorusso: The Structure of Melanins and Melanogenesis. IV. On Some Natural Melanins. Tetrahedron **20**, 1163 (1964).

111. Nordgren, L., H. Rorsman, A.-M. Rosengren, and E. Rosengren: Dopa and Dopamine in the Pigment of Substantia Nigra. Experientia **27**, 1178 (1971).

112. Okun, M., L. Edelstein, N. Or, G. Hamada, and B. Donnellan: Histochemical Studies of Conversion of Tyrosine and Dopa to Melanin Mediated by Mammalian Peroxidase. Life Sciences Part II, **9**, 491 (1970).

113. Okun, M. R., L. M. Edelstein, N. Or, G. Hamada, G. Blumental, B. Donnellan, and J. Burnett: Oxidation of Tyrosine and Dopa to Melanin by Mammalian Peroxidase: The Possible Role of Peroxidase in Melanin Synthesis and Catecholamine Synthesis in Vivo. In: V. Riley: Pigmentation, its Genesis and Biological Control, p. 571. New York: Appleton-Century-Crofts. 1972.

114. Omote, Y., Y. Fujinuma, and N. Sugiyama: Synthesis and Melanogenesis of the DOPA Dimer. Bull. Chem. Soc. Japan. **42**, 1752 (1969).

115. Patel, R. P., M. R. Okun, L. M. Edelstein, and D. Epstein: Biochemical Studies of the Peroxidase-Mediated Oxidation of Tyrosine to Melanin: Demonstration of the Hydroxylation of Tyrosine by Plant and Human Peroxidases. Biochem. J. **124**, 439 (1971).

116. Piattelli, M., E. Fattorusso, and S. Magno: Isolation of Pyrrole-2,3,4-tricarboxylic Acid and Pyrrole-2,3,4,5-tetracarboxylic Acid from Sepiomelanin Oxidation Products. Tetrahedron Letters 718 (1961).

117. — — — Identificazione del 5,6-diossindolo nell'ossidazione enzimatica della dopa. Rend. Accad. Sci. fis. mat. (Napoli) [4] **28**, 168 (1961).

118. Piattelli, M., E. Fattorusso, S. Magno, and R. A. Nicolaus: The Structure of Melanins and Melanogenesis. II. Sepiomelanin and Synthetic Pigments. Tetrahedron **18**, 941 (1962).

119. — — — — The Structure of Melanins and Melanogenesis. III. The Structure of Sepiomelanin. Tetrahedron **19**, 2061 (1963).

120. Piattelli, M., E. Fattorusso, R. A. Nicolaus, and S. Magno: The Structure of Melanins and Melanogenesis. V. Ustilagomelanin. Tetrahedron **21**, 3229 (1965).

121. Piattelli, M., and R. A. Nicolaus: The Structure of Melanins and Melanogenesis. I. The Structure of Melanin in *Sepia*. Tetrahedron **15**, 66 (1961).

122. Pomerantz, S. H.: Separation, Purification, and Properties of Two Tyrosinases from Hamster Melanomas. J. Biol. Chem. **238**, 2351 (1963).

123. Pomerantz, S. H., and J. P.-C. Li: Tyrosinases (Hamster Melanoma). In: H. Tabor and C. W. Tabor (Eds.): Methods in Enzymology, XVII A, p. 620. New York and London: Academic Press. 1970.

124. Prota, G.: Structure and Biogenesis of Phaeomelanins. In: V. Riley: Pigmentation, its Genesis and Biological Control, p. 615. New York: Appleton-Century-Crofts. 1972.

125. PROTA, G., S. CRESCENZI, G. MISCURACA, and R. A. NICOLAUS: New Intermediates in Phaeomelanogenesis in vitro. Experientia 26, 1058 (1970).

126. PROTA, G., A. SUARATO, and R. A. NICOLAUS: The Isolation and Structure of Trichosiderin B. Experientia 27, 1381 (1971).

127. PRYOR, M. G. M.: Sclerotization. In: M. FLORKIN, and H. S. MASON (Eds.): Comparative Biochemistry, Vol. IV, Part B, p. 371. New York and London: Academic Press. 1962.

128. PUGH, C. E. M., and H. S. RAPER: The Action of Tyrosinase on Phenols. With some Observations on the Classification of Oxidases. Biochem. J. 21, 1370 (1927).

129. PULLMAN, A., and B. PULLMAN: The Band Structure of Melanins. Biochim. Biophys. Acta 54, 384 (1961).

130. QUILICO, A.: I pigmenti neri animali e vegetali. Pavia: Tip. Fusi. 1937.

131. RAPER, H. S.: The Aerobic Oxidases. Physiol. Rev. 8, 245 (1928).

132. RILEY, V. (Ed.): Pigmentation, its Genesis and Biological Control. New York: Appleton-Century-Crofts. 1972.

133. RILEY, V., and J. G. FORTNER (Eds.): The Pigment Cell: Molecular, Biological and Clinical Aspects. Ann. New York Acad. Sci. 100, (1963).

134. ROBSON, N. C., and G. A. SWAN: Studies on the Structure of Some Synthetic Melanins. In: DELLA PORTA, G., and O. MÜHLBOCK (Eds.): Structure and Control of the Melanocyte, p. 155. Berlin-Heidelberg-New York: Springer. 1966.

135. SACCHI, S., G. LANZI, and L. ZANOTTI: Electron Spin Resonance Research on Human Hairs under Varying Physiological and Experimental Conditions. In: W. MONTAGNA, and R. L. DOBSON (Eds.): Advances in Biology of Skin, Vol. IX, Hair Growth, p. 169. Oxford: Pergamon Press. 1969.

136. SENOH, S., and B. WITKOP: Formation and Rearrangements of Aminochromes from a New Metabolite of Dopamine and Some of its Derivatives. J. Amer. Chem. Soc. 81, 6231 (1959).

137. SIZER, I. W.: Oxidation of Proteins by Tyrosinase and Peroxidase. Adv. Enzymology 14, 129 (1953).

138. SMITH, P. I., and G. A. SWAN: Unpublished work.

139. SNELL, R. S.: Hormonal Control of Pigmentation in Man and other Mammals. In: W. MONTAGNA, and F. HU (Eds.): Advances in Biology of Skin, Vol. VIII, p. 447. Oxford: Pergamon Press. 1967.

140. SWAN, G. A.: Chemical Structure of Melanins. Ann. New York Acad. Sci. 100, 1005 (1963).

141. — Some Studies on the Formation and Structure of Melanins. Rend. Accad. Sci. fis. mat. (Napoli) [4] 31, 1 (1964).

141a. SWAN, G. A.: Current Knowledge of Melanin Structure. In: V. J. MCGOVERN and P. RUSSELL (Eds.): Pigment Cell, Vol. 1: Mechanisms in Pigmentation, p. 151. Basel: Karger. 1973.

142. SWAN, G. A., and A. WAGGOTT: Studies Related to the Chemistry of Melanins. Part VI. Syntheses of 3-Carboxypyrrole-2-acetic Acid, 3,5-Dicarboxypyrrole-2-acetic Acid, and Related Compounds. J. Chem. Soc. (London) (C) 285 (1970).

143. — — Studies Related to the Chemistry of Melanins. Part X. Quantitative Assessment of Different Types of Units present in Dopa-melanin. J. Chem. Soc. (London) (C) 1409 (1970).

144. SWAN, G. A., and D. WRIGHT: A Study of the Evolution of Carbon Dioxide during Melanin Formation, including the Use of 2-(3:4-Dihydroxyphenyl) [1-^{14}C]- and 2-(3:4-Dihydroxyphenyl) [2-^{14}C]-ethylamine. J. Chem. Soc. (London) 381 (1954).

145. SWAN, G. A., and D. WRIGHT: A Study of Melanin Formation by Use of 2-(3:4-Dihydroxy-[3-^{14}C]phenyl)-, 2-(3:4-Dihydroxy[4-^{14}C]phenyl)-, and 2-(3:4-Dihydroxy[5-^{14}C]phenyl)-ethylamine. J. Chem. Soc. (London) 1549 (1956).

146. Takahashi, H., and T. B. Fitzpatrick: Large Amounts of Deoxyphenylalanine in the Hydrolysate of Melanosomes from Harding-Passey Mouse Melanoma. Nature **209**, 888 (1966).
147. Thathachari, Y. T., and M. S. Blois: Physical Studies on Melanins. II. X-Ray Diffraction. Biophys. J. **9**, 77 (1969).
148. Thomas, M.: Melanins. In: K. Paech, and M. V. Tracey (Eds.): Modern Methods of Plant Analysis, Vol. IV, p. 661. Berlin-Göttingen-Heidelberg: Springer. 1955.
149. Thomson, R. H.: Melanins. In: M. Florkin, and H. S. Mason (Eds.): Comparative Biochemistry, Vol. III, Part A, p. 727. New York and London: Academic Press. 1962.
150. — Some Naturally Occurring Black Pigments. In: T. S. Gore, B. S. Joshi, S. V. Sunthankar, and B. D. Tilak (Eds.): Recent Progress in Chemistry of Natural and Synthetic Colouring Matters and Related Fields, p. 99. New York and London: Academic Press. 1962.
151. Tollin, G., and C. Steelink: Biological Polymers Related to Catechol: Electron Paramagnetic Resonance and Infrared Studies of Melanin, Tannin, Lignin, Humic Acid, and Hydroxyquinones. Biochim. Biophys. Acta **112**, 377 (1966).
152. Van Woert, M. H.: Reduced Nicotinamide-Adenine Dinucleotide Oxidation by Melanin: Inhibition by Phenothiazines. Proc. Soc. Exp. Biol. Med. **129**, 165 (1968).
153. — Activation of Tyrosinase by Chlorpromazine. In: V. Riley (Ed.): Pigmentation, its Genesis and Biological Control, p. 503. New York: Appleton-Century-Crofts. 1972.
154. Van Woert, M. H., K. N. Prasad, and D. C. Borg: Spectroscopic Studies of *Substantia Nigra* Pigment in Human Subjects. J. Neurochem. **14**, 707 (1967).
155. Vercauteren, R., and L. Massart: Model Oxygenases and Theoretical Considerations on the Activation of Oxygen. In: O. Hayaishi (Ed.): Oxygenases, p. 355. New York and London: Academic Press. 1962.
155a. Wassermann, H. P.: Melanin Pigmentation and the Environment. In: Essays on Tropical Dermatology: Excerpta Medica Monograph, p. 7, 1969.
156. Waters, W. A.: Comments on the Mechanism of One-electron Oxidation of Phenols: A Fresh Interpretation of Oxidative Coupling Reactions of Plant Phenols. J. Chem. Soc. (London) (B) 2026 (1971).
157. White, L. P.: Melanin: A Naturally Occurring Cation Exchange Material. Nature **182**, 1427 (1958).
158. Witz, D. F., E. J. Hessler, and T. L. Miller: Bioconversion of Tyrosine into the Propylhygric Acid Moiety of Lincomycin. Biochemistry **10**, 1128 (1971).
159. Yasunobu, K. T.: Mode of Action of Tyrosinase. In: M. Gordon (Ed.): Pigment Cell Biology, p. 583. New York: Academic Press. 1959.
160. Yasunobu, K. T., E. W. Peterson, and H. S. Mason: The Oxidation of Tyrosine-containing Peptides by Tyrosinase. J. Biol. Chem. **234**, 3291 (1959).

Addendum

Okun, Edelstein et al. have strengthened their case for hydroxylation of tyrosine in mammalian melanogenesis by peroxidase (see II.3), and have shown that a "tyrosinase" from melanoma, although able to oxidize dopa, had no ability to oxidize tyrosine (*161*). Moreover the author of this review has repeated certain experiments of Patel, Okun and Edelstein (private communication) and agrees that the results imply that tyrosine can be hydroxylated by peroxidase in the presence of hydrogen peroxide and dopa *in vitro*.

161. Patel, R. P., M. R. Okun, W. A. Yee, G. F. Wilgram, and L. M. Edelstein: Inability of Murine Melanoma "Tyrosinase" (Dopa oxidase) to Oxidize Tyrosine in the Presence or Absence of Dopa or Dihydroxyfumarate Cofactor. J. Invest. Dermatol. **61**, 55 (1973).

(Received February 9, 1973)

Mechanisms of Corrin Dependent Enzymatic Reactions

By G. N. Schrauzer, La Jolla, California, U.S.A.

Contents

I. Introduction

The mechanisms of corrin and of coenzyme B_{12} dependent enzymatic reactions are complex and have been the subject of intense experimental effort, discussion and controversy. The difficulties preventing the understanding of the enzymatic function of corrinoid coenzymes were initially caused by the lack of information on the inorganic and organo-metallic reactions of cobalt in tetradentate, strong-field complexes. Since the discovery and study of vitamin B_{12} model compounds a wealth of experimental and theoretical data on the chemistry of cobalt in complexes related to vitamin B_{12} became available, which is now recognized to be relevant to the chemistry of vitamin B_{12} itself. Indeed, most of the known reactions of vitamin B_{12} have been duplicated with simpler model complexes, and vice versa. The detailed investigation of the nonenzymatic reactions of vitamin B_{12} and of vitamin B_{12} model compounds subsequently led to the formulation of mechanisms of some of the known corrin dependent enzymatic reactions and to a number of surprising qualitative and quantitative experimental correlations between nonenzymatic and enzymatic data.

In the present review we will first describe some of the more important properties and reactions of corrins and of vitamin B_{12} model compounds. This will be followed by an outline and a discussion of the mechanisms of corrin dependent enzymatic reactions. Coverage of the material is not comprehensive. It is intended to give a summary of the reactions of corrins in enzymes from a fundamental chemical point of view. Details concerning the isolation, properties and sources of the enzymes in question will not be given unless such information is mechanistically relevant. As a general policy, author's names are not included in the text. This should in no way be considered as an attempt to diminish the significance or priority of discoveries of any worker in this field. The literature is covered approximately up to October, 1972, but some later work is also quoted.

II. Nomenclature

An abbreviated nomenclature will be used for corrinoid coenzymes as well as vitamin B_{12} model compounds. In most equations, corrins will be represented by [Co], usually without including the axial base component. Coenzyme B_{12} is α-(5,6-dimethylbenzimidazolyl)-Co-5-deoxy-adenosylcobamide (Chart 1). Vitamin B_{12a} is hydroxocobalamin; vitamin B_{12r} and B_{12s} are the Co(II) and Co(I) derivatives of cobalamin, respectively. Cobaloximes are derivatives of bis-dimethylglyoximatocobalt and will be abbreviated by (Co); axial base components will only be shown where necessary. All other abbreviations and nomenclature will be explained in the Text.

III. Properties and Reactions of Corrins and Related Compounds

1. Vitamin B_{12} Chemistry

The biological essentiality of corrins is intimately linked to the presence of cobalt in a special ligand environment which endows the metal with a maximum reactivity and unique catalytic activities. The cobalt atom in corrins possesses the remarkable property of forming stable organometallic derivatives. One of the naturally occurring organocobalamins is coenzyme B_{12} (Chart 1) (1, 2). Although compounds of this type are often described as Co(III) complexes with carbanionic ligands (3), this formulation is rather unrealistic, since most organocorrins exhibit the behavior typical of covalent organometallics. The Co-C bond is resistant to hydrolysis in simple alkylcobalt derivatives. While certain substituted alkylcobalamins undergo Co-C bond cleavage on interaction with acids or bases, these compounds have little in common with organomagnesium halides and should not be considered as "biological" Grignard Reagents" (4). Cleavage of the Co-C bond is observed on reaction with reductants, oxidants, on thermolysis and light-irradiation. Strong reducing agents convert Co(III) and Co(II) derivatives of vitamin B_{12} into the Co(I) form, vitamin B_{12s} (5, 6). The cobalt ion in vitamin B_{12s} is an exceedingly powerful nucleophile, presumably the most powerful biogenic nucleophile of all (7—9). The cobalt(I) ion owes

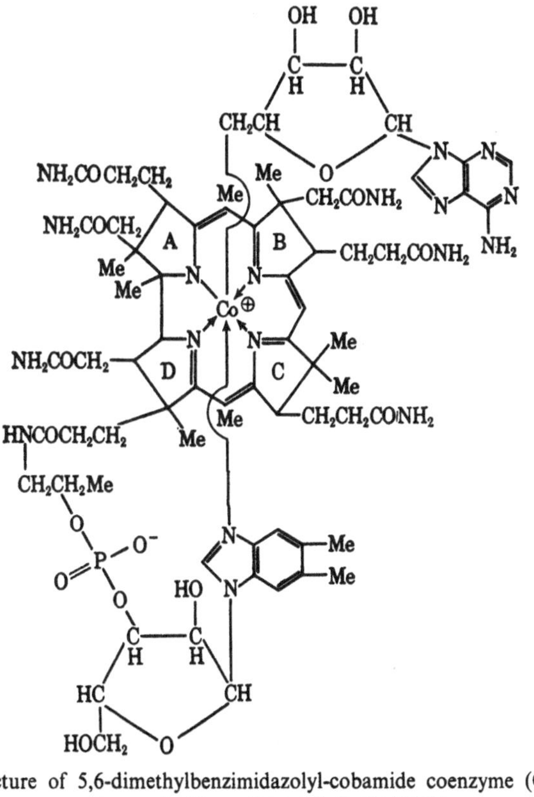

Chart 1. Structure of 5,6-dimethylbenzimidazolyl-cobamide coenzyme (Coenzyme B_{12}). Accounts Chem. Research 1, 97 (1968)

its "supernucleophilicity" to the presence of two electrons in the weakly antibonding $3d_{z2}$ orbital. This generates regions of high electron density perpendicular to the plane of the corrin. As a consequence of it high nucleophilicity, the Co(I) ion is readily alkylated (7—10).

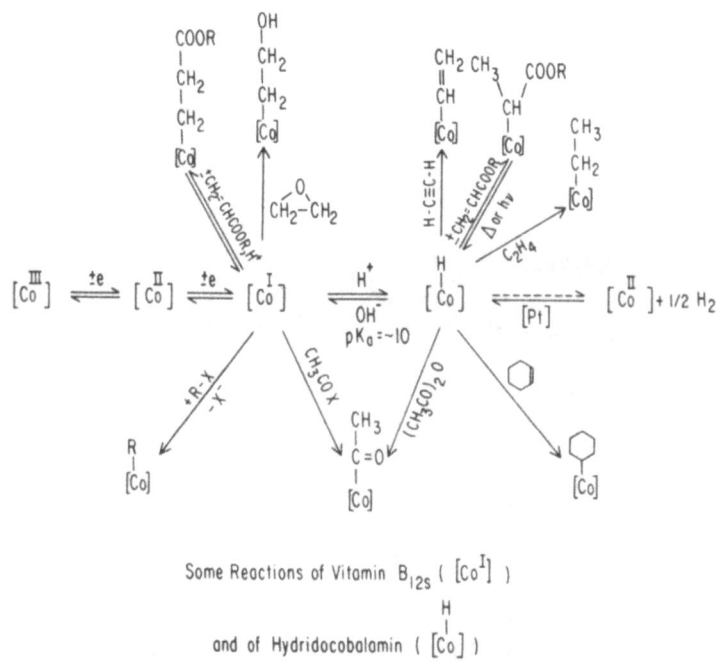

The Brønsted acid of the Co(I) nucleophile (hydridocobalamin) may be generated by reducing aquocobalamin with Zn dust in glacial acetic acid (11). In contrast to the free Co(I) nucleophile, hydridocobalamin reacts with normal, nonactivated olefins (11). Some of the reactions of both vitamin B_{12s} and of hydridocobalamin are summarized in Chart 2.

Some Reactions of Vitamin B_{12s} ($[Co^I]$)

and of Hydridocobalamin ($[\overset{H}{Co}]$)

Chart 2

It is instructive to compare corrins with cobalt porphyrins. The latter do not occur in nature and possess properties substantially different from corrins. They cannot be reduced to the Co(I) state in aqueous solution due to their higher Co(II)/Co(I) reduction potential, in contrast to vitamin B_{12}. This is presumably due to the presence of the additional methine bridge in the porphyrin skeleton. In the corrins, the rings A and D are joined directly, causing a deviation from quadratic coordination geometry, and a change of the effective charge of the ligand system from –2 in the porphyrin to –1 in the corrin. Both effects lower the coordination power of the corrin ligand relative to the porphyrin.

Nature utilizes cobalt most probably because of its exceptionally high nucleophilicity in the univalent state and its ability to form organometallic derivatives. Of other elements only rhodium rivals cobalt to some extent, although the nucleophilicity of Rh(I) is lower than that of Co(I) in identical ligand environments (12). A Rh-B_{12}, if capable of existence should have little or no biological activity.

(2)   (3)

The cobalt-free corrin ligand system is synthesized by several microorganisms, e.g. photosynthetic bacteria as well as by *Streptomyces olivaceus* grown in a cobalt deficient medium (13, 14). *Pseudomonas denitrificans*, an efficient vitamin B_{12} producing microorganism does not yield Rh-B_{12} if grown in a medium containing rhodium and cobalt in the ratio of 100:1. It appears that the enzymes responsible for assembly or the insertion of cobalt into the corrin ring are highly specific and are incapable of utilizing rhodium (15).

2. Vitamin B_{12} Model Compounds

The cobalt ion acquires properties typical of cobalt in vitamin B_{12} in planar tetradentate ligand fields of sufficient strength to stabilize Co(I) ($=d^8$) in the spin-paired state. The nature of ligands and details of their electronic structure are of secondary importance; a relatively large number of cobalt chelates of unrelated ligands are known which to some extent qualify as vitamin B_{12} models (Chart 3), but similarities with vitamin B_{12} are most pronounced in the cobaloximes derived from bis-dimethylglyoximato cobalt (16, 17). Cobaloximes mimick most of the reactions of the cobalt ion of vitamin B_{12} qualitatively, and many also quantitatively. Differences are noted in the rates of axial ligand exchange in Co(III) derivatives, which are exceedingly rapid in Co(III) derivatives of corrins such as hydroxocobalamin (18), whereas cobaloximes are more inert *(19, 20)*. Lability with respect to axial ligand exchange is not a specific property of corrins, however, and is also observed in cobalt- and other metal porphyrins, for example *(21)*. It appears that high

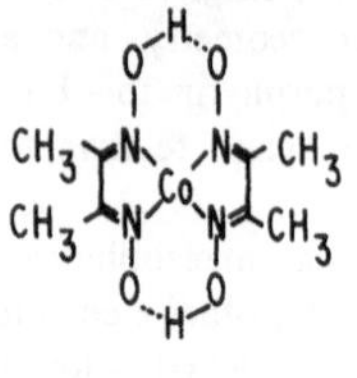

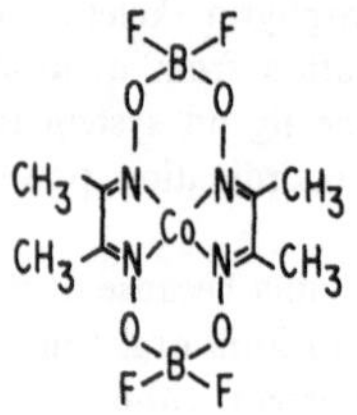

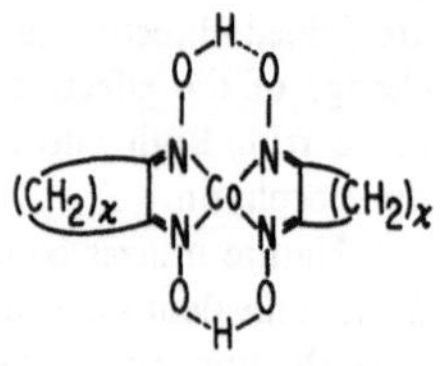

Chart 3. Some vitamin B_{12} model compounds

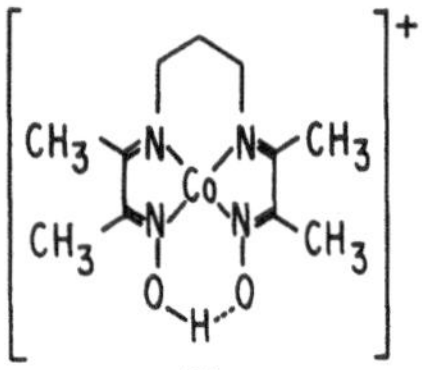

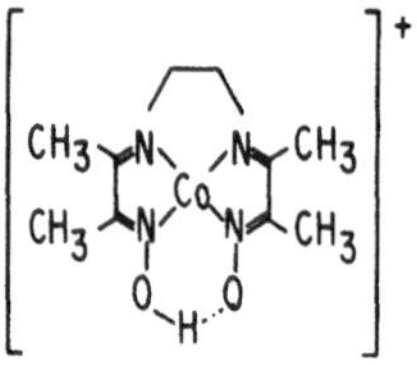

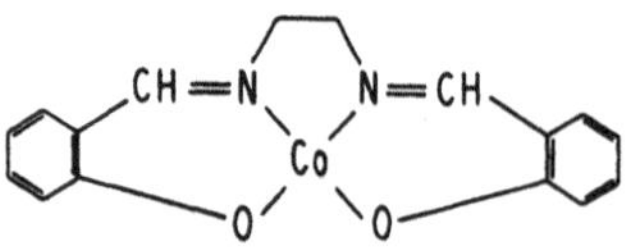

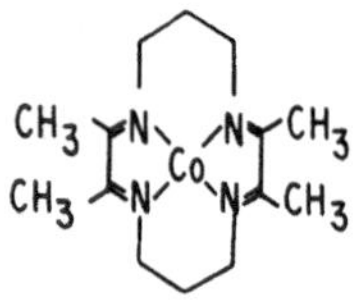

Table 1. *Nucleophilic Reactivity Constants Toward Methyl Iodide*[a]

Nucleophile	
CH_3OH	0.00
Cl^-	4.37
NH_3	5.50
Br^-	5.79
I^-	7.42
$(n-C_4H_9)_3P$	8.69
$S_2O_3^{2-}$	8.95
$(C_6H_5)_3Sn^-$	~ 11.5
$(C_6H_5)_3Ge^-$	~ 12
Cobaloxime$_s$·$P(n-C_4H_9)_3$	13.3
Cobaloxime$_s$·pyridine	13.8[b]
Cobaloxime$_s$(aqua)	14.3[b]
Vitamin B_{12s}	14.4

[a] Except for the last four entries, data taken from *(23)*.
[b] Calculated from relative rates of reaction with n-propyl chloride.

rates of metal ligation are characteristic of complexes with highly π-delocalized ligand systems. The ESR spectra of Co(II) derivatives of cobaloximes are very similar to those of vitamin $B_{12\,r}$, indicating that the unpaired electron is localized predominantly in the $3d_{z2}$ orbital (22). The nucleophilicity (23) of cobaloxime(I) derivatives depends on the nature of the axial base component (8—10). With nitrogen bases, OH^- or water occupying the axial position the nucleophilicity of cobaloximes is virtually identical with vitamin B_{12s} (Table 1).

Other cobalt chelates resemble the corrins less well and many exhibit properties not shared by them. For example, chelate (7) forms uncharged bis-alkylderivatives (24), which thus far could not be obtained in the corrin series. In other complexes (e. g. in 9) the Co(II)/Co(I) reduction potentials are too high to permit generation of the Co(I) nucleophiles under conditions optimal for preparation of vitamin $B_{12\,s}$. Most of the observed differences are caused by variations of in-plane ligand strengths, steric effects, or effective charge on cobalt. The surprising similarity of cobaloximes with corrins is due to a fortuitous combination of variables causing the effective ligand-strength of the bis-dimethylglyoximato-moiety to become virtually equal to that of the corrin system. Molecular orbital calculations furthermore indicate that the molecular orbitals involved in axial bonding in cobaloximes and cobalamins have very similar eigenvalues and eigenvectors (25).

3. Coenzyme B_{12} Reactions

Coenzyme B_{12} reactions will be discussed separately in view of their importance in enzymatic reactions. Coenzyme B_{12} is formally a 2-substituted alkoxyethylcobalt derivative. Compounds of this type undergo Co-C bond cleavage on interaction with acid (26, 27):

$$\begin{array}{c} \overset{\displaystyle OR}{\underset{\displaystyle CH_2}{\mid}} \\ \overset{\displaystyle \mid}{\underset{\displaystyle CH_2}{}} \\ \overset{\displaystyle \mid}{(Co)} \end{array} \quad \xrightarrow[-ROH]{+H^+} \quad (Co^{III}) + CH_2 = CH_2$$

This reaction may proceed via an intermediate complex of the olefin with the (Co^{III}) complex or an equivalent species, as evidenced from solvolysis experiments with carboxyethylderivatives (28, 29). Coenzyme B_{12} is irreversibly decomposed by acids, affording hydroxocobalamin, adenine and 2,3-dihydroxybutenal (8) (30):

$$\text{adenine} + CH_2 = CHCH - CHCH = O + B_{12a} + X^-$$

$$(8)$$

Photolysis of the enzyme causes initial homolysis of the Co-C bond to yield vitamin B_{12r} and the 5'-deoxyadenosyl radical (9). The latter terminates in various ways depending on the reaction conditions. Aerobically it is oxidized mainly to 5'-adenosine aldehyde (**10**) (*31*). Anaerobically, the cyclic adenosine (**11**) is formed (*32*). In the presence of thiols or of organic compounds containing aliphatic hydrogen, 5-deoxyadenosine, (**12**) is produced (*33, 34*).

The only reversible free-radical reaction of coenzyme B_{12} is the recombination of the deoxyadenosyl radical with the Co(II) ion of the corrin; it occurs only if the organic radical remains in the coordination sphere of cobalt. Homolytic Co-C bond cleavage of coenzyme B_{12} has been suggested as the possible mechanism of coenzyme activation in enzymatic reactions (*35, 36*). This is unlikely, however, for reasons to be outlined in Chapter IV, 1—3.

The Co-C bond in coenzyme B_{12} is cleaved in alkali by β-elimination, affording 4'-5'-didehydro-5'-deoxyadenosine (**13**) and the Co(I) nucleophile (*37, 38*):

adenine

HO—
HO—   O

CH_2 + OH$^-$ ⇌ [Co^I]$^-$ +   adenine HO— HO— O

$\overset{H}{\underset{H}{\diagdown}}$ + H$_2$O

[Co]

(**13**)

This reaction yields the Co(I) nucleophile in the absence of reducing agent, is reversible, and provides an attractive model of coenzyme activation and regeneration in enzymatic systems.

Coenzyme B_{12} biosynthesis occurs in many microorganisms. The enzymes utilize ATP as the source of the coenzyme adenosyl group and operates under reducing conditions (*39, 40*). The reaction has not yet been duplicated nonenzymatically. Although coenzyme B_{12} is accessible on a preparative scale by the reaction of 5'-adenosyltosylate-derivatives with vitamin B_{12s} (*6, 41, 42*), ATP, ADP, AMP and 3',5'-cyclic AMP do not react with the Co(I) nucleophile non-enzymatically to form a Co-C bond (*43*). The formation of tripoly-phosphate in coenzyme B_{12} biosynthesis suggests that the enzyme specifically activates the adenosyl moiety in the 5'-position:

ATP + [B_{12s}]-Enzyme ⟶ Coenzyme B_{12} + Tripolyphosphate

The 5'-hydrogen atoms at the adenosyl moiety of coenzyme B_{12} are reversibly transferred to substrates, intermediates or products in co-enzyme B_{12} dependent enzymatic reactions (*44*). Nonenzymatically, the $5'$-protons of coenzyme B_{12} are nonexchangable, indicating a specific activation of the coenzyme on interaction with the apoenzymes. A possible mechanism of the hydrogen exchange reactions will be discussed in Chapter IV.

IV. Coenzyme B_{12} Dependent Enzymatic Reactions

Presently known coenzyme B_{12} dependent enzymes are given in Table 2.

1. Dioldehydrases

Isolated from bacterial sources such as *Aerobacter aerogenes* (DL-propanediol hydrolase) (*45*), but also from *Lactobacillus* (glycerol dehydrase) (*46*), these enzymes catalyze the conversion of vic. diols to aldehydes. They show a specific requirement for coenzyme B_{12} and a univalent alkali metal ion, usually potassium. The overall reaction is:

$$R-\underset{CH_2-OH}{\overset{CH-OH}{|}} \xrightarrow{\text{Coenzyme-Enzyme Complex, K}^+} R-\underset{CH=O}{\overset{CH_2}{|}} + H_2O \quad (1)$$

The substrate specificity is relatively high; propandiol dehydrase utilizes ethylene glycol as well, but not butanediols, mercaptoethanol, erythritol and a variety of other potential substrates, which often act as inhibitors of the reaction (*45, 47, 48*). The net reaction involves a displacement of a hydroxyl group by hydrogen, e. g. from C-2 of propanediol, affording a hydrate of propionaldehyde. Depending on whether the substrate is (S) or (R) propanediol, oxygen at the C-2 position does, or does not, appear in the product propionaldehyde, respectively (*49*). When coenzyme tritiated at the C-5′ position is used for the assay, tritium appears in the product propionaldehyde, while the coenzyme acquires hydrogen in the C-5′ position from the substrate (*44*). This observation has led to the postulate that the coenzyme catalyzes the dehydration of the diols by abstracting hydrogen from the substrate, and subsequently donates it back to the product. Difficulties arise, however, if this interpretation of the H-T transfer phenomena is used to formulate a mechanism of dioldehydrase action, especially since incubation of propionaldehyde with dioldehydrase holoenzyme containing T-labeled coenzyme also leads to H-T exchange (*44*). For this reason detailed model studies were carried out to provide a chemically more plausible mechanism.

Table 2. Coenzyme B_{12} Dependent Enzymes

Enzyme:	Source(s):	Reaction catalyzed:	Mol. Wt.:	Cofactors:
Dioldehydrase	*Aerobacter aerogenes*	$CH_3CH(OH)CH_2OH \rightarrow CH_3CH_2CHO + H_2O$	?	$K^+(NH_4^+)$
Glyceroldehydrase Ethanolamine deaminase	*Lactobacillus sp.* *Clostridium sp.*	$CH_2(OH)CH(OH)CH_2OH \rightarrow CH_2(OH)CH_2CHO + H_2O$ $CH_2(NH_2)CH_2OH \rightarrow CH_3CHO + NH_3$	188,000 520,000	$K^+(NH_4^+)$ $K^+(NH_4^+, Rb)$
Ribonucleotide reductase	*Lactobacillus leichmannii* (and other organisms)	$R(SH)_2 + GTP \rightarrow R(S_2) + dGTP + H_2O$	70,000	Effector (e.g. dGTP)
Methylmalonyl-CoA Mutase	*Propionibacterium shermanii* Sheep liver	$HO_2CCH(CH_3)COCoA \rightleftharpoons HO_2CCH_2CH_2COCoA$	56,000 165,000	
Glutamate Mutase	*Clostridium tetanomorphum* (and other organisms)	$HO_2C(CH_2)_2\overset{NH_3^+}{CH}CO_2H \rightleftharpoons HO_2C\overset{CH_3}{CH}-\overset{NH_3^+}{CH}COOH$	17,000	
α-Methylene-glutarate	*Clostridium barkeri*	$HO_2C(CH_2)_2\overset{CH_2}{\underset{\parallel}{C}}-CO_2H \rightleftharpoons HO_2C\overset{NH_3^+}{CH}-\overset{NH_3^+}{C}-COOH$	ca. 170,000	
L-β-lysine aminomutase Mutase	*Clostridium sticklandii*	$\overset{NH_3^+}{CH_2}CH_2CH_2\overset{NH_3^+}{CH}CH_2CO_2^- \rightleftharpoons CH_3-\overset{NH_3^+}{CH}\ CH_2\overset{NH_3^+}{CH}CH_2CO_2^-$	(I) ca. 60,000 (II) 160,000	(a) $K^+(Rb^+), Mg^{++}(Mn^{++})$. ATP, pyruvate
D-α-lysine aminomutase	*Clostridium sticklandii*	$\overset{NH_3^+}{CH_2}(CH_2)_3\overset{NH_3^+}{CH}CO_2^- \rightleftharpoons CH_3-CH(CH_2)_2\overset{NH_3^+}{CH}CO_2^-$	(I) ca. 60,000 (II) 160,000	$K^+(Rb^+), Mg^{++}(Mn^{++})$. ATP, pyridoxal phosphate
Ornithine aminomutase	*Clostridium sticklandii*	$\overset{NH_3^+}{CH_2}(CH_2)_2\overset{NH_3^+}{CH}CO_2^- \rightleftharpoons CH_3\overset{NH_3^+}{CH}CH_2\overset{}{CH}CO_2^-$	(I) ca. 60,000 (II) 160,000	none (?)

(a) Two components; protein II binds the coenzyme.

A model reaction of possible relevance to the catalytic process of
dioldehydration was observed in a study of the reactions of 2-hydroxy-
alkyl-cobaloximes (26). These compounds are readily available from
the reaction of active glycol derivatives such as olefin oxides or
2-haloalcohols, and decompose in alkaline medium to yield aldehydes or
ketones depending on the structure of the hydroxyalkyl residue attached
to cobalt. Acetaldehyde is formed from 2-hydroxyethylcobaloximes (14)
in alkali:

$$
\begin{array}{ccc}
\underset{\text{(14)}}{\underset{\text{(Co)}}{\overset{CH_2-OH}{\underset{|}{\overset{|}{\underset{CH_2}{}}}}}} & \xrightarrow[-H_2O]{+OH^-} & \underset{\text{(Co)}}{\overset{H}{\overset{|}{\underset{CH_2}{\overset{CH\overset{\frown}{-}O^-}{|}}}}} & \longrightarrow & CH_3-CH=O+(Co^I)^-
\end{array}
$$

The driving force for this reaction appears to be the formation of the
C=O bond from the 2-hydroxyethylate anion, which induces a 1,2-hydride
shift or equivalent process, leading to the formation of acetaldehyde and
the cobaloxime(I) nucleophile. In the presence of excess glycol derivative
the reaction becomes catalytic, since the Co(I) nucleophile may be
re-alkylated, and is regenerated in the process of aldehyde formation.
Hydroxyethylcobalamin and -cobinamide behave similarly, providing
an initial basis for the understanding of the mechanism of dioldehydrase
action. Several extensions are required, however, to accommodate eq. 2
with the available enzymological evidence. First, it is not at all clear
how the Co(I) nucleophile could form from the coenzyme in the
absence of a reducing agent. Secondly, eq. 2 does not account for the
fact that glycols as such do not react with the Co(I) nucleophiles unless
activated. Thirdly, eq. 2 does not provide any clue as to how tritium
from the 5′-position of labeled coenzyme could possibly be transferred
to substrate or product, and vice versa.

If the coenzyme is to show catalytic activity, its Co-C bond must be
broken at the active site of the holoenzyme, and this process must be
reversible, i.e. the coenzyme is regenerated after completion of the
reaction. A mechanism by which this could occur is the base-induced
β-elimination described in III,3. This would yield the Co(I) nucleophile
plus enzyme-bound 4′,5′-didehydro-5′-deoxyadenosine (13) as such
or in a protonated form. Assumption of this process of "coenzyme activation" leads to an interpretation of the H-T exchange phenomena occurring
on incubation of the enzyme with T-5′-C-labeled coenzyme. The
exchange would have to occur between the product aldehyde and the
reactive adenosine derivative at the active site prior to equilibration

with either water or product outside the enzyme (B is an unspecified Brønsted base in the vicinity of the coenzyme active site (*37, 38*).

The exchange reaction has been simulated by model experiments, and the extent of tritium uptake of the coenzyme from labeled substrate was calculated on the basis of this mechanism with excellent agreement with reported experimental data (*37, 38*).

The inability of glycols to undergo nucleophilic displacement reactions with Co(I) nucleophiles under nonenzymatic conditions necessitates the assumption of activation of the substrate by the enzyme.

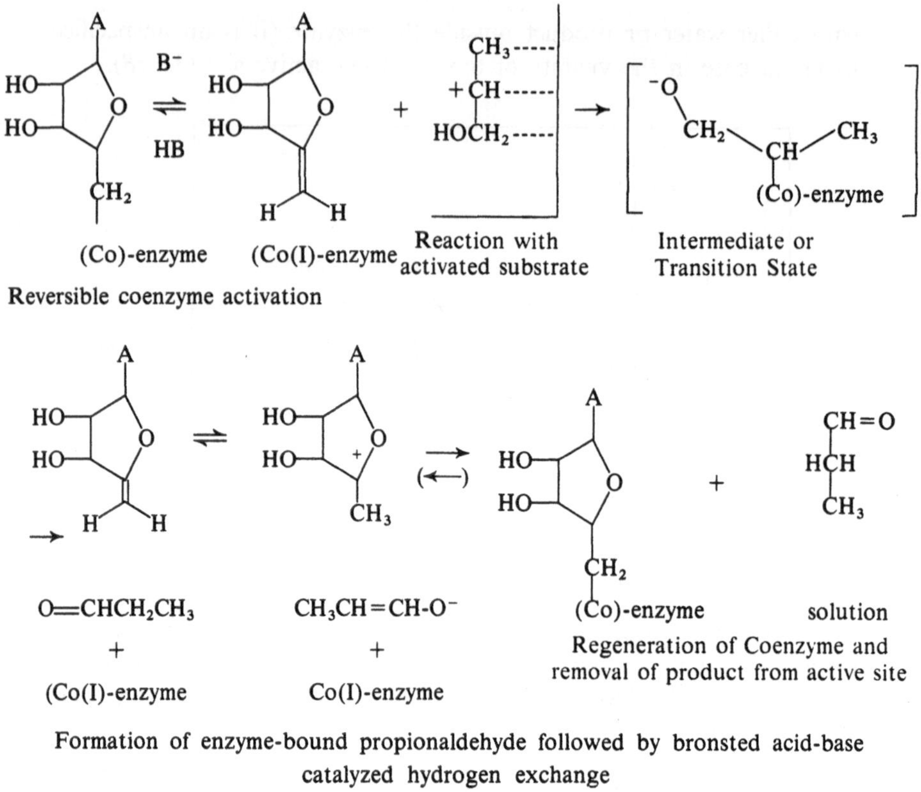

Reversible coenzyme activation

(Co)-enzyme   (Co(I)-enzyme   Reaction with activated substrate   Intermediate or Transition State

$O=CHCH_2CH_3$ $CH_3CH=CH-O^-$ (Co)-enzyme   solution

+   +

(Co(I)-enzyme   Co(I)-enzyme   Regeneration of Coenzyme and removal of product from active site

Formation of enzyme-bound propionaldehyde followed by bronsted acid-base catalyzed hydrogen exchange

Chart 4. Proposed mechanism of the enzymatic dehydration of propanediol *(37, 38)*

Since (R)- or (S)-propanediol behave differently, this activation must be highly specific and was postulated to lead to the formation of an insipient carbonium ion on either C-1 or C-2 of the (R) or (S) propane-diol substrate *(38)*.

The dotted lines symbolize enzyme-substrate binding interactions. Eqs. a and b are consistent with the established stereochemical course of the reaction on C-1 of 1,2-propanediol and the results of ^{18}O distribution experiments, which indicate that ^{18}O from C-2 of propane-diol appears in the aldehyde group only if the initial configuration of substrate is R *(49)*.

A direct SN 2 displacement of the 2-OH group of propanediol by the enzyme-bound Co(I) nucleophile is possible in principle, but is considered less likely in view of the observed net inversion of con-figuration at C-2 and the ^{18}O isotope distribution results. Retention of configuration, although a planar carbonium ion is formed, has been explained in other cases by the enzyme holding the carbonium ion tightly enough to prevent rotation *(38)*.

References, pp. 621—629

A schematic representation of the essential steps of the enzymatic diol dehydration is given in Chart 4. Apart from the available evidence from model studies, this mechanism is supported by the observed inactivation of dioldehydrase holoenzyme by N_2O (*50*). The latter is a specific oxidant for the Co(I) nucleophile of corrins and vitamin B_{12} model compounds, oxidizing them to the Co(II) derivatives (*51*). The degree of N_2O inactivation depends on the source and age of the enzyme preparation. Pre-exposure of the enzyme to air leads to disappearance of N_2O inactivation and also reduces oxygen sensitivity of the holoenzyme, while no significant loss of enzymatic activity is observed (*52*). The pre-exposure of the apoenzyme to air apparently causes oxidation of sulfhydryl groups to disulfide, which produces a less oxygen sensitive and N_2O insensitive modification of holoenzyme. Oxidized apoenzyme may be partially reduced back into the sulfhydryl form by reaction with dithiols such as dihydrolipoate (*52*). It appears that both O_2 and N_2O inactivate the reduced holoenzyme rapidly during the initial lag-period after incubation of the coenzyme with the apoenzyme. In this phase of the reaction the coenzyme is bound to the active site and converted into the functional form by Co-C bond cleavage.

Evidence for the presence of sulfhydryl groups in the vicinity of the active site was independently obtained by other workers (*53, 54*). The SH groups are apparently in some way connected with coenzyme binding but not involved in enzymatic catalysis. The N_2O sensitive holoenzyme also appears to be inactivated by carbon monoxide (*53*). These observations demonstrate the presence of the Co(I) nucleophile in one active modification of the holoenzyme; the air-exposed apoenzyme gives rise to a holoenzyme in which the Co(I) nucleophile is protected against inactivation by gaseous oxidants, presumably due to a conformational change in the apoenzyme as a result of $S-S-$ bond formation. The presence of $4',5'$-didehydro-$5'$-deoxyadenosine in active holoenzyme has not yet been demonstrated, but a number of unidentified nucleosides were noted in oxygen-inactivated holoenzyme after denaturation (*55*). Detection of the unsaturated nucleoside may be difficult since it is likely to be present in a reactive protonated form. Attempts to reconstitute active holoenzyme by incubating $4',5'$-didehydro-$5'$-deoxyadenosine with vitamin B_{12s} and apoenzyme have thus far been unsuccessful (*56*). It would be of interest to study the effect of coenzyme tritiated in the 4-position. This could, but need not, give rise to a kinetic isotope effect or increased lag period. A further test of the new mechanism of dioldehydrase action would be possible by synthesizing $4'$-substituted coenzyme B_{12} derivatives. The attempted synthesis of $4'$-fluoro-deoxyadenosylcobalamin was unsuccessful; although reaction between vitamin B_{12s} and the iodide of $4'$-fluoroadenosine occurred, the product is unstable and immediately decomposes by elimination of fluoride and cleavage of the Co—C bond (57, 58).

Several authors currently favor a free radical mechanism of diol-dehydrase action or alternative processes involving hypothetical organocobalamin intermediates (35, 36). These invoke energetically unfavorable homolytic or heterolytic C—H abstraction steps, as well as reactions of intermediate organocorrins which are not observed in appropriate model systems (59).
not observed in appropriate model systems (59).

Glycerol dehydrase exhibits ESR signals under certain experimental conditions (36). This has been cited in support of the free radical mechanisms, but the nature of the radical species, whose concentration is usually quite low, has not been elucidated. The appearance of the ESR signals is coupled with the formation of vitamin B_{12r} (36) but this evidence is insufficient to postulate Co(II) corrins as the catalytic form of the coenzyme. Ribonucleotide reductase from *Lactobacillus leichmannii* shows an ESR signal identical to that observed in glyceroldehydrase (60). In the absence of substrate, ribonucleotide reductase holoenzyme decomposes slowly with formation of vitamin B_{12r}. Ribonucleotide reductase contains sulfhydryl groups in the vicinity of the coenzyme binding site (61). These may be oxidized to disulfide and do not seem to be involved in the process of ribonucleotide reduction.

From the observed $\langle g \rangle$ value of 2.030 of the ESR signals it may be that the unpaired electron is localized on sulfur and that the paramagnetic species is possibly a form of the apoenzyme intermediate between the fully reduced (—SH) and oxidized (—S—S—) modification. Radicals of the type $R—SS—R^-$ have been characterized (62).

2. Ethanolamine Ammonia Lyase (Ethanolamine Deaminase)

Ethanolamine deaminase, isolated from *Clostridium* sp. (63, 64) catalyzes the coenzyme B_{12} dependent conversion of ethanolamine to acetaldehyde and ammonia (65—74). The apoenzyme is well characterized (MW = 520,000; exposure to 5 M guanidine hydrochloride causes dissociation into inactive subunits of MW 51,000) (66). The carbinol carbon of ethanolamine becomes the aldehyde carbon of product. The overall reaction is:

$$\begin{array}{c} CH_2—OH \\ | \\ CH_2NH_2 \end{array} \xrightarrow{\text{Coenzyme-Enzyme Complex}} CH_3—CH=O + NH_4^+$$

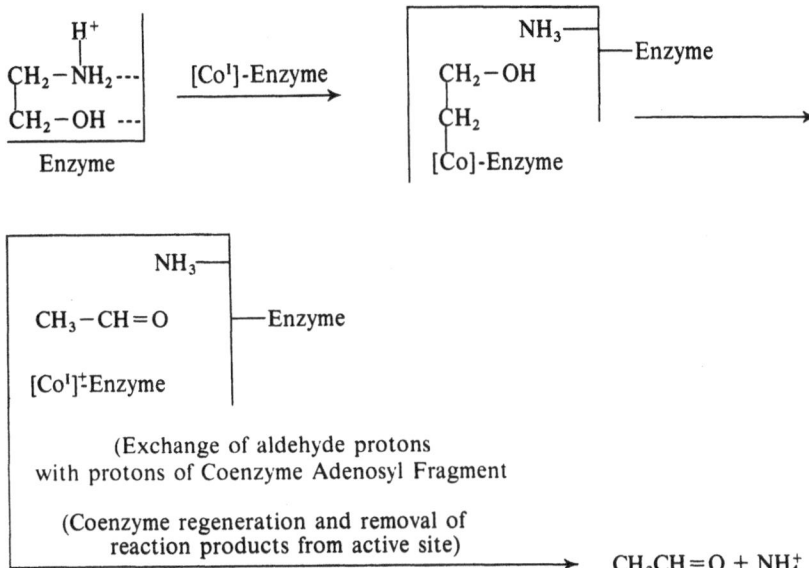

Chart 5. Essential steps of a proposed ionic mechanism of ethanolamine deaminase

Incubation of $1\text{-}^3\text{H}\text{-}2$-aminoethanol with the holoenzyme causes the transfer of tritium to the 5′-position of the adenosyl moiety of coenzyme B_{12}; the reaction occurs without exchange with protons from the aqueous medium (*71*). Ethanolamine deaminase has many features in common with dioldehydrases. In fact, the enzyme has been found to utilize ethylene glycol as a stoichiometric substrate, causing its conversion to acetaldehyde, but with concommittant inactivation of the enzyme (*70*). In view of these observations the mechanism of ethanolamine deaminase action must be similar to that of dioldehydrase, differing only with respect to the initial substrate binding- and activating step, and perhaps details of product removal from the active site. A plausible scheme is given in Chart 5.

Other authors have favored a free radical mechanism and have reported results of ESR measurements, including those with spin-labeled coenzyme, in apparent support of such a mechanism (*74, 75*). It was also observed that 5′-deoxyadenosine is formed on denaturation of holoenzyme (*72*). This adenosine derivative was subsequently assumed to play the part of the intermediate hydrogen carrier in the catalytic process. Since the methyl protons in 5′-deoxyadenosine are inert to exchange by polar mechanisms, this introduces severe mechanistic difficulties which cannot be remedied by the assumption of homolytic $C-H$ cleavage reactions as part of the catalytic process. Several reactions of coenzyme B_{12}

are known which lead to the formation of 5'-deoxyadenosine under
nonenzymatic conditions, i.e. homolysis of the Co-C bond in the presence of hydrogen donors or reductants, or directly on reductive Co-C
bond cleavage of the coenzyme with thiols (33). Since the highest yields
of 5'-deoxyadenosine were observed when the holoenzyme was dena-
tured in the presence of n-propanol (71), and sulfhydryl groups may be
present in the vicinity of the coenzyme binding site, formation of
the saturated nucleoside by any of these mechanisms is possible and thus
seems to be a product of a noncatalytic reaction. Recently, 5'-deoxy-
inosine was obtained by incubating 5'-deoxyinosylcobalamin with
dioldehydrase (76). This coenzyme analog is not active but in the
presence of substrate is evidently reductively cleaved by the apoenzyme. Since the presence of sulfhydryl groups in the vicinity of the
coenzyme binding site of dioldehydrase is well established, this interpretation of the 5'-deoxyinosine formation appears plausible. The effect
of N_2O on functional ethanolaminedeaminase has not yet been investigated.

3. Ribonucleotide Reductase

The summary reaction catalyzed by the coenzyme B_{12} dependent
ribonucleotide reductase of *Lactobacillus leichmannii* consists in the
conversion of ribonucleotides to the corresponding 2'-deoxyribonucleo-
tides. The enzyme utilizes thioredoxins or dithiols such as dihydrolipoate
or dithioerythritol as electron donors (77).

$$XTP + R(SH)_2 \xrightarrow{\text{Coenzyme-Enzyme Complex}} dXTP + R\left<{S \atop S}\right. + H_2O$$

Cell-free extracts of *Rhizobium meliloti* as well as of *Lactobacillus
acidophilus* and of other microorganisms also contain cobamide dependent
ribonucleotide reductases which are presumably similar or identical
with the enzyme from *L. leichmannii* (78, 79).

The enzyme exhibits no selectivity for the base portion of the nucleo-
tides, although GTP is probably most rapidly reduced. The substrate
binding site is highly specific for the ribose moiety, however. Certain
deoxyribonucleotides stimulate the reduction of ribonucleotide substrates and thus are considered effectors of the enzymatic catalysis (80).
The effector action of deoxyribonucleotides is less ribose-specific,
suggesting that the effector binding site or sites are different from the
substrate binding site. Incubation of ribonucleotide with 5'-T labeled

coenzyme causes release of tritium into water. At low enzyme concentrations the amount of tritium transferred to H_2O is stoichiometrically equivalent to the amount of ribonucleotide reduced (81, 82). Incubation of the enzyme with unlabeled coenzyme in tritiated water causes tritium incorporation into both coenzyme and product. The hydroxyl group at $C'2$ is displaced by hydrogen with net retention of configuration. Incubation of coenzyme B_{12} with excess ribonucleotide reductase apoenzyme and a nucleotide triphosphate results in the slow formation of vitamin B_{12r} (83). The ESR signals observed exhibit exceptionally high resolution of the ^{14}N superhyperfine splitting pattern (84). This suggests that the Co(II)-corrin is present at the active site and firmly attached to the protein, causing a diminution of thermal vibrations of the corrin ligand which normally diminish the resolution of the ESR spectra.

Incubation of the enzyme with hydroxocobalamin, dGTP as effector, and dihydrolipoate as reductant causes reduction of the enzyme-bound corrin. The ESR signal of the Co(II) corrin resembles that of vitamin B_{12r} in the absence of enzyme, i.e. the ^{14}N superhyperfine splitting is resolved only partially. However, upon addition of 5′-deoxyadenosine, 4′,5′-didehydro-5′-deoxyadenosine, or adenosine, respectively, the resolution of the ESR signal is significantly improved. This suggest that a special conformation of the corrin-enzyme complex is formed only in the presence of components resembling the coenzyme adenosyl moiety (84). The appearance of ^{14}N superhyperfine splitting indicates that 5,6-dimethylbenzimidazole or another nitrogen base is coordinated to cobalt in this form of the enzyme, but this does not mean that the nitrogen base remains coordinated to cobalt during all stages of enzymatic ribonucleotide reduction (84). Ribonucleotide reductase contains thiol groups in the vicinity of the coenzyme binding site which do not appear to be directly involved in the actual catalytic process (61). Under certain conditions the holoenzyme exhibits an ESR signal identical to that observed with glycerol dehydrase (60). As in the case of the latter, the origin of this signal has not yet been elucidated, it was suggested above that the signal may be due to a paramagnetic species in which the unpaired electron is localized on sulfur; possibly on the sulfur atoms near the coenzyme binding site.

The enzyme catalyzes the reduction of ribonucleotides and utilizes dithiols as the electron donors. As a consequence, mechanistic information for this process can be obtained from a study of the

$$\begin{array}{c} R \\ | \\ (Co) \end{array} + RSH \xrightarrow[-H^+]{} \begin{array}{c} R \\ | \\ (Co) \\ | \\ RS^- \end{array} \xrightarrow[]{+H^+} RH + \begin{array}{c} SR \\ | \\ (Co) \end{array}$$

reactions of organocorrins or -cobaloximes with thiols. Thiols react with organocobalt derivatives by reductive cleavage of the Co-C bond (33, 85).

Further details of this reaction will be outlined in the chapter on methane biosynthesis; it is cited in this context as the model reaction by which corrinoid coenzymes may act as catalysts of the reduction or organic substrates. The fact that a reaction of this type is relevant to ribonucleotide reductase is indicated by the observed correlation of the relative rates of methane evolution from methylcobaloximes with ribonucleotide reductase activity as a function of the buffer concentration. Ribonucleotide reductase activity is dependent on both the nature and concentration of buffers present in the assay mixture (86). The buffers have previously been assumed to exert a chaotropic effect on the enzyme in the vicinity of the active site. However, since a virtually identical buffer dependence is observed in the unrelated nonenzymatic reductive demethylation of methylcobaloximes by dithiols (33), it is clear that the effect of the buffers is in some way associated with a reductive Co-C bond cleavage reaction. Presumably, the buffers influence the dissociation of the dithiol reductant either prior or during its interaction with the corrin or cobaloxime cobalt atom (33). On the basis of this evidence it may be concluded that the reduction of ribonucleotides occurs via organocorrin intermediates, that the dithiol reductant interacts with the corrin cobalt atom and induces its reductive cleavage to a carbanionic species, which is subsequently protonated to form the product deoxyribonucleotide (Chart 6) (33, 34):

The mechanism in Chart 6 incorporates essential features of the mechanism of other coenzyme B_{12} dependent reactions, invoking specific binding and activation of the substrate (in this case at C-2$'$ of the nucleoside triphosphate) and the postulated reversible activation-regeneration of the coenzyme as discussed in Chapter IV,1. In contrast to dioldehydrase and ethanolamine deaminase the protonated fragment of the adenosyl moiety of the coenzyme is accessible to protons of the medium, causing the release of hydrogen from the 5$'$-position into water.

Ribonucleotide reductase holoenzyme is not very oxygen sensitive and not inactivated by N_2O, suggesting that the active coenzyme site is protected against interaction with gaseous oxidants (50). The reductive cleavage of methylcobaloximes by thiols is also virtually unaffected by N_2O. The presence of 4$'$,5$'$-didehydro-5$'$-deoxyadenosine in the holoenzyme has not been established, but 5$'$-deoxyadenosine is formed in the presence of effector and absence of substrate (87). This reaction is unlikely to be part of the catalytic process and indicates that reductive cleavage of the coenzyme may occur under these conditions.

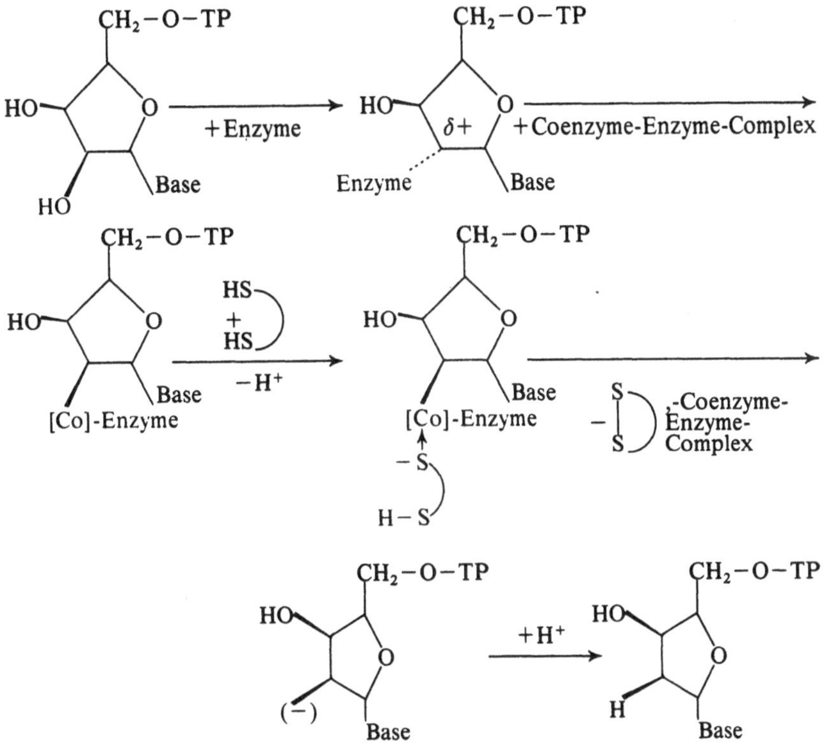

Chart 6. Essential catalytic reaction steps in ribonucleotide reductase

Ribonucleotide reductase activity is stimulated by certain effectors, e.g., dGTP. The function of these effectors is as yet unknown. They appear to bind at a site different from that of the substrate, presumably inducing a conformational change of the enzyme favorable for the catalytic process (*88*).

4. The Coenzyme B$_{12}$ Dependent Mutases

Coenzyme B$_{12}$ dependent enzymes catalyze a number of skeletal rearrangement reactions which are quite unusual and whose mechanisms pose as yet unresolved questions. Before discussing the mechanisms or possible mechanisms of these reactions the individual enzymes are described separately.

a) Glutamate Mutase

The enzyme was discovered in extracts of *Clostridium tetanomorphum* grown in the presence of glutamate and is also found in several other organisms (*89—94*). The overall reaction consists in the reversible conversion of glutamate into methylaspartate. The position of equilibrium is approximately 10.5:1 in favor of glutamate (95):

1
2 COOH
3 |
   CH—NH$_3^+$
   | Coenzyme-Enzyme Complex
4 CH$_2$ $\rightleftarrows$
   |
5 CH$_2$
   |
   COOH

1
2 COOH
   |
   CH—NH$_3^+$
3 4
   CH$_3$—CH
   5 |
   COOH

Assays are usually conducted in the presence of mercaptoethanol which appears to have a stabilizing effect on the enzyme. The net reaction consists in the migration of the glycine moiety from C-3 to C-4 and the migration of hydrogen from C-4 to C-3, respectively. Studies with deuterated methylaspartate reveals that transfer of deuterium from C-3 to C-4 occurs with inversion of configuration on C-4 (95—97). As in other coenzyme B$_{12}$ dependent enzymatic reactions the hydrogen at C-4 of glutamate at one point becomes equivalent with the C-5′ hydrogens of the coenzyme adenosyl moiety. The average kinetic isotope effect for the conversion of methylaspartate to glutamate, $k_H/k_D = 7.5$, suggesting that the cleavage of a C-H bond or an equivalent process is rate determining.

b) Methylmalonyl-CoA Mutase

Methylmalonyl-CoA mutase catalyses the conversion of methylmalonyl-CoA to succinyl-CoA and was discovered first in extracts of propionibacteria (98—100). The enzyme is also present in mammalian systems; B$_{12}$ deficient rats and other animals excrete anomalously high levels of methylmalonic acid. Cases of methylmalonic aciduria have been observed in children receiving a normal dietary supply of vitamin B$_{12}$. Since plasma vitamin B$_{12}$ levels were normal and symptoms of pernicious anemia were absent in these cases, the apparent inability of these children to metabolize methylmalonate suggests a genetic defect in the synthesis of the mutase enzyme or coenzyme B$_{12}$ biosynthesis (101—103). The reaction catalyzed by methylmalonyl-CoA mutase is:

CO—CoA
|
CH$_3$—CH   Coenzyme-Enzyme Complex
| $\rightleftarrows$
COOH

CO—CoA
|
CH$_2$—CH$_2$
|
COOH

In the reaction the CO-CoA moiety and one hydrogen of the methyl-group of methylmalonyl-CoA exchange places; there is no exchange with protons from water during the isomerase reaction, but the migrating hydrogen atom becomes at one point equivalent with the $5'$-protons of the coenzyme adenosyl moiety (104—106). The equilibrium succinyl-CoA/methylmalonyl-CoA is 23.1; K_m values for succinyl-CoA and methylmalonyl-CoA are $3.45 \cdot 10^{-5}$ and $8 \cdot 10^{-5}$ respectively (105).

c) α-Methyleneglutarate Mutase

Cell extracts of a *Clostridium* species contain a coenzyme B_{12} dependent enzyme which catalyzes the conversion of α-methyleneglutarate to β-methylitaconate (107). This enzyme is involved in the enzymatic degradation of nicotinic acid under anaerobic conditions and has been partially purified, MW ca. 170,000 (108). The only known substrates are α-methyleneglutarate and methylitaconate, respectively. Mesaconate, itaconate and succinate are competitive inhibitors of the reaction. The reaction shows similarities to those of other mutases:

$$
\begin{array}{l}
\text{COOH} \\
|\\
\text{CH}_2 \\
|\\
\text{CH}_2 \\
|\\
\text{C}=\text{CH}_2 \\
|\\
\text{COOH}
\end{array}
\quad
\xrightleftharpoons[\text{Coenzyme-Enzyme Complex}]{}
\quad
\begin{array}{l}
\text{COOH} \\
|\\
\text{CH}-\text{CH}_3 \\
|\\
\text{C}=\text{CH}_2 \\
|\\
\text{COOH}
\end{array}
$$

The equilibrium constant for the conversion of methyleneglutarate to methylitaconate is approximately 0.23 at 34° (pH 7.9). Methylitaconate formed is converted to dimethylmaleate by a second coenzyme B_{12}-independent enzyme, which is difficult to remove from the mutase, although the separation was successful on a small scale (107, 108).

There is no exchange of substrate or coenzyme protons with protons of water during reaction, but hydrogen is transferred between the $5'$-position of coenzyme and substrate, as evidenced from experiments with $5'$-T labeled coenzyme. Acrylate and 1-methyl-1,2-cyclopropane-dicarboxylate (*cis* as well as *trans*) do not exchange with the enzyme-bound substrate or derivatives therefrom, nor do they function as enzymic substrates (109).

d) Mechanism of Mutase Action

The known mutases have the following features in common:
1. High substrate specificity
2. Requirement for coenzyme B_{12}
3. Inactivity of other corrins

4. Functionality under anaerobic conditions
5. Exchange of substrate protons with the 5′-deoxyadenosyl moiety of the coenzyme
6. Net attachment/detachment of the CH_2-CH_2-COOH and $CH_3-CH-COOH$ moiety
7. Net migration of hydrogen between two adjacent carbon atoms
8. Lack of exchange with protons of water solvent during enzyme catalysis.

The following discussion of the mechanism of the reactions must be viewed under the aspect that no model reactions duplicating the complete rearrangement of any of the substrates has yet been discovered. However, there are at least two model reactions which may be of possible relevance for the catalytic process. As in the case of other coenzyme B_{12} dependent enzymatic reactions, it may be stated with a reasonable degree of certainty that the Co-C bond of the coenzyme is broken in the mutase holoenzymes and that the cobalt ion is present in a reduced state [either Co(I) or Co(II)]. The exchange of protons from the substrate, product or intermediates of the enzymatic rearrangement with protons at the 5′-position of the coenzyme adenosyl moiety is furthermore likely to occur by an ionic rather than free radical mechanism. This suggests that the mechanism of coenzyme activation is similar to that proposed for the other coenzyme B_{12} dependent enzymes, and that the Co(I) nucleophile of the corrin is the actual catalytic species. The lack of reactivity of virtually all substrates with cobalamins or cobaloximes in various valence states indicates that the enzyme specifically binds and activates the substrates for reaction with the coenzyme. It finally appears reasonable to assume organocorrin intermediates as part of the catalytic process. A potentially relevant model reaction was observed in the cobaloxime series. Ethylacrylate reacts with the Co(I) nucleophile to yield β-carboxyethylcobaloxime (85). With hydridocobaloximes the α-isomer is formed (85, 110). On thermolysis ethylacrylate and hydridocobaloxime are formed from both isomers:

$$
\begin{array}{ccc}
\begin{array}{c}
COOR \\
| \\
CH_2 \\
| \\
CH_2 \\
| \\
(Co)
\end{array}
\;\rightleftharpoons\;
&
\begin{array}{c}
H \\
| \\
(Co) \\
\Big\updownarrow \\
H^+ + (Co^{I-})
\end{array}
+ CH_2{=}CH{-}COOR \;\underset{\Delta}{\rightleftharpoons}\;
&
\begin{array}{c}
CH_3 \;\diagup COOR \\
\diagdown CH \diagup \\
| \\
(Co)
\end{array}
\end{array}
$$

A hypothetical mechanism of mutase action could be based on the assumption that the substrates are activated on interaction with the enzyme, causing cleavage or labilisation of the C-C bond between the propionyl moiety and the migrating group. Formation of the proper organocorrin intermediate would result if the reaction occurred by a heterolytic process, affording an actual or insipient propionyl carbonium ion and a nucleophilic species derived from the migrating moiety, designated R^-

$$
\begin{array}{l}
\underset{\substack{|\\CH_2\\|\\CH_2\\|\\R}}{COOH} \quad \overset{[Enzyme]}{\rightleftharpoons} \quad \underset{\substack{|\\CH_2\\|\\+CH_2\\\vdots\\R^-}}{COOH} \quad \overset{\pm[Co^I]}{\underset{\mp R^-}{\rightleftharpoons}} \quad \underset{\substack{|\\CH_2\\|\\CH_2\\|\\[Co]}}{COOH} \quad \rightleftharpoons \quad \underset{[Co]}{CH_3 \diagdown \overset{H}{\underset{|}{C}} \diagup COOH} \quad \overset{\mp R^-}{\underset{\pm[Co^I]^-}{\rightleftharpoons}}
\end{array}
$$

$$
\underset{\substack{\vdots\\R^-}}{CH_3-\overset{\displaystyle COOH}{\underset{|}{CH^+}}} \quad \overset{[Enzyme]}{\rightleftharpoons} \quad CH_3-\overset{\displaystyle COOH}{\underset{|}{\underset{R}{CH}}}
$$

The mechanism of the C-C bond labilization of the substrates is unknown and represents a weakness of the scheme. It is assumed to be induced by the enzyme, possibly by way of a cooperative interaction with the Co(I) nucleophile and/or prosthetic groups. The ultimate understanding of the mechanism of mutase action will undoubtedly require a detailed study of the role of the enzyme in the processes of substrate binding and -activation. It is presently apparent that the coenzyme cannot possibly catalyze the skeletal rearrangements without prior modification of the substrates.

The identical isomerization step involving the corrin intermediate was postulated in one of the earliest hypothetical mechanisms of methylmalonyl-CoA mutase, on the basis of the known hydro-formylation of ethylacrylate with cobalt carbonyls as catalysts, which yields succinate and methylmalonate (111). In this mechanism the corrin was assumed to act in a manner similar to carbonylation catalysts. The study of acylcobalamins and acylcobaloximes revealed, however, that corrins unfortunately do not catalyze decarbonylation reactions common in metal carbonyl chemistry:

$$
\underset{(Co)}{\overset{\displaystyle R}{\underset{|}{C}}=O} \quad - - - \cancel{\longrightarrow} \quad \underset{(Co)}{\overset{\displaystyle R}{\underset{|}{}}} + CO
$$

Alternative mechanisms of methylmalonyl-CoA mutase were postulated which invoke three-membered ring intermediates, e. g. (4, 85):

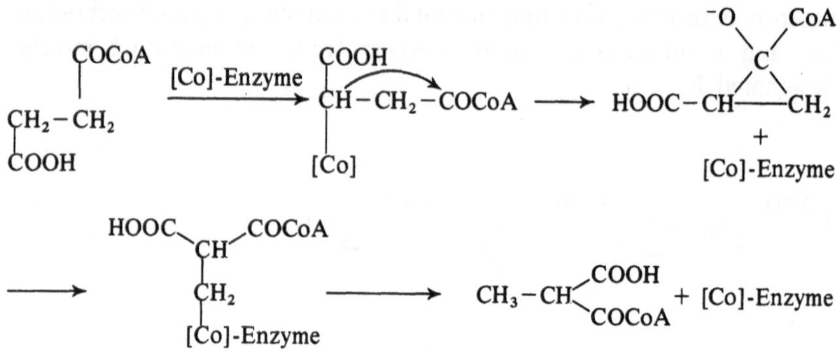

(The valence states of "[Co]-Enzyme" are not shown.) There is some evidence from model experiments with cobaloximes which supports this mechanism (112):

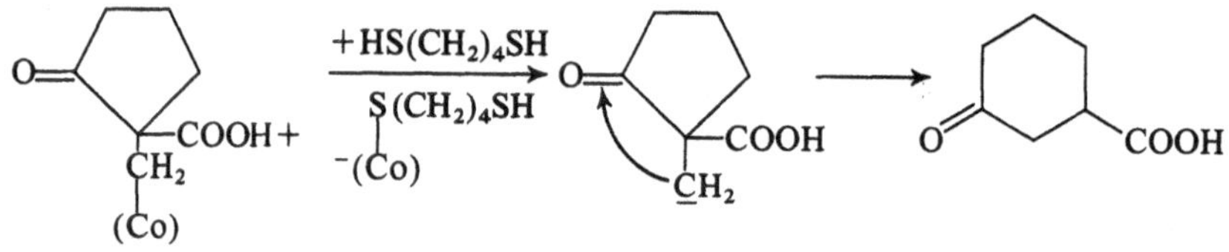

The conversion of (15) to (17) has been achieved in low yields by reductively cleaving to Co-C bond with 1,4-butanedithiol. The reaction occurs via the carbanion (16). On the other hand, a reaction of this type cannot be readily modified to account for the reverse process, or the conversion of methylmalonyl-CoA to succinyl-CoA, and suffers from the fact that, at least in the model system, a thiol is stoichiometrically oxidized. Methylmalonyl-CoA mutase does not show a stoichiometric requirement for a reductant, however. It is also difficult to formulate a plausible mechanism by which the postulated organocorrin intermediates are formed, and how carbanionic species are generated from the intermediates without the stoichiometric requirement of a reducing agent.

Free radical mechanisms of mutase action have also been proposed. These invoke 5'-deoxyadenosine as the intermediate hydrogen carrier, a fact which makes them unlikely just as in dioldehydrase or ethanolamine deaminase (see Chapters IV, 1, 2).

5. Aminomutases

Extracts of *Clostridium sticklandii* grown on lysine, ariginine or ornithine, respectively, contain coenzyme B_{12} dependent enzymes catalyzing the reversible isomerization of these acids in the manner shown for the specific case of β-lysine (*113—119*):

$$CH_2-CH_2-CH_2-CH-CH_2COO^- \quad \underset{Complex}{\overset{Coenzyme-Enzyme-}{\rightleftharpoons}} \quad CH_3-CH-CH_2-CH-CH_2-COO^-$$

$$\underset{NH_3^+}{|} \qquad \underset{NH_3^+}{|} \qquad\qquad\qquad \underset{NH_3^+}{|} \qquad \underset{NH_3^+}{|}$$

The analogous reactions of ariginine and ornithine similarly lead to exchange of the terminal amino group to the adjacent carbon atom accompanied by migration of hydrogen in the reverse direction. All three enzymes consist of two subunits, a corrin-binding protein of MW 160,000, and a thioprotein of MW 60,000; both units must be present to achieve enzymatic activity. L-β lysine aminomutase and D-lysine aminomutase require the corrinoid coenzyme plus a thiol for maximum activity, as well as a monovalent cation, a divalent cation (either Mg^{++} or Mn^{++}), and ATP (*117*). Pyridoxal phosphate is furthermore required for D-lysine aminomutase, but pyruvate for β-lysine aminomutase. The function of ATP appears to be that of an allosteric activator of D-lysine aminomutase, since it is not hydrolyzed by the purified enzymes (*118*). A pyridoxal-dependent exchange of a proton on C-6 of lysine with the solvent was observed (*119*), suggesting interaction of this cofactor with the migrating amino group in D-lysine aminomutase. The pyruvate carbonyl group in β-lysine aminomutase conceivably also interacts with the migrating amino group. On the other hand, ornithine aminomutase appears to require no cofactors other than coenzyme B_{12}. Transfer of hydrogen was demonstrated from substrate to the 5′-carbon of the adenosyl moiety of the coenzyme; in L-β-lysine aminomutase the transfer of hydrogen from C-5 to C-6 is stereospecific (*120*).

The rearrangement of an amine in analogy to the enzymatic process does not occur under simple nonenzymatic conditions. The pyridoxal phosphate or pyruvate cofactor requirement suggests that the migrating amino group is first activated to form a Schiff-base type compound, but the nature of the subsequent reactions is as yet unknown. It is obvious, however, that the rearrangement is accomplished in several steps rather than by way of a simple concerted transfer of H and NH_2. Following the specific binding and activation of the substrate by the enzyme the reaction is probably initiated by the removal of the amino group and the formation of an organocorrin intermediate of the substrate residue.

Isomerisation of the latter e.g. by way of hydridocorrin elimination and subsequent addition to the olefinic intermediate may be considered to be the essential step involving the coenzyme. This reaction is adequately supported by model experiments, especially the demonstrated reaction of hydridocobalamin with nonactivated olefins (*11*). In the following step or steps the amino group is added back on to yield the rearranged product. For lysine such a mechanism may be represented as follows (details of the presumably pyruvate or pyridoxal dependent deamination are not shown):

$$
\begin{array}{ccccccc}
\begin{array}{l}CH_2CO_2^-\\ CH-NH_3^+\\ CH_2\\ CH_2\\ CH_2-NH_3^+\end{array}
&
\begin{array}{c}\pm[Co^I]^-\\ \dashrightarrow\\ \dashleftarrow\\ \mp[NH_3]\end{array}
&
\begin{array}{l}CH_2CO_2^-\\ CH-NH_3^+\\ CH_2\\ CH_2\\ CH_2\\ [Co]\end{array}
& \rightleftharpoons &
\begin{array}{l}CH_2CO_2^-\\ CH-NH_3^+\\ CH_2\\ CH\quad H\\ CH_2\;[Co]\end{array}
& \rightleftharpoons &
\begin{array}{l}CH_2CO_2^-\\ CH-NH_3^+\\ CH_2\\ CH-CH_3\\ [Co]\end{array}
\end{array}
$$

$$
\begin{array}{cc}
\begin{array}{c}\pm[NH_3]\\ \dashrightarrow\\ \dashleftarrow\\ \mp[Co^I]^-\end{array}
&
\begin{array}{l}CH_2CO_2^-\\ CH-NH_3^+\\ CH_2\\ CH-CH_3\\ NH_3^+\end{array}
\end{array}
$$

Attractive features of this mechanism are that it invokes the coenzyme in its reactive forms [i.e. the Co(I) nucleophile and its hydride] and that the assumed coenzyme dependent steps are ionic rather than free radical reactions.

V. Corrin Dependent Enzymatic Reactions

Presently known corrin dependent enzymatic reactions are summarized in Table 3.

1. Methane Biosynthesis

Cell extracts of *Methanosarcina barkeri* form methane from various C_1 precursors including CO_2, formate, formaldehyde methanol, 5-methyltetrahydrofolic acid and methylcorrins (*121*). The enzyme responsible for methane production has been obtained in a partially purified state and was found to contain a corrin. The overall reaction may thus be represented as follows:

$$C_1 \text{ precursors} \xrightarrow{\; +[Co]\text{-Enzyme} \;} \overset{\displaystyle CH_3}{\underset{\displaystyle [Co]\text{-Enzyme}}{|}} \xrightarrow[\; -[Co]\text{-Enzyme} \;]{\text{biogenic reductant, } -H^+} CH_4$$

Extracts of Methanobacterium M.O.H. also contain a methane-forming system which requires stoichiometric amounts of ATP and catalytic amounts of corrin for functionality (122). Molecular hydrogen is used as the source of electrons but a thioprotein or dithiol is probably the terminal reductant in the system. There is also evidence for the requirement of a low-molecular weight cofactor, which, if removed from the cell extracts by dialysis, prevents methane formation. This cofactor ("Coenzyme M") is incompletely characterized but can apparently by methylated and demethylated (123). Apart from typical C_1 precursors, cell extracts of Methanobacterium M.O.H. also utilize methylcobaloximes in the presence of ATP and catalytic amounts of a corrin (123, 124).

Methylcobalamin and methylcobaloximes are reductively cleaved by thiols to form methane in neutral or weakly acidic aqueous solution (124). This reaction involves the initial *trans*-attack of the cobalt ion by the thiol reductant which is followed by the rate-determining cleavage of the Co-C bond (33). This gives rise to a methyl carbanion which is protonated to form methane. The mechanism is supported by the observed inhibition of reductive cleavage by bases competing with the thiol for the axial cobalt binding site, and the demonstrated formation of acetate if the reductive demethylation is performed in the presence of CO_2 (33):

$$\underset{\displaystyle B}{\overset{\displaystyle CH_3}{\underset{\textstyle |}{(\dot{C}o)}}} \underset{\mp B}{\overset{\pm RSH}{\rightleftharpoons}} \underset{\displaystyle R^{\diagup S}{}^{\diagdown}{}_H}{\overset{\displaystyle CH_3}{\underset{\textstyle |}{(\dot{C}o)}}} \rightleftharpoons \underset{\displaystyle R^{\diagup S^-}}{\overset{\displaystyle CH_3}{\underset{\textstyle |}{(\dot{C}o)}}} + H^+ \xrightarrow[\underset{{}^-(\dot{C}o)}{SR}]{\text{rate}} (CH_3^-) \xrightarrow{H^+} CH_4$$

If the biogenic reductant is a thiol or thioprotein, which is very likely, this reaction could represent the terminal step of corrinoid dependent methane biosynthesis. The fact that methylcobaloximes are demethylated by cell extracts of *Methanobacillus* M.O.H. only if catalytic amounts of a corrin are present (123) suggests that either the corrin is essential to produce or maintain a catalytically active enzyme capable of generating methane directly from the methylcobaloximes, or that the cobaloxime-bound methyl groups are first transferred to the corrin cofactor and subsequently demethylated. The relative rates of nonenzymatic demethylation of methylcobaloximes depend on the nature of the axial base component (124). A similar dependence is also observed in the enzymatic demethylation, however, suggesting that the biogenic reductant demethylates the cobaloximes directly. This interpretation is substantiated by a statistical correlation between the observed enzymatic and nonenzymatic rates of methane production (125). If the methyl groups were first transferred to the corrin cofactor and subsequently demethylated enzymatically, a positive correlation between the enzymatic and nonenzymatic rates would not be possible; all cobaloximes should be demethylated at the same or similar rates, since the demethylation of the corrin would be rate determining.

The transfer of cobaloxime-bound methyl groups to the corrin has been studied under nonenzymatic conditions (33). The reaction is rapid,

$$\begin{array}{c}CH_3 \\ | \\ (Co) \end{array} + [Co^I]^- \rightleftharpoons \begin{array}{c}CH_3 \\ | \\ [Co] \end{array} + \begin{array}{c}(Co^I)^- \\ \uparrow \\ B \end{array}$$

reversible but does not correlate with the observed enzymatic rates of methane production (33). Although a transfer of methyl groups according to such a process could occur as a side-reaction, it is unlikely to represent the catalytic pathway. The mechanistic relevance of "Coenzyme M" will become clearer after its purification and identification; it may be suspected that present preparations are contaminated (33). ATP is not consumed in the process of catalytic methane production and may be an allosteric effector (126).

The available evidence indicates that methylated corrins are the terminal acceptors of enzymatic methane biosynthesis, from which methane is produced by reductive cleavage of the Co-C bond on interaction with the biogenic reducing agent.

As in other corrin dependent reactions, methane biosynthesis was also formulated by invoking homolytic cleavage of the Co-C bond of the methylated corrin, but this mechanism is essentially ruled out by the model experiments, and the arguments in favor of a reaction involving methyl radicals are not convincing.

Table 3. *Corrin Dependent Enzymes*

Enzyme:	Source(s):	Reaction catalyzed:	Cofactors:
		(a)	
Methane Synthetase	e.g. *Methanosarcina barkeri*	$C_1 \cdots\to CH_4$	reductant
	Methanobacterium MOH	$C_1 \cdots\to CH_4$	ATP, reductant, "coenzyme M"
Acetate Synthetase	*Clostridium thermoaceticum*	$C_1 + CO_2 \cdots\to CH_3CO_2^-$	reductant
Methylarsine Synthetase	Methanobacterium MOH	$C_1 + As^{+3} \cdots\to (CH_3)_nAsH_{3-n}$	reductant, ATP
Methionine Synthetase	*Escherichia Coli*	$5\text{-}CH_3\text{-}THF + HS(CH_2)_2CH(NH_2)CO_2^- \to$ $THF + CH_3S(CH_2)_2CH(NH_2)CO_2^-$	Adenosylmethionine, reducing system

(a) C_1 denotes one-carbon precursors, including $5CH_3\text{-}THF$.

Propylation of the corrin inhibits methane biosynthesis (127). This is consistent with the reductive Co-C bond cleavage mechanism, since higher alkylcobaloximes and -cobalamins are generally dealkylated at significantly lower rates than the corresponding methylcobalt derivatives.

With methylcobalamin as the terminal acceptor of the methyl group in methane biosynthesis, the mechanism of demethylation is plausibly formulated as follows (the enzyme-bound corrin is shown to be in the Co(I) state after completion of the reductive cleavage, and the thio-protein reductant is oxidized to disulfide; the Co(I) ion can be realkylated to start a new catalytic cycle (33, 124):

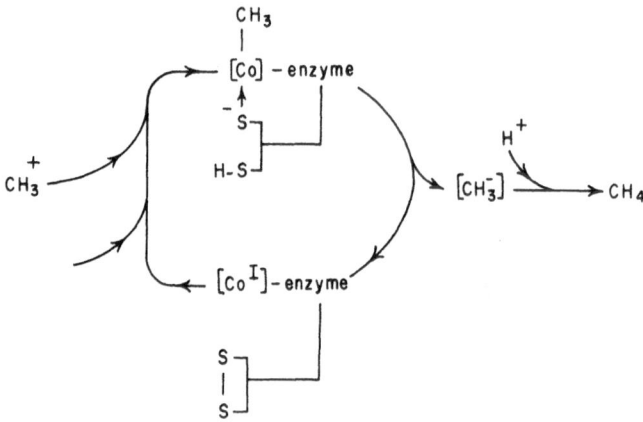

This mechanism is supported by the fact that thiols are known to reduce corrins via mercaptide intermediates which are at equilibrium with the Co(I) nucleophile. Addition of CH_3I to solutions of a corrin in the presence of a thiol at pH 7 produces the methylcorrin (17).

2. Acetate Biosynthesis

Cell extracts of *Clostridium thermoaceticum* convert the methyl group of added methylcobalamin to acetate in the presence of CO_2 (128). Acetate is also formed from C_1-precursors or carriers such as 5'-methyltetrahydrofolic acid and CO_2, in the presence of corrinoid cofactors (129—131). The formation of acetic acid is inhibited by propylation of the corrin, indicating that the methyl groups are first transferred to the corrin and subsequently carboxylated (131). Acetate biosynthesis is inhibited by sulfhydryl blocking reagents, suggesting their participation as reducing agents in this system.

Methylcobalamin and methylcobaloximes are resistant to CO_2 under nonenzymatic conditions and normally do not behave as biogenic Grignard Reagents. Acetate is also not produced from methyl radicals and CO_2, a fact which eliminates a mechanism of acetate biosynthesis involving homolysis of the Co-C bond. However, acetate is formed from the methyl groups of methylcobaloximes or methylcobalamin in the presence of CO_2 and a reductant such as 1,4-butanedithiol or 1,4-dithioerythritol (*132*). This reaction occurs by *trans*-attack of the cobalt ion, followed by the reductive Co-C bond cleavage and generation of a krypto carbanion, which is subsequently carboxylated (*33, 132*):

$$
\begin{array}{c}
CH_3 \\
| \\
(Co) + RSH
\end{array}
\rightleftharpoons
\begin{array}{c}
CH_3 \\
| \\
(Co) \\
\uparrow \\
S \\
R \quad H
\end{array}
\xrightarrow[-SR, -H^+]{}
\begin{array}{c}
(CH_3^-) \\
(Co)
\end{array}
\xrightarrow{CO_2}
CH_3 - CO_2^-
$$

Investigation of the fate of the methyl group during acetate biosynthesis using methylcobalamin-d_3 as the substrate revealed that the methyl group is transformed into acetate as a whole (*133*). This appears not to be the case when 5'-methyltetrahydrofolate is the donor of the methyl group, since 5-CD_3-THF gives rise to the formation of acetate containing varying concentrations of H. The transfer of the methyl group from the tetrahydrofolate to cobalt is a more complicated process and will be discussed in the chapter on methionine biosynthesis. Nevertheless, the mechanism of acetate formation from methyl corrins and CO_2 is now understood and adequately demonstrated by model reactions. The conversion of the enzyme-bound methylcorrin to acetate is expected to occur at a locally aprotic site to prevent methane formation by protonation of the kryptocarbanion formed in the process. It would be of interest to determine the extent of methane production (if any) in the absence of CO_2, from acetate producing cell extracts of *Clostridium thermoaceticum*, and to investigate if methane biosynthesis by cell extracts of methanogenic bacteria may be diverted to yield acetate in the presence of CO_2, since both methane and acetate formation appear to be related processes. Cell extracts of methanogenic bacteria, on the other hand, are also known to convert acetate into methane (*121*). In this case decarboxylation preceeds the formation and reductive cleavage of methylcorrins. The decarboxylation step is unlikely to involve the corrin cofactor, however, and there is no experimental support for a reaction between corrins or cobaloximes and acetate to yield methylcobalt derivatives and CO_2.

In an earlier version of the mechanism of acetate formation, methylcorrins were assumed to be converted to carboxymethylderivatives (*134*):

$$CH_3\text{–}[Co] + CO_2 \longrightarrow CH_2\text{–}COO^-\text{–}[Co] + H^+ \xrightarrow[-[Co]]{\text{reduction}} CH_3\text{–}COO^-$$

This mechanism is unlikely as it requires activation of a C-H bond of the methylcorrin. Nonenzymatically, the cobalt-bound methyl protons show no kinetic lability (they are not abstracted by bases to form a carbanionic species, for example). This mechanism is also eliminated by the labelling experiments demonstrating the conversion of the methyl group as a whole to acetate (*133*).

The terminal stages of acetate biosynthesis are (*33, 132*):

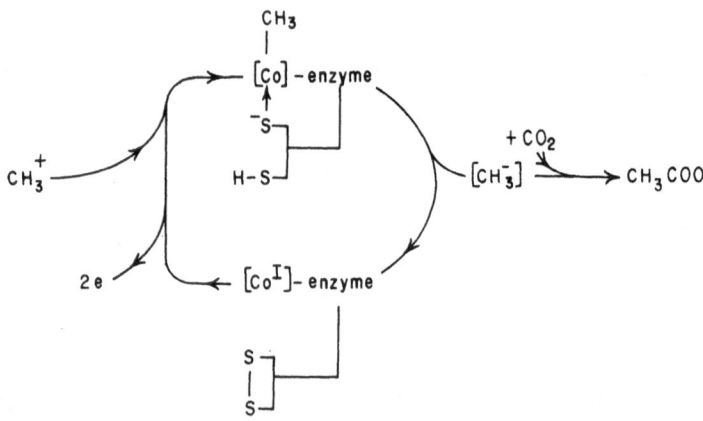

3. Methylarsine Biosynthesis

Cell extracts and whole cells of *Methanobacterium* strain M.O.H. convert methylate arsenate and arsenite to methylarsines under reducing conditions. Arsenate is reduced to arsenite first and subsequently methylated, methylcorrins serve as the methylating agents (*135*). The composition of the methylated arsines has not been rigorously established, but it appears that dimethylarsine is the principal product. The formation of methylarsine is always accompanied by the production of methane. Incubation of ethylcobalamin does not give rise to ethylarsines. Selenite inhibits methylarsine formation and also appears to be methylated (*135*). Methylation of arsenite and arsenate occurs nonenzymatically in the presence of thiols and of other reductants, with methylcobaloximes or methylcobalamin as donors of the methyl group (*33*):

$$\underset{[\text{Co}]}{\overset{\text{CH}_3}{|}} + \text{RSH} \xrightarrow[\;-[\text{Co}^{\text{II}}],\; -1/2\,\text{RSSR},\; -\text{H}^+\;]{} [\text{CH}_3^-] \xrightarrow{+\text{As}^{+3}}$$

$$\text{CH}_3-\text{As}^{++} \xrightarrow[([\text{Co}^{\text{I, II}}])]{+2e,\; +2\text{H}^+} \text{CH}_3-\text{AsH}_2$$

Methyl- and dimethylarsine and traces of trimethylarsine are formed in addition to methane and arsine; traces of ethylarsine result in the corresponding reaction of ethylcobaloxime with arsenite, but not with ethylcobalamin (*33*). Selenite inhibits the nonenzymatic methylation of arsenite and is converted to methylselenide (*33*). These experiments establish that methane, acetate, and methylarsine biosynthesis are related processes involving the reductive cleavage of methylcorrins by a biogenic reductant which is most likely a thioprotein or dithiol (*33*):

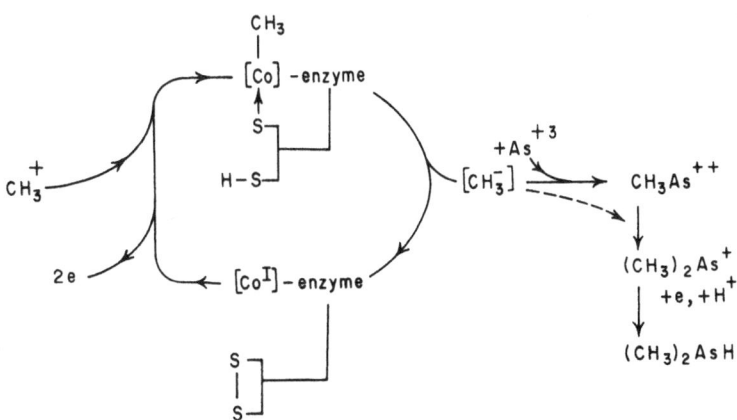

4. Methylmercury Formation

Methane biosynthesis by cell extracts of *Methanobacterium* M.O.H. is inhibited by mercuric salts. The mercury is converted to methylmercuric ion and dimethylmercury in the process (*136*). The reaction occurs with methylcobalamin as the donor of the methyl group, and occurs readily under nonenzymatic conditions (*137, 138*). Methylcobalamin reacts with mercuric salts much more rapidly than methylcobinamide, indicating a rate-enhancing effect due to the presence of the 5,6-dimethylbenzimidazole ligand. This is attributed to an increase of the electron density on cobalt by the coordination of the nitrogen base to cobalt (*138*); the overall reactions thus are:

$$\overset{CH_3}{\underset{}{|}}\ [Co] + Hg^{++} \longrightarrow [Co^{III}] + CH_3Hg^+$$

$$2\,CH_3Hg^+ \rightleftharpoons (CH_3)_2Hg + Hg^{++}$$

Methylmercury is also formed from mercuric salts and methylcobaloximes. The methylation of Hg^{++} by methylcorrins has been claimed to be of importance with regard to problems of mercury pollution, providing a mechanism for the formation of highly toxic dimethylmercury. The uncontrolled accumulation of methylmercury derivatives under anaerobic reducing conditions is actually unlikely. Thus, both dimethylmercury and methylmercuric halides react with reduced corrins and with the cobaloxime(I) nucleophile nonenzymatically to yield the corresponding methylcobalt derivatives (*139, 140*). It may therefore be expected that dimethylmercury is demethylated by anaerobic organisms, ultimately yielding methane and elemental mercury.

5. Methionine Biosynthesis

Animal tissues or *Escherichia coli* methylate homocysteine to methionine with a cobalamin dependent enzyme, utilizing 5-methyltetrahydrofolic acid ($5\text{-}CH_3THF$) as the donor of the methyl group. The enzyme requires adenosylmethionine (AMe) and a reductant as cofactors (*141—147*).
Incubation of the enzyme with methyl-group-labeled AMe indicates that the enzyme-bound cobalamin is methylated by it, and that the

$$5 - CH_3 - THF + HSCH_2CH_2 - \overset{NH_2}{\underset{}{|}}CH - COOH \xrightarrow[AMe]{Cobalamin\text{-}Enzyme\text{-}Complex}$$

$$THF + CH_3 - S - CH_2CH_2\overset{NH_2}{\underset{}{|}}CH - COOH$$

References, pp. 621—629

rate of cobalamin methylation by AMe is faster than the methylation by 5-CH$_3$THF. In the presence of unlabeled 5-CH$_3$THF, methyl-labeled enzyme is formed during the first 30 seconds of reaction. Subsequently, the labeled methyl groups disappear, suggesting a reversible methylation-demethylation process (*148*). It is known that methylated enzyme also transfers the methyl group to THF, and that AMe is also capable of reacting with THF to form 5-CH$_3$THF (*148*). The terminal step of methionine biosynthesis from methylcobalamin is:

$$\overset{\displaystyle CH_3}{\underset{\displaystyle \vert}{[Co]}}\text{-enzyme} + HSCH_2CH_2\overset{\displaystyle NH_2}{\underset{\displaystyle \vert}{CH}}-COOH \longrightarrow [Co^I]\text{-enzyme} +$$

$$CH_3S-CH_2CH_2\overset{\displaystyle NH_2}{\underset{\displaystyle \vert}{CH}}-COOH$$

With methylcobaloxime, this reaction has been demonstrated under nonenzymatic conditions (*150*). The transfer of the methyl group of methylcobalamin to homocysteine occurs nonenzymatically as well, but initial reports suggested that a methyl radical is transferred (*151*). However, the reaction occurs by a nucleophilic attack involving the $^-S-CH_2CH_2\overset{\displaystyle NH_2}{\underset{\displaystyle \vert}{CH}}-COO^-$ ion. (In the experiments with methylcobaloximes, the reaction was shown to occur in the dark in alkaline solution, but not on irradiation, i.e. conditions favoring the production of methyl radicals. In the presence of thiols, methyl radicals generated from methylcobaloxime or -cobalamin are known to terminate mainly by methane formation; alkylation of sulfur is not observed.)

The methylation of the Co(I) nucleophile by AMe is also well understood and occurs nonenzymatically (AH = adenosylhomocysteine (*17, 150*).

$$[Co^I]^- + AMe \xrightarrow[(+H^+)]{} \overset{\displaystyle CH_3}{\underset{\displaystyle \vert}{[Co]}} + AH$$

The Co(I) nucleophiles are also known to react with methylsulfonium salts in similar fashion (*150*).

A fundamental and as yet unresolved problem in methionine bio-synthesis as well as in other biological methyl group transfer reactions (i.e. methane and acetate biosynthesis) concerns the methylation of enzyme-bound corrins by 5-CH$_3$THF. This reaction has not been demonstrated convincingly under nonenzymatic conditions. Nonenzymatically, tertiary amines do not react with the Co(I) nucleophiles. This indicates that 5-CH₃THF must be activated in some way before it can transfer its methyl group to cobalt. Certain quaternary ammonium salts have recently been shown to alkylate the cobaloxime(I) nucleophile (*153*). This suggests that the 5-nitrogen of MeTHF may be quaternized in some fashion on interaction with the enzyme, but the details of this activation of the methyl group are still unknown. Methionine bio-synthesis is inhibited by propylation of the corrin, but enzymatic activity is restored on photolysis of the propylated holoenzyme (*154*). Methylation of the corrin produces an essentially light-insensitive form of the enzyme (*155, 156*). This has been interpreted to suggest that the corrin is coordinated to an imidazole moiety of the enzyme, since the rate of photolysis of imidazole-coordinated methylcobinamide is slower than that of methylcobinamide itself (*157*). However, this conclusion is not warranted, since the photolysis rates of methyl-corrins and methylcobaloximes become exceedingly slow under strictly anaerobic conditions. The virtual light-insensitivity of corrinmethylated methionine synthetase thus is more likely due to the fact that the corrin is in an essentially oxygen protected environment. The efficiency of the reverse reaction of Co-C bond homolysis is very high under strictly anaerobic conditions, and in the enzyme may be further en-hanced by the inability of the CH_3 radical to diffuse out of the coordination sphere of the corrin cobalt ion (*25, 30*).

$$\underset{[\text{Co}]}{\overset{\text{CH}_3}{|}} \underset{}{\overset{h\nu}{\rightleftarrows}} \quad [\text{Co}^{II}] + \text{CH}_3$$

Formaldehyde reacts with the Co(I) nucleophile of cobaloximes in the presence of nitrogen bases to form an organocobaloxime which may be cleaved reductively to form substituted methylamines (R = phenyl) (*157*):

$$(\text{Co}^I)^- + \text{CH}_2 = \text{O} + \text{NH}_2 - \text{R} \longrightarrow \underset{(\text{Co})}{\overset{\text{RNH}}{\underset{|}{\overset{|}{\text{CH}_2}}}} \xrightarrow[-(\text{Co}^I)^-]{\text{reduction}} \text{CH}_3\text{NHR}$$

A similar reaction with thiols affords methylthioethers, the thiols reacting as intrinsic reductants in this process; organocobaloxime inter-mediates were not isolated in this case (*157*):

$$\text{CH}_2 = \text{O} + \text{HSR} \xrightarrow[(\text{Co}^I)^-]{2\text{H}^+} \text{CH}_3 - \text{S} - \text{R} + \text{H}_2\text{O}$$

Although this reaction may be unrelated to any known form of methionine biosynthesis it represents a corrin or cobaloxime catalyzed process by which C_1 precursors at the oxidation level of formaldehyde are reduced to N or S-bound methyl groups. This reaction is cited to create interest in studies involving $-CH_2$-donors (e. g. 5,10-methylenetetrahydrofolate) as potential precursors of the methionine methyl group, and to elicit reconsideration of the question of the role of 5-CH_3-THF in trans-methylation reactions. The detailed role of S-adenosylmethionine in cobalamin-dependent methionine biosynthesis is also not clear. There is some indication for the existence of modifications of the *E. coli* enzyme which are functional in the absence of S-adenosylmethionine (*158, 159*).

References

1. BARKER, H. A., R. D. SMYTH, H. WEISSBACH, J. I. TOOHEY, J. N. LADD, and B. E. VOLCANI: Isolation and properties of crystalline cobamide coenzymes containing benzimidazole or 5,6-dimethylbenzimidazole. J. Biol. Chem. **235**, 480 (1960).

2. LENHERT, P. G., and D. CROWFOOT HODGKIN: Structure of the 5,6-dimethylbenzimid-azolyl-cobamide coenzyme. Nature **192**, 937 (1961).

3. LENHERT, P. G.: The structure of vitamin B_{12}. VII. The X-ray analysis of the vitamin B_{12} coenzyme. Proc. Roy. Soc. A **303**, 45 (1968).

4. INGRAHAM, L. L.: B_{12} coenzymes: Biological Grignard reagents. Ann. N. Y. Acad. Sci. **112**, 713 (1964).

5. SCHINDLER, O.: Reduktionsprodukte von Vitamin B_{12}. Helv. Chim. Acta **34**, 1356 (1951).

6. MÜLLER, O., and G. MÜLLER: Synthesen auf dem Vitamin-B_{12}-Gebiet, XIV. Biochem. Z. **336**, 299 (1962).

7. SCHRAUZER, G. N., R. J. WINDGASSEN, and J. KOHNLE: Die Konstitution von Vitamin B_{12s}. Chem. Ber. **98**, 3324 (1965).

8. SCHRAUZER, G. N., E. DEUTSCH, and R. J. WINDGASSEN: The nucleophilicity of vitamin B_{12s}. J. Am. Chem. Soc. **90**, 2441 (1968).

9. SCHRAUZER, G. N., and E. DEUTSCH: Reactions of Co(I) supernucleophiles. The alkylation of vitamin B_{12s}, cobaloximes (I) and related compounds. J. Am. Chem. Soc. **91**, 3341 (1969).

10. SCHRAUZER, G. N.: The chemistry of Co(I) derivatives of vitamin B_{12} and of related compounds. Ann. N. Y. Acad. Sci. **158**, 526 (1969).

11. SCHRAUZER, G. N., and R. J. HOLLAND: Hydridocobalamin and a new synthesis of organocobalt derivatives of vitamin B_{12}. J. Am. Chem. Soc. **93**, 4060 (1971).

12. WEBER, J. H., and G. N. SCHRAUZER: Bisdimethylglyoximatorhodium derivatives. Analogs of cobaloximes. J. Am. Chem. Soc. **92**, 726 (1970).

13. TOOHEY, J. I.: A vitamin B_{12} compound containing no cobalt. Proc. Natl. Acad. Sci. U.S. **54**, 934 (1966).

14. SATO, K., S. SHIMIZU, and S. FUKUI: A cobalt-free corrinoid in *Streptomyces olivaceus*. Biochem. Biophys. Res. Comm. **39**, 170 (1970).

15. BIRNBAUM, J.: Fermentation of *Pseudomonas denitrificans* in media containing rhodium salts. Unpublished, Merck, Sharp and Dohme Research Laboratories, Rahway, New Jersey.

16. SCHRAUZER, G. N., und J. KOHNLE: Coenzym-B_{12}-Modelle. Chem. Ber. **97**, 3056 (1964).

17. Schrauzer, G. N.: Organocobalt chemistry of vitamin B_{12} model compounds. Accounts Chem. Research 1, 97 (1968).
18. Randall, W. C., and R. A. Alberty: Kinetics of ligand binding to aquocobalamin. Biochemistry 6, 1520 (1967).
19. Tsiang, H. G., and W. K. Wilmarth: Rate and equilibrium studies of the displacement of water in trans-sulfitoaquobis(dimethylglyoximato)cobalt(III) by various nucleophiles. Inorg. Chem. 7, 2535 (1968).
20. Hague, D. N., and J. Halpern: Kinetics of some substitution reactions of trans-bis-dimethylglyoximato)-cobalt(III) complexes. Inorg. Chem. 6, 2059 (1967).
21. Fleischer, E. B., S. Jacobs, and L. Mestichelli: The kinetics of the reaction of Cobalt(III) and Iron(III) hematoporphyrin with cyanide and thiocyanate. Evidence for a dissociative mechanism. J. Am. Chem. Soc. 90, 2527 (1968).
22. Schrauzer, G. N., and L. P. Lee: The molecular and electronic structure of vitamin B_{12r}, cobaloximes(II) and related compounds. J. Am. Chem. Soc. 90, 319 (1968).
23. Pearson, R. G., H. Sobel, and J. Songstad: Nucleophilic reactivity constants toward methyl iodide and trans-$[Pt(py)_2Cl_2]$. J. Am. Chem. Soc. 90, 319 (1968).
24. Costa, G., G. Mestroni, T. Licari, and E. Mestroni: New σ-bonded bisalkyl and alkyl-aryl cobalt complexes of bis(diacetylmonoxime-imino)propane-1,3. Inorg. Nucl. Chem. Letters 5, 561 (1969).
25. Schrauzer, G. N., L. P. Lee, and J. W. Sibert: Alkylcobalamins, alkylcobaloximes: Electronic spectra and mechanism of photodealkylation. J. Am. Chem. Soc. 92, 2997 (1970).
26. Schrauzer, G. N., and R. J. Windgassen: On hydroxyalkylcobaloximes and the mechanism of a cobamide dependent diol dehydrase. J. Am. Chem. Soc. 89, 143 (1967).
27. Hogenkamp, H. P. C., J. E. Rush, and C. A. Swenson: Observations on the organo-metallic bond of the corrinoid coenzymes. J. Biol. Chem. 240, 3641 (1965).
28. Golding, B. T., H. L. Holland, U. Horn, and S. Sakrikar: Solvolysis of β-acetoxy-alkyl-bis(biacetyldioximato)pyridinecobalt: Evidence for a novel intermediate olefinic complex. Angew. Chem. Internat. Edit. 9, 959 (1970).
29. Silverman, R. B., D. Dolphin, and B. M. Babior: A model for the mechanism of action of coenzyme B_{12} dependent enzymes. Evidence for σ-π-rearrangements in cobaloximes. J. Am. Chem. Soc. 94, 4028 (1972).
30. Dolphin, D., A. W. Johnson, and R. Rodrigo: Some reactions of the vitamin B_{12} coenzyme and its alkyl analogues. Ann. N. Y. Acad. Sci. 112, 590 (1964).
31. Hogenkamp, H. P. C., J. N. Ladd, and H. A. Barker: The identification of a nucleo-side derived from coenzyme B_{12}. J. Biol. Chem. 237, 1950 (1962).
32. Hogenkamp, H. P. C., and H. A. Barker: Nucleoside photolysis products of coenzyme B_{12}. Fed. Proc. 21, 470 (1962).
33. Schrauzer, G. N., J. A. Seck, R. J. Holland, T. M. Beckham, E. M. Rubin, and J. W. Sibert: Reductive dealkylation of alkylcobaloximes, alkylcobalamins and related compounds: Simulation of corrin dependent reductase and alkylation reactions. Bioinorganic Chem. 2, 93 (1972).
34. Schrauzer, G. N.: Recent advances in the chemistry of vitamin B_{12} and vitamin B_{12} model compounds. Reductive cobalt-carbon bond cleavage reactions. Pure and Applied Chem. 33, 545 (1973).
35. Babior, B. M., T. H. Moss, and D. C. Gould: The mechanism of action of ethanolamine ammonia lyase, a B_{12}-dependent enzyme. J. Biol. Chem. 247, 4389 (1972).
36. Cockle, S. A., H. A. O. Hill, R. J. P. Williams, S. P. Davies, and M. A. Foster: The detection of intermediates during the conversion of propane-1,2-diol to propionaldehyde by glyceroldehydrase, a coenzyme B_{12} dependent reaction. J. Am. Chem. Soc. 94, 275 (1972).

37. SCHRAUZER, G. N.: Model Studies in nitrogen fixation and cobalamin chemistry. Advances in Chemistry Series **100**, 1 (1971).

38. SCHRAUZER, G. N., and J. W. SIBERT: Coenzyme B_{12} and coenzyme B_{12} model compounds in the catalysis of the dehydration of glycols. J. Am. Chem. Soc. **92**, 1022 (1970).

39. PETERKOFSKY, A., and H. WEISSBACH: Enzymatic synthesis of coenzyme B_{12}. Ann. N. Y. Acad. Sci. **112**, 622 (1964).

40. PAWELKIEWICZ, J., B. BARTOSINSKI, and W. WALERYCH: Enzymic synthesis of corrin coenzymes. Ann. N. Y. Acad. Sci. **112**, 638 (1964).

41. SMITH, E. LESTER, L. MERVYN, P. W. MUGGLETON, A. W. JOHNSON, and N. SHAW: Chemical Routes to coenzyme B_{12} and analogues. Ann. N. Y. Acad. Sci. **112**, 565 (1964).

42. MÜLLER, O., and K. BERNHAUER: Synthesis and biological behavior of additional corrinoid coenzymes. Ann. N. Y. Acad. Sci. **112**, 575 (1964).

43. SCHRAUZER, G. N., and R. J. HOLLAND: Reactions of vitamin B_{12s} with nucleoside phosphates. Unpublished.

44. FREY, P. A., M. K. ESSENBERG, and R. H. ABELES: Studies on the mechanism of hydrogen transfer in the cobamide coenzyme-dependent dioldehydrase reaction. J. Biol. Chem. **242**, 5369 (1967).

45. ABELES, R. H., and H. A. LEE, JR.: Purification and properties of dioldehydrase, an enzyme requiring a cobamide coenzyme. J. Biol. Chem. **238**, 2367 (1963).

46. SMILEY, K. L., and M. SOBOLOV: A cobamide-requiring glycerol dehydrase from an acrolein-forming lactobacillus. Arch. Biochem. Biophys. **97**, 538 (1962).

47. TORAYA, T., and S. FUKUI: Coenzyme B_{12} dependent propanediol dehydrase system. Ternary complex between apoenzyme, coenzyme and substrate analog. Biochim. Biophys. Acta **284**, 536 (1972).

48. HOGENKAMP, H. P. C.: Enzymatic reactions involving corrinoids. Ann. Reviews Biochem. **37**, 225 (1968).

49. RÉTEY, J., A. UMANI-RONCHI, J. SEIBL und D. ARIGONI: Zum Mechanismus der Propandioldehydrase-Reaktion. Experientia **22**, 502 (1966).

50. SCHRAUZER, G. N., R. J. HOLLAND, and J. SECK: The mechanism of coenzyme B_{12} action in dioldehydrase. J. Am. Chem. Soc. **93**, 1503 (1971).

51. HENDERSON, R. G. S., and J. M. PRATT: Reactions of nitrous oxide with some transition metal complexes. Chem. Commun. **1967**, 387.

52. SCHRAUZER, G. N., J. SECK, and R. J. HOLLAND: The mechanism of coenzyme B_{12} action in dioldehydrase: New observations concerning the effects of O_2, N_2O and CO. Z. Naturforsch. **28 c**, 1 (1973).

53. TORAYA, T., M. KONDO, Y. ISEMURA, and S. FUKUI: Propanediol dehydrase system. Nature of cobalamin binding and some properties of apoenzyme-coenzyme B_{12} analog complexes. Biochemistry **11**, 2599 (1972).

54. TORAYA, T., Y. SUGIMOTO, Y. TAMAO, S. SHIMIZU, and S. FUKUI: Role of monovalent cations in binding of vitamin B_{12} coenzyme or its analogs to apoenzyme. Biochemistry **10**, 3475 (1971).

55. WAGNER, O. W., H. A. LEE, P. A. FREY, and R. H. ABELES: Studies of the mechanism of action of cobamide coenzymes. Chemical properties of the enzyme-coenzyme complex. J. Biol. Chem. **241**, 1751 (1966).

56. SCHRAUZER, G. N., and R. J. HOLLAND: Attempted reconstitution of coenzyme B_{12} in dioldehydrase. Unpublished experiments. 1972.

57. HOLLAND, R. J.: Reaction of 4'-fluoroadenosyl-5'-iodide with vitamin B_{12s}. Unpublished, 1972.

58. — Studies in vitamin B_{12} chemistry. Thesis, University of California, San Diego.

59. SCHRAUZER, G. N., R. J. HOLLAND, and W. J. MICHAELY: Concerning the mechanism of coenzyme B_{12} action in dioldehydrase: Synthesis and reactions of proposed intermediates. J. Am. Chem. Soc. **95**, 2024 (1973).

60. Hamilton, J. A., and R. L. Blakley: Electron spin resonance studies of ribonucleotide reduction catalyzed by ribonucleotide reductase of *Lactobacillus leichmannii*. Biochim. Biophys. Acta **184**, 224 (1969).

61. Vitols, E., H. P. C. Hogenkamp, C. Brownson, R. L. Blakley, and J. Connellan: Reduction of a disulphide bond of ribonucleotide reductase by the dithiol substrate. Biochem. J. **104**, 58c (1967).

62. Hoffman, M. Z., and E. Hayon: One-electron reduction of the disulfide linkage in aqueous solution. Formation, protonation and decay kinetics of the RSSR⁻ radical. J. Am. Chem. Soc. **94**, 7950 (1972).

63. Bradbeer, C.: The clostridial fermentations of choline and ethanolamine. I. Preparation and properties of cell free extracts. J. Biol. Chem. **240**, 4669 (1965).

64. — The clostridial fermentations of choline and ethanolamine. II. Requirement for a cobamide coenzyme by an ethanolamine deaminase. J. Biol. Chem. **240**, 4675 (1965).

65. Babior, B. M.: The mechanism of action of ethanolamine deaminase. I. Studies with isotopic hydrogen and oxygen. J. Biol. Chem. **244**, 449 (1969).

66. — The mechanism of action of ethanolamine deaminase, II. The spectrum of the ethanolamine deaminase-coenzyme B_{12} complex during the act of catalysis. Biochim. Biophys. Acta **178**, 406 (1969).

67. — The mechanism of action of ethanolamine deaminase. III. Inhibition by coenzyme B_{12} analogues. J. Biol. Chem. **244**, 2917 (1969).

68. — The mechanism of action of ethanolamine deaminase. IV. Cobamide dependent binding of substrate to ethanolamine deaminase. J. Biol. Chem. **244**, 2927 (1969).

69. — The mechanism of action of ethanolamine deaminase. V. The photolysis of enzyme-bound alkylcobalamins. Biochemistry **8**, 2662 (1969).

70. — The mechanism of action of ethanolamine deaminase. VI. Ethylene glycol, a quasi-substrate for ethanolamine deaminase. J. Biol. Chem. **245**, 1755 (1970).

71. — The mechanism of action of ethanolamine ammonia lyase, a B_{12} dependent enzyme. VII. The mechanism of hydrogen transfer. J. Biol. Chem. **245**, 6125 (1970).

72. Weisblat, D. A., and B. M. Babior: The mechanism of action of ethanolamine ammonia-lyase, a B_{12}-dependent enzyme. VIII. Further studies with compounds labeled with isotopes of hydrogen: Identification and some properties of the rate-limiting step. J. Biol. Chem. **246**, 6064 (1971).

73. Carty, T. J., B. M. Babior, and R. H. Abeles: The mechanism of action of ethanolamine ammonia-lyase, a B_{12}-dependent enzyme. IX. Interaction of the enzyme-coenzyme complex with reaction products. J. Biol. Chem. **246**, 6313 (1971).

74. Babior, B. M., T. H. Moss, and D. C. Gould: The mechanism of action of ethanolamine ammonia-lyase. X. A study of the reaction by electron spin resonance spectrometry. J. Biol. Chem. **247**, 4389 (1972).

75. Law, P. Y., D. G. Brown, E. L. Lien, B. M. Babior, and J. M. Wood: Synthesis and catalytic activity of spin-labeled cobamide coenzymes. Biochemistry **10**, 3428 (1971).

76. Jayme, M., and J. H. Richards: Mechanism of action of coenzyme B_{12}. Release of 5′-deoxyinosine on incubation of deoxyinosylcobalamin, 1,2-propanediol and propanediol dehydrase. Biochem. Biophys. Research Comm. **43**, 1329 (1971).

77. Blakley, R. L.: Cobamides and ribonucleotide reduction. J. Biol. Chem. **240**, 2173 (1965).

78. Blakley, R. L., R. K. Ghambler, P. F. Nixon, and E. Vitols: The cobamide dependent ribonucleotide triphosphate reductase of *Lactobacilli*. Biochem. Biophys. Res. Comm. **20**, 439 (1965).

79. Gleason, F. K., and H. P. C. Hogenkamp: Ribonucleotide reductase from *Euglena*

gracilis, a deoxyadenosylcobalamin-dependent enzyme. J. Biol. Chem. **245**, 4894 (1970).

80. FOLLMANN, H., and H. P. C. HOGENKAMP: Interaction of ribonucleotide reductase with ribonucleotide analogs. Biochemistry **10**, 186 (1971).

81. ABELES, R. H., and W. S. BECK: The mechanism of action of cobamide coenzyme in the ribonucleotide reductase reaction. J. Biol. Chem. **242**, 3589 (1967).

82. BECK, W. S., R. H. ABELES, and W. G. ROBINSON: Transfer of hydrogen from cobamide coenzyme to water during enzymatic ribonucleotide reduction. Biochem. Biophys. Res. Comm. **25**, 421 (1966).

83. HAMILTON, J. A., R. L. BLAKLEY, F. D. LOONEY, and M. E. WINFIELD: Formation of a cobamide containing divalent cobalt by the ribonucleotide-reductase of *Lactobacillus Leichmannii.* Biochim. Biophys. Acta **177**, 374 (1969).

84. HAMILTON, J. A., R. YAMADA, R. L. BLAKLEY, H. P. C. HOGENKAMP, F. D. LOONEY, and M. E. WINFIELD: Cobamides and ribonucleotide reduction. VII. Cob(II)alamin as a sensitive probe for the active center of ribonucleotide reductase. Biochemistry **10**, 347 (1971).

85. SCHRAUZER, G. N., and R. J. WINDGASSEN: Cobalamin model compounds. Preparation and reactions of substituted alkyl- and alkenylcobaloximes and biochemical implications. J. Am. Chem. Soc. **89**, 1999 (1967).

86. JACOBSEN, D. W., and F. M. HUENNEKENS: Ion-dependent activation and inhibition of ribonucleotide reductase from *Lactobacillus leichmannii.* Biochem. and Biophys. Res. Comm. **37**, 793 (1969).

87. YAMADA, R., Y. TAMAO, and R. L. BLAKLEY: Degradation of 5′-deoxyadenosylcobalamin by ribonucleotidetriphosphate reductase and binding of degradation products to the active center. Biochemistry **10**, 3959 (1971).

88. VITOLS, E., C. BROWNSON, W. GARDINER, and R. L. BLAKLEY: Cobamides and Ribonucleotide reduction. V. A kinetic study of the ribonucleotide triphosphate reductase of *Lactobacillus leichmannii.* J. Biol. Chem. **242**, 3035 (1967).

89. BARKER, H. A., F. SUZUKI, A. IODICE, and V. ROOZE: Glutamate mutase reaction. Ann. N. Y. Acad. Sci. **112**, 644 (1964).

90. BARKER, H. A., H. WEISSBACH, and R. D. SMYTH: A coenzyme containing pseudo-vitamin B₁₂. Proc. Natl. Acad. Sci. **44**, 1093 (1958).

91. BARKER, H. A., R. D. SMYTH, R. M. WILSON, and H. WEISSBACH: The purification and properties of β-methylaspartase. J. Biol. Chem. **234**, 320 (1959).

92. BARKER, H. A., V. ROOZE, F. SUZUKI, and A. A. IODICE: The glutamate mutase system. J. Biol. Chem. **239**, 3260 (1964).

93. IODICE, A. A., and H. A. BARKER: The glutamate isomerase reaction: Studies on the incorporation of solvent hydrogen. J. Biol. Chem. **238**, 2094 (1963).

94. OHMORI, H., H. ISHITANI, K. SATO, S. SHIMIZU, and S. FUKUI: Vitamin B₁₂ dependent glutamate mutase activity in photosynthetic bacteria. Biochem. Biophys. Res. Comm. **43**, 156 (1971).

95. EAGAR, R. G., JR., B. G. BALTIMORE, M. M. HERBST, H. A. BARKER, and J. H. RICHARDS: Mechanism of action of coenzyme B₁₂. Hydrogen transfer in the isomerization of β-methylaspartate to glutamate. Biochemistry **11**, 253 (1972).

96. SWITZER, R. L., B. G. BALTIMORE, and H. A. BARKER: Hydrogen transfer between substrates and deoxyadenosylcobalamin in the glutamate mutase reaction. J. Biol. Chem. **244**, 5263 (1969).

97. SPRECHER, M., and D. B. SPRINSON: The stereochemistry of the glutamate mutase reaction. Ann. N. Y. Acad. Sci. **112**, 655 (1964).

98. STJERNHOLM, R., and H. G. WOOD: Methylmalonyl isomerase. II. Purification and properties of the enzyme from propionibacteria. Proc. Nat. Acad. Sci. U. S. **47**, 303 (1961).

99. Eggerer, H., E. R. Stadtman, P. Overath und F. Lynen: Zum Mechanismus der durch Cobalamin-Coenzym katalysierten Umlagerung von Methylmalonyl-CoA in Succinyl-CoA. Biochem. Zeitschr. **333**, 1 (1960).

100. Kellermeyer, R. W., and H. G. Wood: Methylmalonyl isomerase: A study of the mechanism of isomerization. Biochemistry **1**, 1124 (1962).

101. Williams, D. L., G. H. Spray, G. E. Newman, and J. R. O'Brien: Dietary depletion of vitamin B_{12} and the excretion of methylmalonic acid in the rat. Brit. J. Nutrition **23**, 343 (1969).

102. Mudd, S. H., H. L. Levy, and R. H. Abeles: A derangement in B_{12} metabolism leading to homocystinemia, cystathioninemia and methylmalonic aciduria. Biochem. Biophys. Res. Comm. **35**, 121 (1969).

103. Rosenberg, L. E., A. C. Lilljequist, and Y. E. Hsia: B_{12}-dependent methylmalonic-aciduria: Defective B_{12} metabolism in cultured fibroblasts. Biochem. Biophys. Res. Comm. **37**, 607 (1969).

104. Miller, W. W., and J. H. Richards: Mechanism of action of coenzyme B_{12}: H-Transfer in the isomerization of methylmalonyl-CoA to succinyl-CoA. J. Am. Chem. Soc. **91**, 1498 (1969).

105. Wood, H. G., R. W. Kellermeyer, R. Stjernholm, and S. H. G. Allen: Metabolism of methylmalonyl-CoA and the role of biotin and B_{12} coenzymes. Ann. N. Y. Acad. Sci. **112**, 661 (1964).

106. Erfle, J. D., J. M. Clark, Jr., and B. Connor Johnson: Direct hydrogen transfer in the conversion of methylmalonyl-CoA to succinyl-CoA. Ann. N. Y. Acad. Sci. **112**, 684 (1964).

107. Kung, H. F., S. Cederbaum, L. Tsai, and T. C. Stadtman: Nicotinic acid metabolism. V. A cobinamide coenzyme dependent conversion of α-methyleneglutarate to dimethyl-malonate. Proc. Natl. Acad. Sci. U.S. **65**, 978 (1970).

108. Kung, H. F., and T. C. Stadtman: Nicotinic acid metabolism. VI. Purification of α-methylene-glutarate mutase (B_{12}-dependent) and methylitaconate isomerase. J. Biol. Chem. **246**, 3378 (1971).

109. Kung, H. F., and L. Tsai: Nicotinic acid metabolism. VII. Mechanism of action of clostridial α-methylene glutarate mutase (B_{12}-dependent) and methylitaconate iso-merase. J. Biol. Chem. **246**, 6436 (1971).

110. Schrauzer, G. N., and R. J. Holland: Hydridocobaloximes. J. Am. Chem. Soc. **93**, 1505 (1971).

111. Whitlock, H. W.: Mechanistic possibilities for some coenzyme B_{12} catalyzed reactions. Ann. N. Y. Acad. Sci. **112**, 721 (1964).

112. Lowe, J. N., and L. L. Ingraham: A model of the methylmalonyl isomerase reaction. J. Am. Chem. Soc. **93**, 3801 (1971).

113. Tsai, L., and T. C. Stadtman: Anaerobic degradation of lysine. IV. Cobamide coenzyme dependent migration of an amino group from carbon 6 of β-lysine to carbon 5, etc. Arch. Biochem. Biophys. **125**, 210 (1968).

114. Dekker, E. E., and H. A. Barker: Identification and cobamide coenzyme dependent formation of 3,5-diaminohexanoic acid. An intermediate in lysine fermentation. J. Biol. Chem. **243**, 3232 (1968).

115. Stadtman, T. C., and L. Tsai: A cobamide coenzyme dependent migration of the ε-amino group of D-lysine. Biochem. Biophys. Res. Comm. **28**, 920 (1967).

116. Dyer, J. K., and R. N. Costilow: 2,4-Diaminovaleric acid: Intermediate in the anerobic oxidation of ornithine by *Clostridium sticklandii*. J. Bacteriol. **101**, 77 (1970).

117. Morley, C. G. D., and T. C. Stadtman: Studies on the fermentation of D-β-lysine. Purification and properties of an ATP regulated B_{12} coenzyme dependent D-β-lysine mutase complex from *Clostridium sticklandii*. Biochemistry **9**, 4890 (1970).

118. Stadtman, T. C.: Vitamin B_{12}. Science **171**, 859 (1971).

119. STADTMAN, T. C., and P. RENZ: Anaerobic degradation of lysine. V. Some properties of the cobamide coenzyme dependent β-lysine mutase of *Clostridium sticklandii*. Arch. Biochem. Biophys. **125**, 226 (1968).
120. RÉTEY, J., F. KUNZ, T. C. STADTMAN und D. ARIGONI: Zum Mechanismus der β-Lysin-Mutase-Reaktion. Experientia **25**, 801 (1969).
121. STADTMAN, T. C.: Methane Fermentation. Ann. Rev. Microbiol. **21**, 121 (1967).
122. MCBRIDE, B. C., and R. S. WOLFE: A new coenzyme of methyl transfer: Coenzyme M. Biochemistry **10**, 2317 (1971).
123. MCBRIDE, B. C., J. M. WOOD, J. W. SIBERT, and G. N. SCHRAUZER: Methylcobalt derivatives of vitamin B$_{12}$ model compounds as substrates in enzymatic methane formation. J. Am. Chem. Soc. **90**, 5276 (1968).
124. SIBERT, J. W., and G. N. SCHRAUZER: Enzymatic and nonenzymatic demethylation of methylcobalamin and of abiogenic cobaloxime model substrates. Methane biosynthesis by *Methanobacillus omelianskii*. J. Am. Chem. Soc. **92**, 1421 (1970).
125. SCHRAUZER, G. N., J. A. SECK, and T. M. BECKHAM: Reductive Co-C bond cleavage of alkylcorrins and of vitamin B$_{12}$ model compounds by alkaline CO, S$_2$O$_4^{--}$ and stannite. Bioinorganic Chemistry **2**, 211 (1973).
126. ROBERTON, A. M., and R. S. WOLFE: ATP requirement for methanogenesis in cell extracts of *Methanobacterium strain M. o. H*. Biochim. Biophys. Acta **192**, 420 (1969).
127. WOOD, J. M., and R. S. WOLFE: Alkylation of an enzyme in the methane-forming system of *Methanobacillus omelianskii*. Biochem. Biophys. Res. Comm. **22**, 119 (1966).
128. BARKER, H. A., and M. D. KAMEN: Carbon dioxide utilization in the synthesis of acetic acid by *Clostridium thermoaceticum*. Proc. Natl. Acad. Sci. U.S. **31**, 219 (1945).
129. LENTZ, K., and H. G. WOOD: Synthesis of acetate from formate and carbon dioxide by *Clostridium thermoaceticum*. J. Bacteriol. **69**, 645 (1955).
130. POSTON, J. M., K. KURATOMI, and E. R. STADTMAN: Methyl-vitamin B$_{12}$ as a source of methyl groups for the synthesis of acetate by cell-free extracts of *Clostridium thermoaceticum*. Ann. N.Y. Acad. Sci. **112**, 804 (1964).
131. GHAMBEER, R. K., H. G. WOOD, M. SCHULMAN, and L. LJUNGDAHL: Total synthesis of acetate from CO$_2$. III. Inhibition by alkylhalides of the synthesis from CO$_2$, methyltetrahydrofolate and methyl-B$_{12}$ by *Clostridium thermoaceticum*. Arch. Biochem. Biophys. **143**, 471 (1970), and previous papers cited therein.
132. SCHRAUZER, G. N., and J. W. SIBERT: Acetate synthesis from carbon dioxide and methylcorrinoids. Simulation of the microbial carbon dioxide fixation reaction in a model system. J. Am. Chem. Soc. **92**, 3509 (1970).
133. PARKER, D. J., H. G. WOOD, R. K. GHAMBEER, and L. G. LJUNGDAHL: Total synthesis of acetate from carbon dioxide. Retention of deuterium during carboxylation of trideuteriomethyltetrahydrofolate or trideuteriomethylcobalamin. Biochemistry **11**, 3074 (1972).
134. LJUNGDAHL, L., and H. G. WOOD: Total synthesis of acetate from CO$_2$ by heterotrophic bacteria. Ann. Rev. Microbiol. **23**, 515 (1970).
135. MCBRIDE, B. C., and R. S. WOLFE: Biosynthesis of dimethylarsine by *Methanobacterium*. Biochemistry **10**, 4312 (1971).
136. WOOD, J. M., F. S. KENNEDY, and C. G. ROSEN: Synthesis of methylmercury compounds by extracts of a methanogenic bacterium. Nature (London) **220**, 173 (1968).
137. HILL, H. A. O., J. M. PRATT, S. RIDSDALE, F. R. WILLIAMS, and J. R. P. WILLIAMS: Kinetics of substitution of coordinated carbanions in Co(III) corrinoids. Chem. Commun. **1970**, 341.
138. SCHRAUZER, G. N., J. H. WEBER, T. M. BECKHAM, and R. K. Y. HO: Alkyl group transfer from cobalt to mercury: The reaction of alkylcobalamins, alkylcobaloximes and of related compounds with mercuric acetate. Tetrahedron Letters **1971**, 275.

139. SCHRAUZER, G. N., und G. KRATEL: Organometallderivate des Bisdimethylglyoximato-kobalts. Chem. Ber. **102**, 2392 (1969).

140. SCHRAUZER, G. N., and E. M. RUBIN: Formation of Methylcobalamin from dimethyl-mercury and vitamin B_{12s}. Unpublished results, 1972.

141. BUCHANAN, J. M., H. L. ELFORD, R. E. LOUGHLIN, B. M. McDOUGALL, and S. ROSEN-THAL: The role of vitamin B_{12} in methyl transfer to homocysteine. Ann. N. Y. Acad. Sci. **112**, 756 (1964).

142. GUEST, J. R., S. FRIEDMAN, M. J. DILWORTH, and D. D. WOODS: Methylcobalamin as a source of the methyl group of methionine. Ann. N. Y. Acad. Sci. **112**, 774 (1964).

143. DICKERMAN, H. W., B. G. REDFIELD, J. G. BIERI, and H. WEISSBACH: Studies on the role of vitamin B_{12} for the synthesis of methionine in liver. Ann. N. Y. Acad. Sci. **112**, 791 (1964).

144. LARRABEE, A. R., S. ROSENTHAL, R. E. CATHOU, and J. M. BUCHANAN: A methylated derivative of tetrahydrofolate as an intermediate of methionine biosynthesis. J. Am. chem. Soc. **83**, 4094 (1961).

145. KERWAR, S. S., J. H. MANGUM, K. G. SCRIMGEOUR, and F. M. HUENNEKENS: Function of methylcobalamin in methionine synthesis. Biochem. Biophys. Res. Comm. **15**, 377 (1964).

146. TAYLOR, R. T., and H. WEISSBACH: N^5-Methyltetrahydrofolate-homocysteine trans-methylase. J. Biol. Chem. **242**, 1502 (1967).

147. BLAKLEY, R. L.: The biochemistry of folic acid and related pteridines. Amsterdam: North Holland Publishing Co. 1969. Chapter 9, 332.

148. WEISSBACH, H., and R. T. TAYLOR: Roles of vitamin B_{12} and folic acid in methionine synthesis. Vitamins and Hormones **28**, 415 (1970).

149. STAVRIANOPOULOS, J., und L. JAENICKE: Reaktionsschritte der Methioninsynthese bei *E. coli*. Eur. J. Biochem. **3**, 95 (1967).

150. TAYLOR, R. T., and H. WEISSBACH: *Escherichia Coli B* N^5-methyltetrahydrofolate-homocysteine methyltransferase. Sequential formation of bound methylcobalamin with S-adenosyl-L-methionine and N^5-methyltetrahydrofolate. Arch. Biochem. Bio-phys. **129**, 728 (1969).

151. — — *Escherichia Coli B* N^5-methyltetrahydrofolate-homocysteine methyltransferase: Activation with S-adenosyl-L-methionine and the mechanism for methyl group transfer. Arch. Biochem. Biophys. **129**, 745 (1969).

152. SCHRAUZER, G. N., and R. J. WINDGASSEN: On cobaloximes with cobalt-sulfur bonds and some model studies related to cobamide dependent methyl group transfer reactions. J. Am. Chem. Soc. **89**, 3607 (1967).

153. BIED-CHARRETON, C., et A. GAUDEMER: Acoylation de complexes du cobalt(I) par les sels d'ammonium quaternaires. Compt. Rend. Sc. Paris **272**, 1241 (1971).

154. TAYLOR, R. T.: Methylcobalamin as a substrate at a separate site on *Escherichia Coli B* N^5-methyltetrahydrofolate-homocysteine cobalamin methyltransferase. Arch. Bio-chem. Biophys. **144**, 352 (1971).

155. TAYLOR, R. T., and H. WEISSBACH: *Escherichia Coli B* N^5-methyltetrahydrofolate-homocysteine vitamin B_{12} transmethylase: Formation and photolability of a methyl-cobalamin enzyme. Arch. Biochem. Biophys. **123**, 109 (1968).

156. PAILES, W. H., and H. P. C. HOGENKAMP: The photolability of Co-alkylcobinamides. Biochemistry **7**, 4160 (1968).

157. SCHRAUZER, G. N., and R. J. WINDGASSEN: Synthesis of methyl groups catalysed by vitamin B_{12}, *in vitro*. Nature (London) **214**, 492 (1967).

158. RÜDIGER, H., and L. JAENICKE: Methionine synthetase: Existence and interconversion of two enzyme species. Europ. J. Biochem. **16**, 92 (1970).

159. — — On the role of S-adenosylmethionine in the vitamin dependent methionine synthesis. Europ. J. Biochem. **10**, 557 (1969).

(Received January 10, 1973)

Namenverzeichnis. Author Index

Kursiv gedruckte Seitenzahlen beziehen sich auf Literaturverzeichnisse

Page numbers printed in *italics* refer to References

Nespiak, A. 64, *116*
Neubert, M. E. *56, 59*
Neumann, H. *463*
Neville, G. A. *374*
Newman, G. E. *627*
Newberne, P. M. *151*
Nicolaus, R. A. 522, 523, 535, 537, 544, 545, 548, 555, 557, 558, 561, 570, *579, 580, 581*
Niebergall, P. J. *55, 56*
Niemann, G. J. 187
Nieschlag, H. *210*
Nigam, I. C. *374*
Nilsson, B. *114*
Nilsson, E. 187
Nishiie, K. *53*
Nishikawa, Y. *212*
Nishimura, H. *519*
Nixon, P. F. *625*
Nógrádi, M. 199, *208, 210, 212, 214, 215*
Nölken, E. *282*
Nominé, G. *466*
Nonhebel, D. C. *276*
Nordgren, L. *580*
Norin, T. *373, 460*
Norpoth, K. *116*
Norris, G. L. F. *115*
Novak, L. *59*
Nozoe, S. *116*
Numasaki, Y. *57*
Numata, M. *59*

O'Brien, D. H. *465*
O'Brien, J. R. *627*
Ocken, P. R. *464*
Oekedon, D. J. 188
Oelrichs, P. J. 190, *212*
Oesterlin, R. *53, 54*
Oettlé, A. G. *150*
Ogasawara, K. *519*
Ogiso, A. 176, *213, 374*
Ogunkoya, L. 551, *578*
Ohloff, G. *460*
Ohmori, H. *626*
Ohnsorge, U. *149*
Ohta, N. *213*
Ohta, T. *213*
Ohta, Y. 304, *371, 374*
Ohumura, T. *519*
Oishi, T. *518*
O'Kelly, J. *149, 150, 151*
Okigawa, M. *213*
Okuchi, M. *116*
Okun, M. R. 532, *580*

Olechnowicz-Stepien, V. *212*
Oliver, L. K. *520*
Ollis, W. D. *209, 280, 519*
Omote, Y. *580*
Oppolzer, W. *62*
Or, N. *580*
Ord, W. O. *149*
Orris, L. *466*
Ostermayer, J. *210*
O'Sullivan, W. I. *280*
Overath, P. *627*
Overend, W. G. *208*
Overman, L. E. 370, *376*
Oxley, P. *464*
Oyamada, T. 199, *213*

Paal, C. 420, *464*
Paal, Z. *208*
Pacheco, H. 198, *207, 209, 213*
Paech, K. *582*
Pailes, W. H. *629*
Pake, G. E. 563, *576*
Paknikar, S. K. *374*
Pallanza, R. *55*
Pallenbach, D. *56, 59*
Pallos, L. 198, *208, 214*
Panetta, C. A. *56, 57*
Paris, R. R. 180, 187, 188, 189, *208, 209, 213*
Parker, D. J. *628*
Parker, W. *374*
Parry, D. R. *148*
Partyka, A. *59*
Paschal, J. W. *58*
Patel, R. P. *580*
Pathak, M. A. *577*
Paukstelis, J. V. *464*
Paul, D. *463, 464*
Paul, I. C. *372*
Pawelkiewicz, J. *624*
Pearson, M. J. *54, 59*
Pearson, R. G. *623*
Pechan, Z. 529, *577*
Pechmann, H. v. *276*
Pelt, J. G. van *150*
Penfold, A. R. *374*
Pennington, P. A. *54*
Percheron, F. *214*
Percival, A. 550, 551, 567, *575, 576, 578, 580*
Pereira, S. *213*
Perkin, A. G. 179
Perlin, A. S. 162, 166, *209*
Perlman, I. D. *57*
Perron, Y. G. *59, 61*

Sachverzeichnis. Subject Index

Von · By

A. SIEGEL, Wien

Fortschritte der Chemie organischer Naturstoffe
Progress in the Chemistry of Organic Natural Products

Further volumes see next page

Springer-Verlag Wien · New York

Volume 20: 33 figures. XIII, 509 pages. 1962.
Cloth DM 117,—

Cumulative Index / Generalregister 1–20. 1938–1962. XVI, 369 pages. 1964.
Cloth DM 85,—

Volume 21: 14 figures. VII, 362 pages. 1963.
Cloth DM 88,—

Volume 22: 8 figures. VII, 370 pages. 1964.
Cloth DM 103,—

Volume 23: 58 figures. VIII, 397 pages. 1965.
Cloth DM 109,—

Volume 24: 25 figures. VIII, 475 pages. 1966.
Cloth DM 130,—

Volume 25: 25 figures. VII, 348 pages. 1967.
Cloth DM 92,—

Volume 26: 97 figures. IX, 456 pages. 1968.
Cloth DM 144,—

Volume 27: 47 figures. VIII, 412 pages. 1969.
Cloth DM 130,—

Volume 28: 14 figures. XII, 503 pages. 1970.
Cloth DM 155,—

Volume 29: 18 figures. VIII, 554 pages. 1971.
Cloth DM 181,—

Volume 30: 28 figures. VIII, 666 pages. 1973.
Cloth DM 225,—

Contents: **M. J. Cormier, J. E. Wampler, K. Hori,** Bioluminescence: Chemical Aspects. — **L. Jaenicke, D. G. Müller,** Gametenlockstoffe bei niederen Pflanzen und Tieren. — **J. Polonsky,** Quassinoid Bitter Principles. — **B. Franck, H. Flasch,** Die Ergochrome (Physiologie, Isolierung, Struktur und Biosynthese. — **H. D. Locksley,** The Chemistry of Biflavanoid Compounds. — **W. Keller-Schierlein,** Chemie der Makrolid-Antibiotica. — **R. Tschesche, G. Wulff,** Chemie und Biologie der Saponine. — Author Index / Namenverzeichnis. — Subject Index / Sachverzeichnis.

Price reduction for subscribers / Preisermäßigung für Subskribenten: 10%.

Special price reduction (20% of the list price) for the set Vols. 1—20 plus Cumulative Index / Vorzugspreis (20% Nachlaß) bei Bezug der Bände 1—20 inklusive Generalregister.